The Retinal Pigment Epithelium

THE RETINAL PIGMENT EPITHELIUM

Keith M. Zinn and Michael F. Marmor, Editors

PART ONE

Structure and Function

EDITED BY

Michael F. Marmor, M.D.

PART TWO

Pathophysiology and Disease

EDITED BY

Keith M. Zinn, M.D.

HARVARD UNIVERSITY PRESS
Cambridge, Massachusetts
and London, England
1979

Printed in the United States of America

Library of Congress Cataloging in Publication Data

Main entry under title:

The Retinal pigment epithelium.

Includes index.
CONTENTS: pt. 1. Structure and function.—
pt. 2. Pathophysiology and disease.
1. Retina. 2. Visual pigments. 3. Epithelium.
I. Zinn, Keith M. II. Marmor, Michael F.
[DNLM: 1. Pigment epithelium of eye. 2. Retinal
diseases. WW270 R4385]
QP479.R47 617.7'3 79–21
ISBN 0–674–76684–9

Preface

RECENT YEARS have seen a surge of interest in the retinal pigment epithelium (RPE) and an outpouring of scientific data as investigators and clinicians have become aware of the importance of the RPE to retinal function and disease. Our goal is to provide individuals who work with the RPE—either in the laboratory or in patient care—with a comprehensive review of the latest basic research and clinical information. The book is divided into two sections. The first deals with structure and function and represents a core of essential information about the normal RPE. The second is concerned with pathophysiology and with the responses available to the RPE in various disease states. Whenever possible, pathogenetic mechanisms of disease are correlated with RPE structure and function. We hope that the juxtaposition of basic and clinical material will stimulate the researcher to find clinical correlations and will make the clinician more aware of the significance of disease.

The order of the editors' names on this volume has been maintained from an agreement when the project was initiated and does not signify either seniority or relative contribution to the book. This project was begun several years ago, but all of the chapters have been reviewed by their respective authors and updated to ensure that the material is current. We have tried to be reasonably comprehensive in our choice of topics, but new work is already appearing that we could not anticipate. We have also tried to minimize redundancy, but believe the reader will benefit from viewing a few topics from more than one vantage point.

We thank the contributors for their fine efforts in writing these chapters and for their cooperation in revising and updating the material to maintain the cohesiveness and timeliness of the book. We should point out that the contributors are an unusually distinguished group of scientists whose work is to a large degree responsible for our current understanding of the RPE. We also thank Mrs. Adele Siegal and Mrs. Lorinda Cowan for their tireless efforts, without which this volume would not have been possible. Finally, we would like to thank the members of the Harvard University Press for their invaluable help on this project, especially Dr. William Bennett, who recognized its importance, Mr. William Patrick, who has maintained support for it, and Mr. Harry Foster, whose editorial expertise is evident in this volume.

KEITH M. ZINN
MICHAEL F. MARMOR

Contributors

GUSTAVO AGUIRRE, V.M.D.,
Associate Professor of Ophthalmology,
University of Pennsylvania School of Veterinary Medicine

DANIEL M. ALBERT, M.D.,
Professor of Ophthalmic Pathology,
Harvard Medical School

JANICE V. BENJAMIN-HENKIND,
Department of Ophthalmology,
Montefiore Hospital and Medical Center, Bronx, New York

ELAINE R. BERMAN, PH.D.,
Professor of Experimental Ophthalmology,
Hebrew University–Hadassah Medical School, Jerusalem, Israel

DEAN BOK, PH.D.,
Professor of Anatomy,
University of California, Los Angeles, School of Medicine and Jules Stein Eye Institute

BETH BURNSIDE, PH.D.,
Associate Professor of Anatomy,
University of California, Berkeley

NED BUYUKMIHCI, V.M.D.,
Assistant Professor of Ophthalmology,
University of Tennessee College of Veterinary Medicine

RONALD E. CARR, M.D.,
Professor of Ophthalmology,
New York University School of Medicine

ALFRED J. COULOMBRE, PH.D.,
Head, Experimental Embryology Section, Laboratory of Vision Research,
National Eye Institute, National Institutes of Health

THOMAS B. FITZPATRICK, M.D., PH.D.,
Wigglesworth Professor and Chairman, Department of Dermatology,
Harvard Medical School

RAUL I. GARCIA,
Department of Oral Biology,
Harvard University School of Dental Medicine

PAUL HENKIND, M.D., PH.D.,
Professor and Chairman, Department of Ophthalmology,
Albert Einstein College of Medicine

ABRAHAM L. KORNZWEIG, M.D.,
Emeritus Clinical Professor of Ophthalmology,
Mount Sinai School of Medicine

INGRID KREISSIG, M.D.,
University Eye Clinic, Bonn, West Germany

TOICHIRO KUWABARA, M.D.,
Head, Section on Experimental Pathology,
National Eye Institute, National Institutes of Health

ALAN M. LATIES, M.D.,
Irene Heinz Given and John LaPorte Given Professor of Ophthalmology,
University of Pennsylvania School of Medicine and Scheie Eye Institute

MATTHEW M. LAVAIL, PH.D.,
Associate Professor of Anatomy,
University of California, San Francisco, School of Medicine

HARVEY LINCOFF, M.D.,
Clinical Associate Professor of Ophthalmology,
Cornell University Medical College

MARK LURIE, PH.D.,
Research Associate, Division of Ophthalmology,
Stanford University School of Medicine

MICHAEL F. MARMOR, M.D.,
Assistant Professor of Surgery (Ophthalmology),
Stanford University School of Medicine,
Chief, Ophthalmology Section,
Veterans Administration Medical Center, Palo Alto

SHELDON S. MILLER, PH.D.,
Assistant Professor of Physiological Optics,
University of California, Berkeley, School of Optometry

DONALD A. MORRIS, M.D.,
Associate Clinical Professor of Ophthalmology,
Albert Einstein College of Medicine

MICHAEL L. MUND, M.D.,
Assistant Clinical Professor of Ophthalmology,
College of Medicine and Dentistry of New Jersey

MERLYN M. RODRIGUES, M.D.,
Section Chief, Clinical Eye Pathology,
National Eye Institute, National Institutes of Health

LIONEL RUBIN, V.M.D.,
Professor of Ophthalmology,
University of Pennsylvania School of Veterinary Medicine

IRWIN M. SIEGEL, PH.D.,
Professor of Experimental Ophthalmology,
New York University School of Medicine

ROY H. STEINBERG, M.D., PH.D.,
Professor of Physiology and Ophthalmology,
University of California, San Francisco, School of Medicine

GEORGE SZABÓ, PH.D.,
Senior Associate in Oral Biology (Anatomy),
Harvard University School of Dental Medicine

MARK O. M. TSO, M.D.,
Professor of Ophthalmology and Director of Georgiana Theobold Ophthalmic Pathology Laboratory,
University of Illinois College of Medicine

IRMGARD WOOD, M. T.,
Research Associate, Department of Ophthalmology,
University of California, San Francisco, School of Medicine

RICHARD W. YOUNG, PH.D.,
Professor of Anatomy,
University of California, Los Angeles, School of Medicine and Jules Stein Eye Institute

HANAN ZAUBERMAN, M.D.,
Professor and Chairman, Department of Ophthalmology,
Hadassah University Hospital, Jerusalem, Israel

KEITH M. ZINN, M.D.,
Associate Clinical Professor of Ophthalmology,
Mount Sinai School of Medicine

Contents

PART TWO. PATHOPHYSIOLOGY AND DISEASE OF THE RETINAL PIGMENT EPITHELIUM

PART ONE

Structure and Function of the Retinal Pigment Epithelium

1 Anatomy of the Human Retinal Pigment Epithelium

Keith M. Zinn
Janice V. Benjamin-Henkind

THE RETINAL PIGMENT epithelium is one of the most important cell layers in the visual system and it is vital to the integrity of the rod and cone outer segments. It has many complex functions (see Table 1.1), such as absorption of light, storage and conversion of vitamin A esters, formation of an acid-mucopolysaccharide complex that surrounds the outer segments of rods and cones, as well as phagocytosis of rod and cone outer segments. The structure of the RPE is uniquely adapted to many of its diverse functions. This chapter will delineate the normal anatomical features of the human RPE and wherever possible correlate its cytoarchitecture with specific physical, metabolic, biochemical, optical, and transport functions. Structural variations that occur in the various animal species will be covered in Chapter 5.

History

At the turn of the eighteenth century, European anatomists called the choroidal capillary layer, pigment epithelium, and Bruch's membrane the *tunica Ruyschiana* after the work of Frederick Ruysch, a Dutch physician and professor of anatomy at Amsterdam. These anatomists considered the pigment epithelium and Bruch's membrane to be a part of the choroid. However, in 1861 Rudolf von Kölliker demonstrated that these structures were derived embryologically from the outer layer of the secondary optic vesicle and therefore the pigment epithelium was, in reality, a part of the retina (Duke-Elder 1961). The term "retina" was derived from the Latin *rete,* which means net. In the eleventh century, Avicenna (Duke-Elder 1961), translating into Arabic the works of the Greek anatomists Herophilos (Dobson 1925) and Galen, termed this structure *rescheth* (net-like). In the twelfth century, Gerard of Cremona, translating Avicenna's work into Latin, used the term *retina,* which has remained with us to the present day (Duke-Elder 1961). In its most comprehensive sense, the retina includes all the tissues derived from the optic vesicle. Von Kölliker's (1861) work indicated that the correct term for the pigmented cell layer was the *retinal* pigment epithelium (RPE).

In 1833 Wharton Jones, at the youthful age of twenty-four, was the first to describe adequately the morphology of the retinal pigment epithelium, showing that it was a single layer of polygonal cells usually hexagonal in shape. Several years later, Bruch (1844) described a "structureless membrane" upon which the RPE rested. This structure was later termed *Bruch's membrane.* Usher (1906) was the first to point out that the retinal pigment epithelium was darker in the macular zone than elsewhere. Later, histological studies revealed that this difference was largely due to the fact that the macular RPE cells were taller and contained more pigment than the RPE elsewhere. Kühne (1879) used the term *fuscin,* later changed to melanin, to denote the "pigment" in the RPE. Raehlmann (1907) described the pigment as consisting of crystalline needles 1.5 μ long. LoCascio (1923) noted lipoidal droplets at the base of the RPE.

In 1877, Kühnt inferred that each RPE cell was ultimately bound to its neighbor by a cement substance that also bound the basal surface of the RPE to Bruch's membrane. Verhoeff (1903), using light

Table 1.1 Selected functions of the retinal pigment epithelium.

Physical

1. Serves as a barrier protecting the neurosensory retina from unwanted molecules in the choroidal circulation.
2. Creates the major force responsible for the adhesion of neurosensory retina to the RPE through the transport (pumping) of specific fluids.
3. Synthesizes some acid mucopolysaccharides that are found in the subretinal space and which may also play a minor role in the adhesion of neurosensory retina to the RPE.
4. The microvilli have intimate structural relationships (through phagocytosis) with outer segments, which may also play a role in the adhesion of the neurosensory retina to the RPE.

Optical

1. Absorbs light energy (melanin granules) and cuts down on light scatter, improving resolution of images.
2. Acts as a pigmented barrier to prevent excess light coming through the sclera, which would decrease resolution of imagery.
3. The melanin granule's absorption spectrum maximizes the absorption of laser (Argon, Ruby, and Krypton) light to produce the photothermal reaction, which is the key to the laser photocoagulation process.

Metabolic-Biochemical

1. Phagocytizes rod and cone outer segments.
2. Metabolizes vitamin A (esterification, storage, and transport).
3. Synthesizes acid mucopolysaccharides which are involved in the exchange of metabolites in the subretinal space.
4. Transports numerous metabolites to and from visual cells and the choroidal circulation.
5. Contains numerous hydrolytic enzymes (lysosomes) for the degradation of "membranes" obtained from photoreceptor outer segments.
6. Contains enzymes for the synthesis and replacement of melanin granules (tyrosinase).

Transport (Ocular dipole; water flow from vitreous → choroid → sclera responsible for adhesion of neurosensory retina to RPE)

1. Ionic pump systems
   - HCO_3^- (carbonic anhydrase)
   - Na^+, K^+-ATPase (ouabain sensitive)
   - Mg^{+2}, Ca^{+2}-activated ATPase
   - Cl^-
2. Amino acid transport systems (neurosensory retina → RPE → choroid)
   - Taurine
   - Methionine
3. Vitamin A transport system (RPE → Photoreceptors)

microscopy, described a different "membrane" along the lateral walls of the RPE, which was later termed *Verhoeff's membrane*. Electron microscopic studies subsequently revealed a series of desmosomal connections along the lateral walls of the RPE, thereby clarifying the nature of Verhoeff's membrane. Kühne was the first to note that the RPE cells had processes that wrapped around the ends of photoreceptor cells (outer segments), but Iwaki (1958) stated that there was no direct contact between these two cell layers.

Kühne (1879) described cytoplasmic structures in the frog retinal pigment epithelium which he termed *myeloid bodies*. Arey (1928) depicted these structures, under the light microscope, as "colorless, colloidal bodies, spherical or cuboidal in shape and of wax-like luster . . ." Porter (1956), using the electron microscope, confirmed their existence and described their lamellar character. Further studies by Porter and Yamada (1960) indicated that the myeloid bodies of Kühne were PAS (periodic acid Schiff)-positive and stained with hematoxylin. These staining characteristics were identical to those of rod and cone outer segments. In polarized light, myeloid bodies were found to be birefringent (Majima 1925). The myeloid bodies of Kühne were, in fact, stacks of rod outer segment membranes and were termed *lamellated lipid bodies* (Porter and Yamada 1960).

On occasion, myeloid bodies were found to be associated with the nuclear envelope, endoplasmic reticulum, and large lipid bodies of the frog retinal pigment epithelium (Porter and Yamada 1960). However, there were species differences, which led Dowling and Gibbons (1962) to suggest that the term *myeloid body* be reserved for nonmammalian species and *lamellar body* for comparable structures in mammalian species. Young and Bok (1969) introduced the term *phagosomes* to denote that "pigment epithelial cell inclusion bodies are fragments of rod outer segments." More specifically, phagosomes are engulfed foreign bodies different in origin from cellular inclusions—for example, pigment granules produced by the RPE cells themselves (Young and Bok 1969). Dowling and Gibbons' (1962) studies suggested that this extensive array of cytoplasmic lamellar bodies, in the mammalian RPE, could be considered as a series of stages ranging from particles with a relatively simple structure to those with a highly complex lamellar structure.

Karli (1954a, b) was the first to show that the RPE could undergo phagocytosis. Droz (1963), using autoradiography, found evidence that the protein component of the rod outer segments was continually renewed and that the RPE was not the source of this protein. Droz's work raised the question of the fate of rod outer segments. Bairati and Orzalesi (1963) suggested that the photoreceptor outer disc segments were phagocytized by the RPE and subsequently digested by these cells. Young's (1967) studies indicated an orderly turnover (renewal and removal) of the outer segment discs and supported the concept of RPE phagocytosis. The highly ordered lamellar inclusions were the predecessors of the more simple and less organized particles. Feeney, Grieshaber, and Hogan (1965), using electron microscopy, reported "myeloid bodies" resembling fragments of the photoreceptor cell outer segment in the human RPE. These authors considered the fragments to be portions of rod and cone outer segments that were engulfed by the RPE. Spitznas and Hogan (1970) demonstrated RPE phagocytosis in the human eye.

Young and Bok (1969) presented unequivocal autoradiographic evidence, in frogs, that the RPE was phagocytizing rod outer segments. They used the term *phagosomes* for the inclusion bodies with rod material, and they noted that each body was surrounded by two separate cell membranes. The first or inner membrane was derived from the photoreceptor outer segment and the second or outer membrane was derived from the RPE. Young's (1971a, b) studies in the rhesus monkey indicated that each rod manufactured 80–90 new outer segment discs every 24 hours and that the number of discs phagocytized by an RPE cell daily was 2,000 in the parafovea, 3,500 in the perifovea, and about 4,000 in the periphery (see Chapter 9).

Other observations on the anatomy of the retinal pigment epithelium include the discovery of an acid mucopolysaccharide substance coating the outer segments of rods and cones in the subretinal space (Zimmerman 1958). Berman and Bach (1968) identified the polysaccharide and found it to be two-thirds hyaluronidase-sensitive, "nonsulfated" chondroitin sulfate and one-third hyaluronidase-resistant, nonsulfated heteropolysaccharide; the term *sialoglycans* was proposed for the latter compound. Streeten (1961) studied the sudanophilic granules of human RPE histochemically and found them to resemble the lipofuscins. These granules were later termed *residual bodies* and are believed to be the end products of phagosomal activity.

Donders (1854) was the first to fully describe drusen of Bruch's membrane in association with the RPE. Since that time, numerous theories have been proposed concerning the origin of these structures. However, only three theories have survived to date. Donders believed that the RPE was directly converted into drusen material; this was termed the *transformation theory*. Müller (1856) proposed the *deposition theory* whereby drusen were considered deposits secreted by an otherwise intact RPE. Friedman, Smith, and Kuwabara (1963) advocated a *vascular theory*, which suggested that drusen originated from constituents of blood. Lerche (1963, 1964) observed that drusen material initially accumulated within the RPE cell's basal infoldings and later filled the inner collagenous zone of Bruch's membrane. This finding supported both the deposition and transformation theories and was in agreement with the studies of Hogan (1967) and Spencer (1965). Farkas, Sylvester, and Archer (1971) confirmed this sequence of drusen formation and suggested that drusen formation was due to uncontrolled activity of lysosomal enzymes, causing cytolysis of the RPE. Drusen were, therefore, colloid bodies that contained lipid accumulations and other granular materials (debris) and represented, presumably, digestive waste products of the RPE (see Chapter 26).

General Topography

Wolff (1938), using the direct ophthalmoscope, was the first to appreciate that the retinal pigment epithelium was a monolayer of hexagonal cells. Since then, numerous workers have studied the anatomy of the RPE using a variety of techniques (Bairati and Orzalesi 1963; Breathnach and Wyllie 1966; Spitznas and Hogan 1970; Hogan, Alvarado, and Weddell 1971; Fine and Yanoff 1972; Lerche 1963a, b; Hogan 1972).

Histologically, the human RPE consists of a continuous monolayer of cells beginning at the edge of the optic nerve head and extending peripherally to the ora serrata. Anterior to the ora serrata, the RPE continues as the pigment epithelium of the pars plana (ciliary body). The RPE has Bruch's membrane as its basal border, with a blood supply derived from the subjacent choriocapillaris. At its apex, the RPE is in intimate contact with the outer segments of the photoreceptor cells. When viewed enface, as in a flat preparation, the RPE cells have a hexagonal shape (Fig. 1.1). They are tightly packed against one another with little extracellular space between neighboring cells.

In cross-section, the RPE cell varies in shape depending on its age (see Chapter 26) and location within the eye (Fig. 1.2). At the posterior pole in young adults the cells tend to be tall (14 μm) and narrow (10–14 μm) (Salzmann 1912; Tso and Friedman 1967), whereas at the ora serrata, the cells are flatter and wider (60 μm) (Hogan, Alvarado, and Weddell 1971), and have a cuboidal configuration. Within the intervening equatorial zone, the RPE cells vary more in size and shape than they do in the posterior pole and ora serrata regions. With aging, the preceding configurations of the RPE undergo change (see Chapter 26). Counts of the total number of RPE cells per human eye vary from 4.2 to 6.1 million (Hogan, Alvarado, and Weddell 1971) with larger numbers in older eyes. RPE cells have a brown color due to their melanin granules. Histological observations confirm that RPE cells are taller and contain more melanin within the macular zone than elsewhere. This accounts, in part, for the clinical observation that the macula has a darker appearance than other retinal tissue. Also contributing to this color differential is the pigment xanthophyl glycol, presumably located in Henle's fiber layer of the macula, which can act as a filter during fluorescein angiography.

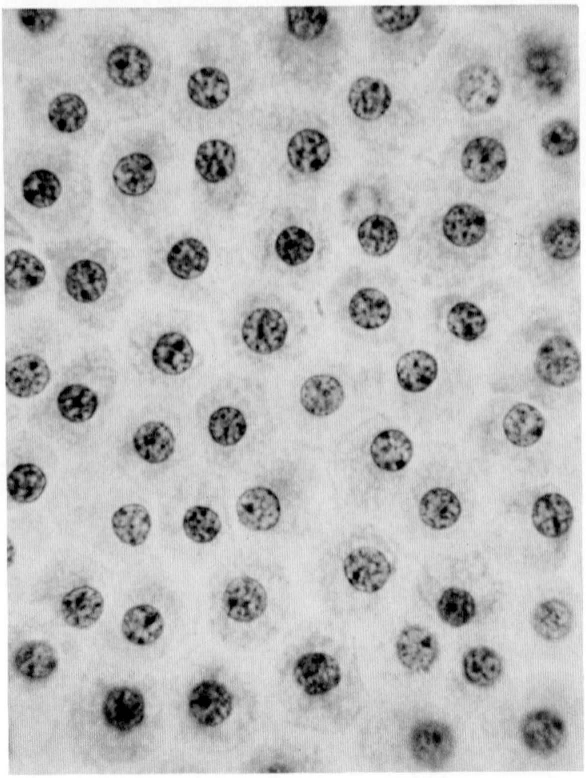

Figure 1.1*A* *Light micrograph of a flat preparation of adult human retinal pigment epithelium. The specimen was obtained from the posterior pole and was bleached with 10 percent hydrogen peroxide and stained with hematoxylin and eosin. Note hexagonal shape of the RPE as well as the round cell nuclei. Nucleoli are also present within the cell nucleus. (Courtesy of Mark O. M. Tso. ×900.)*

Topography of Cell Boundaries and Membranes

BASAL SURFACE. The RPE cell wall is a plasma membrane with the typical trilamellar unit membrane structure formed by two dense zones, each approximately 25Å thick, separated by a middle lucent zone of the same thickness. The basal plasma membrane is markedly convoluted (Figs. 1.3–1.5) and separated from its underlying basement membrane (cuticular layer of Bruch's membrane) by a distance of approximately 1,000Å. The basal infoldings extend 1 μm or more into the RPE cell cytoplasm and are present along the basal surface of the cell. Epithelial cells facing blood vessels often have basal infoldings for absorptive functions. Adjacent RPE cells are connected to each other by desmosome-like attachments (Fig. 1.6) near the basal border (Miki, Bellhorn, and Henkind 1975). Musser and Rosen (1973) found carbonic anhydrase activity in the basal infoldings of the RPE. The work of Steinberg and Miller (1973) suggests that HCO_3^- and Na^+ concentrations can affect the short circuit current of the RPE (see Chapter 12).

The RPE basement membrane has a fibrillar quality, with some fibrils extending into the inner portions of Bruch's membrane (Figs. 1.7 and 1.8). Recently, hemidesmosome-like connections have been observed (Fig. 1.7) between the basal surface of the RPE and its basement membrane (Miki, Bellhorn, and Henkind 1975). It has been suggested that these fine connections may account for the greater cohesion between the RPE and Bruch's membrane than between the RPE and the outer segments of the photoreceptor system. The basement membrane (inner cuticular layer of Bruch's membrane) ends at the optic nerve along with the RPE. How-

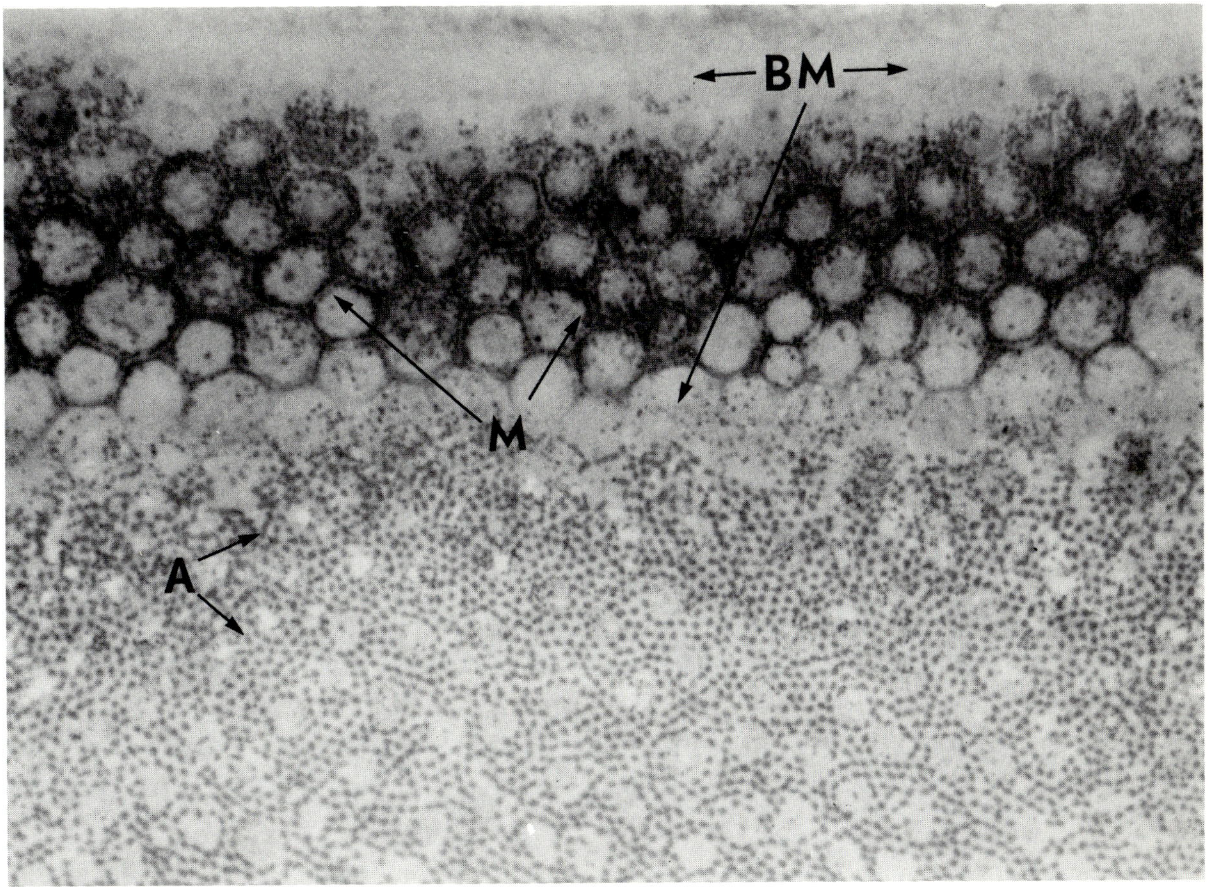

Figure 1.1*B* *Light micrograph of human RPE (macular region) in a frontal plane, demonstrating the hexagonal shape of the cell borders. Melanin granules* (M) *are located within the peripheral cytoplasm adjacent to the cell walls. Bruch's membrane* (BM) *has a relatively homogenous appearance to it. At the apex* (A) *of the RPE cell there is a more diffuse distribution of the melanin granules. The cell nuclei are barely visible in this section. (×570.)*

ever, at the ora serrata, the basement membrane continues as the basement membrane of the pigment epithelium of the pars plana (ciliary body).

BRUCH'S MEMBRANE. Carl Bruch (1844), an anatomist in Zurich, who later became a faculty member at the University of Heidelberg (1855), was the first to demonstrate a "structureless membrane" between the retina and choroid using only a fine brush and a flat knife (Duke-Elder 1961). The "structureless membrane" was later termed Bruch's membrane; it separates the basal surface of the RPE from its blood supply, the choriocapillaris. Early histologists called this layer the *lamina vitrea* because of its resemblance to a glass membrane. However, Wolfrum (1908) noted that Bruch's membrane was composed of a thin inner basal membrane close to the RPE that was separated from an outer elastic layer by a loose layer of collagen. For these reasons Wolfrum did not consider Bruch's membrane to be a true "glass membrane" like Descemet's membrane. Wolter (1955) described the collagen fibers of Bruch's membrane as having a brush-like arrangement, with the fibers radiating from the capillary and intercapillary regions toward the RPE.

The earliest electron microscopic studies of Bruch's membrane were carried out by Sumita (1961), Lerche (1963b, 1965), Garron (1963), and Hogan (1961). These studies have established the pentalaminar structure of Bruch's membrane (layer) as follows (Fig. 1.3):

1. *The basement membrane* of the retinal pigment epithelium is composed of fine filaments

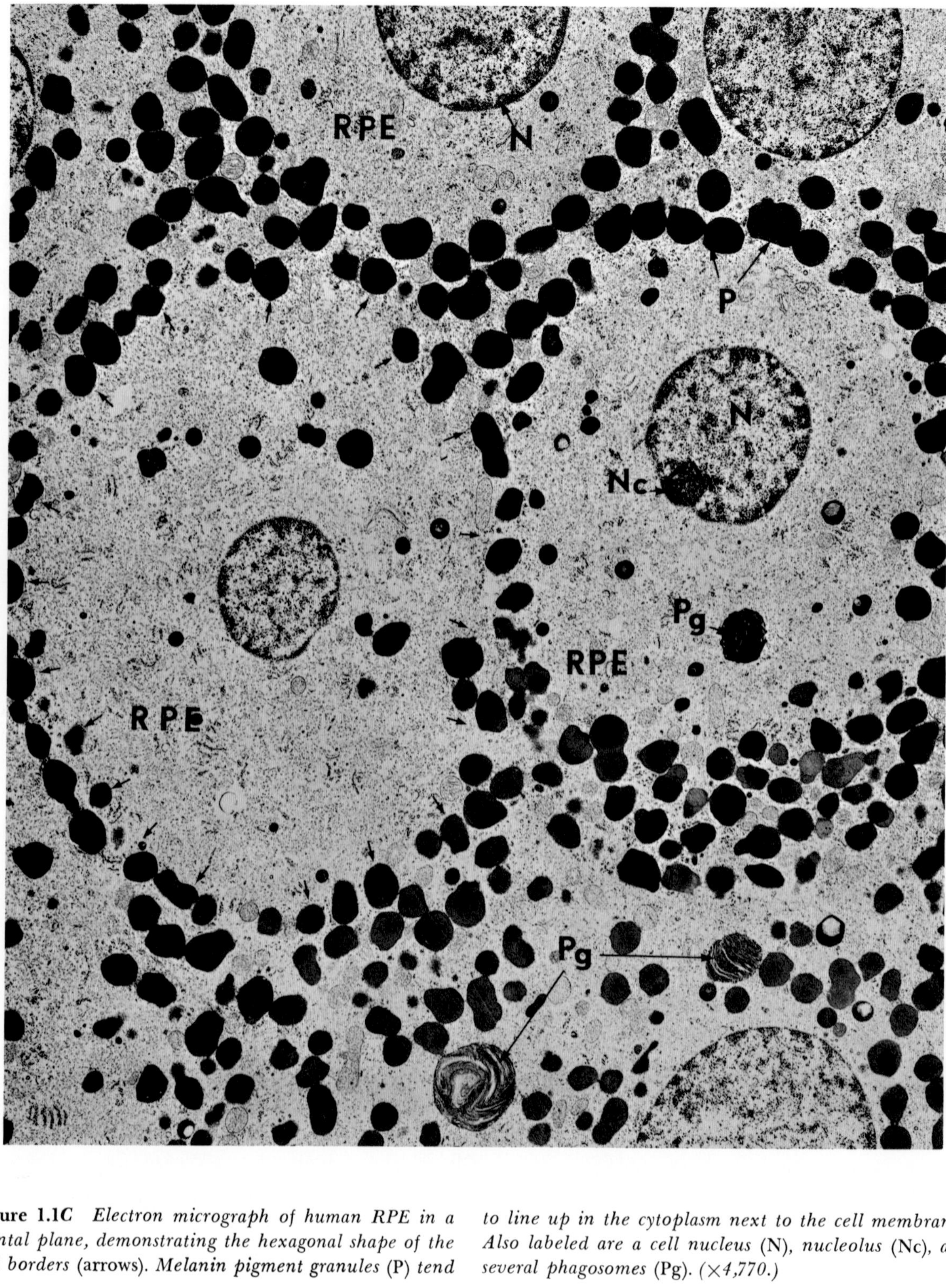

Figure 1.1C *Electron micrograph of human RPE in a frontal plane, demonstrating the hexagonal shape of the cell borders* (arrows). *Melanin pigment granules* (P) *tend to line up in the cytoplasm next to the cell membranes. Also labeled are a cell nucleus* (N), *nucleolus* (Nc), *and several phagosomes* (Pg). *(×4,770.)*

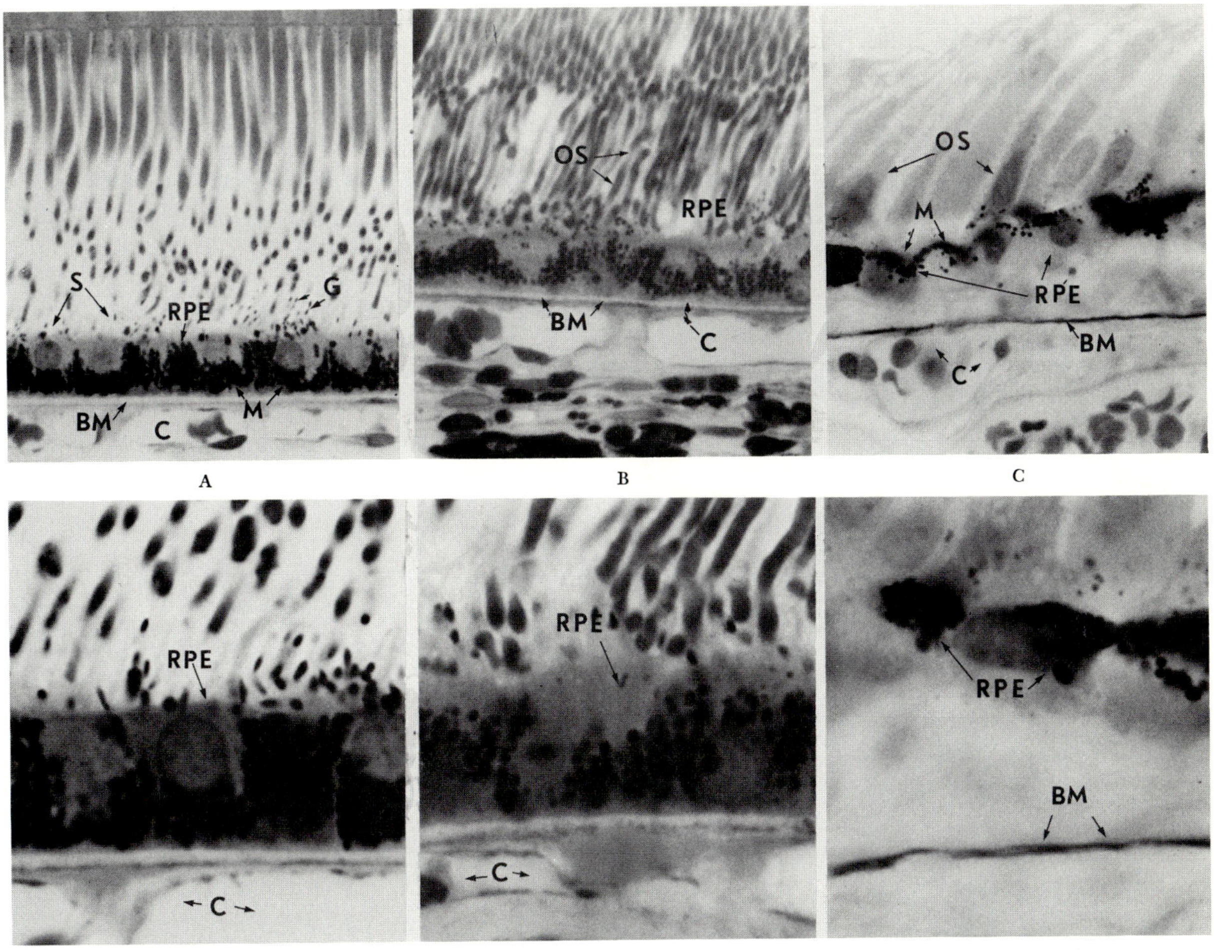

Figure 1.2*A* *Light micrograph of the macular region of adult human RPE in a transverse plane. Melanin granules* (M) *are located mostly at the base of the RPE and along the inner aspect of the cell walls. Note the finely pigmented granules* (G) *in the microvillous processes projecting into the subretinal space* (S) *and surrounding the outer segments of the cone cells. The RPE is resting on Bruch's membrane* (BM). *The choriocapillaris has a very wide lumen* (C) *accommodating several red blood cells at one time.* (Above, ×*570.* Below, ×*1,430.)*

Figure 1.2*B* *Light micrograph of the equatorial region of adult human RPE in a transverse plane. The RPE cells are beginning to lose their highly ordered shape and the cells are shorter than the RPE cells in the macular region. Note the photoreceptor outer segments* (OS), *large lumen of the choriocapillaris* (C) *and Bruch's membrane* (BM). (Above, ×*570.* Below, ×*1,430.)*

Figure 1.2*C* *Light micrograph of the ora serrata region of adult human RPE in a transverse plane. The RPE has a more irregular or amorphous shape in this region and is flatter than in the macular and equatorial regions. The melanin granules* (M) *are clumped together. Note the photoreceptor outer segments* (OS) *abutting obliquely against the apical surface of the RPE and the large lumen choriocapillaris.* (Above, ×*570.* Below, ×*1,430.)*

that extend from its inner and outer surfaces toward the RPE and adjacent collagen layer, respectively. The basement membrane is about 0.3 μm thick (Fig. 1.5) and separated from the basal cell membrane by a space of approximately 1,000Å (Hogan, Alvarado, and Weddell 1971). The basement membrane is elaborated by the RPE and does not enter the cell's basal infoldings (Figs. 1.7 and 1.8).

2. *The inner collagenous zone* is composed of collagen fibrils with a cross-sectional diameter of 600Å and a periodicity of 640Å. The collagen fibrils are interwoven in a loosely arranged meshwork (Figs. 1.3, 1.5, and 1.9). The

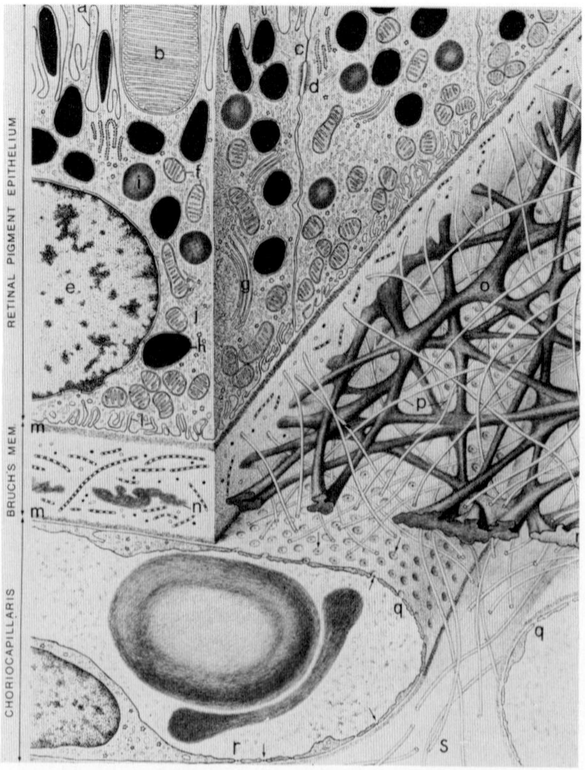

Figure 1.3 *Three-dimensional drawing of the inner choroid and retinal pigment epithelium. The villi of the pigment epithelium* (a) *extend internally to enclose the outer segments of the rods and cones* (b). *The intracellular junctions are characterized by a zonula occludens* (c) *and a desmosome* (d); *otherwise, the cell relations are of the usual type. The cytoplasm of the pigment cells contains a nucleus* (e), *mitochondria* (f), *a Golgi apparatus* (g), *pigment granules* (h), *and phagosomes* (i), *and it is characterized by a large amount of smooth-surfaced endoplasmic reticulum* (j). *The external cell membrane shows complex infoldings* (l) *and a basement membrane* (m). *Bruch's membrane shows an apparently interrupted elastic zone* (n) *in meridianal section, but the elastica is layered and continuous in flat section* (o). *Collagen fibrils* (p) *that form the inner and outer collagenous zones have a random orientation around the elastic zone. The choriocapillaris* (q) *shows fenestrated endothelium internally, laterally, and, to a lesser extent, externally* (r). *The intercapillary zone shows considerable collagen* (s). *The lumen of the capillary contains two red blood cells. (From Hogan, Alvarado, and Weddell 1971.)*

major axes of the collagen fibrils are oriented in a plane parallel to the plane of the outer retina, yet the fibers are running in all directions within this plane. Fine filaments emanating from the basement membrane of the RPE merge with the collagen fibrils and are probably, in part, responsible for the strong adhesion of the RPE to the choroid. The collagen in this layer also appears to be continuous with the collagen in the outer collagenous layer. The fibrils run through the meshwork of the elastic layer. The collagen fibrils appear to be embedded in a matrix or ground substance, which is presumably a mucopolysaccharide-protein complex (Hogan, Alvarado, and Weddell 1971). No cellular elements have been observed in this layer and that includes nerve fibers as well. The average thickness of the inner collagenous zone is about 1.5 μ.

3. *The middle elastic layer* has been termed the "backbone" of Bruch's membrane (Hogan, Alvarado, and Weddell 1971) and is about 0.8 μ thick (Figs. 1.3 and 1.9). The layer is continuous from the optic nerve to the pars plana of the ciliary body. This layer is composed of elastic fibers that vary considerably in size and shape. Generally, the fibers are long and thin, forming a grillwork two to four fibers thick. Collagen fibrils from the inner and outer collagenous zones pass through the polygonal spaces in this interlacing elastic network (Fig. 1.9).
4. *The outer collagenous zone* is similar in structure to the inner collagenous zone. The collagen fibrils have a 600Å periodicity (Figs. 1.3 and 1.9). The fibrils are embedded in a matrix that is presumably made of a mucopolysaccharide protein material. The fibrils have a similar orientation to those found in the inner collagenous zone and communicate with those fibers through spaces in the elastic layer. The outer collagenous zone is 0.7 μ thick and is thus thinner than the inner collagenous zone. In the intercapillary regions, the outer collagenous fibrils become continuous with the choroidal stroma and subcapillary tissue (Fig. 1.10). Under normal conditions, there are rarely any cellular elements, including nerve fibers, present in the layer.
5. *The basement membrane of the choriocapillaris* is the outermost layer of Bruch's membrane and consists of a basement membrane elaborated by the endothelium of the choriocapillaris (Figs. 1.3, 1.9, and 1.10). This layer is 0.14 μm thick (Hogan, Alvarado, and Weddell

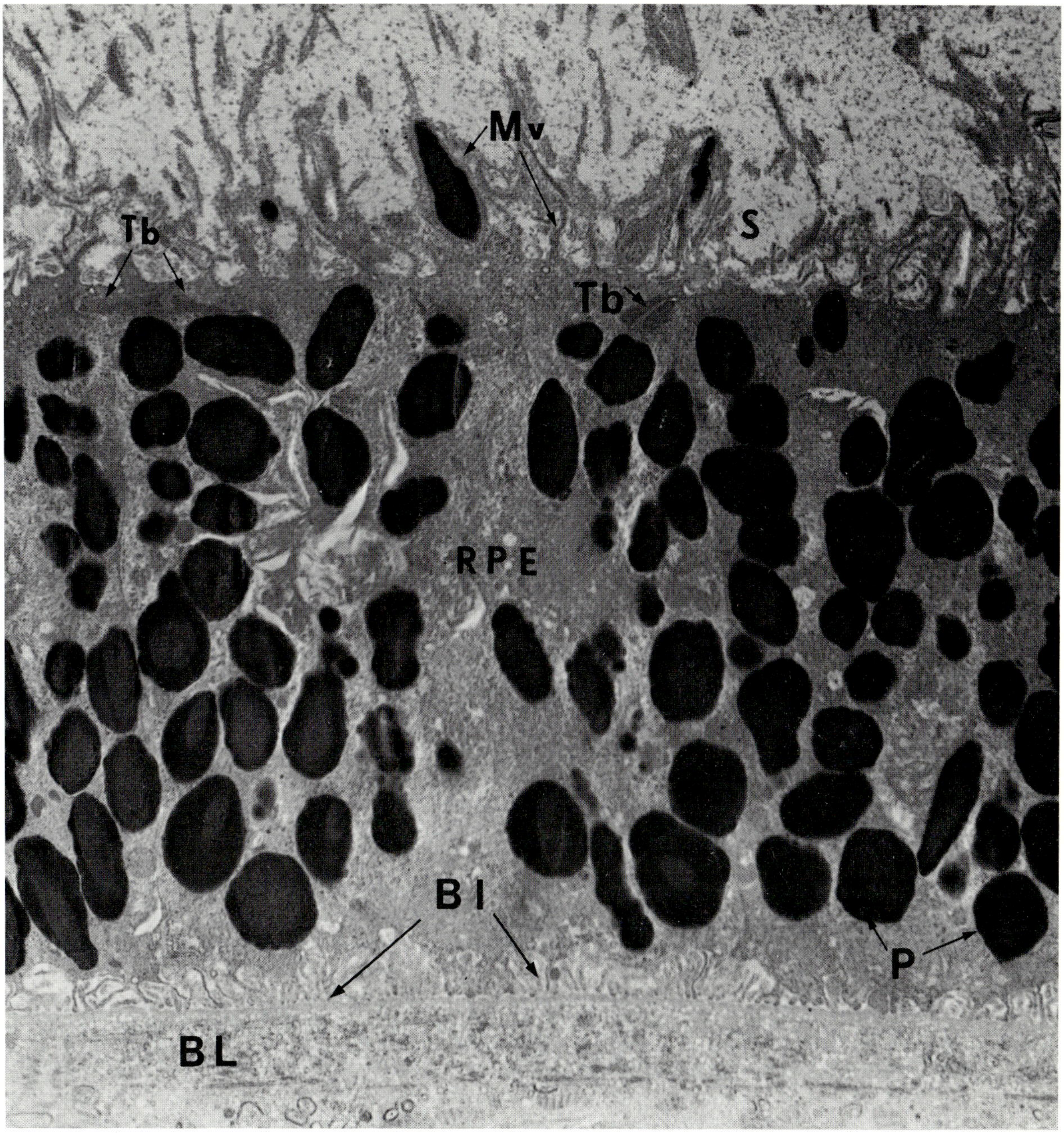

Figure 1.4 *Electron micrograph of the human RPE in a transverse plane. Note the numerous basilar infoldings* (BI) *of the cell membrane at the basilar surface. Also labeled are Bruch's membrane or layer* (BL), *melanin granules* (P), *subretinal space* (S), *microvilli at apex of RPE* (MV), *and the terminal bar* (Tb). *(×8,930.)*

1971) and is therefore thinner than the basement membrane of the RPE. The basement membrane of the choriocapillaris is not a continuous membrane. It is discontinuous at the intercapillary zones (Fig. 1.10).

Bruch's membrane contains open spaces in the collagenous and elastic zones and the basement membrane of the choriocapillaris is discontinuous in the intercapillary zones (Figs. 1.3, 1.5, 1.8, and 1.10). Therefore, Bruch's membrane does not serve as

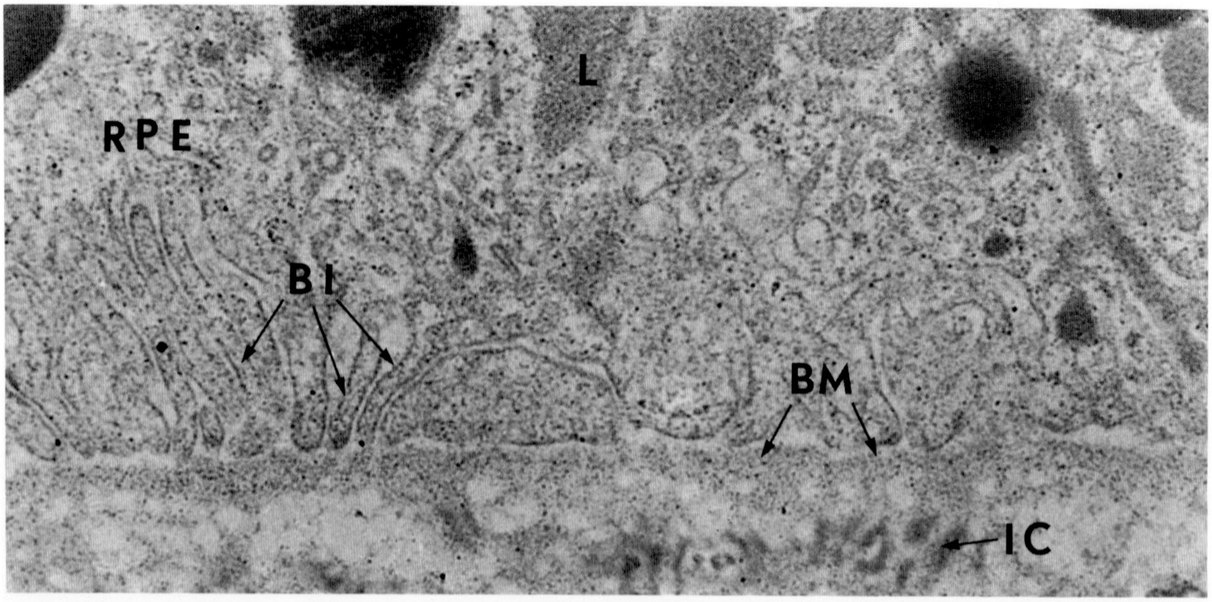

Figure 1.5 *Electron micrograph of the human RPE in a transverse plane, showing the basilar region with highly convoluted basal infoldings* (BI) *resting on a basement membrane* (BM), *which is the innermost zone of Bruch's membrane. Also labeled is the inner collagenous zone* (IC) *of Bruch's membrane and a lysosome* (L). *(×52,500.)*

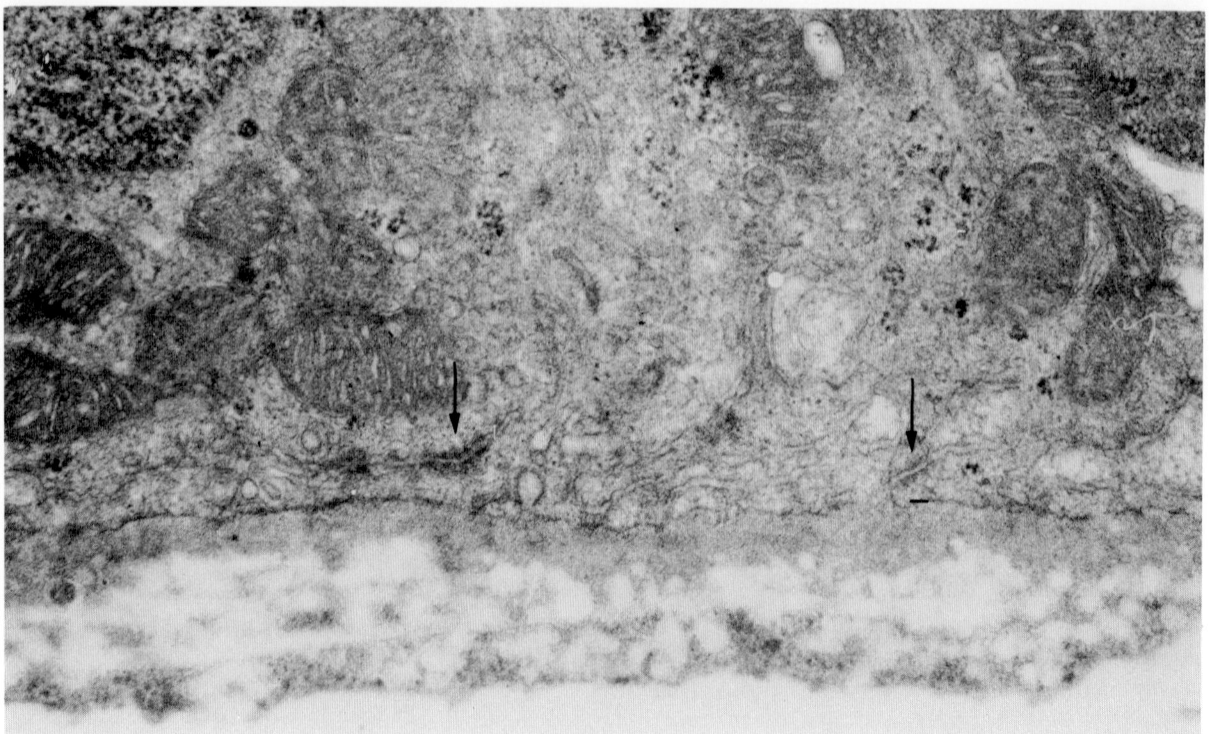

Figure 1.6 *Electron micrograph of the basilar region of the human RPE in a transverse plane. The arrows indicate desmosome connections between adjacent RPE cells in basilar region. (From Miki, Bellhorn, and Henkind, 1975. ×39,000.)*

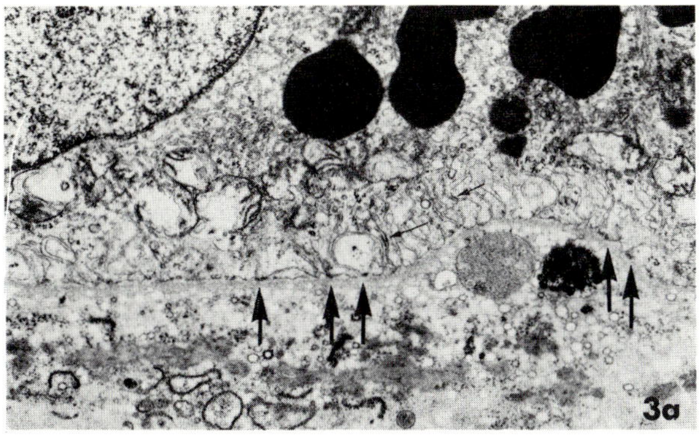

Figure 1.7 *Electron micrograph of the basilar region of the human RPE in a transverse plane. Hemidesmosome connections* (arrows) *indicate spots where the RPE is anchored to the basement membrane (cuticular layer of Bruch's membrane). (From Miki, Bellhorn and Henkind, 1975. ×10,360.)*

a barrier to macromolecules passing between the choriocapillaris and the RPE. Bruch's membrane is thicker (2–4 μ) in the optic nerve region than in the retinal periphery (1–2 μ). With advancing age, there is a progressive alteration of its substructures—that is, collagen and elastic fibers—with retention of the RPE digestive and waste products in Bruch's membrane (Fig. 1.11), leading to increased thickness, PAS-positive staining and drusen formation (see Chapter 26).

LATERAL SURFACE. The lateral cell walls of the RPE are fairly smooth, with only occasional interdigitations compared to the highly convoluted basilar surface (Fig. 1.3). The terminal bar complexes are located at the apical end of the lateral wall and consist of a zonula occludens and zonula adherens. The former structure is more apically

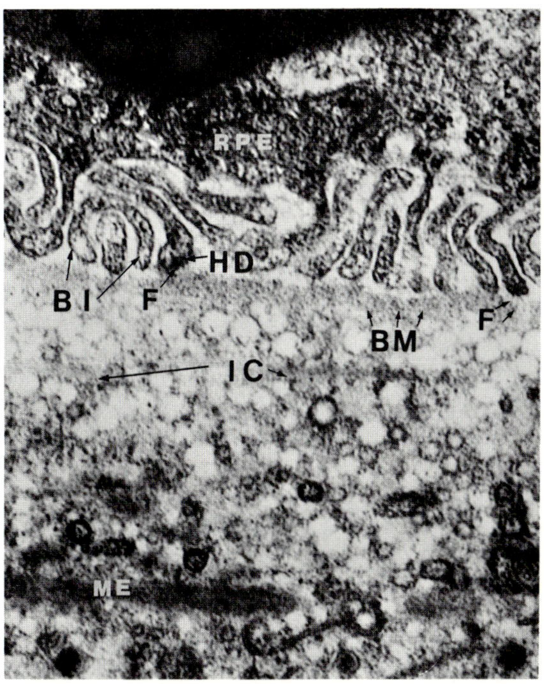

Figure 1.8 *Electron micrograph of the basilar region of the human RPE in a transverse plane. Hemidesmosome connections* (HD) *originate at the basal infoldings* (BI) *and send fine filaments* (F) *into the subjacent basement membrane* (BM) *to help anchor the RPE to Bruch's membrane. Also labeled are the inner collagenous* (IC) *and middle elastic zones* (ME) *of Bruch's membrane. (×38,500.)*

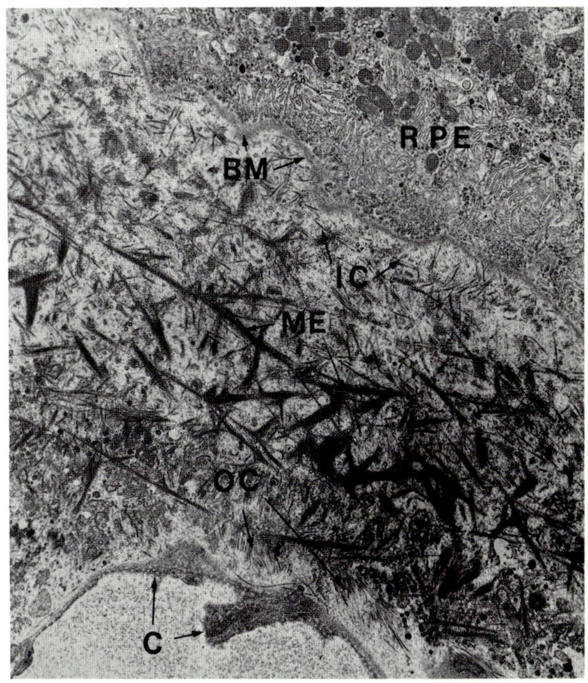

Figure 1.9*A* *Transmission electron micrograph of human RPE in an oblique section of Bruch's membrane, shown at low magnification. Numerous mitochondria* (M) *are present at the base of the cell. Bruch's membrane, or layer, consists of a basement membrane* (BM) *elaborated by the RPE, an inner collagenous zone* (IC), *a middle elastic zone* (ME), *an outer collagenous zone* (OC), *and a basement membrane* (BM_c) *derived from the choriocapillaris endothelium* (C). *(×4,020.)*

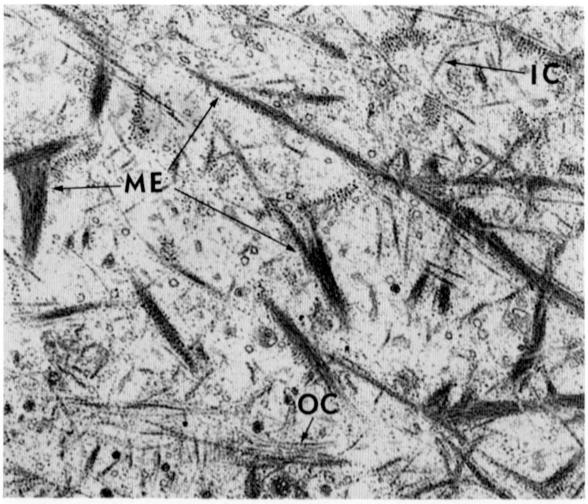

Figure 1.9*B* *Part of Figure 1.9A, shown at high magnification. Inner collagenous zone fibrils* (IC), *middle elastic zone fibers* (ME), *and outer collagenous zone fibrils* (OC) *are present. Open spaces can be seen in the middle elastic zone through which collagen fibrils from the outer and inner collagenous zones traverse and communicate.* (×*16,750.*)

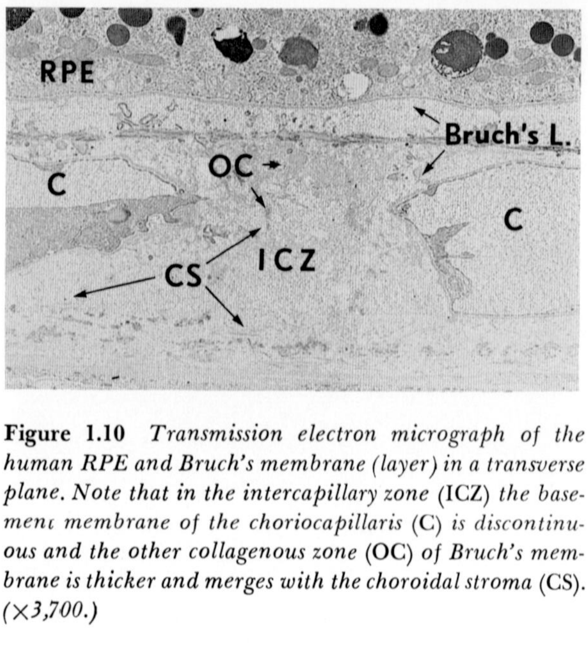

Figure 1.10 *Transmission electron micrograph of the human RPE and Bruch's membrane (layer) in a transverse plane. Note that in the intercapillary zone* (ICZ) *the basement membrane of the choriocapillaris* (C) *is discontinuous and the other collagenous zone* (OC) *of Bruch's membrane is thicker and merges with the choroidal stroma* (CS). (×*3,700.*)

situated than the latter. The terminal bar complex completely encircles the apical portion of the lateral wall and also joins neighboring RPE cells (Figs. 1.12 and 1.13). The zonula occludens is a "tight" intercellular junction characterized by the fusion of adjacent RPE cell membranes, whereas the zonula adherens is an intercellular junction that has a space of approximately 200Å between the cell membranes. In effect, the terminal bar seals off the lateral intercellular space from the apical (subretinal) space (Fig. 1.13) and is responsible for the barrier that prevents the exchange of macromolecules between the photoreceptors and the choroidal circulation, for example, choriocapillaris (Shakib, Rutkowski, and Wise 1972).

Characteristically, "tight" junctions of epithelial cells have a deep zonula occludens that has a substructure in which there are at least five junctional strands bridging the gap, forming a complex mosaic pattern (Claude and Goodenough 1973). Electrically, cells in a single layer held together by tight junctions have a high shunt resistance compared to total cellular resistance (Miller and Steinberg 1977) (see Chapter 12). Accordingly, the RPE fills the criteria for true "tight" junctions within its terminal bar complexes.

Desmosomes may, on occasion, be found along the lateral walls of the RPE (Figs. 1.3, 1.6, and 1.12) and can act as cellular bridges or "spotwelds." Desmosomes have two distinctive features, an electron-dense plate of intercellular material parallel to the cell membrane and numerous tonofilaments in the RPE cell cytoplasm whose long axes are at right angles to the junctional plane within the cytoplasm. The tonofilaments end in the electron-dense cytoplasmic disc or bar that is oriented parallel to the cell membrane. However, desmosomes are not continuous adhesions, but rather discontinuous attachments that have some space in them and therefore do not act as barriers for macromolecules or ions, in contradistinction to terminal bar complexes.

APICAL SURFACE (SUBRETINAL SPACE). The apical surface of the RPE faces the subretinal space in which the outer segments of the photoreceptor cells (rods and cones) are located. (See Chapter 2 for a detailed discussion of the interrelationship between the RPE and the photoreceptor [rod and cone] outer segments in the human.) The subretinal space represents part of the original space within the optic vesicle derived from an early period in embryological development. When there is a neurosensory retinal detachment, excessive fluids collect in the subretinal space. Scanning electron microscopic studies (Sakuragawa and Kuwabara 1976) and transmission electron microscopic studies (Bairati and Orzalesi 1963; Spitznas and Hogan 1970; Hogan, Alvarado, and Weddell 1971; and Steinberg, Wood, and Hogan 1977) have clarified many details about

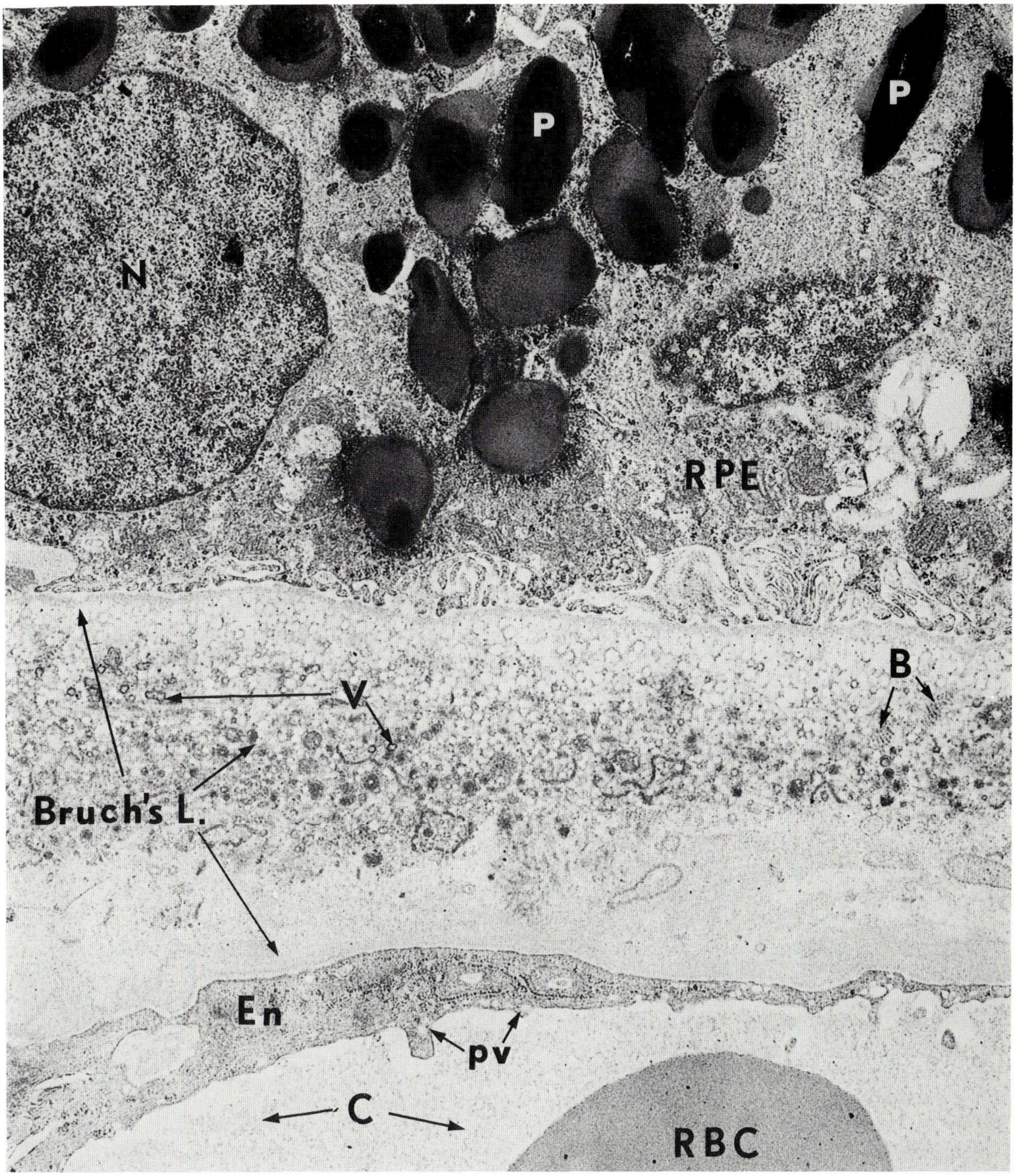

Figure 1.11 *Transmission electron micrograph of the human RPE and Bruch's membrane in a transverse plane. Note the numerous vesicles* (V) *and banded structures* (B) *in the mid-zone of Bruch's membrane (layer), which are digestive waste products of the RPE. There is a red blood cell* (RBC) *in the lumen of the choriocapillaris* (C). *An endothelial cell* (En) *lines the choriocapillaris and some cytoplasmic pinocytotic vesicles* (pv) *are present in it. Also labeled are melanin pigment granules* (P) *and the RPE cell nucleus* (N). *(×13,900.)*

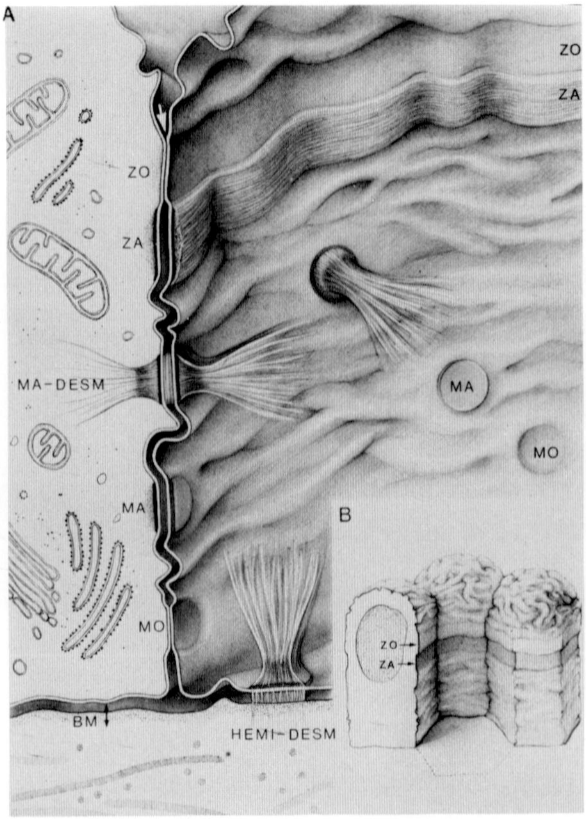

Figure 1.12 *Intercellular junctions. The large view* (A) *shows the intercellular junctions as seen from the interior of the cell. Two beltlike junctions appear near the apex of the cells. One is a zonula occludens* (ZO) *that completely obliterates the intercellular space, and which is characterized by fusion of the outer leaflets of the two contiguous cells* (arrow). *The second is a zonula adherens* (ZA), *in which the intercellular space is unchanged and the contiguous cell membranes are strictly parallel. Cytoplasmic fibrils are found along this junction. A macula adherens, desmosomal-type* (MA-DESM), *contains an intercellular cement substance; numerous tonofilaments are attached to an electron-dense, platelike structure that is closely apposed to the cell membrane. These three junctions are arranged, in the epithelia of many glands, to form a terminal bar or junctional complex. A macula adherens* (MA) *is shown in cross-section and in a "head-on" view. Fine fibrils are associated with it, and it has a buttonlike shape. A macula occludens* (MO) *appears near the bottom of the intracellular space, and since this is a tight junction, the adjacent cell membranes are fused. In a head-on view the macula occludens appears as a round, buttonlike depression. A hemi-desmosome* (HEMI-DESM) *is seen along the basal portion of the cell. Fine fibrils appear to attach the hemi-desmosome to the underlying basement membrane* (BM). *The basement membrane, or basal lamina, is composed of an electron-lucent layer immediately adjacent to the cell membrane and an electron-dense layer consisting of fine filaments. The basement membrane filaments continue into the underlying connective tissue. The insert* (B) *shows that the zonula occludens and the zonula adherens actually form a continuous structure around the circumferences of two cells. (From Hogan, Alvarado and Weddell 1971.)*

the three-dimensional structural relationships between the apical RPE surface and the outer segments of rods and cones in the subretinal space (see Chapter 2).

There are two types of microvilli (Figs. 1.3 and 1.13–1.18) found at the apical surface of the RPE: long microvilli (5–7 μm long), which lie principally between the outer segments of the photoreceptors; and shorter microvilli (3–4 μm long), which form sheaths (sheet-like apical processes) around the ends of the photoreceptor outer segments (Figs. 1.14 and 1.18). These shorter microvilli surround the outer one-third of the outer segment processes of rods and presumably of foveal cones. Rod and foveal cone outer segments are ensheathed at their distal ends by the sheet-like apical processes (Steinberg, Wood, and Hogan 1977). It is these processes that are involved in the phagocytosizing of rod and cone outer segments (Hogan, Wood, and Steinberg 1974). There are no junctional complexes present between the RPE apical processes and the tips of the sheathed outer segments.

The long villous projections (long microvilli) are found to extend between rod outer segments. The extrafoveal cone outer segments are ensheathed by the long villous processes of the RPE (Fig. 1.13*A*). The outer segments of the cone do not reach the apical surface of the RPE. However, as the villous

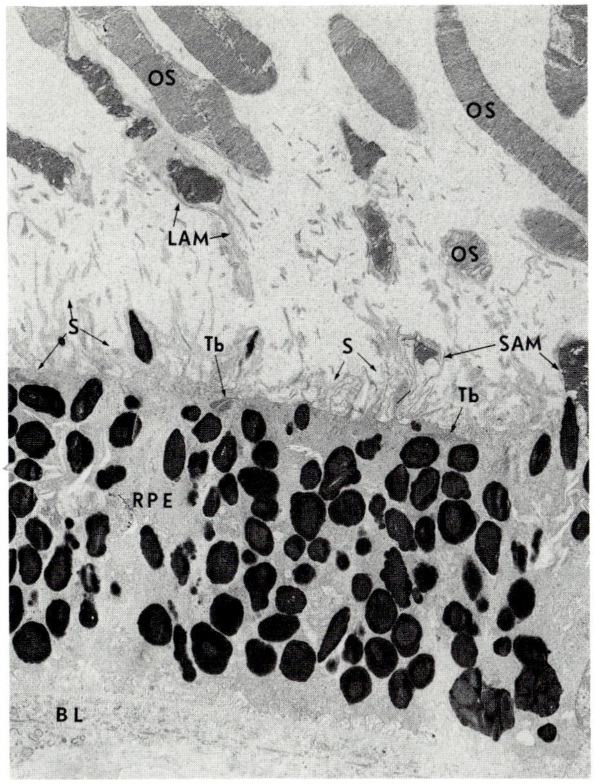

A

Figure 1.13*A* *Transmission electron micrograph of the human RPE: transverse plane through the RPE showing the subretinal space* (S) *at the apex of a cell. The terminal bar* (Tb) *complex is located at the apex of the RPE cell. Outer segments* (OS) *of photoreceptor cells are present. Also labeled are Bruch's Layer* (BL), *long apical microvilli* (LAM), *and short apical microvilli* (SAM). *(×3,500.)*

process of the RPE reaches the cone outer segment, it expands and becomes sheet-like, ensheathing the distal half of the cone (see Chapter 2). Villous processes that surround rods contain smooth endoplasmic reticulum, ribosomes, melanin granules, and possibly actin filaments (see Chapter 10). Villous processes that surround extrafoveal cones vary in length and are usually devoid of intracellular organelles except for pigment granules. On occasion, an individual villous process may contain smooth endoplasmic reticulum, ribosomes, mitochondria and primary lysosomes.

Young (1971a) estimates that 30–45 photoreceptor outer segments engage each RPE cell. In the foveal region, the photoreceptors are exclusively cones, whereas in the retinal periphery they are mostly rods with only 2–4 cones interdigitating with each RPE cell. On the other hand, Steinberg and Wood estimate that as many as 30–40 RPE villous processes are associated with a single cone outer segment (see Chapter 2). In contrast, only several ensheathing villous processes from the apical RPE cell are associated with a single rod outer segment.

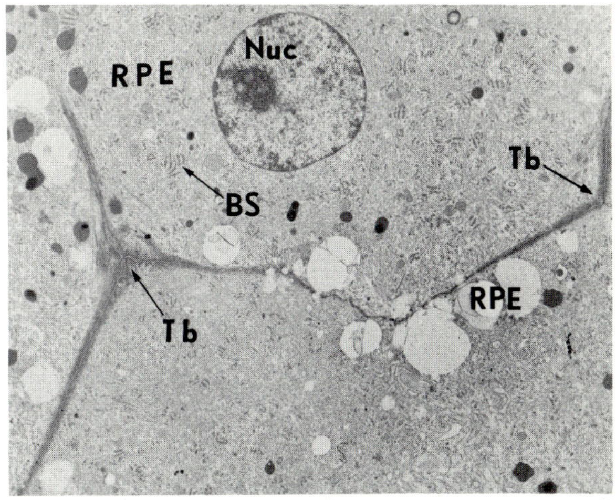

B

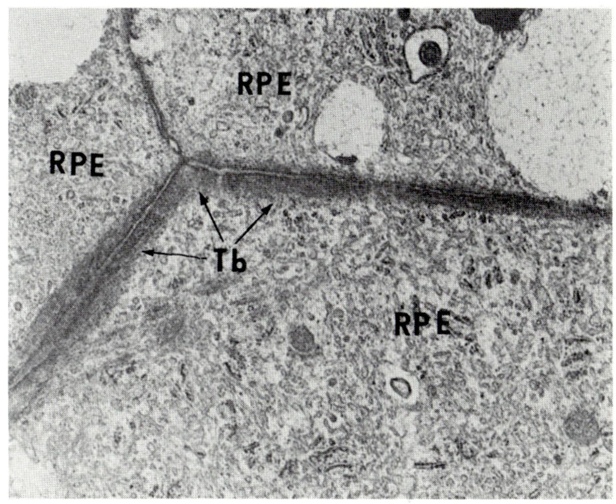

C

Figure 1.13*B* *Transmission electron micrograph of the human RPE: frontal plane through RPE at apex of cell illustrating terminal bar* (Tb) *connections among neighboring cells. Banded structures* (BS) *may be the endoplasmic reticulum, and the cell nucleus is shown* (Nuc). *(×6,200.)*

Figure 1.13*C* *Transmission electron micrograph of the human RPE: frontal plane through RPE at apex of cell illustrating terminal bar connections* (Tb) *between adjacent cells. (×13,300.)*

Sidman (1958) and Zimmerman (1958) noted an acid mucopolysaccharide material coating the apical surface of the RPE as well as the rod and cone outer segments. Berman and Bach (1968) identified the

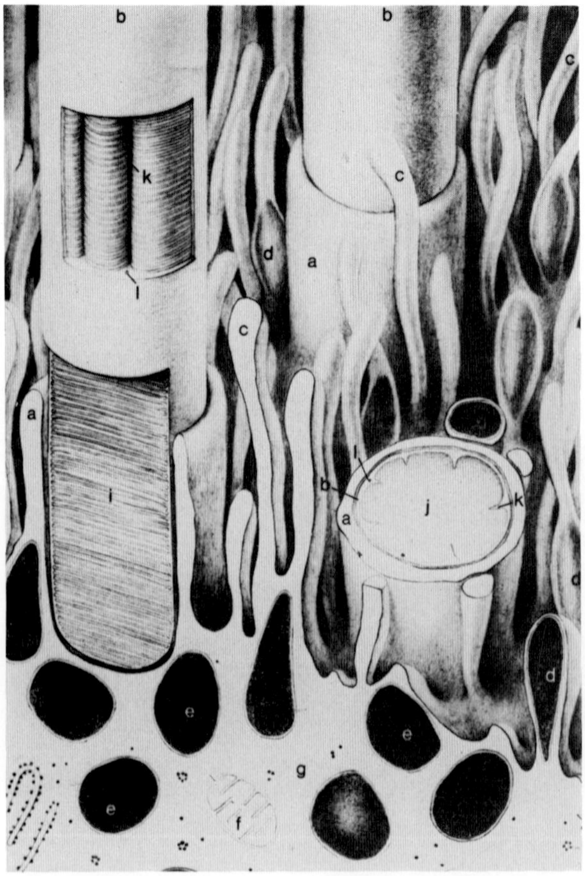

Figure 1.14 *A three-dimensional drawing of the relation of the outer segments of the rods to the retinal pigment epithelial cells. Thick sheaths* (a) *of the pigment epithelium enclose the external portions of the rod outer segments* (b). *Numerous fine finger-like villous processes* (c) *are found between the photoreceptors; they contain pigment granules* (d). *The apical portion of the pigment epithelial layer of cells is seen at the bottom. It contains numerous pigment granules* (e), *mitochondria* (f), *a well-developed, smooth-surfaced, endoplasmic reticulum* (h), *and scattered free ribosomes. The stacks of the rod outer segment discs are depicted in meridianal section at* (i) *and in cross-section at* (j). *There is scalloping of the periphery of the discs* (k). *Microtubules originating in the basal body of the rod cilium extend externally into the outer segment; one such microtubule is shown in cross-section at* (l). *(From Hogan, Alvarado, and Weddell 1971.)*

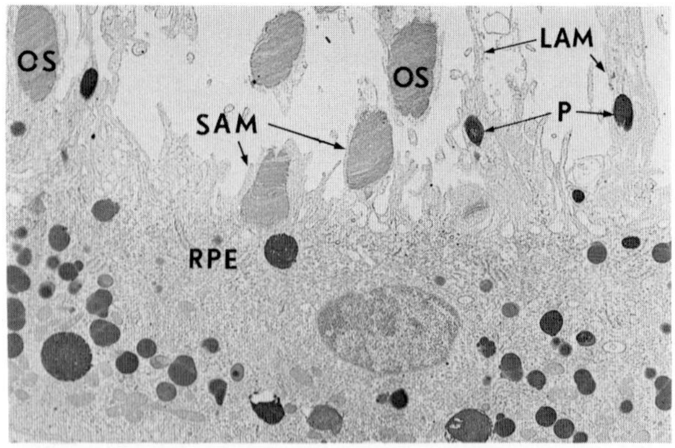

Figure 1.15 *Transmission electron micrograph of the human RPE in a transverse plane. Long, apical microvilli* (LAM) *eventually surround outer segments of photoreceptor cells. Short apical microvilli* (SAM) *form sheaths around the ends of the rod outer segments* (OS). *Also labeled is melanin pigment* (P) *in long apical microvilli. (×2,360.)*

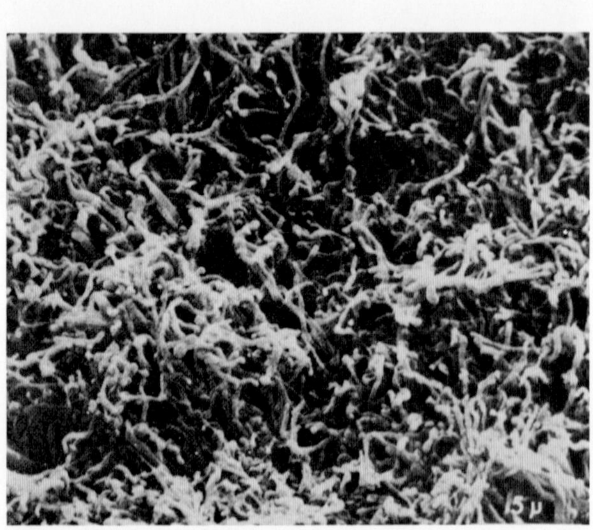

Figure 1.16 *Scanning electron micrograph of apical surface of rhesus monkey RPE. Innumerable fine microvilli from a relatively even surface about five to six cells are present in this area. No individual cells are recognizable. (From Sakuragawa and Kuwabara 1976. ×2,520.)*

material as a polysaccharide, two-thirds of which was hyaluronidase-sensitive and the remainder hyaluronidase-resistant—the former a "half-sulfated" chondroitin sulfate, and the latter a nonsulfated heteropolysaccharide (sialoglycans). Zimmerman and Eastham (1959) demonstrated that the RPE was responsible, in part, for the production of this acid mucopolysaccharide material. Additional studies (Berman 1964; Berman and Bach 1968; Tesoriere et al. 1971; Young 1973) provided detailed biochemical and autoradiographic evidence that the RPE probably contributes to the mucopolysaccharide interphotoreceptor matrix. However, Hall, Ocumpaugh, and Young (1965) showed that the

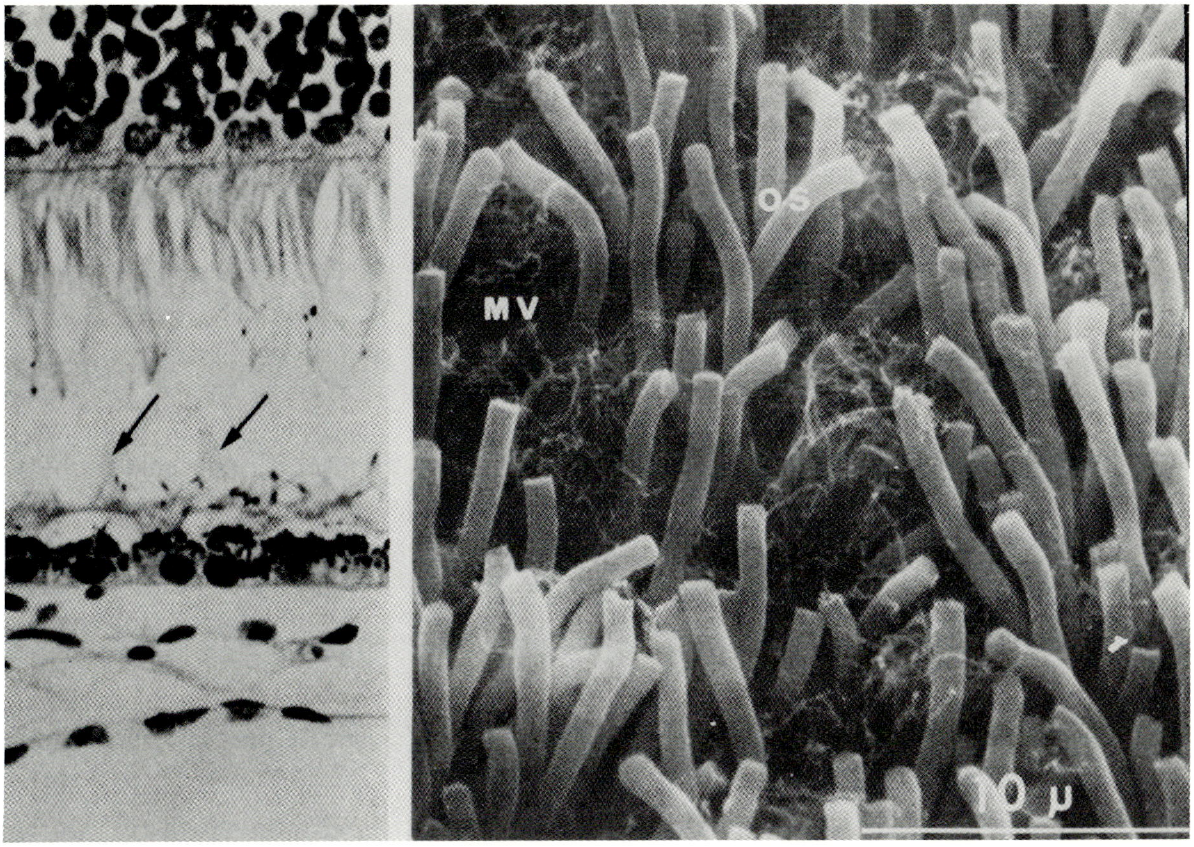

Figure 1.17 *Scanning electron micrograph of rhesus monkey RPE, showing artificial separation by breakage of outer segments.* Left, *broken outer segments in paraffin section* (arrows), *stained with hematoxylin eosin.* Right, *view of surface of pigment epithelium after removal of retina following formalin fixation. Pillar-like outer segments* (OS) *are firmly embedded in microvilli* (MV) *of pigment epithelium. (From Sakuragawa and Kuwabara 1976.* Left, *×280.* Right, *×3,000.)*

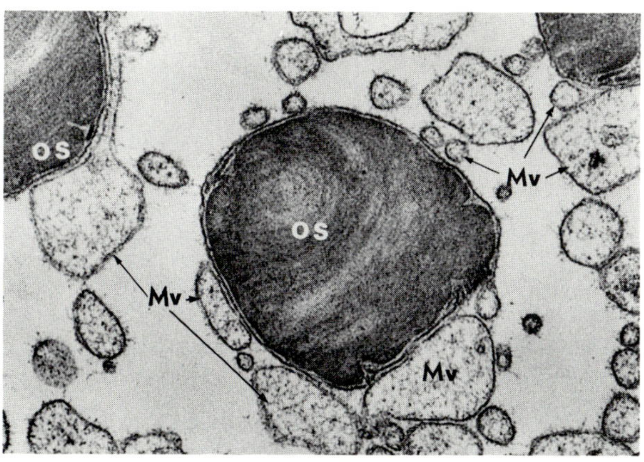

Figure 1.18 *Transmission electron micrograph of the human RPE and outer segment of a photoreceptor cell in a frontal plane. Note the apical microvilli* (Mv) *adjacent to the outer segment* (OS). *(×24,400.)*

inner segments of the photoreceptors could also synthesize sulfated mucopolysaccharides, which, in turn, were secreted to the outer segments (Ocumpaugh and Young 1966), thereby also contributing to the interphotoreceptor matrix.

It was postulated by de Guillebon and Zauberman (1972) that the interphotoreceptor matrix—that is, the acid mucopolysaccharides—might act as a glue to maintain the normal adhesion of the neurosensory retina to the RPE. However, the adhesive forces that might be generated by the viscoelastic properties of the acid mucopolysaccharides are not sufficient to account for the total adhesive force present between neural retina and the RPE (Zauberman, de Guillebon, and Holly 1972; Zauberman and Berman 1969) (see Chapter 11). Fatt and Shantinath (1971) confirmed this finding indirectly by calculating the hydraulic drag of water "flow" from the vitreous through intact neurosensory retina to the choroid and sclera; they found that force sufficient to explain the normal amount of adhesion present between the neurosensory retina and the RPE. However, the interphotoreceptor matrix may serve as a medium for the exchange of metabolites between the outer segments of the photoreceptors and the RPE, but the exact function of the acid mucopolysaccharide material has yet to be discovered (see Chapter 7).

Topography of Intracellular Organelles

NUCLEAR ELEMENTS. The RPE nucleus is located in the basal portion of the cell (Figs. 1.1, 1.3, and 1.19) and is rather large, measuring 8–12 μm in diameter (Hogan, Alvarado, and Weddell 1971). Many of the cells contain several nuclei, especially in the ora serrata region where the ratio of mononucleate to binucleate cells is on the order of 30:1 (Tso and Friedman 1968). Mitoses have rarely, if ever, been observed in the RPE cells. In general, the RPE cell nucleus may contain one or two nucleoli with a diffuse chromatin pattern (Figs. 1.1 and 1.11). The cell nucleus has a double unit membrane with nuclear "pores."

Ribonucleic acid synthesis takes place in the RPE cell nucleus with formation of messenger RNA that flows into the cell cytoplasm to organize protein synthesis in the endoplasmic reticulum, including the production of numerous enzymes. The enzymes will control a variety of metabolic reactions involv-

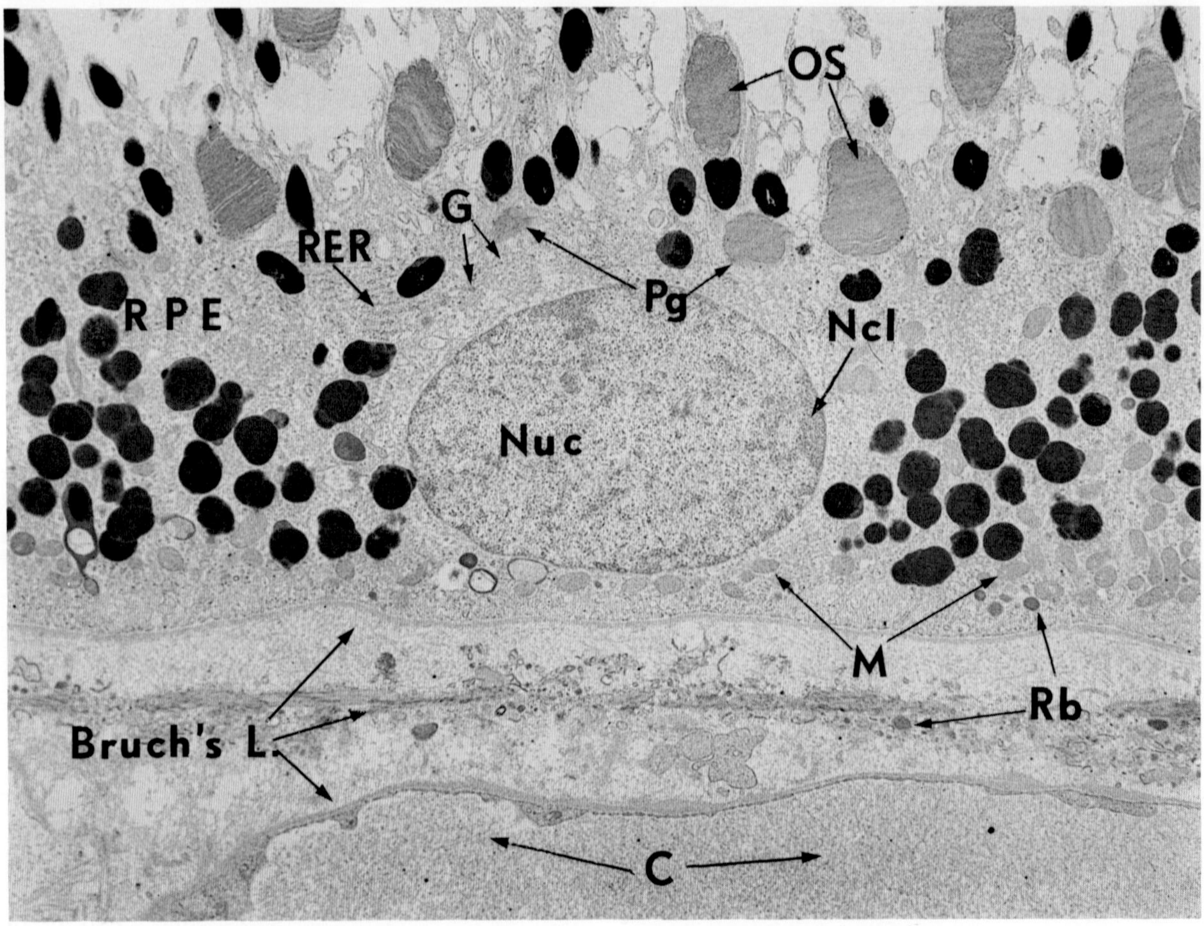

Figure 1.19 *Transmission electron micrograph of the human RPE in a transverse plane. The cell nucleus* (Nuc) *and mitochondria* (M) *are located in the basal portion of the RPE. Chromatin material comprises the nucleolus* (Ncl). *Rough-surfaced endoplasmic reticulum* (RER) *is located at the apex of the cell. Some outer segments* (OS) *of rod photoreceptors are in close apposition to apical microvilli of the RPE. Also labeled are phagosomes* (Pg), *Bruch's membrane* (BL), *the lumen of the choriocapillaris* (C), *residual bodies* (Rb), *and Golgi complexes* (G). *(×7,150.)*

ing lipids, carbohydrates, and other proteins. Protein synthesis is not believed to take place on the surface of the single ribosomes, but only on polyribosome chains or on rough surfaced endoplasmic reticulum.

CYTOPLASMIC ELEMENTS.

Endoplasmic reticulum. The RPE cell cytoplasm contains both smooth- and rough-surfaced endoplasmic reticulum as well as free ribosomes. The smooth endoplasmic reticulum consists of lamellae, vesicular and tubular structures arranged in parallel sheets of 5–10 cisternae (Hogan, Alvarado, and Weddell 1971). Free ribosomes and smooth endoplasmic reticulum tend to be concentrated at the apical region of the RPE (Figs. 1.19 and 1.20). The smooth endoplasmic reticulum is well developed in the RPE within the macular zone, whereas the rough-surfaced endoplasmic reticulum is well developed in the peripheral RPE. RPE cells in the peripheral retina contain more cytoplasmic fibrils, which are responsible for a more compact-looking cytoplasm. There are stacks of annulate lamellae (Fig. 1.20) within the basal cytoplasm of the RPE and they have been identified with active protein synthesis (Hogan, Alvarado, and Weddell 1971).

Golgi complex. The Golgi complex (Fig. 1.20) is also present within the cytoplasm and is believed to be involved in secretory functions as well as the sulfation of mucopolysaccharides (Young 1973) that are secreted to the apical surface of the RPE and eventually coat the outer segments of the photoreceptors. Galactose metabolism, with the formation of galactose-containing lipids, may take place, in part, at the Golgi complex. The Golgi may also play a role in hydrolytic activities by transferring arysulfatase to both lysosomes and phagosomes (see Chapter 6). Acid phosphatase activity has been found in the Golgi complex as well as in lysosomes and phagosomes (Ishikawa and Yamada 1970).

Mitochondria. Mitochondria are located, for the most part, between the cell nucleus and the base of the cell membrane (Figs. 1.19 and 1.20). The mitochondria tend to be small and rod-shaped in longitudinal section, measuring 0.5 to 1.0 μm in length, yet are round in cross-section. The mitochondrial cristae have a transverse orientation and contain enzymes for the Krebs cycle as well as for electron transport particles (ETP). Mitochondria are the centers for

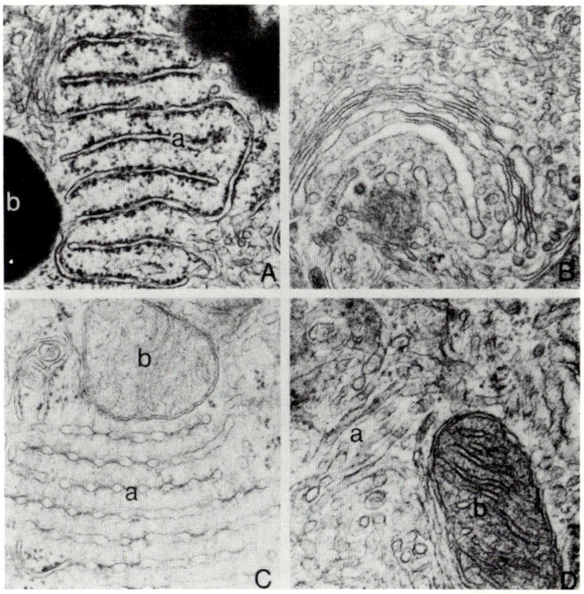

Figure 1.20 *Cytoplasmic organelles in the retinal pigment epithelium.* A, *Rough-surfaced endoplasmic reticulum* (a) *and a pigment granule* (b). *(×25,200.)* B, *Golgi apparatus. (×25,200.)* C, *Annulate lamellae* (a) *and a mitochondrion* (b). *(×37,800.)* D, *Smooth-surfaced endoplasmic reticulum* (a) *and a mitochrondrion* (b). *(From Hogan, Alvarado, and Weddell 1971. ×42,000.)*

oxidative phosphorylation—with the generation of ATP—through the tricarboxylic acid cycle. Other mitochondrial enzyme systems are also present in the RPE (see Table 1.2).

Table 1.2 Retinal pigment epithelial mitochondrial enzymes.

Aerobic (Citric Acid Cycle)
Succinate dehydrogenase
Isocitrate dehydrogenase
Malate dehydrogenase
Electron transport chain enzymes (Cytochrome oxidase, Coenzyme NAD)
β-hydroxybutyrate
Anaerobic
Phosphoglucoisomerase
Lactate dehydrogenase
Phosphogluconate oxidative pathway enzymes (Pentoses, hexoses, NADP)

Note: See Chapters 6 and 7 for details.

Microperoxisomes. Microperoxisomes (Novikoff 1963; Leuenberger and Novikoff 1975) are intracytoplasmic organelles with a granular (Fig. 1.21) matrix

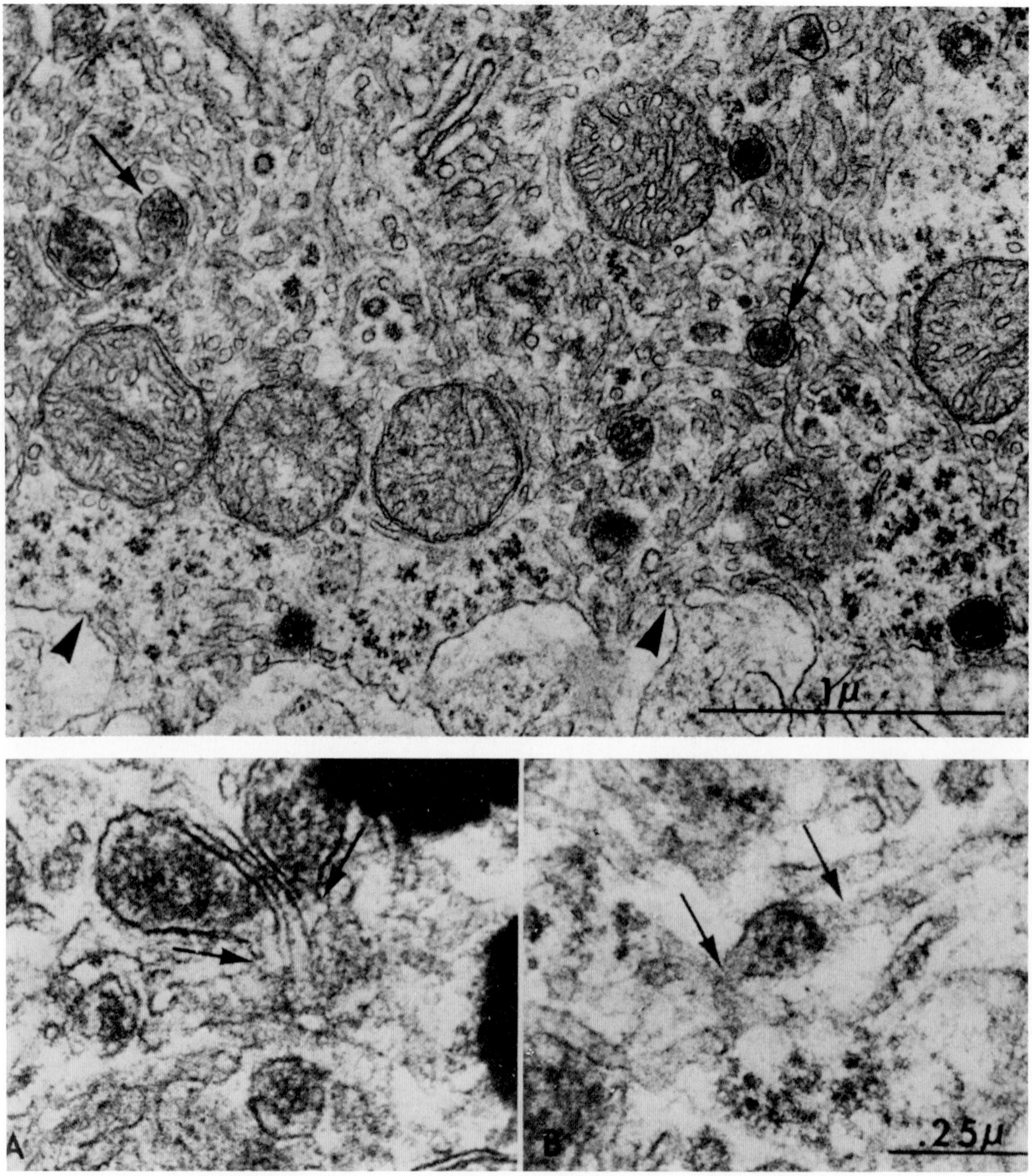

Figure 1.21 *Microperoxisomes.* Above, *retinal pigment epithelium of a 25-year-old human viewed adjacent to the basal infoldings of the plasma membrane* (arrowheads). *Microperoxisomes* (arrows) *histochemically stained for the peroxidatic activity of catalase are closely associated with the smooth endoplasmic reticulum. Mitochondria, coated vesicles, and polyribosomes are numerous.* (×38,000.) Below, *Microperoxisomes in human retinal epithelium showing direct connections* (arrows) *with tubules of the smooth endoplasmic reticulum. The small microperoxisome in micrograph* B *connects with the endoplasmic reticulum on both sides.* (×76,000.) *(From Robison and Kuwabara 1975.)*

ranging in diameter from 0.15 μm to 0.30 μm; they are bound by a single unit membrane, which often has slender connections with smooth endoplasmic reticulum and other microperoxisomes in the RPE (Robison and Kuwabara 1975). Microperoxisomes are distributed throughout the RPE cell cytoplasm, but are more abundant in the basal regions and along the lateral cell boundaries. Microperoxisomes vary in shape from spherical to oval and occur in clusters. They have been thought to be involved in some aspects of lipid metabolism (transport and storage) including the detoxification of H_2O_2 (Leuenberger and Novikoff 1975; Reddy et al. 1974). There is a topographical association of microperoxisomes with lipid droplets and lipofuscin granules (Novikoff and Novikoff 1973; Novikoff et al. 1972). Mitochondria may also be seen near microperoxisome–smooth-ER aggregates (Leuenberger and Novikoff 1975).

RPE microperoxisomes have the following characteristics: catalase is present; nucleoids are absent; there is a tripartite delimiting membrane; and there are numerous slender continuities with smooth ER. In view of the high levels of vitamin A esters present within RPE cells, it is conceivable that the large number of microperoxisomes is related to the metabolic processes that remove vitamin A from the plasma β-lipoproteins and esterify it (Leuenberger and Novikoff 1975).

Lysosomes. Lysosomes are present within the cytoplasmic organelle population of the RPE. There are numerous hydrolytic enzymes contained in these structures—each of which is surrounded by a unit membrane (Fig. 1.19)—including acid lipases, phospholipases and cathepsin D (see Table 1.3). Other hydrolytic enzymes, such as acid phosphatase, are present not only in lysosomes but also in phagosomes and the Golgi complex to a variable extent (see Chapter 7). Cathepsin D, found in lysosomes, may play a role in the turnover of glycoproteins found in the interphotoreceptor matrix, and the release of this agent is thought to be mediated by retinol through altered membrane stability (Dingle 1969; see Chapter 6).

Phagosomes. Phagocytosis is a process by which a cell ingests solid material. Pinocytosis is the ingestion of soluble material. In each case, the material that is ingested is surrounded by a membrane that isolates it from the rest of the cell cytoplasm. The process of phagocytosis has been divided into three phases: (1) recognition and attachment; (2) ingestion; (3) digestion. In the first phase of phagocytosis by RPE cells, the cell recognizes the outer segment of the photoreceptor cell, presumably through recognition sites (see Chapter 9). Interaction with the RPE receptor sites results in the attachment of outer segment material to the RPE (apical) surface membrane, which, in turn, triggers the process of ingestion (Carr 1973; Cohn 1972).

Table 1.3 Hydrolytic enzymes of the retinal pigment epithelium.

Lysosomal fraction
N-Acetyl-β-glucosaminidase
Acid phosphatase
β-galactosidase
Acid lipase
Phospholipases A_1 and A_2
Cathepsin D
Arylsulfatase
β-glucuronidase
Nonlysosomal fraction
Acid phosphatase
N-Acetylglucosaminidase
Naphthylamidase
Cathepsin D
Arylsulfatase
5′-Nucleotidase

Note: See Chapters 6 and 7 for details.

There is some controversy as to the actual mechanism of rod phagocytosis. Spitznas and Hogan (1970) believe that the RPE actively "nibbles away" at the outer segments, whereas Young's (1971b) observations indicate that the rods shed groups of discs that are secondarily engulfed by the RPE.

In any event, the packets of outer segment discs are eventually surrounded by the broad-based apical microvilli of the RPE (Young and Bok 1969) forming a *cytoplasmic phagosome* (Figs. 1.1B, 1.3, and 1.19). These microvilli surround the external 4 μm of the rod outer segments and, in contrast to the long, slim microvilli, appear to participate significantly in the phagocytosis process (see Chapters 2 and 9). Approximately 30–40 discs are sequestered and surrounded by a cell membrane derived from the RPE (Spitznas and Hogan 1970; Young 1971b). The inner membrane of the phagosome, when formed, measures approximately 1.5–2.0 μm in length (Hogan, Alvarado, and Weddell 1971). There is evidence that the number of phagosomes in the peripheral RPE cell is less than in the central (macular) RPE cell (Johnson 1975).

The rough-surfaced endoplasmic reticulum synthesizes the intracellular digestive enzymes, which are then transported to the Golgi complex, where the enzymes are collected in membrane-like vesicles called primary *lysosomes* (DeDuve et al. 1955; Novikoff, Beaufay, and DeDuve 1956). The phagosomes eventually migrate from the apical region of the RPE to the deep regions of the cell, where they encounter lysosomes. The membranes of these two structures presumably fuse (Young 1971b), forming a single "combined vacuole" (see Chapter 9). At this point, the lysosomal hydrolytic enzymes mix with the contents of the phagosome, and the structure is termed either a digestive vacuole or a secondary lysosome. The membrane surfaces on the primary lysosome and the phagosome must have some sort of complimentary recognition sites to allow for membrane attachment and fusion (DeDuve 1964) otherwise the lysosomes could digest other intracellular organelles.

The third phase of phagocytosis is the digestion of the material contained within the digestive vacuole. This is accomplished through the activation of hydrolytic enzymes already present as pH changes from neutral to acid. The rod plasma membrane is probably the first structure to be digested. As digestion proceeds, the following changes take place with regard to the secondary lysosome or digestive vacuole (see Chapter 9): (1) the vacuole's contents become homogenized; (2) there is deformation and disorganization of rod disc membranes; (3) the vacuole proceeds toward the basal region of the RPE; (4) the size of the vacuole diminishes; (5) the shape of the vacuole changes.

Occasionally, indigestible residues remain at the end of the digestion process, and the digestive vacuole with its contents is termed a *residual body*. Structurally, these residual bodies often contain lipids and laminated membrane-like whorls, presumably derived from photoreceptor cell outer segment membranes (Fig. 1.19).

Lipofuscin granules—residual bodies. Streeten (1961), using the light microscope, was the first to document the presence of lipofuscin granules in the human RPE. The granules were found near the cell nucleus and histochemically had sudanophilic staining properties, indicating the presence of unsaturated fatty acids. Zeman and Siakotos (1973) found lipofuscin to be a complex mixture of cross-polymerized materials, consisting mainly of neutral and polar lipids. Initially, lipofuscin granules were considered "wear and tear" pigments, which were also present in heart and brain tissues. The lipofuscin of liver and brain was found to be high in peroxidized lipids and fatty acids (see Chapter 6). The actual chemical composition of lipofuscin in the RPE has not been delineated to date.

Feeney, Grieshaber, and Hogan (1965) demonstrated that lipofuscin granules were part of the lysosomal system. The granules were thought to accumulate in the RPE as one of the end products of phagocytosis (Spitznas and Hogan 1970; Ishikawa and Yamada 1970; Young 1971b; Marshall and Ansell 1971). Feeney (1973) indicated that lipofuscin granules represent incomplete degradation of phagocytized rod outer segments by the lysosomal system (Figs. 1.1, 1.3).

Today there is general agreement that lipofuscin granules are residual bodies (Novikoff, Essner, and Quintana 1964) containing incompletely digested membranes of photoreceptor outer segment discs (Feeney, Grieshaber, and Hogan 1965; Ishikawa and Yamada 1970). Lipofuscin granules essentially lack enzyme activity, indicating that digestion has stopped (see Chapter 9).

Dilley and McConnell (1970) discovered vitamin E (alpha tocopherol) in rod outer segments. Vitamin E is an antioxidant that inhibits peroxidation of polyunsaturated fatty acids in cell membranes (Tappel 1968, 1970). Normally, peroxidation involves the reaction of lipids and oxygen with the formation of free radical substances. The free radicals may cause molecular damage to cell membranes (see Chapter 7) through several mechanisms, one of which involves the cross-linking of proteins (Packer, Deamer, and Heath 1967; Tappel 1970).

Animals on diets deficient in vitamin E (Hayes 1974; Porta and Hartroft 1969) develop large concentrations of lipofuscin in the RPE. Vitamin E protects outer-segment membranes from peroxidation. If, for some reason, these membranes are damaged by peroxidation then they are not completely hydrolyzed by the lysosomal enzyme systems of the RPE. This defect allows the accumulation of lipofuscin granules or true residual bodies. Residual body accumulation may therefore be due, in part, to peroxidation of polyunsaturated lipids in outer segment membranes.

Exocytosis is a process involving the egestion of digestive vacuole residues—for example, residual bodies—and does occur in the retinal pigment epithelium. Hogan (1972) described the gradual accumulation of residual bodies, lipids, and other digestive debris, including mucoproteins, in Bruch's membrane. These undigested waste materials are released at the basal surface of the RPE through exocytosis. Several authors (Hogan and Alvarado 1967; Farkas, Sylvester, and Archer 1971) have concluded that accumulation in Bruch's membrane of residual bodies, along with other digestive materials, contributes to the formation of drusen (see Chapter 26). Chemical analyses (Farkas et al. 1971) of drusen have indicated the presence of PAS-staining complex carbohydrates as well as cerebrosides. The latter were found to be the main lipids in drusen and derived from either RPE lysosomal membranes or from lipofuscin granules. The relationship, if any, between vitamin E and drusen formation remains to be elucidated.

Melanosomes. One of the earliest descriptions of the pigment in the RPE was by Kühne (1879), who termed the pigment *fuscin.* Many years later it was renamed *melanin.* The color of the RPE is due to its melanin content. In older individuals, accumulation of lipofuscin pigment contributes to the color of the RPE. The type of melanin present in the RPE is brown or black eumelanin and not pheomelanin. The latter pigment is responsible for red hair in humans and for the color of animal feathers and fur (see Chapter 8).

Melanin is synthesized from tyrosine through a complex series of reactions, which involve a number of enzymes (such as tyrosinase) and metallic ions (such as Cu^{+}, Cu^{2+} and Zn^{2+}). There is some controversy over the exact biochemical reactions involved (see Chapter 8). The latest evidence indicates that there is a true tyrosinase present that is capable of monophenolase and catecholase activity, which are necessary to get tyrosine → dopa and dopa → dopa quinone, respectively (reviewed by Jimbow et al. 1976).

However, the synthesis of melanin for the melanin granule begins with the formation of the enzyme tyrosinase in the rough-surfaced endoplasmic reticulum. Tyrosinase is then transferred to the Golgi apparatus, where it is stored in membrane-bound vesicles known as *premelanosomes* (Fig. 1.22). At this point, tyrosinase is intimately associated with a finely striated protein framework within the premelanosome (Hogan, Alvarado, and Weddell 1971). Tyrosine is eventually polymerized into melanin on this framework and the vesicle is then termed a melanosome. Melanin synthesis continues, completely filling the granule with melanoprotein, and the structure is then known as a mature melanin granule. At this point the melanin granules do not have any demonstrable tyrosinase activity (see Chapters 3 and 8).

There are two different theories regarding the molecular structure of melanin (see Chapter 8). The first theory (Mason 1948) proposes that melanin is a regular polymer of copolymerized indole 5, 6-quinone subunits. This would mean that melanin is a homopolymer composed of a single type of subunit joined by a single type of linkage. In contrast, Swan (1963, 1974) and others have developed the idea that melanin is a heteropolymer of multiple subunits joined by multiple types of bonding. Presently, melanin is considered a "random polymer derived from the linkage of various different indoles" (Nicolaus and Piatelli 1965; Swan 1974; see Chapter 8).

According to Seiji (1967), intracellular melanin "granules" can be staged as follows:

Premelanosome: A cytoplasmic organelle bounded by a unit membrane, containing oriented fibrils without melanin or tyrosinase activity. These structures may be derived from a specialized region of smooth endoplasmic reticulum (Fig. 1.22) found at the inner surface of the Golgi apparatus (Novikoff, Albala, and Bienpica 1968).

Melanosome: A cytoplasmic organelle having some tyrosinase activity, bounded by a unit membrane with melanin and melanoprotein being present.

Mature Melanin Granule: A cytoplasmic particle that is heavily pigmented with melanoprotein and bounded by a unit membrane (Fig. 1.23). There is no tyrosinase activity in a mature melanin granule.

Melanin granules can be found in RPE cells as early as the fifth week of gestation (Feeney, Grieshaber, and Hogan 1965). Tyrosinase activity was found in melanin granules as late as six months of age in the human fetus (Miyamoto and Fitzpatrick 1957). In the adult, mature melanin granules lack tyrosinase activity (see Chapters 3 and 8).

Melanin granules of the RPE are either ovoid or spherical (Figs. 1.1, 1.3, 1.13, and 1.15). The ovoid

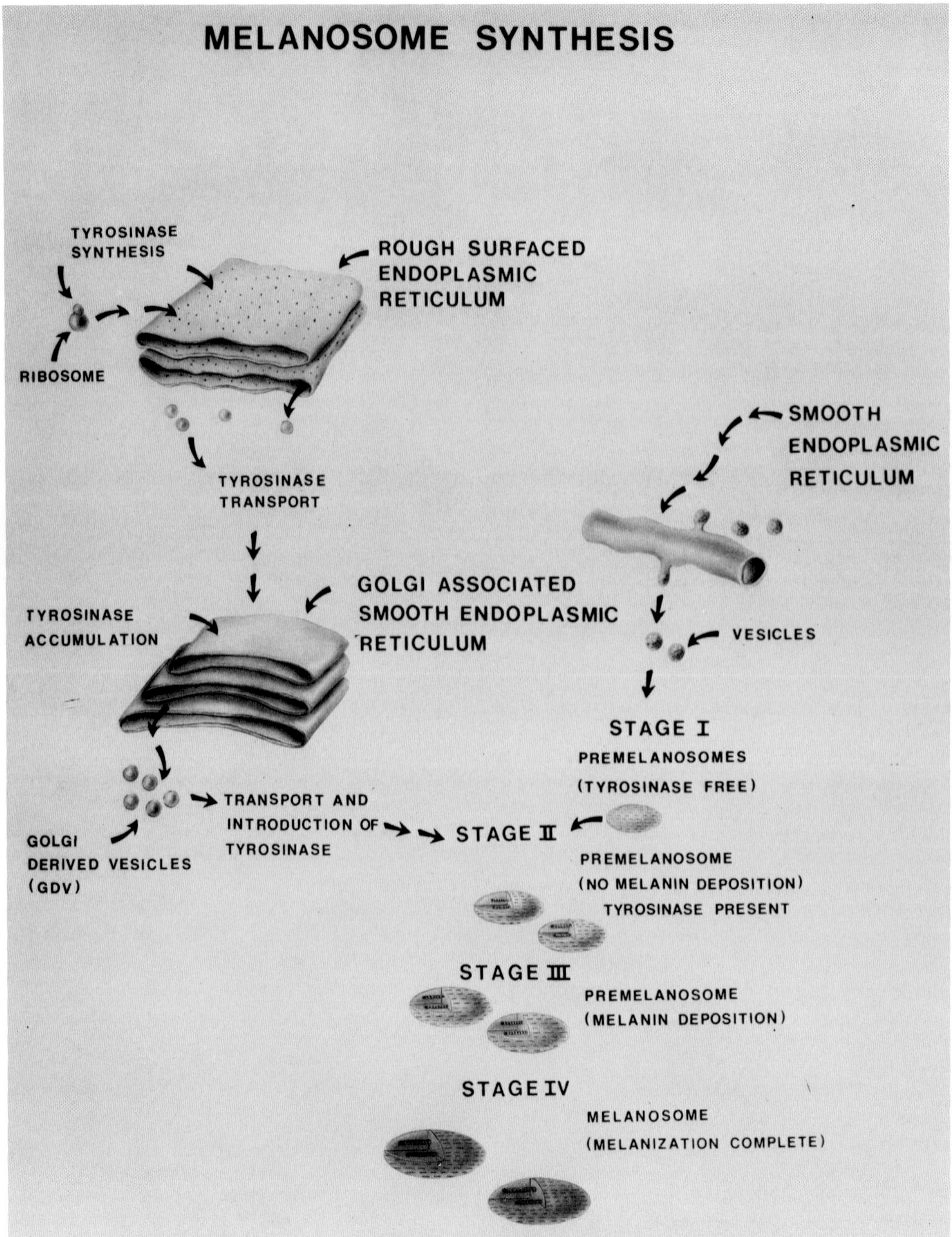
MELANOSOME SYNTHESIS
TYROSINASE SYNTHESIS
ROUGH SURFACED ENDOPLASMIC RETICULUM
RIBOSOME
SMOOTH ENDOPLASMIC RETICULUM
TYROSINASE TRANSPORT
GOLGI ASSOCIATED SMOOTH ENDOPLASMIC RETICULUM
TYROSINASE ACCUMULATION
VESICLES
STAGE I
PREMELANOSOMES (TYROSINASE FREE)
TRANSPORT AND INTRODUCTION OF TYROSINASE
STAGE II
GOLGI DERIVED VESICLES (GDV)
PREMELANOSOME (NO MELANIN DEPOSITION) TYROSINASE PRESENT
STAGE III
PREMELANOSOME (MELANIN DEPOSITION)
STAGE IV
MELANOSOME (MELANIZATION COMPLETE)

← **Figure 1.22** *Schematic Outline of Melanosome Synthesis. The synthesis of melanin begins with the synthesis of tyrosinase on ribosome particles embedded in the cytoplasmic rough-surfaced endoplasmic reticulum of the RPE cell. Tyrosinase is then transported to Golgi-associated smooth endoplasmic reticulum, where it accumulates. Golgi derived vesicles* (GDV) *appear to bud off and carry tyrosinase molecules to the region of premelanosomes. Premelanosomes are derived in some way from smooth endoplasmic reticulum through an intermediary vesicular stage. These premelanosomes (Stage I) do not contain any tyrosinase enzyme, but do have fine filaments (10 nm periodicity). The Golgi-derived vesicles then introduce tyrosinase into the premelanosome, which is then said to be at Stage II. There is no melanin deposition at this point, and the premelanosome contains filaments, with some crosslinking. Then melanin synthesis (Stage III) takes place through a complex series of reactions involving tyrosine, dopa, and dopa → dopa quinone mediated by various enzymes (e.g. tyrosinase) and metallic ions (such as Cu^+, Cu^{2+} and Zn^{2+}). During this process there is a partial obscuration of the fine filaments within the premelanosome by the newly synthesized electron-dense melanin. In Stage IV, the melanosome is now fully melanized, and tyrosinase activity is absent.*

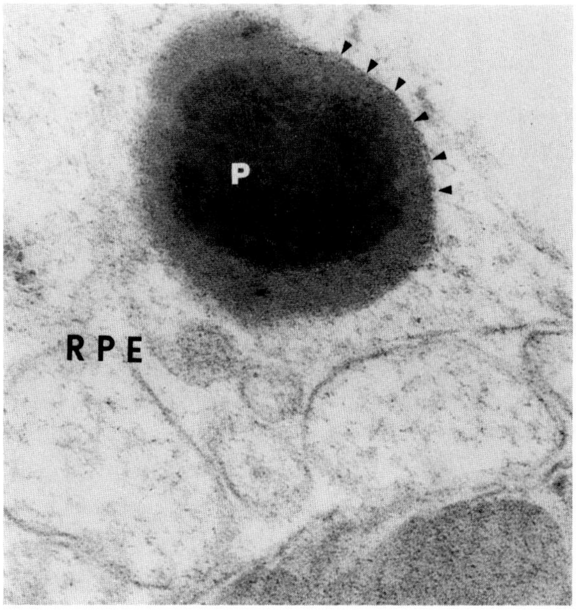

Figure 1.23 *Melanin granule in human RPE in a transverse plane. Mature pigment granule* (P) *containing melanin with a more electron-dense center. A unit membrane surrounds the granule* (arrows). *(×55,900.)*

granules tend to be located within the apical microvilli (Figs. 1.13 and 1.15) whereas the spherical ones are found in the apical to middle portions of the cytoplasm (Figs. 1.3 and 1.4). The basal region of the cell tends to be devoid of melanin granules (Figs. 1.3, 1.4, and 1.13). The pigment granules are 2–3 μm in length, with a cross-sectional diameter of 1 μm. The spherical and oval granules are approximately equal in number in the RPE, in contradistinction to the preponderance of spherical granules in the iris pigment epithelium and ciliary body epithelium (Kozart 1976).

Melanin granules, in adult RPE cells, have a thin unit membrane that appears to be in direct contact with the melanoprotein, which, in turn, fills the entire space of the granule (Figs. 1.3, 1.22, and 1.23). Most of the pigment granules lack any recognizable substructure at the electronmicroscopic level. Melanosomes in different stages of melanization are found in the adult RPE (Fig. 1.22). It has been suggested that lysosomal activity is present in premelanosomes and in melanosomes (Novikoff, Albala, and Bienpica 1968). It is possible that melanin granules are produced, modified, and even degraded by the adult RPE (Bülow 1975).

Melanin is an important cellular biochrome in the RPE and has several known functions (Pathak and Fitzpatrick 1974): (1) melanin decreases scattering of incoming random light waves—from sclera, for example—by absorption; (2) melanin can act as a neutral density filter in the scattering of light waves; (3) melanin absorbs radiant energy in the visible as well as the ultraviolet spectrum, dissipating the absorbed energy in the form of heat; (4) melanin uses the absorbed energy to generate semiquinoid free radicals providing the reactants for oxidation; (5) melanin, when in the form of a stable free-radical, can act as a biologic exchange polymer. Melanin is used as the key pigment in the laser photocoagulation process, whereby absorption of laser energy results in temperature elevations up to several hundred degrees centigrade in a matter of nanoseconds. The heat generated by the melanin absorption turns cytoplasmic H_2O to steam and coagulates nearby tissues, such as the neurosensory retina, retinal blood vessels and the choriocapillaris.

REFERENCES

AREY, L. B. 1928. Visual cells and retinal pigment. In *Special Cytology,* ed. F. V. Cowdry, vol. 2, pp. 889–919. New York: Paul B. Hoeber.

BAIRATI, A., JR., and ORZALESI, N. 1963. The ultrastructure of the pigment epithelium and of the photoreceptor pigment epithelium junction in the human retina. *J. Ultrastruct. Res.* 9:484–496.

Berman, E. R. 1964. The biosynthesis of mucopolysaccharides and glycoproteins in pigment epithelial cells of bovine retina. *Biochim. Biophys. Acta* 83:371–373.

Berman, E. R., and Bach, G. 1968. The acid mucopolysaccharides of cattle retina. *Biochem. J.* 108:75–88.

Breathnach, A. S., and Wyllie, L. M. A. 1966. The ultrastructure of retinal pigment epithelium of the human fetus. *J. Ultrastruct. Res.* 16:584–597.

Bruch, C. L. W. 1844. Untersuchungen zur Kenntniss des Kornigen Pigments der Wirbelthiere. In *Physiologischer und pathologischer Hinsicht,* pp. 1–62. Zürich: Meyer U. Zeller.

Bülow, N. 1975. The retinal pigment epithelium and photoreceptor cells light and electron microscopic studies on monkey eyes. *Cell Tiss. Res.* 161:521–540.

Carr, I. 1963. Ingestion by macrophages-phagocytosis. In *The macrophage: a review of ultrastructure and function,* pp. 61–72. New York and London: Academic Press.

Claude, P., and Goodenough, D. A. 1973. Fractured faces of zonula occludantes from "tight" and "leaky" epithelia. *J. Cell Biol.* 58:390–400.

Cohn, Z. A. 1972. Properties of macrophages. In *Phagocytic mechanisms in health and disease,* ed. R. C. Williams and H. H. Fudenberg, pp. 39–49. New York: Intercontinental Medical Book Corp.

DeDuve, C. 1964. From catases to lysosomes. *Fed. Proc.* 23:1045–1049.

DeDuve, C., Pressman, B. C., Gianetto, R., Wattiank, R. and Appleman, F. 1955. Tissue fraction studies: intracellular distribution patterns of enzymes in rat-liver tissue. *Biochem. J.* 60:604–617.

de Guillebon, H., and Zauberman, H. 1972. Experimental retinal detachment: biophysical aspects of retinal peeling and stretching. *Arch. Ophthalmol.* 87:545.

Dilley, R. A., and McConnell, D. G. 1970. Alphatocopherol in the retinal outer segment of bovine eyes. *J. Membr. Biol.* 2:317–323.

Dingle, J. T. 1969. The extracellular secretion of lysosomal enzymes. In *Lysosomes in biology and pathology,* ed. J. T. Dingle and H. B. Fell, vol. 2, pp. 421–436. Amsterdam and London: North-Holland Publishing Co.

Dobson, J. F. 1925. Herophilus of Alexandria: history. *Proc. R. Soc. Med.* 18:19–32.

Donders, F. C. 1854. Beitrage zur pathologischen Anatomie des Auges. *Graefes Arch. Ophthalmol.* 1:106–118.

Dowling, J. E., and Gibbons, I. R. 1962. The fine structure of the pigment epithelium in the albino rat. *J. Cell Biol.* 14:459–474.

Droz, B. 1963. Dynamic condition of proteins in the visual cells of rats and mice as shown by radio-autography with labelled amino acids. *Anat. Rec.* 145:157–168.

Duke-Elder, S. 1961. *System of ophthalmology.* II. *The anatomy of the visual system,* pp. 137–143, pp. 220–273. London: Henry Kimpton.

Farkas, T. G., Sylvester, V., and Archer, D. 1971. The ultrastructure of drusen. *Am. J. Ophthalmol.* 71:1196–1205.

Farkas, T. G., Sylvester, V., Archer, D., and Altona, M. 1971. The histochemistry of drusen. *Am. J. Ophthalmol.* 71:1206–1215.

Fatt, I., and Shantinath, K. 1971. Flow conductivity of the retina and its role in the retinal adhesion. *Exp. Eye Res.* 12:218–226.

Feeney, L. 1973. The phagolysosomal system of the pigment epithelium: a key to retinal disease. *Invest. Ophthalmol.* 12:635–638.

Feeney, L., Grieshaber, J. A., and Hogan, M. J. 1965. Studies on the human ocular pigment. In *The structure of the eye,* ed. J. W. Rohen, pp. 535–548. Stuttgart: F. K. Schattauer-Verlag.

Fine, B. S., and Yanoff, M. 1972. *Ocular histology: a text and atlas,* pp. 48–107. New York: Harper and Row.

Friedman, E., Smith, T., and Kuwabara, T. 1963. Senile choroidal vascular patterns and drusen. *Arch. Ophthalmol.* 69:220–230.

Garron, L. K. 1963. The ultrastructure of the retinal pigment epithelium with observations on the choriocapillaris and Bruch's membrane. *Trans. Am. Ophthalmol. Soc.* 61:545–588.

Hall, M. O., Ocumpaugh, D. E., and Young, R. W. 1965. The utilization of 35-Sulfate in the synthesis of mucopolysaccharides by the retina. *Invest. Ophthalmol.* 4:322–329.

Hayes, K. C. 1974. Retinal degeneration in monkeys induced by deficiencies of vitamin E or A. *Invest. Ophthalmol.* 13:499–510.

Hogan, M. J. 1961. Ultrastructure of the choroid: its role in the pathogenesis of chorioretinal diseases. *Trans. Pac. Coast Otoophthalmol. Soc.* 42:61–87.

Hogan, M. J. 1967. Bruch's membrane and disease of the macula: role of elastic tissue and collagen. *Trans. Ophthalmol. Soc. U.K.* 87:113–161.

HOGAN, M. J. 1972. Role of the retinal pigment epithelium in macular disease. *Trans. Am. Acad. Ophthalmol. Otolaryngol.* 76:64–80.

HOGAN, M. J., and ALVARADO, J. 1967. Studies on the human macula. IV. Aging changes in Bruch's membrane. *Arch. Ophthalmol.* 77:410–420.

HOGAN, M. J., ALVARADO, J. A., and WEDDELL, J. E. 1971. *Histology of the human eye,* Chap. 9, pp. 393–522. Philadelphia: W. B. Saunders.

HOGAN, M. J., WOOD, I., and STEINBERG, R. H. 1974. Phagocytosis by pigment epithelium of human retinal cones. *Nature* 252:305–307.

ISHIKAWA, T., and YAMADA, E. 1970. The degradation of the photoreceptor outer segment within the pigment epithelial cell of rat retina. *J. Electron Microsc.* 19:85–91.

IWAKI, S. 1958. Electron microscropic study of retinal pigment and the choriocapillaris. *Acta Soc. Ophthalmol. Jap.* 62:995–1001.

JIMBOW, K., QUEVEDO, W. C., JR., FITZPATRICK, T. B., and SZABO, G. 1976. Some aspects of melanin biology: 1950–1975. *J. Invest. Dermatol.* 67:72–89.

JOHNSON, N. F. 1975. Phagocytosis in the normal and ischemic retinal pigment epithelium of the rabbit. *Exp. Eye Res.* 20:97–107.

JONES, W. 1833. Notice relative to the pigmentum nigrum of the eye. *Edin. Med. Surg. J.* 40:77–83.

KARLI, P. 1954*a*. Sur la réaction colloidopexique de l'épithélium pigmentaire au cours d'une rétinose expérimentale. *C. R. Soc. Biol.* (Paris) 148:172–173.

KARLI, P. 1954*b*. Contribution expérimentale à l'étude de la pathogénie des rétinoses pigmentaires. *Ophthalmologica* (Bâle) 128:137–162.

KOZART, D. 1976. Anatomic correlates of the retina. In *Clinical ophthalmology,* ed. T. D. Duane, vol. 3, pp. 1–4. Hagerstown, Maryland: Harper and Row.

KÜHNE, W. 1879. Chemische vorganee in der netzhaut. In *Handbuch der Physiologie,* ed. L. Hermann, F. W. C. Vogel, vol. 3, pp. 235–342.

KÜHNT, H. 1877. Zur Architektonik der retina. *Klin. Monatsbl. Augenheilkd.* (Beilageheft) 15:72–82.

LERCHE, W. 1963*a*. Elektronenmikroskopische untersuchungen zur differenzierung des pigmentepithels und der ausseren kornerzellen (Sinneszellen) in menschlichen auge. *Z. Zellforsch. Mikrosk. Anat.* 58:953–970.

LERCHE, W. 1963*b*. Elektronenmikroskopische Beobachtungen an der Bruchschen Membran des Menschlichen. *Ber. Dtsch. Ophthalmol. Ges.* 65:384–393.

LERCHE, W. 1965. Electron microscopic studies on age related changes in Bruch's membrane in man. *Anat. Gesell. Verh.* 60:123–132.

LERCHE, W. 1965. Electron microscopic observations on the histogenesis of Bruch's membrane in man. *Z. Zellforsch. Mikrosk. Anat.* 65:163–175.

LEUENBERGER, P. M., and NOVIKOFF, A. B. 1975. Studies on microperoxisomes. VII. Pigment epithelial cells and other cell types in the retina of rodents. *J. Cell Biol.* 65:324–334.

LOCASCIO, G. 1923. Sulla Fine costituzione della retina: ricerche chimiche e citologiche. *Ann. Ottal.* 51:653–678.

MAJIMA, K. 1925. Studien über die Struktur der Sehzellen und der Pigmentepithelzellen der Froschnetzhaut. *Arch. Ophthalmol.* 115:286–304.

MARSHALL, J., and ANSELL, P. L. 1971. Membranous inclusions in the retinal pigment epithelium: phagosomes and myeloid studies. *J. Anat.* (Lond.) 110:91–104.

MASON, H. S. 1948. The chemistry of melanin. VII. Mechanism of the oxidation of dihydroxyphenylalanine by tyrosinase. *J. Biol. Chem.* 172:83–99.

MIKI, H., BELLHORN, M. B., and HENKIND, P. 1975. Specializations of the retinochoroidal juncture. *Invest. Ophthalmol.* 14:701–707.

MILLER, S. S., and STEINBERG, R. H. 1977. Passive ionic properties of frog retinal pigment epithelium. *J. Membr. Biol.,* in press.

MIYAMOTO, M., and FITZPATRICK, T. B. 1957. On the nature of the pigment in retinal pigment epithelium. *Science* 126:449–450.

MÜLLER, H. 1856. Untersuchungein uber die Glashaute des Auges, insbesondere die Glaslamelle der Choroidea und ihre senilen Veränderungen. *Graefes Arch. Ophthalmol.* 2:1–65.

MUSSER, G. L., and ROSEN, S. 1973. Localization of carbonic anhydrase activity in the vertebrate retina. *Exp. Eye Res.* 15:105–119.

NICOLAUS, R. A., and PIATELLI, M. 1965. Progress in the chemistry of natural black pigments. *Rend. Acad. Sci. Fis. Math.* (Naples) 32:1–17.

NOVIKOFF, A. B. 1963. Lysosomes in the physiology and pathology of cells contributions of staining methods. *Lysosomes Ciba Found. Symp.* 36–73.

NOVIKOFF, A. B., ALBALA, A., and BIENPICA, L. 1968. Ultrastructural and cytochemical observations on B-16 and Harding-Passey mouse melanomas. The origin of premelanosomes and compared melanosomes. *J. Histochem. Cytochem.* 16:299–319.

Novikoff, A. B., Beaufay, H., and DeDuve, C. 1956. Electron microscopy of lysosome-rich fractions from rat liver. *J. Biophys. Biochem. Cytol.* 2:179–184. (suppl.)

Novikoff, A. B., Essner, E., and Quintana, N. 1964. Golgi apparatus and lysosomes. *Fed. Proc.* 23:1010–1022.

Novikoff, A. B., and Novikoff, P. M. 1973. Microperoxisomes. *J. Histochem. Cytochem.* 21:963–966.

Novikoff, A. B., Novikoff, P. M., Davis, C., and Quintana, N. 1972. Studies on microperoxisomes. II. A cytochemical method for light and electron microscopy. *J. Histochem. Cytochem.* 20:1006–1023.

Ocumpaugh, D. E., and Young, R. W. 1966. Distribution and synthesis of sulfated mucopolysaccharides in the retina of the rat. *Invest. Ophthalmol.* 5:196–203.

Packer, L., Deamer, D. W., and Heath, R. L. 1967. Regulation and deterioration of structure in membranes. *Adv. Gerontol. Res.* 2:77–120.

Pathak, M. A., and Fitzpatrick, T. B. 1974. The role of natural photoprotective agents in human skin. In *Sunlight and man: normal and abnormal photobiologic responses,* ed. T. B. Fitzpatrick, M. A. Pathak, L. C. Harber, M. Seiji, and A. Kutita, pp. 725–750. Japan: Univ. of Tokyo Press.

Porta, E. A., and Hartroft, W. S. 1969. Lipid pigments in relation to aging and dietary factors (lipofuscins). In *Pigments in Pathology,* ed. M. Wolman, pp. 191–235. New York: Academic Press.

Porter, K. R. 1956. The submicroscopic morphology of protoplasm. *Harvey Lect.* 51:175–228.

Porter, K. R., and Yamada, E. 1960. Studies on the endoplasmic reticulum. V. Its form and differentiation in pigment epithelial cells of the frog retina. *J. Biophys. Biochem. Cytol.* 8:181–205.

Raehlmann, E. 1907. Zur anatomie und physiologie des pigment-epithels der netzhaut. *Z. Augen. Heilkunde.* 17:1–25.

Reddy, J. K., Azarnoff, D. L., Svoboda, D. J., et al. 1974. Nafenopin induced hepatic microbody (peroxisome) proliferation and catalase synthesis in rats and mice: absence of sex difference in response. *J. Cell Biol.* 61:344–358.

Robison, W. G., Jr., and Kuwabara, T. 1975. Microperoxisomes in retinal pigment epithelium. *Invest. Ophthalmol.* 14:866–872.

Sakuragawa, M., and Kuwabara, T. 1976. The pigment epithelium of the monkey topographic study by scanning and transmission electron microscopy. *Arch. Ophthalmol.* 94:285–292.

Salzmann, M. 1912. *The anatomy and histology of the human eyeball in the normal state: its development and senescence,* auth. transl. E. V. L. Braun, pp. 1–232. Chicago: Univ. of Chicago Press.

Seiji, M. 1967. Subcellular particles and melanin formation of melanocytes. In *Advances in biology of the skin,* ed. W. Montagna and Funan Hu, vol. 8, pp. 189–222. Oxford, London: Pergamon Press.

Shakib, M., Rutkowski, P., and Wise, G. N. 1972. Fluorescein angiography and the retinal pigment epithelium. *Am. J. Ophthalmol.* 74:206–218.

Sidman, R. L. 1958. Histochemical studies of photoreceptor cells. *Ann. N.Y. Acad. Sci.* 74:182–195.

Spencer, W. H. 1965. Pathogenesis of macular degeneration. *Trans. Am. Acad. Ophthalmol. Otolaryngol.* 69: 662–667.

Spitznas, M., and Hogan, M. J. 1970. Outer segments of photoreceptors and the retinal pigment epithelium. Interrelationship in the human eye. *Arch. Ophthalmol.* 84:810–819.

Steinberg, R. H., and Miller, S. 1973. Aspects of electrolyte transport in frog pigment epithelium. *Exp. Eye Res.* 16:365–372.

Steinberg, R. H., Wood, I., and Hogan, M. J. 1977. Pigment epithelial ensheathment in human retina. *Philos. Trans. R. Soc. Lond.* [Biol. Sci.] 277:459–474.

Streeten, B. W. 1961. Sudanophilic granules of the human retinal pigment epithelium. *Arch. Ophthalmol.* 66:391–398.

Sumita, R. 1961. Electron microscopic study of the choroid. I. The fine structure of Bruch's membrane (lamina vitrea) in the choroid. *Acta Soc. Ophthalmol. Jap.* 65:1188–1203.

Swan, G. A. 1963. Chemical structure of melanins. *Ann. N.Y. Acad. Sci.* 100:1005–1019.

Swan, G. A. 1974. Structure, chemistry and biosynthesis of the melanins. *Fortschr. Chem. Org. Naturst.* 31:521–532.

Tappel, A. L. 1968. Will antioxidant nutrients slow aging processes. *Geriatrics* 23:97–105.

Tappel, A. L. 1970. Biological antioxidant protection against lipid peroxidation damage. *Am. J. Clin. Nutr.* 23:1137–1139.

Tesoriere, G., Vento, R., Magistro, D., and Dones, F. 1971. Biosynthesis of UDP N-acetylhexoseamine in bovine retina. *Eur. J. Biochem.* 19:289–293.

Tso, M. O., and Friedman, E. 1967. The retinal pigment epithelium. I. Comparative histology. *Arch. Ophthalmol.* 78:641–649.

TSO, M. O., and FRIEDMAN, E. 1968. The retinal pigment epithelium. III. Growth and development. *Arch. Ophthalmol.* 80:214–216.

USHER, C. H. 1906. A note on the choroid at the macular region. *Trans. Ophthalmol. Soc. U.K.* 26:107–110.

VERHOEFF, F. H. 1903. A hitherto undescribed membrane of the eye and its significance. *R. Lond. Ophthalmol. Hosp. Reports* 15:309–319.

VON KÖLLIKER, R. 1861. *Entwicklungsgeschichte des Menschen, und der höheren Thiere,* pp. 1–468. Leipzig: W. Engelmann.

WOLFF, E. 1938. Can the network formed by the retinal pigment epithelium be seen by the ophthalmoscope? *Proc. R. Soc. Med.* 31:1104.

WOLFRUM, M. 1908. Beiträge zur Anatomie und Histologie der Aderhaut beim Menschen und bei höheren Wirbeltieren. *Graefes Arch. Ophthalmol.* 67:307–359.

WOLTER, J. R. 1955. Histologic character of the connection between Bruch's membrane and the choriocapillaries of the human eye. *Arch. Ophthalmol.* 53:208–210.

YOUNG, R. W. 1967. The renewal of photoreceptor cell outer segments. *J. Cell Biol.* 33:61–72.

YOUNG, R. W. 1971*a*. The renewal of rod and cone outer segments in the rhesus monkey. *J. Cell Biol.* 49:303–318.

YOUNG, R. W. 1971*b*. Shedding of discs from rod outer segments in the rhesus monkey. *J. Ultrastruct. Res.* 34:190–203.

YOUNG, R. W. 1973. The role of the Golgi complex in sulfate metabolism. *J. Cell Biol.* 57:175–189.

YOUNG, R. W., and BOK, D. 1969. Participation of the retinal pigment epithelium in the rod outer segment renewal process. *J. Cell Biol.* 42:392–403.

ZAUBERMAN, H., and BERMAN, E. R. 1969. Measurement of the adhesive forces between the sensory retina and the pigment epithelium. *Exp. Eye Res.* 8:276–283.

ZAUBERMAN, H., DE GUILLEBON, H., and HOLLY, F. J. 1972. Retinal traction in vitro. *Invest. Ophthalmol.* 11:46–55.

ZEMAN, W., and SIAKOTOS, A. N. 1973. The neutral ceroid-lipofuscinoses. In *Lysosomes and storage diseases,* ed. H. G. Hers and F. Van Hoof, pp. 519–551. New York/London: Academic Press.

ZIMMERMAN, L. E. 1958. Applications of histochemical methods for the demonstration of acid mucopolysaccharides to ophthalmic pathology. *Trans. Am. Acad. Ophthalmol. Otolaryngol.* 62:697–703.

ZIMMERMAN, L. E., and EASTHAM, A. B. 1959. Acid mucopolysaccharides in the retinal pigment epithelium and visual cell layer of the developing mouse eye. *Am. J. Ophthalmol.* 47:488–499.

ACKNOWLEDGMENTS

The authors wish to thank Dr. Paul Henkind for reviewing the manuscript and Mr. Moses Furman for his technical assistance.

2 The Relationship of the Retinal Pigment Epithelium to Photoreceptor Outer Segments in Human Retina

Roy H. Steinberg
Irmgard Wood

THE APICAL SURFACES of mammalian retinal pigment epithelial cells face the rods and cones, and they are densely covered with processes that are apposed to the outer segments. The processes vary in form from species to species and can also vary within individual retinas. Some of them are thin, finger-like projections, which can be appropriately termed microvilli. Others are wider, having a sheet-like form, and are called lamellae (Bairati and Orzalesi 1963). These lamellae tend to wrap around the outer-segment tips (Bairati and Orzalesi 1963; Spitznas and Hogan 1970). In some retinas, such as the cat retina (Steinberg and Wood 1974) and rat retina (Burnside 1976), the lamellae taper as they extend away from the outer segment and have a remarkable leaf-like form. Finally, some of the villous processes in human retina actually expand to become sheet-like when they reach the tips of cones.

In human retina both the rod and cone outer segments are ensheathed at their tips by the sheet-like apical processes (Spitznas and Hogan 1970; Steinberg, Wood, and Hogan 1977). It is these processes that participate in the phagocytosis of the terminal portions of both the rod and cone outer segments. In this chapter we will only briefly describe the well-known ultrastructure of the rod-RPE relationship and will not provide any description of rod phagocytosis, which is fully covered in Chapter 9. The less well-known cone-RPE association, as well as the recently discovered shedding and phagocytosis of human cones, will be described more completely.

Rods

The relationship between human rods and the RPE apical processes has been carefully described in a number of ultrastructural studies (Bairati and Orzalesi 1963; Spitznas and Hogan 1970; Hogan, Alvarado, and Weddell 1971; Fine and Yanoff 1972). The rods abut the apical surface and are capped or ensheathed there by apical processes for several microns (Figs. 2.1, 2.2). The ensheathment is very similar in the rhesus monkey (Young 1971). It is the consensus of these studies that there are no junctional complexes between the apposed RPE apical processes and the rod outer segment membranes (the same is true for cones). In both human and monkey, several ensheathing processes may be concentrically arranged around the outer segment tip (Fig. 2.2). There are also typical villous projections. These are a few microns longer than the unsheathing processes that phagocytose the terminal portions of the rods (Spitznas and Hogan 1970).

Human apical processes have been observed to contain smooth endoplasmic reticulum, ribosomes, and pigment granules (Bairati and Orzalesi 1963). They probably also contain a backbone of regularly arranged intermediate-size filaments similar to the ones that appear in the RPE apical processes in association with rat rods (Burnside 1976; see Chapter 10); and which are also observed in the processes surrounding cat cones (Steinberg and Wood 1974).

Cones

The relation of the apical processes to human cone outer segments has been only recently described through electron microscopy. A very early light-microscopic study of this relationship by Walls (1934) described its most interesting feature—the long apical processes, which ensheath the short outer segments of extrafoveal cones (Figs. 2.3, 2.4).

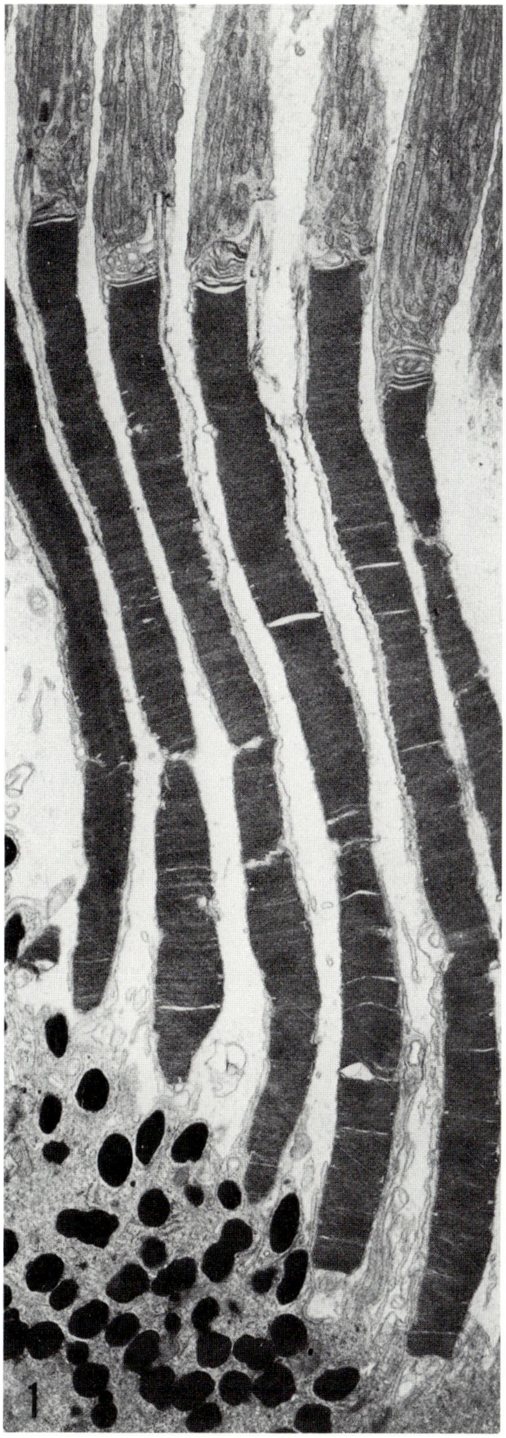

Figure 2.1 *Electron-micrograph montage of a longitudinal section through rod outer and inner segments and the apical portion of the pigment epithelium. The rod outer-segment tips abut against the apical surface and are closely associated with the apical processes. Sixty-year-old retina. (×3,605.)*

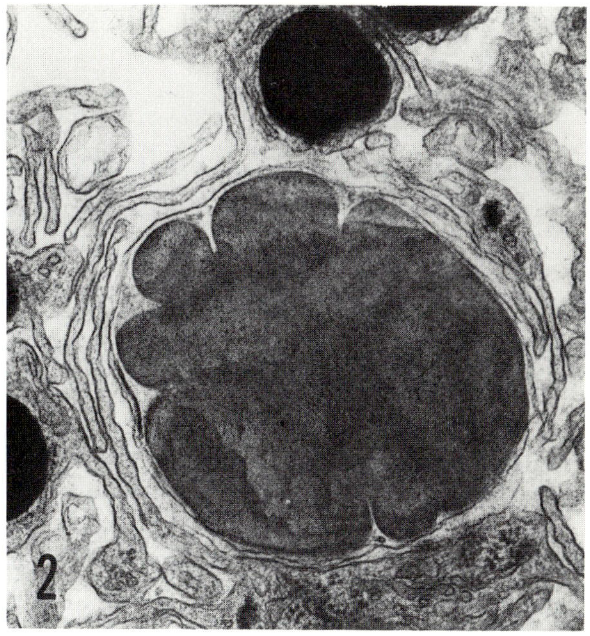

Figure 2.2 *Electron micrograph of a transverse section through the tip of a rod outer segment. Several thin, sheet-like, apical processes concentrically ensheath the rod. A melanin granule appears to the left in a villous-like apical process. Sixty-year-old retina. (×29,685.)*

Unfortunately, Walls' observations were ignored or forgotten, as was a related light-microscopic observation by Eichner (1958). (The poor resolution of these structures with the light microscope may have made it difficult for other investigators to accept the finding.) Even with the electron microscope, it is difficult to describe the relationship, since it cannot be viewed unless the apical processes and cone outer segments are well aligned in the same longitudinal plane. Early electron-microscopic descriptions were obtained by Fine and Yanoff (1972) and Hogan and Wood (1973), and a detailed description was recently published for extrafoveal cones (Steinberg, Wood, and Hogan 1977). There has been, as yet, no detailed description of this relationship for foveal cones.

Although the outer segments of extrafoveal cones do not reach the apical surface, the RPE cells seek out the cones (Figs. 2.3, 2.4). The apical processes are long—in a region where the cone outer segments had an average length of 19 μ, the processes also were 19 μ in length; they usually have a villous or finger-like shape in the space above the cone outer segment (supracone space) (Figs. 2.5, 2.6). Counts of villous profiles in transverse sections through the supracone

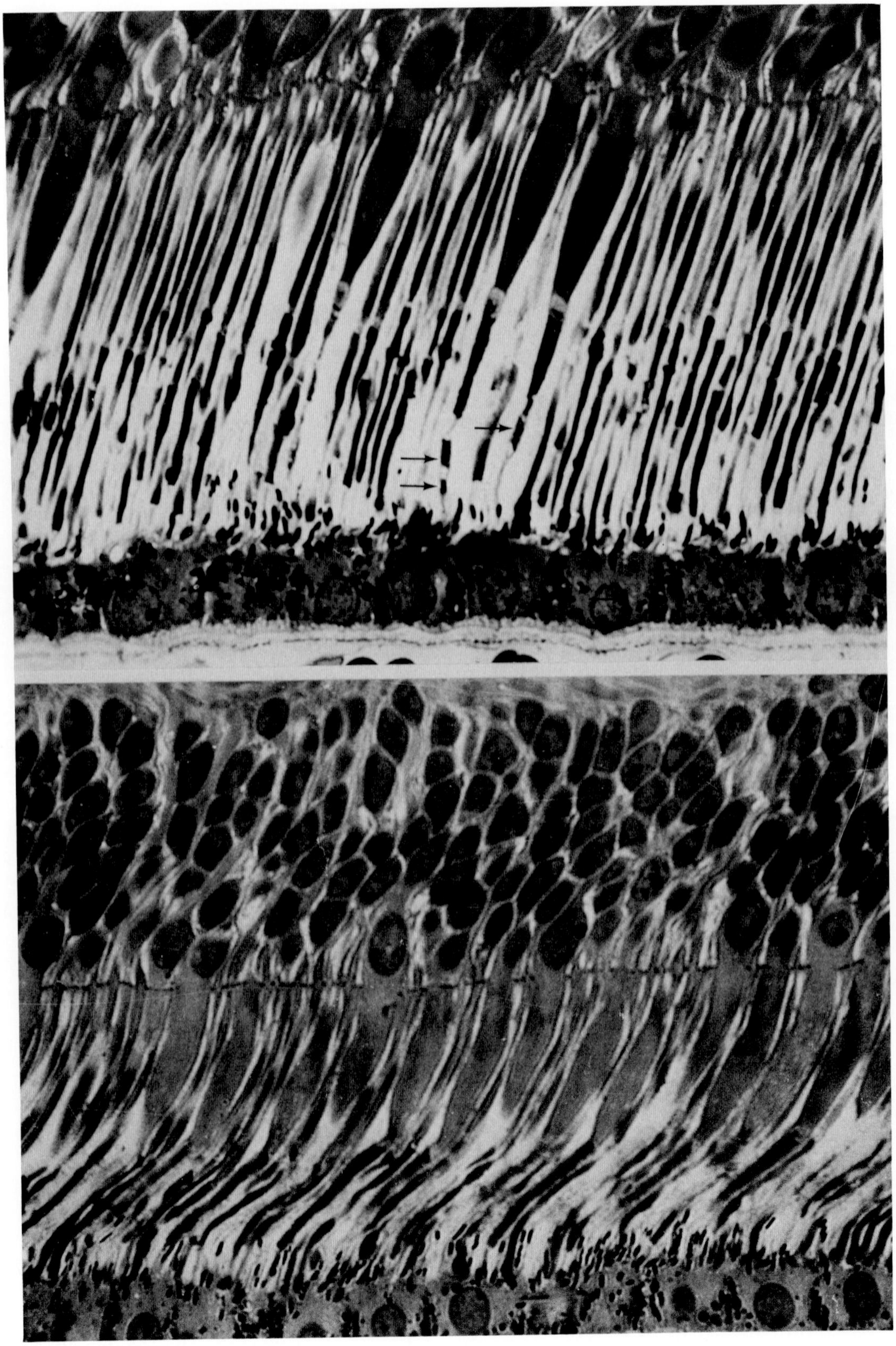

← **Figure 2.3** *Light micrograph of a longitudinal section through the outer portion of the retina, from the outermost part of the outer nuclear layer (top) to the pigment epithelium (bottom). The section is from a region where the rods greatly outnumber the cones. The cone outer segments fall far short of the apical surface of the pigment epithelium, but are associated with long apical processes. The arrows point to inclusions within the apical processes, which are presumed to be phagosomes. Five-year-old retina. (×906.)*

← **Figure 2.4** *Light micrograph, similar to Figure 2.3, except that it shows a section from the parafoveal retina, where the cones are much more numerous than in the region shown in Figure 2.3. The cone outer segments nearly reach the apical surface of the pigment epithelium. Five-year-old retina. (×906.)*

space show that there can be as many as 30–40 processes associated with a single cone (Fig. 2.7). They vary in length, however, so that only a few of them reach the outer segment. They usually are rather empty of intracellular organelles; pigment granules, near the apical surface, are the most prominent organelles. Sometimes, however, an individual process appears enlarged and contains a much richer variety of organelles: smooth endoplasmic reticulum, ribosomes, a variety of vesicle types—some resembling Novikoff's (1973) primary lysosomes—and even mitochondria (Fig. 2.13). We speculate that in such cases the process has enlarged to accommodate a cone phagosome (see below) and that RPE cytoplasm spilled into the enlarged process after the phagosome had passed into the soma of the cell.

Upon reaching the cone, villous apical processes become sheet-like and ensheath the outer segment for as much as one-half of its length (Figs. 2.9–2.12). The tip of the villous process appears to expand upon reaching the outer segment in order to cap the tip of the cone; villous plasma membrane then flows down the sides of the cone to form the sheath around it (Fig. 2.11). A single process may completely surround the outer segment (Fig. 2.9), or multiple processes may wrap around each other concentrically while surrounding the cone (Fig. 2.10). In this way, human extrafoveal cone outer segments, although far removed from the apical surface, are as closely associated with RPE cell membrane as are human rods.

This relationship was found to be essentially the same for extrafoveal cones of three human retinas of different ages: 5, 45, and 60 years (Steinberg, Wood, and Hogan 1977). Recently, additional observations have been made on the central cones from the macula, exclusive of the fovea, in the five-year-old retina (Figs. 2.14–2.19). The relationship of the apical processes to the outer segments was essentially the same in these macular cones as in more peripheral cones. Of course, as the cone outer segments have lengthened, the apical processes are somewhat shorter. One minor difference is the presence of pigment granules in the processes at the tips of the macular cones, or alongside the tips (Figs. 2.14, 2.15). It is probable that the granules extend no farther into the processes in the macular cones than in the more peripheral cones, but that the tips of the longer macular cones are, simply, closer to the granules. It seems reasonable to assume that outer segments associate with their apical processes in the same manner in foveal cones and in macular cones, and this assumption is supported by the photograph of Fine and Yanoff (1972).

SHEDDING OF CONE DISCS AND PHAGOCYTOSIS BY THE RPE. It now has been firmly established that mammalian cones shed their terminal discs in packets, which are phagocytized by the RPE (Anderson and Fisher 1975, 1976; Hogan, Wood, and Steinberg 1974; Steinberg, Wood, and Hogan 1977). Since cones had not been previously thought to shed disc packets (Young 1971b), it was important to establish the normality of this event in the human retina. The existence and normality of cone disc shedding and phagocytosis in the human received strong support from Anderson's and Fisher's (1976) detailed description of these events in the retinas of arboreal and terrestrial squirrels. Furthermore, cone shedding and phagocytosis were observed in all three human retinas studied by Steinberg, Hogan, and Wood (1977) and corroborated in a fourth human retina examined by Fisher and Anderson (personal communication).

Strongly in favor of cone shedding and phagocytosis as a normal event in the three human retinas studied by Steinberg, Hogan, and Wood (1977) were the absence of retinal pathology, particularly in the photoreceptors and RPE; in addition, *rod* disc shedding and phagocytosis were the same as had been previously described in the human retina (Spitznas and Hogan 1970), and in the retinas of other mammals (Young 1971a; Anderson and Fisher 1975, 1976). It is possible, however, that an abnormal condition may have been produced in the cones or RPE

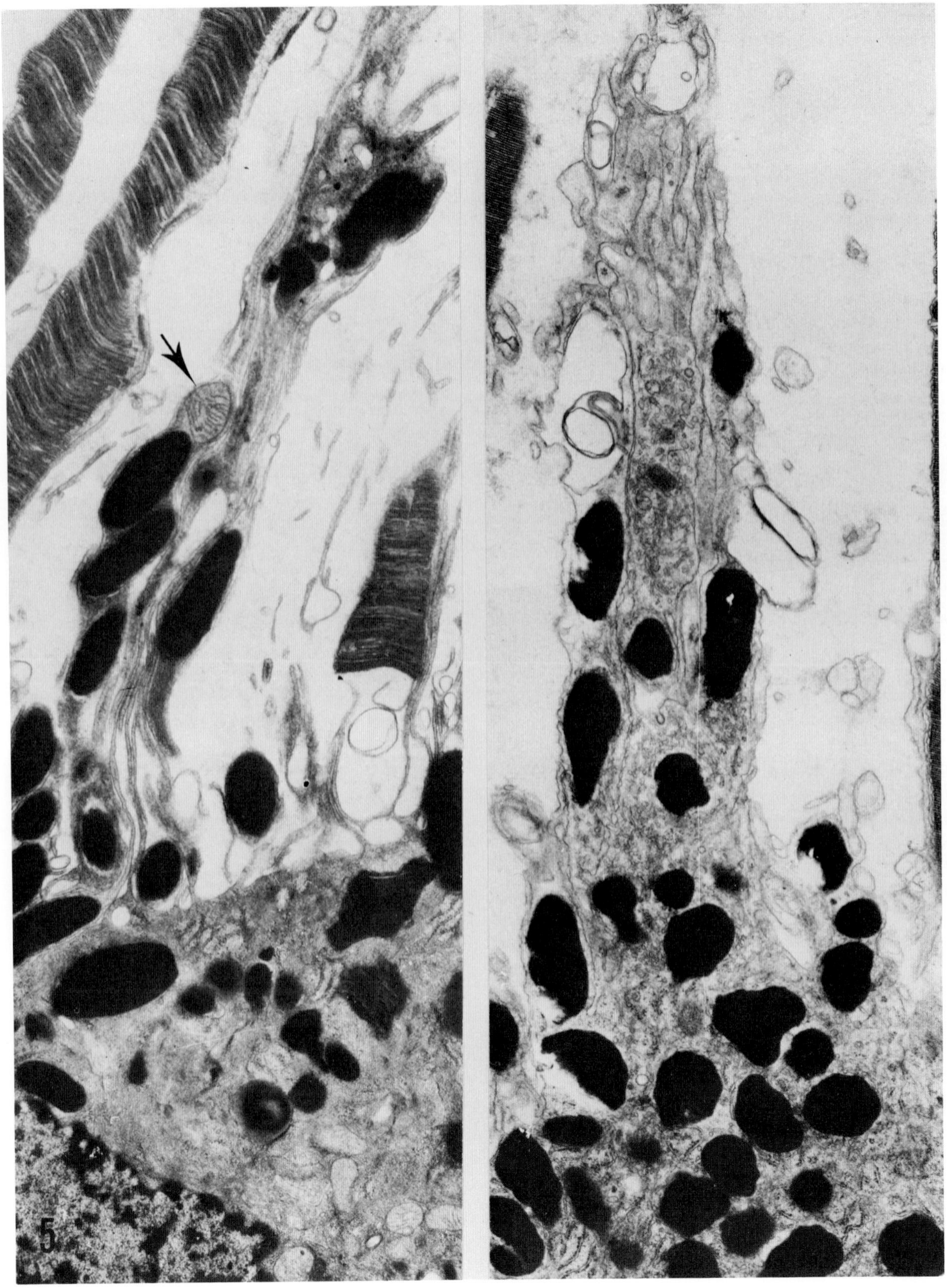
5

← **Figure 2.5** *Electron micrograph of a longitudinal section through pigment epithelial processes at their apical origin. A mitochondrion appears below one of the pigment granules (arrow) and there are several dense bodies further vitreal in the processes. The apical one-half of the pigment epithelial soma appears at the bottom of the micrograph, and two rod phagosomes, having a lamellar structure, appear in the cell above a rod. Five-year-old retina. (From Steinberg, Wood, and Hogan 1977. ×10,075.)*

← **Figure 2.6** *Electron micrograph of a longitudinal section through pigment epithelial apical processes at their apical origin. Sixty-year-old retina. (From Steinberg, Wood, and Hogan 1977. ×10,075.)*

by the light exposure received in preoperative ophthalmoscopic examinations. The same events were present, however, in a rhesus monkey retina, which definitely had no prior history of unusual light exposure (Anderson, Fisher, and Steinberg 1978). Recently, disc shedding from cones has also been observed in nonmammalian retinas (O'Day and Young 1978; Young 1977a,b).

The events of cone disc shedding and phagocytosis in human retina closely resemble those previously described for human rods (Spitznas and Hogan 1970) and monkey rods (Young 1971a). Often, curling of the edges of the terminal packet of discs is the earliest event observed by electron microscopy (Figs. 2.16, 2.17), and in that case, as with monkey rods, there is then no indication of apical-process participation. In contrast, and as described for human rods (Spitznas and Hogan 1970), apical processes sometimes are observed to extend pseudopodia across the tip of the cone outer segment, isolating a terminal packet, whose discs have not curled (Steinberg, Wood, and Hogan 1977). We suggested that this may represent a rapid RPE response to a signal from the outer segment that has as yet no ultrastructural image; while disc curling, in the absence of RPE participation, indicates a relatively sluggish RPE response to the same signal (Steinberg, Wood, and Hogan 1977).

In the next stage, a terminal disc-packet appears within an apical process positioned above the cone outer-segment tip. The discs are enclosed within cone plasma membrane, which, in turn, is surrounded by RPE plasma membrane (Fig. 2.17). A phagosome can consist of a few discs or as many as 60, and the disc membranes are more electron-dense than those of the intact outer segment. Often, the phagosome turns 90° to present a smaller diameter to the apical process (Fig. 2.8). In subsequent stages phagosomes are observed at higher and higher levels above the cone outer segment, and there can be more than one phagosome within the processes above a single cone. The cone phagosomes have some of the appearances that are associated with degeneration and with digestion by lysosomal enzymes—blurring of disc edges, distortion and compression of disc membrane, shrinkage and irregularity of outer membranes—and these changes are more pronounced in the phagosomes that are closest to the cell body of the epithelial cell (Figs. 2.18, 2.19). This strongly suggests that fusion of the phagosome with primary lysosomes occurs in the apical process, and that digestion proceeds as the phagosome moves up the process toward the cell body. Fusion of a vesicle, perhaps a primary lysosome, with a cone phagosome in an apical process has been observed in human retina (Anderson, Fisher, and Steinberg 1978).

As the phagosome ascends in the apical process, a space appears between the phagosome and the cone tip; the process now becomes sheath-like rather than villous-like above the outer segment (Fig. 2.17). Individual apical processes may contain more than one phagosome. This is consistent with the observation that an apical process often continues to ensheath the outer-segment tip after it has phagocytized a packet of discs; it is, therefore, ready to phagocytize the next group. Sometimes the next phagosome appears in a different apical process, indicating that the process that phagocytized the first packet may have come off the cone or have been actively displaced by the new process.

The events of cone disc shedding and phagocytosis are similar in the macular cones and in the more peripheral cones. Since the tips of macular cones are closer to the RPE cell soma, the phagosomes must spend less time in the apical processes, which may explain why macular phagosomes in the processes show less evidence of degeneration. Macular cone phagosomes can be seen entering the soma from the processes, and can also be seen well within the soma above the cone tips (Figs. 2.18, 2.19). Although shedding and phagocytosis are difficult to quantify, they seemed more active in the macular region of the one retina examined, which was from a five-year-old. Multiple phagosomes were common above individual cones, and these were more variable in size than they were more peripherally. Some rather large phagosomes were observed in the macula, which contained as many as 100 discs (Fig. 2.19). The events of cone disc shedding and phagocytosis are summarized in Figure 2.20.

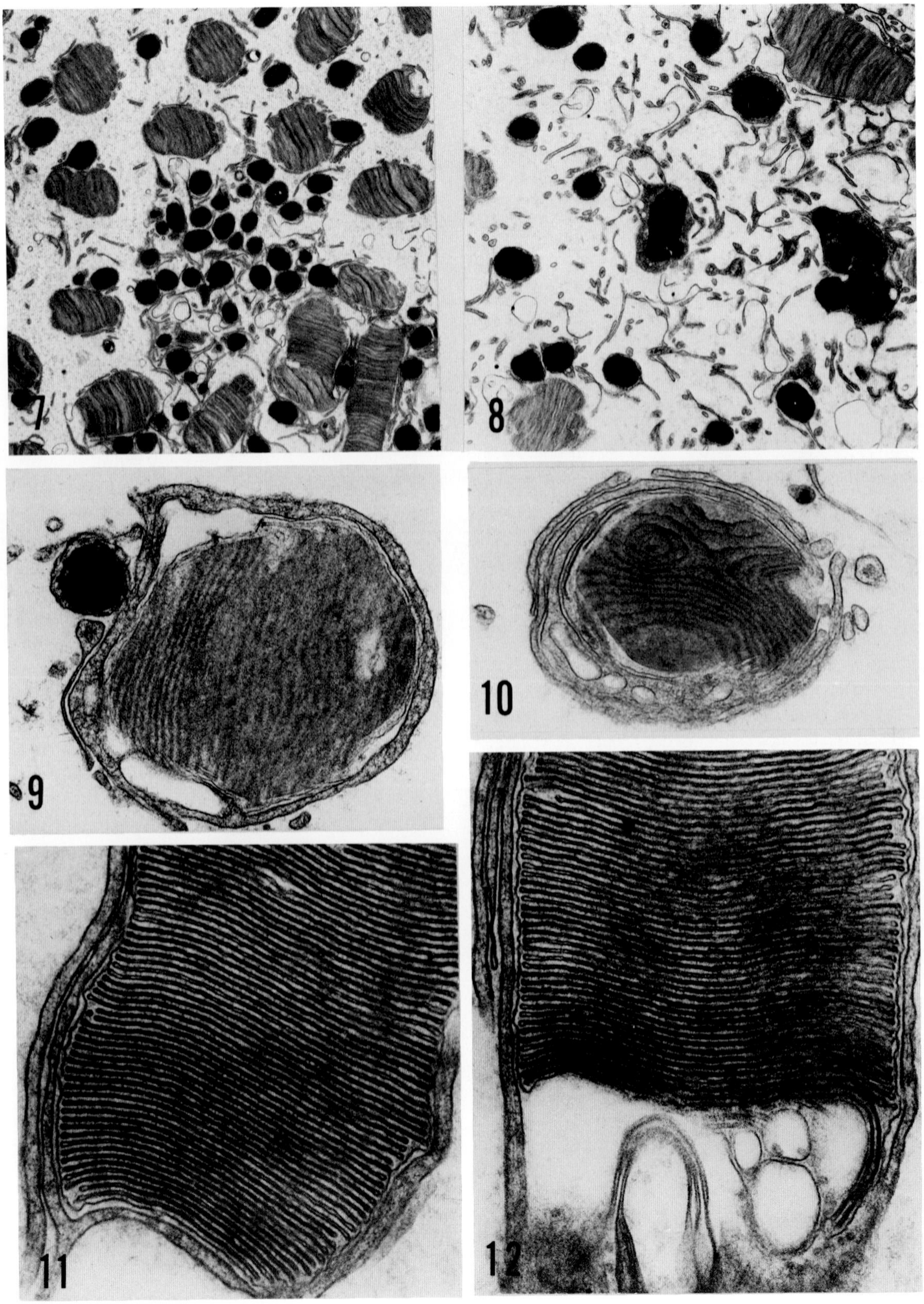
7
8
9
10
11
12

Figure 2.7 *Electron micrograph of a transverse section through a supracone space. The space is surrounded by rod outer segments. It contains apical processes that assume a variety of thinly shaped profiles, and many of them contain pigment granules. Five-year-old retina. (From Steinberg, Wood, and Hogan 1977. ×3,125.)*

Figure 2.8 *Electron micrograph of a transverse section through a supracone space. The space is surrounded by rods. An apical process that contains a phagosome occurs in the center of the space and it is surrounded by other processes that contain pigment granules. Five-year-old retina. (From Steinberg, Wood, and Hogan 1977. ×4,485.)*

Figure 2.9 *Electron micrograph of a transverse section through a cone outer segment, somewhere along its outer one-third. One pigment epithelial process completely ensheaths the outer segment. There also is a circular profile of a process that contains a dense body. Five-year-old retina. (From Steinberg, Wood, and Hogan 1977. ×26,845.)*

Figure 2.10 *Electron micrograph of a transverse section through a cone outer segment, somewhere within its outer one-third. The outer segment is ensheathed by multiple pigment epithelial processes. Five-year-old retina. (From Steinberg, Wood, and Hogan 1977. ×21,240.)*

Figure 2.11 *Electron micrograph of a longitudinal section through a cone outer segment at its tip. Two pigment epithelial apical processes on the left and one on the right extend alongside the outer segment. The inner process on the left bifurcates at the outer-segment tip, and one portion of it extends along the top surface of the outer segment. Five-year-old retina. (From Steinberg, Wood, and Hogan 1977. ×37,760.)*

Figure 2.12 *Electron micrograph of a longitudinal section through a cone outer segment at its tip. Three apical processes on the left and one on the right extend alongside the outer segment. Five-year-old retina. (From Steinberg, Wood, and Hogan 1977. ×37,205.)*

Conclusion

We have seen that the rods and cones of human retina associate with the RPE apical processes in very similar ways. The tips of the outer segments of both photoreceptor types are ensheathed by these processes. This similarity in form reflects a similarity in function—the shed terminal disc packets of both the rods and cones are phagocytized by the ensheathing apical processes. Even though the cone outer segments are farther from the apical surface than the rods, it is clear that the ensheathing apical processes provide the cones with as close an anatomical association to the pigment epithelium as the rods have.

Other functions may depend upon the association between cone outer segments and apical processes and be similar for rods and cones—for example, the rapid interchange of metabolites, such as vitamin A (Dowling 1960) and amino acids (Young 1969; Miller and Steinberg 1976).

It is now certain that any disruption of this relationship, whether for rods or cones, will have pathological consequences. A retinal detachment, for example, can be thought of, as far as the photoreceptors are concerned, as a separation of the outer segment from its apical processes. After reattachment, return of function should depend, to some extent, upon a restoration of this anatomical relationship. It is obvious that this may be more difficult for the cones than for the rods. Each cone will not be reapposed to its apical processes at the original site and other or new apical processes will have to find each cone outer segment. It has not yet been ascertained that the normal relationship can recover, although it has been shown that cone outer segments, which degenerate following detachment, can regrow (Kroll and Machemer 1969).

Figures 2.13–2.20 will be found on pp. 40–43.

Figure 2.13 *Electron-micrograph montage of a longitudinal section through an extramacular cone outer segment, portion of inner segment, and pigment epithelial apical processes. One of the processes is thick and contains a variety of intracellular organelles, including a mitochondrion (arrow). A thick process extends along the left side of the outer segment. Five-year-old retina. (From Steinberg, Wood, and Hogan 1977. ×2,567.)*

Figure 2.14 *As in Figure 2.13, but from the macular region, where cone outer segments end much closer to the pigment epithelium. Pigment granules in the apical processes almost reach the tip of the outer segment. Five-year-old retina. (×2,954.)*

Figure 2.15 *As in Figure 2.13, but from the parafoveal region, nearer to the fovea (area centralis) than Figure 2.14. In this region the cone outer segments almost reach the apical surface of the retinal pigment epithelium. Five-year-old retina. (×2,656.)*

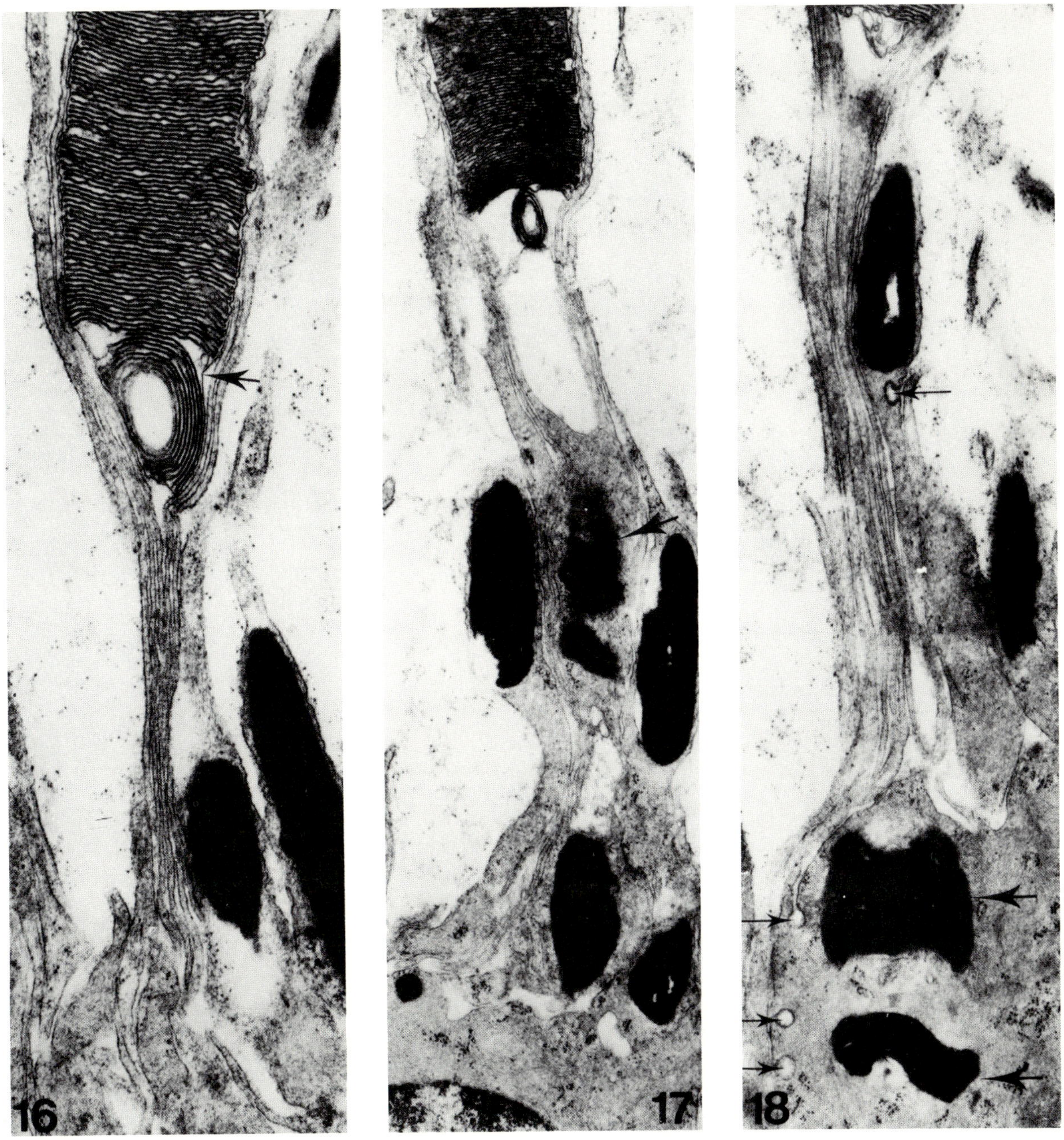

Figure 2.16 *Electron micrograph of a longitudinal section through the tip of a parafoveal cone outer segment and the pigment epithelial apical processes. The outer-segment tip is ensheathed by the apical processes. A terminal packet of nine discs has curled into a whorl, and forms a bulge at the tip of the outer segment. The plasma membrane has begun to pinch in on one side (arrow). Five-year-old retina. (×25,700.)*

Fig. 2.17 *As in Figure 2.16, but from a different parafoveal cone. Here, a packet of three discs has whorled at the tip. A phagosome appears within the apical processes (arrow) on its way to the interior of the retinal pigment epithelial cell. Five-year-old retina. (×21,020.)*

Figure 2.18 *As in Figure 2.16, but from a different parafoveal cone. Two phagosomes have entered the soma of the cell (large arrows) and are positioned near the apical membrane. Note the coated vesicles situated in the apical membrane at various points (small arrows). Five-year-old retina. (×22,090.)*

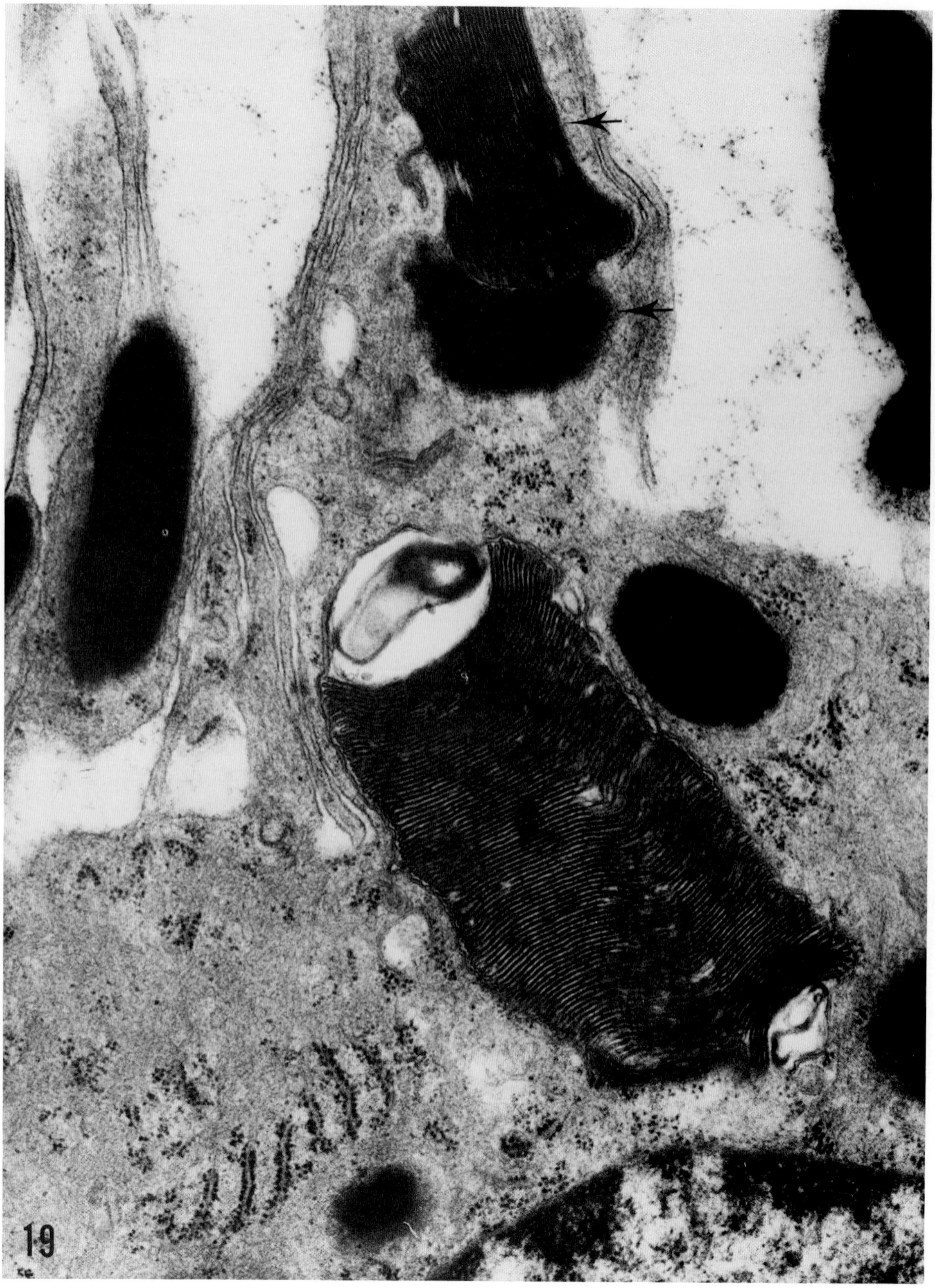
19

← **Figure 2.19** *Electron micrograph of a longitudinal section through pigment epithelial apical processes associated with a parafoveal cone at the apical border of the retinal pigment epithelium. Two phagosomes (arrows), positioned within an apical process, are about to enter the soma. A very large cone phagosome already has entered the soma. This phagosome contains about 100 discs. Five-year-old retina. (×44,460.)*

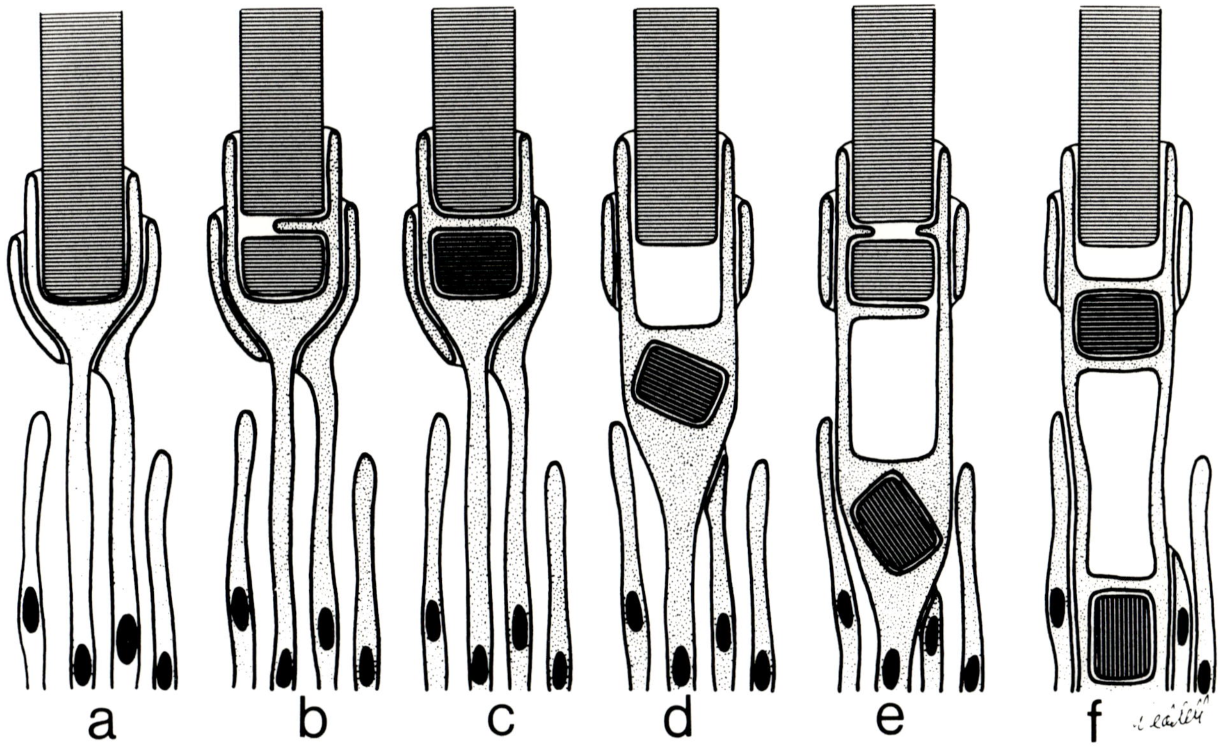

Figure 2.20 *Summary diagram of phagocytosis. Long villous-like pigment epithelial apical processes reach toward the cone from the pigment epithelium. The processes that reach the cone (two are arbitrarily shown) expand to ensheath a portion of the outer segment* (a). *The sequence* a–f *shows the successive phagocytosis of two packets of discs by one apical process. We assumed that separation of the packet and resealing of outer-segment-plasma membrane occurred before phagocytosis, in* b *and* e. *The forms assumed by apical process pseudopodia in* b *and* e *were arbitrarily selected to illustrate several possibilities. The discs in the packets are shown to become more osmiophilic than those in the outer segment, after they have been phagocytized. Note that as the first phagosome ascends in the process* d–f *the sheath-like region of the process lengthens. Phagocytosis of the second packet of discs might also occur by a different process that somehow displaces the first process. (Diagram drawn by J. Weddell. From Steinberg, Wood, and Hogan 1977.)*

cilia extend from the outer neuroblasts towards the RPE. Microvilli are observed surrounding the outer segments of the photoreceptors after 26 weeks of gestation.

Lipofuscin

Lipofuscin, the "wear-and-tear" pigment, which is a residue of activity, was initially noted in the RPE at 2 years of age by Feeney, Grieshaber, and Hogan (1965) and at 16 months of age by Streeten (1961). In early stages, up to 10 months of age, we did not find lipofuscin in any of the epithelial layers. However, we did observe definite lipofuscin in the choroid and iris stroma at 40 weeks gestation. It is interesting to speculate on the significance of this material at birth if it is truly related to aging.

Novikoff et al. (1973) have demonstrated close spatial relations of lipofuscin granules with smooth endoplasmic reticulum and microperoxisomes in human hepatocytes. Microperoxisomes have been found in the RPE of rodents (Leuenberger and Novikoff 1975) and rarely in Müller's cells and vascular endothelium. These organelles occur in cells involved with lipid transport, metabolism and storage.

Bruch's Membrane

Bruch's membrane develops by stages: At 7 weeks of gestation, the age of the earliest specimen studied by electron microscopy, a basement membrane of the retinal pigment epithelium can be identified, and a variable space between it and the choriocapillaris is seen by electron microscopy.

At 9 weeks of gestation, the basement membrane of the pigment epithelium is more dense, the space between it and the choriocapillaris is more uniform, and a delicate basement membrane starts to appear around the choriocapillaris. Amorphous granular material occupies the space between the planes of the two basement membranes.

At 11 weeks of gestation, Bruch's membrane is first recognizable by light microscopy, being most easily seen with the aid of the Masson trichrome stain.

At 12 weeks gestation, scattered collagen fibrils are first observed within the plane of the now-uniform space, which is filled with amorphous material lying external to the plane of the pigment epithelial basement membrane. The basement membrane of the segments of the choriocapillaris is also well developed (Fig. 3.8). Thus, Bruch's membrane is formed by a layer of collagen between the basement membranes of the retinal pigment epithelium and the choriocapillaris. The choriocapillaris portion is less complete than the RPE portion, because of the interrupted pattern of the choriocapillaris.

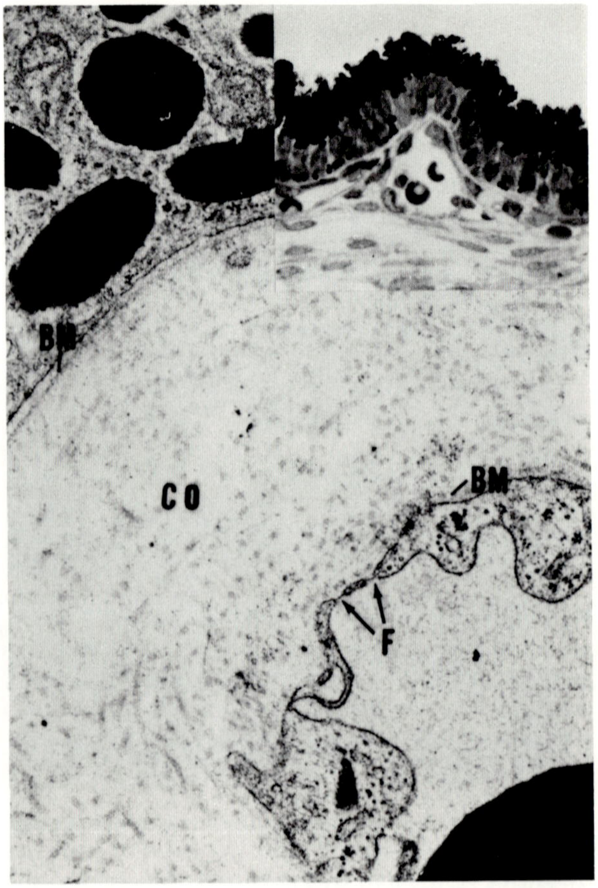

Figure 3.8 *By 12 weeks of gestation, the zone of Bruch's membrane is more highly organized into its characteristic planes. The "collagen layer" now contained more recognizable collagen fibrils* (CO). *The basement membranes of the pigment epithelium and the choriocapillaris* (BM) *are present. Fenestrae* (F) *are clearly visible in the capillary endothelium. (AFIP Neg. 71–1178–11. ×36,600.) Inset shows similar region (also at 12 weeks of development) of pigment epithelium and anterior choroid by light microscopy. (Toluidine blue. AFIP Neg. 70–10623. ×366.)*

The choriocapillaris is first seen at 7 weeks as an incomplete chain of capillaries outside the retinal pigment epithelium. Fenestrae are not consistently demonstrated, but have been seen as early as 12 weeks of gestation. As the fetus matures, the capillaries appear closer together, particularly posteriorly, and the basement membrane around each one becomes more distinct.

REFERENCES

BREATHNACH, A. S., and WYLLIE, L. M. 1966. Ultrastructure of retinal pigment epithelium of the human fetus. *J. Ultrastruct. Res.* 16:584–597.

COULOMBRE, A. J., STEINBERG, S. M., and COULOMBRE, J. L. 1963. The role of intraocular pressure in the development of the chick eye. V. Pigmented epithelium. *Invest. Ophthalmol.* 2:83–89.

DUKE-ELDER, S., ed. 1963. System of ophthalmology, vol. 3, part 1. London: Kimpton.

FEENEY, L. 1973. The phagolysomal system of the pigment epithelium: a key to retinal disease. *Invest. Ophthalmol.* 12:635–638.

FEENEY, L., GRIESHABER, J. A., and HOGAN, M. J. 1965. Studies on human ocular pigment. In *Eye Structure, II Symposium,* ed. J. W. Rohen, p. 535. Stuttgart: Schattauer-Verlag.

FINE, B. S. 1961. Limiting membranes of sensory retina and pigment epithelium. *Arch. Ophthalmol.* 66:847–860.

FITZPATRICK, T. B., MIYAMOTO, M., and ISHIKAWA, K. 1967. The evolution of concepts of melanin biology. *Arch. Dermatol.* 96:305–323.

FONT, R. L., ZIMMERMAN, L. E., and ARMALY, M. F. 1974. The nature of the orange pigment over choroidal melanoma. *Arch. Ophthalmol.* 91:359–362.

FRANCOIS, J., RABAEY, M., and LAGASSEE, A. 1963. Electron microscopic observations on choroid, pigment epithelium and pecten of the developing chick in relation to melanin synthesis. *Ophthalmologica* 146:415–431.

GARRON, L. K. 1963. The ultrastructure of the retinal pigment epithelium with observations on the choriocapillaris and Bruch's membrane. *Trans. Am. Ophthalmol. Soc.* 61:545–588.

HOGAN, M. J. 1972. Role of the retinal pigment epithelium in macular disease. *Trans. Amer. Acad. Ophthalmol. Otolaryngol.* 76:64–80.

HOLLENBERG, M. J., and SPIRA, A. W. 1972. Early development of the human retina. *Can. J. Ophthalmol.* 7:472–491.

ISHIKAWA, T., and YAMADA, E. 1970. The degradation of the photoreceptor outer segment within the pigment epithelial cell of rat retina. *J. Electron. Microsc.* 19:85–99.

KOLB, H., and GOURAS, P. 1974. Electron microscopic observations of human retinitis pigmentosa, dominantly inherited. *Invest. Ophthalmol.* 13:487–498.

KROLL, A. J., and MACHEMER, R. 1968. Experimental retinal detachment in the owl monkey. III. Electron microscopy of retina and pigment epithelium. *Am. J. Ophthalmol.* 66:410–427.

KUWABARA, T., and GORN, R. A. 1968. Retinal damage by visible light. *Arch. Ophthalmol.* 79:69–78.

LAVAIL, M. M., SIDMAN, R. L., and O'NEIL, D. 1972. Photoreceptor-pigment epithelial cell relationships in rats with inherited retinal degeneration. Radioautographic and electron microscope evidence for a dual source of extra lamellar material. *J. Cell Biol.* 53:185–209.

LERCHE, W., and WULLE, K. G. 1967. Uber die genese der melaningranula in der embryonalen menschlichen retina. *Z. Zellforsch* 76:452–457.

LEUENBERGER, P. M., and NOVIKOFF, A. B. 1975. Studies on microperoxisomes VII. Pigment epithelial cells and other cell types in the retina of rodents. *J. Cell Biol.* 65: 324–334.

MANN, I. 1969. The development of the human eye. 3rd ed. New York: Grune and Stratton.

MEIER-RUGE, W. 1968. The pathophysiological morphology of the pigment epithelium and its importance for retinal structure and functions. *Mod. Probl. Ophthalmol.* 8:32.

MOYER, F. 1961. Electron microscopic observations on the origin, development and genetic control of melanin granules in the mouse eye. In *Structure of the Eye,* ed. G. K. Smelser, p. 469. New York: Academic Press.

MUND, M. L., RODRIGUES, M. M., and FINE, B. S. 1972. Light and electron microscopic observations on the pigmented layers of the developing human eye. *Am. J. Ophthalmol.* 73:167–182.

NAKAO, F. 1968. Electron microscopic studies of the retinal pigment epithelium. *Acta Soc. Ophthalmol. Jap.* 72:1049.

NOVIKOFF, A. B., NOVIKOFF, P. M., QUINTANA, N., and DAVIS, C. 1973. Studies in microperoxisomes. IV. Interrelations of microperoxisomes, endoplasmic reticulum and lipofuscin granules. *J. Histochem. Cytochem.* 21:1010.

RODRIGUES, M. M., SPAETH, G. L., WEINREB, S., and SIVALINGAM, E. 1976. The spectrum of trabecular pigmentation. *Trans. Am. Acad. Ophthalmol. Otolaryngol.* 81:258–276.

SELIGMAN, A. M., HANKER, J. S., WASSERKRUG, H. DMOCHOWSKI, H., and KATZOFF, J. 1965. Histochemical demonstration of some oxidized macromolecules with thiocarbohydrazide (TCH) or thiosemicarbazide (TSC). *J. Histochem. Cytochem.* 13:629–639.

SHIELDS, J. A., RODRIGUES, M. M., TASMAN, W., SARIN, L., and ANNESLEY, W. 1976. Lipofuscin pigment over benign and malignant choroidal tumors. *Trans. Amer. Acad. Ophthalmol. Otolaryngol.* 81:871–881.

SMITH, L. T., and IRVINE, A. 1973. Diagnostic significance of orange pigment accumulation over choroidal tumors. *Am. J. Ophthalmol.* 76:212–216.

SPENCER, W. H. 1972. Renaissance of the retinal pigment epithelium. *Arch. Ophthalmol.* 88:1.

STREETEN, B. W. 1961. The sudanophilic granules of the human retinal pigment epithelium. *Arch. Ophthalmol.* 66:391–398.

TSO, M. O. M. 1973. Photic maculopathy in rhesus monkey: a light and electron microscopic study. *Invest. Ophthalmol.* 12:17.

TSO, M. O. M., and FRIEDMAN, E. 1967. The retinal pigment epithelium. I. Comparative histology. *Arch. Ophthalmol.* 78:641–649.

TSO, M. O. M., and FRIEDMAN, E. 1968. The retinal pigment epithelium. III. Growth and development. *Arch. Ophthalmol.* 80:214–216.

WALLOW, I. H. L., and TSO, M. O. M. 1972. Proliferation of the retinal pigment epithelium over malignant choroidal tumors: a light and electron microscopic study. *Am. J. Ophthalmol.* 73:914–926.

WEIDMAN, T. A., and KUWABARA, T. 1968. Postnatal development of the rat retina. *Arch. Ophthalmol.* 79:470–484.

YAMADA, E., and ISHIKAWA, T. 1965. Some observations on the submicroscopic morphogenesis of the human retina. In Eye Structure, II Symposium, ed. J. W. Rohen, p. 5. Stuttgart: Schattauer-Verlag.

4 Roles of the Retinal Pigment Epithelium in the Development of Ocular Tissues

Alfred J. Coulombre

THE EMBRYONIC retinal pigmented epithelium (RPE) is a partner with other emerging tissues of the developing eye in a sequence of tissue interactions. The RPE is not only the target of influences from adjoining tissues that channel its progressive differentiation, it is also the source of influences that elicit the differentiation of other ocular tissues, control their growth in a coordinated fashion, and regulate their morphogenesis in such a manner that the organ they construct is suitably organized and shaped for optical function. Understandably, the steps by which the RPE matures and plays its sequence of diverse roles in development has been elucidated largely by work with nonhuman embryos. In exploring briefly how this simple epithelium exerts or responds to influences as it makes its place in the community of ocular tissues, we will consider its origin, its growth, and some of the functional roles it plays as it becomes more and more highly differentiated. Finally, we will examine the ability of RPE cells to give up their state of specialization and express some other state of differentiation, a phenomenon referred to as metaplasia or transdifferentiation.

Embryological Origins

The developmental ancestry of the RPE has been traced to cells in the retinal field located anterior to the head process in the open neural plate (Adelmann 1929; Woerdeman 1929). The retinal field is induced in the neural ectoderm by the prechordal mesoderm (Adelmann 1930, 1937; Mikami 1941; Waechter 1953). At this early stage, the precursor cells are not differentiated as RPE; they are not even committed definitively as RPE; they are simply destined, in the course of normal development, to fall under influences that will assure their maturation as RPE cells. Each retinal field evaginates from the forebrain as an optic vesicle that is the anlage of the neural retina and of the RPE.

At about this time, an event peculiar to mammals occurs. In female mammals, the sex chromosomal constitution is XX. In each cell, one of the X chromosomes is inactivated relatively early in embryonic development. Since there seems to be equal probability in any given cell that the maternally derived or the paternally derived X chromosome will become inactive, each tissue in such females becomes a mosaic of clones each derived from a progenitor cell with either an active, maternally derived X chromosome or an active, paternally derived X chromosome. Deol and Whitten (1972) elegantly demonstrated this phenomenon in mice heterozygous for an X-linked gene for albinism. In animals of this constitution, the phenotypic expression of random X-chromosomal inactivation became visible later in development, when pigmentation occurred, as clonal patches of pigmented and unpigmented cells in the RPE.

Contact between the tip of the optic vesicle and the lateral head ectoderm initiates the reciprocal induction of lens in the head ectoderm and of neural retina at the apex of the optic vesicle (Mikami 1941; Lopashov 1963). The presumptive neural retina will become the inner wall of the optic cup when the optic vesicle invaginates. The rest of the optic vesicle, a cuff of neural ectoderm continuous with the presumptive neural retina distally and with the brain wall proximally, is brought into contact with periocular mesenchyme. This contact induces the RPE (Lopashov 1963), which becomes the outer wall of the optic cup following invagination of the optic vesicle.

The proliferative activity of the RPE suddenly and markedly decreases following its induction by periocular mesenchyme. This withdrawal from mitosis occurs at 3–3.5 days of incubation in the chick embryo (Coulombre 1955; Zimmerman 1975). Although the decrease in mitotic activity is sudden enough to approach a synchronous withdrawal, and while it is complete enough to affect almost the entire population of RPE cells, two points must be made. First, a small number of mitoses occur in this population throughout the growth of the eye. Furthermore, although the RPE cells no longer divide actively in untreated embryos, they have not lost the ability to do so, for they divide rapidly when they are induced to transdifferentiate in vivo (Coulombre and Coulombre 1965, 1970) or when they are explanted into tissue cultures (Cahn and Cahn 1966; Coon and Cahn 1966; Newsome et al. 1974).

Growth

The profound and early reduction in mitotic rate in the RPE has at least two consequences. First, at the time this reduction occurs, the RPE is relatively minute in area. During the subsequent growth of the eye, the RPE must expand appreciably in area to match the growth of the neural retina, to which it remains intimately apposed throughout development. This apposition of the two retinal layers is maintained because the accumulation of vitreous substance within the globe effects the expansion of the RPE sheet so that it is, at all times following the closure of the choroid fissure, equal in area to the growing neural retina (Coulombre et al. 1963). The consequent manifold expansion in the area of the RPE is effected predominantly by increase in the area of its cells, and very little by the addition of new cells (Coulombre et al. 1963; Tso and Friedman 1968). Second, because the neural retina continues to expand by mitotic addition of cells at its margin until a much later stage in development than does the RPE, the neural retina and the RPE must remain free to slip over one another; otherwise foldings and other distortions would occur at the edge of the optic cup. In fact, the two layers of the optic cup can glide over one another until the RPE cell processes and the outer segments of the visual cells develop, interdigitate, and lock them together. This event occurs after the last cells are added to the neural retina by mitosis (Coulombre 1955).

Concomitantly with their withdrawal from mitosis, the RPE cells begin to synthesize tyrosinase (Harrison 1951), to elaborate premelanosomes (Moyer 1961), to synthesize collagen, which is deposited beneath the basal surfaces (Newsome and Kenyon 1972), and to synthesize melanin, which is deposited in the melanosomes (Zimmerman 1975).

Developmental Roles of the RPE

The RPE plays a succession of roles during embryonic development. These include inductive interaction, phagocytic activity, trophic influence, transport function, and retinomotor response, among others.

The RPE induces the scleral coat in periocular mesenchyme, which is largely neural crest (Reinbold 1968; Newsome 1972). The available evidence suggests that this induction is mediated by a basement membrane deposited beneath the RPE at an early stage in development (Newsome 1976). This inductive potency of the RPE is transitory and is lost at intermediate stages in development. It is still untested whether the other concentric coats of the eye wall (such as Tenon's capsule or the choroid coat) are similarly induced, the outermost first in development and the innermost last. If such a sequence of inductions occurs, it is likely that each is mediated by a separate basement membrane produced by and deposited beneath the basal surface of the RPE. This would involve the production, in proper order, of a series of qualitatively different basement membranes by the developing RPE. Presumably, such a sequence is terminated by the deposition of Bruch's membrane.

Uptake (Young 1971) and breakdown (Ishikawa and Yamada 1970) of apical segments of the rod outer segments is a well-established function of the mature RPE. Phagocytic function, more broadly defined, is present from early developmental stages, long before the appearance of the rod outer segments or the RPE cell processes, since, at the optic-cup stage, the RPE cells of the frog take up melanin granules that have been ejected from the cells of the primitive neural retina (Hollyfield 1972). Thus, the phagocytic function of the RPE develops early, is not confined in its specificity to the uptake of rod outer segment material, and is (at least in early development) independent of the presence of the RPE cell processes.

At least one trophic function has been demonstrated for the embryonic RPE. The outer segments of the differentiating visual cells remain rudimentary when the neural retinal cells are not apposed, apex to apex, to the cells of the RPE (Johnson 1935; Hollyfield and Witkovsky 1974). The basis for this trophic interaction has not been elucidated.

Interposed, as it is, between the vascular bed of the choroid coat and a neural retina that is avascular in most submammalian vertebrates, the RPE would be expected to engage in active transport. Transport of retinol, for example, would be especially important because of the central role played by vitamin A in visual photochemistry. A receptor for retinol is present at least as early as 11 days of incubation in the RPE of the chick embryo (Wiggert and Chader 1975). It is present in the soluble fraction, binds retinol reversibly and noncovalently, and is a protein of approximately 19,000–20,000 daltons. Its sedimentation properties differ from the retinol-binding protein that is present in the serum of the blood. This receptor, which is probably involved in transport of retinol across the RPE, is specific since its interaction with retinol is affected little by the presence of retinal or retinoic acid. Retinol receptors in the RPE appear to differ in different species, both qualitatively and in the number of types present (Wiggert et al. 1976).

Metaplastic Transformations of the RPE

Although the RPE cells become differentiated melanocytes, they are not immediately and irreversibly locked into this particular differentiated state (Dragomirow 1932). In at least some species, RPE cells retain, for a time following their differentiation, the capacity to transdifferentiate, in response to appropriate challenge, to form neural retinal tissue, lens cells, or iris epithelium. Among urodeles, the embryonic pigmented epithelium forms neural retina when it is brought into contact with nasal placode or with otocyst (Dragomirow 1936; Ikeda 1937; Detwiler and van Dyke 1953, 1954). The metaplastic formation of neural retina from RPE can be initiated by simple physical separation of the embryonic RPE from the neural retina (Orts Llorca and Génis Gálvez 1960; Coulombre and Coulombre 1965, 1970). This metaplastic capacity tends to be lost with age. Early in embryonic development the RPE can give rise to neural retina in fish (Dabaghian 1959), birds (Dorris 1938; Reinbold 1968; Orts Llorca and Génis Gálvez 1960; Coulombre and Coulombre 1965, 1970) and mammals (Stroeva 1960). In anurans this capacity is retained until metamorphosis (Sato 1933; Lopashov 1949). In some urodeles the capacity persists throughout life (Wachs 1920; Ikeda 1937; Sperry 1944, 1945; Stone 1950a, 1950b; Reyer 1956; Stone and Steinitz 1957; Hasegawa 1958). In these forms (many of the Salimandridae) removal of the neural retina elicits from the RPE a metaplastic regeneration of an entire neural retina. Axons from the ganglion cells of such regenerates grow to and reinervate appropriate centers in the brain in such a way that the topographic mapping of the retina in the brain is reconstructed normally and that vision is restored. This phenomenon suggests that the topographic map of the neural retina is duplicated in the RPE.

Conclusion

Failure of the developing RPE to exert influences on neighboring tissues, to respond appropriately to influences from adjoining tissues, or to successfully complete steps in its pathway of differentiation, has more or less serious consequences for subsequent development of the eye. Failure of induction of the eye field by prechordal mesenchyme results in anophthalmus, since, in this circumstance, the neural and pigmented portions of the retina (prime movers in subsequent ocular morphogenesis) fail to develop. Failure of the eye field to separate and lateralize results in cyclopia or in different degrees of synophthalmus. X-chromosomal inactivation in mammalian females leads to a mosaic of focal defects in the RPE when the remaining active X chromosome bears a defective gene that is expressed in the RPE. Failure of the mechanism by which the enlarging vitreous body effects expansion of the RPE, so that its area remains tailored to that of the neural retina, has several consequences. For example, when vitreous substance drains from the developing eye through a colobomatous defect, the globe fails to enlarge (microphthalmia), the RPE fails to increase normally in area, and the neural retina continues to expand in area, becomes folded within the small eye, and degenerates in those regions that fold away from the RPE and the underlying choroid vascular bed. In some submammalian forms this condition is attended by focal metaplastic formation of neural retina from patches of RPE not in contact with the folded neural retina. Local failure of the RPE to induce the sclerotic coat could lead to colobomatous defects (usually associated with retinal cysts) outside the region of the choroid fissure. Failure of proper development of phagocytic function in the RPE is responsible for a category of inherited retinal dystrophies. Should the trophic interaction between the RPE and the visual cells fail to occur, mature outer segments do not develop. Defects in the development of the transport functions of the RPE would be expected to seriously affect the neural retina.

The complex developmental roles played by the RPE assure, when they come into play normally and on schedule, the normal development and function of neighboring tissues. Conversely, when the RPE fails to play its roles normally and on schedule, a range of developmental ocular anomalies and functional deficits results.

REFERENCES

ADELMANN, H. 1929. Experimental studies on the development of the eye. II. The eye forming potencies of the median portions of the urodelan neural plate (*Triturus taeniatus* and *Amblystoma punctatum*). *J. Exp. Zool.* 54: 219–317.

ADELMANN, H. 1930. Experimental studies on the development of the eye. II. The effect of substrate on the heterotopic development of median and lateral strips of the anterior end of the neural plate of *Amblystoma*. *J. Exp. Zool.* 57:223–281.

ADELMANN, H. 1937. Experimental studies on the development of the eye. IV. The effect of the partial and complete excision of the precordal substrate on the development of the eyes of *Amblystoma punctatum*. *J. Exp. Zool.* 75:199–227.

CAHN, R., and CAHN, M. 1966. Heritability of cellular differentiation: clonal growth and exprėssion of differentiation in retinal pigment cells *in vitro*. *Proc. Natl. Acad. Sci. U.S.A.* 55:106–144.

COON, H., and CAHN, R. 1966. Differentiation *in vitro*. Effects of Sephadex fraction of embryo extract. *Science* 153:1116–1119.

COULOMBRE, A. 1955. Correlation of structural and biochemical changes in the developing retina of the chick. *Am. J. Anat.* 96:153–189.

COULOMBRE, A., COULOMBRE, J., and STEINBERG, S. 1963. The role of intraocular pressure in the development of the chick eye. V. Pigmented epithelium. *Invest. Ophthalmol.* 2:83–89.

COULOMBRE, J., and COULOMBRE, A. 1965. Regeneration of neural retina from the pigmented epithelium in the chick embryo. *Dev. Biol.* 12:79–92.

COULOMBRE, J., and COULOMBRE, A. 1970. Influence of mouse neural retina on regeneration of chick neural retina from chick embryonic pigmented epithelium. *Nature* 228:559–560.

DABAGHIAN, H. 1959. Regulative properties of the eye in the embryos of *Acipenseridae*. *Dokl. Akad. Nauk. SSSR* 125:938–940.

DEOL, M., and WHITTEN, W. 1972. Time of X chromosome inactivation in retinal melanocytes of the mouse. *Nature* [New Biol.] 238:159–160.

DETWILER, S., and VAN DYKE, R. 1953. The induction of neural retina from the pigment epithelial layer of the eye. *J. Exp. Zool.* 122:367–384.

DETWILER, S., and VAN DYKE, R. 1954. Further observations on retinal inductions. *J. Exp. Zool.* 126:135–155.

DORRIS, F. 1938. Differentiation of the chick eye *in vitro*. *J. Exp. Zool.* 78:385–415.

DRAGAMIROW, N. 1932. Uber die Entwicklung von Augenbechern aus transplantierten Stuckchen des embryonalen Tapetums. *Arch. Entwicklungsmech.* 126:636–662.

DRAGOMIROW, N. 1936. Uber Inducktion sekundarer Retina im transplantierten Augenbecher bei Triton und Pelobates. *Arch. Entwicklungsmech.* 134:716–737.

HARRISON, J. 1951. *In vitro* analysis of differentiation of retinal pigment in the developing chick embryo. *J. Exp. Zool.* 118:209–241.

HASEGAWA, M. 1958. Restitution of the eye after removal of the retina and lens in the newt, *Triturus pyrrhogaster*. *Embryologia* 4:1–32.

HOLLYFIELD, J. 1972. Egg pigment: early elimination from the developing retina of the frog. *Experientia* 28: 68–69.

HOLLYFIELD, J., and WITKOVSKY, P. 1974. Pigmented retinal epithelium involvement in photoreceptor development and function. *Exp. Zool.* 189:357–377.

IKEDA, Y. 1937. Uber die Bildung akzessorischer Retina aus dem Tapetum bei Hynovius. *Arch. Entwicklungsmech.* 136:676–680.

ISHIKAWA, T., and YAMADA, E. 1970. The degradation of the photoreceptor outer segment within the pigment epithelial cell of the rat retina. *J. Electron Microsc.* 19: 85–99.

JOHNSON, M. 1935. Visual cells of the amphibian retina in the absence of the epithelial pigment layer. *Anat. Rec.* 63:53–75.

LOPASHOV, G. 1949. On the significance of different processes in the restoration of amphibian eyes. *Dokl. Akad. Nauk SSSR* 69:865–868.

LOPASHOV, G. 1963. *Developmental mechanisms of vertebrate eye rudiments,* trans. Jean Medawar. New York: Macmillan.

MIKAMI, Y. 1941. Experimental studies on the development of the eye in Urodele, *Triturus pyrrhogaster* (Boie), *Jap. J. Zool.* 9:303–319.

MOYER, F. 1961. Electron microscope observations on the origin, development and genetic control of melanin granules in the mouse eye. In *The structure of the eye,* ed. G. Smerser, pp. 469–486. New York: Academic Press.

NEWSOME, D. 1972. Cartilage induction by retinal pigmented epithelium of chick embryo. *Dev. Biol.* 27:575–579.

NEWSOME, D. 1976. *In vitro* stimulation of cartilage in embryonic chick neural crest cells by products of retinal pigmented epithelium. *Dev. Biol.* 49:496–507.

NEWSOME, D., FLETCHER, R., ROBISON, W., KENYON, K., and CHADER, G. 1974. Effects of cyclic AMP and Sephadex fractions of chick embryo extract on cloned retinal pigmented epithelium in tissue culture. *J. Cell Biol.* 61:369–382.

NEWSOME, D., and KENYON, K. 1973. Collagen production *in vitro* pigmented retinal epithelium of the chick embryo. *Dev. Biol.* 32:387–400.

ORTS-LLORCA, F., and GÉNIS-GÁLVEZ, J.-M. 1960. Experimental production of retinal speta in the chick embryo: differentiation of pigment epithelium into neural retina. *Acta Anat.* (Basel) 42:31–70.

REINBOLD, R. 1968. Rôle du tapétum dans la differenciation de la sclérotique chez l'embryon de poulet. *J. Embryol. Exp. Morphol.* 19:43–47.

REYER, R. 1956. Lens regeneration from homoplastic and heteroplastic implants of dorsal iris into the eye chamber of *Triturus viridescens* and *Amblystoma punctatum. J. Exp. Zool.* 133:145–189.

SATO, T. 1933. Uber die Determination des fetalen Augenspalts bei *Triton taeniatus. Arch. Entwicklungsmech.* 128:342–377.

SPERRY, R. 1944. Optic nerve regeneration with return of vision in anurans. *J. Neurophysiol.* 7:57–70.

SPERRY, R. 1945. Restoration of vision after crossing of optic nerves and after contralateral transplantation of eye. *J. Neurophysiol.* 8:15–28.

STONE, L. 1950*a*. Neural retina degeneration followed by regeneration from surviving retinal pigment cells in grafted adult salamander eyes. *Anat. Rec.* 106:89–109.

STONE, L. 1950*b*. The role of retinal pigment cells in regenerating neural retinal of adult salamander eyes. *J. Exp. Zool.* 113:9–13.

STONE, L., and STEINITZ, H. 1957. Regeneration of neural retina and lens from retina pigment cell grafts in adult newts. *J. Exp. Zool.* 135:301–318.

STROEVA, O. 1960. Experimental analysis of the eye morphogenesis in mammals. *J. Embryol. Exp. Morphol.* 8:349–368.

TSO, M., and FRIEDMAN, E. 1968. The retinal pigment epithelium. III. Growth and development. *Arch. Ophthalmol.* 80:214.

WACHS, H. 1920. Restitution des Auges nach Extirpation von Retina und Linse bei *Tritonen. Arch. Entwicklungsmech.* 46:328–390.

WAECHTER, H. 1953. Die Inductionsfahigkeit der Gehirnplatte bei Urodelen und ihr medianlaterales Gefalle. *Arch. Entwicklungsmech.* 146:201–274.

WIGGERT, B., BERGSMA, D., and CHADER, G. 1976. Retinol receptors of the retina and pigment epithelium: further characterization and species variation. *Exp. Eye Res.* 22:411–418.

WIGGERT, B., and CHADER, G. 1975. A receptor for retinol in the developing retina and pigment epithelium. *Exp. Eye Res.* 21:143–151.

WOERDEMAN, M. 1929. Experimentelle Untersuchungen uber Lage und Bau der augenbildenden Bizerke in der Medularplatte beim Axolotl. *Arch. Entwicklungsmech.* 116:220–241.

YOUNG, R. 1971. Shedding of discs from rod outer segments in rhesus monkey. *J. Ultrastruct. Res.* 34:190–203.

ZIMMERMAN, J. 1975. The initiation of melanogenesis in the chick retinal pigment epithelium. *Dev. Biol.* 44:102–118.

5 Species Differences in the Retinal Pigment Epithelium

Toichiro Kuwabara

CLOSE EXAMINATION reveals that the retinas of different species are considerably different, although the basic structures and presumably the functions are not too different. The common variations among animals are in vascularity, in macular specialization, in the type of photoreceptor cells, and in the distribution of certain cells such as the horizontal cell. On the other hand, there appears to be little variation among animals in the structure of the retinal pigment epithelium (RPE). The RPE cells form a single cell layer covering the choroidal vascular tissue, the basic structure of which is identical in all animals. However, studies of the fine structure have revealed that the RPE cell has small but definite differences between species and the differences appear to have some relationship to those in the structure of the retina.

Although the fine structure of sensory retinas of various animals has been described extensively, detailed description of the RPE of nonhuman animals is rather scarce. The purpose of this chapter is to present a comparative study of the fine structure of the RPE in animals that may be commonly used in various biomedical experiments. No attempt has been made to cover all species.

The general structure of the RPE of the human has been described in detail by several authors (Bairati and Orzalesi 1963; Breathnach and Wyllie 1966; Spitznas and Hogan 1970; Hogan et al. 1971; Fine and Yanoff 1972; Lerche 1963; and Hogan 1972), but only a few studies on animals are found in the literature. A report on the sheep cell by Leure-DuPree (1968) appears to be one of the most comprehensive, and the RPE of both sheep and cow (Berman et al. 1974) has a structure nearly identical with that in man. The basic structure that has been described in these earlier reports is applicable to all the animals examined in this study. The RPE is a single cell layer forming the outer lining of the subretinal space. The cells are relatively uniform in size and are closely joined by their apicolateral junctions. The apical end of each cell is covered with dense microvilli, which extend into the subretinal space and surround the outer segments. Sakuragawa and Kuwabara (1976) have emphasized the extremely close physical relationship between the retinal outer segments and the numerous microvilli of the RPE. The prominent basement membrane, which forms the inner limit of Bruch's membrane, is found beneath the cell. The structures of the choroidal capillary and Bruch's membrane are strikingly similar in all animals.

The animals that will be described in this chapter were enucleated under light-adapted conditions, generally under ordinary animal-room illumination. Some eyes were removed during the hibernating period of the animal. The enucleated eyes were fixed by immersion in 4 percent glutaraldehyde solution in 0.15 M phosphate buffer (pH 7.2) at room temperature, but prolonged fixation was avoided. The pieces of retina and RPE were postfixed in 1 percent osmium tetroxide in the same buffer solution and embedded in epoxy resin. Thick epoxy sections for light microscopic study were stained with toluidine blue, and ultra-thin sections were stained with uranyl acetate and lead citrate.

Rhesus Monkey (*Macaca mulatta*)

Despite the abundance of retinal studies, the structure of the RPE of the monkey has been described

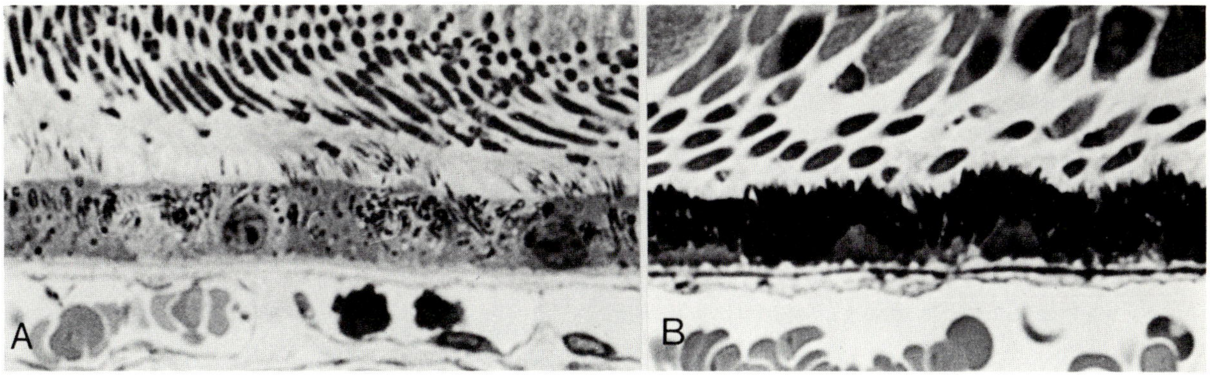

Figure 5.1. *Monkey RPE.* (A) *The posterior zone; the apical surface of the cell is flat.* (B) *The peripheral zone; the cell is bulged and heavily pigmented. (Epoxy sections, toluidine blue staining. ×512.)*

rather sparsely (Sakuragawa and Kuwabara 1976; Young 1971). Cells in the posterior zone of the globe are uniformly hexagonal, 15 μ in diameter and 10 μ in height (Fig. 5.1*A*). The apical cell surface from which innumerable microvilli extend is relatively flat, whereas the cells in the periphery have bulged surfaces. The peripheral cells are short (15–20 μ in diameter and 7–10 μ in height) and heavily pigmented (Fig. 5.1*B*). Also, the microvilli form mounds corresponding to the sites of cone cells that have shorter outer segments. The basal cell membrane is moderately infolded. The lateral cell membrane is smooth and straight. Cells are firmly attached to each other by the junctional complex at the apicolateral portion. The complex consists of zonula occludens and zonula adherens. Conspicuous tonofibrils attach to the lower portion of the junction. This junction complex is termed the apicolateral junction (Figs. 5.2, 5.3). The desmosome is rare on the lateral cell membrane. The microvilli are extremely dense in all cells and are divided into two kinds. Both microvilli extend into the spacious subretinal space with willowy villi, measuring 0.1–0.3 μ in diameter and 5–7 μ long. One type consists of a thick and straight basal portion in which a large pigment granule and some microorganelles are present and transform into fine villi at the tip portion. The other type has no melanin-containing thick portion. The fine villi contain no microorganelles except for fine vesicles and filaments. The tips of the outer segments are closely surrounded by these fine microvilli (Fig. 5.4). The subretinal space is filled with moderately thick fluid, which appears as a fine granular substance in thin sections.

The main cytoplasmic constituent is the smooth endoplasmic reticulum, which is distributed uniformly throughout the cell, except in the basal infoldings and in the fine microvilli. The smooth endoplasmic reticulum is a finely branched saccular membranous structure that contains a weakly stained substance (Figs. 5.2, 5.3). Golgi apparatuses are occasionally found in the para-nuclear zone. A moderate amount of rough endoplasmic reticulum is present among smooth endoplasmic reticulum and it is often arranged into a lamellar pack in the apical portion. The lamellarly arranged rough endoplasmic reticulum is abundant in certain mild pathological conditions, particularly in light-exposed animals. Large mitochondria, cristae of which are well developed, are abundantly present in the lateral and basal regions of the cell. Lysosomes are numerous in the apicolateral and perikaryon zones. Many are primary lysosomes, measuring about 0.1–0.3 μ in diameter. Several secondary lysosomes are also present, depending upon the pathophysiological conditions of the cell. Lamellar bodies, phagocytized outer tips of the photoreceptors, are seen in the cytoplasm. Numerous electron-dense substances that appear to be the degrading product of the lamellar body are regularly present in the cytoplasm. They are also present in thick microvilli (Fig. 5.5). Pigment granules are mainly spindle in shape and are distributed in the apical portion of the cell. Melanin granules consist of homogeneous electron-dense substance. Some large round particles show a striated pattern. However, no structure suggesting melanogenesis is observed in the RPE of the adult monkey. The background cell matrix is electron-lucent and contains ribosomes that frequently form small clusters. Glycogen particles are present in moderate numbers.

The cytoplasm shows considerable variation, depending on the pathophysiological conditions of the

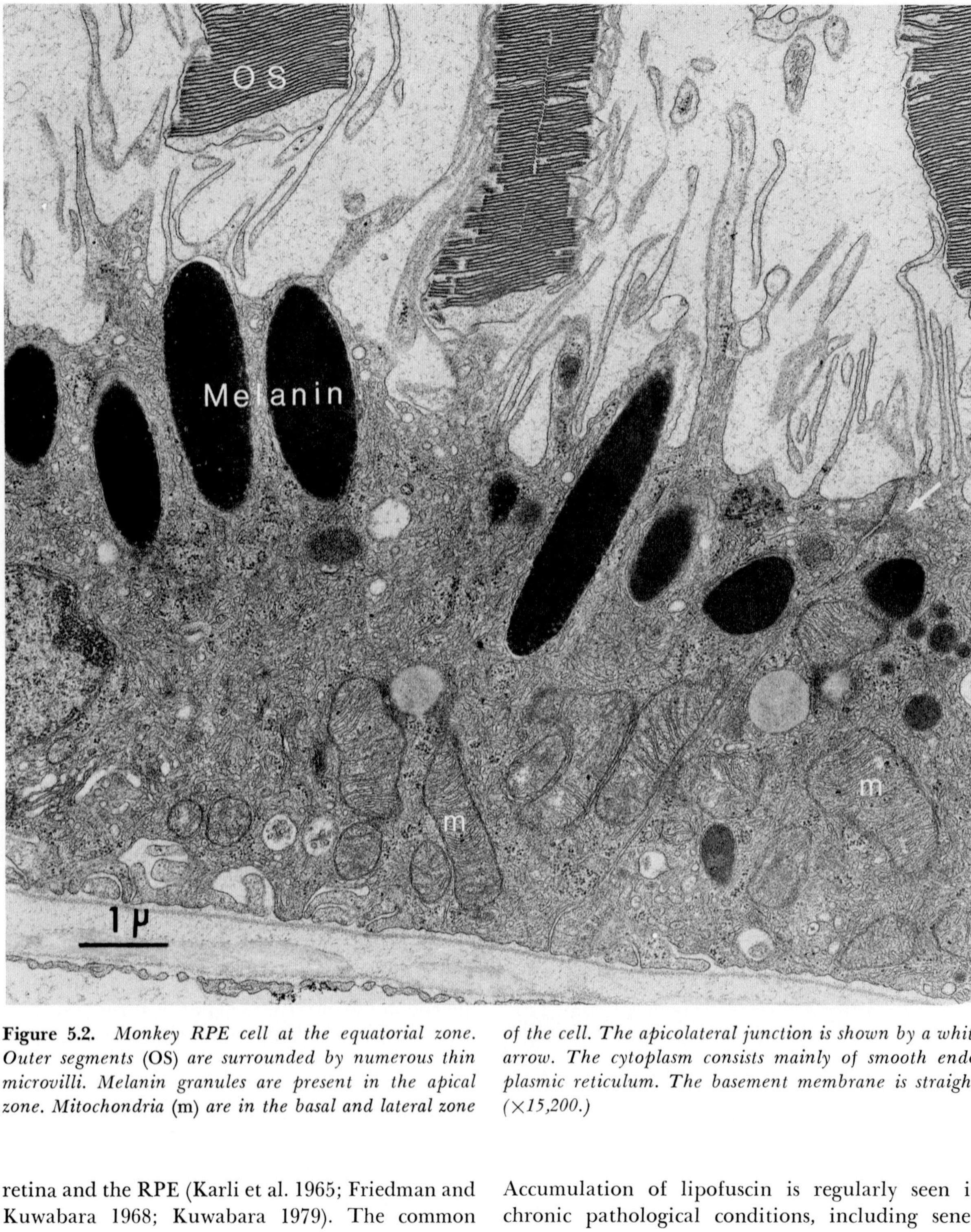

Figure 5.2. *Monkey RPE cell at the equatorial zone. Outer segments* (OS) *are surrounded by numerous thin microvilli. Melanin granules are present in the apical zone. Mitochondria* (m) *are in the basal and lateral zone of the cell. The apicolateral junction is shown by a white arrow. The cytoplasm consists mainly of smooth endoplasmic reticulum. The basement membrane is straight. (×15,200.)*

retina and the RPE (Karli et al. 1965; Friedman and Kuwabara 1968; Kuwabara 1979). The common changes are swelling of mitochondria and accumulation of inclusion bodies. The swollen mitochondria become large vacuoles in nonphysiologic conditions. The smooth endoplasmic reticulum also shows considerable variations, either being compact or loose. Accumulation of lipofuscin is regularly seen in chronic pathological conditions, including senescence (Okuda et al. 1960; Feeney et al. 1965). Although the lateral and basal attachments of the epithelium easily become loose, the apicolateral junctions remain intact even in severely damaged cells. The fluid in the subretinal space becomes

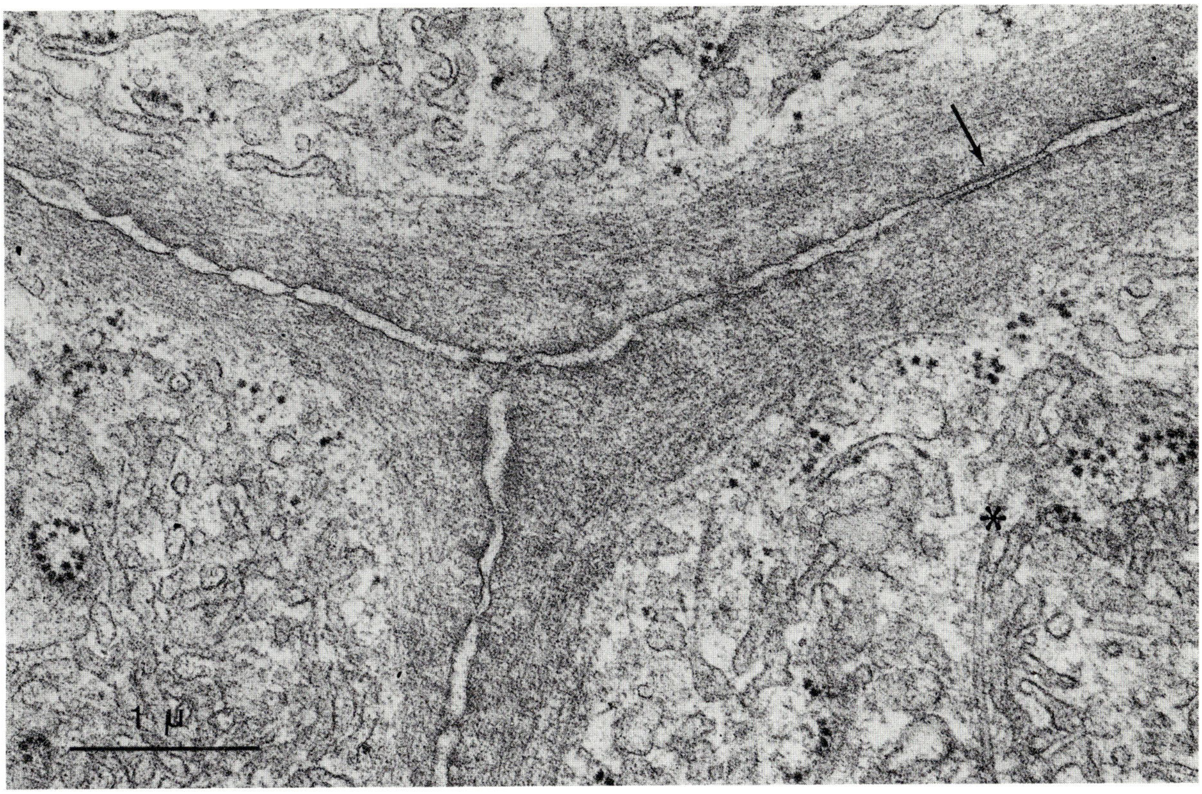

Figure 5.3. *Monkey RPE cell. Flat section at the level of the apicolateral junction. The junction is formed by the chain of desmosomes and gap junctions (arrow). Gap junctions become more numerous in the slightly apical side of this zone. Conspicuous tonofibrils are present along the junction. A microtubule is seen in this photograph (asterisk). (×51,920.)*

Figure 5.4. *Monkey RPE. Scanning electron-microscopic view of the microvilli. Tip portions of the outer segment that have been broken off during the tissue processing are closely surrounded by dense microvilli. (×3,540.)*

thicker in various pathologic conditions; the thicker substance is seen as a denser granular substance. It is most frequently present in the crypts of the microvilli.

Dog and Cat

The normal fine structure of the RPE cells of both the dog and cat have been infrequently described (Smith and Berson 1971; Steinberg and Wood 1974). These animals have the tapetum beneath a portion of the retina, although the tapetum is absent in blue-eyed and Siamese cats. The RPE cells over the tapetum, which occupy the most posterior zone of the fundus, are not pigmented. Fine capillaries distribute between the RPE and the flat inner surface of the tapetum. The sinus-like dense meshwork of the choroidal vessels, which is present in the non-tapetal zone, is missing in this area. Capillaries over the tapetum tissue form impressions into the basal cytoplasm of the RPE. Since the apical surface of these RPE cells is flat, the cell body in this portion is extremely thin (about 1–2 μ) (Figs. 5.6, 5.7). In addi-

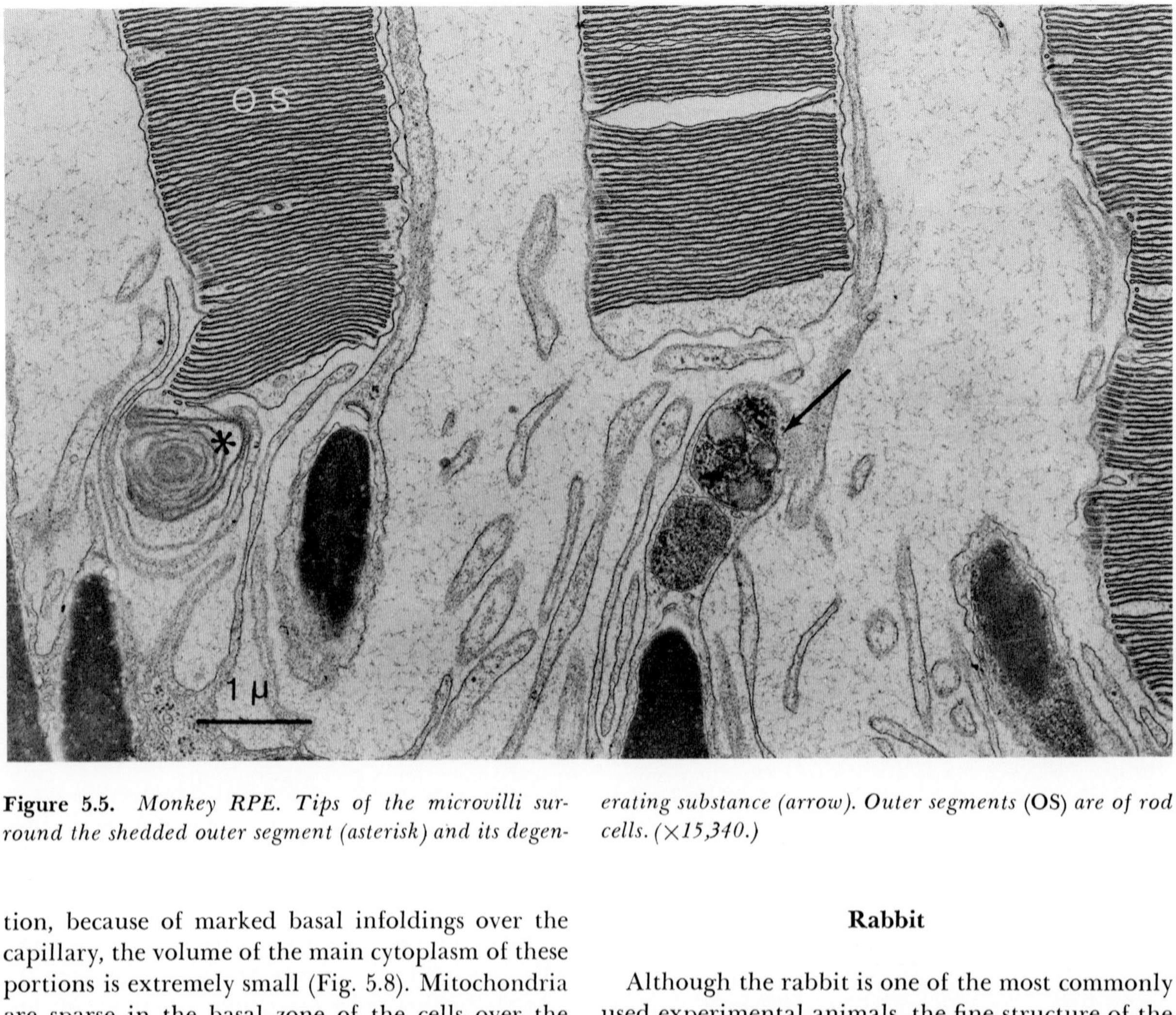

Figure 5.5. *Monkey RPE. Tips of the microvilli surround the shedded outer segment (asterisk) and its degenerating substance (arrow). Outer segments* (OS) *are of rod cells.* (×*15,340.*)

tion, because of marked basal infoldings over the capillary, the volume of the main cytoplasm of these portions is extremely small (Fig. 5.8). Mitochondria are sparse in the basal zone of the cells over the tapetal capillary. The RPE cells, which are directly attached to the tapetal cells, are taller (10 μ in height) and have no basal infoldings (Fig. 5.6). A thin basement membrane is shared by the RPE and the tapetal cells. Bruch's membrane is also irregularly formed. Regular Bruch's membrane with the elastic membrane is present only between the epithelial cells and the capillary. The apicolateral junctions are present at the apical margin. The nature and number of microorganelles are not different from those of the monkey. Microvilli of the cells in the tapetal zone are thin and long. The outer tips of the rod outer segments reach the apical surface of the RPE cell, but the shorter cone outer segments are surrounded by the tip of bundles of extending microvilli. Cells in the nontapetal zone are heavily pigmented. Melanin pigments are large and are mainly distributed in the thick microvilli.

Rabbit

Although the rabbit is one of the most commonly used experimental animals, the fine structure of the RPE has rarely been described (Yamada et al. 1958; Ohkuma 1972). The retina is avascular except in the area of myelinated nerve fibers. The RPE cell of the posterior part of the rabbit eye is considerably thinner than that of the monkey (about 7 μ high). The cells are bulged at the apical center portion and the apicolateral junctions are found in the area not far from the base (Fig. 5.9). The basal infoldings are conspicuous. Large cytoplasmic protrusions and fine microvilli extend into the subretinal space. The nature and distribution of microorganelles are about equal to those of the monkey except for oil droplets. The oil droplets are large (often occupy the whole cell) and randomly distributed within the cytoplasm. They are moderately electron-dense and are surrounded by the cell membrane. In pigmented rabbits, the melanin granules have a distribution similar to those of the monkey cell. Nonpigmented melanosomes of albino animals are seen only in the developing stage.

Guinea Pig

The structure of the RPE of the guinea pig has not been specifically reported. The size and the fine structure are similar to that of the rabbit (Fig. 5.10).

Rat and Mouse

The RPE cells in rats and mice are not significantly smaller than in larger animals. The RPE cells of rats and mice have very similar structures (Yamada et al. 1958; Dowling and Gibbons 1962; Cohen 1960). These cells have well developed basal infoldings, a relatively flat apical surface, and dense fine microvilli (Fig. 5.11). Apicolateral junctions are present near the apical ends of the cell. The cytoplasm consists of fine smooth endoplasmic reticulum, Golgi apparatus, scattered rough endoplasmic reticulum, large mitochondria, and lysosomes. Large oil droplets are also found. Electron-dense inclusion bodies are commonly found in the apical zone in both normal and pathological conditions (Fig. 5.11*B*). The outer tips of the retinal outer segments touch the apical end of the cell in light-adapted conditions. However, with dark adaptation microvilli are disengaged from the outer segment and the distance between the outer segment and the pigment epithelium becomes markedly greater (Fig. 5.12). Since the height of the pigment epithelial cell and length of the microvilli are not appreciably altered, this change appears to be due to contraction of the photoreceptor cell. Melanin granules are absent in albino animals. Only amelanotic melanosomes are present in the developing cell.

Examination of horizontal sections of the RPE cells of the rat has revealed that many of the cells have more than one nucleus. Multinucleated RPE cells begin to develop during the last few prenatal days and become apparent in the first postnatal week. Most of the RPE cells of young rats contain two nuclei of equal size and shape. Occasionally, cells with three or more nuclei are present. There are fewer multinucleated cells in older rats. Multinucleated RPE cells are not uncommon in other animals, such as monkeys and rabbits, and doubly nucleated cells have been reported in human RPE by Friedman and Tso (1968).

Hamster

The RPE of the hamster is similar in structure to that of the rat and the mouse.

Ground Squirrel (*Citellus tridecemlineatus*)

The photoreceptors of this animal are mainly cone cells. The general appearance of the RPE cells in the ground squirrel is not greatly different than the pigmented rat (Hollenberg and Bernstein 1966). The cells have scarce basal infoldings, high positioned apicolateral junctions, and large microvilli (Fig. 5.14). However, the smooth endoplasmic reticulum has a different appearance in the ground squirrel than in other small animals; it is dense but sparsely branched and has a convoluted tubular appearance (Fig. 5.15*A*). A moderate number of mitochondria are present in the lateral zone. Lamellar inclusion bodies are rare, but electron-dense inclusion bodies of various kinds are abundantly scattered in the apical and parakaryon zone.

The photoreceptor cells and the RPE show marked cytologic changes during hibernation (Kuwabara 1975). The outer segments of the cone cell disappear and the dense tubular smooth endoplasmic reticulum of the RPE becomes vacuolar (Fig. 5.15*B*). When the animal is aroused, the recovery process of the outer segments and of the cytologic appearance of the RPE cell is extremely fast.

Bat (*Myotis lucifugus*)

The fine structure of the RPE of this small brown bat has not been heretofore described. The retina and pigment epithelium of the large fruit bat have been described by Pedler and Tilley (1969). The fruit bat is reported to have intracellular reflective bodies. *Myotis lucifugus* has no reflecting fundus.

The eye of this animal is extremely small (2 mm in diameter) but in size and structure, the RPE cells are similar to those of the mouse (Fig. 5.16). However, bundles measuring about 1 μ long and 0.1 μ in diameter and consisting of small tubes (100Å in diameter) are often found among the smooth endoplasmic reticulum. The longitudinal section of the tubular structure gives an appearance similar to the myeloid body of lower animals.

Cells examined during hibernation show marked cytologic changes (Kuwabara 1975). Smooth endoplasmic reticulum becomes sparse and the bundles of tubular structure described above become large and numerous (Fig. 5.16*B*). The ends of the tubules are often continuous to the smooth endoplasmic reticulum. During hibernation, the structure of the outer segment of the retina is not altered, but the number of lamellar phagosomes in the pigment epithelium is reduced.

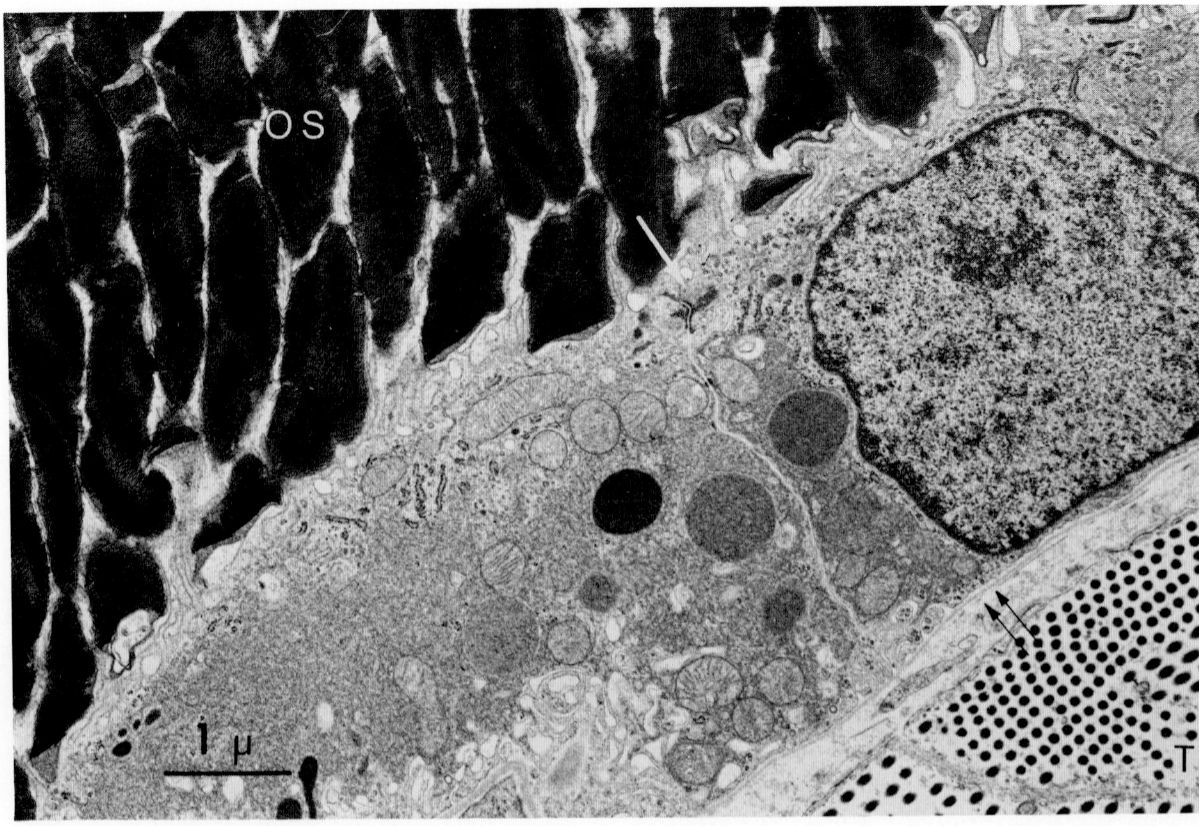

Figure 5.6. *Cat RPE in the tapetal area. The tapetum (T) is seen in the lower right corner. The cell over the tapetum has no melanin pigments. The basal infoldings are absent in the nonvascular area (double arrows). The white arrow indicates the apicolateral junction. (×16,640.)*

Opossum (*Didelphis virginiana*)

The opossum fundus is reflectile and'is thought to have a tapetum. However, instead of having the regular tapetal tissue in the choroid, the reflecting substance (a guanine-like substance) is present within the cytoplasm of the RPE cell. The RPE cell is extremely large and the cytoplasm is filled with the guanine-like substance, which forms membrane-bound bodies of various sizes and shapes (0.1–1.0 μ). These bodies are electron-lucent and have no specific structure (Fig. 5.17). The cytoplasmic matrix is electron-lucent, structureless, and contains almost no microorganelles. Only small numbers of mitochondria and rough endoplasmic reticulum are present, and smooth endoplasmic reticulum and melanin pigment are absent. Lamellar phagosomes are present in the structureless matrix (Fig. 5.17). The microvilli are sparse and the outer tips of the photoreceptors touch the apical end or occasionally slightly invaginate into the cell.

The RPE cells in the nonreflectile zone contain abundant melanin pigment and the cytoplasmic appearance is similar to that of the mouse.

Birds (Chicken and Pigeon)

Structure of the RPE of the bird, especially of the chicken, has been reported by several authors (Yamada et al. 1958; Nishida 1964; Matsusaka 1967).

Although the RPE layer of the bird is thick in appearance, the cell consists of an extremely small soma and long large villi (Figs. 5.18, 5.19). The basal cell membranes are markedly infolded. The apicolateral junctions that are associated with conspicuous tonofibrils are present only a short distance from the basal infoldings, often within one micron. Composition of the junction is slightly different. The portion of the zonula occludens is not clear. The villous processes extend almost to the level of the microvilli of Müller's cells. The main cell body contains relatively sparse smooth endoplasmic reticulum, numerous mitochondria and myeloid bodies, and occasional melanin pigments. The structure of the myeloid

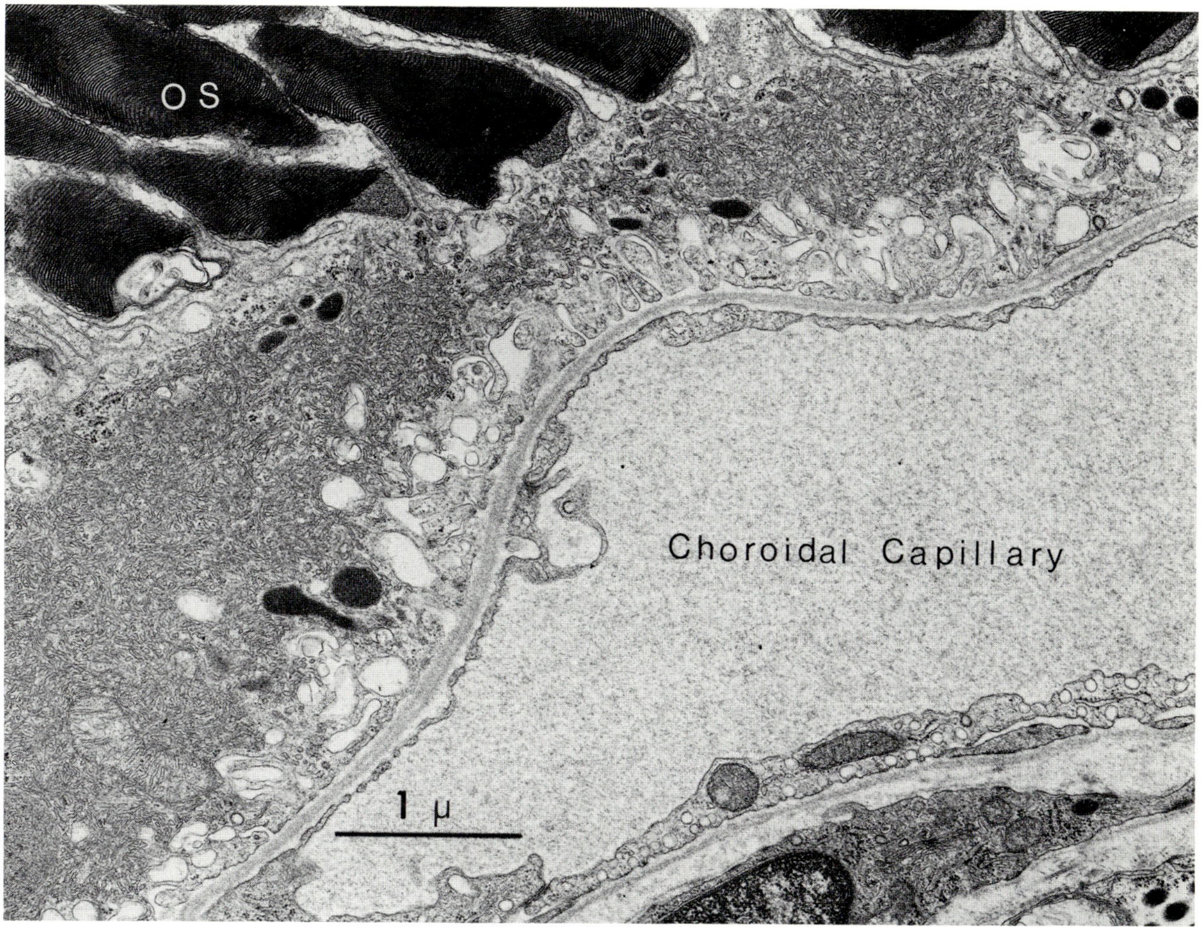

Figure 5.7. *Cat RPE cell over the choroidal capillary in the tapetal area. The cell is thin and has well-developed basal infoldings. The cytoplasm is packed with smooth endoplasmic reticulum, but has no melanin pigments. (×25,600.)*

body is similar to that of the frog. The villi contain smooth endoplasmic reticulum and groups of lamellarly arranged membranes that have an appearance of the myeloid body. Large and elongated melanin pigments are abundant in the villi. Because of the flat shape of the main cell body, the tips of the outer segments are found close to the base of the cell.

Frog (*Rana pipiens*), Toad, and Newt

Only slight differences in the cytology of the RPE are noted between these animals. The retinal structures are also similar. Electron-microscopic studies of their retinas and RPE's have been reported by several authors (Porter and Yamada 1960; Nilsson 1964; Yamada 1960; Steinberg 1973; Keefe 1971).

The RPE of these animals is thick and has several unique features (Fig. 5.20). The basal cell membrane is sparsely infolded. The apicolateral junctions are present close to the base. Lateral cell membranes below the junction are loosely attached to each other and often form large intercellular spaces (Fig. 5.21). The apical cytoplasm is conspicuously large and sends out numerous large microvilli between large rod outer segments. The villi and the rod outer segments are closely attached to each other. The tips of the villi reach the level of the inner segments.

The cytoplasm consists of fine smooth endoplasmic reticula, which are uniformly and densely distributed (Fig. 5.21). The smooth endoplasmic reticula in the large villi are somewhat irregular in size. Numerous myeloid bodies are present throughout the cytoplasm. The relationship between the myeloid bodies and the smooth endoplasmic reticulum has been described in detail by Porter and Yamada (1960). The myeloid bodies are seen as rhomboid shapes of various sizes. Membranes of the myeloid body and

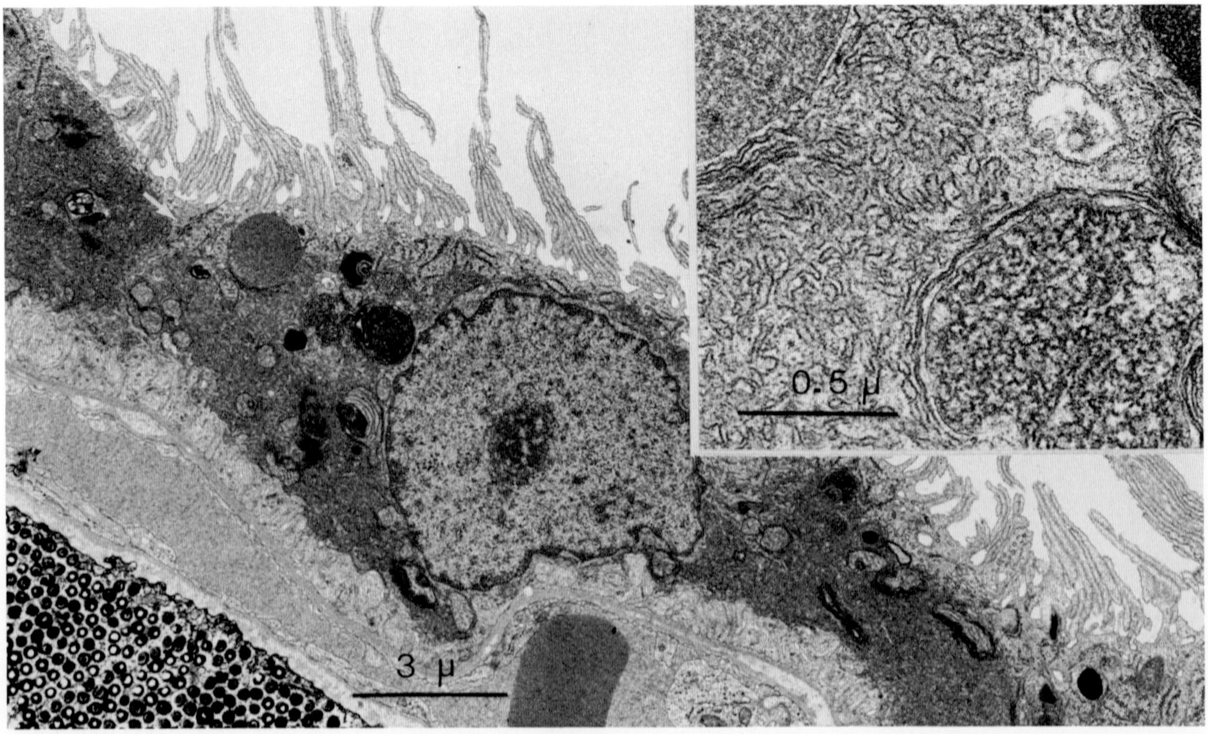

Figure 5.8. *Dog RPE. Cells over the choroidal capillaries in the tapetal zone are thin and have well developed basal infoldings. The cytoplasm contains numerous lysosomes and inclusion bodies. Melanin pigment is absent in this area. The apicolateral junction is indicated by a white arrow. (×7,320.) The inset shows fine smooth endoplasmic reticulum and inclusion bodies. (×42,600.)*

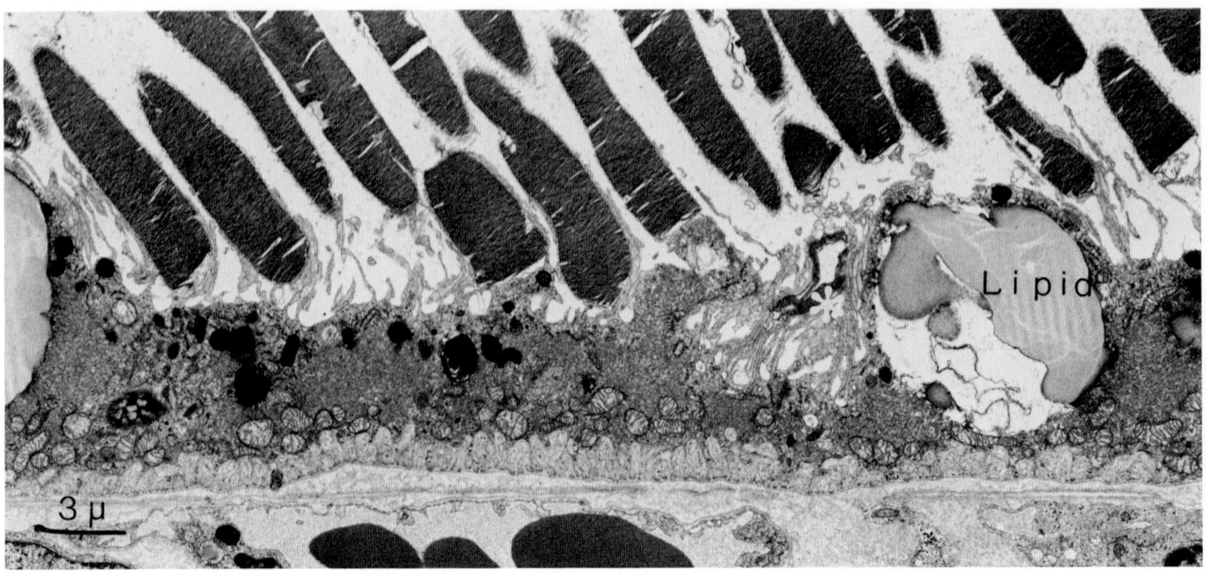

Figure 5.9. *Rabbit RPE cells contain large oil droplets. The basal infoldings are conspicuous. Lamellar membranes are trapped by microvilli (white asterisk). Electron-dense inclusion bodies are abundant in some cells. (×4,130.)*

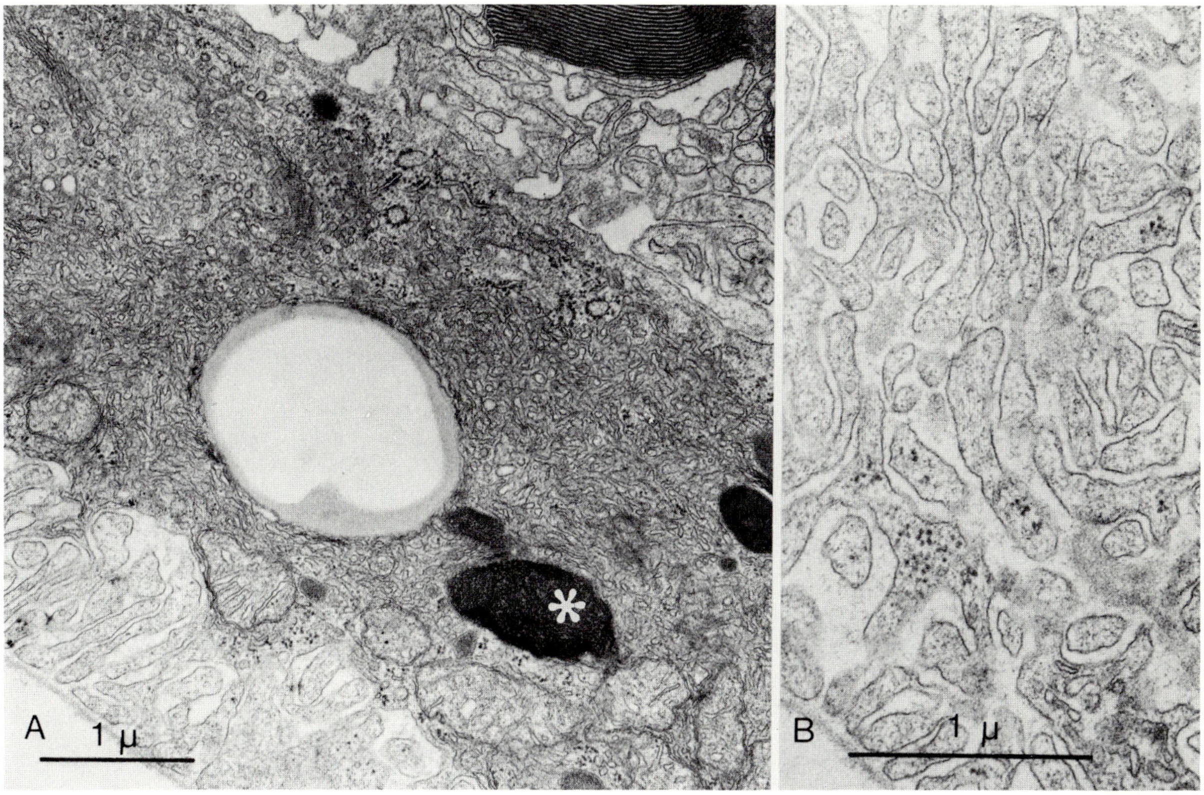

Figure 5.10. *Guinea pig RPE.* (A) *The pigment epithelial cell contains large lipid droplets. Lamellar inclusion bodies are occasionally found (white asterisk). The basal infoldings are conspicuous. (×18,000.)* (B) *A portion of the basal infoldings. A few ribosome granules are present in the infolded cytoplasm. The basement membrane substance is seen among infoldings. (×34,700.)*

the smooth endoplasmic reticulum are continuous in several places (Fig. 5.21*B*). Membranes of some myeloid bodies appear to form a whorled pattern (Fig. 5.21). Mitochondria are present in the basal regions. The lamellar phagosomes are abundant mainly in the apical portion. Lysosomes and dark inclusion bodies are present in the apical cytoplasm. Large membrane-bound oil droplets of various sizes are present in the cytoplasm. The lipid is moderately electron-opaque and structureless. Melanin granules are abundant; they are small and round in the apical cytoplasm and large and elongated in the microvilli. The pigmentation of the newt cell is slightly different from that of the frog. Small and round melanin granules are distributed throughout the cytoplasm, including the basal portion. The subretinal space is compactly filled by villi and outer segments (Steinberg 1973).

During hibernation the RPE cells show a considerable change in the cytoplasm (Kuwabara 1975). The basal infoldings completely disappear, and the smooth endoplasmic reticulum becomes extremely sparse and large. Also, the myeloid bodies become sparse (Fig. 5.22). Smooth endoplasmic reticula transform into tubules that form bundles of various sizes in several locations (Fig. 5.22*B*). During hibernation, the appearance of the retinal outer segments is not altered, but large lamellar bodies that have been phagocytized by the RPE cell appear to remain without further alteration. This finding suggests that the turnover process of the rod outer segment is slowed down considerably. The outer synaptic organs of the retina become smaller and the numbers of vesicles and synaptic ribbons appear to be reduced during hibernation.

Lizard and Turtle

In the lizard and turtle the photoreceptors are mainly cone cells (Yamada 1960; Yamada 1961). Each RPE cell consists of a small main cytoplasm and

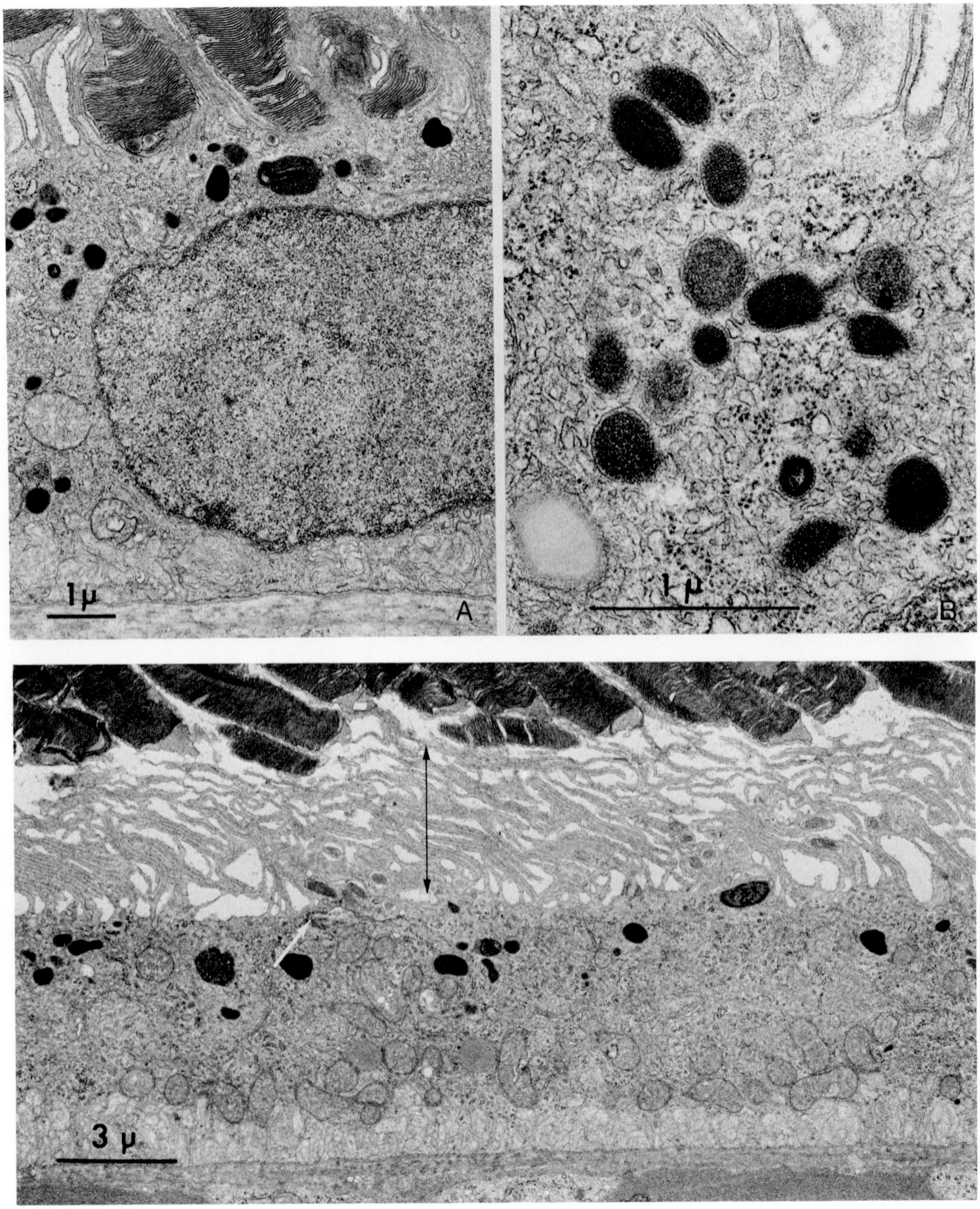
1μ
A
1 μ
B
3 μ

Figure 5.11. *Rat RPE.* (A) *The cytoplasmic appearance is similar to that of other animals. The outer segments touch the apical cell membrane. The cell contains numerous electron-dense particles.* ($\times$*11,560.*) (B) *The upper left portion of* A *is shown in higher magnification. The dark inclusion bodies appear to be lysosomes.* ($\times$*34,700.*)

Figure 5.12. *Dark-adapted rat RPE. The outer segments are disengaged from the microvilli. The increased distance between the outer segment and the pigment epithelium is shown by an arrow. The white arrow indicates the apicolateral junction.* ($\times$*6,840.*)

large, long villous processes (Fig. 5.23). The apicolateral junction is a short distance from the basal cell membrane, which is moderately infolded. The retinal outer segments also intrude into the parakaryon regions of the RPE cells. A large portion of the cell nucleus is often found above the junction line (Fig. 5.24). The apical cytoplasm, including villi, contains smooth endoplasmic reticulum, which is irregularly distributed. Myeloid bodies are large and abundant throughout the cytoplasm, and a few small ones are present in the villi. The cytoplasm contains numerous mitochondria in the basal region, and lysosomes and dense inclusion bodies in the parakaryon zone. Clusters of glycogen granules are found in the matrix. Large spindle-shaped melanin granules are present only in the villi. The villi extend far toward the retina, and their tips often touch the microvilli of the inner segment of the photoreceptor cells. Microvilli of the inner segment of the cone cell are extremely well developed (Fig. 5.25).

Snake

Although the fine structure of the photoreceptor cell of the snake has been described by Ishikawa and Yamada (1969), fine structure of the RPE cells has not been described in detail. The snake retina consists of large cone cells that have stubby outer segments (Fig. 5.26). Abundant lipidic material is found in mitochondria of the ellipsoid. The RPE cells are cuboidal in shape and have large cell bodies and long microvilli. The villi, containing melanin granules and some smooth endoplasmic reticulum, form

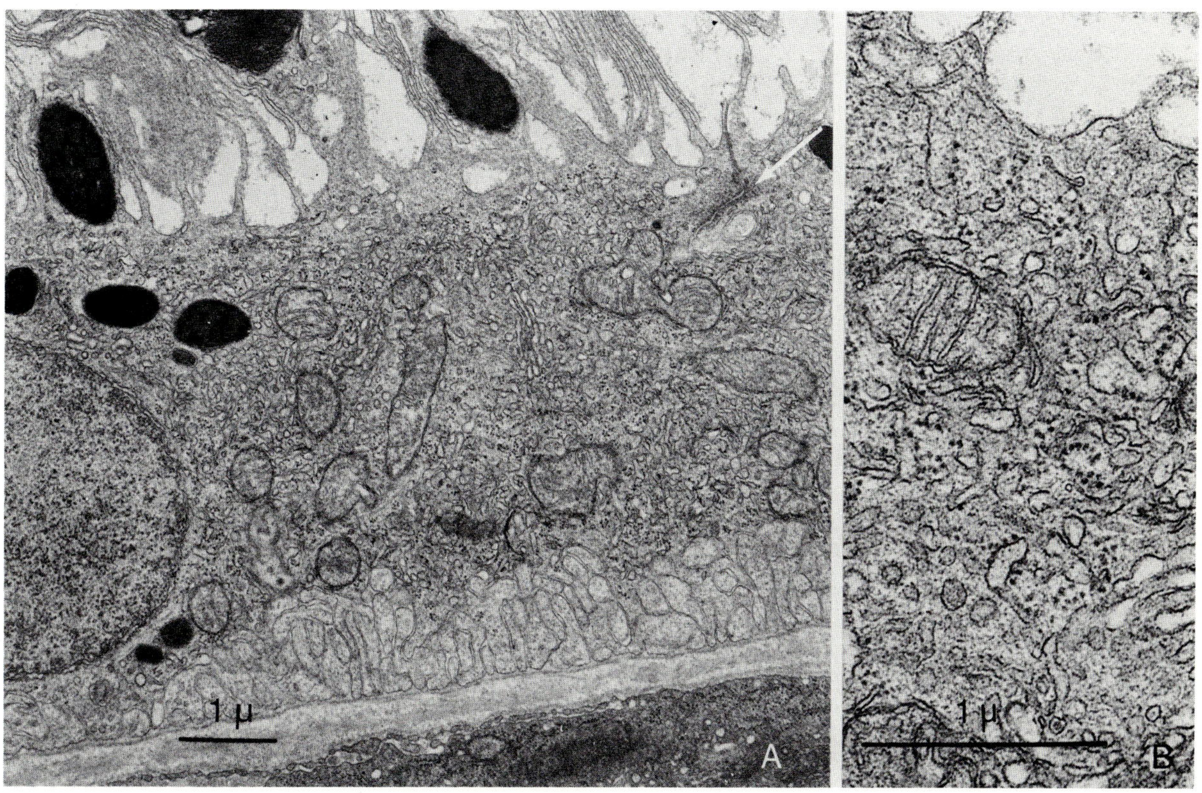

Figure 5.13. *Hamster RPE.* (A) *The cytoplasmic appearance is similar to that of the rat RPE. Golgi apparatus are abundant in this cell. The apicolateral junction is shown by a white arrow.* ($\times$*13,220.*) (B) *Higher magnification.*

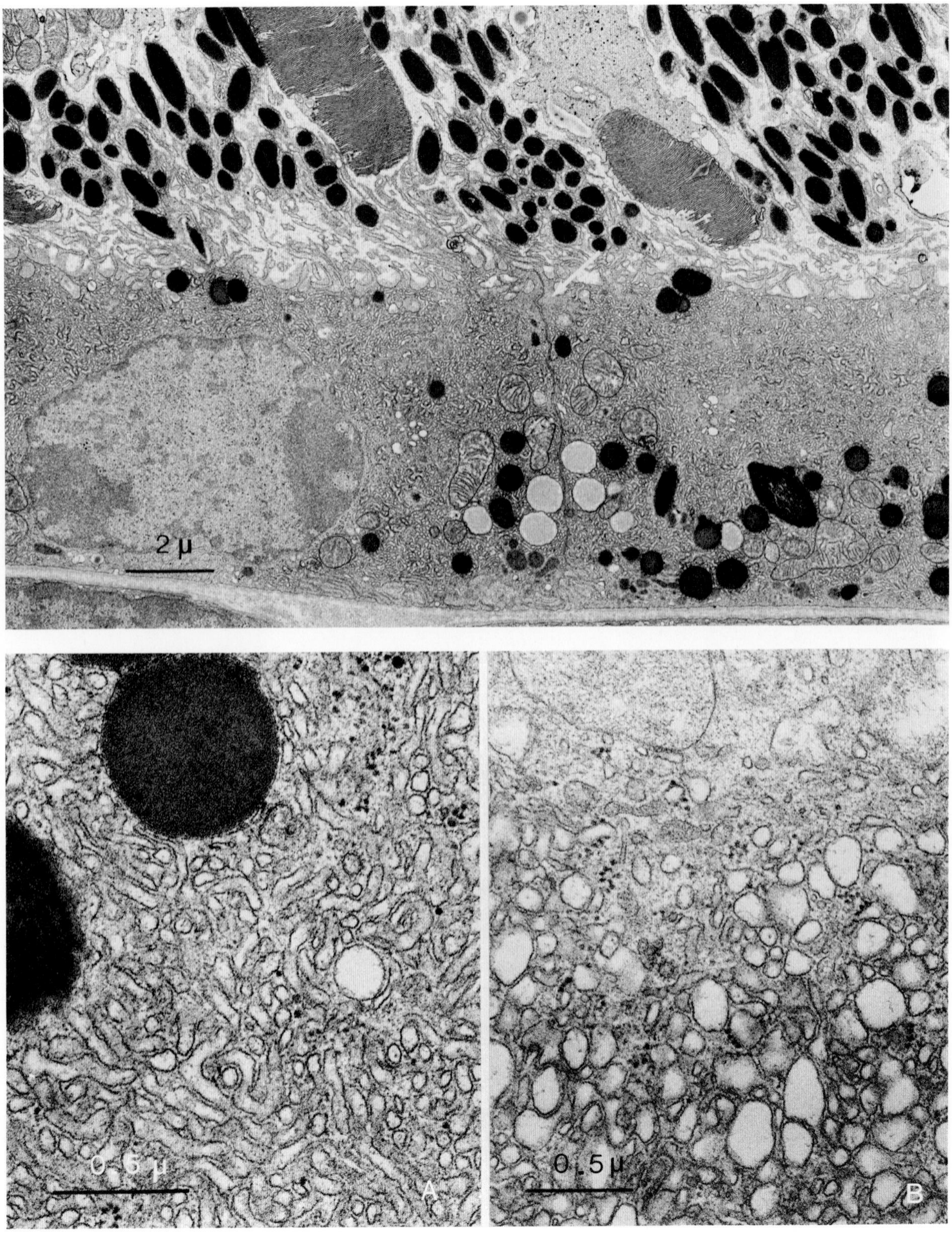
2 μ
0.5 μ
A
0.5 μ
B

Figure 5.14. *Ground squirrel RPE. Photoreceptors are mainly cone cells. Melanin pigment is abundant in the microvilli. The cytoplasm contains electron-dense inclusion bodies and some lipid droplets. The apicolateral junction is shown by a white arrow. (×7,790.)*

Figure 5.15. *Ground squirrel RPE.* (A) *The smooth endoplasmic reticulum shows a convoluted tubular structure. (×41,000.)* (B) *The smooth endoplasmic reticulum becomes vesicular during hibernation. (×37,760.)*

large bundles and extend into the spaces between cone cells. Their tips often reach the microvilli of Müller's cell. Tips of the cone outer segments wedge deep into the epithelium. The basal cell membranes are sparsely infolded. The conspicuous apicolateral junctions are present at a moderate height. The cytoplasm contains abundant smooth endoplasmic reticulum and numerous Golgi apparatus. Myeloid bodies are absent in the two species examined (Fig. 5.27). Small and sparse mitochondria are distributed along the basal and lateral cell membranes. Lamellar inclusion bodies are rare. The distribution and number of other microorganelles appear to be similar to those of other animals. Curiously, the diaphragm of the fenestration of the choroidal endothelial cell consists of two cell membranes (Fig. 5.28).

Fish (Teleost)

The photoreceptors and synaptic organs of the fish retina have been thoroughly described, but the structure of the RPE has been only incompletely described. The RPE cells of the teleost appear to be tall when observed by light microscopy. However,

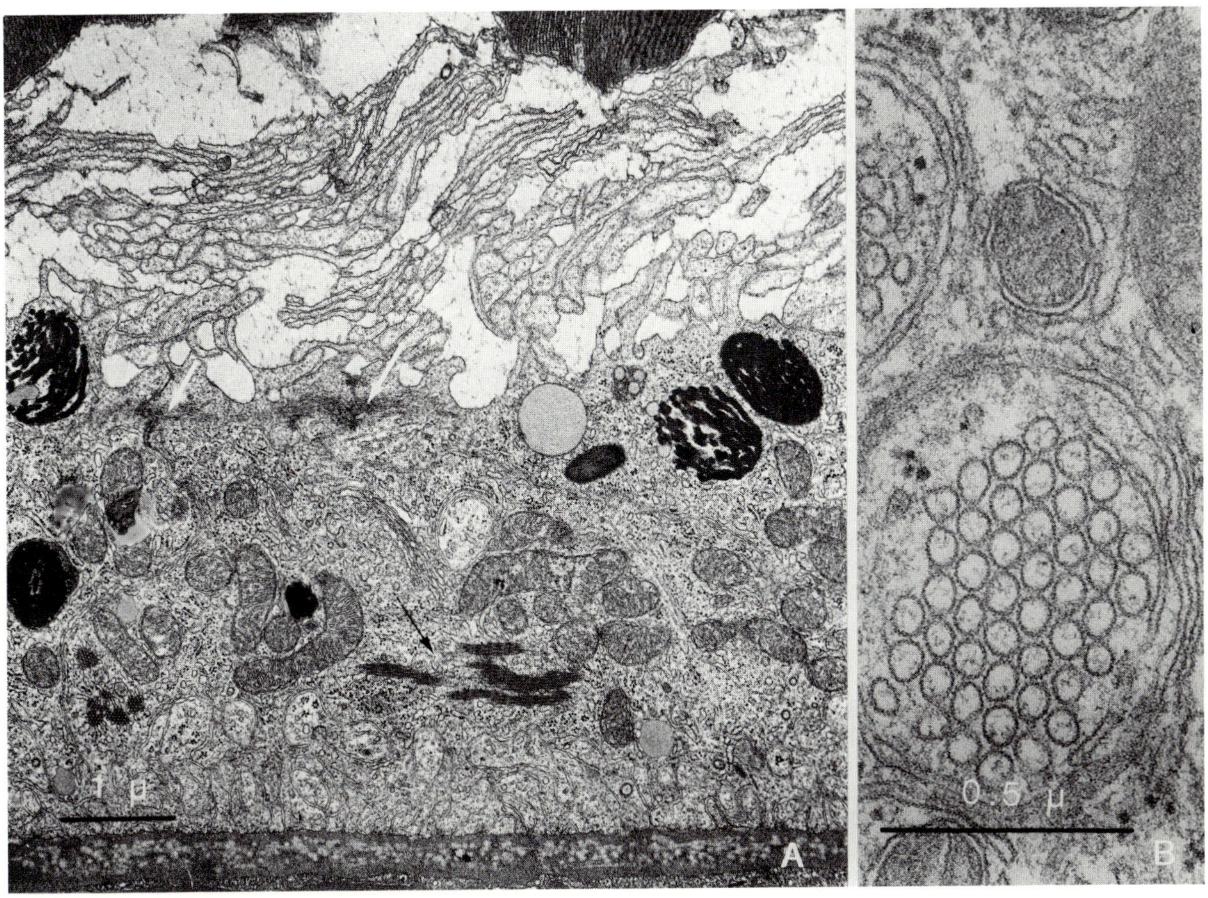

Figure 5.16. *Bat RPE.* (A) *Despite the small size of the globe, the size of the pigment epithelium (and the retina) is not too different from that of other animals. Melanin pigment is present in the apical zone. Fine tubular structure is present in the cytoplasm (arrow). White arrows indicate apicolateral junctions. (×15,340.)* (B) *The tubular structure becomes large and conspicuous during hibernation. The smooth endoplasmic reticulum becomes sparse. (×68,900.)*

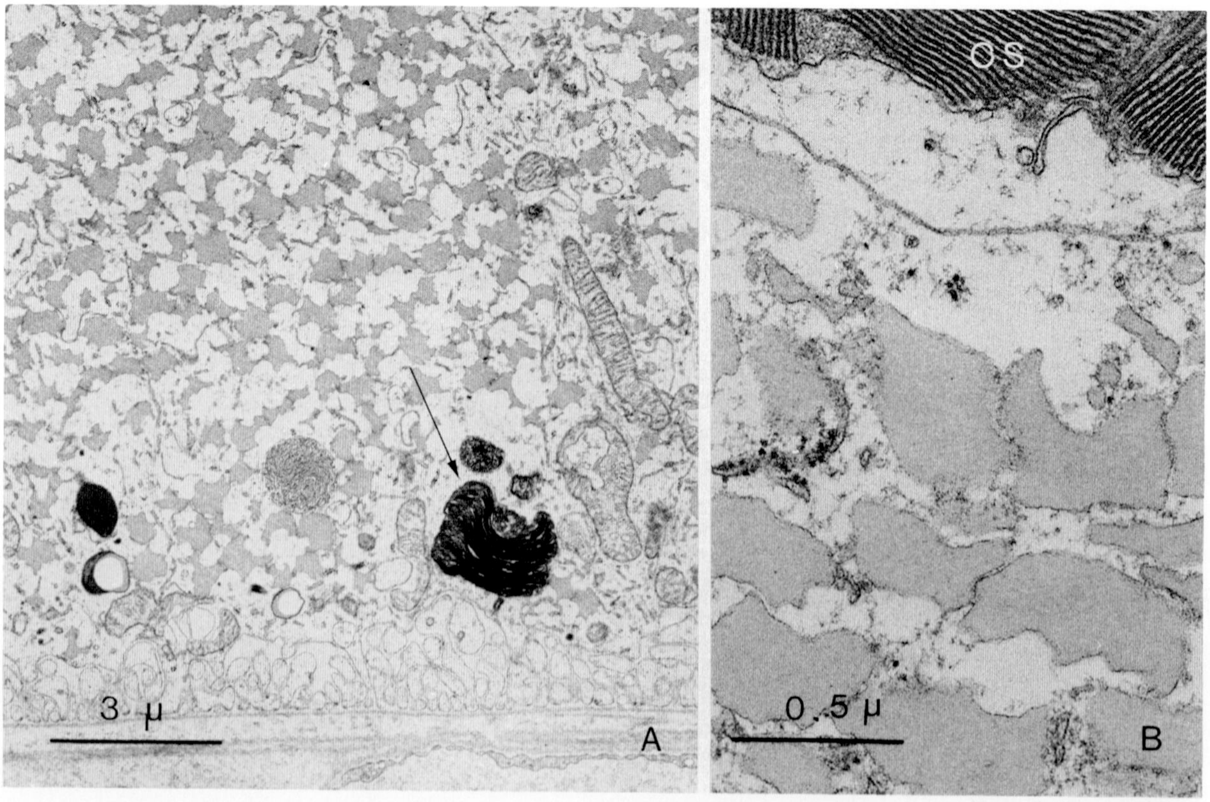
OS
3 μ
A
0.5 μ
B

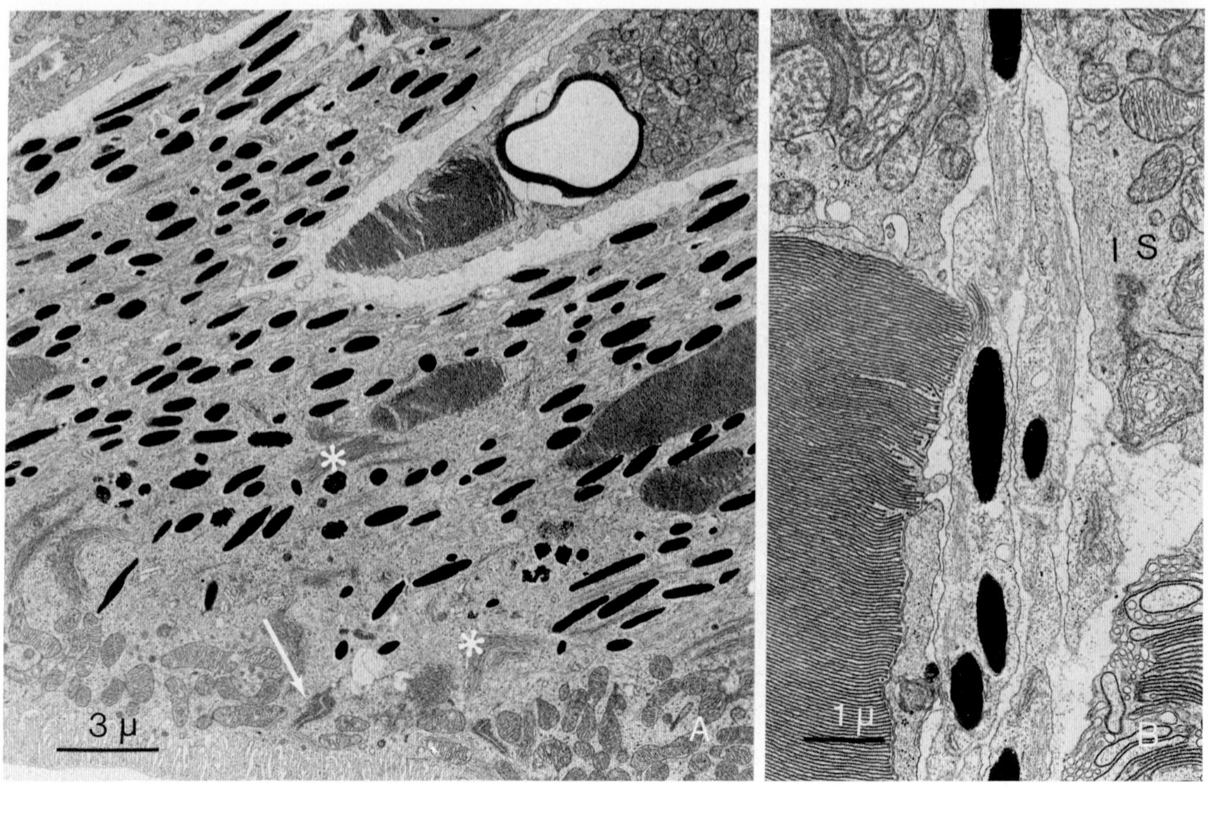
IS
3 μ
A
1 μ
B

Figure 5.17. *Opossum RPE.* (A) *The cell is large and contains abundant guanine-like substance and a moderate number of mitochondria, but no smooth endoplasmic reticulum. Lamellar inclusion bodies are occasionally found (arrow). The basal infoldings are moderate in degree. (×7,790.)* (B) *Higher magnification of the apical portion of the cell. Guanine granules are encased within the cell membrane and irregular in shape. The cytoplasm contains a small amount of rough endoplasmic reticulum but no smooth endoplasmic reticulum. The microvilli are sparse. (47,200.)*

Figure 5.18. *Bird RPE.* (A) *Pigeon. Retinal photoreceptor cells are predominantly cones. The pigment epithelial cells have the apicolateral junctions near the basal membrane (white arrow). The cytoplasm contains numerous myeloid bodies (asterisks). Villi are large and long. They contain large pigment granules. (×4,720.)* (B) *Chicken. Large villi extend to the level of the inner segment. (×10,620.)*

the main cytoplasm is relatively small and large villi occupy the major part of the cell. The villi extensively interdigitate with the outer segments of the photoreceptor cells (Fig. 5.29). The apicolateral junctions are situated near the base of the cell, and the tips of the outer segments also reach the vicinity of the basal cell membrane. It is common to find nuclei of RPE cells among the outer segments. The basal infoldings are sparse and the basal cell membranes form hemidesmosomes. Melanin pigment granules are found abundantly throughout the cytoplasm, including the basal portion. In some fish (trout and goldfish), melanin granules are intermingled with particles that have a structure similar to guanine (Fig. 5.29*B*). The guanine particles are membrane-bound and electron-lucent. They are round in shape and measure 0.5–1.0 μ in diameter. The smooth endoplasmic reticulum is not dense in the paranuclear portion, particularly in the vicinity of the guanine particles, but is abundant in the villi. Myeloid bodies are numerous in the apical cytoplasm and in the villi (Fig. 5.30). These myeloid bodies are smaller than those in the main cell body. Mitochondria are small and scanty. Lamellar inclusion bodies are occasionally present. Other microorganelles are also infrequent. The villi extend to the level of the inner segment of the photoreceptor cells. The crown microvilli of the photoreceptor cell are surrounded by the tips of the villi of the RPE cell. The layer of the choroidal vessels is extremely thick.

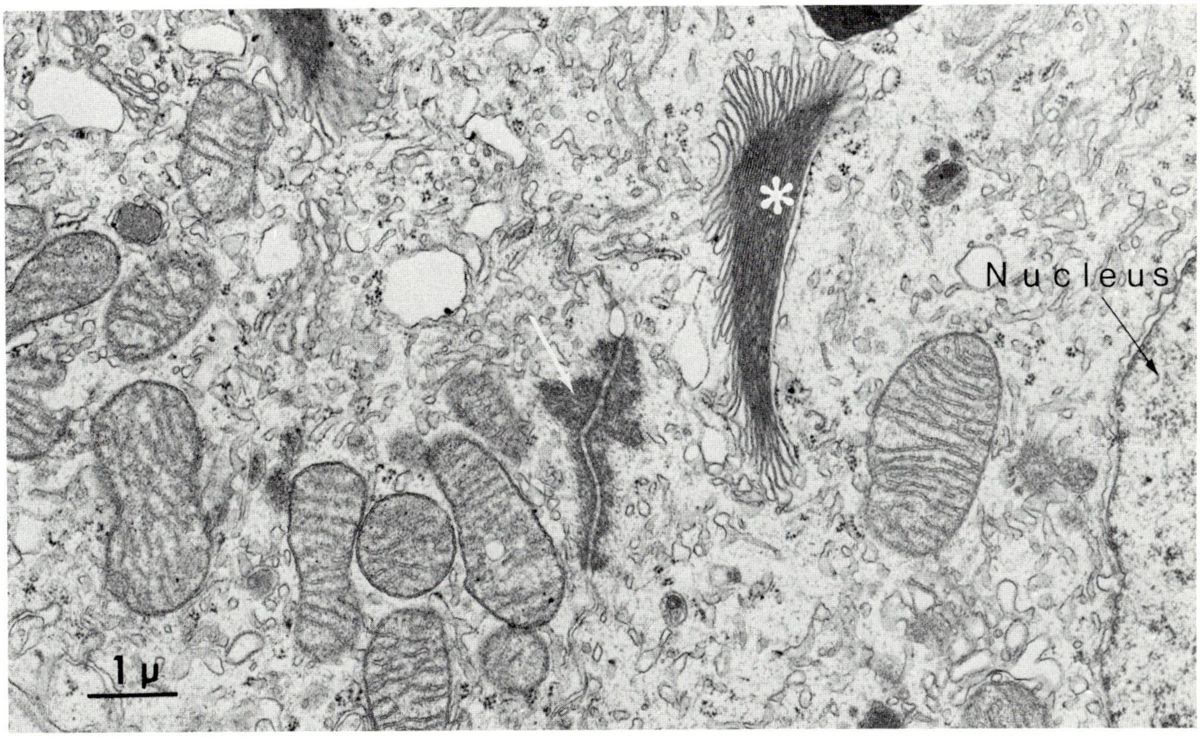

Figure 5.19. *Chicken RPE. Higher magnification of the cytoplasm shows the low-positioned junction (white arrow) and a myeloid body (asterisk) beside the usual microorganelles. (×12,270.)*

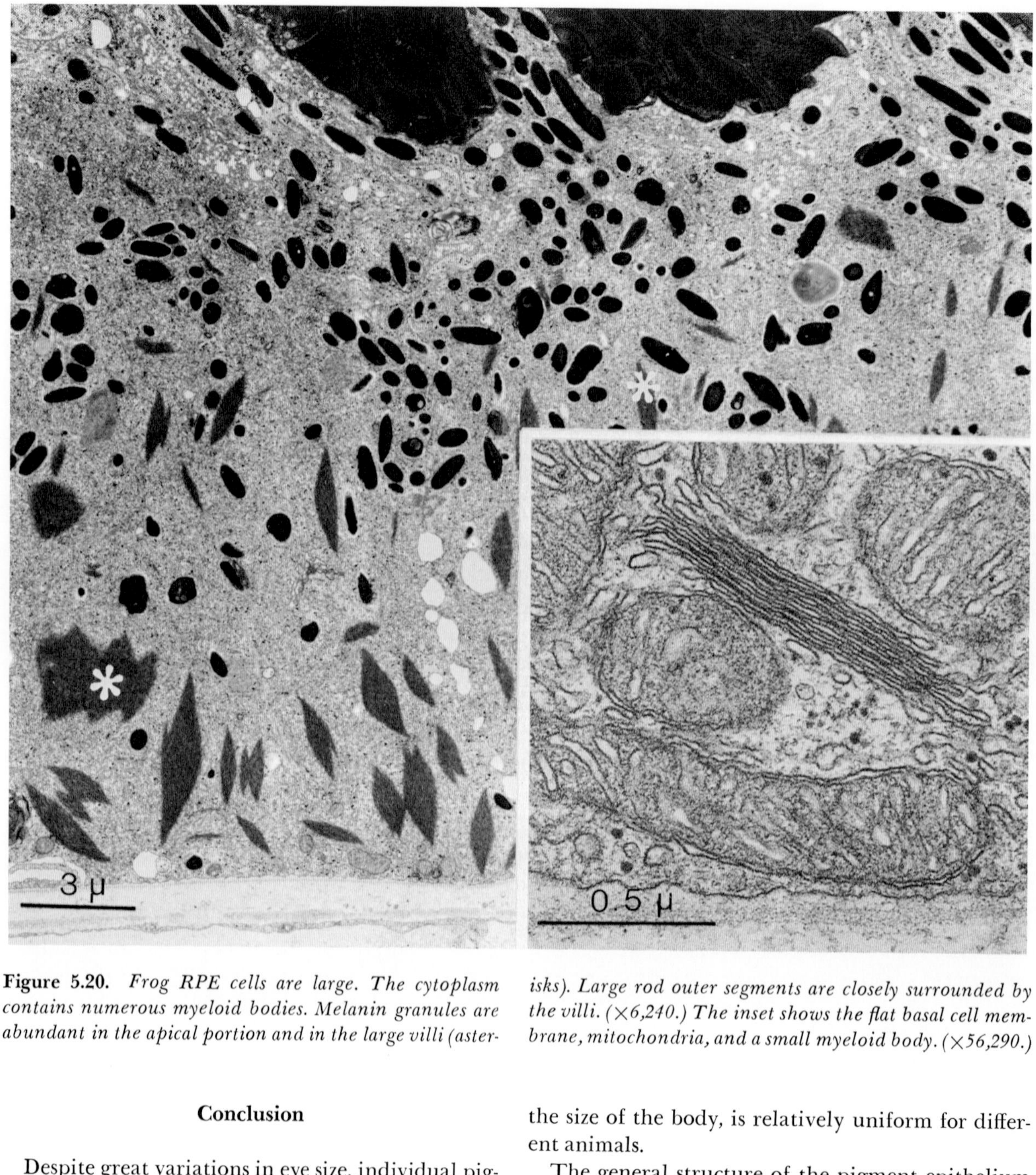

Figure 5.20. *Frog RPE cells are large. The cytoplasm contains numerous myeloid bodies. Melanin granules are abundant in the apical portion and in the large villi (asterisks). Large rod outer segments are closely surrounded by the villi. (×6,240.) The inset shows the flat basal cell membrane, mitochondria, and a small myeloid body. (×56,290.)*

Conclusion

Despite great variations in eye size, individual pigment epithelial cells of different animals show a relatively uniform size. The height of the main cell body in the posterior portion measures 4–7 μ and in the microvilli 5–10 μ. The smallest eye discussed in this chapter is that of the bat, measuring about 1.5 mm in diameter, and the largest is that of the cow, measuring about 45 mm. Duke-Elder (1958) has pointed out that the size of the eye, in proportion to the size of the body, is relatively uniform for different animals.

The general structure of the pigment epithelium is also uniform in all animals. The cells form a single cell layer that lines the outer limit of the subretinal space. They are joined firmly by apicolateral junctions. The apical ends of the epithelial cells are covered with numerous microvilli that stretch into the subretinal space. The main microorganelle is the smooth endoplasmic reticulum, which occupies most of the cytoplasmic area; the distribution of

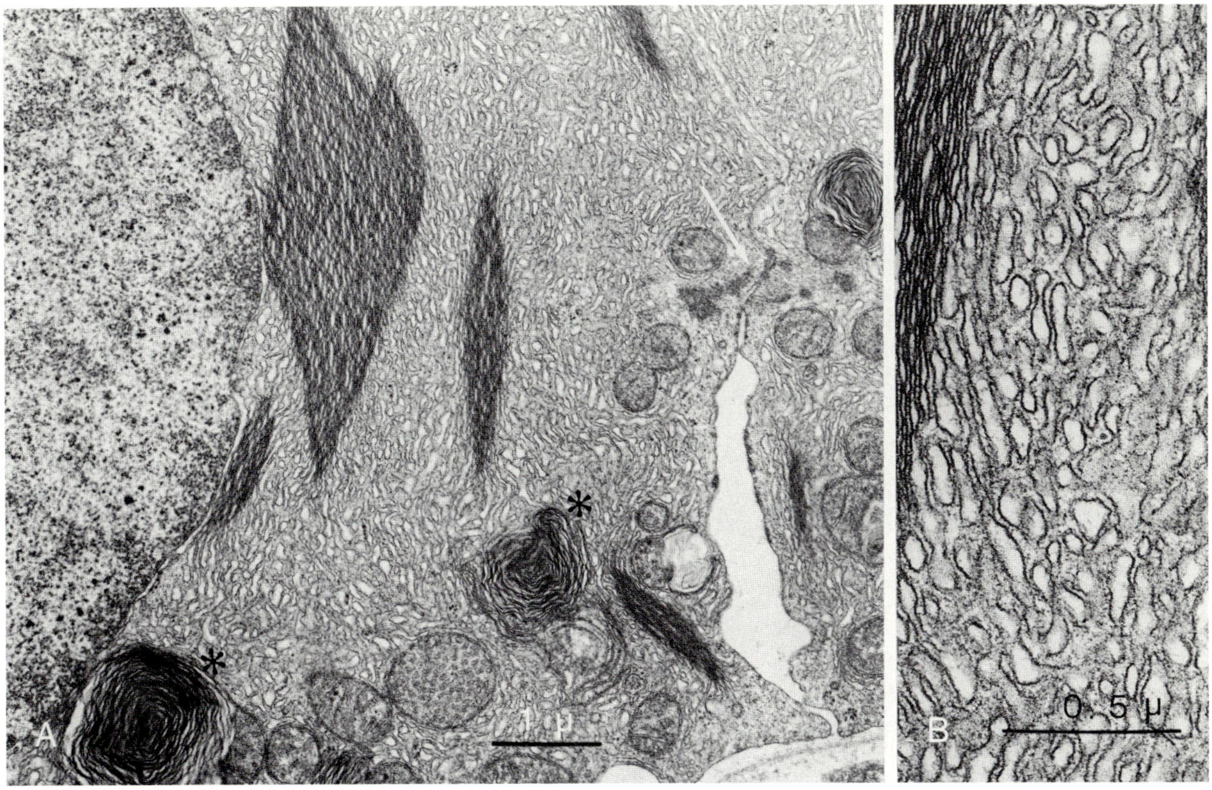

Figure 5.21. *Frog RPE.* (A) *Numerous myeloid bodies are seen in dense smooth endoplasmic reticulum. Some form whorled membranes (asterisks). The apicolateral junction (white arrow) is present near the basal end. Lateral cell membranes below the junction are separated widely. (×14,870.)* (B) *Membranes of the myeloid body are continuous with those of the smooth endoplasmic reticulum. (×47,440.)*

other microorganelles is about equal in all animals. The structure of the underlying Bruch's membrane and choroidal tissue is identical in all animals. The uniformity of the basic structure indicates that the pigment epithelium may have the same function in all animals.

However, there are certain variations among species in shape and cytologic constituents. Although the difference in structure between the mammals and lower animals is often great, the variation is not always related to the evolutionary order of the species. One of the most striking characteristics seen in lower animals is that the microvilli are large and long. The cytoplasm contained by the microvilli of the pigment epithelial cells of lower animals is generally larger in volume than the cytoplasm in the main portion of the cell. Dense and large cell processes surround the outer segments very closely and often extend to the level of the inner segments. The tips of these villi reach the crown villi or the microvilli of the photoreceptor inner segments. In some animals, the villi often make a direct contact with the microvilli of the Müller's cell of the retina. Also it is common to see deep invagination of the outer segment into the cell body of the pigment epithelium. Lower animals apparently require such close physical contact for metabolic and functional performance of both cell elements. Although contraction of the microvilli and migration of the melanin pigment by light adaptation has been discussed, no specific retractile organelles have been demonstrated in the villous processes except for the filamentous structure. In lower animals the melanin granules within the microvilli, as well as the whole microvilli, move toward the scleral direction by dark adaptation. However, in mammals, especially in rats, the main moving part appears to be the photoreceptor cell itself.

In general, the cells of lower animals have low-positioned apicolateral junctions; however, this vari-

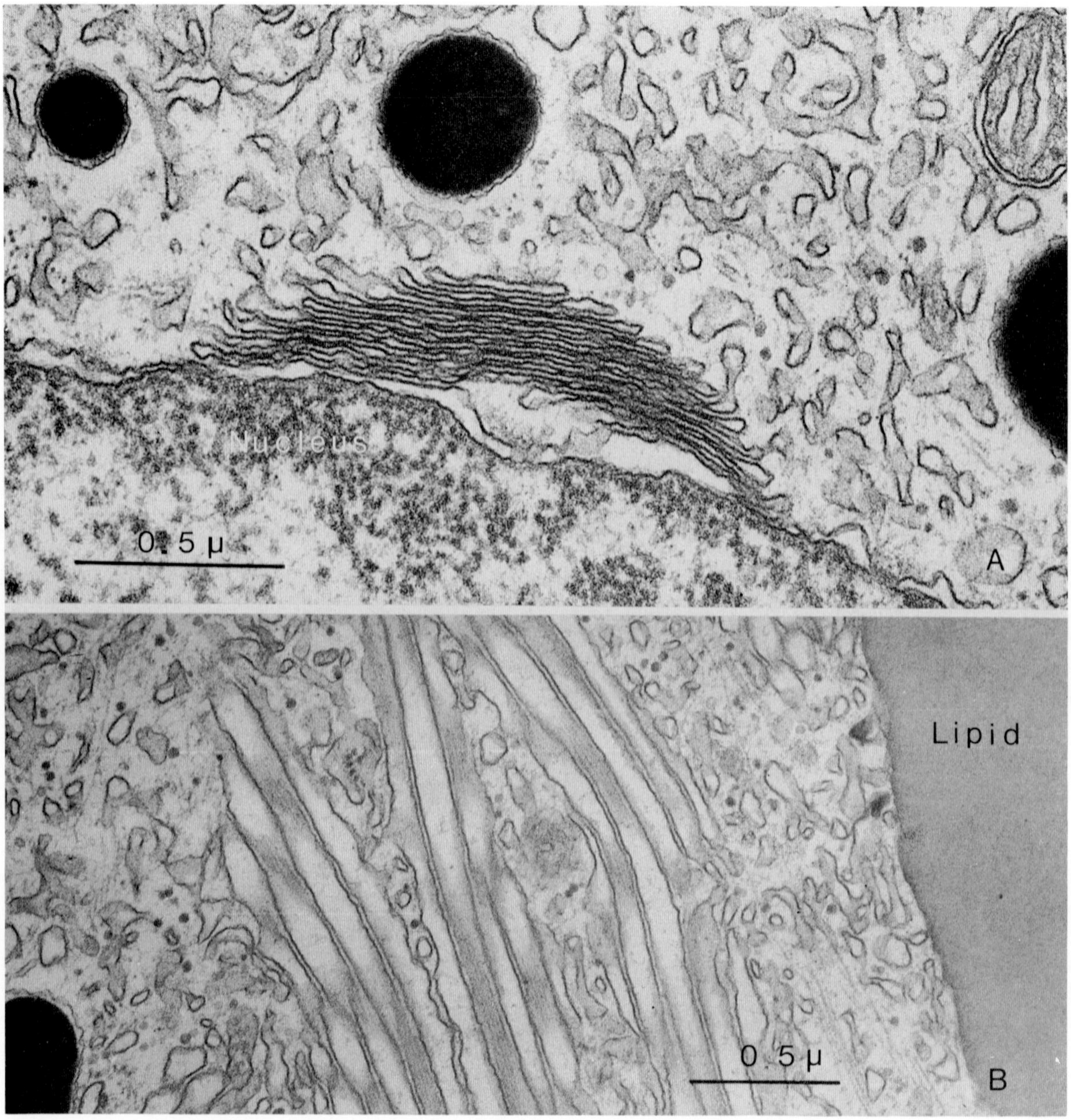

Figure 5.22. *Hibernating frog RPE.* (A) *Smooth endoplasmic reticulum is sparse and the myeloid body is small. This myeloid body is continuous to the nuclear membrane.* (×*65,140.*) (B) *Some smooth endoplasmic reticulum transform into tubules. Lipid droplets remain unchanged.* (×*44,840.*)

ation does not always correspond to the evolutionary position of an animal. The pigment epithelium of the rabbit has junctions near the basal line and has large bulged apical cytoplasm; the RPE cells of the snake have high-positioned junctions.

The pigment epithelial cells of lower animals (birds and below) contain myeloid bodies. The function of these bodies is not clearly known, but structurally they are modified smooth endoplasmic reticulum, as proposed by Porter and Yamada (1960). The regular arrangement of their membrane often suggests certain photosensitive activities, but the functions may be mainly related to nutritional metabolism of the cell. Myeloid bodies become sparse

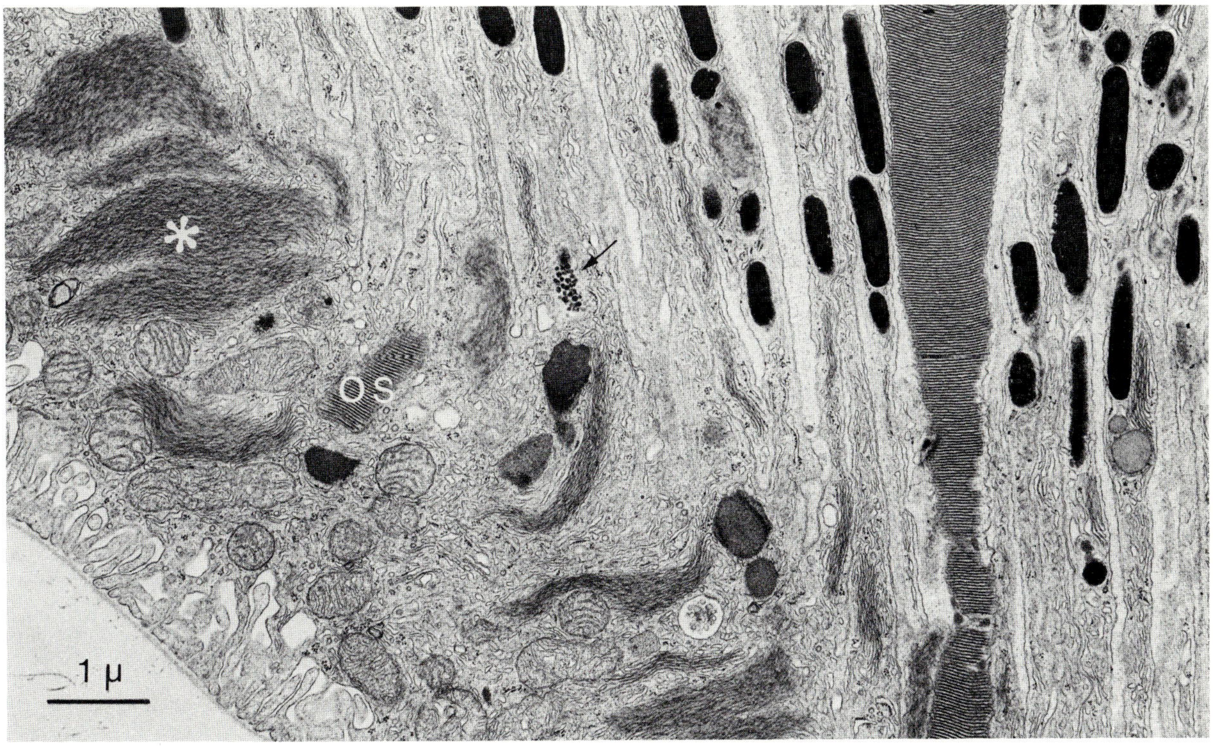

Figure 5.23. *Lizard RPE. The cytoplasm is small but the villi are large and long. Numerous myeloid bodies (asterisk) and dark inclusion bodies are present. Outer segments of cone cells* (OS) *extend deep into the cytoplasm. Clusters of glycogen particles are seen occasionally.* (×*13,690.*)

during hibernation in some animals. Many hibernating animals also lose their photoreceptor outer segments. Since cytologic changes in the photoreceptor cells and pigment epithelium during hibernation are often so great, considerable caution must be taken when animals that naturally hibernate during the winter season are used in experiments. It is especially noteworthy that many animals (ground squirrels, frogs, and snakes) keep their seasonal cycle, including cytologic changes, even when they are being fed normally in a warm laboratory during the winter.

Although the general distribution and fine structure of the choroidal blood vessels are uniform in all animals, the vascularity of the retina is greatly variable. The pigment epithelium of animals with poorly vascularized retinas appears to have well developed infoldings of the basal cell membranes. This converse correlation between retinal vascularity and the basal infolding is most apparent in the guinea pig and rabbit. All other mammals with vascularized retinas have moderate numbers of basal infoldings. However, the correlation does not hold in the lower animals. The frog and snake, which have no blood vessels in the retina, have almost flat basal lines. The pigment epithelial cells of hibernating animals or of animals in which photoreceptor cells have been severely damaged show very little basal infolding. Also, the pigment epithelium over the tapetum cell has no basal infoldings.

The pigment cells of rodents and many lower animals contain lipid droplets. The lipid droplets in the rabbit and guinea pig are large and abundant. The amount of lipid in the pigment epithelium appears to be somewhat correlated to the vascularity of the retinal tissue. The pigment epithelium of animals in which retinal blood vessels are absent contains more lipid droplets. Also, similar lipid droplets accumulate in the abnormal pigment epithelium of monkeys and humans.

Lamellar phagosomes are present more frequently in the cells of animals in which the retina is predominantly made up of rods. For example, cells of the monkey and frog contain numerous lamellar

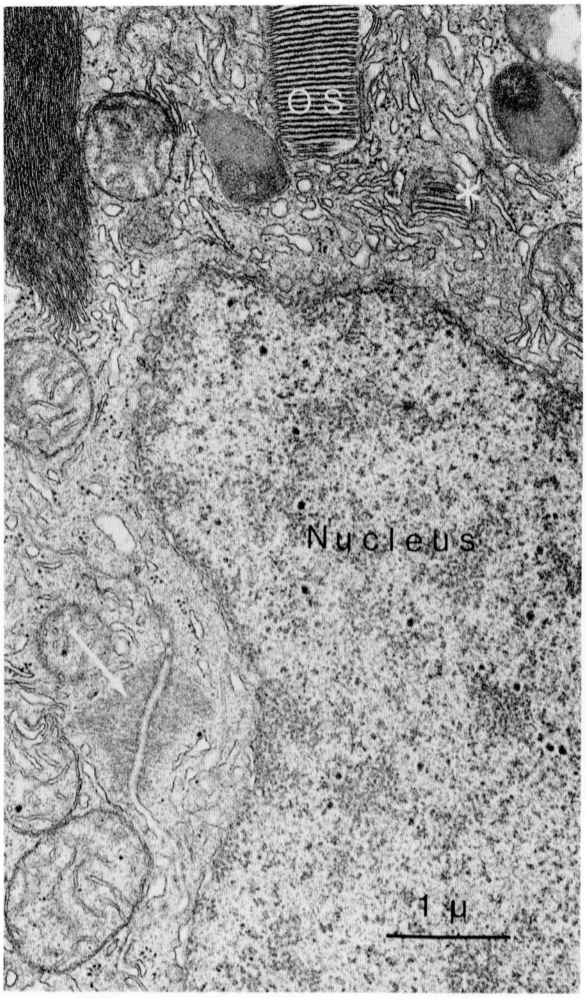

Figure 5.24. *Turtle RPE. The outer segment* (OS) *extends as deep as the nuclear area. Small fragments of cone outer segment are frequently seen in the cytoplasm (asterisk). The nucleus protrudes above the level of the apicolateral junction (white arrow). The cytoplasm is rich in smooth endoplasmic reticulum. A myeloid body is seen in the upper left corner. (×17,250.)*

bodies, whereas those of the ground squirrel and snake contain a very small number or none. However, small particles that are similar to the degrading substances of the lamellar body are equally abundant whether an animal has predominantly rods or cones. The variations in cytologic appearance within an animal often seem to be caused in part by different stages and degrees of degeneration of the lamellar phagosomes, though the basic structure of the cytoplasm itself is relatively standard.

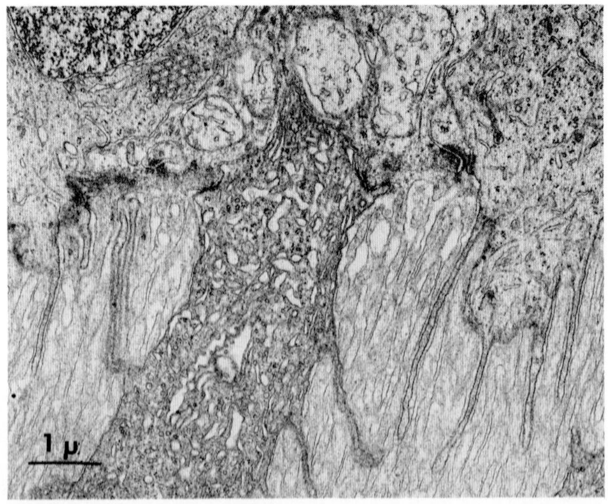

Figure 5.25. *Lizard RPE. Long microvilli extending from the inner segment of a cone cell interdigitate with those of a Müller's cell near the outer limiting membrane. (×9,480.)*

Although each animal has distinguishable characteristics in the cytoplasm of the pigment epithelium, the main cytoplasmic microorganelle is the smooth endoplasmic reticulum, indicating that function of the pigment epithelial cell depends upon this microorganelle. However, it is curious to find that smooth endoplasmic reticulum is sparse or absent in pigment epithelium that contains a reflecting substance.

Figure 5.26. *Snake RPE. This snake had been kept at room temperature, but it was killed for study in midwinter.* (A) *The photoreceptors are mainly cone cells. Outer segments are short. The pigment epithelial cells are large. Epoxy section, toluidine blue. (×472.)* (B) *The inner segment of a cone cell contains lipid droplets in the mitochondria* (IS)*. Long microvilli extend to the level of the inner segment. (×2,830.)*

Figure 5.27. *Snake RPE.* (A) *Cells are large. Numerous microvilli surround the outer segment* (OS) *closely. The apicolateral junction is near the apex of the cell (white arrow). (×8,140.)* (B) *The smooth endoplasmic reticulum is dilated. The appearance is similar to that of a hibernating animal. (×18,170.)*

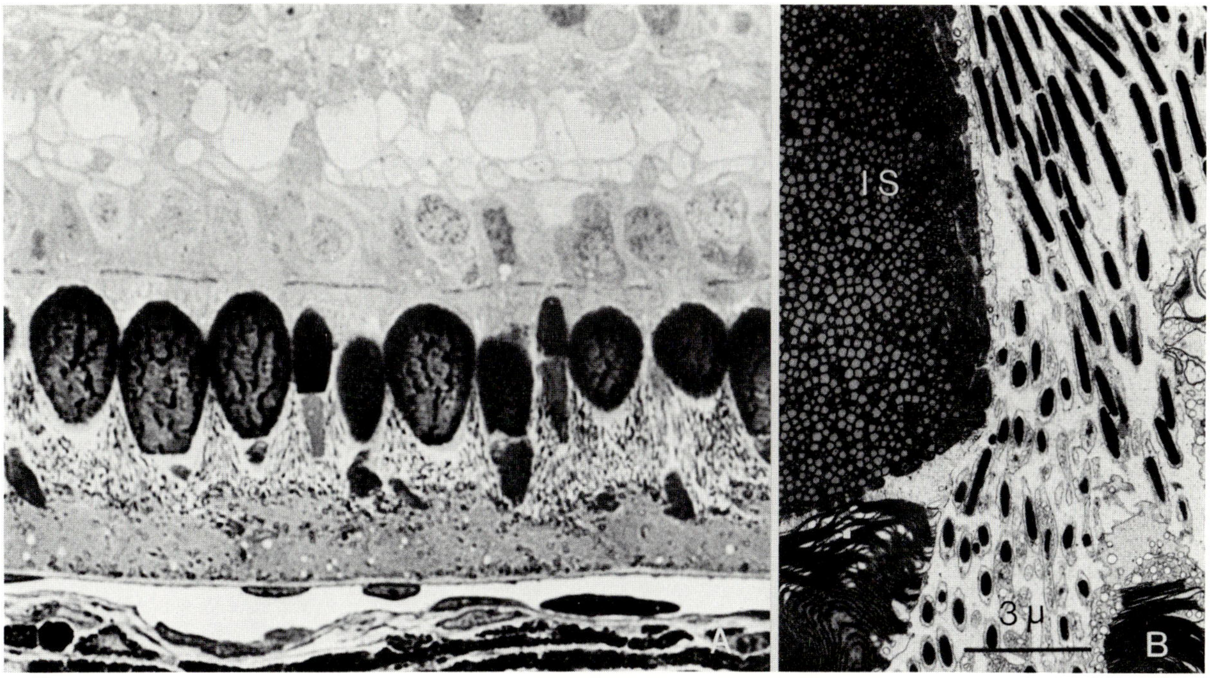
A
IS
3 μ
B

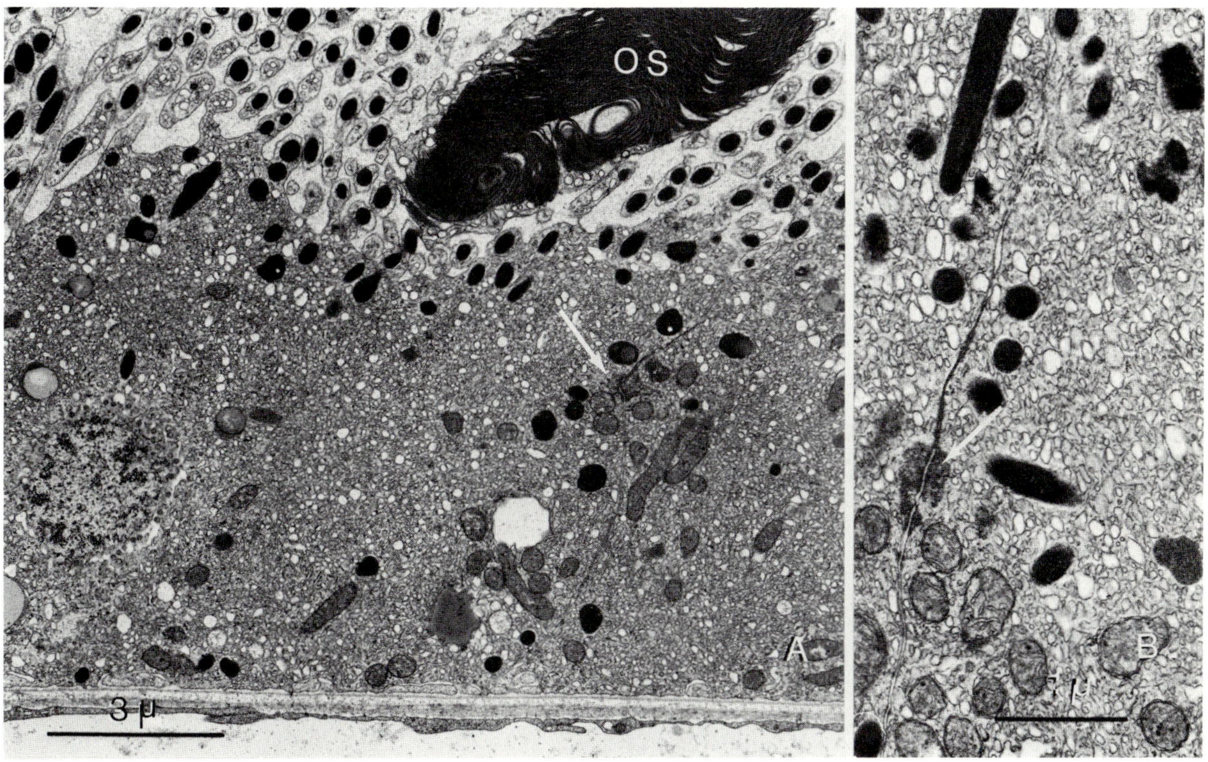
OS
A
3 μ
B

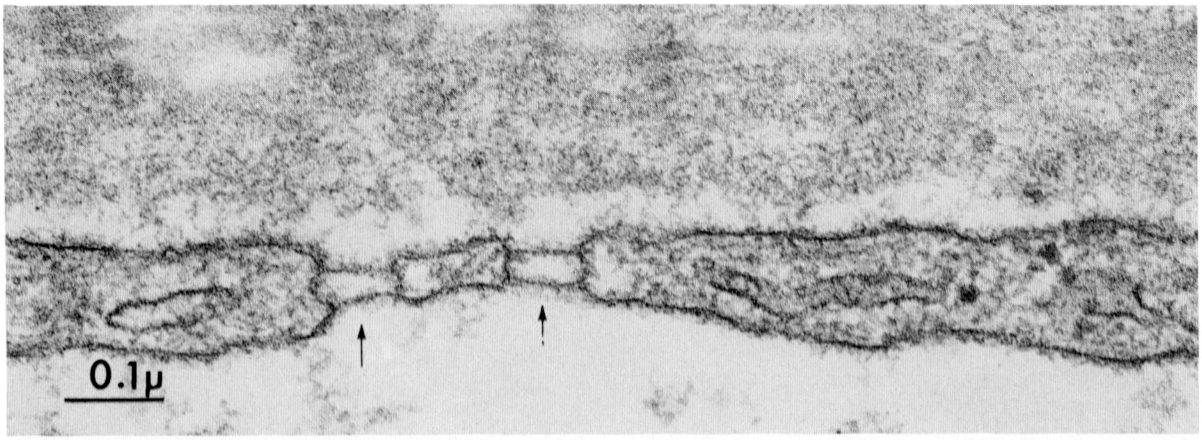

Figure 5.28. *Snake RPE. Fenestrations of the choroidal capillary consist of two-membrane diaphragm (arrows). (×7,550.)*

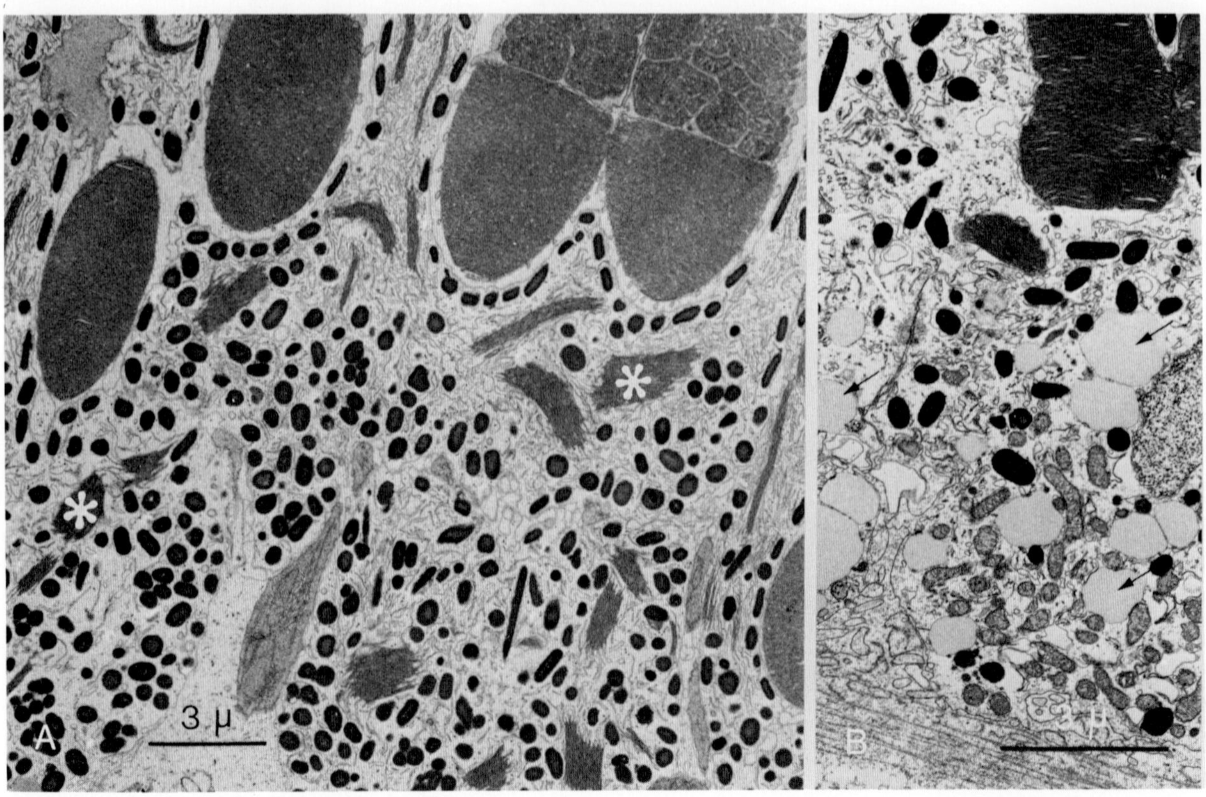

Figure 5.29. *Fish RPE.* (A) *Guppy. Cells are large and heavily pigmented. The main cytoplasm contains sparse smooth endoplasmic reticulum. Myeloid bodies (asterisks) are abundant. Large villi surround photoreceptor elements. Outer segments extend deep into the cell. A double cone is shown in the upper right corner. (×5,430.)* (B) *Goldfish. The cell contains numerous guanine granules (arrows). The apicolateral junction is close to the base of the cell. Bruch's membrane consists of fine fibrils. (×7,550.)*

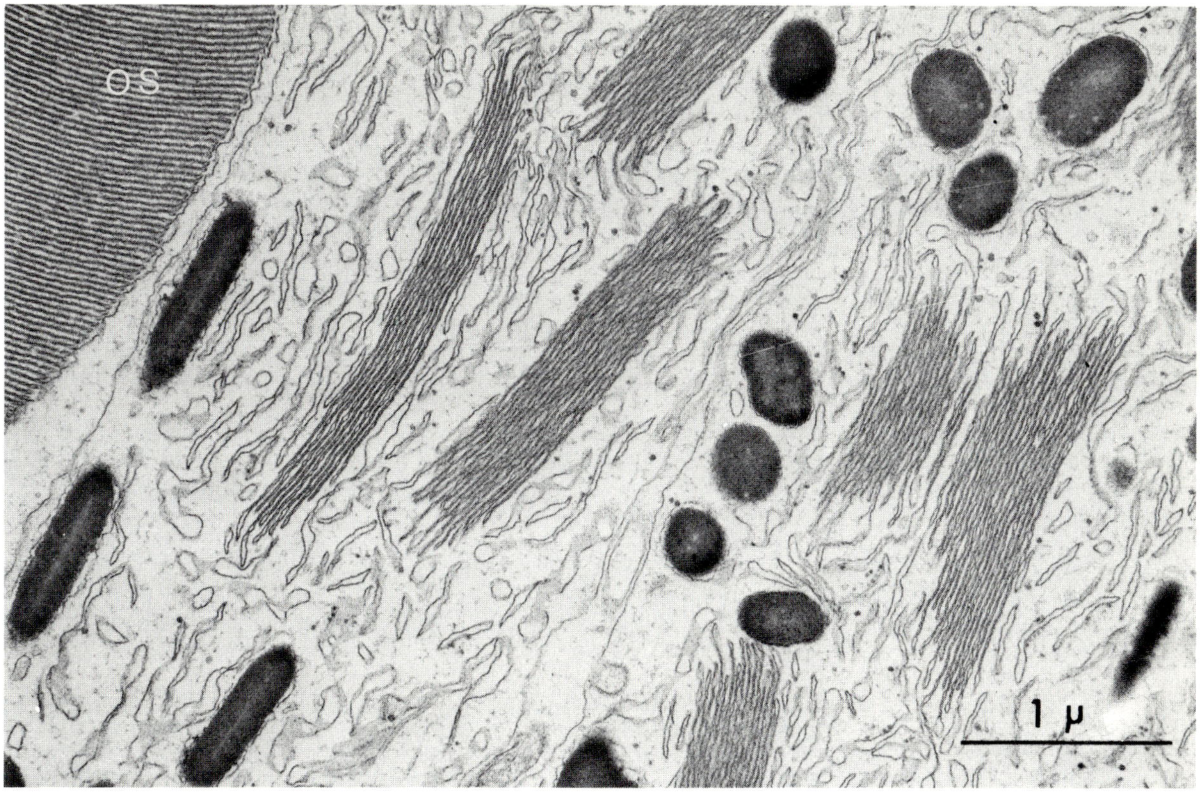

Figure 5.30. *Trout RPE. Large villi extend between outer segments* (OS). *They contain a moderate number of smooth endoplasmic reticulum and numerous small myeloid bodies. (×28,080.)*

REFERENCES

BAIRATI, A., JR., and ORZALESI, N. 1963. The ultrastructure of the pigment epithelium and of the photoreceptor-pigment epithelium junction in the human retina. *J. Ultrastruct. Res.* 9:484–496.

BERMAN, E. R., SCHWELL, H., and FEENEY, L. 1974. The retinal pigment epithelium: chemical composition and structure. *Invest. Ophthalmol.* 13:675–687.

BREATHNACH, A. S., and WYLLIE, L. M. A. 1966. The ultrastructure of retinal pigment epithelium of the human fetus. *J. Ultrastruct. Res.* 16:584–597.

COHEN, A. I. 1960. The ultrastructure of the rods of the mouse retina. *Am. J. Anat.* 107:23–48.

DOWLING, J. E., and GIBBONS, I. R. 1962. Fine structure of pigment epithelium in albino rat. *J. Cell Biol.* 14:459–474.

DUKE-ELDER, S. 1958. *System of ophthalmology. I. The eye in evolution*, p. 449. St. Louis: C. V. Mosby.

FEENEY, L., GRIESHABER, J., and HOGAN, M. J. Studies on human ocular pigment. In *The structure of the eye*, ed. H. W. Rowen, vol. 2, pp. 535–548. Stuttgart: Schattauer-Verlag.

FINE, B. S., and YANOFF, M. 1972. *Ocular histology: a text and atlas*, pp. 48–107. New York: Harper and Row.

FRIEDMAN, E., and KUWABARA, T. 1968. The retinal pigment epithelium. IV. The damaging effects of radiant energy. *Arch. Ophthalmol.* 80:265–279.

FRIEDMAN, E. and TSO, M. O. 1968. The retinal pigment epithelium. Histologic changes associated with age, *Arch. Ophthal.* 79:315–320.

HOGAN, M. J. 1972. Role of the retinal pigment epithelium in macular disease. *Trans. Am. Acad. Ophthalmol. Otolaryngol.* 76:64–80.

HOGAN, M. J., ALVARADO, J. A., and WEDDELL, J. E. 1971. *Histology of the human eye*, pp. 405–420. Philadelphia: W. B. Saunders.

HOLLENBERG, M. J., and BERNSTEIN, M. H. 1966. Fine structure of the photoreceptor cells of the ground squirrel (*Citellus tridecemlineatus*). *Am. J. Anat.* 118:359–363.

ISHIKAWA, T., and YAMADA, E. 1969. Atypical mitochondria in the ellipsoid of the photoreceptor cells of vertebrate retinas. *Invest. Ophthalmol.* 8:302–316.

KARLI, P., STOECKEL, M. E., and PORTE, A. 1965. Dégénérescence des cellules visuelles photoreceptrices et persistance d'un sensibilité de la rétine a la stimulation photique: observations au microscope électronique. *Z. Zellforsch. Mikrosk. Anat.* 65:238–352.

KEEFE, J. R. 1971. The fine structure of the retina in the newt, *Triturus viridescens. J. Exp. Zool.* 177:263–294.

KUWABARA, T. 1975. Cytologic changes in the pigment epithelium during hibernation. *Invest. Ophthalmol.* 14: 457–467.

KUWABARA, T. 1979. Photic and photo-thermal effects on the pigment epithelium. In *The retinal pigment epithelium,* ed. K. M. Zinn and M. F. Marmor. Cambridge: Harvard University Press.

LERCHE, W. 1963. Elekronenmikroskopische Untersuchunger zur Differenzierung des Pigmentepithels und der ausseren Kornerzellen (Sinneszellen) in menschlichen Auge. *Z. Zellforsch. Mikrosk. Anat.* 58:953–970.

LEURE-DUPREE, A. 1968. Ultrastructure of the pigment epithelium in the domestic sheep. *Am. J. Ophthalmol.* 65: 383–398.

MATSUSAKA, T. 1967. The intracytoplasmic channel in pigment epithelial cells of the chick retina. *Z. Zellforsch. Mikrosk. Anat.* 81:100–113.

NILSSON, S. E. G. 1964. An electron microscopic classification of the retinal receptors of the leopard frog (*Rana pipiens*). *J. Ultrastruc. Res.* 10:390–416.

NISHIDA, S. 1964. Electron microscopic study of chicken retina: the ultrastructure in the retinal epithelium of light adapted chicken. *Acta Soc. Ophthalmol. Jap.* 68:1431–1443.

OHKUMA, M. 1972. Transmission and scanning electron microscopic observations of the experimentally detached retina in the rabbit. 1. Early changes of the detached retina. *Acta Soc. Ophthalmol. Jap.* 76:303–311.

OKUDA, K., AONO, H., and YOKOYAMA, A. 1960. Electron microscopic studies on the fine structure and morphogenesis of fuscin granules. *Folia Ophthalmol. Jap.* 11:635.

PEDLER, C., and TILLEY, R. 1969. The retina of a fruit bat (*Pteropus giganteus* Brunnich). *Vision Res.* 9:909–922.

PORTER, K. R., and YAMADA, E. 1960. Studies on the endoplasmic reticulum. V. Its form and differentiation in pigment epithelial cells of the frog retina. *J. Biophys. Biochem. Cytol.* 8:181–205.

SAKURAGAWA, M., and KUWABARA, T. 1976. The pigment epithelium of the monkey: a topographic study by transmitting and scanning electron microscopic study. *Arch. Ophthalmol.* 94:285–292.

SMITH, R. S., and BERSON, E. L. 1971. Acute toxic effects of chloroquine on the cat retina: ultrastructural changes. *Invest. Ophthalmol.* 10:237–246.

SPITZNAS, M., and HOGAN, M. J. 1970. Outer segments of photoreceptors and the retinal pigment epithelium: interrelationship in the human eye. *Arch. Ophthalmol.* 84:810–819.

STEINBERG, R. H. 1973. Scanning electron microscopy of the bullfrog's retina and pigment epithelium. *Z. Zellforsch. Mikrosk. Anat.* 143:451–463.

STEINBERG, R. H., and WOOD, I. 1974. Pigment epithelial cell ensheathment of cone outer segments in the retina of the domestic cat. *Proc. R. Soc. Lond.* [Biol.] 187:461–478.

YAMADA, E. 1960. Observation on the fine structure of photoreceptive elements in the vertebrate eye. *J. Electron Microsc.* 9:1–14.

YAMADA, E. 1961. The fine structure of the pigment epithelium in the turtle eye. *The structure of the eye,* ed. G. K. Smelser, pp. 73–84. New York: Academic Press.

YAMADA, E., TOKUYASU, K., and IWAKI, S. 1958. The fine structure of retina studied with electron microscope. II. Pigment epithelium and capillary of the choriocapillary layer. *J. Electron Microsc.* 6:42–46.

YOUNG, R. W. 1971. The renewal of rod and cone outer segments in the rhesus monkey. *J. Cell Biol.* 49:303–318.

ACKNOWLEDGMENTS

Continuous interest and encouragement of Dr. David G. Cogan are gratefully acknowledged. Thanks are due to Miss Dorothy Drago for her excellent technical assistance. Figure 27 is prepared by Mr. D. Player and Figure 28 is provided by Dr. M. Funahashi.

The initial portion of this investigation was supported by a U. S. Public Health Research Grant EY 00006 from National Eye Institute, National Institutes of Health.

6 Biochemistry of the Retinal Pigment Epithelium

Elaine R. Berman

THERE HAVE BEEN many reviews on the biochemistry of the retina, but none—as far as the author is aware—on the pigment epithelium, apart from a brief survey given before the Association for Research in Vision and Ophthalmology (Berman 1972). Present knowledge of the metabolism of this tissue is highly fragmentary; nevertheless, the biochemical transformations that do take place in the retinal pigment epithelium (RPE) are considered to play an important, if not essential, role in retinal physiology and function. Stimulating reviews on the molecular organization of normal RPE and of ultrastructural changes in the disease state have appeared (Moyer 1969; Spencer 1972; Hogan 1972; Kuwabara 1972). The present chapter brings together the information presently available on RPE biochemistry, incorporating both published literature as well as some recent (unpublished) findings in the author's laboratory.

Techniques for Studying RPE Metabolism

The paucity of information on the biochemistry of the RPE is due in great part to the lack of suitable experimental material. Unlike other ocular tissues, the RPE is not readily available in the amounts needed for accurate biochemical studies. Although the cells can be dissected out and purified from large eyes such as those of cattle and sheep, or even cat, dog, and rabbit, nevertheless, 20–30 eyes are usually needed for a complete metabolic experiment, which, for obvious reasons, can only be carried out in vitro. The amounts of RPE available from small laboratory animals such as the rat, mouse, or frog are so limited in quantity as to preclude conventional biochemical approaches. However, in spite of limited amounts of material, it is possible to measure retinol in albino rats (Dowling 1960; Noell, Delmelle, and Albrecht 1971) and in frogs (Bridges 1975, 1976). More recently, a technique based on cochromatography of unlabeled vitamin A compounds with labeled visual cycle intermediates has been successfully applied to rat eyes (Zimmerman 1974; Zimmerman, Yost, and Daemen 1974). But, for all practical purposes, histochemical and radioautographic techniques, at the level of both the light and electron microscope, are the main methods used for small laboratory animals. In fact, most of our present knowledge is based on experimental findings using these techniques. Other methods such as tissue and organ cultures and standard biochemical techniques are also available now for the study of RPE biochemistry.

HISTOCHEMISTRY AND RADIOAUTOGRAPHY. The elegant ultramicro techniques developed by Lowry and coworkers (Lowry, Roberts, and Lewis 1956; Lowry et al. 1961) for analyzing enzyme activities of individual retinal layers have added considerably to our understanding of the metabolic characteristics of RPE tissue. However, this approach has not been used extensively by other workers, probably because of the technical difficulties involved. During the past decade, radioautography has yielded much valuable information; the fate of labeled precursors is followed by both light and electron microscopy, and time studies provide a basis for kinetic information as well. Electron-microscopic cytochemistry has also been used, but this approach is limited mainly to studies of enzymes such as acid phosphatase (Ishikawa and Yamada 1970; Marshall 1970; Marshall and Ansell 1971; Alvarado, Sarris, and Hogan 1972; Hollyfield and Ward 1974b) or catalase (Leuenberger and Novikoff 1975; Robison and Kuwabara 1975). Histochemistry may be helpful for the identification of certain chemical components (Streeten 1961; Farkas et al. 1971b) and has also been applied to the detection of oxidative enzyme activities (Pearse 1961; Hansson 1970). Although it is considered to give "semiquantitative" results, histochemistry—at its present stage of development—appears to be principally a qualitative approach.

TISSUE AND ORGAN CULTURE. The long-held view that only embryonic pigment epithelial cells can be successfully cultured has been dispelled by several independent groups of investigators. Adult cells from pig (Francois et al. 1971) and rat (Edwards and Szamier 1977) as well as human tissue (Barishak 1960; Albert, Tso, and Rabson 1972; Mannagh, Arya, and Irvine 1973) have been established in culture, with survival times as long as 3 to 6 months. Initially, the cells divide actively and spread, forming monolayers on the floor of the culture vessel. Mitotic activity then ceases, although it can be resumed after subculture. Cultured cells are not only viable (as evidenced by their morphological appearance and their ability to undergo mitosis) but they are also capable of phagocytizing Thorotrast particles (Tso, Albert, and Zimmerman 1973). RPE cells from embryonic chick eyes have also been grown in tissue culture (Newsome et al. 1974). By including in the culture media a low molecular weight protein fraction extracted from chick embryos and purified on Sephadex G-25, well-differentiated confluent sheets of RPE cells were obtained after 70 days. Addition of dibutyryl cyclic 3′,5′-adenosine monophosphate was shown to alter significantly the morphology, growth rate, adhesive properties and pigmentation of these cells. Other details on tissue culture of RPE cells are given in Chapter 16.

Successful organ culture of both cattle and human RPE has been reported by Feeney and Mixon (1976). Tissue explants maintained for 24 hours actively phagocytized latex beads measuring 1.1 μm in diameter. A latent period of about 8 hours was consistently observed but by 20 hours, 90 percent or more of the pigment cells were found to be actively phagocytizing the latex microspheres. As in chick embryo cultures, dibutyryl cyclic 3′,5′-adenosine monophosphate had striking effects on the system, in this case enhancing greatly the rate of incorporation of the latex spheres.

STANDARD BIOCHEMICAL TECHNIQUES. Whole cell suspensions or homogenates of bovine RPE are the most commonly used tissue preparations. The first published method for isolating pigment cells by brushing them from the optic cup after removal of the retina appeared in 1962 (Glocklin and Potts 1962). This procedure has been used not only by the originating investigators (Glocklin and Potts 1965; Gonasun and Potts 1974) but by others as well (Berman 1964, 1971; Shichi 1969, 1973; Vento and Cacioppo 1973; Zimmerman 1976). It is important to point out, however, that most of the preparations used had not been examined systematically under the microscope, hence there was no assurance that they had been free of contamination by fragments originating from adjacent ocular tissues. These suspicions were confirmed in a study (Berman, Schwell, and Feeney 1974) that showed that only under carefully defined conditions of centrifugation in solutions of known density or viscosity can all of the impurities be effectively separated from the cells. The principal contaminants present in unwashed RPE preparations are red blood cells, broken and intact outer segments, and retinal tissue. Bovine RPE cells suspended in .25 M sucrose and centrifuged at 112 *g* for 10 minutes are, after four or five washes, essentially free of extraneous tissue. The appearance of cell preparations isolated in this manner is shown in Figure 6.1. Alternatively, the

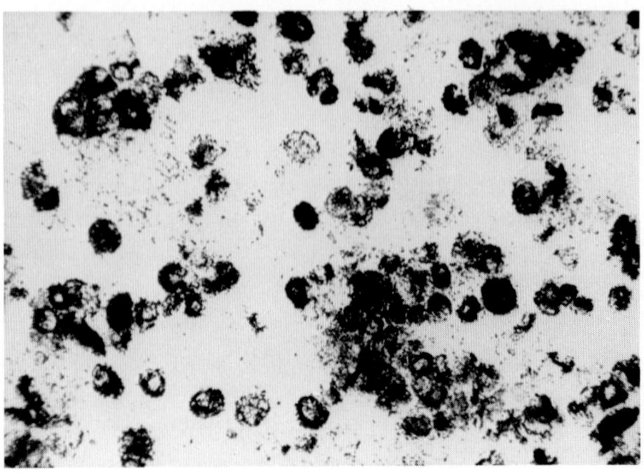

Figure 6.1 *Washed RPE cells prepared as described in the text. Note the absence of rod outer segments and red blood cells. (Stained with toluidine blue. ×35.)*

brushed-out cells may be purified either by filtration on a Nucleopore polycarbonate membrane (12 μm), or by passing the crude suspension through a column of acid-washed glass beads (#100) (Siakotos, Aguirre, and Schuster, 1978). The yield of cells from cattle eyes using the latter method was over 8 mg per eye, and the preparations were essentially free of rod outer segments and red blood cells; however, they did contain some choroidal tissue. The procedure may be modified to obtain RPE cells from small animals such as rats and cats. In this case, the cells are removed under a dissecting microscope with a microspatula and purified on small glass bead columns.

Whatever method is used for the purification of RPE cells, it is now possible to fractionate cell homogenates and to isolate their subcellular organelles. Although either density-gradient centrifugation or differential pelleting can be used for fractionation, the latter method is somewhat simpler and, with the limited amount of material available, is the method of choice. Fractions enriched in melanin granules, nuclei, mitochondria, lysosomes, and microsomes are shown in Figure 6.2. Enzymatic analyses of these subcellular fractions are described later in this chapter (see the section on "Metabolism of the RPE"). It is important to point out that although this method is effective for the isolation of RPE organelles, a recent investigation has shown that cells prepared in this manner lack their soluble cytosol (Saari et al. 1977). Biochemical and morphological examination of both the cells and the wash solutions (especially the first one) showed conclusively that the plasma membranes suffer considerable damage during the brushing-out procedure, and it is this which leads to leakage and loss of the cellular cytosol. Alternative methods for harvesting would be highly desirable, but until this is accomplished, all published studies on the biochemistry of isolated RPE cell preparations must be considered in the light of two major methodological weaknesses. The first, which applies to many early studies, is that of contamination with non-RPE material. The second is that purified preparations of cells, although retaining the subcellular organelles, may have lost most of the soluble cytosol during their removal from the optic cup. The findings to be reviewed in this chapter are taken, for the most part, from reports in which the contribution of these two factors to the published results cannot be accurately assessed.

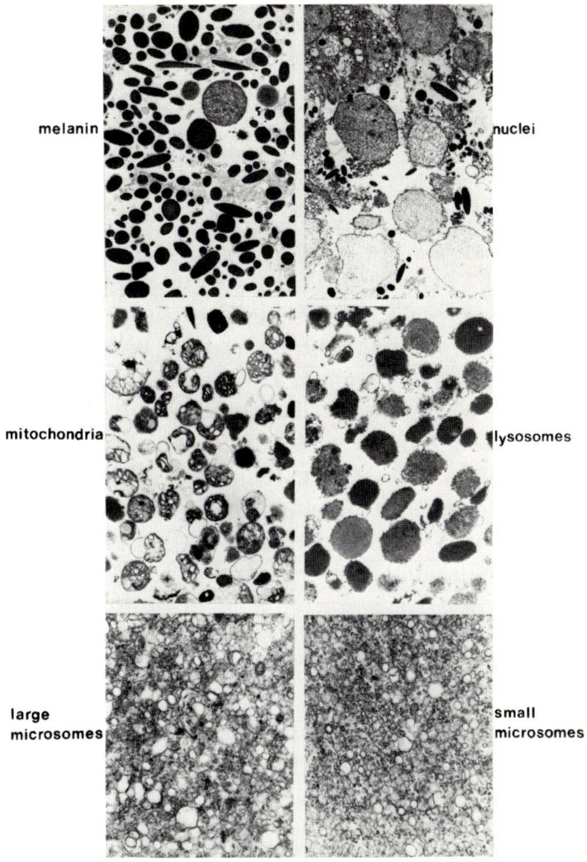

Figure 6.2 *Electron micrographs of subcellular fractions of cattle RPE homogenates isolated by differential centrifugation.*

Chemical Composition of RPE Cells

Table 6.1 summarizes the water, protein, lipid, and DNA contents of RPE cells; the content of retinol in frog, rat, and cattle RPE is shown sepa-

Table 6.1 Comparative chemical compositions of bovine RPE and retina.

	Wet weight (%)		*Dry weight (%)*	
Component	*Pigment epithelium*	*Retina*	*Pigment epithelium*	*Rod outer segments*
Water	81.4	80.6	—	—
Protein	8.1	7.7	43.5	41[a]
Total lipid	3.00	2.74	16.1	39[b]
Phospholipids	1.59	1.67	8.5	31.5[b] 36.0[a]
DNA	0.72	0.79	3.9	—

Source: Adapted from Berman et al. 1974.
a. Nielson et al. 1970.
b. Borggreven et al. 1970.

Table 6.4 Fatty acid composition of phospholipids in RPE and rod outer segments of cattle retina.

	Composition (% of each fatty acid by weight)			
	RPE			*Rod outer segments*
Fatty acid	*Total Phospholipids*[a]	*Phosphatidyl choline*[b]	*Phosphatidyl ethanolamine*[b]	*Total phospholipids*[c]
16:0	30.7	61.3	40.1	34.4
16:1	2.6	1.9	0.3	2.2
18:0	14.9	17.0	20.1	17.5
18:1	13.1	11.1	6.1	11.2
18:2	14.9	10.2	2.6	0.9
20:4	16.6	2.5	11.4	5.6
22:4	Trace	—	—	1.5
22:5	—	—	—	2.3
22:6	Trace	2.7	7.5	23.0

a. Data of Berman et al. 1974.
b. Data of Anderson et al. 1976.
c. Data of Anderson and Sperling 1971.

47 percent of the fatty acids in the various phospholipids (Anderson et al. 1976). In contrast, only small amounts of docosahexaenoate are detectable among the fatty acids of RPE phospholipids (Table 6.4 and Fig. 6.3). When analyzed on an individual basis, the highest concentration of docosahexaenoate (11.7 percent) is in phosphatidylserine (Anderson et al. 1976). Since, however, this phospholipid constitutes only 4 percent of the total phospholipid in RPE cells, the contribution of 22:6 from this component is quite negligible. In the absence of direct experimental data, one can only speculate that polyenoates such as docosahexaenoate, present in the outer segments phagocytized by RPE (Young 1974; Young and Bok 1969, 1970), must be very efficiently metabolized by pigment cells. Degradative pathways may be present, converting long chain polyenoates to 16- and 18-carbon fatty acids. It seems more likely, however, that lipid peroxidations catalyzed by free radicals in the presence of oxygen (Feeney and Berman 1976) may account for the extremely rapid disappearance (LaVail 1976) of phagocytized outer segments. In many tissues, possibly including RPE, the lipid peroxide free radical decomposes into highly reactive fragments such as malonaldehyde, which in turn forms cross-linkages (through Schiff bases) with free amino groups of proteins, phospholipids, and nucleic acids. This may be one of the crucial steps in lipofuscin formation in RPE cells, an aspect that is discussed further in the following section.

OTHER COMPONENTS. Before leaving the subject of the chemical composition of RPE, structural components such as melanin and lipofuscin should also be mentioned. The precise chemical composition of these complex macromolecules is only poorly understood in other cells, and for RPE cells we have exceedingly limited information. In those sources where it has been examined, melanin was found to consist of highly cross-linked polymers formed from tyrosine-derived semiquinone free radicals (Blois 1971). Studies on melanosomes isolated from a variety of mammalian and avian tissues have provided further insight into their chemical composition and structure (Duchon, Borovansky, and Hach 1973). The melanosomes isolated from embryonic chick RPE contained 51.6 percent melanin and 19.3 percent protein; the remaining 29.1 percent of the components could not be identified. The protein was found to consist of about 18 amino acids, the most abundant being Glu, Gly, Ala, and Asp. Traces (0.2 to 0.8 percent) of 3,4-dihydroxyphenylalanine (DOPA) were also present.

Melanosomes have also been isolated from eyes of black mice (Hearing and Lutzner 1973), using the whole eye as starting material. Thus, the chemical analyses performed were presumably on melanosomes originating not only from RPE but from other melanotic tissues as well, the principal ones being choroid and iris. Nevertheless, the findings reported by Hearing and Lutzner contribute significantly to

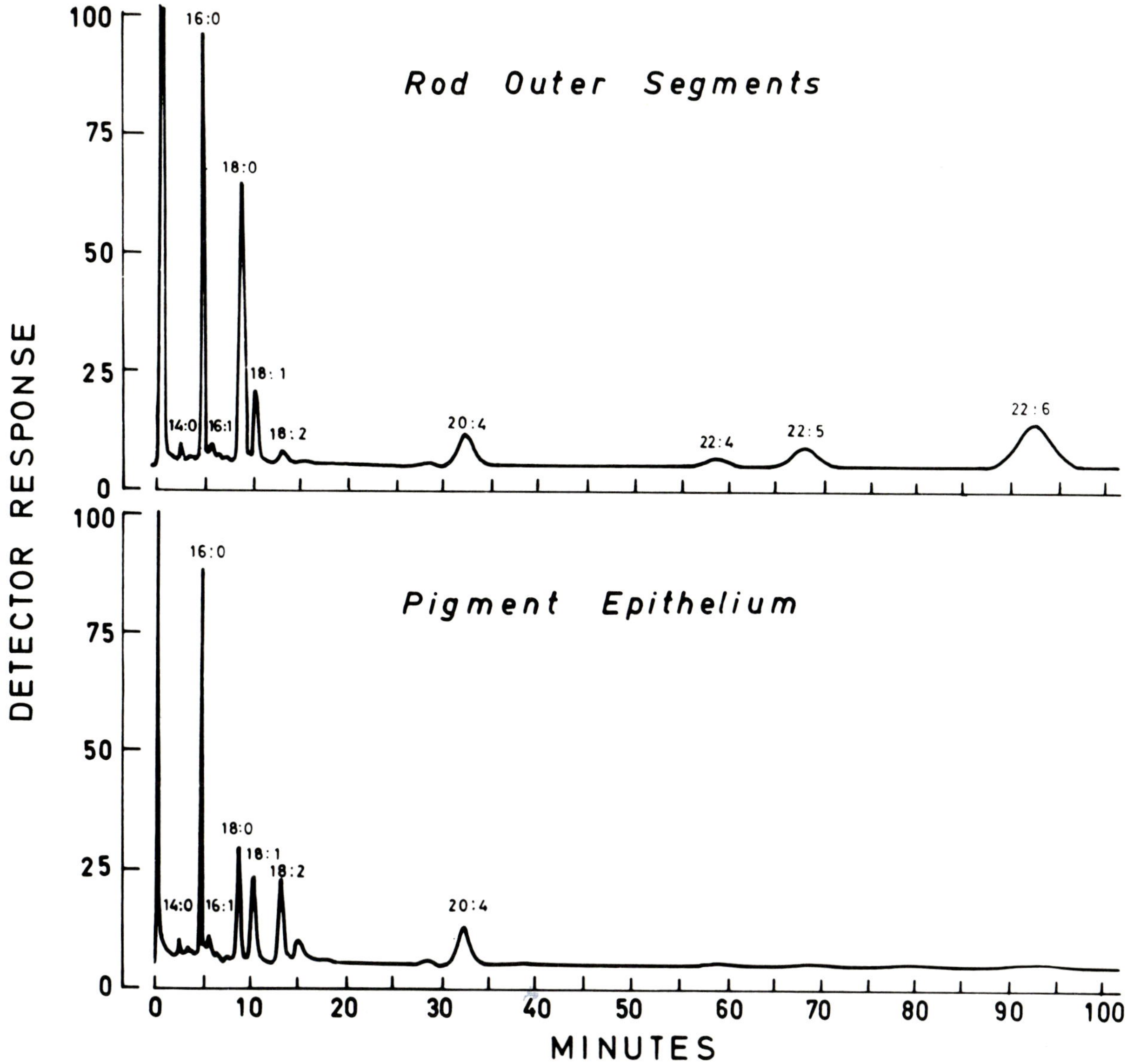

Figure 6.3 *Gas-liquid chromatogram of phospholipid fatty acids from rod outer segments and from RPE. (Reprinted with permission from Berman et al. 1974.)*

our understanding of melanogenesis in general. These authors describe, for the first time, the use of relatively mild methods for the complete solubilization of melanin. Acrylamide gel electrophoresis of urea-treated or sodium-dodecyl-sulfate-treated fractions gave seven protein bands, one of which was considered to represent either the principal structural fibrillar protein or possibly the site of tyrosinase activity in the melanosome particle, or both.

The lipofuscin granules of human RPE were first characterized histochemically by Streeten (1961), using light microscopy, and were later identified as part of the lysosomal system by Feeney, Grieshaber, and Hogan (1965). Lipofuscin (often called "age" pigment) isolated from human brain and other tissues is a complex mixture of polymerized substances, consisting mainly of neutral and polar lipids (Zeman and Siakotos 1973) cross-linked to partially degraded cellular membranes. The granules are recognized both in vivo and in vitro by their characteristic autofluorescent properties. Spectral analyses show excitation and emission peaks at approximately 360 and 450 nm, respectively (Siakotos and Armstrong 1975), which serves to distinguish these particles from vitamin A—whose emission maximum, under the same conditions, is at 495 nm (Csallany and Ayaz 1976). Lipofuscin granules presumably accumulate in the RPE as one of the principal end products of phagocytosis; these particles represent remnants of rod outer segments (and other cellular components) for which enzymatic machinery for further degradation is lacking (Feeney 1973a; Feeney and Berman 1976). Since lipofuscin granules normally are neither excreted nor secreted from RPE cells, once formed, they accumulate in the cell cytoplasm. Hence, it is not surprising that autofluorescent lipofuscin granules are abundant in RPE of senescent individuals, but are rarely observed in young human eyes (Feeney 1978).

Information on other chemical components of human RPE has been deduced indirectly from histochemical analyses of drusen (Farkas et al. 1971b), assuming that the latter represent pathological exudates originating from within the pigment cells. Apart from PAS-staining complex carbohydrates, cerebrosides were also identified on the basis of combined solubility and staining reactions. It was suggested (Farkas 1971a) that drusen represent substances extruded from the RPE cell as a consequence of either degenerative or autophagic processes.

Metabolism of RPE

The most important aspect of the biochemistry of any tissue is its overall (intermediary) metabolism. Unfortunately, for the pigment epithelium our knowledge is so limited that we are unable to deduce anything more than the most superficial kind of biochemical information, and in a fragmentary state, at that.

RETINOL TRANSFORMATIONS. The present discussion will be confined to recent biochemical findings related to the storage, metabolism, and transport of vitamin A by the RPE. Other aspects of the physiology and turnover of vitamin A, including the role of the RPE in the visual cycle, are discussed in the chapter that follows.

Retinol is mobilized from storage depots in the liver as a retinol-RBP (retinol-binding protein) complex (Smith et al. 1973), and in this form, in 1:1 noncovalent linkage to serum prealbumin (Kanai, Raz, and Goodman 1968), it is carried to peripheral tissues for utilization. The molecular weight of serum RBP is approximately 21,000 daltons (Kanai et al. 1968), while that of the prealbumin complex is about 76,000, corresponding to a sedimentation coefficient of 5.2S (Abe, Muto, and Hosoya 1975). On the basal surface and lateral plasma membranes of the RPE cells there are specific receptor sites (Heller 1975; Bok and Heller 1976) for serum RBP which, however, does not enter the cell; its diffusion is blocked by the tight junctions (zonulae occludentes) existing between RPE cells. Nevertheless, injected retinol does in fact find its way into the pigment epithelium cells (Sherman 1970; Hall and Bok 1974); therefore it is reasonable to assume that free retinol injected into the living animal is able to cross the RPE plasma membrane and penetrate into the cell cytoplasm. This process may be mediated by a specific plasma membrane retinol binding protein of molecular weight 14,500 that has recently been isolated from cattle RPE (Maraini et al. 1977).

Retinol taken up from the circulation by the RPE may be handled in several possible ways after entering the cell. The first way, for which there is now considerable evidence, is for it to be bound to a specific receptor, or binding protein. This protein has been isolated from, and characterized in, RPE cytosol from a variety of species. Detailed studies on chick embryo RPE cytosol (Wiggert and Chader 1975; Wiggert, Bergsma, and Chader 1976) have shown that the principal retinol receptor is a protein with a sedimentation constant of 2S that corresponds closely to the retinol-binding protein identified in bovine RPE cytosol, whose molecular weight is 17,000 daltons (Saari et al. 1977). This value for the molecular weight of RPE retinol-binding protein has been corroborated in several laboratories using different, though complementary, experimental techniques. These include sucrose density gradients and Sepharose gel filtration for the chick embryo protein, and Sephadex gel filtration and gel electrophoresis for the bovine RPE retinol-binding protein (Saari et al. 1977). It is highly unlikely that there are any other macromolecular binding proteins for retinol in RPE cytosol. If this is the case, then it is probably not fortuitous that a retinol-binding protein of similar size has been found in cytosol of cattle retina (Futterman, Saari, and Swanson 1976; Saari and Futterman 1976), as well as in retina of a variety of other species (Wiggert and Chader 1975; Wiggert et al. 1976; Wiggert et al. 1977). Immunological studies have shown that in both the neural retina and RPE of embryonic chick, the cytosol retinol-binding proteins are separate and distinct from purified serum retinol-binding protein (Abe et al.

1977). Whether this protein is a "carrier of retinol" between the RPE and the retina remains to be established.

Esterification is another possible way in which the pigment epithelium could transform the retinol entering the cell. All of the vitamin A reaching the RPE from the blood (and possibly some that may enter during light adaptation) is in the form of the free alcohol, yet 95 percent of the retinol stored in the RPE is in the ester form. Krinsky (1958) demonstrated esterifying activity in "pigmented layers" of bovine RPE, and, by analogy with the retinal system (Andrews and Futterman 1964), it is reasonable to assume that a similar esterifying mechanism exists in the RPE. It is therefore not surprising that Hirosawa and Yamada (1976) found considerable amounts of labeled retinol in the endoplasmic reticulum of the RPE after parenteral injection in the mouse. This corroborates well with quantitative data which show that more than two-thirds of the endogenous vitamin A of the RPE is bound to the microsomes as retinyl ester (Berman et al. 1979). Another storage site is in the cytoplasm, where vitamin A is deposited in the form of lipid droplets, the size and number depending on the nutritional status of the animal (Robison and Kuwabara 1977). Taken together, these findings suggest that retinol taken up from the blood may enter both of the metabolic pools of retinyl ester present in the RPE. It is possible that the catalase-containing peroxisomes found in close association with the vitamin-A-storing lipid droplets in mouse RPE (Robison and Kuwabara 1977) play an important role in transporting or converting vitamin A into one of the storage pools.

Another biochemical transformation of retinol that may be occurring in the RPE is isomerization. Retinol from serum (and from the retina during light adaptation) is of all-*trans* configuration, yet a high proportion of the retinol in both bovine (Krinsky 1958) and amphibian (Bridges 1975 and 1976) RPE is the 11-*cis* isomer. In spite of this, the conversion of all-*trans* retinol to the 11-*cis* isomer has in fact only been demonstrated in one study (Hubbard 1955–56). Isomerizing enzymes have not been detected in any species except the frog, either because they are actually not present, or because this important interconversion is taking place in the rod outer segments and not in the RPE.

There are many questions about RPE vitamin A that still remain unanswered. For example, to what fatty acid, or acids, is RPE retinol esterified? How is retinol mobilized from its storage form, as retinyl ester, to the biologically active form required for rhodopsin regeneration, namely retinol or an active derivative of this alcohol. Hydrolytic enzymes were not detected by Krinsky (1958), but it is reasonable to suppose that they should be present. Finally, what is the precise role of 11-*cis* retinol dehydrogenase found in particulate fractions of RPE cells (Zimmerman 1976) and what is its relationship, if any, to the metabolic transformation described above?

HYDROLYTIC ENZYMES. In view of the phagocytic character of RPE cells in virtually every species examined (Young and Bok 1969; Spitznas and Hogan 1970; Hollyfield and Ward 1974a; LaVail 1976), a highly developed lysosomal system containing a variety of hydrolytic enzymes would be anticipated (Feeney 1973a). This in fact appears to be the case, judging from histochemical, electron-microscopic, and biochemical evidence. Acid phosphatase, N-acetylglucosaminidase, naphthylamidase, and cathepsin activities have been demonstrated in rat RPE by light microscopy (Abraham et al. 1969; Burden et al. 1971). These and other acid hydrolases have also been found in cattle RPE by use of standard biochemical procedures (Berman 1971; Vento and Cacioppo 1973; Hayasaka, Hara, and Mizuno 1975a, b; Rothman, Feeney, and Berman 1976). The electron-microscopic studies of Ishikawa and Yamada (1970) showed that, in rat RPE, acid phosphatase activity is localized in lysosomes and phagosomes, as well as in the Golgi complex. Later studies of rat RPE, as well as of frog and pigeon RPE, confirmed, on the whole, the strong reaction in lysosomes and phagosomes; however acid phosphatase activity in the Golgi apparatus was not a consistent finding (Marshall and Ansell 1971; Hollyfield and Ward 1974b). Other hydrolytic enzymes detected by electron microscopic cytochemistry in RPE include arylsulfatase and 5′-nucleotidase (Alvarado et al. 1972).

On the assumption that acid hydrolases play a key role in the degradation of phagocytized rod outer segment discs, the question of delivery of these enzymes to the phagosomal particle is an exceedingly important one, but the precise mechanism still remains obscure. Preliminary evidence (Alvarado et al. 1972) suggests that at least one of the enzymes, arylsulfatase, is transferred from the Golgi apparatus to both lysosomes and phagosomes in a manner similar to that found in other phagocytic cells. More recent studies using tadpoles (Hollyfield and Ward 1974b) have shown that some (though not all) of the polystyrene spheres injected into the interphotoreceptor space and phagocytized by the cells acquire acid phosphatase activity. These findings suggest that "artificial phagosomes," that is, polystyrene spheres, can interact with lysosomes of the RPE.

Direct demonstration of acid hydrolases in homogenates and in subcellular fractions of cattle RPE has been reported (Berman 1971). In these experiments, the highest specific activity was found in a lysosomal-mitochondrial fraction isolated by differential centrifugation. In the course of further investigations, using highly purified RPE cell preparations (Berman and Feeney 1976; Rothman et al. 1976), unexpectedly high levels of acid lipase were found. This enzyme was measured using a synthetic substrate, 4-methylumbelliferyl (MU) palmitate; its specific activity in RPE is about 30 times greater than in retina, and 6 times greater than in rat liver (Fig. 6.4). The acid lipase of RPE has a single peak of activity at pH 5–6, in contrast to retina, where additional peaks occur at pH values of 4 and 5.5. It may also be seen (Fig. 6.4) that both rat liver and retina have, in addition, neutral lipases with optima at pH 7.5 and 8, respectively. This activity appears to be absent in cattle RPE.

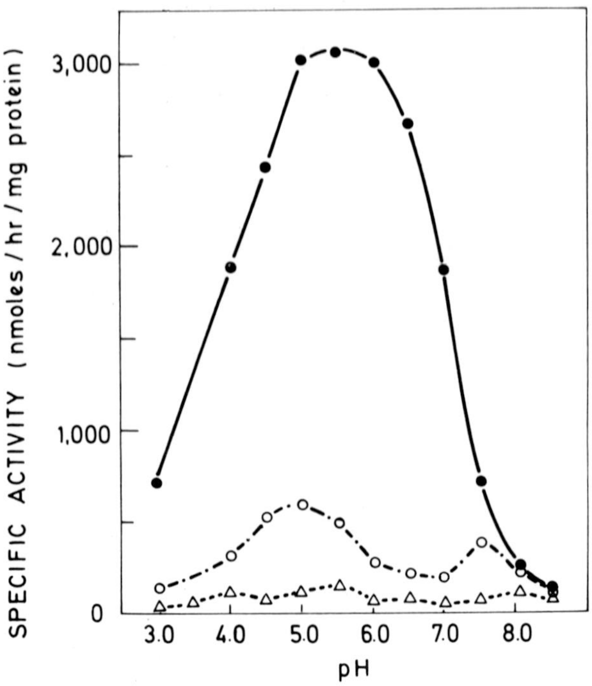

Figure 6.4 *Effect of pH on the hydrolysis of 4-methylumbelliferyl palmitate in homogenates of cattle RPE (●—●), cattle retina (Δ–Δ) and rat liver (o–o). (Reprinted with permission from Rothman et al. 1976.)*

The intracellular distribution in RPE homogenates of several enzymes and of organelle markers such as DNA is shown in Figure 6.5. These fractions had been isolated by differential pelleting; the electron micrographs shown in Figure 6.2 provide a representative picture of the kinds of organelles

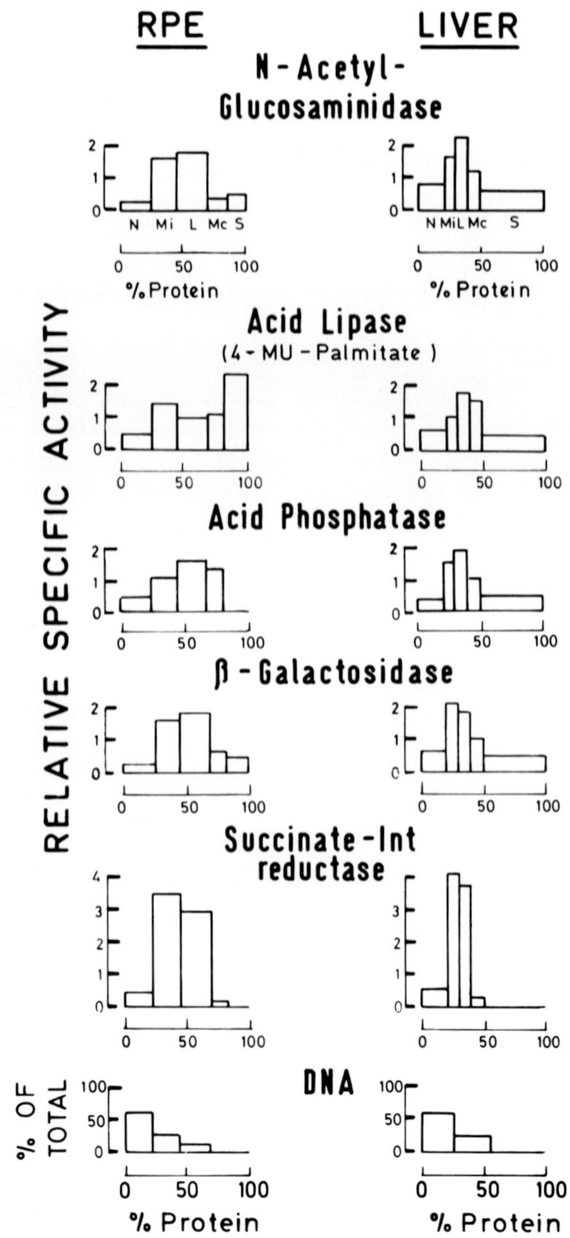

Figure 6.5 *Subcellular distribution of enzymes and DNA after fractionation of RPE and liver homogenates by differential centrifugation. (After Rothman et al. 1976.)*

present. The lysosomal fraction is of major interest when considering localization of hydrolytic enzymes of the RPE. However, N-acetylglucosaminidase, acid phosphatase, and β-galactosidase, the three lysosomal markers used, are present in highest concentration in the mitochondrial as well as lysosomal pellets, in both RPE and liver (Fig. 6.5). It should be pointed out that these organelles cannot always be isolated in a pure form, especially if the amount of starting material is limited, because of the intrinsic heterogeneity in both size and density of the particles. This becomes apparent by careful inspection of electron micrographs of these fractions, shown in Figure 6.2. If the mitochondrial-lysosomal fractions are considered together, however, then the three acid hydrolases used as lysosomal markers are found mainly in these two fractions. On the other hand, close to 50 percent of the total acid lipase activity, using 4 MU-palmitate as substrate, is in the RPE cytosol. This water-insoluble lipid is a highly sensitive and convenient substrate for studying the hydrolytic cleavage of palmityl esters; however, it is by no means a physiological substrate, that is, a naturally-occurring substance in cells. Other studies using glycerol tridecanoate (also a nonphysiological substrate), show that this lipase activity is found mainly in the lysosomal fraction of RPE cells (Hayasaka et al. 1977). The pH for optimum cleavage of this 10-carbon triglyceride is between 4 and 5. Its specific activity is ten times higher in RPE than in retina or liver. Taken together these findings suggest that there must be a family of acid lipases in RPE cells; the detection of the individual enzyme depends entirely on the substrate used in the assay procedure. The very high levels of acid lipases in general in RPE cells, compared to other tissues, supports the view that they may have an important role in lipid turnover, possibly one related to the degradation of phagocytized rod outer segments. More specific information on this question awaits further characterization of the enzymes, which will then allow precise studies of their substrate specificities to be carried out.

Owing to their ubiquitous distribution in nature, specific lipases such as phospholipases should also be present in the pigment epithelium. In the only published study on this subject, phospholipase activity was indeed found in RPE cells, but no details were given as to how the cells were prepared (Swartz and Mitchell 1973). Recent experiments using palmitate-labeled phosphatidylcholine and phosphatidylethanolamine have demonstrated clearly the presence of phospholipase A_1 activity in RPE homogenates, as well as in subcellular fractions (Berman and Kliger, unpublished data). The optimum conditions for demonstrating this activity in homogenates are given in Table 6.5. These data

Table 6.5 Phospholipase A_1 activity in bovine pigment epithelial cell homogenates.

	Fatty acid (nmoles released/hr/mg protein)					
	Phosphatidylethanolamine			*Phosphatidylcholine*		
	Calcium concentration			*Calcium concentration*		
pH	*None*	*5 mM*	*10 mM*	*None*	*5 mM*	*10 mM*
4	12.4	10.3	12.4	29.8	32.5	29.8
9	4.9	19.5	18.7	1.6	7.6	7.6

Note: Incubation mixtures of 0.2 ml contained 100 nmoles of labeled phospholipid, Triton-X-100, 0.15 M buffer and 90 μg protein.

show that phosphatidylcholine is the preferred substrate at pH 4, whereas at pH 9, phosphatidylethanolamine is degraded two to three times faster. Ca^{2+} has no effect on phospholipase activity at pH 4, but at pH 9, the degradation of both phospholipids is stimulated about fourfold in the presence of 5 mM Ca^{2+}. Further studies on isolated cell fractions have revealed high activity in all of the particulate fractions (lysosomes, mitochondria, and microsomes). As in homogenates, the Ca^{2+} effect is only demonstrable for phospholipases whose optimum activity is at pH 9. Similar highly specific effects of Ca^{2+}, as a function of pH, on the degradation of phospholipids have been observed in other cell types. The Ca^{2+} effects may be of special significance in RPE cells for controlling the degradation of the phospholipids contained in phagocytized rod outer segment disks. In this context, it should be noted that RPE cells have an unusually high (15.0 mmole/kg wet weight) endogenous level of calcium (Hess 1975). Elegant investigations using secondary ion mass spectrometry confirmed the presence of this metal ion in RPE, and in addition showed that its localization was both diffuse (throughout the cytosol?) and specific (in the cell nuclei) (Bellhorn and Lewis 1976). Among the many physiological roles attributed to Ca^{2+} in cellular metabolism in general, and as a transmitter in phototransduction in particular, we are justified in adding still another possible function, namely, control of phospholipid metabolism.

Cathepsin D has been isolated from extracts of bovine RPE and purified by ammonium sulfate precipitation and gel filtration on Sephadex G-200 (Hayasaka et al. 1975a). The apparent molecular weight of the enzyme was 60,000, and it showed optimum activity at pH 4.0, in line with most other lysosomal enzymes. Cathepsin D is thought to play an important role in the degradation of the extracellular matrix in connective tissues such as chick cartilage (Dingle 1969). The findings of Hayasaka et al. (1975a, b), thus raise the important question of whether this enzyme may be playing a similar role in the turnover of the proteoglycans and glycoproteins of the interphotoreceptor matrix. These macromolecules consist mainly of undersulfated chondroitin sulfate and sialic-acid-rich glycoproteins (Bach and Berman 1971a, b). Their protein cores could be degraded by proteases such as cathepsin D. Of special relevance here is that the release of cathepsin D from the lysosome is thought to be mediated by retinol (Dingle 1969; Burden et al. 1971) through labilization of the lysosomal membrane. As in other tissues, cathepsin D of the RPE is mainly of lysosomal origin (Hayasaka et al. 1975b), hence its release from this organelle could be controlled by the amount of retinol present in the RPE. Apart from extracellular catabolic activities, cathepsin D may also be playing an important role intracellularly, in the degradation of the opsin portion of the rod outer segment disks phagocytized by RPE cells. Hayasaka et al. (1975a, b) used a synthetic substrate, bovine serum albumin, to determine cathepsin activity in RPE; it would be of great interest to know whether the putative native substrate, opsin, is also degraded by this enzyme.

Although not lysosomal in origin, very high Mg^{2+}-Ca^{2+}-activated ATPase activity has been demonstrated in developing chick RPE and retina (Coulombre 1955). In addition there is evidence suggesting the presence of a ouabain-sensitive Na^{+}-K^{+} pump on the apical RPE membrane, which may play a role in the production of the short-circuit current by these cells (Steinberg and Miller 1973). This topic is discussed in greater detail in Chapter 12.

ENERGY PRODUCTION AND BIOSYNTHETIC ACTIVITIES. There are insufficient data at the present time to assess accurately the energy requirements of the RPE. We do know, however, that several endergonic reactions occur in these cells—for example, esterification of retinol, melanin synthesis, and transport of essential metabolites between the choriocapillaris and the outer retinal layers. The absolute levels of ATP and of pyridine nucleotides have never been examined, but there is little doubt that both must be present. The early experiments of Glocklin and Potts (1962) demonstrated that [^{32}P]phosphate is incorporated into acid-soluble organic phosphorous compounds. This process was strongly inhibited by diaminodiphenoxyheptane (DDH).

Further investigations (Glocklin and Potts 1965) showed that cattle RPE has a very low, albeit measurable, aerobic metabolism, oxidizing (per mg of tissue protein) approximately one-twentieth as much glucose as the adjoining neural retina. Whatever little oxidative metabolism is present in RPE appears to proceed by way of the tricarboxylic cycle. On the other hand, the rate of aerobic glycolysis in RPE, as measured by lactic acid production, is about one-half of that found in retina. Thus the RPE cells glycolyze (that is, convert glucose into lactic acid) both aerobically and anaerobically. Expressed differently, it may be said that in spite of the existence of a functional tricarboxylic acid cycle in RPE cells, the predominant pathway for glucose catabolism, even in the presence of oxygen, is glycolysis. The Pasteur effect (inhibition of glucose consumption and cessation of lactate production in the presence of oxygen) does in fact exist in RPE cells, but can only be demonstrated under certain conditions, for example, when incubations are carried out in bicarbonate medium (Glocklin and Potts 1965). These studies leave us with the impression that most of the ATP produced by RPE cells is generated by the glycolytic breakdown of glucose, a pathway far less efficient than oxidation through the tricarboxylic acid cycle, but apparently sufficient to meet the biosynthetic, and other, needs of the RPE cells.

More direct evidence for the existence of an active tricarboxylic acid cycle has been obtained using succinic dehydrogenase as a mitochrondrial marker (Rothman et al. 1976). Even though the mitochondria could not be isolated as a single clean fraction (Fig. 6.2), the presence of succinic dehydrogenase in mitochondria-enriched fractions (Fig. 6.5) justified the conclusion that oxidative pathways exist in RPE. Their precise quantitation in terms of the cellular regulation of carbohydrate metabolism is an important question requiring further investigation.

Apart from mitochondrial oxidations through well-established pathways, another (nonphosphorylating) oxidative pathway has also been described in cattle RPE. Specifically, this involves a microsomal electron transport system capable of slowly oxidizing NADH (and NADPH), in the presence of ascorbate or other electron acceptors such as cytochrome *c* and 3,4-dihydroxyphenylalanine (Shichi 1969). A microsome-bound hemoprotein has been found that is apparently identical to P_{450} of liver, suggesting that it may have an important role in RPE metabolism, for example, in hydroxylation reactions. In line with this possibility, a specific aryl hydrocarbon hydroxylase acting on benzo(a)pyrene has been described in mouse RPE (Shichi, Atlas, and Nebert 1975). Present as both endogenous and genetically inducible enzyme, it is localized primarily in the endoplasmic reticulum of the cell, as shown by recent studies using cultured RPE preparations (Shichi, Tsunematsu, and Nebert 1976). These and other monooxygenase reactions are mediated by microsomal cytochrome P_{450} systems, and play an important role in drug metabolism and detoxification reactions in the cell. The aryl hydrocarbon hydroxylase enzymes in the endoplasmic reticulum of RPE may have both a protective and destructive influence on the cell. Thus, polycyclic (and other foreign) compounds, which cannot be adsorbed effectively by the melanin granules of the choroid, may enter the RPE (Shichi et al. 1976). Such harmful substances will indeed be "detoxified" by the microsomal hydroxylases, but the products formed—monooxygenated intermediates—may in turn cause as much, or perhaps more, cellular damage than the original foreign substance. For example, hydroxybenzo(a)pyrene, at a concentration of only 2 nmole/mg of tissue, caused the release of about 80 percent of the acid phosphatase of RPE lysosomes (Shichi et al. 1976). Whether such mechanisms may be an underlying cause of certain retinal degenerations remains to be established.

Only a few reports have appeared describing biosynthetic activities of the RPE. Using cell suspensions of cattle RPE, it was shown that [^{14}C]glucose was incorporated into both glycoproteins and mucopolysaccharides, the former being more heavily labeled than the latter (Berman 1964). These findings have recently been confirmed and extended in a series of autoradiographic studies using [^{3}H]fucose that defined more precisely the site of glycoprotein biosynthesis in RPE of mice and rats (Feeney 1973b). Pulse-chase experiments showed that early labeling occurred mainly in the Golgi apparatus of both the inner segments of the photoreceptors and the RPE, suggesting both of these as active sites in the biosynthesis of fucose-containing glycoproteins. The newly synthesized glycoproteins were rapidly incorporated into cell surface components of both the RPE and the outer segments, thereby contributing to the extracellular (or interphotoreceptor) matrix of the subretinal space. Labeled glycoproteins were also found in phagosomes of the RPE and were assumed to have originated from phagocytized rod outer segment disks. Additionally, some of the radioactivity present in phagosomes could have represented lysosomal glycoproteins.

Gonasun and Potts (1974), using suspensions of cattle RPE cells, have shown that labeled amino acids such as phenylalanine, leucine, and lysine are efficiently incorporated into TCA-insoluble proteins. Chloroquine, at a concentration of 10^{-3} M, caused approximately 75 percent inhibition of phenylalanine incorporation, while puromycin and cycloheximide had similar effects at concentrations of 10^{-4} to 10^{-5} M. The site of protein inhibition by chloroquine was not determined, but the data support the view that it occurs at the level of the endoplasmic reticulum.

Protein synthesis from labeled amino acid precursors has also been demonstrated in frog RPE (Young and Bok 1970). This radioautographic study showed that initially the labeling was present in both the nucleus and cytoplasm. Later, some of the radioactivity was transferred to the cytoplasmic processes of the RPE cell. Glycoprotein biosynthesis from labeled galactose was not detected; instead, most of the galactose appeared to be incorporated into complex lipids.

MELANIN SYNTHESIS. Only a limited number of studies have been reported on melanogenesis in RPE, in contrast to other types of pigmented cells, where vast information has become available in recent years. Tyrosinase, which plays a central role in melanin synthesis, has been detected in RPE of chick embryo but initially appeared to be absent from the RPE of both adult chickens and adult mice or humans (Miyamoto and Fitzpatrick 1957). Later studies using "eye homogenates" of pigmented as well as nonpigmented mice (Hearing 1973a) have in fact shown this enzyme activity to be present. Of all the cell fractions examined, the highest specific activities were in the premelanosomal (10,000 *g*) and microsomal (100,000 *g*) fractions. Tyrosinase is a component of both Golgi vesicles and smooth endoplasmic

on the lysosomal enzymes of bovine retina and pigment epithelium. *Exp. Eye Res.* 15:43–49.

Wiggert, B. O., Bergsma, D. R., and Chader, G. J. 1976. Retinol receptors of the retina and pigment epithelium: further characterization and species variation. *Exp. Eye Res.* 22:411–418.

Wiggert, B., Bergsma, D. R., Lewis, M., Abe, T., and Chader, G. J. 1977. Vitamin A receptors. II. Characteristics of retinol binding in chick retina and pigment epithelium. *Biochim. Biophys. Acta* 498:366–374.

Wiggert, B. O., and Chader, G. J. 1975. A receptor for retinol in the developing retina and pigment epithelium. *Exp. Eye Res.* 21:143–151.

Young, R. W. 1974. Biogenesis and renewal of visual cell outer segment membranes. *Exp. Eye Res.* 18: 215–223.

Young, R. W., and Bok, D. 1969. Participation of the retinal pigment epithelium in the rod outer segment renewal process. *J. Cell Biol.* 42:392–403.

Young, R. W., and Bok, D. 1970. Autoradiographic studies on the metabolism of the retinal pigment epithelium. *Invest. Ophthalmol.* 9:524–536.

Zeman, W., and Siakotos, A. N. 1973. The neuronal ceroid-lipofuscinoses. In *Lysosomes and storage diseases,* ed. H. G. Hers and F. van Hoof, p. 519. New York/London: Academic Press.

Zimmerman, W. F. 1974. The distribution and proportions of vitamin A compounds during the visual cycle in the rat. *Vision Res.* 14:795–802.

Zimmerman, W. F. 1976. Subcellular distribution of 11-*cis* retinol dehydrogenase activity in bovine pigment epithelium. *Exp. Eye Res.* 23:159–164.

Zimmerman, W. F., Yost, M. T., and Daemen, F. J. M. 1974. Dynamics and function of vitamin A compounds in rat retina after a small bleach of rhodopsin. *Nature* 250: 66–67.

ACKNOWLEDGMENTS

The author wishes to thank Dr. Lynette Feeney for critical comments and suggestions in the preparation of this manuscript, and for the electron micrographs of subcellular fractions of RPE cells. This work was supported in part by NIH Grant EY-01131 and by the United States-Israel Binational Science Foundation, Jerusalem, Israel.

7 Metabolism of the Retinal Pigment Epithelium

Richard W. Young
Dean Bok

THE PIGMENT EPITHELIUM, interposed between the choroidal capillaries and the visual cells of the retina, is in a strategic position to influence the exchange of metabolites between the visual cells and the bloodstream. This metabolic interaction is facilitated by an intimate anatomical relationship in which cytoplasmic extensions of the pigment epithelium closely envelop the rod and cone outer segments. Recent research has revealed that the pigment epithelium is indeed an indispensable participant in visual-cell activities. Consequently, a discussion of the metabolism of the pigment epithelium necessarily involves its interrelationship with the rods and cones. Melanogenesis, phagocytosis, and intracellular digestion are described in other chapters, and will not be discussed in detail here.

Energy Metabolism

The pigment epithelial cells, which contain numerous mitochondria, are characterized by a vigorous oxidative metabolism. Particularly active are the mitochondrial enzymes of the citric acid cycle, such as succinate, isocitrate, and malate dehydrogenase, and associated electron transport enzymes, including cytochrome oxidase and coenzyme NAD (Lowry et al. 1956; Berkow and Patz 1961; Pearse 1961; Hansson 1970, 1971). β-hydroxybutyrate dehydrogenase, an enzyme that participates in fatty acid oxidation by mitochondria, has also been identified in pigment epithelial cells (Pearse 1961). These aerobic pathways serve to provide the cell with energy in the form of ATP. Phosphoglucoisomerase and lactate dehydrogenase are also active in pigment epithelial cells (Lowry et al. 1956, 1961; Hansson 1970, 1971). These enzymes function in glycolysis, a pathway that generates ATP from glucose under anaerobic conditions. Both glycolysis and the citric acid cycle have been demonstrated in pigment epithelial cells cultured in vitro (Glocklin and Potts 1965). Evidence has also been obtained of activity in the phosphogluconate oxidative pathway (Pearse 1961; Lowry et al. 1961; Hansson 1970, 1971). This sequence does not generate ATP, but it serves as a mechanism for the synthesis and disposal of pentose and sugars and reduces $NADP^+$. Carbonic anhydrase, which may affect acid-base balance, has been detected (Bhattacherjee 1971; Musser and Rosen 1973).

The retina consumes oxygen more rapidly than other tissues (Cohen and Noell 1965), and the activity of some oxidative enzymes is higher in the pigment epithelium than in any other part of the retina, with the possible exception of the visual cell inner segments (Pearse 1961; Hansson 1970). This strongly suggests that the metabolic functions of these cells demand a great deal of energy. As we shall see, although some of this energy is certainly required for the transport of metabolites and ions against concentration gradients, most of it is probably consumed in those reactions by which the cells continually renew their own structure and participate in the renewal of visual cell structure as well.

Vitamin A

Although most of our understanding of the metabolic activities of the pigment epithelium is of recent origin, knowledge of the critical role of the pigment epithelium in retinal vitamin A metabolism can be traced back to 1877, when Franz Boll, in announcing his discovery of a visual pigment in frogs, suggested the carotenoid nature of this substance. He called attention to the gold-colored oil droplets in the pigment epithelium, related them to the yellow color of visual pigment obtained by treatment with acid, and speculated that the substance in the pigment epithelium could be the precursor from which visual pigment decomposed by light is regenerated. This remarkable insight found little acceptance until the identification of vitamin A (retinol) in the retina revived the possibility that visual pigment contained a carotenoid component. The pigment epithelium of the frog proved to contain large stores of vitamin A (Wald 1935) concentrated in lipid droplets (Young and Bok 1970; Bridges 1975; Hirosawa and Yamada 1976; Robison and Kuwabara 1977) (Fig. 7.1). The storage of vitamin A by the pigment epithelium now appears to be a general property of that cell layer, although in albino rats the quantities are relatively small (Zimmerman 1974), except in bright light (Jancso and Jancso 1936; Greenberg and Popper 1941; Dowling 1960). Frogs and certain fish contain two or three times as much retinol in the pigment epithelium as in the visual pigment (where it is contained as the aldehyde, retinal) (Hubbard and Dowling 1962; Bridges and Yoshikami 1970). The pigment epithelium of cattle may contain nearly 70 times as much vitamin A per gram of tissue as the retina (Berman et al. 1974). Goldfish pigment epithelium stores retinol to the extent of 40 or more molar equivalents of the retinal in visual pigment (Bridges 1973).

Retinol is supplied to the pigment epithelium from at least three different sources: from the blood, from fragments of rod outer segments engulfed by phagocytosis, and from the adjacent rod outer segments following a strong bleach of visual pigment. Since vitamin A is not synthesized by vertebrate tissues, the ultimate source is the diet and the bloodstream.

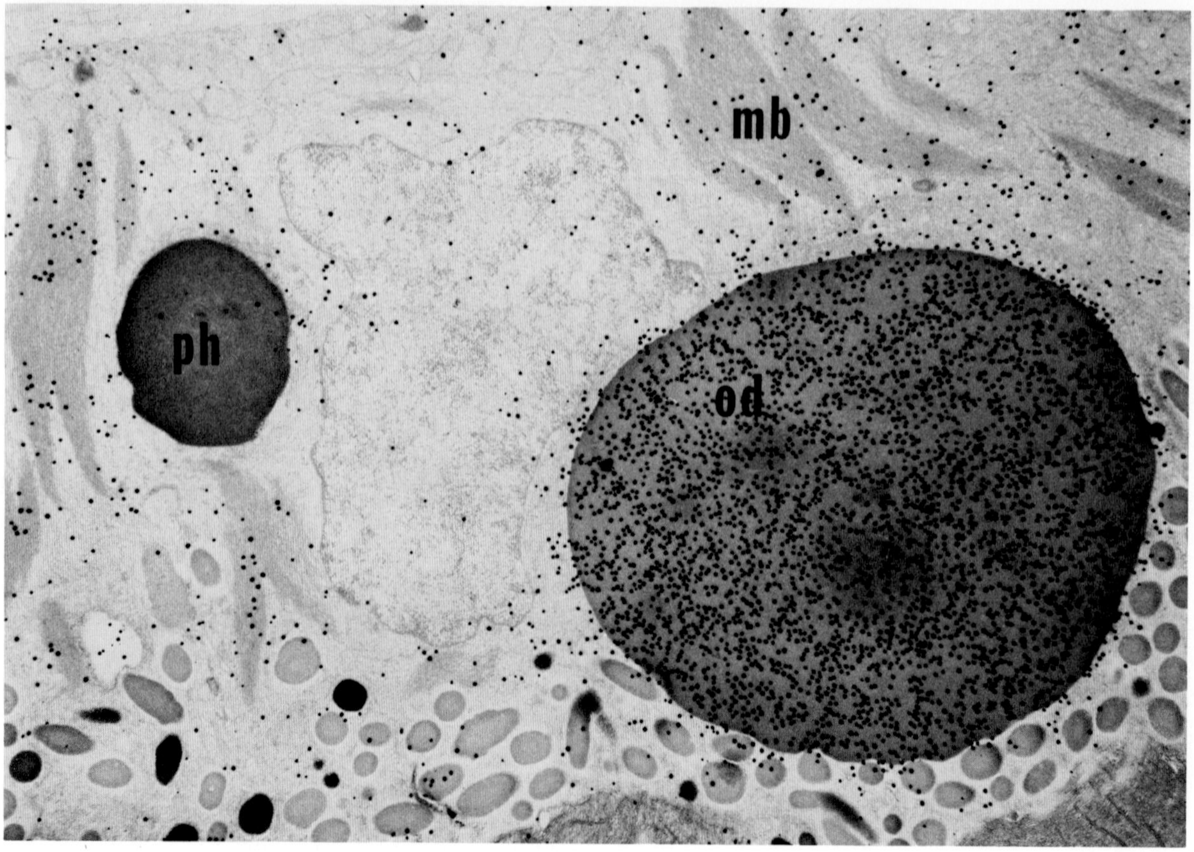

Figure 7.1 *Electron-microscopic autoradiogram of frog pigment epithelium following an injection of [^{3}H] retinol. The oil droplet* (od) *is heavily radioactive. Myeloid bodies* (mb) *and cytoplasm have also bound some of the labeled vitamin A, but the phagosome* (ph) *is nearly free of radioactivity.*

In the blood, vitamin A is transported by retinol-binding protein (RBP), a small protein that firmly binds one molecule of retinol. In the human, RBP is itself bound to one molecule of a larger protein, prealbumin, probably to prevent the loss of the small RBP molecule in the urine (Kanai et al. 1968). Neither of these bonds is covalent. The protein-protein bond is in part dependent upon RBP being charged with retinol, suggesting that removal of retinol leads to a conformational change in the protein (Heller and Horwitz 1973). Several geometric isomers of retinol and retinal (vitamin A aldehyde) will bind to RBP. However, the native compound isolated from RBP in human plasma is entirely all-*trans* retinol (Horwitz and Heller 1973). RBP is joined with retinol in the liver (where the protein is synthesized) and is then released into the blood (Muto et al. 1972). Upon reaching the pigment epithelium, the retinol-RBP complex binds to a specific receptor on the cell membrane. Heller (1975) has estimated that as many as 50,000 retinol-RBP molecules may bind to a single bovine pigment epithelial cell. The binding sites appear to be concentrated on the basal and lateral surfaces of the cell, adjacent to the choroidal capillary network (Heller and Bok 1976; Bok and Heller 1976). The protein has significantly less affinity for the cell-membrane receptor after it has released its molecule of retinol (again suggesting a conformational change), so that it is readily displaced from the membrane by another retinol-RBP molecule (Heller 1975). The process by which retinol is taken into the cell, and the question of whether there is a feedback process by which the pigment epithelium can regulate the amount of retinol taken from the blood, have not yet been explored.[1]

The mechanisms by which vitamin A is transported through the pigment epithelium and exchanged with the visual cells are also unknown. It is considered likely, however, that during such movements retinol is bound to a carrier protein that solubilizes and stabilizes it, just as in the blood. This view is supported by analogy with other lipids, which have been shown to move through the cytoplasm by means of carrier proteins (Wirtz 1974), and by the fact that specific retinol-binding proteins have been found in the cytosol of the pigment epithelium (Wiggert and Chader 1975; Wiggert et al. 1976; Heller 1976). There are indications that peroxisomes are involved in the metabolism of retinol by the epithelial cells (Leuenberger and Novikoff 1975; Robison and Kuwabara 1977).

The pigment epithelium does not readily allow retinol to return to the blood. This is revealed when vitamin A is withheld from the diet. Retinol begins to decline in the retina only when it is no longer detectable in the blood (Dowling and Wald 1958) and other organs are depleted of the vitamin (Popper and Greenberg 1941). Rats lose weight and die with no evidence of visual cell degeneration until shortly before death (Johnson 1943). The pigment epithelium itself shows no morphological evidence of damage when vitamin A is withdrawn from the diet, even at a stage when the rods are beginning to degenerate (Johnson 1943). As the rod outer segments undergo breakdown, the number of phagosomes in the pigment epithelium initially increases (Hayes 1974), but later, after the outer segments are gone, these inclusions disappear entirely (Dowling and Gibbons 1961).

Excess vitamin A is toxic to cells, apparently because of its destructive effects on cell membranes (Vento and Cacioppo 1973). Hypervitaminosis A can lead to death. But even with an abnormally high burden of retinol, the pigment epithelium shows little or no ill effects, apparently storing the excess vitamin in droplets of lipid (Amemiya 1967).

The pigment epithelium is not harmed by its large deposits of vitamin A because the vitamin is stored in the ester form (Wald 1935; Krinsky 1958; Dowling 1960; Bridges 1975), which is not nearly as harmful to membranes (Dingle 1963). Retinal does not occur in the pigment epithelium (Dowling 1960). Vitamin A is esterified with fatty acids (Krinsky 1958), principally palmitic, stearic, and oleic acids (Futterman and Andrews 1964). The site of esterification in the pigment epithelium is unknown, but in the retina, which can also esterify vitamin A, esterification takes place in the "microsome" fraction in vitro. Rod outer segments lack the esterifying enzyme system (Andrews and Futterman 1964). In just what cell these "microsomes" are located is not known. Autoradiograms indicate that the only site of [^{3}H] vitamin A incorporation in the frog neural retina is the photoreceptor outer segments (Hall and Bok 1976; Bok et al. 1977). The inner segments and the other cells in the neural retina show no measurable radioactivity between 30 minutes and 3 weeks after an injection of tritiated vitamin A.

1. The rate of uptake by the pigment epithelium of labeled retinol from the blood is about the same in light- and dark-adapted frogs, although the amount of retinol stored in the cells is higher in the light-adapted (Hall and Bok 1974).

The visual pigment rhodopsin, situated in the membranes of the rod outer-segment discs, consists of the protein opsin combined with the 11-*cis* isomer of retinal. When light is absorbed by a rhodopsin molecule, the 11-*cis* retinal is isomerized to the all-*trans* conformation, which must be reisomerized to 11-*cis* retinal to regenerate the visual pigment (Hubbard and Wald 1952). When rod outer segments are exposed to bright light, a significant amount of retinal is released from opsin in the all-*trans* form. This is reduced to retinol by an oxidoreductase (retinol dehydrogenase) in the outer segment membranes. The retinol, perhaps bound to a carrier protein, then diffuses into the pigment epithelium, from which it may return during dark adaptation (Jancso and Jancso 1936; Greenberg and Popper 1941; Hubbard and Colman 1959; Dowling 1960; Sidman and Dowling 1963; Zimmerman 1974; Zimmerman et al. 1974). In order to understand the role of the pigment epithelium in this molecular exchange, we need to know if the to-and-fro migration of retinol is an absolute requirement in the visual cycle. The critical step in pigment regeneration is the reisomerization of all-*trans* to 11-*cis* retinal. Does this take place in the pigment epithelium?

Apparently not in the rat. No 11-*cis* isomer has been detected in the pigment epithelium of the light-adapted rat. When retinol returns to the retina from the pigment epithelium, it is still in the all-*trans* form (Dowling 1960). However, both cattle and frogs may store some vitamin A ester in the 11-*cis* conformation (Krinsky 1958; Hubbard and Dowling 1962; Bridges 1975), although the retinol isomer supplied to the pigment epithelium by the blood and the visual cells is in the all-*trans* form. This suggests that in these species the pigment epithelium may contain a retinol isomerase. Alternatively, 11-*cis* retinol produced in the visual cells may "leak" back to the pigment epithelium and be stored there (Bridges 1976). Another source of the 11-*cis* isomer might be visual-cell membranes phagocytized by the pigment epithelium (Zimmerman et al. 1975; Nicotra and Livrea 1976). In the bovine eye, when 11-*cis* retinol reaches the outer segments, it is first converted to all-*trans* retinol before being oxidized to retinal (Daemen et al. 1974). Thus, in cattle as in rats, the final all-*trans* to 11-*cis* isomerization does not occur in the pigment epithelium.

Does the isomerization take place in the retina? It has been known for a century that visual pigment can be regenerated in the isolated frog retina from the yellow-orange initial photoproducts of bleaching (Kühne 1878; Wald 1935), which appear prior to the release of retinal from opsin (Hubbard et al. 1965). Later, it was shown that regeneration of rhodopsin might occur in isolated cattle retinas and in retinal homogenates free of pigment epithelium (Wald and Hubbard 1950; Collins et al. 1954). An isomerizing agent was extracted from cattle retinas that could convert all-*trans* to 11-*cis* retinal in the light (Hubbard 1956). After bleaching the visual pigment in the photoreceptors of the isolated toad retina, Brown (1965) observed a significant recovery of the early receptor potential (ERP). Recovery of the ERP is believed to reflect the regeneration of visual pigment. Recent evidence indicated the presence of isomerizing agents in isolated cattle-rod outer segments. These organelles will regenerate a visual pigment from all-*trans* retinal in the dark (Amer and Akhtar 1972; Akhtar et al. 1973; Futterman and Rollins 1973; Futterman 1974), although it is the 9-*cis* not the 11-*cis* isomer that is produced in these systems. However, Cone and Brown (1969) have reported that retinas of rat, frog, and monkey will spontaneously regenerate rhodopsin in the dark, if the isolated retina is incubated in a very small chamber. Frog cones can also regenerate visual pigment without the presence of the pigment epithelium (Goldstein 1970). Nonenzymatic photoisomerization may play a role. Shichi and Somers (1974) have obtained evidence that photoisomerization of retinal complexed with phosphatidyl ethanolamine in rod outer segments can contribute to rhodopsin regeneration. Furthermore, Amer and Akhtar (1973) succeeded in demonstrating the formation of 11-*cis* retinal in the isolated rat retina by partially bleaching the retinas with relatively low levels of white light, prior to incubating them in the dark with all-*trans* retinal.

In summary, there is growing evidence that the isomerization of all-*trans* to 11-*cis* retinal may occur in the retina, probably in the visual cell outer segments. What, then, is the purpose of the observed transfer of retinol to and from the pigment epithelium following a strong (10 percent or more) bleach of rhodopsin?

One possibility is that the outer segments contain an isomerase that is specific for the ester, but lack the esterifying enzyme, which is present in the pigment epithelium (Bridges 1976a). At present, there is no compelling evidence in favor of a particular derivative of vitamin A as the form isomerized (Bridges 1977).

It seems more likely that under conditions of dim illumination, when rods are active and less than 1 percent bleached (Pirenne 1961), the rod outer segment isomerase system is capable of regenerating the small amount of all-*trans* retinal that is produced. In brighter light, however, the capacity of this system is exceeded, and quantities of all-*trans* retinal are freed from their attachment to opsin. This highly reactive aldehyde does not accumulate, however (Dowling 1960). Instead, it is immediately reduced to the transport form, retinol, by the oxidoreductase present in the outer-segment membranes (Futterman 1963). The retinol diffuses from its site of production, perhaps being solubilized and protected from the aqueous environment by a transport protein (Futterman et al. 1976, 1977; Saari and Futterman, 1976; Heller 1976; Wiggert et al. 1976; Bergsma et al. 1977; Saari et al. 1977; Futterman and Saari 1977), and it soon reaches the pigment epithelium, whose cytoplasmic extensions envelop the rod outer segments. The pigment epithelium converts the retinol to its storage form by esterifying it with fatty acids. The major function that the pigment epithelium performs in this process is the conservation of ocular vitamin A (Zimmerman et al. 1974). It prevents the escape into the blood of the retinol, which diffuses away from the rod outer segments in bright light.[2] Thus, the esterification and storage of retinol that reaches the pigment epithelium from any source—the blood, the tissue fluid, or (presumably) phagocytized visual-cell outer-segment fragments—appears to be a general property of this layer of cells, but it is probably not a crucial component of the vitamin A cycle.

The factors that control the transfer of retinol from the pigment epithelium to the visual cells are so far undiscovered. In the visual cell outer segments, the retinol is converted to retinal (the form required for visual pigment) by the same oxidoreductase that catalyzes the reverse reaction following a strong bleach (Wald and Hubbard 1950; Futterman 1963; Daemen et al. 1974).

The pigment epithelium supplies retinol to the visual cells not only for visual pigment regeneration, but also for visual pigment synthesis, which accompanies the continual renewal of the rod and cone outer segment membranes. Rod visual cells repeatedly assemble new outer segment membranes, with about 10 percent of the membrane system being replenished daily in mammals (Young 1967; 1971). The renewal rate in cones is known with less certainty, but it appears to be roughly comparable to that in rods (Young 1977, 1978; O'Day and Young, 1979). The pigment epithelium not only phagocytizes and destroys the used disc membranes that are shed from the rods and cones (Young and Bok 1969; Young 1971b; see Chapter 9), it must supply to the visual cells the molecular precursors, including retinol, from which these membranes are synthesized and assembled. Since new membrane formation occurs under all lighting conditions (Young 1967), although at varying rates (Bridges et al. 1976; Besharse et al. 1977), the pigment epithelium must continually supply retinol to the rods, even when the retina is totally dark-adapted (Bridges and Yoshikami 1969; Hall and Bok 1974).

In certain fish and amphibians, the rod visual pigment protein may at different times be conjugated with either retinal or 3-dehydroretinal (forming rhodopsin or porphyropsin). The change from one chromophore to the other may occur at metamorphosis, or may be induced by long-term light or dark adaptation (Bridges and Yoshikami 1970; Reuter et al. 1971; Bridges 1973, 1974). The pigment epithelium controls these conversions, and the rods accept whatever form of retinol the pigment epithelium provides. In these species the pigment epithelium apparently can interconvert retinol and 3-dehydroretinol.

Fatty Acids

Fatty acids, as we have seen, are used by pigment epithelial cells to esterify retinol. In addition, fatty acids are oxidized in mitochondria to provide energy and are an important constituent of phospholipids, which, in association with protein, make up the various types of cell membranes. The phospholipids in rod outer segment membranes have a remarkably high content of polyunsaturated fatty acids (Borggreven et al. 1970; Nielsen et al. 1970; Poince-

2. If the pigment epithelium is damaged by sodium iodate, it either cannot trap the retinol released by bleaching or is unable to return it to the rods. As a result, bleached visual pigment is not regenerated in the dark (Grignolo et al. 1966). Not all of the retinol is trapped by the normal pigment epithelium. Bleaching of rhodopsin accelerates the loss of retinol from the eye in vitamin A deficiency (Noell et al. 1971).

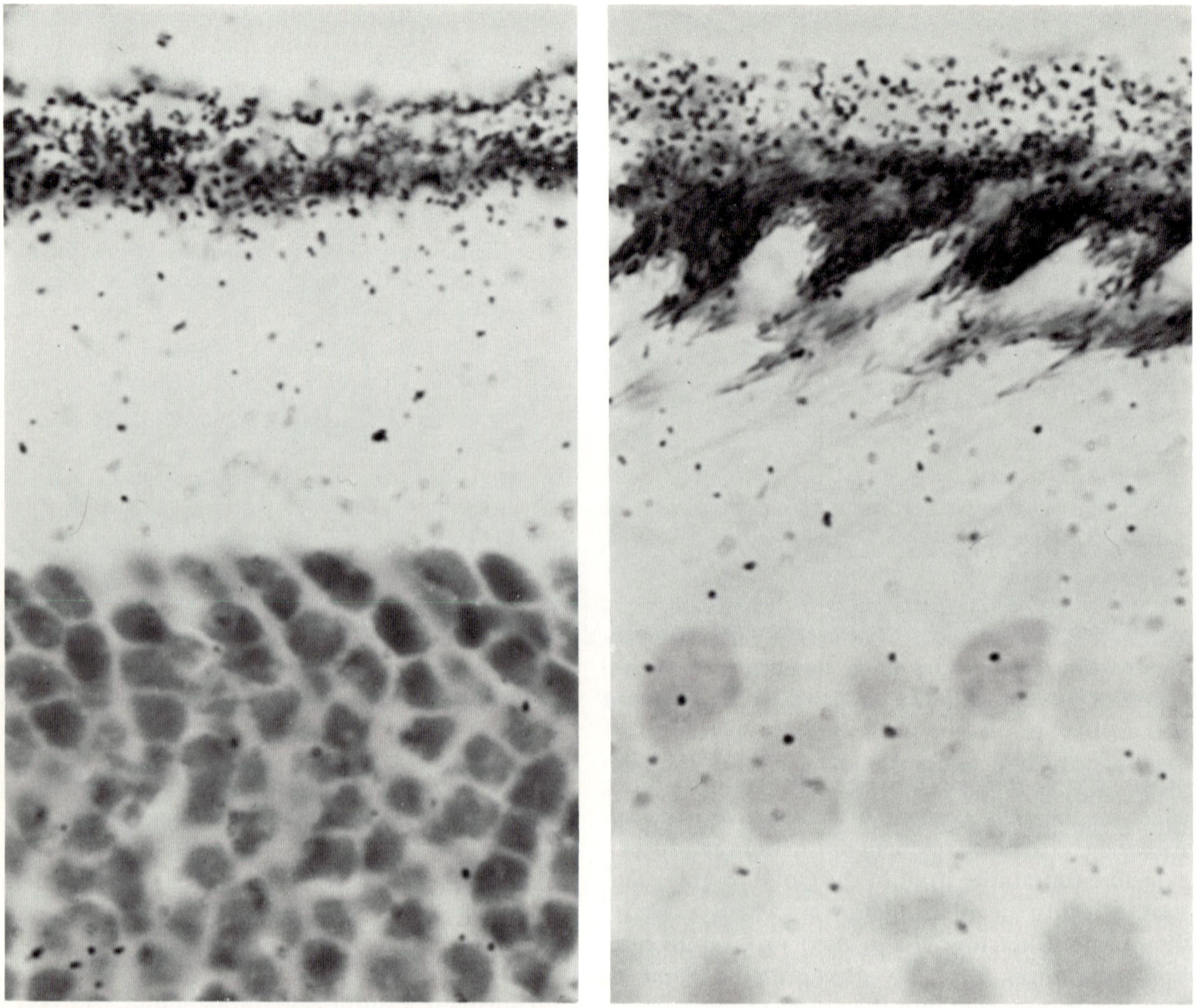

Figure 7.12 *One hour after injection of [^3H] taurine, the pigment epithelium shows heavy concentration of radioactivity in the rat* (left) *and in the frog* (right). *(From Young 1969; copyright © 1969 by the Regents of the University of California; reprinted by permission of the University of California Press.)*

retinal taurine metabolism by actively concentrating exogenous taurine obtained from the bloodstream (Fig. 7.12) (Young 1969; Lake et al. 1975, 1977; Kennedy and Voaden 1976; Voaden et al. 1977; Edwards 1977) and subsequently releasing it to the retina (Young 1969; Lake et al. 1977). There is also evidence that the epithelial cells can clear taurine from the retina by an active transport system (Miller and Steinberg 1976).

Autophagy

The preceding discussion has brought out the fact that pigment epithelial cells take up from the fluids of surrounding tissues, in selectively varying amounts, most of the small molecules which they normally encounter. Some of these, such as retinol and taurine, do not appear to serve as precursors for the synthesis of larger molecules. Others, such as glucose and fatty acids, may in part be degraded to provide a source of energy used in the synthesis of ATP. Still others, such as glycerol, may be retained in relatively low quantities, most of the molecules taken up from the blood being passed on to the visual cells. However, a major proportion of the small molecules incorporated into the cells appear to be used for the synthesis of larger molecules. Some mucopolysaccharide may be secreted, but the overwhelming majority of these newly synthesized molecules seem to remain within the cell.

Except for the DNA, all the major classes of large

molecules in the cell—phospholipids, proteins, glycoproteins, and ribonucleic acid—undergo repeated synthesis. These macromolecules may then be assembled into larger components, such as ribosomes, membranes, lysosomes, and peroxisomes. The continued formation of melanosomes and mitochondria has also been documented in mature pigment epithelial cells (Spitznas 1971; Bülow 1975; Young 1977, 1978a) (Fig. 7.13). Therefore, a significant component of the metabolism of the pigment epithelium is the synthesis of cellular constituents.

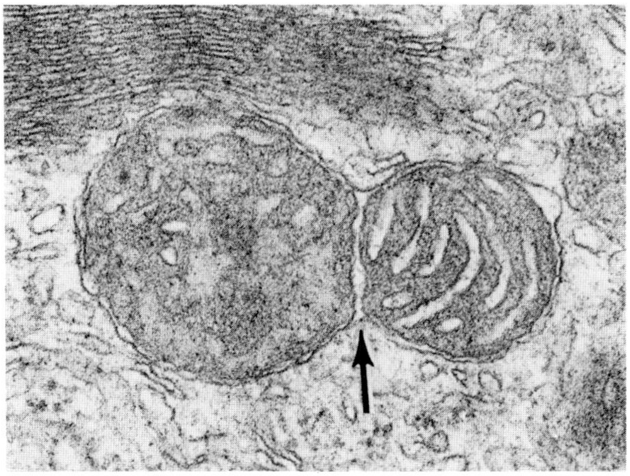

Figure 7.13 *New mitochondria are produced in mature pigment epithelial cells. The organelles grow, then divide in two. This mitochondrion appears to be preparing for division by the formation of a transverse partition* (arrow). *(From Young 1977.)*

Why, then, do the cells fail to increase in size? Evidently, there must be a disappearance of molecules to balance the production of new ones. In radioisotope studies, this is reflected in the progressive disappearance of radioactivity from the cells with the passage of time, although the mechanism of molecular removal is not revealed by these methods.

The discovery that autophagy is a normal part of the metabolism of pigment epithelial cells (Remé 1977; Young 1977) seems to clarify this situation. Autophagy is a process in which the cell isolates part of its own cytoplasm by surrounding it with a membrane, thereby forming a membrane-limited vacuole. Digestive enzymes are introduced into the autophagic vacuole, probably by fusion with lysosomes. The enzymes then degrade the contents of the vacuole into small molecules, which escape through the bounding membrane into the cytoplasm, where they may be used again for synthetic reactions. The vacuolar membrane protects the surrounding cytoplasm from this digestive action. Autophagic degradation is fundamentally similar to the process by which the contents of phagocytic (heterophagic) vacuoles are digested (Holtzman 1976). Indeed, autophagic and phagocytic vacuoles may coalesce.

All of the major cytoplasmic structures in pigment epithelial cells, including endoplasmic reticulum (Figs. 7.14, 7.15), mitochondria (Fig. 7.16), and melanosomes (Remé 1977; Young 1977, 1978a) may be destroyed by autophagy. It is through the degradative process of autophagy that the epithelial cells seem to compensate for the unremitting synthesis of their own structural and metabolic constituents.

In summary, it appears that the pigment epithelial cells destroy their own molecules just as fast as they produce them. Production of new molecules is balanced by the degradation of older molecules, yielding a steady state, in which there is neither growth nor diminution of size. The overall process is called *renewal* (Young 1976). The predominant feature of the metabolism of the pigment epithelium appears to be the renewal of the cell's own structure.

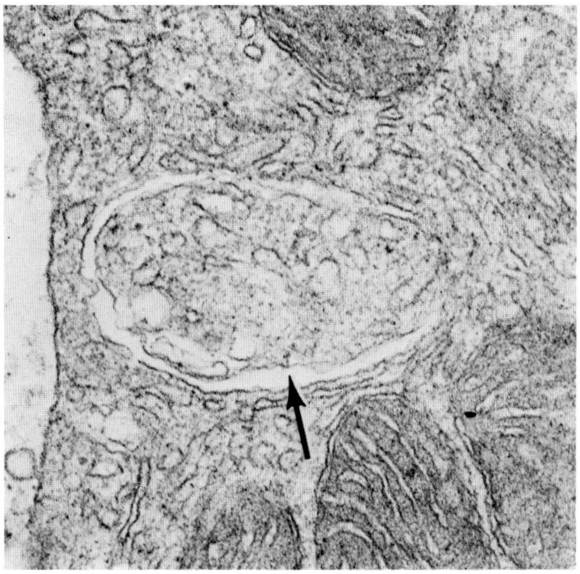

Figure 7.14 *Pigment epithelial cells repeatedly destroy parts of their own cytoplasm by autophagy. In this early stage in formation of an autophagic vacuole, a small area of cytoplasm is being surrounded by double membranes. The arrow indicates the cleft between the two membranes. (From Young 1977.)*

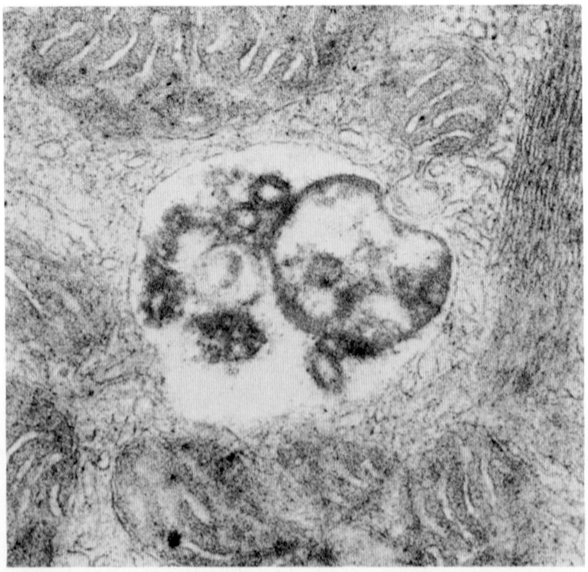

Figure 7.15 *The contents of this autophagic vacuole are in a more advanced stage of degradation. The inner of the two membranes has been destroyed, and the contents are digested beyond recognition. (From Young 1977.)*

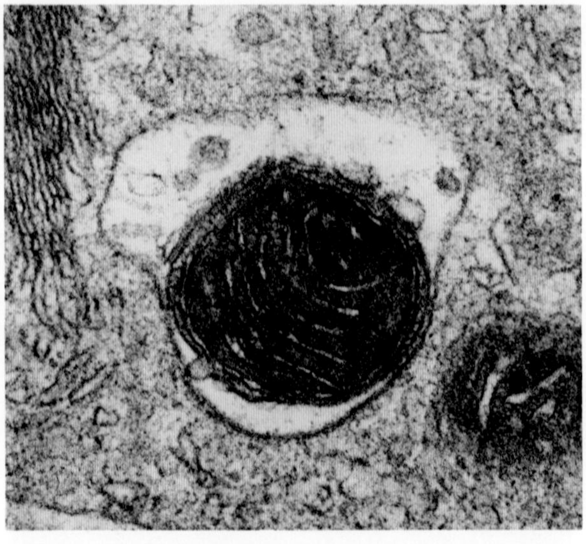

Figure 7.16 *Mitochondria are also degraded by autophagy. The mitochondrion in this autophagic vacuole shows an increased density, which is frequently observed during digestion. Small molecule breakdown products may escape through the vacuole membrane and be used again for synthesis. (From Young 1977.)*

Daily Rhythms

The recent observation that the rods and cones shed used membranes from the ends of their outer segments according to a daily schedule must influence importantly our thinking about the metabolism of the pigment epithelium. It is now documented that in rats (LaVail 1976), as well as in frogs, goldfish (O'Day and Young 1978), and chickens (Young 1978a) *rods* shed photosensitive membranes from their tips early in the morning. Similarly, in goldfish (O'Day and Young 1978), lizards (Young 1977), and chickens (Young 1978a) *cones* shed membranes early in the night (Fig. 7.17). No variations from this alternating rhythmic shedding by rods and cones are known, and there is no reason to expect that the human eye will prove to be an exception to what appears to be a general principle (O'Day and Young 1978; Young 1978b).

As soon as the membranes have been detached from the ends of the visual cells, they are immediately phagocytized and destroyed by the pigment epithelium. Thus, the pigment epithelium participates directly in the renewal of the visual cell outer segments: It maintains the balance between formation and degradation by destroying the membranes repeatedly formed and intermittently cast off by the visual cells.

This repetitive event, which is synchronized by the fluctuation of light in the environment (Young 1978b), imposes a rhythm on the metabolism of the pigment epithelium. Twice each day, at regularly recurring (approximately 12-hour) intervals, the pigment epithelium is presented with a substantial mass of detached rod (or cone) membranes. This initiates the sequence of membrane recognition and ingestion, described elsewhere in this volume (Chapter 9). Degradation of the membranes begins immediately after ingestion. This leads to the efflux of small-molecule breakdown products from the phagocytic vacuoles into the cytoplasm of the pigment epithelium. This, in turn, must soon reach a peak and begin to decline, as the membranes are progressively degraded to smaller and fewer fragments. Within 12 hours the cells have essentially completed the task of degradation, only to receive another batch of discarded visual cell membranes, thereby initiating another sequence of ingestion, digestion, and diffusion of small molecules into the cytoplasm (Fig. 7.17).

Because the rod and cone outer segment membranes are chemically different and are not present in equal amounts (due to the preponderance of one or the other type of visual cell), the chemical events of the dark period are not simply the mirror image

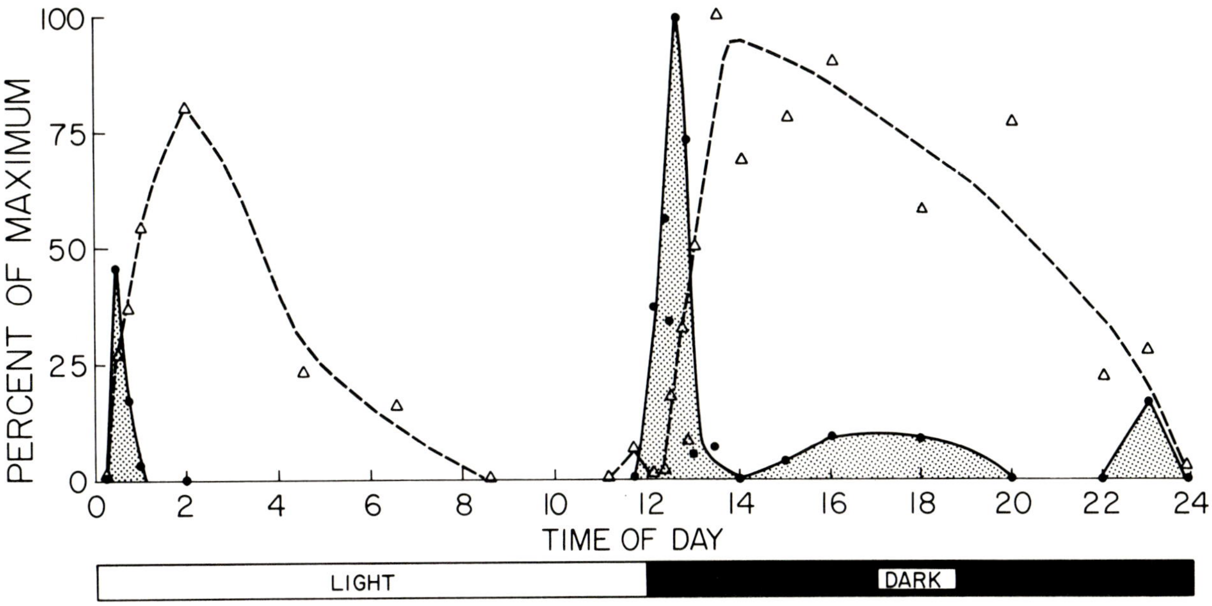

Figure 7.17 *Twice each day the pigment epithelium phagocytizes and destroys groups of membranes shed from the outer segments of rods and cones. This imposes a daily rhythm on the metabolism of the pigment epithelium. The graph shows changes in the number of phagosomes in pigment epithelial cells of the chick retina at different times of the day, expressed as a percentage of the maximum observed concentration. The peak early in the light period is due to shedding from rods. The peak early in the dark period is due to shedding from cones. The shaded curves represent freshly detached membranes. The other curves* (triangles *and* dashed lines) *represent older phagosomes, showing signs of degradation. (From Young 1978a.)*

of those during the light period. Instead, these events must generate *continuously changing* chemical conditions within the pigment epithelial cells, incessantly modifying the chemical environment in a cyclic manner that repeats itself daily. Because all chemical reactions are extremely sensitive to their chemical environment, it has been suggested that the entire metabolism of the pigment epithelium may be rhythmic, for it is difficult to imagine that any reaction or process could continue steadily in continuously changing chemical conditions (Young 1978b, 1978c).

Although this area of research is just beginning to be exploited, other daily rhythms have already been detected in the pigment epithelium, including rhythms of autophagy and the prevalence of small granules in lizards (Young 1977). There is also preliminary evidence for rhythmic changes in mitochondrial form and in the degradation of melanosomes in chickens (Young 1978a). It appears certain that the variable *time of day* must hereafter be given serious consideration in any analysis of the metabolism of the pigment epithelium.

Conclusion

As noted at the beginning of this chapter, there is noteworthy evidence of a remarkably high utilization of energy by the cells of the pigment epithelium. The chemical reactions of the vitamin A cycle, which have received much attention from researchers, do not seem to constitute an important drain on the cells' energy resources. Energy is required to transport small molecules and ions against concentration gradients (Miller and Steinberg 1977), but this too is unlikely to involve an exorbitant energy cost. Discovery of the phagocytic activities of the pigment epithelium brought to attention another need for metabolic energy. Ingestion and digestion of phagocytized material requires the consumption of energy, and as we have pointed out elsewhere in this volume (Chapter 9), there is no cell in the body with greater powers of phagocytosis. Furthermore, the recent recognition that the cells are continually engaged in digesting their own cytoplasmic constituents by autophagy reveals another need for energy, in order to produce vacuolar membranes and a variety of diges-

FEENEY, L., GRIESHABER, J. A., and HOGAN, M. J. 1965. Studies on human ocular pigment. In *The structure of the eye,* ed. J. W. Rohen, vol. 2, pp. 535–548. Stuttgart: F. K. Schattauer-Verlag.

FINE, B. S., and ZIMMERMAN, L. E. 1963. Observations on the rod and cone layer of the human retina. *Invest. Ophthalmol.* 2:446–459.

FREDRICKSON, D. S., and GORDON, R. S., JR. 1958. Transport of fatty acids. *Physiol. Rev.* 38:585–630.

FUTTERMAN, S. 1963. Metabolism of the retina. III. The role of reduced triphosphopyridine nucleotide in the visual cycle. *J. Biol. Chem.* 238:1145–1150.

FUTTERMAN, S. 1974. Recent studies on a possible mechanism for visual pigment regeneration. *Exp. Eye Res.* 18:89–96.

FUTTERMAN, S., and ANDREWS, J. S. 1964. Metabolism of the retina. IV. The composition of vitamin A ester synthesized by the retina. *J. Biol. Chem.* 239:81–84.

FUTTERMAN, S., DOWNER, J. L., and HENDRICKSON, A. 1971. Effect of essential fatty acid deficiency on the fatty acid composition, morphology, and electroretinographic response of the retina. *Invest. Ophthalmol.* 10:151–156.

FUTTERMAN, S., and ROLLINS, M. H. 1973. A catalytic role of dihydroriboflavin in the geometrical isomerization of all-trans retinal. In *Biochemistry and physiology of visual pigments,* ed. H. Langer, pp. 123–129. New York: Springer-Verlag.

FUTTERMAN, S., and SAARI, J. C. 1977. Occurrence of 11-*cis*-retinal-binding protein restricted to the retina. *Invest. Ophthalmol.* 16:768–771.

FUTTERMAN, S., SAARI, J. C., and BLAIR, S. 1977. Occurrence of a binding protein for 11-*cis*-retinal in retina. *J. Biol. Chem.* 252:3267–3271.

FUTTERMAN, S., SAARI, J. C., and SWANSON, D. E. 1976. Retinol and retinoic acid-binding proteins in bovine retina: aspects of binding specificity. *Exp. Eye Res.* 22: 419–424.

GLOCKLIN, V. C., and POTTS, A. M. 1965. The metabolism of retinal pigment cell epithelium. II. Respiration and glycolysis. *Invest. Ophthalmol.* 4:226–234.

GOLDSTEIN, E. B. 1970. Cone pigment regeneration in the isolated frog retina. *Vision Res.* 10:1065–1068.

GREENBERG, R., and POPPER, H. 1941. Demonstration of vitamin A in the retina by fluorescence microscopy. *Am. J. Physiol.* 134:114–118.

GRIGNOLO, A., ORZALESI, N., and CALABRIA, G. A. 1966. Studies on the fine structure and the rhodopsin cycle of the rabbit retina in experimental degeneration induced by sodium iodate. *Exp. Eye Res.* 5:86–97.

GUIDOTTI, A., BADIANI, G., and PEPEU, G., 1972. Taurine distribution in cat brain. *J. Neurochem.* 19:431–435.

HALL, M. O., and BOK, D. 1974. Incorporation of (^{3}H) vitamin A into rhodopsin in light- and dark-adapted frogs. *Exp. Eye Res.* 18:105–117.

HALL, M. O., and BOK, D. 1976. Reduction of the retinal-opsin linkage in isolated frog retinas. *Exp. Eye Res.* 22:595–609.

HALL, M. O., and HELLER, J. 1969. Mucopolysaccharides of the retina. In *The retina,* ed. B. R. Straatsma, M. O. Hall, R. A. Allen, and F. Crescitelli, pp. 211–234. Los Angeles: Univ. of California Press.

HALL, M. O., OCUMPAUGH, D. E., and YOUNG, R. W. 1965. The utilization of ^{35}S-Sulfate in the synthesis of mucopolysaccharides by the retina. *Invest. Ophthalmol.* 4:322–329.

HANSSON, H. A. 1970. A histochemical study of oxidative enzymes in rat retina damaged by visible light. *Exp. Eye Res.* 9:285–296.

HANSSON, H. A. 1971. A histochemical study of cellular reactions in rat retina transiently damaged by visible light. *Exp. Eye Res.* 12:270–274.

HAYES, K. C. 1974. Retinal degeneration in monkeys induced by deficiencies of vitamin E or A. *Invest. Ophthalmol.* 13:499–510.

HAYES, K. C., CAREY, R. E., and SCHMIDT, S. Y. 1975. Retinal degeneration associated with taurine deficiency in the cat. *Science* 188:949–951.

HELLER, J. 1975. Interactions of plasma retinol-binding protein with its receptor: Specific binding of bovine and human retinol-binding protein to pigment epithelium cells from bovine eyes. *J. Biol. Chem.* 250:3613–3619.

HELLER, J. 1976. Intracellular retinol-binding proteins from bovine pigment epithelial and photoreceptor cell fractions: purification of higher molecular weight lipoglycoproteins. *J. Biol. Chem.* 251:2952–2957.

HELLER, J., and BOK, D. 1976. A specific receptor for retinol binding protein as detected by the binding of human and bovine retinol binding protein to pigment epithelial cells. *Am. J. Ophthalmol.* 81:93–97.

HELLER, J., and HORWITZ, J. 1973. Conformational changes following interaction between retinol isomers and human retinol-binding protein and between the retinol-binding protein and prealbumin. *J. Biol. Chem.* 248:6308–6316.

HIROSAWA, K., and YAMADA, E. 1976. Localization of vitamin A in the mouse retina as revealed by radioautography. Structure of the eye, III, ed. E. Yamada and S. Mishima. *Jap. J. Ophthalmol.,* pp. 165–175.

HOLTZMAN, E. 1976. *Lysosomes: a survey.* New York: Springer-Verlag.

HORWITZ, J., and HELLER, J. 1973. Interactions of all-*trans,* 9-, 11-, and 13-*cis*-retinal, all-*trans*-retinal acetate, and retinoic acid with human retinol-binding protein and prealbumin. *J. Biol. Chem.* 248:6317–6324.

HUBBARD, R. 1956. Retinene isomerase. *J. Gen. Physiol.* 39:935–962.

HUBBARD, R., BOWNDS, D., and YOSHIZAWA, T. 1965. The chemistry of visual photoreception. *Cold Spring Harbor Symp. Quant. Biol.* 30:301–315.

HUBBARD, R., and COLMAN, A. D. 1959. Vitamin-A content of the frog eye during light and dark adaptation. *Science* 130:977–978.

HUBBARD, R., and DOWLING, J. E. 1962. Formation and utilization of 11-*cis* vitamin A by the eye tissues during light and dark adaptation. *Nature* 193:341–343.

HUBBARD, R., and WALD, G. 1952. Cis-trans isomers of vitamin A and retinene in the rhodopsin system. *J. Gen. Physiol.* 36:269–315.

JANCSO, N., VON and JANCSO, H. VON 1936. Fluoreszenzmikroskopische Beobachtung der reversiblen Vitamin-A-Bildung in der Netzhaut während des Sehaktes. *Biochem. Z.* 287:289–290.

JOHNSON, M. L. 1943. Degeneration and repair of the rat retina in avitaminosis. *A. Arch. Ophthalmol.* 29:793–810.

KANAI, M., RAZ, A., and GOODMAN, D. S. 1968. Retinol-binding protein: the transport protein for vitamin A in human plasma. *J. Clin. Invest.* 47:2025–2044.

KENNEDY, A. J., and VOADEN, M. J. 1975. Studies on the uptake and release of radioactive taurine by the frog retina. *J. Neurochem.* 27:131–137.

KRINSKY, N. 1958. The enzymatic esterification of vitamin A. *J. Biol. Chem.* 232:881–894.

KUBICEK, R., and DOLÉNEK, A. 1958. Taurine et acids aminés dans la rétine des animaux. *J. Chromatogr.* 1:266–268.

KÜHNE, W. 1878. *On the photochemistry of the retina and on visual purple.* London: Macmillan.

KUWABARA, T., and COGAN, D. G. 1961. Retinal glycogen. *Arch. Ophthalmol.* 66:680–688.

LAKE, N., MARSHALL, J., and VOADEN, M. J. 1975. Studies on the uptake of taurine by the isolated neural retina and pigment epithelium of the frog. *Biochem. Soc. Trans.* 3:524–525.

LAVAIL, M. M. 1976. Rod outer segment disk shedding in rat retina: relationship to cyclic lighting. *Science* 194: 1071–1074.

LEHNINGER, A. L. 1971. *Bioenergetics.* Menlo Park: W. A. Benjamin.

LEUENBERGER, P. M., and NOVIKOFF, A. B. 1975. Studies on microperoxisomes. VII. Pigment epithelial cells and other cell types in the retina of rodents. *J. Cell. Biol.* 65:324–334.

LOWRY, O. H., ROBERTS, N. R., and LEWIS, C. 1956. The quantitative histochemistry of the retina. *J. Biol. Chem.* 220:879–892.

LOWRY, O. H., ROBERTS, N. R., SCHULTZ, D. W., CLOW, J. E., and CLARK, J. R. 1961. Quantitative histochemistry of the retina. II. Enzymes of glucose metabolism. *J. Biol. Chem.* 236:2813–2820.

MEAD, J. F., and HOWTON, D. R. 1960. *Radioisotope studies of fatty acid metabolism.* New York: Pergamon Press.

MILLER, S., and STEINBERG, R. H. 1976. Transport of taurine, L-methionine and 3-o-methol-D-glucose across frog retinal pigment epithelium. *Exp. Eye Res.* 23:177–189.

MILLER, S. S., and STEINBERG, R. H. 1977. Active transport of ions across frog retinal pigment epithelium. *Exp. Eye Res.* 25:235–248.

MUSSER, G. L., and ROSEN, S. 1973. Localization of carbonic anhydrase activity in the vertebrate retina. *Exp. Eye Res.* 15:105–119.

MUTO, Y., SMITH, J. E., MILCH, P. O., and GOODMAN, D. S. 1972. Regulation of retinol-binding protein metabolism by vitamin A status in the rat. *J. Biol. Chem.* 247: 2542–2550.

NICOTRA, C., and LIVREA, M. A. 1976. Alcohol dehydrogenase and retinol dehydrogenase in bovine retinal pigment epithelium. *Exp. Eye Res.* 22:367–376.

NIELSEN, N. C., FLEISHER, S., and MCCONNELL, D. G. 1970. Lipid composition of bovine retinal outer segment fragments. *Biochim. Biophys. Acta* 211:10–19.

NOELL, W. K., DELMELLE, M. C., and ALBRECHT, R. 1971. Vitamin A deficiency effect on retina: dependence on light. *Science* 172:72–76.

NOVIKOFF, A. B., ESSNER, E., and QUINTANA, N. 1964. Golgi apparatus and lysosomes. *Fed. Proc.* 23:1010–1022.

OCUMPAUGH, D. E., and YOUNG, R. W. 1966. Distribution and synthesis of sulfated mucopolysaccharides in the retina of the rat. *Invest. Ophthalmol.* 5:196–203.

O'Day, W. T., and Young, R. W. 1978. Rhythmic daily shedding of outer-segment membranes by visual cells in the goldfish. *J. Cell Biol.* 76:593–604.

O'Day, W. T., and Young, R. W. 1979. The effects of prolonged exposure to cold on visual cells of the goldfish. *Exp. Eye Res.* in press.

Orr, H. T., Cohen, A. I., and Carter, J. A. 1976. The levels of free taurine, glutamate, glycine and γ-amino butyric acid during the postnatal development of the normal and dystrophic retina of the mouse. *Exp. Eye Res.* 23:377–384.

Orr, H. T., Cohen, A. I., and Lowry, O. H. 1976. The distribution of taurine in the vertebrate retina. *J. Neurochem.* 26:609–611.

Packer, L., Deamer, D. W., and Heath, R. L. 1967. Regulation and deterioration of structure in membranes. *Adv. Gerontol. Res.* 2:77–120.

Pasantes-Morales, H., Klethi, J., Urban, P. F., and Mandel, P. 1972. The physiological role of taurine in retina: uptake and effect on electroretinogram. *Physiol. Chem. Phys.* 4:339–348.

Pasantes-Morales, H., Salceda, R., and Lopez Colome, A. M. 1976. The role of taurine in retina: factors affecting its release. In *Taurine,* ed. R. Huxtable and A. Barbeau, pp. 191–200. New York: Raven Press.

Pearse, A. G. E. 1961. Localization of oxidative enzymes in rat and chick retina in various physiological conditions. In *The structure of the eye,* ed. G. K. Smelser, pp. 53–70. New York: Academic Press.

Pirenne, M. H. 1961. Light quanta and vision. *Endeavour* 20:197–209.

Poincelot, R. P., and Abrahamson, E. W. 1970. Fatty acid composition of bovine rod outer segments and rhodopsin. *Biochim. Biophys. Acta* 202:382–385.

Popper, H., and Greenberg, R. 1941. Visualization of vitamin A in rat organs by fluorescence microscopy. *Arch. Pathol.* 32:11–32.

Porta, E. A., and Hartroft, W. S. 1969. Lipid pigments in relation to aging and dietary factors (lipofuscins). In *Pigments in pathology,* ed. M. Wolman, pp. 191–235. New York: Academic Press.

Pourcho, R. G. 1977. Distribution of [^{35}S] taurine in mouse retina after intravitreal and intravascular injection. *Exp. Eye Res.* 25:119–127.

Remé, C. E. 1977. Autophagy in visual cells and pigment epithelium. *Invest. Ophthalmol.* 16:807–814.

Reuter, T. E., White, R. H., and Wald, G. 1971. Rhodopsin and porphyropsin fields in the adult bullfrog retina. *J. Gen. Physiol.* 58:351–371.

Robison, W. G., Jr., and Kuwabara, T. 1977. Vitamin A storage and peroxisomes in retinal pigment epithelium and liver. *Invest. Ophthalmol.* 16:1110–1117.

Saari, J. C., Bunt, A. H., Futterman, S., and Berman, E. R. 1977. Localization of cellular retinol-binding protein in bovine retina and retinal pigment epithelium, with a consideration of the pigment epithelium isolation technique. *Invest. Ophthalmol.* 16:797–805.

Saari, J. C., and Futterman, S. 1976. Retinol-binding protein in bovine retina: isolation and partial characterization. *Exp. Eye Res.* 22:425–433.

Schmidt, S. Y., Berson, E. L., Watson, G., and Huang, C. 1977. Retinal degeneration in cats fed casein. III. Taurine deficiency and ERG amplitudes. *Invest. Ophthalmol.* 16:673–678.

Shichi, H. 1973. Modified rhodopsin in the pigment epithelium? *Vision Res.* 13:477–480.

Shichi, H., and Somers 1974. Possible involvement of retinylidene phospholipid in photoisomerization of all-trans-retinal to 11-cis-retinal. *J. Biol. Chem.* 249:6570–6577.

Sidman, R. L., and Dowling, J. E. 1963. Autoradiographic localization of C^{14}-vitamin A in dark and light-adapted rat eyes. *Anat. Rec.* 145:286.

Spector, A. A., John, K., and Fletcher, J. E. 1969. Binding of long-chain fatty acids to bovine serum albumin. *J. Lipid Res.* 10:56–67.

Spiro, R. G. 1970. Glycoproteins. *Annu. Rev. Biochem.* 39:599–638.

Spitznas, M. 1971. Morphogenesis and nature of the pigment granules in the adult human retinal pigment epithelium. *Zeit. Zellforsch.* 122:378–388.

Starr, M. S., and Voaden, M. J. 1972. The uptake, metabolism and release of ^{14}C-taurine by rat retina *in vitro.* *Vision Res.* 12:1261–1269.

Streeten, B. W. 1961. The sudanophilic granules of the human retinal pigment epithelium. *Arch. Ophthalmol.* 66:391–398.

Tappel, A. L. 1968. Will antioxidant nutrients slow aging processes? *Geriatrics* 23:97–105.

Tappel, A. L. 1970. Biological antioxidant protection against lipid peroxidation damage. *Am. J. Clin. Nutr.* 23:1137–1139.

Tesoriere, G., Vento, R., Magistro, D., and Dones, F. 1971. Biosynthesis of UDP *N*-acetylhexosamine in bovine retina. *Eur. J. Biochem.* 19:289–293.

Vento, R., and Cacioppo, F. 1973. The effect of retinol

on the lysosomal enzymes of bovine retina and pigment epithelium. *Exp. Eye Res.* 15:43–49.

VOADEN, M. J., LAKE, N., MARSHALL, J., and MORJARIA, B. 1977. Studies on the distribution of taurine and other neuroactive amino acids in the retina. *Exp. Eye Res.* 25: 249–257.

WALD, G. 1935. Carotenoids and the visual cycle. *J. Gen. Physiol.* 19:351–371.

WALD, G., and HUBBARD, R. 1950. The synthesis of rhodopsin from vitamin A_1. *Proc. Nat. Acad. Sci. U.S.A.* 36:92–102.

WIGGERT, B. O., BERGSMA, D. R., and CHADER, G. J. 1976. Retinol receptors of the retina and pigment epithelium: further characterization and species variation. *Exp. Eye Res.* 22:411–418.

WIGGERT, B. O., and CHADER, G. J. 1975. A receptor for retinol in the developing retina and pigment epithelium. *Exp. Eye Res.* 21:143–151.

WIRTZ, K. W. A. 1974. Transfer of phospholipids between membranes. *Biochim. Biophys. Acta* 344:95–117.

YOUNG, R. W. 1967. The renewal of photoreceptor cell outer segments. *J. Cell Biol.* 33:61–72.

YOUNG, R. W. 1969. The organization of vertebrate photoreceptor cells. In *The retina,* ed. B. R. Straatsma, M. O. Hall, R. A. Allen, and F. Crescitelli, pp. 177–210. Los Angeles: Univ. of California Press.

YOUNG, R. W. 1971*a*. The renewal of rod and cone outer segments in the rhesus monkey. *J. Cell Biol.* 49:303–318.

YOUNG, R. W. 1971*b*. Shedding of discs from rod outer segments in the rhesus monkey. *J. Ultrastruct. Res.* 34:190–203.

YOUNG, R. W. 1973. The role of the Golgi complex in sulfate metabolism. *J. Cell Biol.* 57:175–189.

YOUNG, R. W. 1976. Visual cells and the concept of renewal. *Invest. Ophthalmol.* 15:700–725.

YOUNG, R. W. 1977. The daily rhythm of shedding and degradation of cone outer segment membranes in the lizard retina. *J. Ultrastruct. Res.* 61:172–185.

YOUNG, R. W. 1978*a*. The daily rhythm of shedding and degradation of rod and cone outer segment membranes in the chick retina. *Invest. Ophthalmol.* 17:105–116.

YOUNG, R. W. 1978*b*. Visual cells, daily rhythms and vision research. *Vision Res.* 18:573–578.

YOUNG, R. W. 1978*c*. Metabolism of the pigment epithelium. In *Proceedings, XXIII International Congress of Ophthalmology,* ed. K. Shimizu. In press. Amsterdam: Excerpta Medica.

YOUNG, R. W., and BOK, D. 1969. Participation of the retinal pigment epithelium in the rod outer segment renewal process. *J. Cell Biol.* 42:392–403.

YOUNG, R. W., and BOK, D. 1970. Autoradiographic studies on the metabolism of the retinal pigment epithelium. *Invest. Ophthalmol.* 9:524–536.

YOUNG, R. W., and DROZ, B. 1968. The renewal of protein in retinal rods and cones. *J. Cell Biol.* 39:169–184.

ZIMMERMAN, L. E. 1961. Acid mucopolysaccharides in ocular histology and pathology. *Proc. Inst. Med. Chi.* 23: 267–277.

ZIMMERMAN, L. E., and EASTHAM, A. B. 1959. Acid mucopolysaccharide in the retinal pigment epithelium and visual cell layer of the developing mouse eye. *Am. J. Ophthalmol.* 47:488–498.

ZIMMERMAN, W. F. 1974. The distribution and proportions of vitamin A compounds during the visual cycle in the rat. *Vision Res.* 14:795–802.

ZIMMERMAN, W. F., LION, F., DAEMEN, F. J. M., and BONTING, S. L. 1975. Biochemical aspects of the visual process. XXX. Distribution of stereospecific retinol dehydrogenase activities in subcellular fractions of bovine retina and pigment epithelium. *Exp. Eye Res.* 21:325–332.

ZIMMERMAN, W. F., YOST, M. T., and DAEMEN, F. J. M. 1974. Dynamics and function of vitamin A compounds in rat retina after a small bleach of rhodopsin. *Nature* 250: 66–67.

ACKNOWLEDGMENTS

Our research is supported by United States Public Health Service Grants EY00095 and EY00444.

8 Molecular and Cell Biology of Melanin

Raul I. Garcia
George Szabó
Thomas B. Fitzpatrick

THE BIOLOGY of the melanocyte provides an opportunity to establish a focal point for different scientific disciplines, ideas, and techniques. By presenting three simple questions one may illustrate the astonishingly wide range of ramifications in the study of melanocyte structure and activity, and of melanogenesis. Why does a tiger have stripes? Because it has a certain specific distribution of melanin-synthesizing cells, melanocytes, in its integument. How do certain animals, such as lizards, change color? Because their melanocytes are under hormonal and nervous control and interact with other integumentary cells on short- and long-term bases. Why do some of us turn brown on the beach, whereas others become afflicted with disfiguring and discomforting skin lesions when exposed to sun? Because in some of us the integumentary melanocytes can be influenced to produce protective melanin, and in others they cannot.

As might be expected, melanin biology is a fertile meeting ground for biochemists, biologists, anthropologists, and clinicians. Pigmentation research allows us to bridge the "resolution gap" between a macroscopic, anthropological study of the phenotypical appearance of the individual and the investigation of tissues and cells on the microscopical level. To a trained eye the examination of electron-microscopical sections of the skin is as informative in regard to the phenotypic appearance, and very often to the racial background, of an individual, as the study of bones was to Cuvier, who prided himself on being able to reconstruct a whole animal from a single fossil bone.

The macroscopically visible coloration of human skin is due to three pigments: hemoglobin, carotene, and melanin. Melanin is formed from the amino acid tyrosine, which, in the presence of molecular oxygen and through the catalytic action of the enzyme tyrosinase, is formed in a specialized cell type called the melanocyte. The final morphological appearance of the melanin within a cell varies across different classes of animals and plants. It is interesting, however, that basically the same enzyme, tyrosinase, is responsible for melanin synthesis in numerous species of animals and plants. Although there is still a controversy over whether one enzyme or two, or more, is involved in melanin synthesis, the great majority of workers in the field hold to the one-enzyme hypothesis and the existence of a true tyrosinase.

Tyrosinase is a copper-containing enzyme widely distributed in plants, animals, and man, and it catalyzes the oxidation of tyrosine to melanin. It causes the cut surface of many fruits and plants to darken. The well-known darkening of potatoes after they are cut and the dark integumental covering of insects are both due to various forms of melanin. Insects contain melanin in their cuticles, and cephalopods utilize melanin in defense against predators. This "ink" was widely used in the ancient world, in the form of sepia, in various artistic endeavors.

This discussion will be confined to the formation and the function of melanin in vertebrates, and in particular to their integumentary melanocytes and the cells of the retinal pigment epithelium.

In vertebrates there are specialized cell types, the melanocytes, which synthesize the enzyme (or enzymes) necessary for the formation of melanin. The melanocytes that are located in the retinal pigment epithelium of the eye derive *in situ* from the optic cup itself, but those located in the integument originate from the neural crest. Consequently, both types of melanocytes are closely related to the nervous system. Retinal pigment cells and integumentary melanocytes share many important features, although they differ in some respects. Both cell types are capable of synthesizing tyrosinase and producing a specific cell organelle, the melanosome, which is the site of melanin formation. Both types of melanocytes are able to move melanosomes from a perinuclear location into cell processes, and both possess a network of microtubules and 10 nm microfilaments thought to be necessary for this purpose.

The major difference between the melanocytes located in the eye and the melanocytes located in the epidermis of the integument is the final fate of the melanosomes. The epidermal melanocytes transfer these pigment granules into the keratinocytes, whereas the cells of the retinal pigment epithelium do not transfer the melanosomes to other cells. There is, however, a slow but regular turnover of melanosomes within the retinal pigment epithelial cells.

It is therefore logical that the retinal pigment epithelium has been used in research on melanin formation. The retinal pigment cells are very useful not only because they are in many respects similar to the epidermal melanocytes in their melanogenic machinery, but also because they are readily obtainable from developing eyes of various vertebrate embryos, may be easily isolated from other cell types, and can be very easily maintained in cell culture systems.

In this chapter we shall first give an account of current knowledge of the biochemical pathway of melanin synthesis and deal with the various levels of the biological control of melanogenesis. We will then turn to the role of melanin inside the cell, and finally we will illustrate how the retinal pigment epithelium cells serve as excellent models in studying these various aspects of melanin biology.

Biochemistry of Melanin Synthesis and Tyrosinase Activity in Vertebrates

Medawar (1953), referring to the work of Fitzpatrick (1952; Fitzpatrick et al. 1950, 1967), Lerner (Lerner and Fitzpatrick 1950; Lerner et al. 1949, 1950), Foster (1963, 1968) and Becker (1927), and others, states that "while it is true that inasmuch as a 'DOPA oxidase' does catalyze the oxidation of tyrosine, and a 'tyrosinase' that of DOPA, one enzyme or the other is probably of supererogatory status. 'Tyrosinase' enjoys the right of prior definition, tyrosine is the prior substrate, and for mainly historical reasons this enzyme having both monophenolase and diphenolase activities, has been termed 'tyrosinase'."

Von Furth, in 1901, first described the enzyme tyrosinase in vertebrate tissues. Bloch (1917) was the first to show that in human skin certain cells of the basal layer of the epidermis, when incubated in a dopa (dihydroxyphenylalanine) solution, become loaded with newly formed pigment in vitro. In 1928 Raper conclusively showed that dopa is the oxidation product of tyrosine in melanin synthesis (Fig. 8.1). Hogeboom and Adams (1942), studying

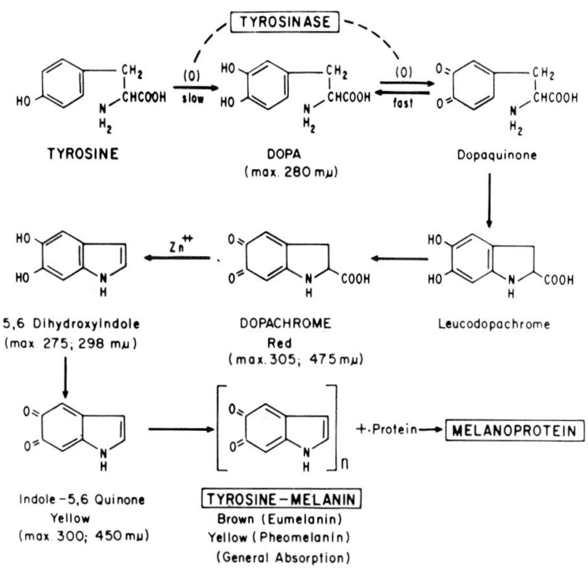

Figure 8.1 *Pathway of eumelanin and pheomelanin synthesis.*

Harding-Passey mouse melanomas, found that both a monophenolase (tyrosine → dopa) and a diphenolase (dopa → dopa quinone) are involved in the formation of melanin, proposing that the activities could be separated and were due to different enzymes. However, Lerner and his group (1949) reported that it was not possible to distinguish between a "tyrosinase" and a "dopa oxidase" activity. They showed that there is a very long lag period before tyrosine is converted into dopa, but that a trace amount of dopa speeds up the reaction.

Although the conflict between the one-enzyme

and two-enzyme hypotheses has not been fully resolved (Hearing et al. 1973, 1973b; Holstein et al. 1973; Okun et al. 1970, 1973; Patel et al. 1971), the great weight of evidence (reviewed by Jimbow et al. 1976b) favors the existence of a true mammalian tyrosinase, capable of both monophenolase and catecholase activity. True tyrosinase, then, is quite an interesting enzyme, catalyzing two apparently different oxidations using molecular oxygen, namely, the hydroxylation of phenols to orthodiphenols (tyrosine → dopa) and the oxidation of catechols to orthoquinones (dopa → dopa quinone).

Tyrosinase is a copper-containing enzyme. Vanneste and Zuberbuhler (1974) have proposed (Fig. 8.2) that the active enzyme would carry copper in the

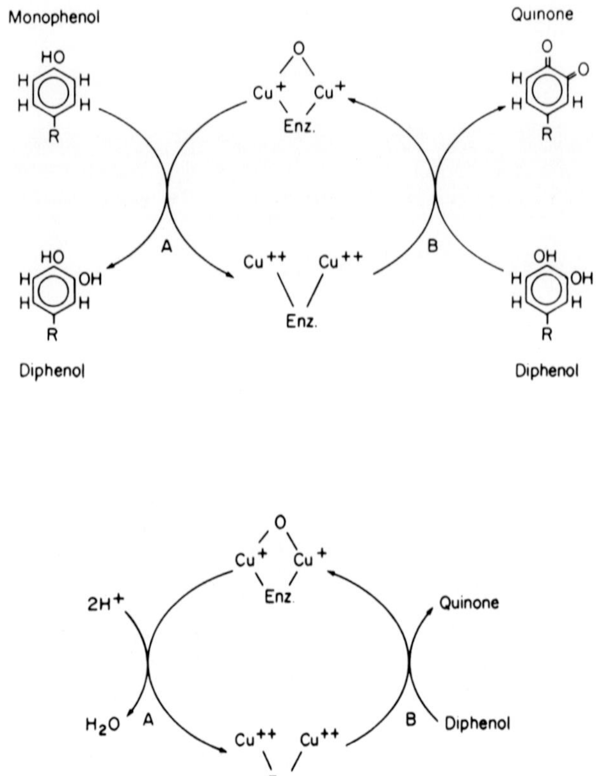

Figure 8.2 *Oxidation state changes of copper in tyrosinase action. (After Vanneste and Zuberbuhler 1974.)*

reduced, or cuprous, state (Cu^{+}). The copper in the resting form of the enzyme is in the oxidized, or cupric, state (Cu^{2+}), but would be reduced to cuprous by reacting with ortho diphenol. In this reaction the ortho diphenol is oxidized to orthoquinone. The Vanneste-Zuberbuhler hypothesis holds that in latent tyrosinase, copper exists in the cupric state and is then reduced during the activation step. This hypothesis is quite valuable in that it cogently explains the catalytic effect of dopa on the tyrosinase-tyrosine reaction. This scheme is discussed by Riley (1974) in his excellent review of the histochemical and biochemical evidence, and he has proposed that there may be physiological circumstances where the amount of reduced enzyme (Cu^{+}) may be insufficient to generate enough diphenol to promote the formation of pigment. This would seem to be the case in white human skin that has not been adequately exposed to sunlight.

Two major hypotheses exist regarding the composition of the melanin polymer. Mason (1948) holds that melanin is a regular polymer of copolymerized indole 5,6-quinone subunits. According to this view, melanin is a homopolymer, involving a single type of subunit joined by a single type of linkage. The other side, however, maintains that the Raper-Mason scheme may be oversimplified. Swan (1963, 1964, 1974), Blois (1967), and Nicolaus and his coworkers (1965, 1968) have developed the concept of melanin structure as a heteropolymer of multiple subunits joined by multiple types of bonding. The work of Nicolaus and Piatelli (1965) with squid melanin has shown that the polymer is composed of 5,6-dihydroxyindole, and 5,6-dihydroxy-indole-2-carboxylic acid moieties at different oxidative levels. These authors, together with Swan (1974), have concluded that melanin is a random polymer derived from the linkage of various different indoles.

The type of melanin we have been discussing, and which is the focus of this chapter, is the brown or black eumelanin. A related type of melanin, pheomelanin, is responsible for the yellow and red coloration of animal fur, feathers, human red hair, and the yellow band of agouti hair. Pheomelanin is formed by the reaction of cysteine with dopa quinone to form the major product of 5-S-cysteinyl dopa, which is then polymerized to form the pheomelanin heteropolymer (Prota and Nicolaus 1967). However, pheomelanin synthesis is restricted to hair follicle and feather germ melanocytes, and there is no evidence for pheomelanin synthesis in epidermal melanocytes or retinal pigment epithelium cells.

Morphological Aspects of Melanogenesis

Actual synthesis of melanin in melanocytes is restricted to a specific organelle, the melanosome, or pigment granule (Seiji et al. 1961, 1963, 1972; Seiji and Iwashita 1963a, b, 1965). The melanin itself is intimately associated with the protein component of the melanosome, the internal matrix, to which it is believed melanin is attached by sulfhydril linkages. Four separate stages in the development of the melanosome have been described (Fitzpatrick and Quevedo 1971), and these hold true for both neural crest-derived melanocytes and retinal pigment epithelium (Toda and Fitzpatrick 1972) (Fig. 8.3).

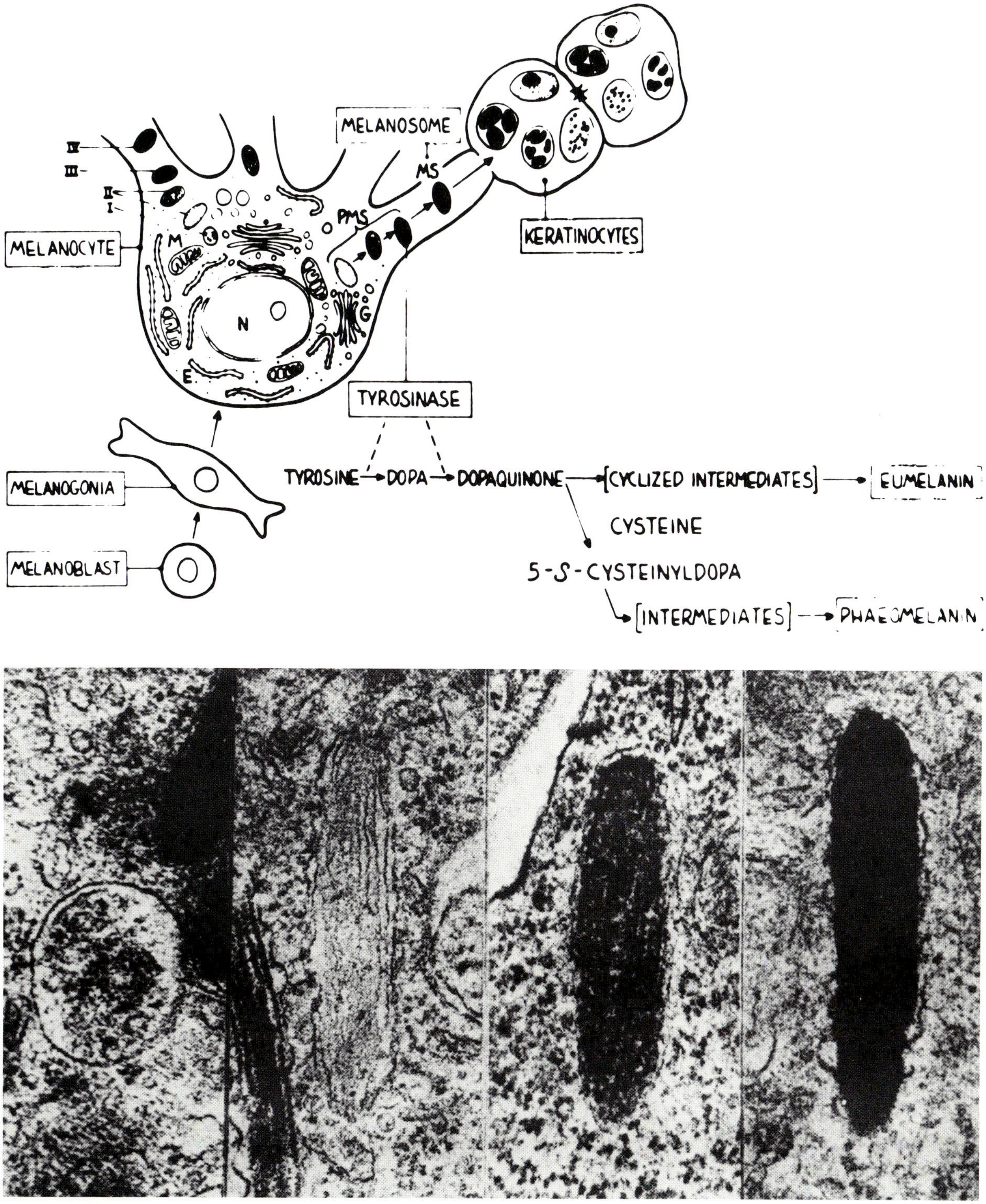

Figure 8.3 *Stages of melanosome development.* (Top, *from Fitzpatrick et al. 1967.* Bottom, *from Toda and Fitzpatrick 1972.)*

The stage I melanosome is a spherical, membrane-delineated vesicle that can be assumed to be a melanosome if: (a) tyrosinase can be demonstrated by autoradiographic and histochemical changes in these vesicles, or (b) it contains filaments that have a distinct periodicity (10 nm). Stage II is an ellipsoid organelle, containing numerous filaments with or without cross-linking, and having a distinct periodicity. At this stage no melanin deposition has yet taken place. Stage III is an ellipsoid organelle in which the internal structure characteristic of stage II has become partly obscured by electron-dense melanin. Stage IV is an organelle in which melanin deposition masks the internal structure characteristic of stages II and III. This final stage is also referred to as the mature melanosome, or the fully melanized melanosome.

The nature of the melanosome matrix, whether filamentous or membranous, has been open to question. However, Breathnach et al.'s (1973) freeze-fracture study of melanosomes in human retinal pigment epithelium and in epidermal and choroidal melanocytes has conclusively demonstrated that the internal organization of the melanosome is not membranous. Moyer (1963) and Lutzner and Lowrie (1972) have suggested that the melanosome internal protein matrix is composed of a regular network of fibrils or filaments, which is consistent with the freeze-fracture observations. Further studies are needed to elucidate the structure and composition of the internal protein framework. (Schroeder 1969; Hearing and Lutzner 1973).

The subcellular events leading to premelanosome formation, and to the subsequent melanin synthesis resulting in the mature melanosome, have not all been clearly identified. According to the "classic" theory of melanosome formation developed by Seiji and his coworkers, tyrosinase is synthesized on the ribosomes and then transported through endoplasmic reticulum to the Golgi area, where the enzyme is concentrated into vesicles. There, on the maturing face, tyrosinase-containing vesicles are released. These then develop into membrane-limited particles (premelanosomes) with a distinctive, ordered, internal matrix. This matrix is progressively obscured during melanin synthesis, as the recently formed melanin covers and blocks off the tyrosinase associated with the matrix, so that in the fully melanized mature melanosome, tyrosinase activity is no longer detected. A key feature of this scheme is that tyrosinase is seen as an integral part of the melanosome matrix from the very beginning of granule synthesis.

Toda and Fitzpatrick's findings (1972) have led them to revise the original Seiji scheme by having the stage I melanosome arise by ballooning out from the smooth endoplasmic reticulum. They agree with Maul (1969) that most premelanosomes are contained in, and develop within, a tubular smooth endoplasmic reticulum that is connected with the Golgi during melanogenesis, but they hold to the view that the tyrosinase is an integral part of the membrane of the developing melanosome from stage I.

Other studies of the role of the Golgi and associated elements in premelanosome formation (Novikoff et al. 1968, 1971; Maul 1969) have led to the suggestion that new premelanosomes are formed from dilated cisternae of smooth endoplasmic reticulum. This endoplasmic reticulum is part of a specialized system of Golgi-associated smooth endoplasmic reticulum (GERL), involved in the packaging of secretory proteins in many cell types (Novikoff et al. 1968, 1971).

Thus, according to the most current concept, the tyrosinase accumulates in the GERL, is packaged in Golgi-derived vesicles, and is then transported by the vesicles to the premelanosomes derived from smooth endoplasmic reticulum. In this scheme the melanosome-protein matrix does not contain tyrosinase as a structural component from the beginning, but rather the tyrosinase is introduced to the premelanosome (at stage II) by its release from the Golgi-derived vesicles into the stage II melanosome (Maul and Brumbaugh 1971; Ide 1972).

This latter scheme of the elaboration of tyrosinase-free premelanosomes and the subsequent introduction of tyrosinase has gained support in part from the work of Eppig (1970a, b) and Eppig and Dumont (1972) on RPE, using the tyrosinase inhibitor PTU (phenylthiourea) in vivo. They elaborated on the earlier work of Mitchell and Szabo (1966), who raised embryos of various cold-blooded vertebrates in PTU and produced unpigmented retinal pigment epithelium that contained melanosomes (stage II) normal in every respect except for lack of melanin.

The matrix of mature mammalian melanosomes has generally been thought to be electron-dense and amorphous after the completion of melanization. Several electron-microscopical studies (Jimbow and Kukita 1971; Durrer and Villiger 1957; Stanka 1971), however, have shown that the completely mature melanosome is not amorphous but contains some unknown structures that are spherical, about 40 nm in diameter, and translucent with regard to electron density. This new structure, called the "vesiculo-globular" body, is found not only in the melanosomes of mammals, but also in those of other vertebrates (Jimbow and Fitzpatrick 1974; Turner et al. 1976) (Fig. 8.4).

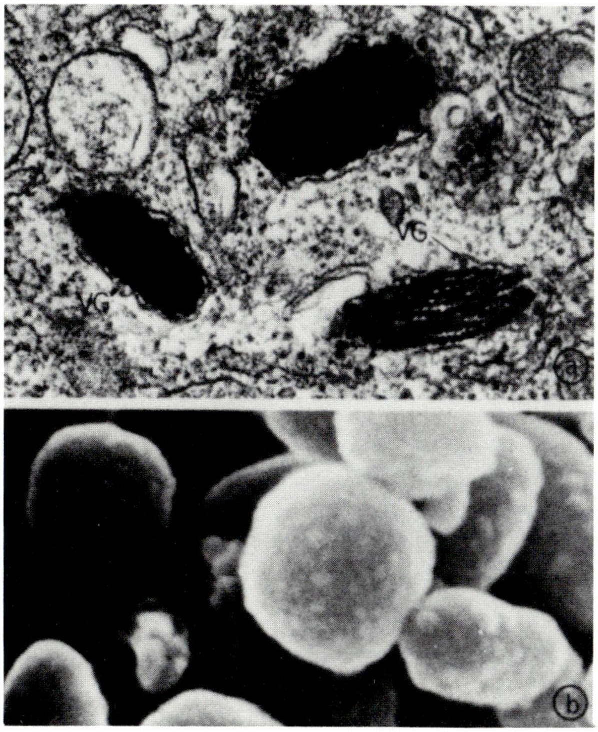

Figure 8.4 *Vesiculo-globular* (VG) *bodies in melanosomes of the retinal pigment epithelium and hair follicles.* (a) *Melanosomes in RPE. Note that the VG bodies are present between the outer membrane and the inner matrix. Each VG body is limited by a membrane.* (b) *Scanning electron microscope view of the VG bodies in the follicular melanosomes of black mice (C57BL/6J). Note that the globular bodies protrude from the surface of the melanosomes; the outer membrane of the melanosome was leached off by treatment with thioglycolic acid and phenol. (From Jimbow et al. 1976, © The William and Wilkins Co.)*

The vesiculo-globular bodies: (1) are present in all stages of melanosomal development; (2) increase in number during developing stages; (3) are attached to the surface of the lamellae of the inner matrices; (4) are covered by osmiophilic fine grains of melanin after melanization of melanosomes; (5) do not undergo internal melanization during melanosomal development; (6) are unchanged in size even though the melanosomes change in size; and (7) are not degraded once they are incorporated into the inner matrices. The roles of these vesiculo-globular bodies are still unknown, but they might, among other things, be the carriers of enzymes (Jimbow and Fitzpatrick 1974).

One tentative conclusion is that different organisms may have developed different mechanisms for melanogenesis: there is apparently more than one way to make a melanosome. It is also clear that the older theories of the nuclear or mitochondrial origin of melanosomes can be dismissed.

Several questions concerning melanin synthesis still remain unanswered. Why does melanin synthesis only take place in the developing melanosome, and not at the site of tyrosinase synthesis? More evidence is needed to explain how tyrosinase is introduced into the melanosome, and how it is activated. Why is the tyrosinase inactive while in the transport vesicles? How is the melanosomal protein matrix constructed, and what is its composition?

Control of Melanogenesis

Various factors may bring about an increase in melanogenesis, and this end result, reflected in higher tyrosinase activities and rates of melanin synthesis, may in turn come about through different intracellular mechanisms (for example, activation of a preexisting inactive tyrosinase, translation of a long-lived tyrosinase mRNA, or transcription of new message). It appears that the extent to which each mechanism is used varies according to the developmental state of the pigment cell and to the melanogenic factor involved.

Functional melanocytes appear early in the development of the mammalian integument, and there is a similarly early onset of melanogenic activity in developing retinal pigment epithelium (Hu and Montagna 1971; Ide 1972). The early commitment of chick embryo retinal pigment epithelium to melanin synthesis has been studied in detail by Zimmerman (1975, 1976; Zimmerman et al. 1974). He has found an interesting correlation between growth and differentiative functions—that is, mitotic activity and melanin synthesis—and has shown (1976) that there is a lapse of 24 hours between the time retinal pigment epithelium in developing chick eye withdraws from mitotic activity and the onset of a high rate of melanogenesis. This sharp increase in melanogenesis is due to newly synthesized messenger

RNA, and not to any long-lived species of mRNA. As withdrawal from mitotic activity was shown to be directly linked to the increase in melanin synthesis, his results suggest that the link occurs at the level of transcription of new mRNA, rather than translation of already existing message.

A number of studies on the control of growth and differentiation in melanocytes have been carried out using retinal pigment cells in various tissue culture systems. Primary cultures of dissociated chick embryo retinal pigment epithelial cells initially undergo a process of dedifferentiation during which they lose melanin synthesizing activity as they enter a period of logarithmic growth. As the mitotic rate drops, especially in cells that reach confluency, melanin synthesis sharply increases as the cells redifferentiate (Werner 1969, 1973; Whittaker 1963, 1965, 1967, 1968a).

The temporary loss of melanotic phenotype by RPE cells in vitro has been described in detail by Whittaker (1963, 1968a, b) and is attributed not to regulation of gene expression but to translational competition. This process of dedifferentiation and subsequent redifferentiation may be considered a modulation, referring to a "reversible fluctuation within the already established range of determination; also the covert maintenance of the potentiality to express an original function" (Weiss, 1949).

Results from studies of cultured mammalian melanoma cells parallel those obtained from RPE. Using B-16 melanoma cells in vitro, Silagi (1969) found a decreased rate of growth in melanotic lines, and believed this to be a significant factor in maintaining differentiated functions. Kitano and Hu (1970, 1971), also working with melanoma cells in vitro, found a similar inverse correlation between melanin synthesis and DNA and protein synthesis.

Control of melanogenesis is exercised at various levels, by neural, hormonal, and environmental factors. However, recent work has shown that all these factors may act in a similar manner through the adenyl-cyclase–cyclic-AMP system of pigment cells (Pawelek et al. 1973). For example, elevated intracellular cyclic AMP levels cause increases in melanization, cessation of growth, and gross morphological changes in cultured melanoma cells (Johnson and Pastan 1972; Wong and Pawelek 1973; Wong et al. 1974; Pawelek 1976).

NEURAL AND HORMONAL FACTORS. Except for the direct innervation of melanophores in some fishes, the neural control of pigment cells is an indirect one. Various adrenergic and cholinergic agents, as well as their blocking agents, exert strong influences on melanocytes. For example, the integumentary melanophores of various reptiles and amphibians contain alpha- and beta-adrenergic receptors on their cell membranes. Alpha receptor stimulation produces an aggregation of melanosomes inside cells and a visible skin lightening, while beta-adrenergic receptor stimulation results in skin darkening due to the intracellular dispersal of melanosomes into dendrites (Lerner 1971).

It has recently become clear that neural and hormonal agents act on pigmentation through the same mechanism (Abe et al. 1969). Beta-adrenergic agonists, together with the hormones MSH and ACTH, act on the cellular membrane and activate the enzyme adenylate cyclase, which converts ATP to cyclic AMP. Intracellular cyclic AMP levels are further controlled by a cyclic nucleotide phosphodiesterase, which inactivates cyclic AMP by converting it to 5′-AMP. The action of phosphodiesterase may, in turn, be blocked by certain compounds, including caffeine and theophylline (methyl xanthines). In fact, exposure of frog skin melanophores to theophylline produces pigment dispersal, a response similar to that obtained using cyclic AMP or MSH (Bagnara and Hadley 1969; Goldman and Hadley 1969). Still, the molecular mechanisms by which an increase in intracellular cyclic AMP results in pigment dispersal in melanophores, and in an increased tyrosinase activity and melanin synthesis in melanocytes, have yet to be elucidated.

GENETIC CONTROL OF MELANOGENESIS. Several excellent reviews of the genetic aspects of pigmentation in mammals are available (Searle 1968; Quevedo 1971, 1973). Dramatic microscopic and visible phenotypic differences in pigment cells and pigmentation may be caused by a change in a single gene locus, as in the homozygous recessive DBL mouse, where such a change produces a gray colored fur that differs markedly from the black fur of the dominant C57BL mouse (Gerson and Szabo 1969).

Tyrosinase-negative albinism is also due to a single gene substitution, resulting in a recessive homozygous at the C locus (Witkop et al. 1971, 1973, 1976). In this condition there is no active enzyme and no melanin synthesis. Premelanosomes, however, are regularly made in integumentary melanocytes and are transferred to keratinocytes (Parakkal 1967). Melanosomes are normally synthesized up to stage II of development, but stop there, as no melanin is deposited due to the inability of the tyrosinase to catalyze melanogenesis.

Genotypic variation is also reflected in the different isoenzyme patterns that are obtained when tyrosinase is analyzed by gel electrophoresis. Multiple forms of active tyrosinase having different electrophoretic mobilities have been resolved using disc electrophoresis. The isoenzyme patterns produced vary according to genotype (Holstein et al. 1967, 1971), and also vary according to disease state, as reported by Voulot and Ortonne (1975), who studied the tyrosinases in normal and diseased human skin (naevi, malignant melanoma, vitiligo), in mouse melanoma (Burnett 1971; Burnett et al. 1967; Miyazaki and Ohtaki 1975), and in human malignant melanoma (Burnett and Seiler 1969).

Differences among various pigmentary genotypes, as well as changes brought about by disease of the pigmentary system, are reflected in melanosomes of altered shape, size, and internal matrix structure (Moyer 1961, 1963, 1966), degree of melanization (Hearing et al. 1973b), and functional life time. Melanosomes that are spherical rather than the normal ellipsoid, smaller than normal, and with an altered internal matrix have been found regularly in human melanomas (Clark et al. 1972), and in cultured retinal pigment cells (Whittaker 1971; Garcia et al. 1976). The giant pigment granules, or macromelanosomes (Jimbow et al. 1973), found in the cutaneous pigmented macules of neurofibromatosis and in other disorders, are a fascinating example of altered melanosomes and the disease effects on melanogenesis (Figs. 8.5, 8.6).

Normal ellipsoidal melanosomes measure approximately 0.7 × 0.3 μm, while the macromelanosomes are approximately 5.3 × 3.5 μm. In neurofibromatosis (Szabo 1959; Benedict et al. 1968; Jimbow et al. 1973) both normal and giant melanosomes have been found in both keratinocytes and melanocytes on the light- and electron-microscopical levels, signifying that transfer of melanosomes takes place. Giant melanosomes are also found in the human disorder nevus spilus (Konrad et al. 1974), multiple lentigines syndrome (Weiss and Zelickson 1977), the Aleutian mink (Lutzner et al. 1968), in dermal melanocytes in the black Long-Evans rat (Glimcher and Szabo 1978), and in the mouse genotype responsible for beige color (Lutzner and Lowrie 1972). The beige mouse has been used as a model of the Chediak-Higashi syndrome by Robison et al. (1975), who found that the giant granules contained both melanin and the enzyme acid-phosphatase. In fact, the large granules in Chediak-Higashi syndrome in humans are the result of the accumulation of melanosomes within large phagosomes or secondary lysosomes (Zelickson et al. 1967; Windhorst et al. 1968).

In the hypomelanotic skin of tuberous sclerosis there is a decrease in the synthesis, degree of melanization, and size of the melanosomes in melanocytes, with the occurrence of aggregates of melanosomes (melanosome complexes), actually secondary lysosomes, in keratinocytes. Melanosome complexes in melanocytes are characteristic of nevus depigmentosus. The associated hypomelanosis is related to decreased melanosome synthesis, and also possibly to an abnormal transfer system (Jimbow et al. 1975).

In mammalian skin, melanosomes are synthesized in the epidermal melanocytes, but normally spend

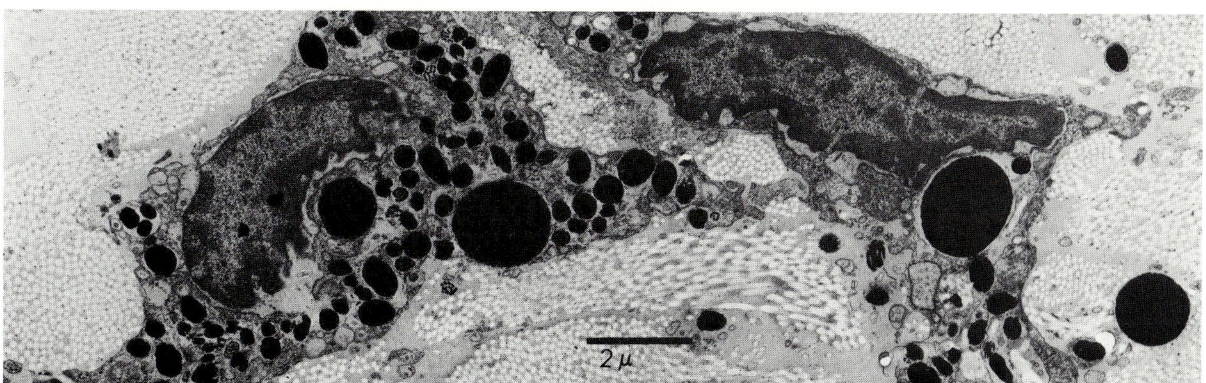

Figure 8.5 *Macromelanosomes in dermal melanocytes of a castrated black Long-Evans rat. Abnormal giant granules are found alongside normal melanosomes.*

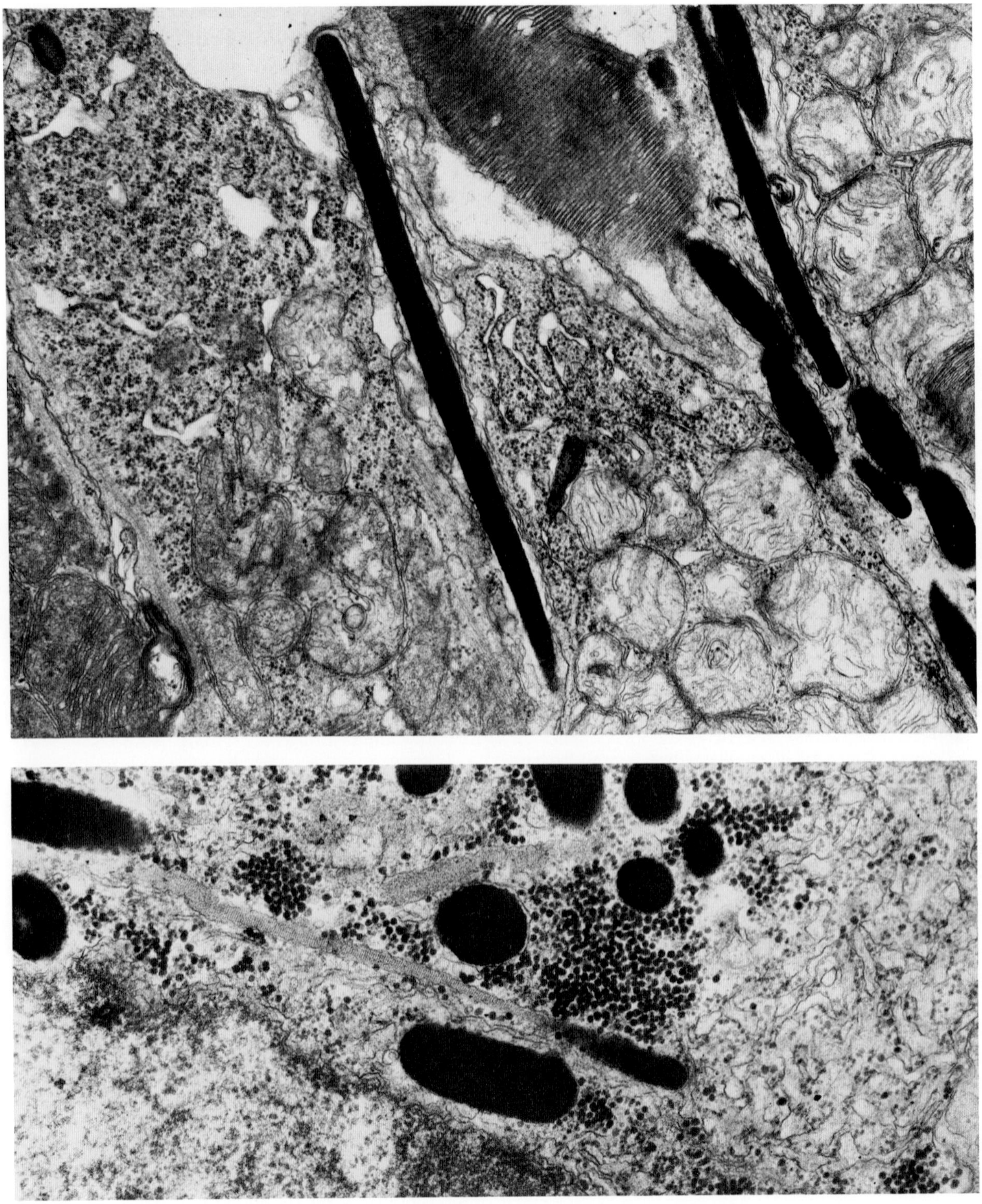

Figure 8.6 *Extra-long melanosomes (stages II and IV) alongside normal-size melanosomes in Fundulus retinal pigment epithelium, a normal finding.*

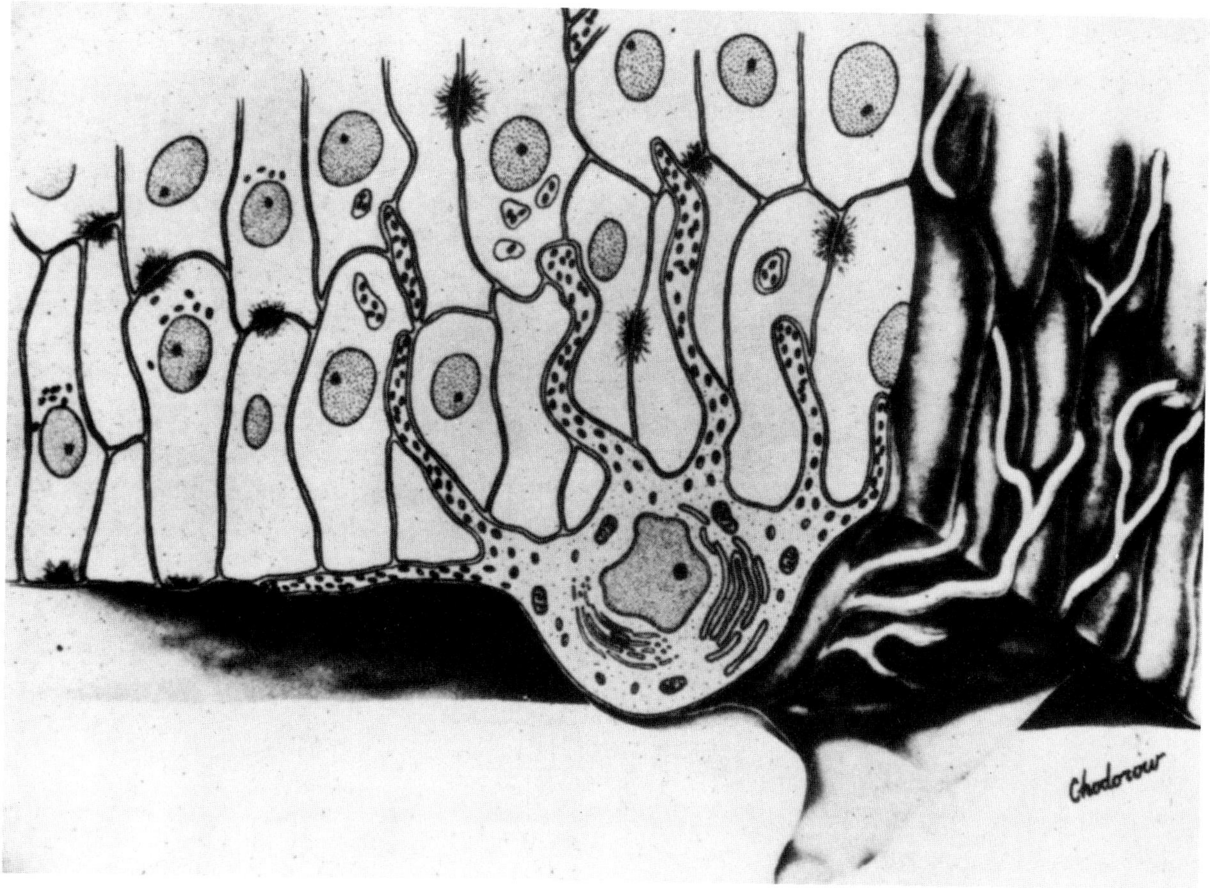

Figure 8.7 *The epidermal melanin unit, consisting of an epidermal melanocyte and its associated keratinocytes. (From Fitzpatrick and Breathnach 1963.)*

only a part of their lives in their mother cells. Melanocytes and keratinocytes are members of a unique symbiotic system, termed the "epidermal melanin unit" (Fitzpatrick and Breathnach, 1963) (Fig. 8.7). Melanosomes are transferred into keratinocytes, which eventually proceed up the epidermis as they keratinize and are eventually sloughed off. This regular upward movement of keratinocytes in the epidermis, taking with them the transferred melanosomes, produces a constant demand on melanocytes for new melanosomes. Following transfer, the melanosomes may be degraded as the keratinocyte undergoes cytomorphosis and passes up the epidermis.

There exist genetic-racial differences in the fate of melanosomes in keratinocytes (Szabo et al. 1969; Rosdahl and Szabo 1976; Garcia et al. 1978) (Fig. 8.8). The long melanosomes in keratinocytes of African and American Negroes and Australian aborigines are found as discrete, single granules. However, in Caucasian and Mongoloid skin, where melanosomes are shorter, they are found as aggregates, or melanosome complexes, which may be lysosomal structures (Hori et al. 1968; Wolff and Honigsmann 1972; Saito and Seiji 1973, 1976). Keratinocytes play an active and important role in producing phenotypical coloration in that they possess a melanosome size-sorting mechanism determining dispersion and degradation of melanosomes following transfer.

Melanosome degradation in mammalian epidermis normally occurs in keratinocytes and not in melanocytes. However, a different situation exists in the retinal pigment epithelium, where the melanosomes produced are permanent residents and are not transferred to other cells. In studies of the retinal pigment epithelium of White Leghorn and Rhode Island Red chicken embryos, there is apparently no biochemically or histochemically demonstrable tyro-

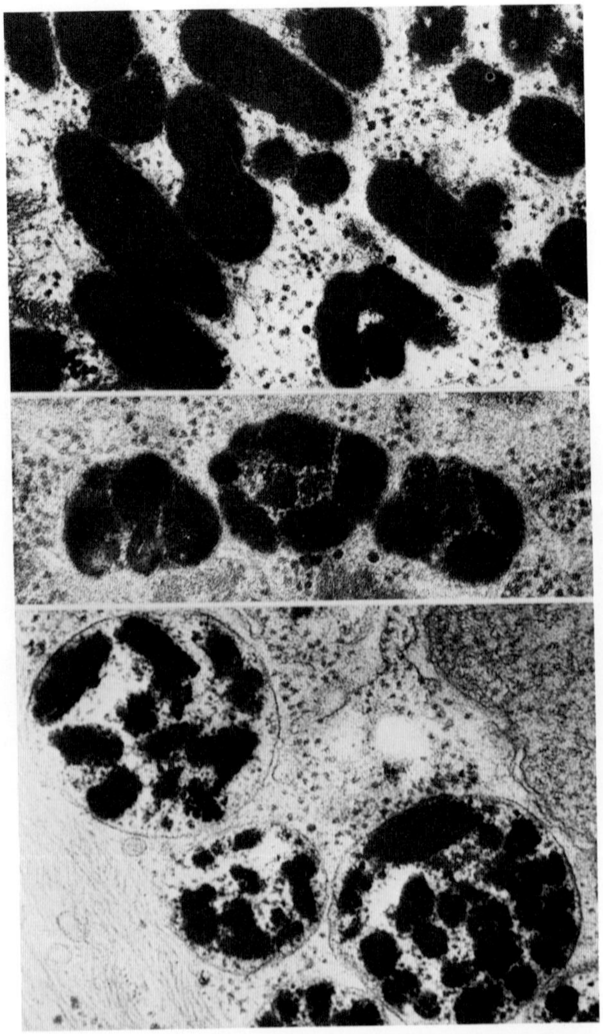

Figure 8.8 *Racial differences in melanosome packaging in human skin. Top to bottom: Negro, Oriental, and Caucasian keratinocytes. Melanosomes are found in large membrane-bound aggregates in Caucasian and Oriental keratinocytes. (From Szabo et al. 1969.)*

sinase activity following the second week in ovo (Miyamoto and Fitzpatrick 1957; Toda and Fitzpatrick 1972; Doezema 1973). However, in retinal pigment epithelium the melanosomes manufactured in ovo or in utero do not last forever. Recently, tyrosinase activity has been demonstrated in adult bovine retinal pigment epithelium (Dryja et al. 1978). There is a definite regular, if slow, turnover of melanosomes in mature retinal pigment epithelium. Evidence for this comes from the occurrence of new premelanosomes in adult retinal pigment epithelium and the association of mature melanosomes with the lysosomal enzyme acid-phosphatase (Toda and Fitzpatrick 1972; Hollyfield and Ward 1974a, b), together with the degradation of old melanosomes by incorporation into autophagosomes and lipofuscin granules (Endo and Hu 1973).

Interestingly, in their study of human ocular pigment, Feeney et al. (1965) found a number of "compound granules" with the characteristics of both melanin and lipofuscin granules. The existence of these intermediate forms supports the view of Endo and Hu (1973) that lipofuscin granules and lysosome-like inclusions in the retinal pigment epithelium are one and the same, representing degenerated and degraded melanin granules in secondary lysosomes or in residual bodies (Fig. 8.9).

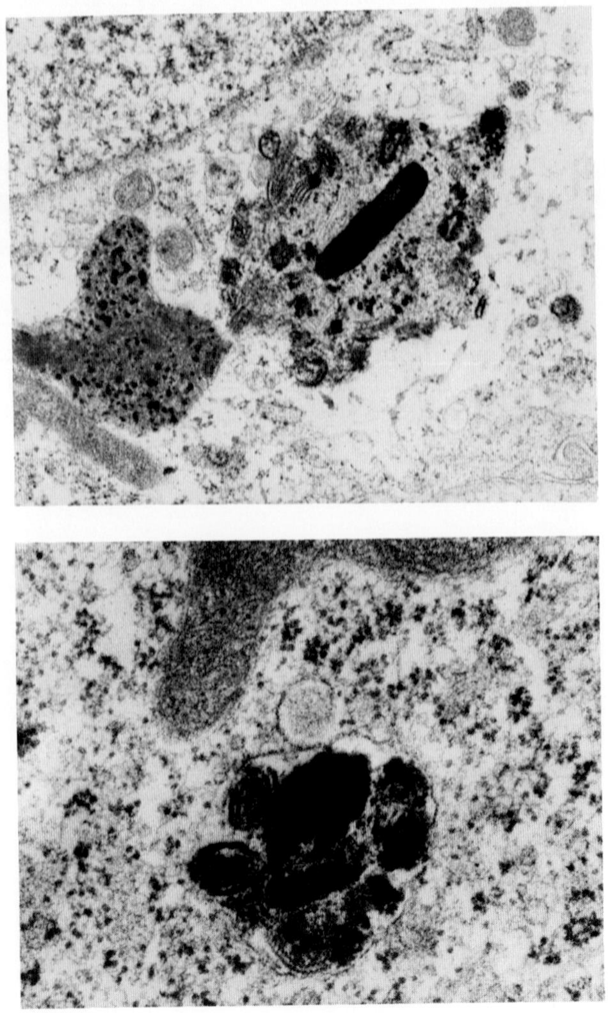

Figure 8.9 *Autophagosomes in cultured chick embryo retinal pigment cells, containing melanosomes in various stages of development. These autophagosomes have been found to be positive for acid phosphatase activity.*

Not only does there appear to be a slow but regular turnover of melanosomes in retinal pigment epithelium, but in certain cases there is a dramatic and wide-spread autophagocytosis of melanosomes. A general feature of these situations appears to be a realization by retinal pigment epithelium cells that they may no longer need their melanosomes. The most marked example is in the albino goldfish where stage II melanosomes are synthesized during early developmental stages, are not melanized, and eventually disappear from the retinal pigment epithelium (Takeuchi and Kajishima 1974). Another case is the amelanotic and intermediate-type cells of bovine retinal pigment epithelium, which have characteristic phagolysosomal particles and autophagic vacuoles containing melanosomes (Berman et al. 1974). Autophagocytosis of melanosomes also occurs in cultured retinal pigment epithelium cells (Takeuchi et al. 1975; Garcia et al. 1976) (Fig. 8.9). For unknown reasons, the phagolysosomal system of the retinal pigment epithelium (Feeney 1973) is used in these cases for the degradation of the retinal pigment epithelial cells' own melanosomes.

Role of Melanin in Cellular Metabolism

Exactly what melanin does inside cells has not yet been clarified. One role that does seem to be universally recognized, although not completely understood, is that melanin acts as a neutral density filter in scattering of light and, in so doing, protects organisms in a variety of ways. Examples include the squid's use of melanin as a "smoke screen" against predators, the protection of vertebrate skin against ultraviolet light by epidermal melanin, and the light-filtering role of melanin in the eye. In many invertebrates and vertebrates there is, additionally, migration of melanosomes in cell processes of retinular cells and of retinal pigment epithelium during light adaptation (see Chapter 10).

In the case of retinal pigment epithelium, Robison et al. (1975) have reported that a defect in retinal melanin may result in abnormal susceptibility to light damage. In Chediak-Higashi patients, hypopigmentation may contribute (together with a defect in the rod-renewal process) to this increased susceptibility, as the ocular tissues "that would normally provide protection from photons are lightly pigmented."

It is now generally accepted that melanin in human epidermis protects the skin from the damaging effects of ultraviolet light (UVL) (Szabo 1975). In experiments with isolated human epidermis an inverse correlation has been found between epidermal transmittance of UVL and the epidermal melanin content. Although thickness of the stratum corneum is a factor in shielding skin from UVL, epidermal melanin content is by far the more important factor (Thomson 1955). In dark-skinned individuals, for example, there is a lower incidence of skin cancers, including melanomas. Also, the minimum amount of UVL needed to produce an erythemal response in darkly pigmented human skin is much greater than that needed to produce a similar response in lightly pigmented skin. This difference is most marked when comparing Negroes with Caucasians. Nature has also provided an ideal experimental situation in the case of dark-skinned persons suffering from vitiligo. Following exposure to solar radiation, the exposed hypomelanotic vitiliginous macules are invariably erythematous in contrast to the adjacent pigmented skin, which rarely shows any evidence of erythema (Pathak and Fitzpatrick 1974).

Irradiation of normal human skin with UVL ($\geqq$ 290 nm), either from an artificial source or from sunlight, produces three responses, each with a different time course. Immediately, there is a darkening of the melanin already present in the skin. This phenomenon of immediate pigment darkening (IPD) is believed to be due to light-induced oxidation of 5,6-dihydroxyindole subunits in the melanin heteropolymer (Fig. 8.10). A second response, with

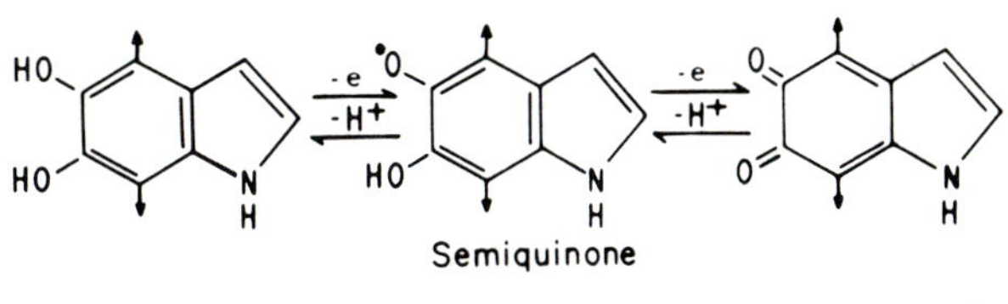

Figure 8.10 *Diagram illustrating the oxidation of the monomeric unit (5, 6-dihydroxyindole) of melanin during immediate pigment darkening (IPD) induced by radiation. It is suggested that this oxidation takes place through an intermediate stage of semiquinone formation (free radical state). (After Pathak 1967.)*

a slower time course, is erythema. Third is a melanogenic response, slower and longer-lasting than the first two responses, with the resultant skin hyperpigmentation (sun tan) due to an increase in melanogenesis and newly synthesized melanin.

The mechanism by which UVL induces this increase in melanogenesis is not known. It is not clear whether UVL acts directly on the epidermal melanocytes, or indirectly through the keratinocytes or by affecting the dermis underlying the melanocyte. There is much evidence supporting an indirect mechanism of action. For example, the irradiation of human skin with UVL produces sunburn, an erythemal response. It has been suggested that this inflammatory response is mediated by prostaglandins, and in fact inhibitors of prostaglandin synthesis act as antiinflammatory agents in human skin (Snyder and Eaglestein 1974). In turn, prostaglandins (specially PGE_1 and PGE_2) have been shown to induce melanogenesis in cultured melanoma cells and also in cultured retinal pigment cells (Garcia et al. in press).

Evidence against a direct UVL melanogenic effect comes from experiments with melanoma cell cultures where UVL irradiation of these melanocytes did not induce hyperpigmentation. Interestingly, though, it was found that the less pigmented cells were killed by UVL, showing that cellular melanin content afforded protection from UVL (Kitano and Hu 1969). Further evidence for an intracellular protective role for melanin comes from London et al. (1976), who have studied the UVL-mediated formation of thymine dimers in melanotic and amelanotic strains of Cloudman mouse melanoma cells in vitro. They find that the rate of formation of thymine dimers in the nonpigmented cells is nearly twice that in the pigmented cells. Their data demonstrate the photoprotective property of intracellular melanin in shielding cells from one type of photomediated injury to DNA.

Darkly pigmented human skin in regions of high UVL intensities is selectively advantageous because of the protection it affords against erythema and, over a longer time period, skin cancers and melanoma. In fair-skinned individuals the incidences of squamous and basal cell carcinomas are related to the degree and intensity of solar exposure, which vary according to latitude. The susceptibility to carcinomas of the solar exposed skin regions is related to lightness of skin and the inability to tan uniformly. However, a resistance to cancer of the exposed surfaces is seen among pigmented Caucasoids (East Indians), American Indians, and other races such as Orientals (Pathak and Fitzpatrick 1974).

Most interestingly, in certain circumstances pigmentation may actually be detrimental. It has been suggested that pigmented skin may be maladaptive in cold and that this might account for the observed racial differences and geographic distribution of human skin color (Post et al. 1975). In experiments on the induction of cold damage to skin, using black and white spotted (piebald) guinea pigs, it was found that the severity of frostbite was much greater in the pigmented areas than in the nonpigmented areas, and that this cold damage was directly related to the melanin content of epidermal cells (Post 1971). In this case the presence of melanin in the tissue did not afford protection, but actually resulted in a toxic response to the cold exposure. Work by Johnson et al. (1970) has shown that in vitro pigmented melanoma cells are considerably more sensitive to UVL at 8 to 16 times the minimal erythemal dose than are nonpigmented cells. They show that, while at lower doses of UVL irradiation, melanin may have a protective effect on the pigmented cell, at higher doses the melanin content results in a cytotoxic response.

It has become clear that cellular melanin is not an inert, unreactive substance. Rather, it appears to play a vital role in the metabolism of the pigmented cell. Whereas normally melanin affords protection, under certain special circumstances, where its functions are interfered with or overloaded, it may become cytotoxic. The consistent appearance of melanin in areas where energy conversion occurs, as in the skin, retina, and inner ear, supports the suggestion of a biological role for melanin. Combining biological data and the known amorphous semiconductor properties of the melanins, Proctor et al. (1974) have developed a hypothesis on the role of melanin as an agent in energy transfer and conversion. Accordingly, melanin may indicate the presence of electronically activated species in the pigmented cell. Under certain conditions of high energy input, melanins may act either as intermediaries in the dissipation of excited state energy or in the creation of toxic agents. Thus, excited state, or free radical, energy transferred to the melanosome may be converted into a form that is cytotoxic. Alternately, at low energy input rates, melanin directs a conversion to "innocuous phonovibrational modes" (Proctor et al. 1974).

It is clear from the phenomenon of immediate pigment darkening (IPD) that melanin is not biologically inert, but chemically active. The clinical and photobiologic data presented by Pathak (1967; Pathak and Stratton 1968; Pathak et al. 1968) suggest that IPD is an oxidation reaction involving the generation of unstable semiquinone-like free radicals in the melanin polymer. In this scheme, 5,6-dihydroxyindole monomer units of the melanin heteropolymer undergo reversible oxidation-reduction. The formation of semiquinone-like radicals during IPD is the result of the detachment of protons caused by absorption of light of wavelengths between 320 nm and 700 nm (Fig. 8.10). It is not yet known whether IPD also occurs in the melanosomes of retinal pigment cells. However, IPD seems closely related to the photochemical and photomechanical responses of melanin described by Moyer (1969) in the retinal pigment epithelium (Rosenberg 1962; Sever et al. 1962; Cope et al. 1963). Studies of the contribution of melanin to the photostable fast photovoltage of the eye clearly point out melanin as a "photosensitive" agent in retinal pigment epithelium (Brown 1965; Brown and Crawford 1967; Zanen and Debecker 1971; Debecker and Zanen 1975; see Chapter 13).

Mason and his coworkers (1960) have described the role of melanin in biological systems: "Melanin may act in some organisms as a biological electron-exchange polymer, able by means of its capacity for oxidation and reduction, and its free radical state, to protect a melanin-containing tissue or associated tissues against reducing or oxidizing conditions which might otherwise set free within living cells reactive free radicals capable of disrupting metabolism." This protective function ascribed to melanin —to serve as a radical trap—is in part analogous to the protective role played by beta carotenes in photosynthetic systems, where they may act as quenchers of excited state species of oxygen (Krinsky 1968, 1971).

The presence of melanin is not always beneficial, however, and the phenomenon of drug-binding to melanin is a case where, in the extreme, cellular melanin content leads to cytotoxicity. The association of melanin with drugs is relatively harmless in the skin where there is a continuous turnover of the melanin granules, whereas it may be harmful in the eye where the compounds may be retained by the melanin granules for a considerable time, the turnover of the captor melanosomes being so very slow (Szabo 1968). For example, it is now well established that the affinity of chloroquine and the phenothiazines for melanin is the primary factor determining their ocular toxicity (Dencker et al. 1973). Studying drug-induced chronic otic lesions, Dencker et al. (1973) have concluded that inner-ear melanin is responsible for the drug accumulation. These authors believe that the damage to the receptor cells in both the eye and the inner ear is secondary to damage in the melanin-containing cells in the sensory structures. Using bovine retinal pigment epithelium melanin in vitro, they have shown that "kanamycin had the highest melanin affinity (89%), followed by chloroquine (85%), quinine (68%), dihydrostreptomycin (65%), streptomycin (60%), and viomycin (46%). Salicylic acid and the tuberculostatics, isoniazid and paraaminosalicylic acid, showed no melanin affinity."

This drug affinity for melanin appears to be unique, as no other tissue accumulates these compounds in a similar manner, and no accumulation is seen in albino animals. That melanin is capable of binding or collecting certain drugs indicates that it may bind or collect other, more physiologic, substances. This action of melanin may be an important part of the functioning of the retinal pigment epithelium (Hague 1968).

A possible mechanism by which melanin drug-accumulation results in damage to the pigment cell is by interference with the normal functions of melanin in the pigment cell. This shutting-off or reduction of normal melanin function may result, in turn, in the observed tissue damage. However, a more likely possibility is that "after the drug has reached the tissue and is there in adequately high concentration, it must exert a biochemical effect other than its combination with melanin to actually cause tissue damage" (Potts 1968). The nature of the molecular interaction of chloroquine and melanin has been studied and the available data support the latter view (cf. Rubin 1968).

In summary, the role played by melanin in cellular metabolism, including the protection it affords against light and other agents, is accomplished by the following physical and chemical properties of the biochrome: (a) melanin acts as a neutral density filter, (b) melanin in the melanosomes attenuates the impinging radiation by scattering, (c) melanin absorbs the radiant energy in the ultraviolet and visible spectrum and dissipates the absorbed energy as heat, (d) melanin can utilize the absorbed energy and undergo immediate oxidation through the generation of semiquinonoid free radicals in the polymer, (e) melanin, as a stable free radical, has a capacity to act as a biologic electron exchange polymer (Pathak and Fitzpatrick 1974).

Retinal Pigment Epithelium Cell as a Model in Melanocyte Biology

The retinal pigment epithelium is an excellent model for the study of melanogenesis. Its most beneficial qualities for this purpose are the compactness of the tissue in terms of high pigment-cell population density, and also the intensive melanin synthesis associated with the retinal pigment epithelium during a limited period in its development. Appreciation of these qualities has led to the experimental creation of unpigmented retinal pigment epithelium in vivo by raising embryos of *Fundulus, Rana pipiens,* and other animals in the presence of phenylthiourea, an inhibitor of tyrosinase (Mitchell and Szabo 1966). Retinal pigment epithelium is relatively easy to separate and isolate from other tissues and cell types, and it has been used to establish pure cultures of melanin-producing cells (Trinkaus 1963; Cahn and Cahn 1966; Mannagh et al. 1973; Tso et al. 1973; Edwards 1977).

Although partially pure cultures of nonmalignant mammalian epidermal melanocytes have been established in some laboratories (Prunieras 1965; Prunieras et al. 1976; Riley 1975), no specific technique is available for the consistent isolation of purified populations of such cells. The use of pure cultures of melanoma cells is also of limited value if one's goal is to study the biology of normal melanin-producing cells. The great advantage, then, in the use of retinal pigment cells lies in the facility of establishing pure cultures of normal nonmalignant melanin-producing cells.

Several investigators have used a culture system of embryonic chick retinal pigment cells to study the relation of cell growth and DNA synthesis to melanogenesis, and, more generally, the processes of growth and differentiation (Coon and Cahn 1966; Newsome et al. 1974). In terms of differentiation, melanin is a highly specific, morphologically and biochemically identifiable macromolecular product characteristic of the differentiated state. The complex process of melanogenesis, involving the synthesis, intracellular transport, and activation of tyrosinase, together with the macromolecular synthesis of the melanosome, is quite sensitive to factors that inhibit differentiation. In fact, chick embryo retinal pigment cells, when cultured, have been observed to undergo a process of dedifferentiation or modulation (Whittaker 1963). Werner and Szabo (1973) have shown that differentiation in chick embryo retinal pigment cells in vitro is suppressed by bromodeoxyuridine (BrdU). Epithelial morphology and melanogenesis (that is, the cells' expression of differentiation) are more sensitive to BrdU than is their growth rate. While addition of 10^{-5} M BrdU to cultured cells inhibits both DNA and melanin synthesis, 10^{-6} M BrdU inhibits only melanin synthesis and tyrosinase activity and does not disturb general growth functions. The separation of growth and differentiative functions in this system produced by BrdU provides an experimental model where the differences between growth and differentiation may be studied at the transcriptional and translational levels in pigment cells (Wrathall et al. 1973; Garcia et al. in press).

The process of melanogenesis is alike in both epidermal melanocytes and retinal pigment cells. Several factors that stimulate melanogenesis in epidermal melanocytes and melanoma cells in vitro (for example, dibutyryl cyclic AMP, theophylline, prostaglandin E_1) have also been found to increase tyrosinase activity and melanin synthesis in cultured chick embryo retinal pigment cells (Garcia et al. 1975, in press; Redfern et al. 1976). It is not known, though, whether retinal pigment cells have the same or similar hormone receptors on their cell membranes as epidermal melanocytes, which are responsive to MSH and ACTH.

Retinal pigment cells are also valuable in studying the effects of ultraviolet light (290 nm–320 nm) on isolated nonmalignant melanocytes. It is known that melanogenesis is stimulated by prostaglandin E_1, which is also involved in the skin erythemal response induced by UVL. Oikawa and Nakayama (1974) have found that cholecalciferols (vitamin D) stimulate melanogenesis in cultured melanoma cells. Vitamin D is synthesized in skin by the action of UVL on 7-dehydrocholesterol. UVL may thus affect melanocytes indirectly through one of these mechanisms.

Working with melanoma cell cultures, Kitano and Hu (1969) were unable to stimulate melanin synthesis in UVL-irradiated cultures. It does not appear that the culture environment is inhibiting a melanogenic response, as epidermal melanocytes in organ culture of skin are stimulated by UVL in vitro (Glimcher et al. 1978). One idea is that there may be something unique about the tissue milieu in situ, and experimental evidence suggests that UVL acts on melanocytes indirectly through this milieu. However, recent work on UVL irradiation of cultured chick embryo retinal pigment cells has demonstrated a direct UVL stimulation of melanogenesis (Garcia et al. in press).

MELANOSOME MIGRATION AND PIGMENT DONATION. While RPE fulfills a useful role in pigmentation research, certain important aspects cannot be fully studied in RPE. For the sake of completeness, we outline some characteristics that can best be studied in melanocytes other than RPE. Masson, in 1948, described the epidermal melanocytes as "unicellular glands," which by "cytocrine" function secrete and selectively deposit their products, the melanosomes, into the keratinizing cells. This transfer of melanosomes has been observed in vitro by several investigators (Cruickshank 1964; Cohen and Szabo 1968; Prunieras 1969). Donation has been observed in cultures of melanocytes and keratinocytes from the same animal, and also in mixed cultures of melanocytes taken from black guinea pigs and keratinocytes taken from albino guinea pigs (Prunieras et al. 1976a, b). Although the term "melanocyte" may be generally applied to cover any and all melanin-producing cells, it is generally reserved for the melanin-producing cells of vertebrate epidermis and oral mucosa. What makes these cells remarkable and quite unique is their cytocrine activity. Although similar to melanocytes (in the strict sense) in numerous ways, retinal pigment cells differ from them in this one major characteristic.

One characteristic shared by retinal pigment cells and melanocytes is intracellular migration of melanosomes (see Table 8.1). (This appears to be restricted to poikilotherms. Melanosome migration in mammalian RPE has not been demonstrated.) In numerous vertebrates the cells of the RPE have long, thin processes that extend between the outer segments. The melanosomes move reversibly in the RPE cell processes in response to light, and this serves to screen the adjacent photoreceptors (cf. Chapter 10). This phenomenon is known to occur in many fishes and amphibians (cf. Prosser and Brown 1961; Ali 1959, 1971; Burnside 1976). Extension of rods into the layer of the RPE in response to light occurs in the catfish Ameiurus (Welsh and Osborn 1937) and Ictalurus (Adomian and Sjostrand 1973). The general observation in those species studied is that more pigment granules are found around the rods in the light adapted retina than in the dark adapted retina.

The mechanism by which melanosomes migrate within pigment cells is not known. However, in those pigment cells where melanosome movement is known to take place, melanosomes are found associated with microtubules, microfilaments, or both. There is an extensive literature on pigment granule movement in invertebrate photoreceptor structures and the relationship of pigment granule migration to light adaptation (Kleinholz 1961; Fahrenbach 1969; Vernon 1974; Schonenberger 1977). Discussing this phenomenon in Crustacea, Kleinholz (1961) states that "the chief function of such movement appears to be to screen the sensory component of the ommatidium, the rhabdom, in bright light, and to uncover the rhabdom in darkness or in light of low intensity."

Table 8.1 Comparison of properties of melanocytes.

Melanocyte	*Secretion of granules*	*Movement of granules*	*Association with other cell types*	*Cytocrinia*	*UVL excitation*
Retinal pigment epithelium	−	⇌	+	−	?
Epidermal melanocyte[a]	+	→	+	+	+
Dermal melanocyte[b]	−	?	−	−	+
Melanophore[c]	?	⇌	?	?	?
Squid (ink gland)	+	−	−	−	?

Note: + = affirmative data available; − = negative data available; ? = problematical; ⇄ = aggregation and dispersion; → = unidirectional along dendrites.

a. Epidermal melanocytes are melanin-producing cells with a special and unique relationship with the adjacent keratinizing Malpighian cells.

b. Dermal melanocytes manufacture and contain melanosomes and are found in dermis of homeotherms; they most likely do not have cytocrine or effector functions.

c. Melanophores are unique in their ability to disperse and aggregate melanosomes; they constitute the pigmentary effector system of poikilotherms.

Miller and Cawthon (1974) have shown that pigment granule movement in Limulus photoreceptors is associated with radially oriented microtubules in the retinular cell. Colchicine treatment results in pigment granule translocation to the light-adapted position and inhibits the return of pigment granules to the dark-adapted position. This association between microtubules and pigment granule movement also exists in the dermal melanophores of fishes, amphibians, and reptiles. Here, centripetal migration of melanosomes produces skin lightening, while centrifugal dispersion of melanosomes into melanophore dendrites produces skin darkening. Dispersion is inhibited by cytochalasin, which disrupts microfilaments, while centripetal migration is colchicine-sensitive.

In the frog RPE, melanosome migration appears to be associated with 10 nm microfilaments which have been tentatively identified as actin (Murray and Dubin 1975). These actin-like microfilaments are quite similar to the cytochalasin-sensitive microfilaments found in the frog dermal melanophores, in regard to their distribution and association with melanosomes. Actin filaments have also been identified in primate (squirrel monkey) RPE cell processes (Burnside and Laties 1976). Melanosome migration is not believed to occur in mammalian RPE and the role of actin filaments in these species is not known.

Melanosome migration does take place in mammalian epidermal melanocytes, however, and this occurs in the process of pigment donation to keratinocytes. Melanosomes migrate out into melanocyte dendrites from a perinuclear location and are subsequently transferred to keratinocytes. This process is stimulated by light and involves the association of melanosomes with microfilaments in human epidermal melanocytes (Jimbow et al. 1974). The immediate tanning reaction in human skin is produced by UV-A (320–380 nm) and visible light and involves a rapid translocation of melanosomes into melanocyte dendrites together with a change in the distribution of microfilaments. Following irradiation, bundles of 10 nm microfilaments are found in and running parallel to the long axis of dendrites, and associated with melanosomes (Jimbow et al. 1976a).

REFERENCES

ABE, K., BUTCHER, R. W., NICHOLSON, W. E., BAIRD, C. E., LIDDLE, R. A., and LIDDLE, G. W. 1969. Adenosine 3′, 5′-monophosphate (cyclic AMP) as the mediator of the actions of MSH and norepinephrine on the frog skin. *Endocrinology* 84:362–368.

ADOMIAN, G. E., and SJOSTRAND, F. S. 1973. Morphology of myoid elongation and shortening. *J. Cell Biol.* 59:2a.

ALI, M. A. 1959. Retinomotor responses in fishes. *Can. J. Zool.* 37:965–996.

ALI, M. A. 1971. Les responses rétinomotrices: caractères et mécanismes. *Vision Res.* 11:1225–1288.

BAGNARA, J. T., and HADLEY, M. E. 1969. Control of bright colored pigment cells of fishes and amphibians. *Am. Zool.* 9:465–478.

BECKER, S. W. 1927. Melanin pigmentation: a systematic study of the pigment of the human skin and upper mucous membrane with special consideration of pigmented dendritic cells. *Arch. Dermatol.* 16:259–290.

BENEDICT, P. H., SZABO, G., FITZPATRICK, T. B., and SINESI, S. J. 1968. Melanotic macules in Albright's syndrome and in neurofibromatosis. *J. Am. Med. Assoc.* 205: 618–622.

BERMAN, E. R., SCHWELL, H., and FEENEY, L. 1974. The retinal pigment epithelium: chemical composition and structure. *Invest. Ophthalmol.* 13:675–687.

BLOCH, B. B. 1917. Chemische Untersuchungen uber das spezifische pigmentbildende Ferment der Haut, die Dopaoxydase. *Hoppe Seylers Z. Physiol. Chem.* 98:226–254.

BLOIS, M. S. 1967. A note on the problem of melanin structure. In *Advances in biology of skin,* ed. W. Montagna and F. Hu. Vol. 8. *The pigmentary system,* pp. 319–321. New York: Pergamon Press.

BREATHNACH, A. S., GROSS, M., and MARTIN, B. 1973. Freeze-fracture replication of melanocytes and melanosomes. *J. Anat.* 116:303–320.

BROWN, K. T. 1965. An early potential evoked by light from the pigment epithelium-choroid complex of the eye of the toad. *Nature* 207:1249–1253.

BROWN, K. T., and CRAWFORD, J. M. 1967. Melanin and the rapid light-evoked responses from pigment epithelium cells of the frog eye. *Vision Res.* 7:165–178.

BURNETT, J. B. 1971. The tyrosinases of mouse melanoma: isolation and molecular properties. *J. Bio. Chem.* 246:3079–3091.

BURNETT, J. B., SEILER, H., and BROWN, I. V. 1967. Separation and characterization of multiple forms of tyrosinase from mouse melanoma. *Cancer Res.* 27:880–889.

BURNETT, J. B., and SEILER, H. 1969. Multiple forms of tyrosinase from human melanoma. *J. Invest. Dermatol.* 52:199–203.

BURNSIDE, B. 1976. Possible roles of microtubules and actin filaments in retinal pigmented epithelium. *Exp. Eye Res.* 23:257–275.

BURNSIDE, B., and LATIES, A. M. 1976. Actin filaments in apical projections of the primate pigmented epithelial cell. *Invest. Ophthalmol.* 15:570–575.

CAHN, R. D., and CAHN, M. B. 1966. Heritability of cellular differentiation: clonal growth and expression of differentiation in retinal pigment cells *in vitro*. *Proc. Natl. Acad. Sci. U.S.A.* 55:106–114.

CLARK, W. H., JR., TEN HEGGELER, B., and BRETTON, R. 1972. Electron microscopic observations of human cutaneous melanomas correlated with their biologic behavior. In *Melanoma and skin cancer,* pp. 121–141. Sydney, Australia: International Union against Cancer and Australian Cancer Society.

COHEN, J., and SZABO, G. 1968. Study of pigment donation *in vitro*. *Exp. Cell Res.* 50:418–433.

COON, H. G., and CAHN, R. D. 1966. Differentiation *in vitro:* effects of sephadex fractions of chick embryo extract. *Science* 153:1116–1119.

COPE, F. W., SEVER, R. J., and POLIS, B. D. 1963. Reversible free radical generation in the melanin granules of the eye by visible light. *Arch. Biochem. Biophys.* 100:171–177.

CRUICKSHANK, C. N. D., and HARCOURT, S. A. 1964. Pigment donation *in vitro*. *J. Invest. Dermatol.* 42:183–185.

DEBECKER, J., and ZANEN, A. 1975. Intensity functions of the early receptor potential and of the melanin fast photovoltage in the human eye. *Vision Res.* 15:101–106.

DENCKER, L., LINDQUIST, N. G., and ULBERG, S. 1973. Mechanism of drug-induced chronic otic lesions. Role of drug accumulation on the melanin of the inner ear. *Experientia* 29:1362–1364.

DOEZEMA, P. 1973. Tyrosinase from embryonic chick retinal pigment epithelium. *Comp. Biochem. Physiol.* 46B:509–517.

DRYJA, T. P., O'NEIL-DRYJA, M., PAWELEK, J. M., and ALBERT, D. M. 1978. Demonstration of tyrosinase in the adult bovine uveal tract and retinal pigment epithelium. *Invest. Ophthalmol.* 17:511–514.

DURRER, H., and VILLIGER, W. 1957. Bildung der Schillerstruktur beim Glanzstar. *Z. Zellforsch. Mikrosk. Anat.* 81:445–456.

EDWARDS, R. B. 1977. Culture of rat retinal pigment epithelium. *In Vitro* 13:301–304.

ENDO, H., and HU, F. 1973. Pigment cell development in Rhesus monkey eyes: an electron microscopic and histochemical study. *Dev. Biol.* 32:69–81.

EPPIG, J. J., JR. 1970*a*. Melanogenesis in amphibians. I. A study of the fine structure of the normal and PTU-treated pigmented epithelium in *Rana pipiens* tadpole eyes. *Z. Zellforsch. Mikrosk. Anat.* 103:238–246.

EPPIG, J. J., JR. 1970*b*. Melanogenesis in amphibians. II. Electron microscopic studies of the normal and PTU-treated pigment epithelium of developing *Notophtalmus viridescens* eye. *J. Embryol. Exp. Morphol.* 24:447–454.

EPPIG, J. J., JR., and DUMONT, J. N. 1972. Cytochemical localization of tyrosinase activity in pigmented epithelial cells of Rana pipiens and Xenopus laevis larvae. *J. Ultrastruct. Res.* 39:397–410.

FAHRENBACH, W. H. 1969. The morphology of the eyes of Limulus. II. Ommatidia of the compound eye. *Z. Zellforsch. Mikrosk. Anat.* 93:451–483.

FEENEY, L. 1973. The phagolysosomal system of the pigment epithelium: a key to retinal disease. *Invest. Ophthalmol.* 12:635–638.

FEENEY, L., GRIESHABER, J. A., and HOGAN, M. J. 1965. Studies on human ocular pigment. In *Structure of the eye. II. Symposium,* ed. J. W. Rohen, pp. 535–548. Stuttgart: Schattauer-Verlag.

FITZPATRICK, T. M. 1952. Human melanogenesis. *A.M.A. Arch. Dermatol. Syphilol.* 65:379–391.

FITZPATRICK, T. B., BECKER, S. W., JR., LERNER, A. B., and MONTGOMERY, H. 1950. Tyrosinase in human skin: demonstration of its presence and of its role in human melanin formation. *Science* 112:223–225.

FITZPATRICK, T. B., and BREATHNACH, A. S. 1963. Das epidermale Melanin-Einheit-System. *Dermatol. Wochenschr.* 147:481–489.

FITZPATRICK, T. B., MIYAMOTO, M., and ISHIKAWA, K. 1967. The evolution of concepts of melanin biology. In *Advances in biology of skin: the pigmentary system,* ed. W. Montagna and F. Hu, vol. 8, pp. 1–30. New York: Pergamon Press.

FITZPATRICK, T. B., and QUEVEDO, W. C., JR. 1971. Biological processes underlying melanin pigmentation and pigmentary disorders. In *Modern trends in dermatology,* ed. P. Borrie, pp. 122–149. London: Butterworth.

FOSTER, M. 1963. Some physiological genetic aspects of mammalian melanogenesis. *Ann. N. Y. Acad. Sci.* 100:743–748.

FOSTER, M. 1968. Mammalian pigment genetics. *Adv. Genet.* 13:311–339.

FURTH, V. O., and SCHREIDER, H. 1901. Uber tierische Tyrosinasen und ihre Beziehungen zur Pigmentbildung. *Hofmeister's Beitr. Chem. Physiol. Pathol.* 1:241–242.

Garcia, R. I., Flynn, E. A., and Szabo, G. 1976. Autophagocytosis of melanosomes in cultured retinal pigment cells. *In Vitro* 12:294–295.

Garcia, R. I., Fuentes, M. A., Kornhauser, A., and Szabo, G. 1975. Comparison of the effects of cyclic AMP and embryo extract on DNA and melanin synthesis in chick embryo retinal pigment cells *in vitro*. *J. Cell Biol.* 67:129a.

Garcia, R. I., Szabo, G., Mitchell, R. E., and Bloom, J. 1978. Variation in ultrastructure of skin pigmentation in Solomon Islanders. *Am. J. Phys. Anthropol.* 48:396–397.

Garcia, R. I., Werner, I., and Szabo, G. in press. Effect of prostaglandins, 5-bromodeoxyuridine, and ultraviolet light on growth and differentiation of retinal pigment cells *in vitro*. In *Pigment cell,* ed. S. Klaus, vol. 4, Basel: S. Karger.

Gerson, D. E., and Szabo, G. 1969. The effects of a single gene substitution on the mammalian melanocyte system: a qualitative and quantitative histological study in the C57BL and DBL mice. *Am. J. Phys. Anthropol.* 31:363–382.

Glimcher, M. E., Garcia, R. I., and Szabo, G. 1978. Organ culture of mammalian skin and the effects of ultraviolet light and testosterone on melanocyte morphology and function. *J. Exp. Zool.* 204:229–238.

Glimcher, M. E., and Szabo, G. 1978. Giant pigment granules in dermal melanocytes of rat scrotal skin. *Experientia* 34:1079–1080.

Goldman, J. M., and Hadley, M. E. 1969. The beta adrenergic receptor and 3′,5′-adenosine monophosphate: possible roles in the regulation of melanophore response of the spadefoot toad, *Scaphiopus couchi. Gen. Comp. Endocrinol.* 13:151–163.

Hague, E. B. 1968. Opening remarks: symposium on prolonged pharmocotherapy and the eye. *Dis. Nerv. Syst.* 29:5–9 (suppl.).

Hearing, V. J. 1973. Mammalian melanogenesis: tyrosinase versus peroxidase involvement and activation mechanisms. *Arch. Biochem. Biophys.* 158:720–725.

Hearing, V. J., and Ekel, T. M. 1975. Involvement of tyrosinase in melanin formation in murine melanoma. *J. Invest. Dermatol.* 64:80–83.

Hearing, V. J., and Lutzner, M. A. 1973. Mammalian melanosomal proteins: characterization by polyacrylamide gel electrophoresis. *Yale J. Biol. Med.* 46:553–559.

Hearing, V. J., Phillips, P., and Lutzner, M. A. 1973*b*. The fine structure of melanogenesis in coat color mutants of the mouse. *J. Ultrastruct. Res.* 43:88–106.

Hogeboom, G. H., and Adams, M. H. 1942. Mammalian tyrosinase and dopaoxidase. *J. Biol. Chem.* 145:273–279.

Hollyfield, J. G., and Ward, A. 1974*a*. Phagocytic activity in the retinal pigment epithelium of the frog, *Rana pipiens*. I. Uptake of polystyrene spheres. *J. Ultrastruct. Res.* 46:327–338.

Hollyfield, J. G., and Ward, A. 1974*b*. Phagocytic activity of the pigmented retinal epithelium. III. Interaction between lysosomes and ingested polystyrene spheres. *Invest. Ophthalmol.* 13:1016–1023.

Holstein, T. J., Burnett, J. B., and Quevedo, W. C., Jr. 1967. Genetic regulation of multiple forms of tyrosinase in mice: action of a and b loci. *Proc. Soc. Exp. Biol. Med.* 126:415–420.

Holstein, T. J., Quevedo, W. C., Jr., and Burnett, J. B. 1971. Multiple forms of tyrosinase in rodents and langomorphs with special reference to their genetic control in mice. *J. Exp. Zool.* 177:173–184.

Holstein, T. J., Stowell, C. P., Quevedo, W. C., Jr., Zacaro, R. M., and Bienieki, T. C. 1973. Peroxidase "protyrosinase," and multiple forms of tyrosinase in mice. *Yale J. Biol. Med.* 46:560–571.

Hori, Y., Toda, K., Pathak, M. A., Clark, W. H., Jr., and Fitzpatrick, T. B. 1968. A fine-structure study of the human epidermal melanosome and its acid phosphatase activity. *J. Ultrastruct. Res.* 25:109–120.

Hu, F., and Montagna, W. 1971. The development of pigment cells in the eyes of Rhesus monkeys. *Am. J. Anat.* 132:119–132.

Ide, C. 1972. The development of melanosomes in the pigment epithelium of the chick embryo. *Z. Zellforsch. Mikrosk. Anat.* 131:171–186.

Jimbow, K., Davison, P. F., Pathak, M. A., and Fitzpatrick, T. B. 1976*a*. Cytoplasmic filaments in melanocytes. In *Pigment cell,* ed. V. Riley, vol. 3, pp. 13–32. Basel: S. Karger.

Jimbow, K., and Fitzpatrick, T. B. 1974. Characterization of a new melanosomal structural component—the vesiculo-globular body—by conventional transmission, high voltage, and scanning EM. *J. Ultrastruct. Res.* 48: 269–283.

Jimbow, K., Fitzpatrick, T. B., Szabo, G., and Hori, Y. 1975. Congenital circumscribed hypomelanosis: a characterization based on electron microscopic study of tuberous sclerosis, nevus depigmentosus, and piebaldism. *J. Invest. Dermatol.* 64:50–62.

Jimbow, K., and Kukita, A. 1971. Fine structure of pigment granules in the human hair bulb: ultrastructure of pigment granules. In *Biology of normal and abnormal melanocytes,* ed. T. Kawamura, T. B. Fitzpatrick, and M. Seiji, pp. 171–193. Tokyo: Univ. of Tokyo Press.

JIMBOW, K., PATHAK, M. A., SZABO, G., and FITZPATRICK, T. B. 1974. Ultrastructural changes in human melanocytes after ultraviolet radiation. In *Sunlight and man,* ed. T. B. Fitzpatrick, M. A. Pathak, L. C. Haber, M. Seiji, and A. Kukita, pp. 195–216. Tokyo: Univ. of Tokyo Press.

JIMBOW, K., QUEVEDO, W. C., JR., FITZPATRICK, T. B., and SZABO, G. 1976*b*. Some aspects of melanin biology, 1950–1975. *J. Invest. Dermatol.* 67:72–89.

JIMBOW, K., SZABO, G., and FITZPATRICK, T. B. 1973. Ultrastructure of giant pigment granules (macromelanosomes) in the cutaneous pigmented macules of neurofibromatosis. *J. Invest. Dermatol.* 61:300–309.

JOHNSON, B. E., MANDELL, G., and DANIELS, F., JR. 1970. Melanin and cell reactions to ultraviolet radiation. *J. Invest. Dermatol.* 54:89.

JOHNSON, G. S., and PASTAN, I. 1972. N^6,O^2-Dibutyryl adenosine 3′, 5′-monophosphate induces pigment production in melanoma cells. *Nature* [New Biol.] 237:267–268.

KITANO, Y., and HU, F. 1969. The effects of ultraviolet light on mammalian pigment cells *in vitro. J. Invest. Dermatol.* 52:25–31.

KITANO, Y., and HU, F. 1970. Proliferation and differentiation of pigment cells *in vitro. J. Invest. Dermatol.* 55:444–451.

KITANO, Y., and HU, F. 1971. Melanin vs. protein synthesis in melanocytes *in vitro. Exp. Cell Res.* 64:83–88.

KLEINHOLZ, L. H. 1961. Pigmentary effectors. In *The physiology of crustacea,* ed. T. H. Waterman, vol. 2, pp. 133–169. New York: Academic Press.

KONRAD, K., WOLFF, K., and HONIGSMANN, H. 1974. The giant melanosome: a model of deranged melanosome-morphogenesis. *J. Ultrastruct. Res.* 48:102–123.

KRINSKY, N. I. 1968. The protective function of carotenoid pigments. In *Photophysiology,* ed. A. C. Giese, vol. 3, pp. 123–195. New York: Academic Press.

KRINSKY, N. I. 1971. Function of carotenoids. In *Carotenoids,* ed. O. Isler, pp. 669–716. Basel: Birkhauser-Verlag.

LERNER, A. B. 1971. Neural control of pigment cells. In *Biology of normal and abnormal melanocytes,* ed. T. Kawamura, T. B. Fitzpatrick, and M. Seiji, pp. 3–13. Tokyo: Univ. of Tokyo Press.

LERNER, A. B., and FITZPATRICK, T. B. 1950. Biochemistry of melanin formation. *Physiol. Rev.* 30:91–126.

LERNER, A. B., FITZPATRICK, T. B., CALKINS, E., and SUMMERSON, W. H. 1949. Mammalian tyrosinase: preparation and properties. *J. Biol. Chem.* 178:185–195.

LERNER, A. B., FITZPATRICK, T. B., CALKINS, E., and SUMMERSON, W. H. 1950. Mammalian tyrosinase: the relationship of copper to enzymatic activity. *J. Biol. Chem.* 187:793–802.

LONDON, D. A., CARTER, D. M., and CONDIT, E. S. 1976. Effect of pigment on photomediated production of thymine dimers in cultured melanoma cells. *J. Invest. Dermatol.* 67:261–264.

LUTZNER, M. A., and LOWRIE, C. T. 1972. Ultrastructure of the development of the normal black and beige melanin granules in the mouse. In *Pigmentation: its genesis and biologic control,* ed. V. Riley, pp. 89–105. New York: Appleton-Century-Crofts.

LUTZNER, M. A., TIERNEY, J. H., and BENDITT, E. P. 1968. Giant granules and widespread cytoplasmic inclusions in a genetic syndrome of Aleutian mink. *Lab. Invest.* 14:2063–2079.

MANNAGH, J., ARYA, D. V., and IRVINE, A. R., JR. 1973. Tissue culture of human retinal pigment epithelium. *Invest. Ophthalmol.* 12:52–64.

MASON, H. S. 1948. The chemistry of melanin. III. Mechanisms of the oxidation of dihydroxyphenylalanine by tyrosinase. *J. Biol. Chem.* 172:83–99.

MASON, H. S., INGRAM, D. J. E., and ALLEN, B. 1960. The free radical property of melanins. *Arch. Biochem. Biophys.* 86:225–230.

MASSON, P. 1948. Pigment cells in man. In *Biology of melanomas,* ed. R. W. Miner and M. Gordon. New York: New York Academy of Sciences.

MAUL, G. G. 1969. Golgi-melanosome relationship in human melanoma *in vitro. J. Ultrastruct. Res.* 26:163–176.

MAUL, G. G., and BRUMBAUGH, J. A. 1971. On the possible function of coated vesicles in melanogenesis of the regenerating fowl feather. *J. Cell Biol.* 48:41–48.

MEDAWAR, P. 1953. The microanatomy of the mammalian epidermis. *Q. J. Microscop. Sci.* 94:481–506.

MILLER, W. H., and CAWTHON, D. F. 1974. Pigment granule movement in *Limulus* photoreceptors. *Invest. Ophthalmol.* 13:401–405.

MITCHELL, B. S., and SZABO, G. 1966. The effect of phenylthiourea on melanogenesis in the embryo of *Fundulus heteroclitus. Biol. Bull.* 131:398–399.

MIYAMOTO, M., and FITZPATRICK, T. B. 1957. On the nature of the pigment in the retinal pigment epithelium. *Science* 126:449–450.

MIYAZAKI, K., and OHTAKI, N. 1975. Tyrosinase as glycoprotein. *Arch. Dermatol. Forsch.* 252:211–216.

MOYER, F. H. 1961. Electron microscopic observations on the origin, development, and genetic control of melanin granules in the mouse eye. In *The structure of the eye*, ed. G. Smelser, pp. 469–486. New York: Academic Press.

MOYER, F. H. 1963. Genetic effects on melanosome fine structure and ontogeny in normal and malignant cells. *Ann. N.Y. Acad. Sci.* 100:584–606.

MOYER, F. H. 1966. Genetic variation in the fine structure and ontogeny of mouse melanin granules. *Am. Zool.* 6:43–66.

MOYER, F. H. 1969. Development, structure, and function of the retinal pigmented epithelium. In *The retina*, ed. B. R. Straatsma et al., pp. 1–30. Berkeley: Univ. of California Press.

MURRAY, R. L., and DUBIN, M. W. 1975. The occurrence of actinlike filaments in association with migrating pigment granules in frog retinal pigment epithelium. *J. Cell Biol.* 64:705–710.

NEWSOME, D. A., FLETCHER, R. T., ROBISON, W. G., JR., KENYON, K. R., and CHADER, G. J. 1974. Effects of cyclic AMP and sephadex fractions of chick embryo extract on cloned retinal pigmented epithelium in tissue culture. *J. Cell Biol.* 61:369–382.

NICOLAUS, R. A. 1968. Melanins. In *Chemistry of natural product series*, ed. E. Lederer. Paris: Hermann.

NICOLAUS, R. A., and PIATELLI, M. 1965. Progress in the chemistry of natural black pigments. *Rend. Accad. Sci. Fis. Mathemat.* (Naples). 32:1–17.

NOVIKOFF, A. B., ALBALA, A., and BIEMPICA, L. 1968. Ultrastructure and cyto-chemical observations on B16 and Harding-Passey mouse melanomas: the origin of premelanosomes and compound melanosomes. *J. Histochem. Cytochem.* 16:299–319.

NOVIKOFF, P. M., NOVIKOFF, A. B., QUINTANA, N., and HAUW, J. 1971. Golgi apparatus, GERL, and lysosomes of neurons in rat dorsal root ganglia studied by thick section and thin section cytochemistry. *J. Cell Biol.* 50:859–886.

OIKAWA, A., and NAKAYAMA, M. 1974. Stimulation of melanogenesis in cultured melanoma cells by calciferols. *F.E.B.S. Lett.* 42:32–35.

OKUN, M., EDELSTEIN, L., OR, N., HAMADA, G., and DONELLAN, B. 1970. The role of peroxidase vs. the role of tyrosinase in enzymatic conversion of tyrosine to melanin in melanocytes, mast cells, and eosinophils: an autoradiographic-histochemical study. *J. Invest. Dermatol.* 55:1–12.

OKUN, M. B., EDELSTEIN, L. M., PATEL, R., and DONNELLAN, B. 1973. A revised concept of mammalian melanogenesis: the possible synergistic functions of aerobic Dopa-oxidase and peroxidase. A review. *Yale J. Biol. Med.* 46:535–540.

PARAKKAL, P. F. 1967. The transfer of premelanosomes into the keratinizing cells of albino hair follicles. *J. Cell Biol.* 35:473–477.

PATEL, R., OKUN, M., EDELSTEIN, L., and EPSTEIN, D. 1971. Biochemical studies of the peroxidase-mediated oxidation of tyrosine to melanin: demonstration of the hydroxylation of tyrosine by plant and human peroxidases. *Biochem. J.* 124:439–441.

PATHAK, M. A. 1967. Photobiology of melanogenesis: biophysical aspects. In *Advances in biology of skin*, ed. W. Montagna and F. Hu. Vol. 8. *The pigmentary system*, pp. 397–420. New York: Pergamon Press.

PATHAK, M. A., and FITZPATRICK, T. B. 1974. The role of natural photoprotective agents in human skin. In *Sunlight and man: normal and abnormal photobiologic responses*, ed. T. B. Fitzpatrick, M. A. Pathak, L. C. Harber, M. Seiji, and A. Kukita, pp. 725–750. Tokyo: Univ. of Tokyo Press.

PATHAK, M. A., and STRATTON, K. 1968. Free radicals in human skin before and after exposure to light. *Arch. Biochem. Biophys.* 123:468–476.

PATHAK, M. A., STRATTON, K., and HORI, Y. 1968. Light-induced immediate pigment darkening reaction in human skin. *Clin. Res.* 16:258.

PAWELEK, J. M. 1976. Factors regulating growth and pigmentation of melanoma cells. *J. Invest. Dermatol.* 66:201–209.

PAWELEK, J., WONG, G., SANSONE, M., and MOROWITZ, J. 1973. Molecular controls in mammalian pigmentation. *Yale J. Biol. Med.* 46:430–443.

POST, P. W. 1971. Pigmentation and its role in human adaptation. Ph.D. dissertation, Columbia University. Ann Arbor, Michigan: University Microfilms #71–23, 613.

POST, P. W., DANIELS, F., JR., and BINFORD, R. T., JR. 1975. Cold injury and the evolution of "white" skin. *Hum. Biol.* 47:65–80.

POTTS, A. M. 1968. Agents which cause pigmentary retinopathy. *Dis. Nerv. Syst.* 29:16–18 (suppl.).

PROCTOR, P., MCGINNESS, J., and CORRY, P. 1974. A hypothesis on the preferential destruction of melanized tissues. *J. Theor. Biol.* 48:19–22.

PROSSER, C. L., and BROWN, F. A., JR. 1961. *Comparative Animal Physiology*, p. 350. Philadelphia: W. B. Saunders.

PROTA, T., and NICOLAUS, R. A. 1967. On the biogenesis of pheomelanins. In *Advances in biology of skin*, ed. W. Montagna and F. Hu. Vol. 8. *The pigmentary system*, pp. 323–332. New York: Pergamon Press.

PRUNIERAS, M. 1965. *La culture de l'épiderme de mammifère adulte.* Paris: Société Parisienne d'Expansion Immobilière, Département Médical.

PRUNIERAS, M. 1969. Interactions between keratinocytes and dendritic cells. *J. Invest. Dermatol.* 52:1–17.

PRUNIERAS, M., DELESCLUSE, C., and REGNIER, M. 1976*a*. The culture of skin: a review of theories and experimental methods. *J. Invest. Dermatol.* 67:58–65.

PRUNIERAS, M., MORENO, G., DOSSO, Y., and VINZENS, F. 1976*b*. Studies on guinea pig skin cell cultures. V. Cocultures of pigmented melanocytes and albino keratinocytes: a model for the study of pigment transfer. *Acta Derm. Venereol.* (Stockh.) 56:1–9.

QUEVEDO, W. C., JR. 1971. Genetic regulation of pigmentation in mammals. In *Biology of normal and abnormal melanocytes,* ed. T. Kawamura, T. B. Fitzpatrick, and M. Seiji, pp. 99–115. Tokyo: Univ. of Tokyo Press.

QUEVEDO, W. C., JR. 1973. Genetic control of melanin metabolism within the melanin unit of mammalian epidermis. *J. Invest. Dermatol.* 60:407–417.

RAPER, H. S. 1928. The aerobic oxidases. *Physiol. Rev.* 8:245–282.

REDFERN, N., ISRAEL, P., BERGSMA, D., ROBISON, W. G., JR., WHIKEHART, D., and CHADER, G. 1976. Neural retinal and pigment epithelial cells in culture: patterns of differentiation and effects of prostaglandins and cyclic AMP on pigmentation. *Exp. Eye Res.* 22:559–568.

RILEY, P. A. 1974. Melanin and melanocytes. In *The physiology and pathophysiology of the skin,* ed. A. Jarrett, pp. 1101–1130. London: Academic Press.

RILEY, P. A. 1975. Growth inhibition in normal mammalian melanocytes *in vitro. Brit. J. Dermatol.* 92:291–304.

ROBISON, W. G., JR., KUWABARA, T., and COGAN, D. G. 1975. Lysosomes and melanin granules of the retinal pigment epithelium in a mouse model of the Chediak-Higashi syndrome. *Invest. Ophthalmol.* 14:312–317.

ROSDAHL, I., and SZABO, G. 1976. Ultrastructure of the human melanocyte system in the newborn, with special reference to 'racial' differences. In *Pigment cell,* ed. V. Riley, vol. 3. Basel: S. Karger.

ROSENBERG, B. 1962. Electronic charge transport in carotenoid pigments and a primitive theory of the electroretinogram. *Photochem. Photobiol.* 1:117–129.

RUBIN, M. 1968. The antimalarials and the tranquilizers. *Dis. Nerv. Syst.* 29:67–76 (suppl.).

SAITO, N., and SEIJI, M. 1973. Epidermal lysosomes and the degradation of melanosomes. *Acta Dermatol. Venereol.* (Stockh.) 53:69–74 (suppl. 73).

SAITO, N., and SEIJI, M. 1976. Degradation of melanosomes *in vitro* and *in vivo.* In *Pigment Cell,* ed. V. Riley, vol. 3, pp. 384–392. Basel: S. Karger.

SCHONENBERGER, N. 1977. The fine structure of the compound eye of *Squilla mantis (Crustacea, Stomatopoda). Cell Tissue Res.* 176:205–233.

SCHROEDER, H. E. 1969. Melanin containing organelles in cells of the human gingiva. I. Epithelial melanocytes. *J. Periodont. Res.* 4:1–18.

SEARLE, A. G. 1968. *Comparative genetics of coat colour in mammals.* London: Logos Press.

SEIJI, M., FITZPATRICK, T. B., and BIRBECK, M. S. C. 1961. The melanosome: a distinctive subcellular particle of mammalian melanocytes and the site of melanogenesis. *J. Invest. Dermatol.* 36:243–252.

SEIJI, M., ITAKURA, H., and IRIMAJIRI, T. 1972. Tyrosinase in the membrane system of mouse melanoma. In *Pigmentation: its genesis and biologic control,* ed. V. Riley, pp. 525–542. New York: Appleton-Century-Crofts.

SEIJI, M., and IWASHITA, S. 1963*a*. Intracellular localization of tyrosinase in melanocytes. *J. Biochem.* 54:103–106.

SEIJI, M., and IWASHITA, S. 1963*b*. On the site of melanin formation in melanocytes. *J. Biochem.* 54:465.

SEIJI, M., and IWASHITA, S. 1965. Intracellular localization of tyrosinase and site of melanin formation in melanocytes. *J. Invest. Dermatol.* 45:305–314.

SEIJI, M., SHIMAO, K., BIRBECK, M. S. C., and FITZPATRICK, T. B. 1963. Subcellular localization of melanin biosynthesis. *Ann. N.Y. Acad. Sci.* 100:497–533.

SEVER, R. J., COPE, F. W., and POLIS, B. D. 1962. Generation by visible light of labile free radicals in the melanin granules of the eye. *Science* 137:128–129.

SILAGI, S. 1969. Control of pigment production in mouse melanoma cells *in vitro. J. Cell Biol.* 43:263–274.

SNYDER, D. S., and EAGLESTEIN, W. H. 1974. Intradermal anti-prostaglandin agents and sunburn. *J. Invest. Dermatol.* 62:47–50.

STANKA, P. 1971. Elektronmikroskopische Untersuchungen über die Prëmelanosomenentstehung am retinalen Pigmentepithel von Huhnerembroyonen. *Z. Zellforsch. Mikrosk. Anat.* 112:120–128.

SWAN, G. A. 1963. Chemical structure of melanins. *Ann. N.Y. Acad. Sci.* 100:1005–1019.

SWAN, G. A. 1964. Some studies on the formation and structure of melanin. *Rend. Accad. Sci. Fisiche Math.* (Naples) 31:212–231.

SWAN, G. A. 1974. Structure, chemistry and biosynthesis of the melanins. *Fortschr. Chem. Org. Naturst.* 31:521–582.

SZABO, G. 1959. Quantitative histological investigations on the melanocyte system of the human epidermis. In *Pigment cell biology,* ed. M. Gordon, pp. 549–561. New York: Academic Press.

SZABO, G. 1968. Melanin pigmentation. *Dis. Nerv. Syst.* 29:58–62 (suppl.).

SZABO, G. 1975. Skin as an adaptive organ. In *physiological anthropology*, ed. A. Damon, pp. 39–58. New York: Oxford Univ. Press.

SZABO, G., GERALD, A. B., PATHAK, M. A., and FITZPATRICK, T. B. 1969. Racial differences in the fate of melanosomes in human epidermis. *Nature* 222:1081–1082.

TAKEUCHI, I. K., and KAJISHIMA, T. 1974. Pigment cell differentiation in albino goldfish. I. Ultrastructural differentiation of retinal pigment epithelium. *Cell Tissue Res.* 155:383–398.

TAKEUCHI, Y. K., TAKEUCHI, I. K., and KAJISHIMA, T. 1975. Inhibition of melanosome transformation in embryonic chick pigment retina cultured *in vitro. J. Exp. Zool.* 192:391–400.

THOMSON, M. L. 1955. Relative efficiency of pigment and horny layer thickness in protecting the skin of Europeans and Africans against solar ultraviolet radiation. *J. Physiol.* (Lond.) 127:236–246.

TODA, K., and FITZPATRICK, T. B. 1972. Ultrastructural and biochemical studies of the formation of melanosomes in the embryonic chick retinal pigment epithelium. In *Pigmentation: its genesis and biologic control,* ed. V. Riley, pp. 125–141. New York: Appleton-Century-Crofts.

TRINKAUS, J. P. 1963. Behavior of dissociated retinal pigment cells in heterotypic cell aggregates. *Ann. N.Y. Acad. Sci.* 100:413–434.

TSO, M. O. M., ALBERT, D., and ZIMMERMAN, L. E. 1973. Organ culture of human retinal pigment epithelium and choroid: a model for the study of cytologic behavior of RPE *in vitro. Invest. Ophthalmol.* 12:554–566.

TURNER, W. A., JR., TAYLOR, J. D., and CHEN, T. T. 1976. The role of multivesicular bodies in melanosome formation. In *Pigment cell,* ed. V. Riley, pp. 33–45. Basel: S. Karger.

VANNESTE, W. H., and ZUBERBUHLER, A. 1974. Copper-containing oxygenases. In *Molecular mechanisms of oxygen activation,* ed. O. Hayaishi, pp. 371–404. New York: Academic Press.

VERNON, J. E. 1974. Physiological color changes in Odonata eyes: a comparison between eye and epidermal chromatophore pigment migrations. *J. Insect. Physiol.* 20:1491–1505.

VOULOT, C., and ORTONNE, J.-P. 1975. The tyrosinases of normal and diseased human skin. *Arch. Dermatol. Res.* 254:205–214.

WEISS, L. W., and ZELICKSON, A. S. 1977. Giant melanosomes in multiple lentigines syndrome. *Arch. Dermatol.* 113:491–494.

WEISS, P. 1949. Differential growth. In *The chemistry and physiology of growth,* ed. A. K. Parpart, pp. 135–186. Princeton: Princeton Univ. Press.

WELSH, J. H., and OSBORN, C. M. 1937. Diurnal rhythm in fish retina. *J. Comp. Neurol.* 66:349–359.

WERNER, I. 1969. Growth and differentiation of chicken embryo retinal pigment cells in culture. Unpublished Ph.D. thesis, Brown University.

WERNER, I. 1973. Control of pigment synthesis in culture. *Yale J. Biol. Med.* 46:444–463.

WERNER, I., and SZABO, G. 1973. The effect of 5-Bromodeoxyuridine on growth and melanogenesis in cultured pigment cells. *J. Cell Biol.* 59:362a.

WHITTAKER, J. R. 1963. Changes in melanogenesis during the dedifferentiation of chick retinal pigment cells in cell culture. *Dev. Biol.* 8:99–127.

WHITTAKER, J. R. 1965. A relationship between increased protein synthesis and loss of melanin synthesis in monolayer cultures of chick retinal pigment cells. *Am. Zool.* 5:643.

WHITTAKER, J. R. 1967. Loss of melanotic phenotype *in vitro* by differentiated retinal pigment cells: demonstration of mechanisms involved. *Dev. Biol.* 15:553–574.

WHITTAKER, J. R. 1968*a*. Translational competition as a possible basis of modulation in retinal pigment cell cultures. *J. Exp. Zool.* 169:143–160.

WHITTAKER, J. R. 1968*b*. The nature and probable cause of modulations in pigment cell cultures. In *Results from problems in cell differentiation,* ed. H. Ursprung. Vol. 1. *The stability of the differentiated state,* pp. 25–36. Berlin: Springer-Verlag.

WHITTAKER, J. R. 1971. Melanin granule shape in tissue-cultured embryonic pigment cells. *Experientia* 27:1464–1465.

WINDHORST, D. B., ZELICKSON, A. S., and GOOD, R. A. 1968. A human pigmentary dilution based on a heritable subcellular structural defect—the Chediak-Higashi syndrome. *J. Invest. Dermatol.* 50:9–18.

WITKOP, C. J., JR., HILL, C. W., DESNICH, S., THIES, J. K., THORN, H. U., JENKINS, M., and WHITE, J. G. 1973. Ophthalmologic, biochemical, platelet, and ultrastructural defects in the various types of oculocutaneous albinism. *J. Invest. Dermatol.* 60:443–456.

WITKOP, C. J., JR., KING, R. A., and CREEL, D. J. 1976. The abnormal albino animal. In *Pigment cell,* ed. V. Riley, vol. 3., pp. 201–210. Basel: S. Karger.

WITKOP, C. J., JR., WHITE, J. G., NANCE, W. E., and UMBER, R. E. 1971. Mutations in the melanin pigment system in man resulting in features of oculocutaneous albinism. In *Pigmentation: its genesis and biologic control,* ed. V. Riley, pp. 359–377. New York: Appleton-Century-Crofts.

WOLFF, K., and HONIGSMANN, H. 1972. Are melanosome complexes lysosomes? *J. Invest. Dermatol.* 59:170–176.

WONG, G., and PAWELEK, J. 1973. Control of phenotypic expression of cultured melanoma cells by melanocyte stimulating hormones. *Nature* (New Biol.) 241:213–215.

WONG, G., PAWELEK, J., SANSONE, M., and MOROWITZ, J. 1974. Response of mouse melanoma cells to melanocyte stimulating hormone. *Nature* 248:351–354.

WRATHALL, J. P., OLIVER, C., SILAGI, S., and ESSNER, E. 1973. Suppression of pigmentation in mouse melanoma cells by 5-bromodeoxyuridine: effects on tyrosinase activity and melanosome formation. *J. Cell Biol.* 57:406–423.

ZANEN, A., and DEBECKER, J. 1971. Visual pigments and melanin contributions to the fast photo-voltage of the human eye. *Vision Res.* 11:169–172.

ZELICKSON, A. S., WINDHORST, D. B., WHITE, J. G., and GOOD, R. A. 1967. The Chediak-Higashi syndrome: formation of giant melanosomes and the basis of hypopigmentation. *J. Invest. Dermatol.* 49:575–581.

ZIMMERMAN, J. 1975. The initiation of melanogenesis in the chick retinal pigment epithelium. *Dev. Biol.* 44: 102–118.

ZIMMERMAN, J. 1976. Synthesis of messenger RNA involved in increased levels of melanogenesis. *J. Cell Biol.* 70:145a.

ZIMMERMAN, J., BRUMBAUGH, J. A., BIEHL, J., and HOLTZER, H. 1974. The effect of 5-bromodeoxyuridine on the differentiation of chick embryo pigment cells. *Exp. Cell Res.* 83:159–165.

ACKNOWLEDGMENTS

The research reported here was supported by USPHS Grant AM-20669 from the National Institute of Arthritis, Metabolism, and Digestive Diseases. We thank Drs. M. A. Fuentes, A. Kornhauser, and I. Werner for their advice and criticism, and gratefully acknowledge the assistance of Michael Agostino, Wayne Colin, Evelyn Flynn, Ann Hawthorne, Mark Glimcher, Elsie Rosenfield, and Susannah Swihart.

9 Phagocytic Properties of the Retinal Pigment Epithelium

Dean Bok
Richard W. Young

DURING RECENT YEARS, it has increasingly been recognized that the retinal pigment epithelium carries out several functions that are crucial for the normal operation of the visual system. One of these important roles, appreciated for about a decade, is the phagocytosis of rod outer segment debris. This scavenging activity goes on daily at an impressive rate in the normal retina. It can be accelerated to extraordinary levels when outer segments are damaged. Disruption of this phagocytic function may underlie a variety of clinical disorders, some of which result in blindness. In this chapter, we will review the research that ultimately led to an awareness of these phagocytic properties, current concepts regarding the general process of phagocytosis, the present understanding of the phagocytic process in the pigment epithelium, and the possible clinical and pathological implications of these properties. The related process of autophagy within the pigment epithelium is reviewed in Chapter 7.

Early Evidence of Phagocytosis by the Pigment Epithelium

The first indication that the pigment epithelium was capable of phagocytosis was presented by Karli (1954a, b). In rabbits injected with sodium iodoacetate, he observed that the pigment epithelial cells migrated away from their attachments to Bruch's membrane and engulfed necrotic debris that resulted from the poisoning effect of the drug. Trypan blue was ingested by migrating and partially detached pigment epithelial cells, but did not appear to be phagocytized by cells that remained firmly attached to Bruch's membrane.

Kühne (1879) had described cytoplasmic structures in the frog pigment epithelium, which he termed myeloid bodies. Porter (1956) confirmed their existence, using the electron microscope, and described their lamellar character. However, when the pigment epithelium was examined in several different vertebrate species, two types of lamellar structures could be distinguished. One of these matched Porter's description of myeloid bodies, whereas the other type did not (Yamada et al. 1958; Bernstein and Pease 1959). Confusion existed over the relationship between the two forms, and the term "myeloid body" continued to be used for both.

Porter and Yamada (1960) then made a definite distinction between myeloid bodies and other lamellar structures. They described the myeloid body in the frog as a component of the smooth endoplasmic reticulum (Figs. 9.1, 9.20), and distinctly set it apart from the other components, which they called "lamellated lipid bodies." At the same time, Cohen (1960) stated that the mouse pigment epithelium lacked myeloid bodies, although "fragments of receptor outer segments, when partially enveloped by retinal pigment cells, gave a superficial appearance of myeloid elements." He later mentioned the rare occurrence of degenerated rod fragments in association with the pigment epithelium in the rhesus monkey, but did not state whether or not he considered them to be intracellular (Cohen 1961).

In 1961, Dowling and Gibbons published their classic work on the relationship between vitamin A deficiency and retinal ultrastructure. They observed that the "myeloid bodies" in the rat pigment epithelium disappeared during severe vitamin A defi-

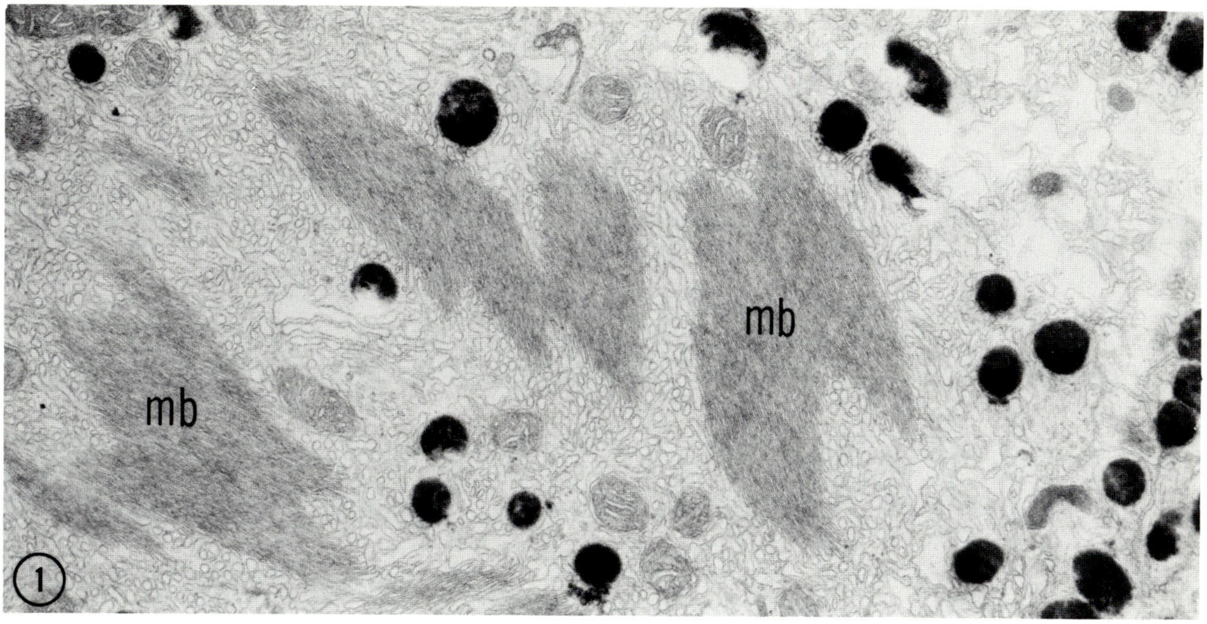

Figure 9.1 *Electron micrograph of myeloid bodies* (mb) *in the pigment epithelium of the desert night lizard,* Xantusia vigilis. *These organelles are usually angular in shape and represent highly differentiated portions of the smooth endoplasmic reticulum.* (×22,790.)

ciency, but reappeared within three days when vitamin A was restored to the diet. Subsequently (1962), they stated that the "myeloid bodies" of the normal rat pigment epithelium were not the same structures as the myeloid bodies of lower vertebrates, and correctly classified the former as inclusion bodies. The essential distinction was that in the rat the lamellar bodies were surrounded by a membrane, rather than being continuous with the smooth endoplasmic reticulum. Dowling and Gibbons (1962) also showed that a series of stages could be constructed, ranging from particles with little internal structure to highly ordered lamellar particles (Fig. 9.2). This was the first study in which it was clearly suggested that the extensive array of lamellar inclusion bodies in the mammalian pigment epithelium was morphologically and, perhaps, ancestrally related.

In 1963, in an important and insightful paper, Bairati and Orzalesi reported investigations of the human pigment epithelium in which they offered a hypothesis concerning the significance of the lamellar inclusions. Their observations led them to predict that discs might detach in small groups from the photoreceptor outer segment and be phagocytized by the pigment epithelium. They further speculated that these fragments might be digested by the pigment epithelium, so that the highly ordered lamellar inclusions could be considered to be the predecessors of the disorganized ones. Of equal importance was their prediction that a loss of discs from the apex of the outer segment would require a compensatory addition of new discs at the base.

These authors had at their disposal a new and crucial experimental observation: Droz (1963) had just presented autoradiographic evidence indicating that the protein component of rod outer segments was continually renewed. Droz demonstrated that the primary center for protein synthesis in retinal rods is located in the inner segment, and that proteins are transported from there to the outer segment. It was his impression that the protein became randomly distributed throughout the outer segment in 4 to 7 days.

Droz's kinetic analysis of protein metabolism in the pigment epithelium indicated that this cell layer was not a source of protein for the photoreceptor outer segment. Earlier investigators had suggested that the pigment epithelium might produce the layers of outer segment membrane in much the same way that the Schwann cell forms the myelin sheath. In this view, now known to be incorrect, inclusion bodies in the pigment epithelium were

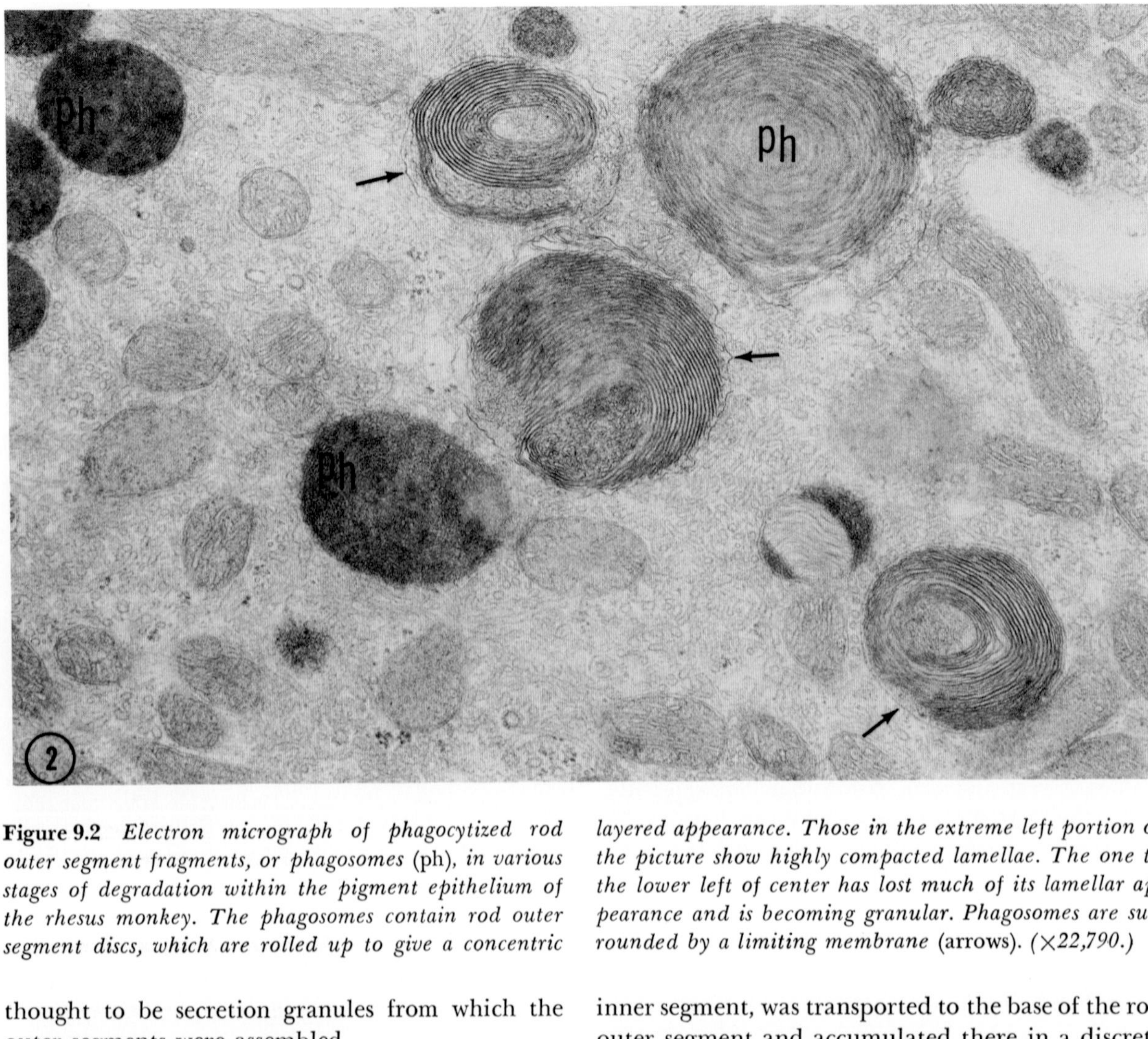

Figure 9.2 *Electron micrograph of phagocytized rod outer segment fragments, or phagosomes* (ph), *in various stages of degradation within the pigment epithelium of the rhesus monkey. The phagosomes contain rod outer segment discs, which are rolled up to give a concentric layered appearance. Those in the extreme left portion of the picture show highly compacted lamellae. The one to the lower left of center has lost much of its lamellar appearance and is becoming granular. Phagosomes are surrounded by a limiting membrane* (arrows). *(×22,790.)*

thought to be secretion granules from which the outer segments were assembled.

Shortly after the appearance of Droz's important publication, Cohen (1963) and Nilsson (1964) suggested on morphological grounds that some rod outer segment discs might continue to form after the cell had completed its differentiation.

"Myeloid bodies" that resembled fragments of visual cell outer segments were then observed by electron microscopy in the human pigment epithelium by Feeney, Grieshaber, and Hogan (1965), who considered them to be "broken or detached portions of outer rod and cone segment engulfed by the pigment epithelial cell."

The autoradiographic studies by Young (1965, 1966, 1967) on rat, mouse, and frog retinas firmly established the concept of rod outer segment renewal. He demonstrated a sequence of events in which newly synthesized protein, formed in the rod inner segment, was transported to the base of the rod outer segment and accumulated there in a discrete "reaction band" of radioactive protein disposed at right angles to the long axis of the outer segment. The reaction band was then gradually displaced along the outer segment toward the pigment epithelium. When the labeled protein reached the tip of the rod outer segment (after approximately 9 days in the mouse and rat, and 6 weeks in the frog), it disappeared. Regardless of the species and transit time, the reaction band neither increased in width nor decreased in intensity. Young concluded that it was not simply the protein which was undergoing renewal. The entire outer segment disc structure was being replaced by a mechanism involving the continual addition of new membranous discs at the base of the rod outer segment and the balanced removal of material from the apex of the cell (where the labeled protein "disappeared"). Since that part of the outer segment is in contact with the pigment epithelium, Young (1967) suggested that the pigment epithelium "may participate actively in the disposal phase of photoreceptor outer segment renewal."

At about this time, Kroll and Machemer (1968) reported a series of studies on experimental retinal detachment in monkeys. Detachment of the retina caused the disintegration of photoreceptor outer segments and the associated disappearance of "lamellar inclusion bodies" from the pigment epithelium. Next, they observed (1969) that when the experimentally detached retina was reattached, there was a rapid regeneration of damaged outer segments and a reappearance of "lamellar inclusion bodies" in the cytoplasm of the pigment epithelium. (These results were similar to those obtained by Dowling and Gibbons, cited above, during vitamin A deficiency and recovery.) Kroll and Machemer concurred with earlier workers that the lamellar bodies might represent broken or detached portions of photoreceptor outer segments engulfed by the pigment epithelium.

Further evidence of phagocytic activity by the pigment epithelium was derived from experiments in which retinal degeneration was induced by visible light. In 1966, it was reported (Noell et al. 1966) that intense, unfocused visible light applied for less than an hour could cause irreversible photoreceptor damage in albino rats. Electron-microscopic studies showed severe outer segment degeneration and a massive population of lamellar structures in the pigment epithelium (Kuwabara and Gorn 1968). Although these structures were not recognized as being of outer segment origin at the time, a subsequent report (Grignolo et al. 1969) correctly identified these inclusions as the result of "widespread phagocytosis . . . of degenerated outer segments."

Direct Autoradiographic Evidence of Phagocytosis by the Pigment Epithelium

Direct and unequivocal evidence for the phagocytosis of rod outer segment discs by the pigment epithelium was finally provided by the autoradiographic studies of Young and Bok (1969). Using frogs in which a group of rod outer segment discs had been labeled with radioactive amino acids (Figs. 9.3, 9.4), these authors observed that when the intensely labeled discs disappeared from the end of the rod outer segment, highly radioactive inclusion bodies suddenly appeared in the cytoplasm of the pigment epithelium (Figs. 9.5–9.10). Examination of the radioactive inclusions in the electron microscope revealed their close resemblance to rod outer segments (Figs. 9.9, 9.10). None of the pigment epithelial inclusion bodies showed this intense concentration of radioactivity during the several weeks prior to the arrival of radioactive discs at the ends of the outer segments (Figs. 9.3, 9.4), in spite of the fact that radioactive precursors had been made available to the pigment epithelium as well as the photoreceptor cells. Thus, it was directly demonstrated that the inclusion bodies were not synthesized by the pigment epithelium, but were ingested by phagocytosis. In order to denote this clearly, the "lamellated lipid bodies" or "inclusion bodies" of earlier authors were termed *phagosomes* to emphasize the fact that they were engulfed foreign bodies.

Labeled phagosomes were first observed in the apical cytoplasm of the pigment epithelium, at which time the resemblance of their contents to outer segments was apparent (Figs. 9.9, 9.10). Next, they were displaced to more basal positions, where they took on a variety of forms similar to those observed by earlier investigators. It was apparent that the pigment epithelium destroyed the material in the phagosomes, because (a) there was no increase in their number; (b) the unlabeled phagosomes disappeared following detachment of the labeled rod discs; and (c) the intensely radioactive phagosomes themselves subsequently disappeared (Young and Bok 1969).

Quantitative Aspects of Rod Outer Segment Phagocytosis by the Pigment Epithelium

The rapid rate of assembly of new outer segment disc membranes by rod visual cells presents the pigment epithelium with a quantitatively significant disposal task. Using autoradiography and electron microscopy, Young (1971a) has determined that in the rhesus monkey each rod assembles 80 to 90 new discs daily; that is, one every 16 to 18 minutes. Assuming that no net growth in length is occurring, the number of discs produced daily by each rod at the base of the outer segment must be balanced by the loss of a comparable number of discs at the apical end of the cell. Having determined that the outer segments of from 24 to 45 rods were inserted in the cytoplasm of each pigment epithelial cell, depending on the location within the retina, Young estimated that the number of discs phagocytized and destroyed daily was about 2,000

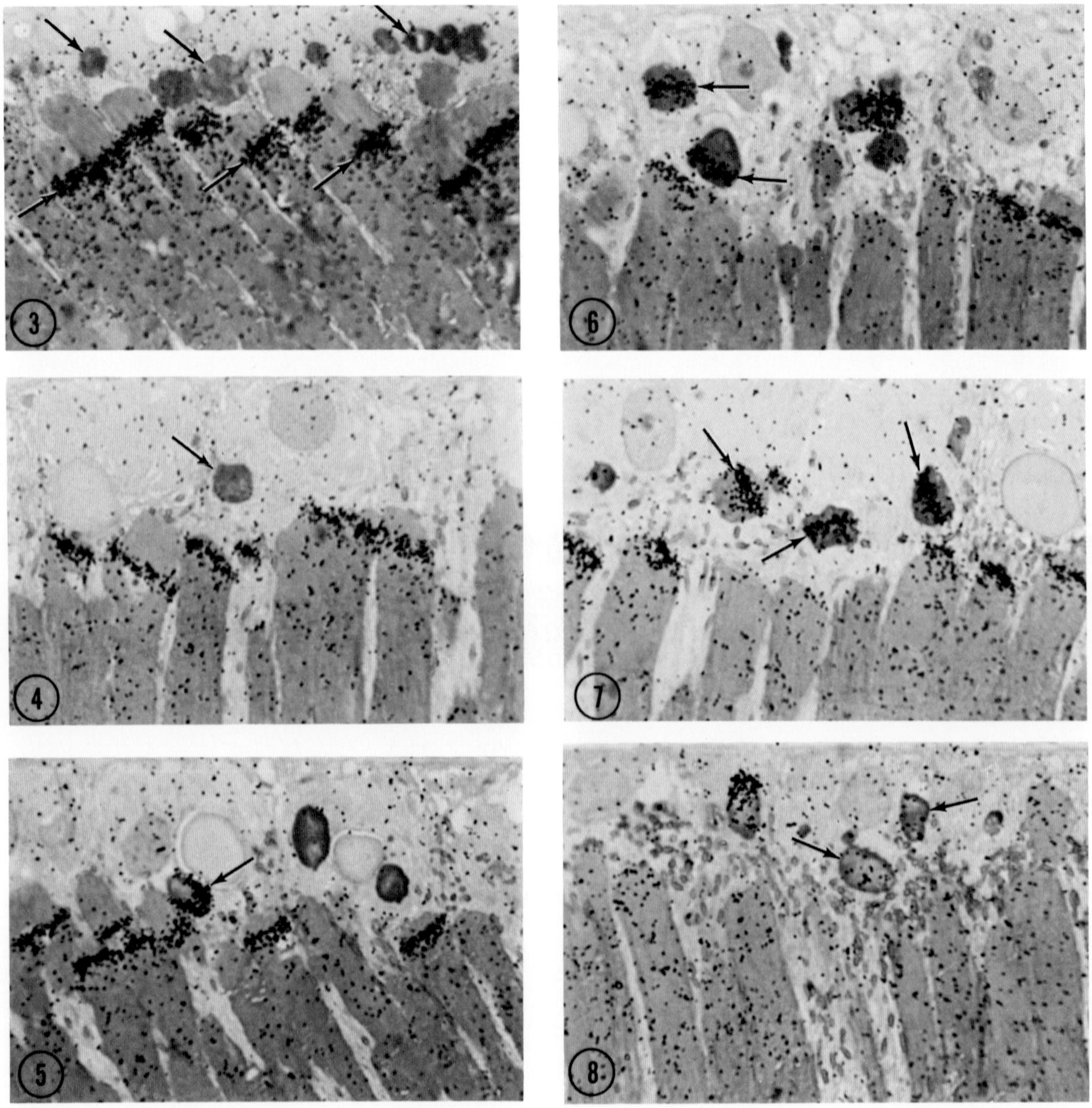

Figures 9.3–9.8 *Light-microscopic autoradiograms of rod outer segment—pigment epithelial junction in the frog retina. The frogs were injected with a mixture of [^{3}H] leucine and [^{3}H]phenylalanine and sacrificed between 47 and 68 days after injection. Rod outer segment discs that were assembled shortly after the injection have been displaced along the outer segment. Some of the discs are being shed from the ends of the outer segment. Others have been engulfed by the pigment epithelium. (After Young and Bok 1969. ×1,210. See individual captions at right.)*

in the parafovea, 3,500 in the perifovea, and nearly 4,000 in the periphery. The disc diameter is about 1.7 μ in peripheral rods, so that the disc area is approximately 2.27 μ^2. This represents an area of some 9080 μ^2 of doubled membrane consumed daily by a pigment epithelial cell in the periphery.

An approximate calculation can be made in the rat, where each rod outer segment contains on the order of 950 discs. These are renewed every 9 to 10 days (Young 1967), representing a daily disc output of roughly 100 discs per rod. In the rat, the pigment epithelial cells are very broad, apparently because

Figure 9.3 *Forty-seven days after injection, radioactive discs* (lower arrows) *are approaching the tips of the outer segments. Discs that were assembled prior to the injection of radioactive amino acids are not radioactive. Phagosomes* (upper arrows) *in the pigment epithelium are not radioactive.*

Figure 9.4 *Fifty days after injection, radioactive discs are close to the outer segment tips. A single nonradioactive phagosome* (arrow) *is visible.*

Figure 9.5 *Fifty-nine days after injection. One of the phagosomes in the field is radioactive* (arrow).

Figure 9.6 *Sixty-one days after injection. As more radioactive fragments are shed from the rod outer segments, the population of radioactive phagosomes increases. Those that have been recently engulfed* (arrows) *retain the "banding" pattern observed initially in the outer segment.*

Figure 9.7 *Sixty-one days after injection. The three outer segments in the center of the field have shed their radioactive discs. The resulting phagosomes* (arrows) *are visible in the cytoplasm of the pigment epithelium.*

Figure 9.8 *Sixty-eight days after injection. As the most intensely radioactive phagosomes are digested, they are replaced by phagosomes with lower levels of radioactivity* (arrows).

they are binucleate (Young 1966; Ishikawa and Yamada 1970), and two nuclei can support a larger volume of cytoplasm. In the Long-Evans rat, each epithelial cell is in contact with some 250 to 300 rods (Young 1966). This represents an astounding daily consumption of 25,000 to 30,000 rod outer segment discs!

In frogs raised at 22.5°C, discs are formed at the rate of 36 per day in red rods, and there are up to five red rods per pigment epithelial cell, representing 180 daily discs. The production of discs is approximately doubled when the temperature is raised by 10°C (Young 1967) so that in frogs raised at a mammalian body temperature of 37.5°C, the red rods would produce about 120 discs per day, providing each pigment epithelial cell (associated with five red rods) with 600 discs to destroy. The diameter of these discs is large, averaging about 7μ, which represents an estimated area of 23,100 μ^2 of double membrane digested daily.

In the frog, there are two types of rods, and they produce new outer segment discs at different rates (Young 1967). Red rods produce (and therefore shed) discs nearly 50 percent faster than green rods (36 vs. 25 per day at 22.5°C: Young and Droz 1968). Because red and green rods exist side by side and may project into the same pigment epithelial cell, it is apparent that a delicately balanced system controls disc production and disposal among these retinal cells.

The quantitative estimates cited above, while only simplified estimates (see next section on light and circadian effects), confirm the impression that a very substantial amount of intracellular digestion of membranes is performed at all times in the normal pigment epithelium. Nevertheless, the phagocytic *capacity* in these cells is much greater than that which is normally operative. Kuwabara and Gorn (1968), Kuwabara (1970), and Grignolo et al. (1969) all observed a massive increase in the number of phagosomes in the pigment epithelium of rats suffering outer segment damage from intense visible light. Marshall (1970) destroyed outer segments with a ruby laser in rabbits, in which each pigment epithelial cell normally contains from 20 to 30 phagosomes. In the laser-irradiated tissue, 250 to 400 of these inclusions were present, in many cases occupying over 70 percent of the cell's volume. Clearly, the pigment epithelium retains a very considerable phagocytic reserve which is available for emergencies.

Effects of Light and Circadian Factors on Rod Outer Segment Disc Shedding

Recent evidence has demonstrated that delivery of the phagocytic burden to the pigment epithelium, rather than being uniform throughout the day, proceeds according to a daily rhythm that is synchronized by light. LaVail (1976a) recently reported that rats maintained on a lighting cycle of 12 hours light and 12 hours darkness shed their rod discs shortly after the lights were turned on. However, this synchronized shedding process occurred on schedule over a period of 3 days in the absence of a light stimulus, indicating that the phenomenon might follow a light-entrained circadian rhythm (LaVail 1976b). Frogs (Basinger, Hoffman, and Matthes 1976; and Hollyfield, Besharse, and Rayborn 1976), goldfish (O'Day and Young 1977) and chickens (Young 1977b) also shed their rod discs in response to a light stimulus. It was initially claimed that disc shedding in frogs was totally dependent on light stimulation and did not follow a circadian rhythm (Basinger et al. 1976). However, Besharse, Hollyfield, and Rayborn (1977b) have suggested that rod disc

Digestive vacuoles occur in a variety of different sizes and forms, depending on the size and nature of the digested material and the state of digestion of the contents. They often exhibit internal structure such as granular matrix, membrane fragments, and myelin figures. When the digestive process is complete, if indigestible residues remain, the vacuole and its contents are termed a *residual body*. These generally appear as irregular, dense masses of variable size, often containing lipid and laminated whorls. Protozoa expel such indigestible residues, but egestion (exocytosis) is said to occur rarely, if at all, in the cells of higher animals (Vernon-Roberts 1972; Hirsch 1974; DeDuve and Wattiaux 1966; Steinman and Cohn 1974).

Figure 9.11 *Electron-microscopic autoradiogram of a normal albino rat retina that received an injection of radioactively labeled outer segment debris in the extracellular space between pigment epithelium and photoreceptors. The rat was sacrificed 12 hours after injection. The pigment epithelium* (pe) *readily phagocytized nonradioactive outer segment fragments from neighboring damaged photoreceptors, as evidenced by the presence of nonradioactive phagosomes* (ph). *It rarely ingested the injected radioactive debris* (rd). *(After Custer and Bok 1975. ×13,900.)*

Figure 9.12 *Electron-microscopic autoradiogram of normal albino rat pigment epithelium* (pe) *with a history identical to that shown in Figure 9.11. Exogenous radioactive outer segment debris* (rd) *remains in the extracellular space. ×13,900.)*

Mechanisms of Phagocytosis in Pigment Epithelial Cells

RECOGNITION AND ATTACHMENT. As we have seen, one general characteristic of phagocytosis is selectivity, determined by recognition sites on the outer surface of the cell membrane. The retinal pigment epithelial cells manifest such discrimination.

In the frog, during embryogenesis, the developing pigment epithelium scavenges egg pigment expelled from the apical ends of the retinal neuroepithelial cells (Hollyfield 1972), and it engulfs polystyrene beads at all ages (Hollyfield and Ward 1974a). The intact rat pigment epithelium will phagocytize India ink particles (Custer and Bok 1975; Reich-D'Almeida and Hockley 1975a), intact monkey pigment epithelium will ingest polystyrene spheres (Funahashi, Okisaka, and Kuwabara 1976) and partially detached rabbit pigment epithelium will engulf trypan blue (Karli 1954a). The human pigment epithelium will ingest particles of thorotrast (Tso, Albert, and Zimmerman 1973) and polystyrene spheres (Feeney and Mixon 1976). In all of these cases, the pigment epithelium manifests its general phagocytic character, because other phagocytic cells and even some nonphagocytic ones also commonly ingest particles with nonspecific recognition sites, such as gold, carbon, polystyrene beads, and vital dyes such as trypan blue.

On the other hand, frog pigment epithelium refuses to phagocytize pasteurized and native micrococcus, although other phagocytic cells in the frog will ingest these organisms (Hollyfield and Ward 1974b; Hollyfield 1976). Apparently, the appropriate recognition sites are not present on the pigment epithelial cell outer membrane. This is dramatically illustrated when micrococci and polystyrene spheres are provided simultaneously to the pigment epithelium (Hollyfield 1976). Almost without exception the polystyrene is engulfed while the micrococci are excluded. This is in keeping with the observations of Griffin and Silverstein on mouse macrophages (1974) in which it was noted that recognition and ingestion of one particle does not induce general phagocytosis of unrelated particles in the general vicinity. To put it in popular terms, a phagocytic cell cannot, by virtue of a "feeding frenzy," ingest everything in its path.

Even fragments of rod outer segments may not be recognized under certain circumstances. When such fragments are isolated by sucrose flotation without precautions to preserve their integrity and then presented to normal, intact pigment epithelium, they are frequently not phagocytized (Figs. 9.11, 9.12), although the same pigment epithelium voraciously engulfs fragments of its native retinal outer segments that are damaged by the injection procedure (Custer and Bok 1975). This is, no doubt, due to the fact that the plasma membrane is stripped away, thereby removing complementary recognition sites that are present on the plasma membrane of the outer segment but lacking on the disc membranes themselves. When outer segments are isolated by gentler methods and fed to normal, cultured rat pigment epithelium, they are readily phagocytized (Edwards and Szamier 1977).

Whether or not the pigment epithelium can phagocytize other types of retinal debris deserves study. Marshall (1970) states that no increase in

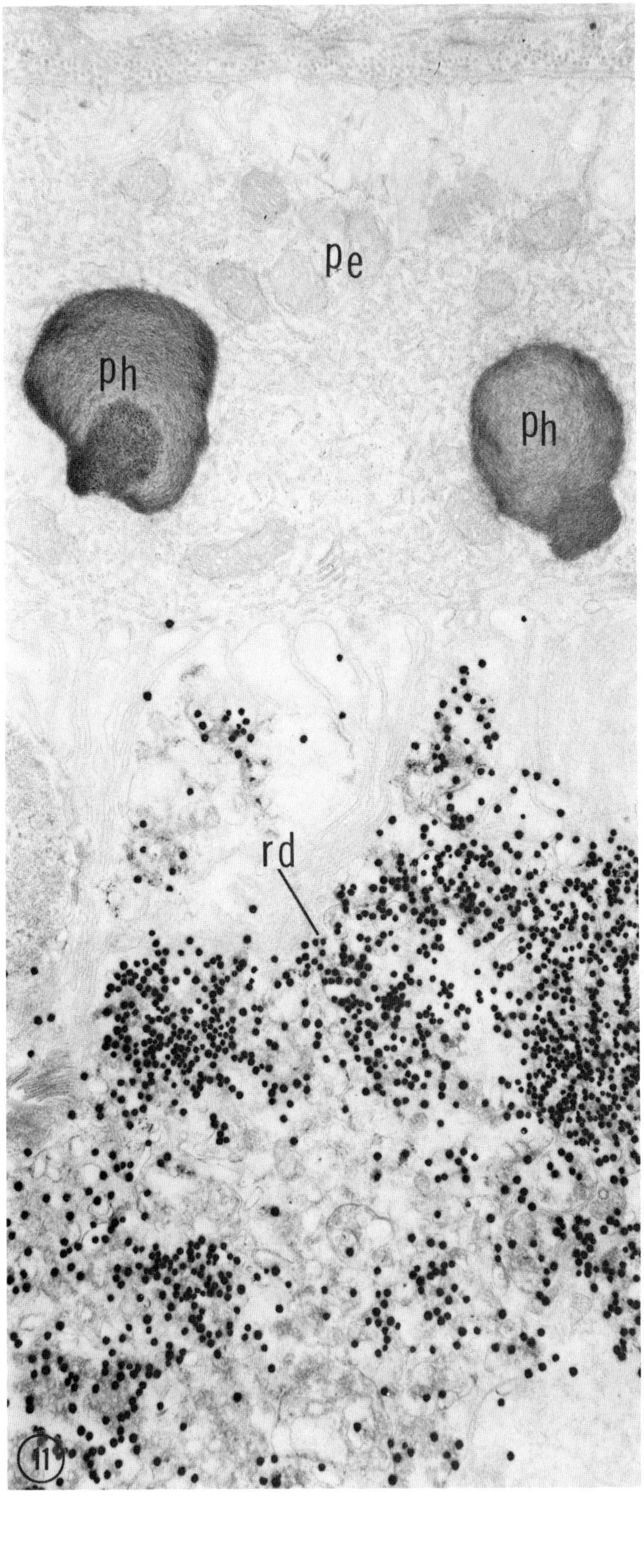
pe
ph
ph
rd
11

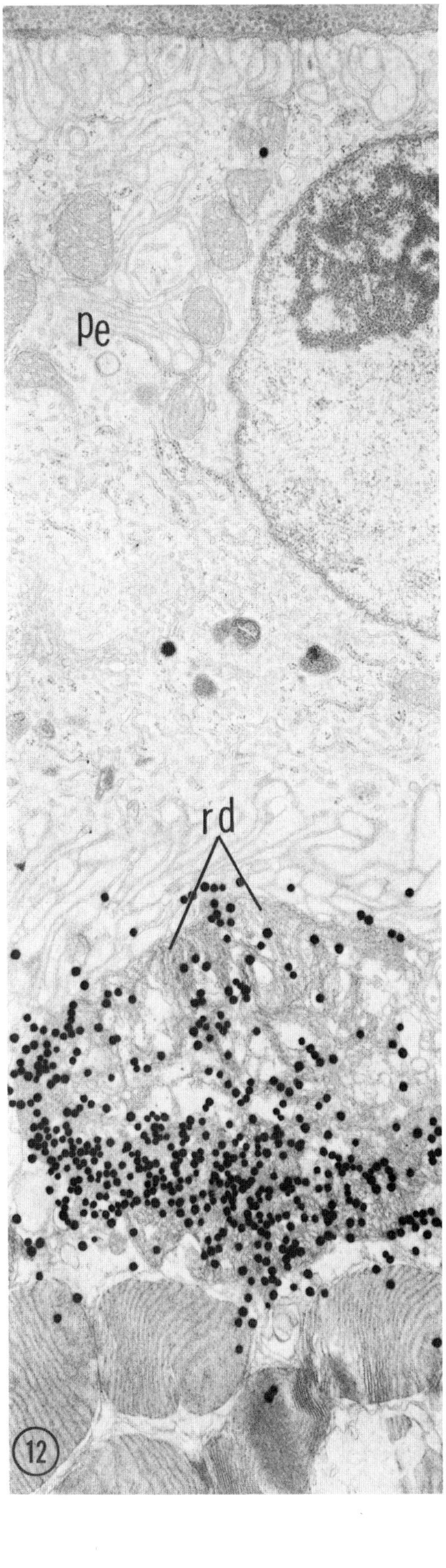
pe
rd
12

phagosomes is observed when the inner retina is damaged under conditions that spare the receptor outer segments, but in this case the epithelial cells may not have been directly exposed to the debris. The phagocytes which appear following severe retinal trauma are commonly considered to be derived from the blood (Gloor 1974).

The attachment phase of phagocytosis, which is believed to trigger ingestion, has not been studied in the pigment epithelium. However, Ishikawa and Yamada (1970) found that when they made flat preparations of the pigment epithelium in the rat, fragments of rod outer segments adhered more firmly to the pigment epithelium than did intact outer segments. This might be an indication that the first stage of rod outer segment phagocytosis involves an intimate attachment of the particulate material to recognition sites on the cell membrane.

INGESTION. Young and Bok (1969) were the first to note that disc formation in rods is continuous, and that detachment of old discs is intermittent and occurs in groups of variable size. A given packet of ejected discs may contain the product of a few hours (Young 1971b) or more than a day's continuous disc production (Young and Bok 1969). This observation is strengthened by recent reports that only about one quarter of frog rods undergo rhythmic shedding when the light and dark periods are about equally spaced (Basinger et al. 1976; Besharse et al. 1977b; Hollyfield et al. 1977).

In the human, between 30 or 40 discs are shed at one time, according to Spitznas and Hogan (1970), although Bairati and Orzalesi's pictures (1963) show that ten or fewer discs may be shed (cf. Young 1969, Fig. 89). In the rhesus monkey, the rods commonly shed discs in groups of eight to thirty, although smaller and larger clusters can be observed (Young 1971b). Although the disc assembly and shedding rates can be readily altered in heterothermic animals like the frog by manipulation of light (Besharse et al. 1977a, 1977b) and temperature (Hollyfield et al. 1977), there does appear to be an upper limit to phagosome size. This, according to Hollyfield et al. (1977) rarely exceeds 150 discs. In frogs, there is a population of small phagosomes (5–30 discs each) that is derived from a low-grade shedding of rod discs which is independent of the lighting conditions but increased at elevated temperatures (Hollyfield, Besharse, and Rayborn 1976, 1977). This phagosome population accounts for about 20 percent of the outer segment disc production.

On the basis of electron microscopic observations, two different mechanisms have been proposed for the detachment and ingestion of rod outer segment discs:

Spitznas and Hogan (1970) interpreted their electron microscopic analysis of the human retina to indicate that the epithelium "nibbles away" at the rod outer segment tips. The "sheath" of apical cytoplasm that encloses the tip of the outer segment was described as forming "cytoplasmic protrusions which extend inward from all sides between 30 or 40 of the discs of the outer segment. The protrusions then fuse in the middle, separating the selected group of discs from the outer segment of the receptor." In other words, the apical processes were thought to constrict around the outer segment tips and "bite off" a packet of discs and its surrounding outer membrane.

Young's (1971b) examination of this process in the rhesus monkey led to a different conclusion. According to his observations, the discs are not pulled or bitten off. Instead, the rods themselves shed groups of discs, which are only secondarily engulfed by the pigment epithelium.

In the earliest recognizable stage of the disc-shedding process observed by Young, the edges of the discs at the end of the outer segment become bent or curled, with the convex side commonly facing the remaining discs in the outer segment (Fig. 9.13). (This was also observed by Bairati and Orzalesi in the human retina.) Next, the membrane that surrounds the rod outer segment begins to fold into the cytoplasm separating the group of deformed discs from the remainder of the stack (Figs. 9.14, 9.15). This probably occurs rapidly, because behind the infolded membrane a gap is momentarily created that appears empty of the material filling the rest of the extracellular space. The cytoplasmic extensions of the pigment epithelium show no involvement in these changes. The progressive infolding of the outer membrane effectively dissects the group of discs from the top of the stack and seals off the end of the outer segment (Figs. 9.14–9.16). The infolded membrane fuses with itself on the other side, so that the cytoplasm is not exposed to the extracellular space, and the isolated discs are not only fully detached but also completely surrounded by plasma membrane (Fig. 9.16).

According to Young's (1971b) description, "as soon

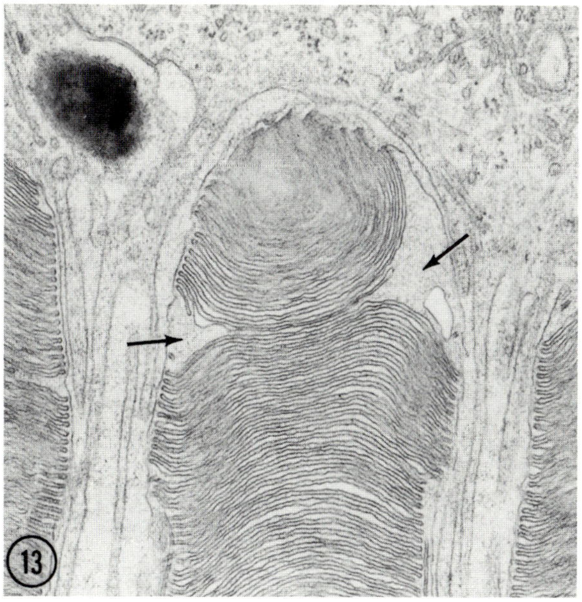

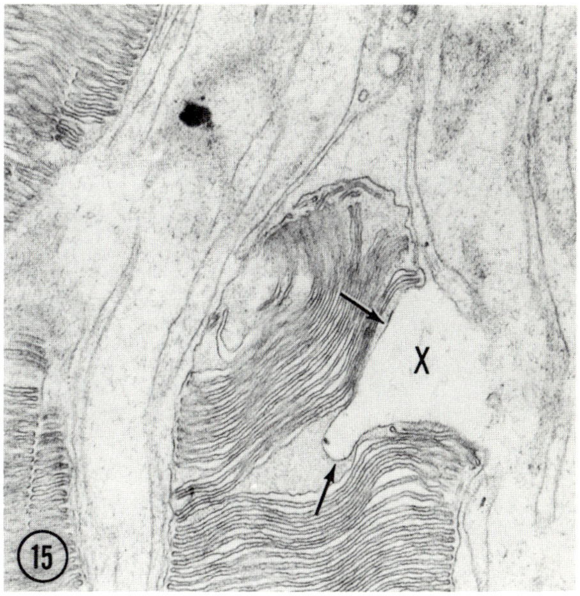

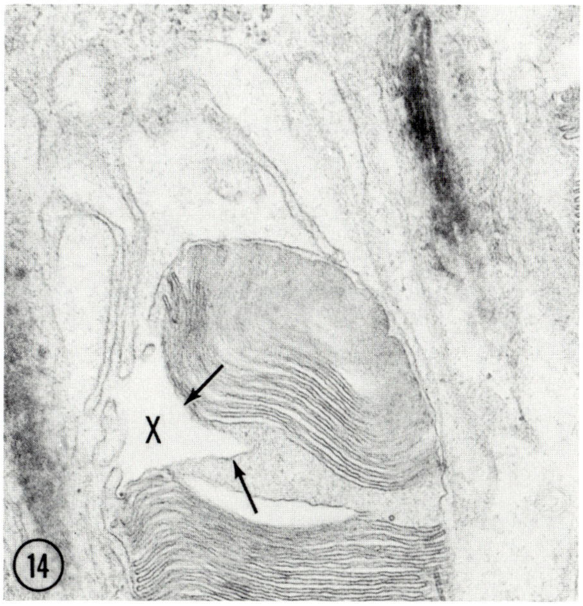

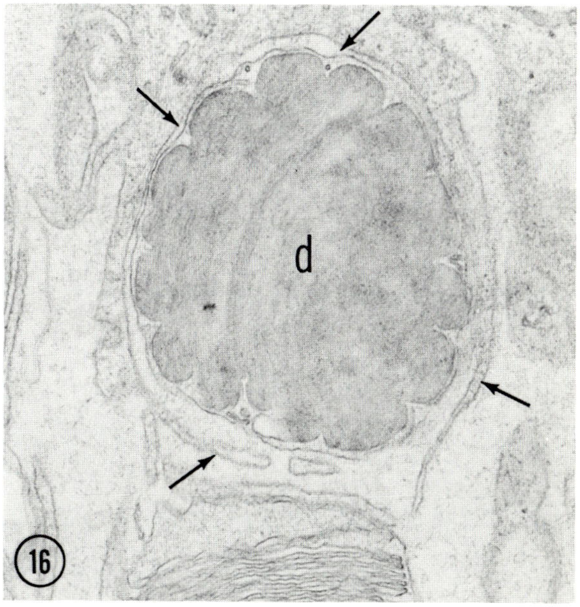

Figures 9.13–9.16 *Electron micrographs of rod outer segment disc shedding and phagocytosis in the rhesus monkey. (Figures 9.14–9.16 after Young 1971. ×32,570.)*

Figure 9.13 *In the earliest stage of shedding, the edges of the discs at the outer segment tip frequently become bent or curled. Cytoplasm* (arrows) *is displaced into the region vacated by the discs.*

Figure 9.14 *After a stack of discs becomes deformed and separated from those that are to remain behind, the outer segment plasma membrane* (arrows) *folds into the cytoplasm separating the two groups of discs. A space* (x), *temporarily lacking in extracellular matrix, is created in the wake of the invaginating membrane.*

Figure 9.15 *As disc shedding progresses, the invagination of the plasma membrane* (arrows) *increases in depth. The gap* (x) *in the extracellular matrix remains.*

Figure 9.16 *A group of discs* (d) *has detached completely from the rod outer segment and has turned sideways so that the stack is cut in cross section. Progressive infolding and fusion of the outer segment plasma membrane has sealed the end of the outer segment. For the first time, microvilli from the pigment epithelium* (lower arrows) *surround and begin to enclose the rejected packet of discs. Part of the outer segment plasma membrane* (upper arrows) *is shed with the discs.*

as the discs are shed into the extracellular space (or perhaps slightly before they are fully detached) the adjacent cytoplasmic extensions of the pigment epithelium begin to surround them. Slender, watery fingers of cytoplasm coming from opposite sides of the detached discs meet and fuse, thereby enclosing the ejected packet of outer segment membranes in a cytoplasmic phagosome, close to the tip of the rod from which they were shed [Fig. 9.17]. Next, the phagosome is withdrawn into the main body of the pigment epithelial cell [Fig. 9.18]."

The differences between these two concepts are

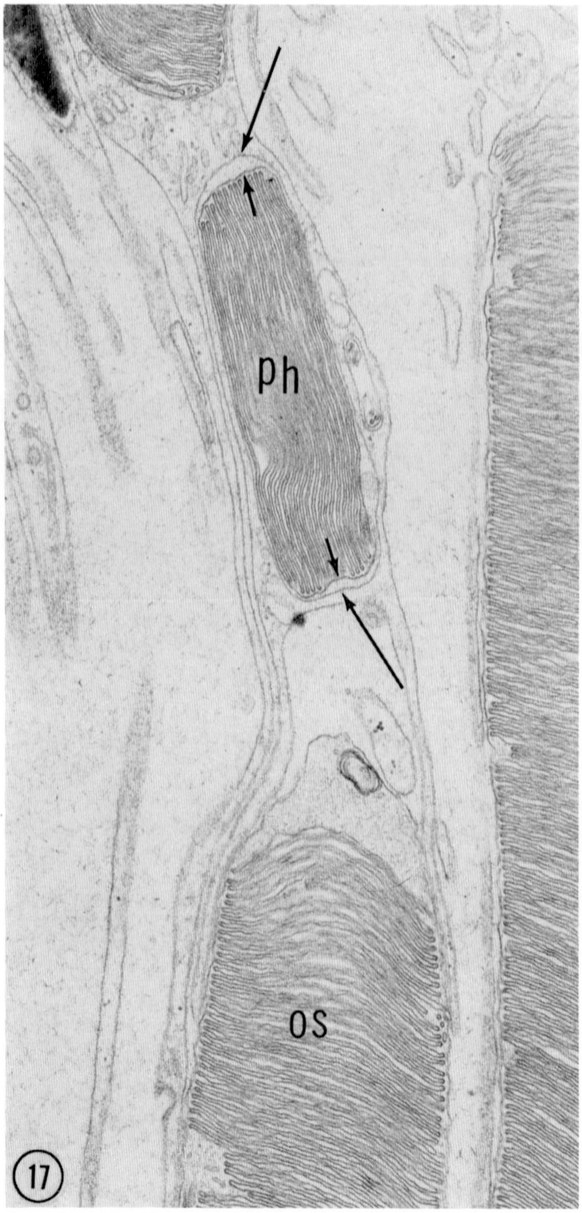

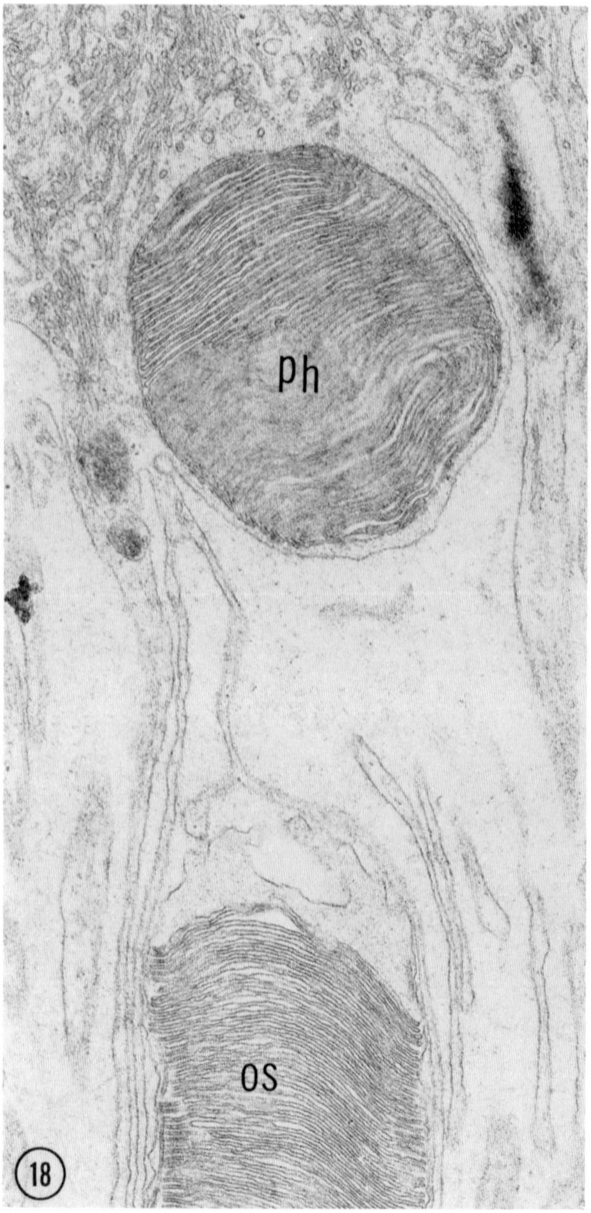

Figures 9.17–9.18 *Electron micrographs of rod outer segment disc shedding and phagocytosis in the rhesus monkey. (After Young 1971. ×27,430.)*

Figure 9.17 *An outer segment fragment has been shed and is now present as a phagosome* (ph) *within the cytoplasm of the pigment epithelium. The rejected plasma membrane of the outer segment* (short arrows) *and the limiting membrane of the phagosome* (long arrows) *are both visible.*

Figure 9.18 *The newly formed phagosome* (ph) *is drawn into the cell body of the pigment epithelium. The outer segment* (os) *from which it is derived is also visible.*

not trivial. In one, the pigment epithelium is the active agent in removing the discs. In the other, the epithelial cells scavenge discs actively shed from the rods.

Although the resolution of this controversy must depend upon further research, we tend to favor the latter alternative for several reasons. First of all, there is, to our knowledge, no other example in any other cell system of the process postulated by Spitznas and Hogan: that is, the "biting off" of part of a living cell by a phagocyte. Second, the detachment of discs by infolding of the outer membrane accounts for the wrapping of the shed discs in plasma membrane, and, more importantly, explains how the end of the cell remains sealed off during the process. Patency of the outer membrane over the tip of the rod outer segment is not accounted for by a mechanism in which the end of the rod is pulled off by the pigment epithelium. Also, infolding of the plasma membrane is a known feature of the rod visual cells (it represents the mechanism by which the discs are formed). In the parietal eyes of lizards, and even more prominently in the pineal photoreceptors of frogs and lampreys, sloughed tips of outer segments have been observed (Eakin 1973). There are no cells in contact with these outer segments. However, there are free macrophages in the lumen of the eye containing stacks of membranes which are almost certainly outer segment fragments (Petit 1968; Bunt and Kelly 1971). In this system, the discs appear to be exfoliated first, then phagocytized. Additionally, we have unpublished evidence of what appears to be disc shedding from frog rods incubated in vitro without the presence of the pigment epithelium. Finally, we have observed numerous examples of groups of discs in various stages of detachment in the rhesus monkey without any apparent involvement of the apical processes of the pigment epithelium (Figs. 9.13–9.15).

What perturbations initiate disc detachment? Perhaps lengthening of the outer segment due to repeated disc assembly introduces a progressive instability into the system. The gradually increasing separation between the tip of the outer segment and the metabolic centers in the inner segment, coupled with the microenvironmental effects of proximity to the pigment epithelium, might combine until a critical event such as light stimulation or light deprivation activates the disc shedding mechanism. Alterations in certain humoral titers that follow circadian patterns might also contribute to this phenomenon. If the pigment epithelium is the active agent, possibly it is the increasing pressure of the lengthening outer segment that initiates the process. Are there age changes in the membrane analogous to those that enable macrophages to identify erythrocytes as effete? Or are terminal glycoprotein sugar residues such as galactose or fucose added to visual pigment in the photoreceptor plasma membrane as recognition sites that allow the pigment epithelium to initiate phagocytosis (O'Brien 1976)? These are among the numerous questions that remain to be answered.

All observers are agreed that when the packet of discs is first drawn into the pigment epithelial cell cytoplasm it is enclosed by two membranes (Ishikawa and Yamada 1970; Spitznas and Hogan 1970; Young 1971b; Bairati and Orzalesi 1963). The inner membrane is the rod plasma membrane, and the outer membrane is a portion of the epithelial cell plasma membrane that was pinched off when the phagosome was internalized (Fig. 9.17). Evidently, during ingestion, the pigment epithelium plasma membrane adheres to the rod outer membrane, but does not fuse with it. It does, however, fuse with itself, just as the rod outer membrane readily fuses with itself (during rod disc formation, shedding, and in vesicle formation when the membranes are damaged). The selective fusion of membranes is as important to the phagocytic process as is the selective recognition and attachment of membranes.

DIGESTION. Digestion of the contents of the phagocytic vacuole probably begins very soon after the membranes are ingested. Phagosomes that retain the initial double outer membrane, just described, are rare, and when present are located in the apical cytoplasm, suggesting that the rod plasma membrane is one of the first structures to be destroyed. As digestion progresses, the contents of the vacuole bear less and less resemblance to disc membranes, although layered myelin figures are common, even in residual bodies. Characteristic changes include: deformation and compaction of the discs; rolling up of the discs to yield a concentric, layered appearance; disorganization of the membranes with homogenization of the contents or the formation of reticulated or vesicular structures; and, appearance of a granular matrix and dense cores (Fig. 9.2). The smaller, more disorganized masses tend to be located nearer to Bruch's membrane, indicating a progressive movement of the phagosome toward the basal aspect of the cell as digestion proceeds. Several authors have shown the variety of appearance of

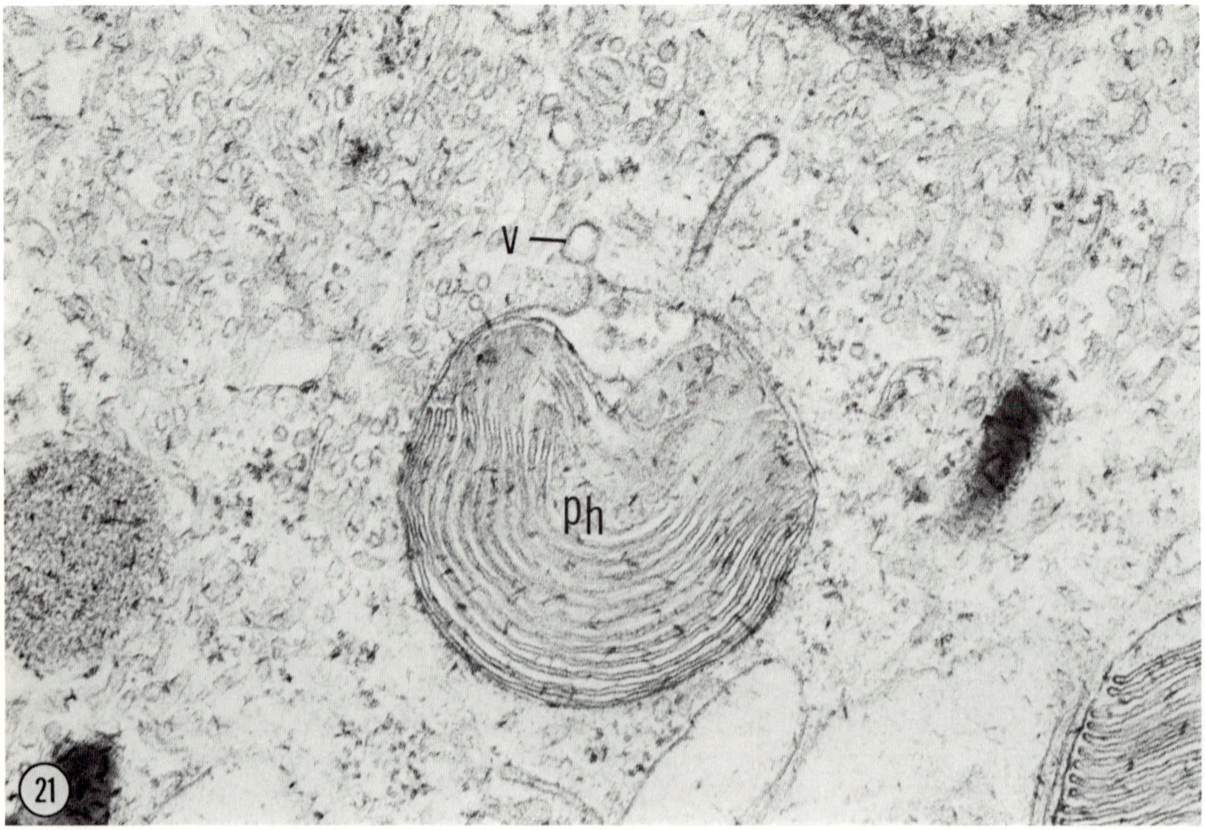

Figure 9.21 *Electron micrograph of phagosome* (ph) *in the rhesus monkey pigment epithelium. A small vesicle* (v) *has fused with the limiting membrane of the phagosome suggesting, perhaps, the release of hydrolytic substances into that structure.* (×*54,000.*)

does occur in the pigment epithelium, and that this process may be associated with aging, which will now be considered.

Age Changes

The pigment epithelium acquires phagocytic powers early during retinal development, at least in the frog, where it phagocytizes melanin extruded from the differentiating neuroepithelium and will ingest polystyrene spheres (Hollyfield 1972; Hollyfield and Ward 1974a).

In the rat and mouse, phagocytosis of fragments of rod outer segments, signaled by the appearance of phagosomes in the pigment epithelium, begins shortly after these structures begin to differentiate, and well before they reach their adult proportions (Dowling and Gibbons 1962; Ishikawa and Yamada 1970; Braekevelt and Hollenberg 1970; Bok and Hall 1971). This appears to be the case in the human retina as well (Hollenberg and Spira 1973).

LaVail (1973) has clearly demonstrated in the mouse that the detachment and phagocytosis of discs is relatively slow during the early growth period (7–13 days), when the rate of rod outer segment disc production is higher than it is in the adult. The ingestion of disc membranes by the pigment epithelium increases to 70 percent of the adult level by 13–15 days of age and then gradually rises until a steady-state is achieved between the rates of disc membrane assembly and disposal, thereby terminating net outer segment growth.

Membranes containing a high proportion of polyunsaturated fatty acids—such as those of rod outer segments—are particularly susceptible to peroxidation. This renders them resistant to degradation (Tappel 1970), so that instead of the contents of phagocytic vacuoles being completely digested to small molecules that diffuse into the cytoplasm, undigested residues may remain as residual bodies. In the pigment epithelium, as in many other types of cells, residual bodies gradually accumulate as "lipofuscin granules."

Streeten (1961) observed by means of cytochemical methods that in the human pigment epithelium lipofuscin granules are not present in the fetus, but begin to appear about 16 months after birth and continue to increase in number thereafter. These age changes were subsequently confirmed by electron microscopy (Feeney et al. 1965; Kolb and Gouras 1974). Leuenberger (1975) is of the opinion that there is a concomitant decrease in acid phosphatase levels in the aging rat pigment epithelium.

Hogan (1972) has described the gradual accumulation of vesicles, residual bodies, partially digested phagosomes, lipid crystals, and mucoprotein in Bruch's membrane of the human retina; he attributes this accumulation largely to extrusion from the pigment epithelium. Apparently, the pigment epithelium attempts to rid itself of its increasing burden of residual bodies and other detritus by exocytosis from the basal surface. This process, which is rare or negligible in short-lived phagocytic cells, may be an adaptation of the pigment epithelium to a long life span.

A number of authors (Hogan and Alvarado 1967; Farkas, Sylvester, and Archer 1971) have reached the conclusion that congestion of Bruch's membrane by residual bodies and other extrusions from the pigment epithelium contributes to the formation of drusen. However, drusen vary in their clinical presentation and there may be several varieties with different pathological origins. Even with regard to a specific type of drusen, this concept, although intriguing, will require more direct experimental evidence before it can be confirmed.

Clinical-Pathological Implications

Any biological process, including the renewal of rod outer segment membranes and their phagocytosis by the pigment epithelium, has implications for clinical medicine, because each process inherently contains the potential for derangement, resulting in a pathological condition. With regard to phagocytosis alone, the possibilities for disturbance are apparent in recognition-attachment, ingestion, and digestion. One may well imagine, for example, that the progressive accumulation of undigested residues might underlie some of the retinopathies of old age (Hogan 1972; Feeney 1973).

Some evidence has already been obtained of derangement of the phagocytic mechanism in pigment epithelial cells. A number of investigators have now shown that the pigment epithelium in a strain of rats that has the autosomal recessive gene for retinal dystrophy does not phagocytize outer segment discs (Bok and Hall 1969, 1971; Herron et al. 1969; LaVail et al. 1972). This is the same strain of rats in which Dowling and Sidman (1962) had earlier demonstrated the accumulation of rhodopsin-rich lamellar debris at the pigment epithelium apical surface. Why does the pigment epithelium fail to phagocytize outer segment debris in these animals? Is it because the pigment epithelium is defective, or is there something aberrant in the photoreceptor membranes that causes them to be ignored by the pigment epithelium? Herron et al. (1969) termed this "a pigment epithelial disease," although the necessary experimental evidence for this conclusion was lacking at that time. Custer and Bok (1975) and Reich-D'Almeida and Hockley (1975b and 1976) injected India ink particles between the neural retina and pigment epithelium, and showed that in the dystrophic rat, the retinal pigment epithelium will engulf particles of ink with vigor (Figs. 9.22, 9.23). The resulting phagosomes undergo normal fusion with acid-phosphatase-positive lysosomes (Reich-D'Almeida and Hockley 1976). Therefore, the ingestion and digestion phases of phagocytosis appear to be functional in these mutant rats. The most satisfying alternative, in our opinion, is that there is a failure in recognition and attachment between the plasma membrane of the photoreceptor and the pigment epithelium. The best evidence for this was recently presented by Mullen and LaVail (1976). These investigators fused the blastulae from pink-eyed RCS animals with blastulae from a normal, pigmented strain of rats. In the resulting tetraparental offspring, rod visual cells with associated whorls of degenerating outer segment membranes were always observed adjacent to patches of albinotic, mutant pigment epithelium. In contrast, the visual cells appeared normal when they were in contact with melanotic pigment epithelium from the normal animals (see Chapter 20 for further details).

Some investigators have suggested that the pigment epithelial lysosomes can become leaky under certain conditions. This has been postulated as the mechanism responsible for retinal dystrophy in rats (Burden et al. 1971). These authors suggest that the enveloping membranes of primary lysosomes in the pigment epithelium of these animals are defective. As a result, hydrolytic enzymes leak into the extracellular space. This allegedly leads to the observed

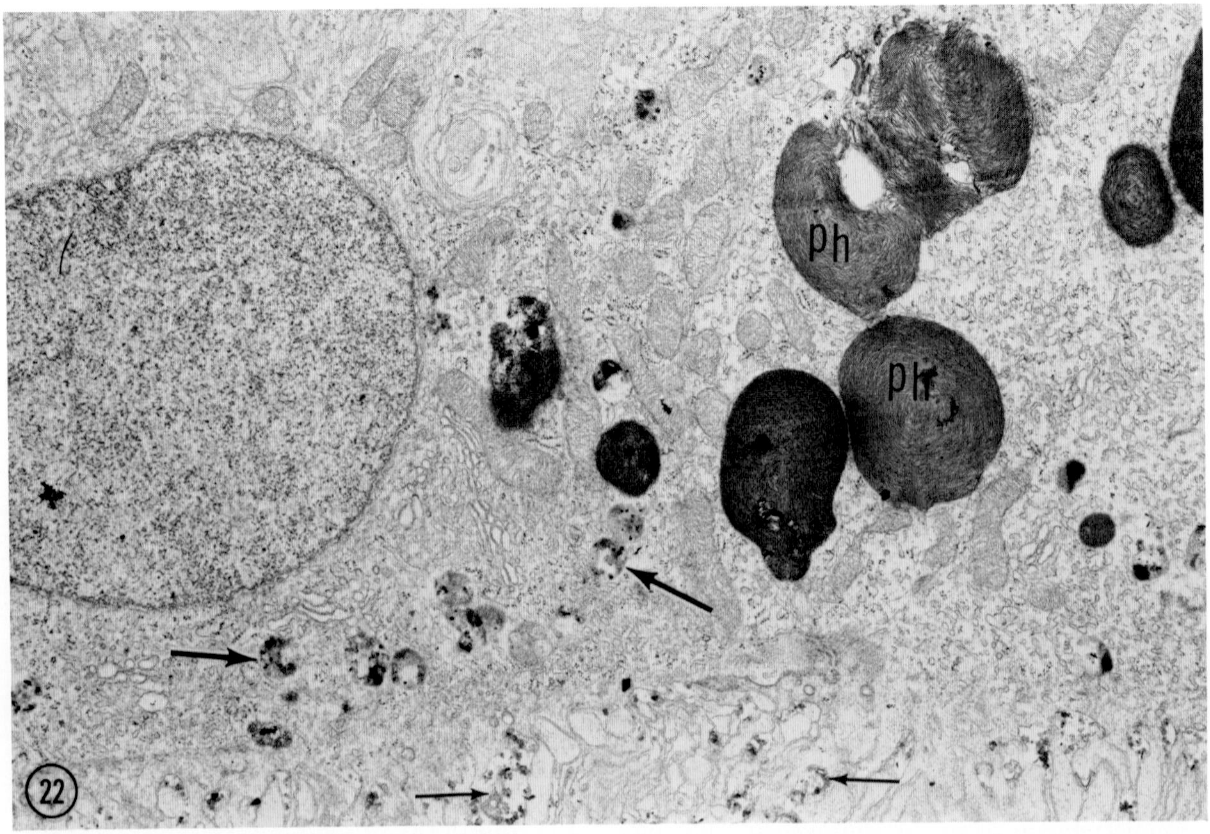
ph
ph
22

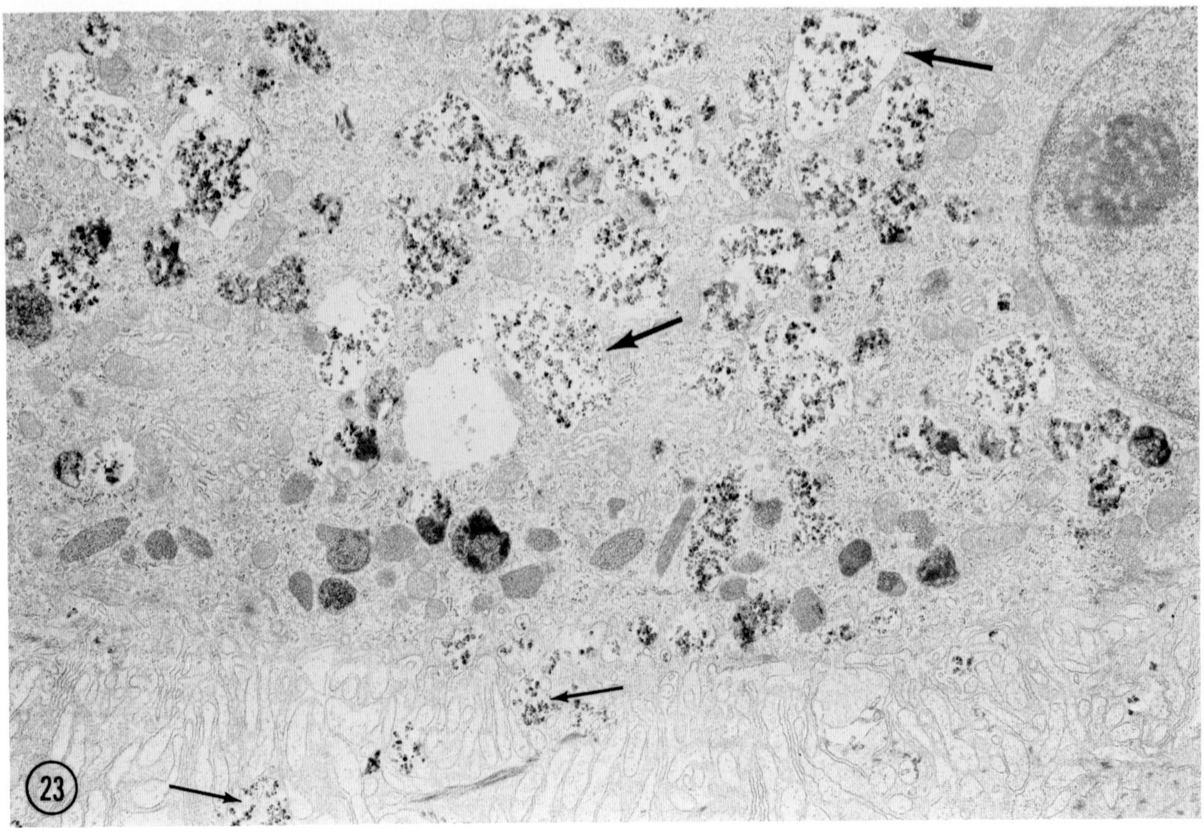
23

Figure 9.22 *Electron micrograph of normal albino rat pigment epithelium 12 hours after an injection of carbon particles into the extracellular space between pigment epithelium and photoreceptors. Carbon is observed in the extracellular space between microvilli of the pigment epithelium* (small arrows) *and in phagosomes in the cytoplasm* (large arrows). *Phagosomes derived from outer segments* (ph) *are also visible. (×10,530.)*

Figure 9.23 *Electron micrograph of pigment epithelium from adult dystrophic rat retina 12 hours after an injection of carbon particles into the extracellular space. Carbon is observed between microvilli* (small arrows) *and in phagosomes in the cytoplasm* (large arrows). *The ingestion phase of phagocytosis is functional in the pigment epithelium of these mutant rats. However, the pigment epithelium does not appear to recognize debris from adjacent photoreceptors. (After Custer and Bok 1975. ×14,040.)*

disruption of rod outer segments. If this is the case, it is difficult to understand why the pigment epithelium is not also destroyed, since the enzymes would first have to traverse the cytoplasm. Vento and Cacciopo (1973) lent support to this idea when they showed that retinol can damage lysosomes obtained from the retina and pigment epithelium of calf eyes, and suggested that this might happen in the rats affected with retinal dystrophy. Ansell and Marshall (1974) have recently presented cytochemical evidence to support the theory of Burden et al. (1971) that hydrolytic enzymes are released into the extracellular space in the dystrophic rat retina. They observed acid phosphatase activity in the extracellular debris, but no activity was found either in lysosomes or free within the cytoplasm of the pigment epithelium. This is in contrast to recent evidence obtained by Essner, Gorin, and Griewski (1977) and evidence from one of our laboratories (Bok). These investigators have demonstrated abundant acid-phosphatase- and arylsulfatase-positive lysosomes in the pigment epithelium of the dystrophic rat retina, with no evidence for leakage from these structures, nor was there any evidence for enzyme activity in the extracellular debris.

These dystrophic rats, in which the phagocytosis mechanism fails, have often been considered to be a model for the human disease retinitis pigmentosa. However, a recent examination of the eyes of a patient with autosomally dominant retinitis pigmentosa (Kolb and Gouras 1974) indicates important differences, at least in this form of the disease. Except for the foveal cones, the visual cells were gone. In the foveal pigment epithelium there were enormous quantities of lipofuscin, strong evidence of an intact mechanism for ingesting photoreceptor membranes.

It can be anticipated that genetic defects that result in the deletion of one or more of the lysosomal enzymes will also affect the pigment epithelium. The beige mouse, a mutant of the C57 strain, is supposedly a model for the Chediak-Higashi syndrome in humans. Individuals with this syndrome have impaired intracellular digestion in leukocytes and macrophages. Robison et al. (1975) have observed unusually large acid-phosphatase-positive melanin granules in the pigment epithelia of these mice, suggesting the possibility of abnormal intracellular digestion.

Exceptionally large accumulations of undigested residues have been recorded in the pigment epithelium of monkeys that were maintained on a diet deficient in vitamin E (Hayes 1974). Because vitamin E (alpha tocopherol) is an antioxidant, it appears that peroxidation of the rod outer segment membranes rendered them indigestible by the cells' hydrolytic enzymes.

Conclusion

Although awareness of the phagocytic properties of the retinal pigment epithelium is very recent, it is already apparent that these cells manifest all of the classical properties of phagocytic cells, including discrimination between particles of different types, ingestion of particulate matter into vacuoles bounded by plasma membrane, production of a large variety of hydrolytic enzymes, a system for packaging these enzymes and delivering them to the phagocytic vacuoles, and the capacity to digest completely the engulfed particles under most circumstances, or, when required, to store undigestible material in residual bodies.

Previous research on phagocytosis has largely concerned neutrophilic leukocytes and macrophages. These are sometimes considered the "professional phagocytes" of the body, because phagocytosis is their primary, perhaps even their sole function. As the other chapters in this volume abundantly document, phagocytosis is only one of the several important functions of the pigment epithelium.

However, the vast majority of the "professionals" are never called upon to manifest their potentialities during their short life span and, when induced to display them, may exhaust their powers in a single phagocytic ordeal. In striking contrast, the pigment epithelial cells of the retina begin to perform their phagocytic role before embryological development has been completed. They manifest a rate of intake and digestion of particulate material that under normal conditions is enormous, yet can be accelerated to nearly incredible levels in emergencies. They carry out these activities continuously, without ceasing during a lifetime that, in man, may extend over many decades. Because of these properties, we suggest that the retinal pigment epithelial cells are the most highly developed phagocytes in the body.

REFERENCES

Abraham, R., Hume, M., and Smith, J. 1969. A histochemical study of the lysosomal enzymes in the retina of the rat. *Histochemie* 18:195–201.

Anderson, D. H., and Fisher, S. K. 1975. Disc shedding in the rod-like and cone-like photoreceptors of tree squirrels. *Science* 187:953–955.

Anderson, D. H., and Fisher, S. K. 1976. The photoreceptors of diurnal squirrels: outer segment structure, disc shedding and protein renewal. *J. Ultrastruct. Res.* 55:119–141.

Ansell, P. L., and Marshall, J. 1974. The distribution of extra-cellular acid phosphatase in the retinas of retinitis pigmentosa rats. *Exp. Eye Res.* 19:273–279.

Bairati, A., and Orzalesi, N. 1963. The ultrastructure of the pigment epithelium and of the photoreceptor-pigment epithelium junction in the human retina. *J. Ultrastruct. Res.* 9:484–496.

Basinger, S., Hoffman, R., and Matthes, M. 1976. Photoreceptor shedding is initiated by light in the frog retina. *Science* 194:1074–1076.

Berman, E. R. 1971. Acid hydrolases of the retinal pigment epithelium. *Invest. Ophthalmol.* 10:64–68.

Bernstein, M. H., and Pease, D. C. 1959. Electron microscopy of the tapetum lucidum of the cat. *J. Biophys. Biochem. Cytol.* 5:35–53.

Besharse, J. C., Hollyfield, J. G., and Rayborn, M. E. 1977*a*. Photoreceptor outer segments: accelerated membrane renewal in rods after exposure to light. *Science* 196:536–538.

Besharse, J. C., Hollyfield, J. G., and Rayborn, M. E. 1977*b*. Turnover of rod photoreceptor outer segments. II. Membrane addition and loss in relationship to light. *J. Cell Biol.* 75:507–527.

Bok, D., and Hall, M. O. 1969. The etiology of retinal dystrophy in RCS rats. *Invest. Ophthalmol.* 8:649 (abstr.).

Bok, D., and Hall, M. O. 1971. The role of the pigment epithelium in the etiology of inherited retinal dystrophy in the rat. *J. Cell Biol.* 49:664–682.

Borovjagin, V. L., and Ivanina, T. A. 1973. An ultrastructual study of the frog retinal rod photoreceptor membranes phagocytized by pigment epithelium cells after aldehyde fixations and organic solvents treatments. *Vision Res.* 13:753–757.

Braekevelt, C. R., and Hollenberg, M. J. 1970. Development of the retinal pigment epithelium, choriocapillaris and Bruch's membrane in the albino rat. *Exp. Eye Res.* 9:124–131.

Bunt, A. H., and Kelly, D. E. 1971. Frog pineal photoreceptor renewal: preliminary observations. *Anat. Rec.* 171:99–116.

Burden, E. M., Yates, C. M., Reading, H. W., Bitensky, L., and Chayen, J. 1971. Investigation into the structural integrity of lysosomes in the normal and dystrophic rat retina. *Exp. Eye Res.* 12:159–165.

Carr, I. 1973. *The macrophage,* chap. 5. New York: Academic Press.

Cohen, A. I. 1960. The ultrastructure of the rods of the mouse retina. *Am. J. Anat.* 107:23–48.

Cohen, A. I. 1961. The fine structure of the extrafoveal receptors of the Rhesus monkey. *Exp. Eye Res.* 1:128–136.

Cohen, A. I. 1963. Vertebrate retinal cells and their organization. *Biol. Rev.* 38:427–459.

Cohn, Z. A. 1972. Properties of macrophages. In *Phagocytic mechanisms in health and disease,* ed. R. C. Williams and H. H. Fudenberg, pp. 39–49. New York: Intercontinental Medical Book Corp.

Custer, N. V., and Bok, D. 1975. Pigment epithelium-photoreceptor interactions in the normal and dystrophic rat retina. *Exp. Eye Res.* 21:153–166.

DeDuve, C. 1963. General properties of lysosomes: the lysosome concept. In *Ciba Foundation Symposium on lysosomes,* ed. A. V. S. de Reuck and M. P. Cameron, pp. 1–35. Boston: Little Brown.

DeDuve, C. 1964. From cytases to lysosomes. *Fed. Proc.* 23:1045–1049.

DeDuve, C., Pressman, B. C., Gianetto, R., Wattiaux, R., and Applemans, F. 1955. Tissue fraction studies, intracellular distribution patterns of enzymes in rat-liver tissue. *Biochem. J.* 60:604–617.

DeDuve, C., and Wattiaux, R. 1966. Functions of lysosomes. *Annu. Rev. Physiol.* 28:435–492.

Ditto, M. 1975. A difference between developing rods

and cones in the formation of outer segment membranes. *Vision Res.* 15:535–536.

Dowling, J. E., and Gibbons, I. R. 1961. The effect of vitamin A deficiency on the fine structure of the retina. In *The structure of the eye,* ed. G. K. Smelser, pp. 85–99. New York: Academic Press.

Dowling, J. E., and Gibbons, I. R. 1962. The fine structure of the pigment epithelium in the albino rat. *J. Cell Biol.* 14:459–474.

Dowling, J. E., and Sidman, R. L. 1962. Inherited retinal dystrophy in the rat. *J. Cell Biol.* 49:664–682.

Droz, B. 1963. Dynamic condition of proteins in the visual cells of rats and mice as shown by radioautography with labeled amino acids. *Anat. Rec.* 145:157–168.

Eakin, R. M. 1973. *The third eye,* chap. 3. Berkeley: Univ. of California Press.

Edwards, R. B., and Szamier, R. B. 1977. Defective phagocytosis of isolated rod outer segments by RCS rat retinal pigment epithelium. *Science* 197:1001–1003.

Eichner, D. 1955. Zur frage der Fermentlokalization in der Netzhaut des Rindes. *Z. Zellforsch. Mikrosk. Anat.* 41:493–508.

Eichner, D. 1958. Zur Histologie und Topochemie der Netzhaut des Menschen. *Z. Zellforsch. Mikrosk. Anat.* 48: 137–186.

Elsbach, P. 1974. Phagocytosis. In *The inflammatory process,* ed. B. W. Zweifach, L. Grant, and R. T. McCluskey, 2d ed., pp. 363–408. New York: Academic Press.

Essner, E., Gorin, M., and Griewski, R. A. 1978. Localization of lysosomal enzymes in retinal pigment epithelium of rats with inherited retinal dystrophy. *Invest. Ophthalmol. and Visual Science* 17:278–288.

Farkas, T. G., Sylvester, V., and Archer, D. 1971. The ultrastructure of drusen. *Am. J. Ophthalmol.* 71:1196–1205.

Feeney, L. 1973. The phagolysosomal system of the pigment epithelium: a key to retinal disease. *Invest. Ophthalmol.* 12:636–638.

Feeney, L., Grieshaber, J. A., and Hogan, M. J. 1965. Studies on human ocular pigment. In *The structure of the eye,* ed. J. W. Rohen, pp. 535–548. Stuttgart: F. K. Schattauer-Verlag.

Feeney, L., and Mixon, R. 1976. An *in vitro* model of phagocytosis in bovine and human retinal pigment epithelium. *Exp. Eye Res.* 22:533–548.

Funahashi, M., Okisaka, S., and Kuwabara, T. 1976. Phagocytosis by the monkey pigment epithelium. *Exp. Eye Res.* 23:217–225.

Gloor, B. P. 1974. On the question of the origin of macrophages in the retina and the vitreous following photocoagulation (autoradiographic investigations by means of ^{3}H-thymidine). *Graefes Arch. Klin. Ophthalmol.* 190:183–194.

Godfrey, A. J. 1973. The photoreceptors of the chick. Ph.D. dissertation, Dept. of Anatomy, Univ. of California, Los Angeles.

Griffin, F. M., and Silverstein, S. C. 1974. Segmental response of the macrophage plasma membrane to a phagocytic stimulus. *J. Exp. Med.,* 139:323–336.

Grignolo, A., Orzalesi, N., Castellazzo, R., and Vittone, P. 1969. Retinal damage by visible light in albino rats. *Ophthalmologica* 157:43–59.

Hayes, K. C. 1974. Retinal degeneration in monkeys induced by deficiencies of vitamin E or A. *Invest. Ophthalmol.* 13:499–510.

Herron, W. L., Riegel, B. W., Myers, O. E., and Rubin, M. L. 1969. Retinal dystrophy in the rat: a pigment epithelial disease. *Invest. Ophthalmol.* 8:595–604.

Hirsch, J. G. 1972. The digestive tract of phagocytic cells. In *Phagocytic mechanisms in health and disease,* ed. R. C. Williams and H. H. Fudenberg, pp. 23–35. New York: Intercontinental Medical Book Corp.

Hirsch, J. G. 1974. Neutrophil leukocytes. In *The inflammatory process,* ed. B. W. Zweifach, L. Grant, and R. T. McCluskey, 2d ed., pp. 411–447. New York: Academic Press.

Hogan, M. J. 1972. Role of the retinal pigment epithelium in macular disease. *Trans. Am. Acad. Ophthalmol. Otolaryngol.* 76:64–80.

Hogan, M. J., and Alvarado, J. 1967. Studies on the human macula. IV. Aging changes in Bruch's membrane. *Arch. Ophthalmol.* 77:410–420.

Hogan, M. J., Wood, I., and Steinberg, R. H. 1974. Phagocytosis by pigment epithelium of human retinal cones. *Nature* 252:305–307.

Hollenberg, M. J., and Spira, A. W. 1973. Human retinal development: ultrastructure of the outer retina. *Am. J. Anat.* 137:357–386.

Hollyfield, J. G. 1972. Egg pigment: early elimination from the developing retina of the frog. *Experientia* 28: 68–69.

Hollyfield, J. G. 1976. Phagocytic capabilities of the pigment epithelium. *Exp. Eye Res.* 22:457–468.

Hollyfield, J. G., Besharse, J. C., and Rayborn, M. E. 1976. The effect of light on the quantity of phagosomes in the pigment epithelium. *Exp. Eye Res.* 23:623–635.

Hollyfield, J. G., Besharse, J. C., an
1977. Turnover of rod photoreceptor o
Membrane addition and loss in relatio
ture. *J. Cell. Biol.,* 75:490–506.

Hollyfield, J. G., and Ward, A. 1
activity in the retinal pigment epitheliu
pipiens. I. Uptake of polystyrene spher
Res. 46:327–338.

Hollyfield, J. G., and Ward, A. 1
activity in the retinal pigment epithelium
pipiens. II. Exclusion of *Sarcina subfla*
Res. 46:339–350.

Hollyfield, J. G., and Ward, A. 1
activity of the pigmented retinal epith
actions between lysosomes and inge
spheres. *Invest. Ophthalmol.* 13:1016–10

Ishikawa, T., and Yamada, E. 1970.
of the photoreceptor outer segment wit
epithelial cell of rat retina. *J. Electron*

Jensen, M. S., and Bainton, D. F.
changes in pH within the phagocytic vac
morphonuclear neutrophilic leukocyte.
379–388.

Johnson, N. F. 1975. Phagocytosis in
ischaemic retinal pigment epithelium of
Eye Res. 20:97–107.

Karli, P. 1954*a*. Sur la réaction col
l'épithelium pigmentaire au cours d'une
mentale. C. R. *Soc. Biol.* (Paris) 148:172–

Karli, P. 1954*b*. Contribution experin
de la pathogenie des rétinoses pigmen
mologica 128:137–162.

Kolb, H., and Gouras, P. 1974. Elec
observations of human retinitis pigmen
inherited. *Invest. Ophthalmol.* 13:487–49

Kroll, A. J., and Machemer, R. 196
detachment in the owl monkey. III. Ele
of retina and pigment epithelium. *Am.*
66:410–427.

Kroll, A. J., and Machemer, R. 196
retinal detachment in the owl monkey. V.
copy of the reattached retina. *Am. J.*
117–130.

Kühne, W. 1879. Chemische Vorgange
In *Handbuch der Physiologie,* ed. L. Her
235–342. F. W. C. Vogel.

Kuwabara, T. 1970. Retinal recovery f
light. *Am. J. Ophthalmol.* 70:187–198.

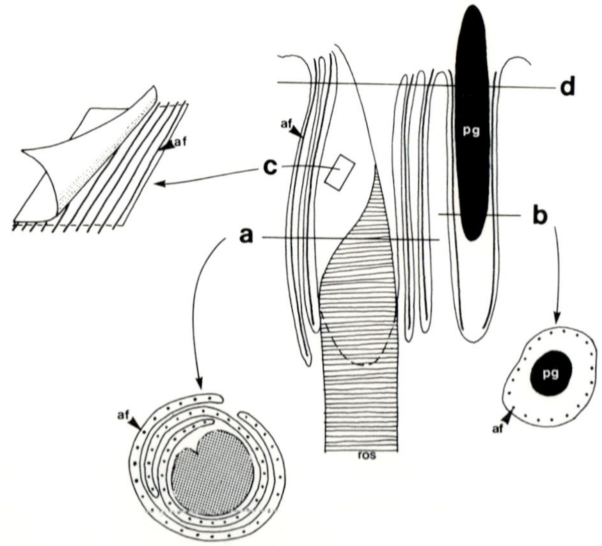

Figure 10.3 *Schematic illustration of apical projections in RPE cells of normal rat. Parallel rows of actin filaments* (af) *line the cytoplasmic face of the plasma membranes bounding the apical projections.* a *and* b *are transverse sections at the level indicated;* c *illustrates a three-dimensional view of a leaflike projection.* ros = *rod outer segments,* p = *pigment granule. (From Burnside 1976.)*

ments in the RPE are attached in evenly spaced parallel longitudinal rows to the cytoplasmic faces of the membranes of the apical projections (Burnside 1976; Burnside and Laties 1976). Though the function of this elaborate arrangement has not yet been demonstrated, several possibilities are apparent. The arrangement is appropriate to provide a structurally stable membrane skeleton, one that could take part episodically in engulfment. It should be remembered that the wandering phagocyte is free—it can migrate over a foreign particle that is engulfed—while the RPE is fixed; it must reach out to enclose the distal outer segment. Shed outer segment discs must thereafter be transported within the delicate apical projections. It should also be mentioned that in mammals the attachment between RPE and photoreceptors is conveyed by a relatively short (4–5 μ) apposition of rod outer segments with RPE apical projections; the membrane-bound rows of actin filaments may lend necessary structural stability to the delicate apical projections not provided by cell membranes alone.

Pigment Migration

As early as 1877, Boll and Kühne independently observed that the pigment of the frog RPE moved in response to light. They found that the pigment dispersed almost to the outer limiting membrane in the light and retracted to the base of the RPE cell in the dark. From these initial observations grew an enormous literature over the next century. A host of workers investigated retinal pigment migration in anything that crawled, flew, or swam (see reviews by Ali 1971, 1975; Arey 1915, 1916a; Detwiler 1943; Garten 1907; Nicol 1963; Parker 1932; Walls 1942). After numerous skirmishes it was generally agreed that dramatic retinal pigment migration is restricted to teleost fish, amphibians (anurans more than urodeles), and some birds. Reptiles in general exhibit slight pigmentary movement. In mammals, as well as anyone has been able to tell, pigment is immobile.

Generally, when pigment migration occurs, it is accompanied by coincident lengthening or shortening of the photoreceptors (Fig. 10.5). In the light, cones shorten while rods elongate to seclude themselves in the advancing pigment. In darkness, the opposite happens: cones elongate while rods shorten and thus extricate their outer segments from the retreating pigment. Movements of pigment granules and of photoreceptors are generally called photomechanical movements or retinomotor responses (Ali 1971, 1975; Walls 1942).

The function of retinal pigment migration has been the subject of some debate. Garten (1907) suggested that the expanded pigment served to absorb light refracted by the photoreceptors (especially their ellipsoids and oil droplets) and thereby prevent light scatter within the retina. Later, Walls (1942) argued that photomechanical movements served to modulate the amount of light allowed to strike the photoreceptor outer segments. Thus, he espoused the theory of Exner and Januschke (1906) that photomechanical movements are a mechanism for adapting the eye for day and night vision. Supporting this view is the observation that with a few exceptions (particularly birds) photomechanical movements take the place of pupillary responses; where one occurs, the other is absent.

It has subsequently been shown that, in fact, pigment migration can mask outer segments from light. By relating rhodopsin concentration to position of pigment in the RPE in the frog retina, Bäck and coworkers (1965) demonstrated that the pigment reduces the bleaching effect of light on rod outer segments by about one-third. It is consistent with these observations that the photic spectral sensitivity

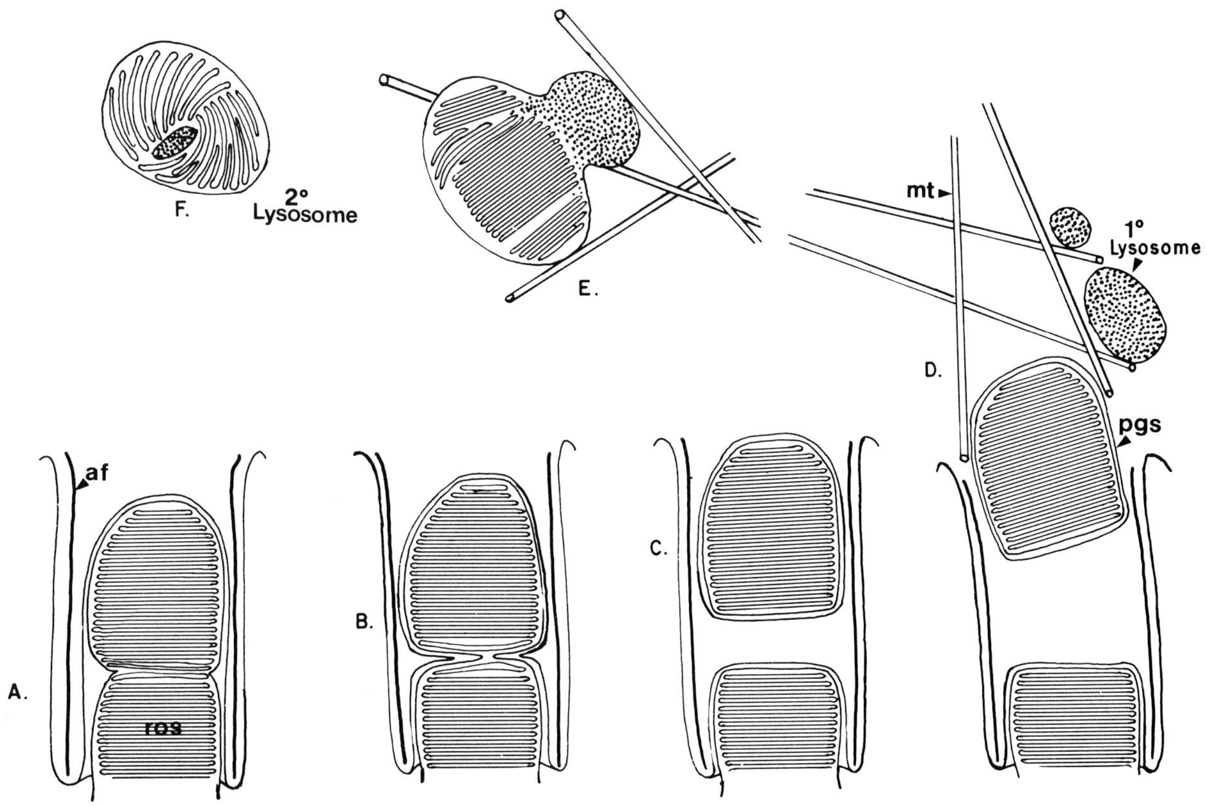

Figure 10.4 *Phagocytosis of effete rod outer segments by RPE.* A, *leaflike RPE projection with membrane-bound rows of actin filaments ensheaths tip of rod outer segments* (ros). B–C, *detached, effete rod-outer-segment tip is enclosed by membranes of the RPE projection.* D, *newly formed phagosome of rod-outer-segment discs is transported into apical cytoplasm, where fusion with primary lysosome takes place* (E); *a secondary lysosome is formed* (F), *and digestion is initiated. Actin filaments on apical projections may participate in engulfment and/or transport, while microtubules present in the apical cytoplasm are appropriately arranged for transporting phagosomes to lysosomes. These hypothesized roles for actin filaments and microtubules are based on morphological dispositions of the organelles; inhibitor studies have not yet been done. (From Burnside 1976.)*

curve of the albino trout differs from that of its pigmented relative (Kobayashi and Ali 1971).

A relationship between pigment migration and rod function is suggested by other observations. First, during development in fish, the initial appearance of pigment migration coincides with the differentiation of rods (Ali 1959, 1961; Blaxter and Staines 1970); in younger larvae the retina has only cones and the RPE is stationary. Second, in several species, the time required for regeneration of visual pigment parallels the time required for RPE contraction; for this reason some workers have suggested a relationship between pigment migration and visual pigment regeneration (Bäck, Donner, and Reuter 1965).

Photomechanical pigment migration has long been thought to result from movement of pigment granules within the confines of fixed apical projections of RPE cells (Parker 1932). This hypothesis arose, in part, from comparison with pigment migration in dermal melanophores (Parker 1932) and in part from examination of paraffin-sectioned material by light microscopy (a difficult matter since apical projections in many species are exceedingly fine and stain poorly) (Parker 1932). On the other hand, since the attachment between neural retina and RPE, so tenacious in the light-adapted state, lessens in darkness (Czerny 1867; Owczarek, Marak, and Pilkerton 1975), it was also reasoned that the decreased adhesion between the two layers in darkness might result from partial retraction of the apical projections of the RPE. However, such an argument is faulted on two grounds. First, as clearly seen in dissociated light- and dark-adapted fish RPE cells, apical projections do in fact remain extended in the dark-

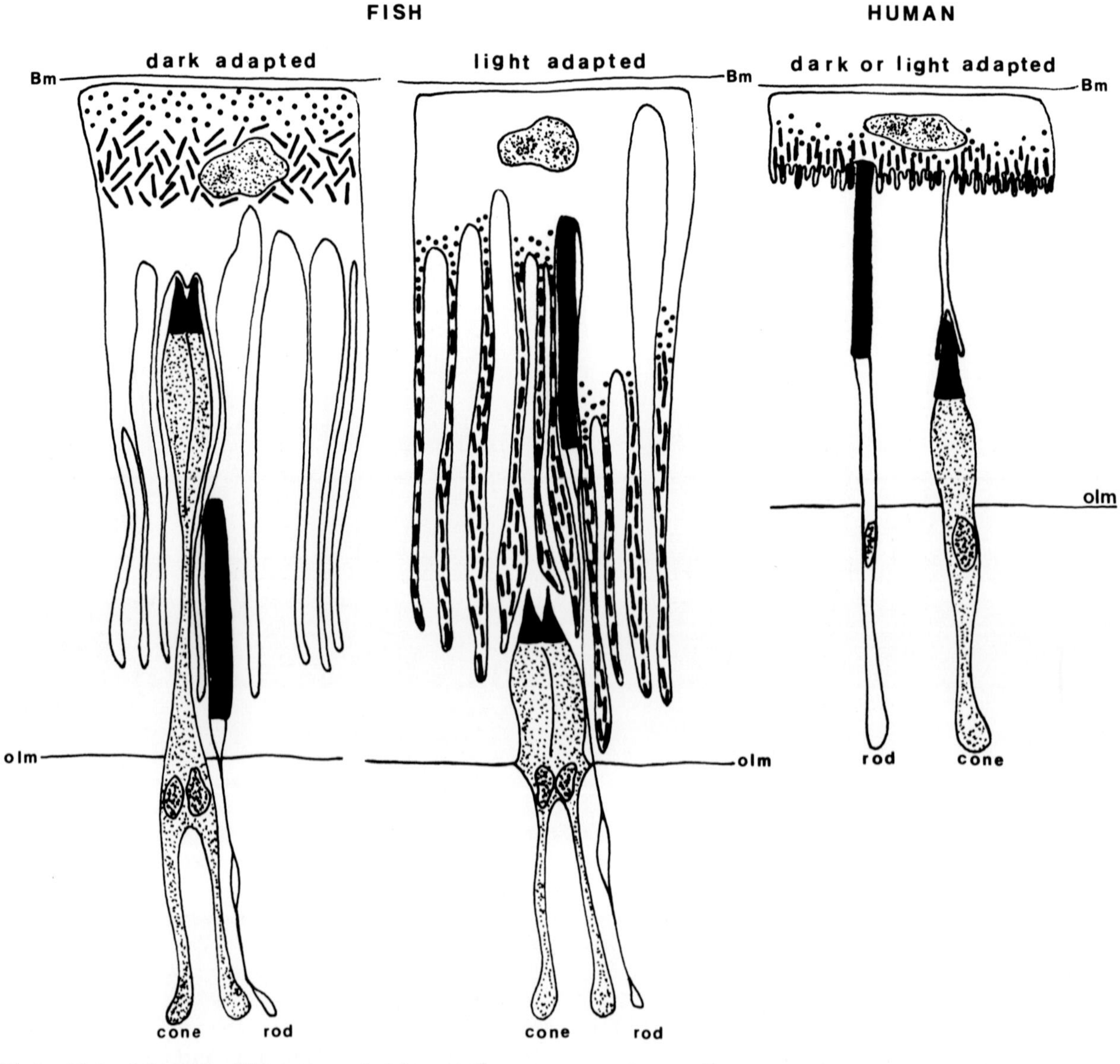

Figure 10.5 *Schematic illustration of fish and human RPE interrelationship with photoreceptors. Fish RPE cells possess long delicate apical projections that extend almost to the outer limiting membrane* (olm), *while the mammalian RPE has both short apical projections, some of which ensheath the tips of the rod outer segments, and some long delicate projections that ensheath the tips of cone outer segments. In the fish (grey snapper), pigment moves into the projections in light and reaggregates to the base of the cell in darkness. As far as is known, no light-induced movement occurs in mammalian retina.*

adapted state (Fig. 10.6) (Burnside and Nachmias, unpublished observations). Figure 10.6 illustrates that RPE cells from dark-adapted grey-snapper retinas possess numerous slender apical projections, some more than 100 μ in length. This observation corroborates the assumption of earlier workers that pigment granules move within fixed processes. Second, a difference in tenacity of adhesion of neural retina to RPE has been observed in species (rabbit) that display no evidence of melanosome motility (Owczarek, Marak, and Pilkerton 1975). Though a light-dark change in adhesion between RPE and

Figure 10.6 *Light micrographs of isolated RPE cells dissociated from light-adapted* (a,b) *and dark-adapted* (c,d) *grey snapper retinas. The long delicate apical projections* (arrows) *are filled with pigment in light and devoid of it in the dark.* a,b: *Nomarski optics;* c,d: *phase optics.* (×*1,400.*)

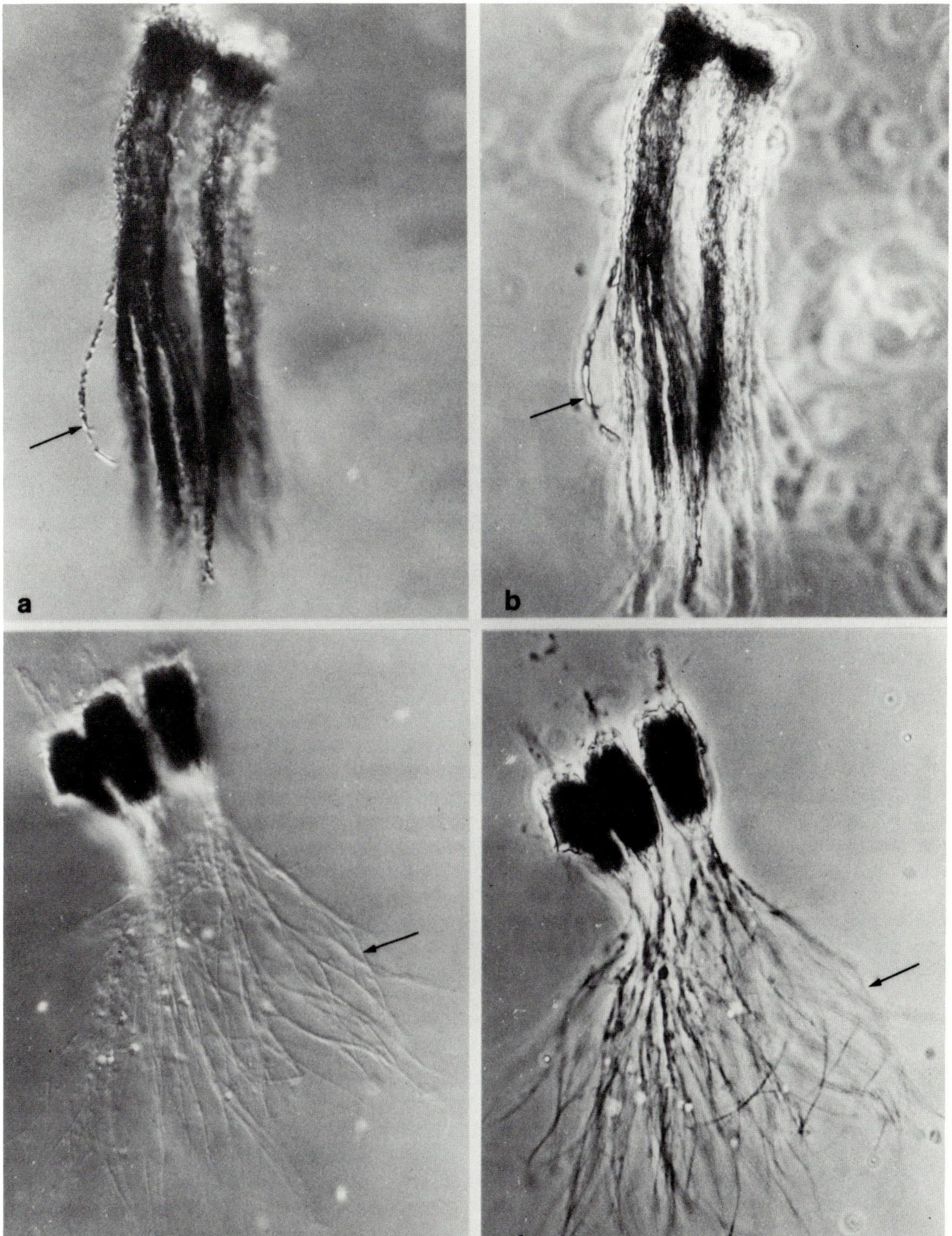
a
b
c
d

neural retina seems certain, it does not appear to result from retraction of the RPE apical projections during pigment aggregation.

MORPHOLOGY. The presence of long, relatively stiff, apical projections in isolated RPE cells (Fig. 10.6) demonstrates that the cytoskeletal mechanism responsible for maintaining the shape of the projections resides within the projections themselves and does not depend upon the presence of the neural retina. Studies of cellular extensions in other cells have revealed that elaboration and maintenance of the extension may depend on microtubules (Burnside 1975), bundles of laterally associated actin filaments (Tilney et al. 1973; Yamada et al. 1970) or both (Yamada et al. 1970). Microtubules are often thought to have a stabilizing role in large projections, while actin bundles are more often associated with very fine projections. A single structure can serve to illustrate both points: in outgrowing neuronal projections, actin filaments stabilize the delicate microspikes of the growing tip while microtubules stabilize the main body of the large newly forming neural projection; both are required for outgrowth (Yamada et al. 1970). Preliminary results from colchicine studies using isolated RPE cells suggest that microtubules are required for maintenance of the apical projections (Burnside and Nachmias, unpublished observations). It seems reasonable that a further search be made for microtubules in species in which they have not yet been observed.

In certain fish species the RPE cells contain reflecting granules as well as pigment granules; such fish are said to have a retinal tapetum (Arnott, Maciolek, and Nicol 1970; Arnott, Nicol, and Querfeld 1971; Parker 1932; Wunder 1926). In the sea trout these granules are composed of the triglyceride tridocosahexaenoin (Arnott, Nicol, and Querfeld 1971). In some species, the small round tapetal granules and the pigment granules move in opposite directions during dark- and light-adaptation (Garten 1907). Similar bidirectional streaming of particles has been observed in nerve axons, in cytoplasmic strands in plant cells, and in several other cell types (Rebhun 1972). Bidirectional streaming suggests that transport of pigment granules is selective and may not be explicable simply in terms of bulk cytoplasmic flow.

The RPE cells of amphibians, birds, fish, and reptiles contain lamellated stacks of membranous sacs, which have been called myeloid bodies because of regular spacing of membranes in the central portion of the stacks (see Rodieck 1973; although Rodieck does not report myeloid bodies in fish, Burnside has seen them in each of ten species examined). Because of the morphological similarity between these bodies and outer segment saccules, Porter and Yamada (1960) suggested that myeloid bodies might be sites of photoreception within the RPE cells which serve to trigger photomechanical pigment migration. It is consistent with this hypothesis that myeloid bodies are found in all animal groups known to exhibit photomechanical movements, while in mammals both pigment movement and myeloid bodies are absent. An inconclusive test resulted when, by microspectrophotometry, Liebman and coworkers (Liebman, Carroll, and Laties 1969; Liebman and Entine 1968) described inclusions in the frog retina that contained an unbleached visual pigment with absorption maximum of 500 nm. Since this value coincides with that of rod outer segments, it is not possible to differentiate myeloid bodies with certainty from ingested rod outer segment phagosomes.

Characteristics of Pigment Migration

Pigment migration is affected by many factors: rate and extent vary from species to species, while external conditions such as previous light exposure have a decided influence on dark-adaptive movements (Ali 1971, 1975; Parker 1932). Moreover, different regions of the same retina may display different extents of pigment migration (Hess 1910). The combination of species differences and location variability was no doubt responsible for the elaborate controversies that arose in the nineteenth-century literature as to the nature and occurrence of retinal pigment migration.

Taking into account these variations, most authorities agree that pigment migration in the RPE is relatively slow. The time required for complete displacement of pigment ranges from 20 to 70 minutes under various conditions (Ali 1975; Nicol 1963; Parker 1932). Thus, the rate of pigment granule movement is approximately 0.5–5 μ/min. This rate is slow compared to particle movements in other cell types, most of which are in the order of 1 μ/sec (Rebhun 1972). This slower rate is similar to that of chromosomal movement in mitosis (Rebhun 1972).

LIGHT INTENSITY. Both rate and extent of pigment expansion are influenced by the intensity of incident light (Ali 1971, 1975; Smith 1927). At low light intensities movement is slow and the RPE may reach a steady state that is only partially dispersed. In intense light, movement is fast, and pigment granules may pile up in the distal tips of the RPE projections near the outer limiting membrane.

The rate of pigment retraction in the dark is influenced by previous light exposure; the higher the light intensity to which the animal was exposed before the dark period, the slower the subsequent pigment retraction (Ali 1971, 1975). Thus the rates of retraction reported by various investigators are not comparable without full specification of experimental conditions.

The threshold of light intensity required to set off dispersion or retraction of pigment has been examined in several species and found to show considerable variation among species (Ali 1971, 1975; Nicol 1963). In general, the pigment begins to adopt a dark position when illumination falls below one foot-candle (Nicol 1963). This level of illumination is experienced during twilight or in a landscape illuminated by the half moon (Nicol 1963). In addition, the threshold for "dark-adaptation" movements (pigment retraction) drops during the night in some fishes (Tamura 1957).

SPECTRAL SENSITIVITY. Liebman and coworkers (Liebman, Carroll, and Laties 1969) have shown that the action spectrum for pigment migration in the frog corresponds to that of red rods (rhodopsin). They conclude that triggering depends on the red rods rather than on rhodopsin-containing receptors in the RPE cell itself, because the observed action spectrum corresponds to that of rods, rather than to a theoretical RPE action spectrum calculated for the RPE. Such a calculation takes into account light preadsorbed by the rhodopsin of the overlying rods before reaching hypothetical receptors in the RPE (Liebman and Entine 1968).

TEMPERATURE. The effects of temperature on the pigment migration are complex and somewhat ambiguous. Arey (1916a) found that in several fishes, both in light and darkness, the retinal pigment exhibits greater expansion at low than at high temperatures. In frogs, on the other hand, Arey (1916a) and later Detwiler and Lewis (1926) found different temperature effects on light- and dark-adapted retinas: in darkness the RPE pigment expanded at low and high temperatures (0°–14° C and 19°–33° C), remaining contracted at normal temperatures (14°–19° C), while no temperature effect was found in the light-adapted retina. One way to account for this effect would be light dominance. Another explanation is that the expansion of pigment in dark-adapted eyes at temperature extremes may result from stress, since adrenalin has been shown to produce pigment expansion in dark-adapted frog eyes (Fujita 1912; Vilter 1946). Detwiler and Lewis (1926) found that dark-induced contraction of pigment in previously light-adapted eyes was less sensitive to temperature than was maintenance of the contracted state, retraction occurring between 18°–23° C. In salmon retina, temperature had no effect on light-adapted retinas, but in dark-adapted retinas, higher temperatures (23° C) produced a slight expansion (Ali 1975). In fishes, the rate of pigment migration increases with higher temperatures up to 20° C (Ali 1964).

DIURNAL RHYTHMS. Welsh and Osborn (1937) were the first to note the imposition of circadian rhythm on photomechanical responses. They observed that the extreme of dark-adaptation (completely retracted pigment) is found only in eyes fixed during the night. Furthermore, they saw that in animals kept in constant darkness, the RPE approaches normal light-adapted morphology (expanded pigment) at the time of sunrise outside, and it returns to the dark-adapted state at sunset. In contrast, the RPE in constant light retains the expanded position throughout; thus light appears to dominate the rhythmic influence. Ali (1964) and John et al. (1967) have examined other fish and found evidence of species variation: some are rhythmic in total darkness while others are not.

HORMONAL CONTROL. Despite a number of observations relating neurohumors to retinal photomechanical movement, some dating back 120 years (Mueller 1856), it is just now becoming possible to undertake critical studies of the roles of neurotransmitters and neurohormones in the control of pigment migration. Because of recent changes in our understanding of neurohormone action and because of the lack of harmony in many previous reports, we must interpret past results with great care. Nonetheless, the results of prior work serve as an aid in setting out those questions which clarify the role of chemical neurotransmitters and neurohormones in the control of retinal photomechanical movements.

Dermal melanocytes, which are at once more readily visible and more accessible than retinal pigment epithelial cells, have been far more intensively investigated with regard to their reactivity to neurotransmitters and neurohormones. The wide and often contrasting variation in response observed among different fish and amphibian species clearly indicates that dermal melanocytes are responsive to a broad variety of agents: to biogenic amines, to acetylcholine, to MSH, and at times to ACTH. Further, for biogenic amines such as adrenalin, receptors of both the α- and β-type have been reported (for brief review, see Lerner 1971).

Although the eye does not afford quite the same advantages for study as does the skin, it is clear that photomechanical movements take place in response to a number of biological agents (Detwiler 1943). For instance, there is substantial evidence that catecholamines, especially adrenalin, can bring about photomechanical movement: adrenalin has been shown to produce the light-adapted (expanded) state of the retinal pigment eipthelium in dark-adapted frogs and eels (Fujita 1912; Vilter 1946). There is also evidence that indoleamines, especially melatonin and serotonin, have cell motility effects as well. In fact, in teleost fish it has recently been shown that melatonin and serotonin have opposite effects with respect to pigment migration or dark adaptation (see Cheze and Ali 1976, fig. 4). In the frog, melatonin apparently has two effects: it brings about contraction of cone myoids in *Xenopus* tadpole (Quay and McLeod 1968), and it alters the dark-adaptational pigmentary migration (Kraus-Ruppert and Lembeck 1965). For photoreceptors there is also some negative evidence: acetylcholine and β-amino butyric acid apparently have little or no effect on their electrical activity (Murakami, Ohtsu, and Ohtsuka 1972).

The evidence for the participation of polypeptide hormones in photomechanical movements is much less up-to-date and much less complete than is that for biogenic amines. There are a few references concerning work with MSH (Intermedin); although they are not recent, they make a reasonable case for the participation of MSH in pigment retraction in the dark-adapting eel (Vilter 1946). However, this work was done before the characterization of two different forms of MSH, α and β, and before it was realized just how similar in structure both forms of MSH were to ACTH (ACTH and β-MSH apparently share a common prohormone).

The whole of neuropharmacology has been turned topsy-turvy by the discovery of cyclic AMP. The successive findings that hormones affect cells by activation of membrane-affiliated adenyl cyclase systems (Sutherland, Rall, and Berthet 1957) and that, in turn, cyclic nucleotides have protein kinases as their site of action (Walsh et al. 1971) have completely refocused neuropharmacology toward postsynaptic and intracellular events. To date, retinal photomechanical movements have not been studied in this regard. However, there is abundant reason to do so: in dermal melanocytes, which are responsive to the same neurotransmitters as are retinal pigment epithelial cells, adenyl cyclase clearly acts as a transmembrane cellular switching device. Alterations of intracellular cyclic AMP levels in dermal melanocytes have been recorded: (a) after the application of biogenic amines—both catecholamines such as epinephrine and indoleamines such as melatonin; (b) after the application of acetylcholine (Moellman, Lerner, and Hendee, Jr., 1974); and (c) after the application of polypeptides such as MSH (Bitensky, Keirns, and Keirns 1975).

THE RESPONSE TO LIGHT. In vertebrates, the pigmentary response to light can take place in an enucleated eye (Hamburger 1889), indicating local origin. Further, since pigmentary migration has an action spectrum with a single minimum, it must be initiated by the photoreceptors (Liebman, Carroll, and Laties 1969). That this is so is supported by the recent demonstration that the pigmentary response is lost when the retina is detached (Snyder and Zadunaisky 1976).

The nature of the signal from photoreceptors to pigment epithelial cell is incompletely known. In vertebrates a role for calcium ion has been adduced (Snyder and Zadunaisky 1976), but it is not presently understood whether this role is primary and an intrinsic part of photoreception or secondary and hence part of a distinct cell-to-cell signal.

In case of an invertebrate, the squid, in which pigment migration actually takes place within the photoreceptor cell, there is evidence for an electrochemical dependence: when isolated slices of squid eye are bathed in sodium-free Ringer's solution (that is, one in which choline+ is substituted for Na+), pigment migration in response to light is stilled (Hagins and Liebman 1962).

EFFECTS OF DRUGS AND ANESTHETICS. A variety of agents have been shown to arrest pigment migration in the RPE. Among them are cocaine (Lodato 1895; Ovio 1895), ether (Arey 1916a), and carbon dioxide (Arey 1916a). Increase or decrease of oxygen by nitrogen substitution appeared to have no effect so long as the fish studied remained alive (Arey 1916a): thus the effect of carbon dioxide appears to be independent of oxygen deprivation. However, since all of these agents have been applied to the whole animal it is difficult to interpret the immediate effect on RPE migration. No effect on migration was produced by choretone, urethane (Arey 1916a), strychnine, or quinine (Fujita 1912).

Possible Motile Mechanisms for Pigment Movement

Any movement of pigment granules in the RPE requires energy. When an animal is exposed to carbon dioxide, pigment movement stops and granules remain wherever they have stopped (Arey 1916a). Thus, neither the expanded nor retracted states represent a "relaxed state" to which granules return without energy expenditure.

In species with photomechanical movement, pigment granules move within long, fixed apical projections. Thus, one might expect to find within the projections the cellular machinery responsible for movement. Unfortunately, the cytoplasm of these RPE projections is especially difficult to fix for electron microscopy; consequently, in most species these projections appear to be empty except for the indestructible pigment granules (Arnott, Maciolek, and Nicol 1970; Rosenkranz 1973; Young and Droz 1968). When, on occasion, cytoplasmic detail within the apical projection is retained, longitudinal microtubules (Burnside 1976) and bundles of filaments (Burnside 1976; Murray and Dubin 1975) are observed. The longitudinal filaments are 60–80 Å in diameter; that they are actin filaments is shown by the cytochemical procedures of myosin subfragment-1-binding in fishes (Burnside 1976) and heavy meromyosin-binding in frogs (Murray and Dubin 1975). On this basis Murray and Dubin (1975) have suggested that the actin bundles are responsible for pigment movement, for in their preparations microtubules were not observed.

Results from recent studies with intraocularly injected colchicine and cytochalasin-B in fishes (Burnside, Adler, and Sandu, unpublished results with the blue-striped grunt) are consistent with Murray and Dubin's suggestion that actin filaments rather than microtubules mediate pigment migration in the RPE cell processes. Microtubules are, however, necessary for movements within the RPE cell body. Cytochalasin blocks light-induced outward migration of pigment (dispersion) but fails to inhibit aggregation in the dark. Thus it seems likely that actin filaments participate only in outward migration. The normal mechanism of inward migration (aggregation) is not clear; it seems to function independently of actin or microtubules.

Mammalian Pigment

It is generally held that photomechanical movements and rapidly moving pupils are mutually exclusive, presumably because prompt iris sphincter and dilator action renders pigmentary migration and rod-cone movements superfluous. Whether or not this explanation is correct, the absence of photomechanical movement in species with rapidly reactive iris muscles is widely recognized (Ali 1971; Duke-Elder 1958). Since we have considered at length the motile aspects of photomechanical RPE, we examine here several adaptations of RPE cells in species lacking photomechanical movements; these adaptations may relate to motile or cytoskeletal aspects of these RPE cells.

First of all, the apical projections of the RPE cells are generally much shorter in species that lack photomechanical movements (Figs. 10.5, 10.7). For instance, in a typical fish, apical processes may extend some 60 microns to ensheath fully the rod outer segments (Fig. 10.6). In contrast, in the retina of a macaque monkey, the entire rod outer segment is never fully sheathed; instead, RPE apical processes form a short distal cap, wrapping the terminal 4 microns of a 28-micron rod outer segment. For cones this relationship is yet to be fully worked out. However, there is evidence that in the human only the distal one-half of the cone is wrapped, while in the cat and in the squirrel the entire cone outer segment can be surrounded by pigment epithelial apical processes (Anderson and Fisher 1976; Hogan, Alvarado, and Weddell 1971; Steinberg and Wood 1974).

Second, in the monkey and in most other mammalian species the melanosomes themselves are

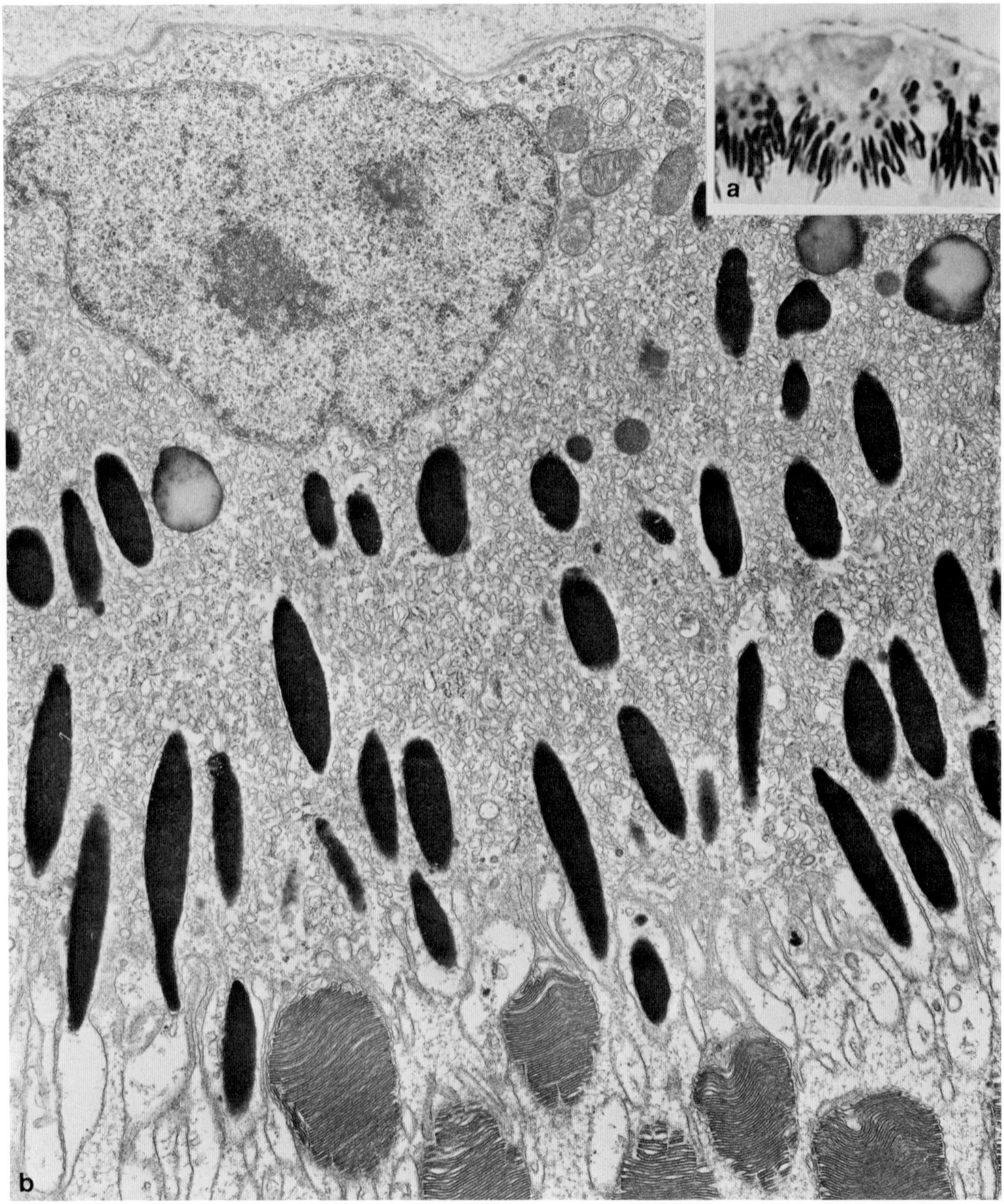

Figure 10.7 *Squirrel monkey RPE cell.* a, *light micrograph showing apical distribution and parallel alignment of rod-shaped pigment granules.* b, *electron micrograph of cell similar to* a: *rod-shaped granules often extend into the apical projections while round granules are more basally situated.* (a: *×2,990.* b: *×12,650.)*

stratified to form a band across the apical region of the RPE cell, such that a continuous line is seen when histological sections are viewed at low power. At higher magnifications, the black line is often found to consist of two types of melanosomes—spherical and elongate—the two types being more or less segregated (Fig. 10.7). Thus, elongate melanosomes generally hold an apical position in the band, while spherical ones are more basal. In some species such as the squirrel, the division is so sharp that the two types of melanosomes are truly separated from one another by pigment-free cytoplasm (Fig. 10.8). Elongate melanosomes have, in addition, two special qualities: they are, for the most part, vertically oriented and thus lie parallel one to the other; and they are frequently found just at the root of an apical process or, in some species, within it (Feeney, Grieshaber, and Hogan 1965) (Figs. 10.7, 10.8). Lastly, it appears to be a general property of melanosomes that they do not touch each other; instead they are separated from one another by a variable but definite intervening space (Figs. 10.7, 10.8).

All of these observations point to a complex mechanism for distributing melanosomes. From a cytoskeletal point of view there are many questions to answer: how are melanosomes confined to a given cytoplasmic level; how are the two types of melanosomes held separate from one another within the cytosol, and, lastly, by what mechanism are elongate melanosomes held in a vertical position.

The apical region of the RPE cell contains an elaborate network of filaments associated with the junctional complexes that provide attachment between cells. Numerous cytoplasmic filaments extend into the cytoplasm both from the occasional desmosomes and also from the *zonula adherens* that forms a continuous band around the cell near the apex. This apical filamentary complex (occupying the same part of the cell as the pigment granules) corresponds to the "terminal web" described in many epithelia by classical light microscopy. Whether this terminal web is sufficient to orient melanosomes and fix them within the cytosol, or whether an additional and distinct filamentary rigging is also required, is not known.

Thus, at present, our knowledge of the internal structure of the apical part of the retinal pigment epithelium is incomplete. Some help may come from critical study of comparative cellular pathology. It would be of great interest to know the arrangement of the terminal web or the disposition of actin filaments in cases where defects in the cytoplasmic disposition of melanin granules have been noted. For instance, clumping and basal displacement of melanosomes occurs in the pigment epithelium of mice with *d*-locus (Maltese dilution) (Moyer 1966). Also, basal displacement of melanin granules in retinal pigment epithelial cells has been recognized in at least one tissue specimen of dominant human retinitis pigmentosa (Kolb and Gouras 1974). In each case, appropriate cytoskeletal studies are yet to be done.

Whether or not the mechanism by which melanosomes are held in place in the cytosol is fully elucidated, the nature of their disposition still has important meaning. Arranged as they are in a spaced array, melanosomes are geographically disposed to play serial complimentary optical roles. First, light that directly impinges on melanosomes after traversing the neural retina can be absorbed. Since the absorption spectrum for melanin is essentially flat over a broad range, absorption is in good measure independent of wave length (Liebman, Carroll, and Laties 1969). Second, light that is not absorbed by melanin is likely to be susceptible to Mie scattering; that is, in accordance with electromagnetic theory, both the size and separation of melanin granules are such as to scatter an incoming beam of light in an asymmetric fashion, a preponderance of light being scattered in a forward direction. As a practical result of this, light that has already gone past the photoreceptors to enter the pigment epithelium tends, if not absorbed by the melanin pigment itself, to be scattered forward toward the next layer of pigment in the choroid. Lastly, those few photons that do finally reach the sclera and are then reflected back toward the neural retina not only face a second hazardous trip through the same two layers of melanin-containing cells but also, because of the elongate and oriented nature of the most apical melanosomes of the pigmented epithelium, have an increasing chance of being absorbed as their approach becomes more oblique to the elongate melanosomes. The solid angle through which light can reach the rod outer segments has been decreased. In summary then, because of the way they are held within the cell and how they are shaped, melanosomes act in double series to minimize glare and thus to promote the optical efficiency of the eye.

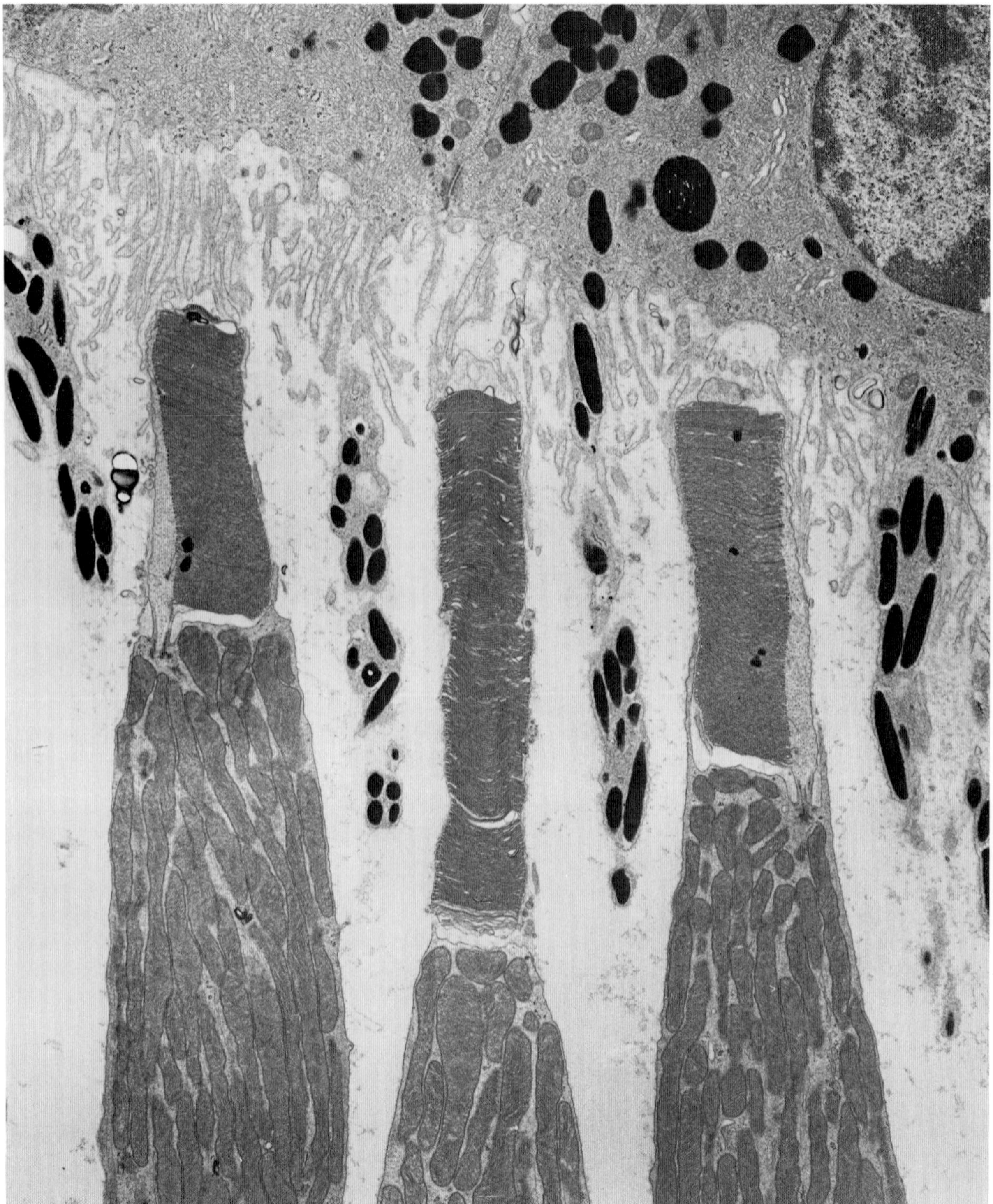

Figure 10.8 *RPE and photoreceptors of the 13-lined ground squirrel. Rod-shaped pigment granules are primarily situated in apical projections, while rounded pigment granules are found in the apical cytoplasm. (From Anderson and Fisher 1976. ×8,390.)*

REFERENCES

ALI, M. A. 1959. The ocular structure, retinomotor and photobehavioral responses of juvenile Pacific salmon. *Can. J. Zool.* 37:965–996.

ALI, M. A. 1961. Histophysiological studies on the juvenile Atlantic salmon (*Salmo salar*) retina. II. Responses to light intensities, wavelengths, temperatures, and continuous light or dark. *Can. J. Zool.* 39:551–562.

ALI, M. A. 1964. Uber den einfluss der temperatur auf die geschwindigkeit der retinomotorischen reacktionen des lachses (*Salmo salar*). *Naturwissenschaften* 51:471.

ALI, M. A. 1971. Les responses rétinomotrices: caractères et mécanismes, *Vision Res.* 11:1225–1288.

ALI, M. A. 1975. Retinomotor responses. In *Vision in fishes,* ed. M. A. Ali, series A, pp. 313–355. NATO Advanced Study Institute.

ANDERSON, D. H., and FISHER, S. K. 1976. The photoreceptors of diurnal squirrels: outer segment structure, disc shedding, and protein renewal. *J. Ultrastruct. Res.* 55:119–141.

AREY, L. B. 1915. The occurrence and significance of photomechanical changes in the vertebrate retina: an historical survey. *J. Comp. Neurol.* 25:525–554.

AREY, L. B. 1916. The movements in the visual cells and retinal pigment of the lower vertebrates. *J. Comp. Neurol.* 26:121–201.

ARNOTT, H. J., MACIOLEK, N. J., and NICOL, J. A. C. 1970. Retinal tapetum lucidum: a novel reflecting system in the eye of teleosts. *Science* 169:478–480.

ARNOTT, H. J., NICOL, J. A. C., and QUERFELD, C. W. 1971. Reflecting spheres in the eyes of weakfishes (*Sciaenidae*). *Nature* 233:130–133.

BÄCK, I., DONNER, K. O., and REUTER, T. 1965. The screening effect of the pigment epithelium on the retinal rods in the frog. *Vision Res.* 5:101–111.

BITENSKY, M. W., KEIRNS, C. N., and KEIRNS, J. J. 1975 The amphibian melanocyte: microtubules, cyclic amp, and organelle translocation. *Ann. N.Y. Acad. Sci.* 253: 685–691.

BITENSKY, M. W., MIKI, N., KEIRNS, J. J., KEIRNS, M., BARABAN, J. M., FREEMAN, J., WHEELER, M. A., LACY, J., and MARCUS, F. R. 1975. Activation of photoreceptor disk membrane phosphodiesterase by light and ATP. In *Advances in cyclic nucleotide research,* ed. G. I. Drummond, P. Greengard, and G. A. Robinson, vol. 5, pp. 213–240. New York: Raven Press.

BLAXTER, J. H. S., and STAINES, M. 1970. Pure-cone retinae and retino-motor responses in larval teleosts. *J. Mar. Biol. Ass. U.K.* 50:449–460.

BOK, D., and YOUNG, R. W. 1979. Phagocytic properties of the retinal epithelium. In *The retinal pigment epithelium,* ed. K. M. Zinn and M. F. Marmor. Cambridge: Harvard Univ. Press.

BOLL, F. 1877. Zur anatomie und physiologie der retina. *Arch. Anat. Physiol.* 4:783–787.

BURNSIDE, B. 1975. The form and arrangement of microtubules: an historical primarily morphological review. *Ann. N.Y. Acad. Sci.* 253:14–26.

BURNSIDE, M. B. 1976. Possible roles of microtubules and actin filaments in retinal pigmented epithelium. *Exp. Eye Res.* 23:257–275.

BURNSIDE, B., and LATIES, A. M. 1976. Actin filaments in apical projections of the primate pigmented epithelial cell. *Invest. Ophthalmol.* 15:570–575.

CHEZE, G., and ALI, M. A. 1976. Rôle de l'épiphyse dans la migration du pigment epithelial rètinien chez quelques teleosteens. *Can J. Zool.* 54:475–481.

COPELAND, M. 1974. The cellular response to cytochalasin-B: a critical review. *Cytologia.* 39:709–727.

CZERNY, V. 1867. Uber Blendung der Netzhaut durch Sonnenlicht. *Sitzungsber. Akad. Wiss. Wien., Math-Naturwiss. Kl.* (abst. 2) 56:409–428.

DETWILER, S. R. 1943. *Vertebrate photoreceptors,* p. 184. New York: Macmillan.

DETWILER, S. R., and LEWIS, R. W. 1926. Temperature and retinal pigment migration in the eyes of the frog. *J. Comp. Neurol.* 41:153–169.

DUKE-ELDER, S. 1958. *System of ophthalmology.* I. *The eye in evolution.* St. Louis: C. V. Mosby.

EXNER, S., and JANUSCHKE, H. 1906. Die Stabchenwanderung im Auge von *Abramis brama* bei Lichtveranderungen. *Ber. Akad. Wiss. Wien, Math-Naturwiss. Kl.* 115:269–280.

FEENEY, I., GRIESHABER, J. A., and HOGAN, M. J. 1965. Studies on human ocular pigment. In *The structure of the eye.* II. *Symposium,* ed. J. W. Rohen, pp. 535–548. Stuttgart: F. K. Schattauer-Verlag.

FUJITA, H. 1912. Pigmentbewegung und Zapfenkontraktion im Dunkelauge des Frosches bei Einwirkung verschiedener Reize. *Arch. Vergl. Ophthalmol.* 2:164–179.

GARTEN, S. 1907. Die Veranderungen der Netzhaut durch Licht. *Graefe-Saemisch Handbuch gesamten Augenheilkunde* (Leipzig) 2:250–280.

GOLDMAN, R., POLLARD, T., and ROSENBAUM, J., ed. 1976. *Cell motility*. Cold Spring Harbor Laboratory. 1367 pp.

HAGINS, W. A., and LIEBMAN, P. A. 1962. Light induced pigment migration in the squid retina. *Biol. Bull.* 123:498.

HAMBURGER, D. J. 1889. Dorsnijding van der nervus opticus bij Kikvorschen, in Verband met de Beweging van Pigment en Kegels in het Netvlies, onder der Invloed van Licht en Duister. *Onderzoekingen Utrechtsche. Hoogeschol., Reeks 3.* 11:58–67.

HARTWIG, J., and STOSSEL, T. P. 1976. Interactions of actin, myosin, and an actin-binding protein of rabbit pulmonary macrophages. III. Effects of cytochalasin-B. *J. Cell Biol.* 71:295–303.

HESS, C. 1910. Beitrage zur Kenntnis regionarer Verschiedenheiten der Netzhaut und des Pigment epithels in der Wirbeltierreihe. *Arch. Vergl. Ophthalmol.* 1:413–442.

HOGAN, M. J., ALVARADO, J. A., and WEDDELL, J. E. 1971. *Histology of the human eye*. Philadelphia: W. B. Saunders Co.

HUXLEY, H. E. 1963. Electron microscopic studies on the structure of natural and synthetic protein filaments from striated muscle. *J. Molec. Biol.* 7:281–308.

JOHN, K. J., SEGALL, M., and ZAWATSKY, L. 1967. Retinomotor rhythms in the goldfish, *Carassius auratus*. *Biol. Bull.* 132:200–210.

KOBAYASHI, H., and ALI, M. A. 1971. Electroretinographic determination of spectral sensitivity in albino and pigmented brook trout (*Salvelinus fontinalis,* Mitchell). *Can. J. Physiol. Pharmacol.* 49:1030–1037.

KOLB, H., and GOURAS, P. 1974. Electron microscopic observations of human retinitis pigmentosa, dominantly inherited. *Invest. Ophthalmol.* 13:487–498.

KRAUS-RUPPERT, R., and LEMBECK, F. 1965. Melatonin auf die Pigmentzellen der Retina. *Pflugers Arch. Gesamte Physiol.* 284:160–168.

KÜHNE, W. 1878. Fortgesetzte Untersuchungen uber die Retina und die Pigmente des Auges. *Untersuch. Physiol. Inst. Univ. Heidelberg* 2:89.

LERNER, A. B. 1971. Neurocontrol of pigment cells. In *Biology of normal and abnormal melanocytes*, ed. M. Fitzpatrick, T. Kawaneura, and M. Seiji, p. 3–16. Baltimore: University Park Press. Also Tokyo: Univ. of Tokyo Press.

LIEBMAN, P. A., CARROLL, S., and LATIES, A. 1969. Spectral sensitivity of retinal screening pigment migration in the frog. *Vision Res.* 9:377–384.

LIEBMAN, P. A., and ENTINE, G. 1968. Visual pigments of frog and tadpole (*Rana pipiens*). *Vision Res.* 8:761–775.

LIEBMAN, P. A., RICE, R., CARROLL, S., ENTINE, G., and LATIES, A. 1967. Triggers for C wave and pigment migration. *Invest. Ophthalmol.* 6:214.

LODATO, G. 1895. Richerche sullo fisiologia dello strato neuroepitheliale della retina. *Arch. Ottalm.* 3:141–148.

MOELLMAN, G., LERNER, A. B., and HENDEE, J. R., JR. 1974. The mechanism of frog skin lightening by acetylcholine. *Gen. Comp. Endocrinol.* 23:45–51.

MOYER, F. H. 1966. Genetic variations in the fine structure and ontogeny of mouse melanin granules. *Am. Zoologist.* 6:43–66.

MUELLER, H. 1856. Anotomisch-physiologische Untersuchungen uber die Retina bei Menschen und Wirbelthieren. *Z. Wiss. Zool.* 8:1–122.

MURAKAMI, M., OHTSU, K., and OHTSUKA, T. 1972. Effects of chemicals on receptors and horizontal cells in the retina. *J. Physiol.* 227:889–913.

MURRAY, R. L., and DUBIN, J. W. 1975. The occurrence of actin-like filaments in association with migrating pigment granules in frog retinal pigment epithelium. *J. Cell Biol.* 64:705–710.

MURRAY, J. M., and WEBER, A. 1974. The cooperative action of muscle proteins. *Scient. Amer.* 230:58–71.

NICOL, J. A. C. 1963. Some aspects of photoreception and vision in fishes. *Adv. Mar. Biol.* 1:171–208.

OVIO, G. 1895. Di una speciale azione della Cocaine sulle funzione visiva. *Ann. Ottalm.* 24:23.

OWCZAREK, F. R., MARAK, G. E., and PILKERTON, A. R. 1975. Retinal adhesion in light and dark-adapted rabbit. *Invest. Ophthalmol.* 14:353–358.

PARKER, G. H. 1932. The movements of retinal pigment. *Ergeb. Biol.* 9:239.

PORTER, K. R., and YAMADA, E. 1960. Studies on the endoplasmic reticulum. V. Its form and differentiation in pigment epithelial cells of the frog retina. *J. Biophys. Biochem. Cytol.* 8:181–205.

QUAY, W. B., and MCLEOD, R. W. 1968. Melatonin and photic stimulation of cone contraction in the retina of larval *Xenopus laevis*. *Anat. Rec.* 160:491.

REBHUN, L. I. 1972. Polarized intracellular particle transport: saltatory movements and cytoplasmic streaming. *Int. Rev. Cytol.* 39:93–137.

RODIECK, R. W. 1973. *The vertebrate retina: principles of structure and function*. San Francisco: W. H. Freeman and Co.

ROSENKRANTZ, J. 1973. New results on the ultrastructure of frog rod outer segments. *Z. Zellforsch. Mikrosk. Anat.* 143:45–52.

SMITH, D. C. 1927. The relation of light intensity to retinal pigment migration in the frog. *J. Comp. Neurol.* 42:457–473.

SNYDER, W. Z., and ZADUNAISKY, J. A. 1976. A role for calcium in the migration of retinal screening pigment in the frog. *Exp. Eye Res.* 22:377–388.

STEINBERG, R. H., and WOOD, I. 1974. Pigment epithelial cell ensheathment of cone outer segments in the retina of the domestic cat. *Proc. R. Soc. Lond.* [Biol.] 187:461–478.

STEPHENS, R. E., and EDDS, K. T. 1976. Microtubules: structure, chemistry, and function. *Physiol. Rev.* 56:709–777.

STOSSEL, T. P. 1974. Phagocytosis. *New Eng. J. Med.* 290:833–839.

SUTHERLAND, E. W., RALL, T. W., and BERTHET, J. 1957. The relationship of epinephrine and glucagon to liver phosphorylase. IV. Effect on epinephrine and glucagon or the reactivation of phosphorylase in liver homogenase. *J. Biol. Chem.* 224:463–475.

TAMURA, T. 1957. On the relation between the intensity of illumination and the shifting of cones in the fish retina. *Bull. Jap. Soc. Sci. Fish.* 22:742–746.

TILNEY, L. G., HATANO, S., ISHIKAWA, H., and MOOSEKER, M. S. 1973. The polymerization of actin: its role in the generation of the acrosomal process of certain echinoderm sperm. *J. Cell. Biol.* 59:109–126.

VILTER, V. 1946. Déterminisme intrinsique de migrations pigmentaires dans la rétine. *C. R. Soc. Biol.* (Paris) 140:277–278.

WALLS, G. L. 1928. An experimental study of the retina of the brook lamprey, *Entosphenus appendix* (De Kay). *J. Comp. Neurol.* 46:465–473.

WALLS, G. L. 1942. *The vertebrate eye and its adaptive radiation,* p. 785. Cranbrook Institute of Science Bull. 19, Bloomfield Hill, Michigan.

WALSH, P. A., PERKINS, J. P., BRONSTROM, C. O., HO, E. S., and KREBS, E. G. 1971. Catalysis of the phosphorylase-kinase activation reaction. *J. Biol. Chem.* 246:1968–1976.

WEIHING, R. R. 1976. Cytochalasin-B inhibits actin-related gelation of HeLa cell extracts. *J. Cell Biol.* 71:303–307.

WELSH, J. H., and OSBORN, C. M. 1937. Diurnal changes in the retina of the catfish, *Ameiurus nebulosus. J. Comp. Neurol.* 66:349–359.

WILSON, L., ANDERSON, K., and CHEN, D. 1976. Non-stoichiometric poisoning of microtubule polymerization. In *Cell motility,* ed. R. D. Goldman, T. Pollard, and J. Rosenbaum. Cold Spring Harbor Laboratory. 1367 pp.

WUNDER, W. 1926. Die Bediutung des Adaptationsczustandes fur das Verhalten der Schelemente und des Pigmentes in der Netzhaut von Knochenfischen. *Z. Vergl. Physiol.* 3:595–614.

YAMADA, K. M., KENNETH, M., SPOONER, B. S., and WESSELLS, N. K. 1970. Axon growth: roles of microfilaments and microtubules. *Proc. Natl. Acad. Sci. U.S.A.* 66:1206–1212.

YOUNG, R. W., and DROZ, B. 1968. The renewal of protein in retinal rods and cones. *J. Cell Biol.* 39:169–184.

11 Adhesive Forces Between the Retinal Pigment Epithelium and Sensory Retina

Hanan Zauberman

THE SEPARATE LAYERS of neural retina and pigment epithelium come into apposition during embryonic life. Simultaneously, the potential space between them, the interphotoreceptor matrix, becomes filled with a thin layer of colloidal iron-staining mucopolysaccharides (Sidman 1958; Zimmerman and Eastham 1959; Fine and Zimmerman 1963). The villous processes of the pigment epithelium interdigitate with the tips of the outer segments. The villi are of two main types: slender and finger-like villi projecting into the interphotoreceptor matrix, and broad, rampart-like villi encompassing the tips of the outer segments. Both types contain melanin pigment granules (Spitznas and Hogan 1970). Membrane-bound actin filaments have been recently described in the monkey and human villous processes. It has been suggested that they provide the support necessary to maintain the elongated shape of the villous processes of the pigment epithelium and may play a role in retinal adhesion (Burnside and Laties 1976). With the exception of the ora serrata and the optic nerve, no anatomic connections have ever been demonstrated between these layers. Hence, the adhesion between the photoreceptor cells and the pigment epithelium is thought to be a very tenuous one throughout most of the retina. For example, in post-mortem eyes, the retina is often detached by the time the globe is sectioned, simply because it is more adherent to the shrinking vitreous than it is to the pigment epithelium. Also, in life, detachment of the sensory retina from the pigment epithelium may occur as a consequence either of vitreous traction or of exudative chorioretinopathy. In view of these observations, and others, it is not surprising that, until recently, it was generally conceded that little or no adhesion existed between the sensory retina and the pigment epithelium (Duke-Elder and Dobree 1967).

Despite this prevailing view, there have been a number of clinical observations suggesting that a firm retinal adhesiveness does in fact exist. For example, mechanical traction by the vitreous on the retina can often induce retinal tears, as noted in 5–10 percent of normal eyes (Okun 1961; Rutnin and Schepens 1967). However, long-term studies of individuals with retinal tears have shown that only infrequently do they lead to detachment (Byer 1972; Neumann and Hyams 1972). This suggests that some type of adhesion is present in the area surrounding the tear that "protects" the retina from detaching. We could speculate that progression from the tear stage to detachment is the consequence of a diffuse loss of this native cellular adhesion. Another observation supporting the notion of adhesive forces in the retina is the rarity of retinal detachment among Africans (Av-Shalom et al. 1967a), even though the incidence of retinal breaks is similar to that found in Caucasians (Av-Shalom et al. 1967b). It is possible that adhesion of the retina is intrinsically stronger in the former population.

This chapter will describe a number of biophysical approaches that have shed some light on the mechanism of retinal adhesion at the cellular and tissue levels, as well as histochemical and biochemical research that has helped to clarify some hitherto unknown structural properties of the subretinal space.

Cellular Adhesion

A full discussion of the theoretical basis of this important concept is beyond the scope of the present chapter, but a brief summary of current thinking in this field is nevertheless relevant. "Cellular adhesion" in vitro is usually quantitated by measuring the rate at which single cells are able to form aggregates; either fresh suspensions or cells cultured from embryonic sources are used for this purpose. Unfortunately, these model systems are far from the photoreceptor–pigment-epithelium interaction under discussion here. Even when retinal tissue is used, the results are applicable in only a limited sense. However, since glycoproteins and glycosyl transferases play important roles in determining cell-surface interactions in embryonic chick retina (Roth, McGuire, and Roseman 1971; Hausman and Moscona 1975), it is tempting to speculate that the glycoproteins found on the cell coat of mature mammalian photoreceptor cells (Bach and Berman 1970, 1971a; Feeney 1973b, c) have a similar function.

Of equal importance are membrane lipids, especially phospholipids. For example, treating embryonic chick retina with phospholipase A_2, or with the product of this phospholipase action, lysolecithin, decreases the adhesiveness of the cells (Curtis, Campbell, and Shaw 1975). Resynthesis of the phospholipids has been shown to restore cellular adhesion in these cultured cells.

Another parameter commonly used to measure cell adhesion is the ease with which cells growing in culture can be detached from the surface of the flasks in which they are cultured (Shields and Pollock 1974). Among the many agents tested, cyclic nucleotides are of greatest interest. It has been postulated that certain areas on the cell surface ("adhesion patches") are, after their formation, linked to the cytoplasmic contractile system of the cell. The latter, consisting of microtubules and microfilaments, could be responsible for the stimulatory effect on adhesion found in the presence of either cAMP or cGMP. Cytochalasin B, surprisingly, also causes increased adhesion of the cells.

Divalent cations such as Ca^{+2} may also play a role in cellular adhesion, as shown by recent studies using EGTA as the chelating agent (Shields and Pollock 1974). It is thought that calcium ions on or near the cell surface are able to cross-link the membrane with other nearby anionic sites. Removal of Ca^{+2} decreases the adhesiveness of cells, an effect that is canceled at temperatures below 20°C.

It would be highly desirable to apply the techniques used in these model systems to study interaction between the photoreceptors and the pigment epithelium. This could explain, at the molecular level, many of the biological observations to be described below. However, until carefully controlled experiments on isolated systems can be carried out, one can only speculate that in the mammalian eye, there must be many factors, in addition to the aforementioned ones, that contribute to adhesion between these cell layers. For example, van der Waals forces, or other types of short-range interactions between hydrophobic molecules, would be expected to come into play in this area. Spitznas and Hogan (1970) have suggested that capillary forces could intervene in promoting adhesion between the outer segments and the enveloping villous processes of the pigment epithelial cells. It is also possible that the orientation of the rhodopsin molecules in the rod outer segment disc membranes could contribute to the adhesive properties of the photoreceptor cells. Blasie (1972a) has shown by x-ray diffraction studies that although most of the rhodopsin molecule is submerged in a bilipid layer, part of it is exposed to the aqueous phase. Bleaching causes the rhodopsin to sink further into the lipid layer, causing a decrease in net charge and, hence, a reduction in electrostatic interaction between the discs (Blasie 1972b). This may explain the recent observation that the force required to experimentally detach the rabbit retina in vitro is greater by 20 percent in light-adapted eyes than in dark-adapted ones (Owczarek, Marak, and Pilkerton 1975).

In summary, there are many factors that contribute to the adhesive qualities of cells. Most of them are due to specialized chemical components on the cell surface. Others, which may be uniquely adapted to the pigment-epithelium–retinal area, include short-range intermolecular forces, electrostatic forces created by the peculiar orientation of the rhodopsin molecules, and "sticky substances" in the interphotoreceptor matrix.

The Subretinal Space: Structural and Biochemical Properties

The anatomic boundaries of the subretinal space are determined externally by the apical surfaces of the pigment epithelial cells and internally by the outer limiting membrane of the sensory retina. The latter is formed from the zonulae adherentes, which attach each photoreceptor to its neighboring Muller cell.

The subretinal space ends blindly in two circumferential cul-de-sacs. The anterior cul-de-sac occurs at the ora serrata, and the posterior is around the optic nerve head. Within the cul-de-sac itself the external limiting membrane unites with the pigment epithelium to form a tight seal (Fine and Yanoff 1972).

The extracellular space, as seen ultrastructurally, is somewhat wider than the 20 nm interneuronal space (Feeney 1973a), but it could indeed be even thinner in vivo. In terms of our present knowledge, "space" may be a somewhat misleading definition of the area, since, strictly speaking, it is a matrix. As recently suggested by Rohlich (1970), the term interphotoreceptor matrix is especially appropriate. Although this matrix lacks the typical morphology of other extracellular connective tissue matrices, filamentous structures that form a kind of three-dimensional network have been described (Röhlich 1970). Collagen has not been detected. The principal chemical components of the matrix are mucopolysaccharides, first noted histochemically by Sidman (1958) and later confirmed by many investigators (Zimmerman and Eastham 1959; Fine and Zimmerman 1963; Rohlich 1970; Feeney 1973b). The mucopolysaccharides show a positive reaction with a variety of specific stains, the most commonly used being Alcian blue and colloidal iron. In addition, PAS-staining substances, presumably glycoproteins, are especially prominent in (or on the surface of) the photoreceptor outer segments, but are not present in the matrix itself (Fine and Zimmerman 1963).

The mucopolysaccharides of the matrix could not be identified by histochemical studies, but in subsequent investigations, interphotoreceptor matrix was isolated from bovine retinas, and the mucopolysaccharides examined chemically (Berman 1969; Bach and Berman 1971a). Three groups of amino-sugar-containing macromolecules were isolated and identified. The principal components were undersulfated chondroitin sulfates, and the remainder of the mucopolysaccharides consisted of fully sulfated chondroitin sulfate and hyaluronic acid. Approximately 25 percent of the amino sugar-containing macromolecules consisted of sialoglycans—glycoproteins containing a significant amount of sialic acid. The chondroitin sulfate isolated from cattle retina was readily digested by hyaluronidase (Berman 1969), even though earlier histochemical studies suggested that these components were hyaluronidase-resistant (Zimmerman and Eastham 1959). The reasons for this discrepancy are not clear, but it is possible that in the histochemical studies, either the enzyme had not penetrated the tissue or the pH employed for the incubation was not the optimum one (Berman 1969).

It has been postulated (Zimmerman and Eastham 1959) that the mucopolysaccharides of the interphotoreceptor may be acting as "sticky substances" and in vivo could play a role in promoting adhesion between the neural retina and the pigment epithelium. Although this may be true to some extent, no direct proof is as yet available. Arguing against this possibility is the observation that the principal mucopolysaccharides of the interphotoreceptor matrix, the chondroitin sulfates, are neither of high molecular weight nor are they unusually viscous (Bach and Berman 1971b).

The PAS-positive substances identified by Fine and Zimmerman (1963) are probably the same glycoproteins as those isolated from saline extracts of bovine retina (Bach and Berman 1970). Structural studies have shown that they contain sialic acid and fucose, both being linked to the glycoprotein in terminal positions (Bach and Berman 1971b). Subsequent investigations (Feeney 1973b, c) support the view that the fucose-containing glycoprotein is part of the cell coat covering the rod outer segment membrane. In view of recent work on the role of surface glycoproteins as recognition factors regulating the uptake and entry of a variety of substances into cells, it is reasonable to suppose that the glycoproteins of the interphotoreceptor matrix, localized on the photoreceptor cell surface, may have other functions as well. One of these could be the kind of cell-surface interactions discussed above (Roth et al. 1971; Hausman and Moscona 1975).

What controls the amount of fluid in the interphotoreceptor space? The affinity for water of the mucopolysaccharides that form the matrix has not been examined directly, but it is tempting to speculate that these components, being highly polar substances and very soluble in water, may have important hydrophilic properties in situ. This could be related to the ease with which subretinal edema develops in a variety of conditions. It is also possible that active processes mediated by the bordering cell layers play a role in regulating the amount of fluid present in the interphotoreceptor matrix and in maintaining the thinness of the subretinal space. This would explain why subretinal fluid is often easily reabsorbed after closing a retinal tear in retinal detachment surgery (Custodis 1953; Lincoff et al. 1965).

Biophysical Data on Retinal Adhesion

Our present understanding of the physical characteristics of retinal adhesion is based on (a) quantitative measurements, obtained by peeling and traction methods in vitro, and (b) qualitative evaluations of the resistance of the sensory retina to detachment by traction in vivo, as well as post mortem. These studies, and the structural changes produced when the neural retina and the pigment epithelium are separated, are described below.

PEELING EXPERIMENTS. The force needed to separate the retina from the pigment cell layer in freshly enucleated rabbit eyes has been measured in rectangular strips, 6.5 mm wide, of vitreous-free tissue (de Guillebon and Zauberman 1972). Using an instrument specifically designed for this purpose (Fig. 11.1), the edge of the retina is glued to a thin

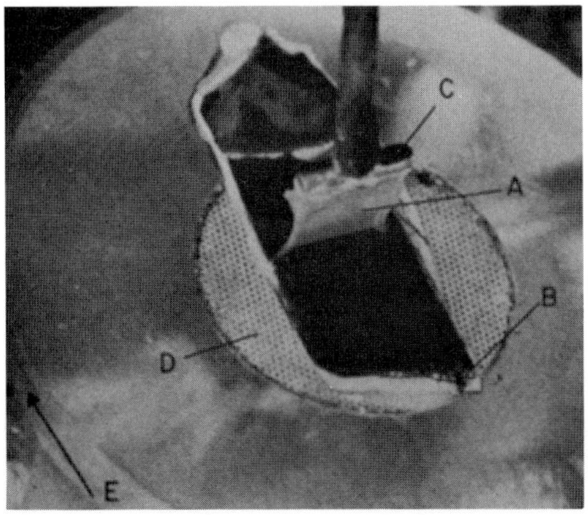

Figure 11.1 *Peeling off the retina to measure adhesive force.*

metallic rod attached in such a way that the retina can be peeled off at rates of 2, 8.5, 42, and 215 mm/min. The force required for separating the retina varies between 40 and 100 mg over most of the strip, using peeling rates ranging from 2 to 42 mm/min (Fig. 11.2). At a faster rate of 210 mm/min, the retina could not be separated because the force, which had been rapidly transferred to the choroid, produced instead a separation between the cho-

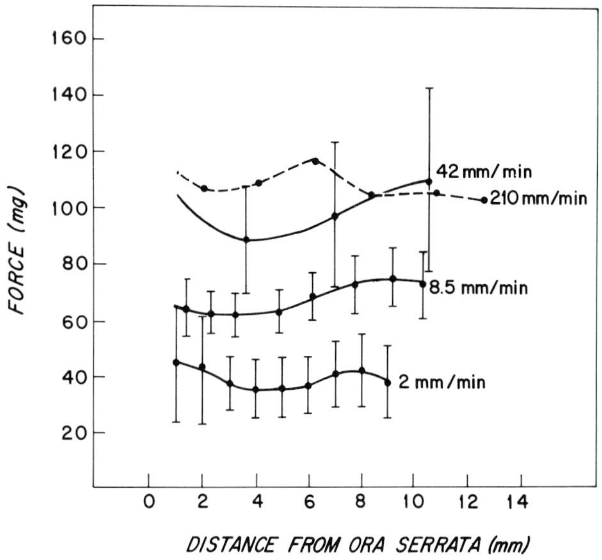

Figure 11.2 *The relationship between peeling force across the retinal strip and peeling rate.*

roid and the sclera. At lower peeling rates, the force needed to separate the retina in rabbit eyes increases with the peeling rate applied; a semilogarithmic relationship exists between peeling rates of 2 and 42 mm/min (Fig. 11.3). In freshly enucleated

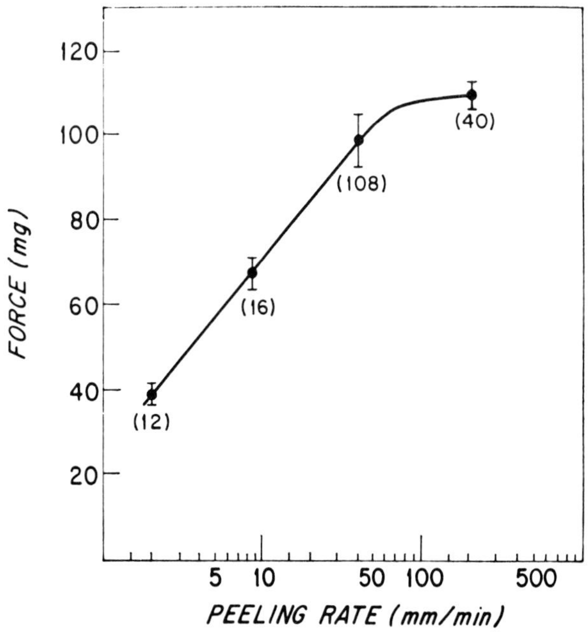

Figure 11.3 *A semilogarithmic relationship between peeling force and peeling rate, at rates of 2 to 42 mm/min. The number of experiments for each peeling rate is shown in brackets.*

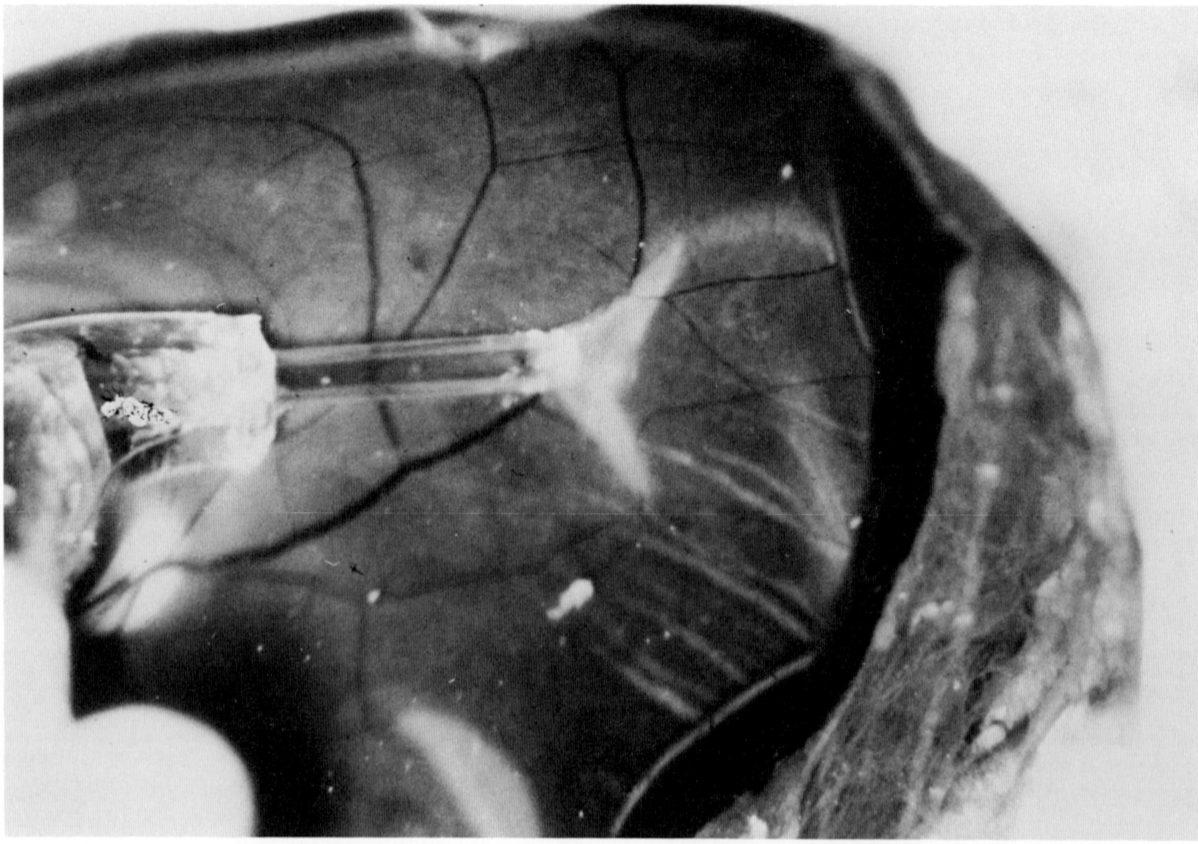

Figure 11.4 *Traction on the retina at 0.2 mm/min. Note developing large detachment.*

cat eyes, a peeling rate of 5 mm/min yields forces similar to those obtained in rabbits using the same rate of peeling (Zauberman and Berman 1969).

TRACTION EXPERIMENTS. The adhesive properties of the retina have been examined in whole, undisturbed, posterior segments of freshly enucleated owl-monkey and rabbit eyes in which the vitreous was not removed. Localized traction exerted on the retina by means of a glued-on fine glass tube connected to a force transducer and recording system showed that lower traction rates result in larger and higher retinal detachment than progressively higher traction rates (Zauberman, de Guillebon, and Holly 1972). Figure 11.4 shows the large detachment and break at a traction rate of 0.2 mm/min, while Figure 11.5 shows the minute detachment and multiple breaks obtained at a rate of 8.5 mm/min using owl-monkey eyes. In these experiments too, very high rates of traction resulted in choroid separation, with the retina remaining attached to the pigment epithelium.

An interesting relationship has been observed between the traction rate applied and the plane of cleavage between the layers of neural retina and pigment epithelium. At a low traction rate of 0.2 mm/min, the cleavage occurs at the interface between the pigment epithelial villous processes and

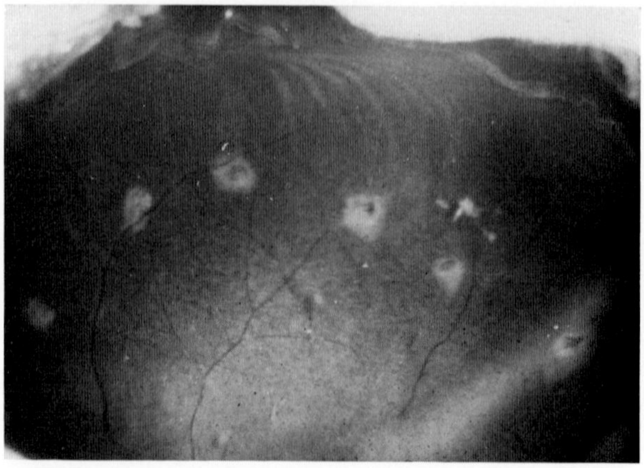

Figure 11.5 *Traction at a rate of 8.5 mm/min. The minute detachment and tear are localized to the area of the attachment of the traction tube.*

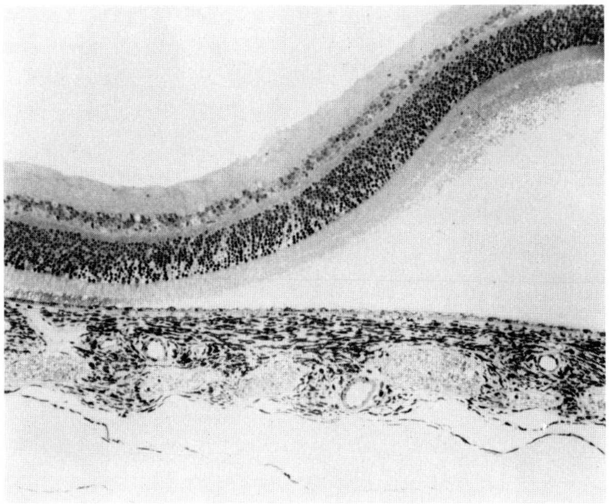

Figure 11.6 *Cleavage of retina between outer segments and pigment epithelium at 0.2 mm/min. (Stained with toluidine blue. ×80.)*

the photoreceptor outer segments (Fig. 11.6), while at 8.5 mm/min, traction induces separation at the level of the apical surface of the pigment epithelium (Fig. 11.7).

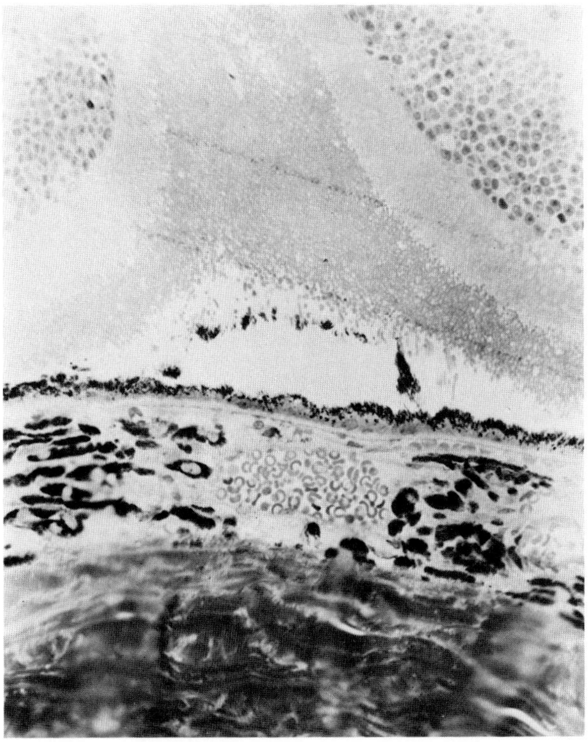

Figure 11.7 *At 8.5 mm/min, traction on the retina results in torn clumps of pigment epithelium. (Stained with toluidine blue. ×308.)*

As already indicated in the peeling experiments, the force recorded when the retina was separated was shown to be progressively higher with increasing traction rates. Figure 11.8 shows that the relationship between the force needed to separate the retina and the traction rate is semilogarithmic.

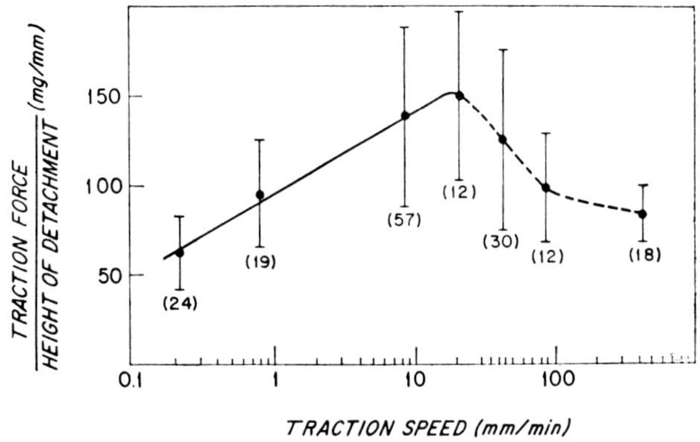

Figure 11.8 *Relationship between rate of traction and ratio of traction force to height of detachment. Numbers indicate number of experiments for each traction rate.*

INTERPRETATION OF PEELING AND TRACTION EXPERIMENTS IN VITRO. Taking into account anatomic, histochemical (Sidman 1958; Zimmerman and Eastham 1959; Fine and Zimmerman 1963) and chemical (Berman 1969; Ocumpaugh and Young 1966; Feeney 1973b) data, the adhesive joint between the villous processes of the pigment epithelium and the outer segments can be likened to the adhesive joint of two solid irregular surfaces mediated by a thin layer of viscous substances (Fig. 11.9). For such a simple model, the time needed to separate

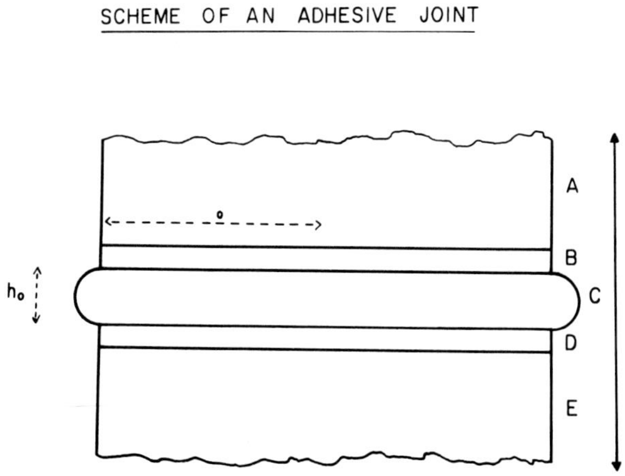

Figure 11.9 *Scheme of adhesive joint:* A *and* E, *solids;* B *and* D, *adherend surfaces;* C, *adhesive layer;* O, *radius of adhesive joint. (Adapted from Bikerman and Yap 1958.)*

the two solids depends both on the mechanical properties of the adhesive—for example, its viscosity or viscoelasticity—and on the geometry of the system. When traction is applied to the system, the behavior of the components will determine the type of cleavage. For the initial stage of pulling—the most important in the process—the following equation has been derived to describe tackiness (Bickerman and Yap 1958):

$$Ft = \frac{\eta^3 \pi a^4}{4}\left(\frac{1}{h_1^2} - \right)\frac{1}{h_2^2}$$

where F is the force applied, t is time (the duration of the traction), η is the viscosity of the adhesive liquid, a is the radius of the specimen, and h_1 and h_2 are the initial and final thicknesses of the adhesive. The equation shows that there is no definite force opposing the breaking of the adhesive joint, because the resistance to separation is dependent on the rate of separation. As h is increased by a force attempting to separate the surfaces, a viscous flow of the adhesive will take place toward the center. On the other hand, the separation will occur within the adhesive material itself if the force applied is small enough to allow for such viscous flow. These rheological phenomena could be invoked to explain the clean separation observed between neural retina and pigment epithelium when low traction rates are applied (Fig. 11.6). If the force applied to separate the surfaces is great, and is exerted over a short period of time, a cohesive failure may take place in the adherent, because such an application of force does not allow the viscous flow to occur. The cleavage observed at the level of the villous processes of the pigment epithelium when high traction rates are used is consistent with this hypothesis (Fig. 11.7).

It is clear from this model that the thinner the layer interposed between the surfaces, the stronger will be the resistance to separation. This is a fundamental point since it is reasonable to assume that in life the subretinal space is thinner than it is after death, and this results in a greater resistance to detachment of the retina.

Influence of Pigment and the State of Light Adaptation on Retinal Adhesion

There is, in some animals, a direct relationship between the amount of pigmentation and the degree of retinal adhesion when measured in vitro. Thus, the force needed to peel off the neural retinal from the pigment epithelium is stronger over the tapetum negrum than over the tapetum lucidum of the cat (Zauberman and Berman 1969). In the frog, two processes—pigmentation and the state of light adaptation—are interrelated in terms of their effects on retinal adhesion (Zauberman 1972a). During light adaptation, melanin pigments have been observed moving toward (and into) the villous process of the pigment epithelium, while during dark adaptation the pigments retreat to the base of the cell. Measurements of the force needed to separate the neural retina from the pigment epithelium in vitro showed it to be about twice as high in the light as in the dark (Zauberman 1972a). A hypothesis put forward to explain this phenomenon is that the pigment granules may be acting like barbs between the individual outer segments.

Pigment migration is not known to occur in mammalian eyes, so that the problems of pigmentation and of light-adaptation—as they relate to retinal adhesion—would have to be approached differently. Owczarek et al. (1975) compared the adhesive forces in vitro in albino, moderately pigmented, and heavily pigmented rabbits and found no statistically significant differences between the three groups. These results are at variance with previous findings (de Guillebon, Tribonniere, and Pomerantzeff 1971) which showed that in rabbits pigmentation does in fact increase the force of adhesion. Light may also play a role in the rabbit eye since in animals exposed to 100 watts of illumination at 50 cm for 5 minutes prior to enucleation, the adhesive forces were slightly (20 percent) greater than in dark-adapted animals (Owczarek et al. 1975). Subtle changes in electrostatic forces between the photoreceptor discs during light adaptation were thought to play a role in promoting greater adhesive forces in this area in the light.

Comparison of Retinal Adhesion in Vivo and Post Mortem

As pointed out above, the force needed to separate the neural retina from the pigment epithelium is quite small in vitro, especially at low peeling or traction rates. In contrast, the components responsible for retinal adhesion, whatever they may be, appear to be stronger in the living animal. This has been measured semiquantitatively by applying traction to localized areas of the retina, using the instrument shown in Figure 11.10 (Zauberman and de Guillebon 1972). Using the albino rabbit, a fine tube was glued to the retina and the tissue was separated

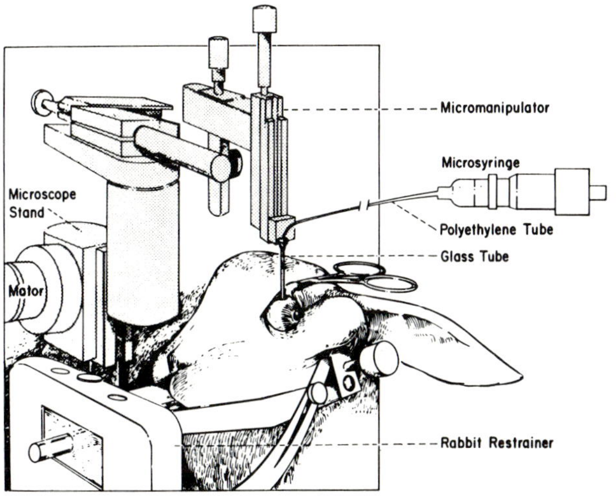

Figure 11.10 *Instrument to exert traction on the retina in vivo.*

by applying traction at a rate of 0.15 mm/min. Under these conditions, the retina was detached and torn off only over an area measuring 0.9 mm in diameter, precisely the radius of the tube that had been glued to the tissue. No detachment was observed beyond this area (Fig. 11.11). Even when the experiment was repeated several times over areas of retina adjacent to a tear, the surface detached was never

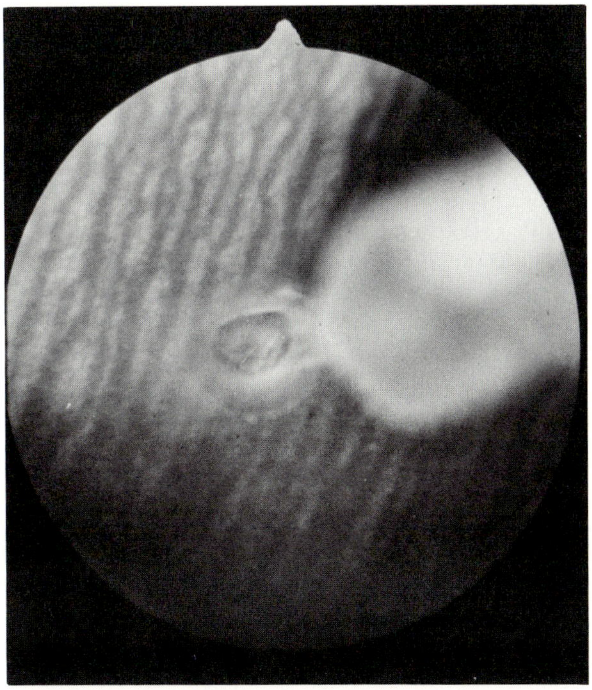

Figure 11.11 *Traction on the retina in vivo at 0.15 mm/min. Note localized tear.*

found to be any wider (Fig. 11.12). These results support the view that retinal adhesion is very firm in vivo and, moreover, that there is still considerable adhesion in the area surrounding an experimentally induced retinal tear.

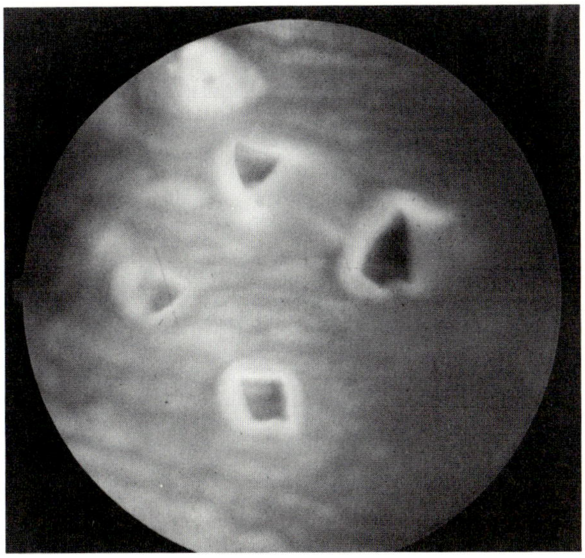

Figure 11.12 *Four adjacent tears produced in vivo at 0.15 mm/min traction rate. Note that each detachment and tear is limited to the area of the tip of the traction element.*

Yet, if the same experiment is carried out shortly after killing the animal with an overdose of anaesthesia, the area of detachment is three to eight times larger than in vivo (Fig. 11.13).

To gain further insight into this finding, the effects of retinal separation by traction on the structure of the plane of cleavage were also examined. Electron micrographs showed that traction on the retina in the living animal results in stretching and disruption of the villous processes of the pigment epithelium (Fig. 11.14), suggesting a strong resistance of the tissues to separation. In contrast, after death no such signs of traction on the apical surface of the pigmented cells could be observed (Fig. 11.15). The explanation for the stronger retinal adhesion in vivo is not entirely clear, but the following possibilities should be considered.

Active Transport Mediated by the Retina or the Pigment Epithelium

It is well known clinically that in many individuals with retinal detachment, sealing of the retinal tear will lead to fluid reabsorption within a few

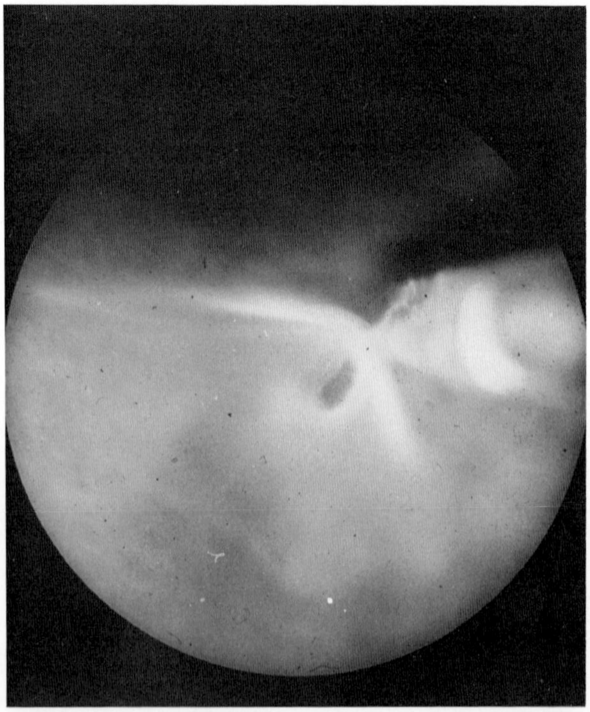

Figure 11.13 *Traction on the retina after death. Rate of traction 0.15 mm/min. Note large area of detachment.*

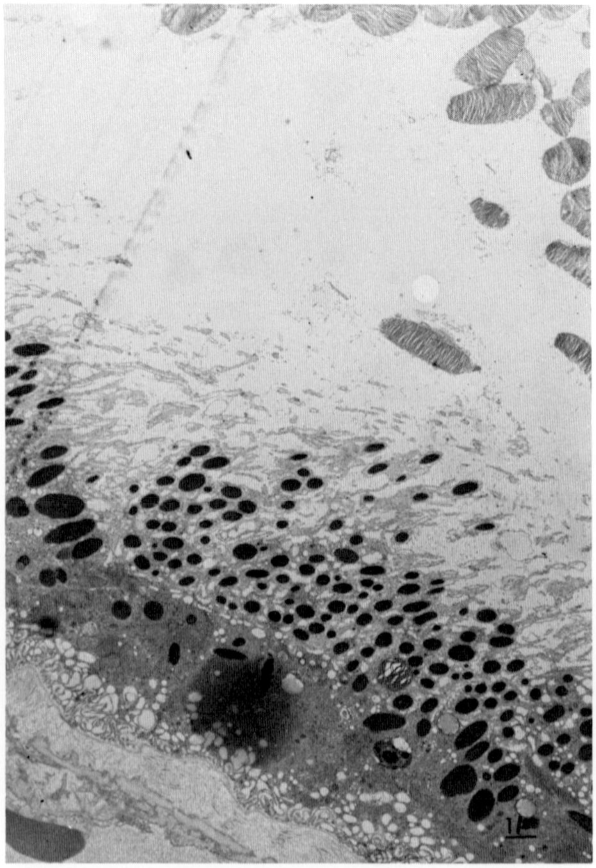

Figure 11.14 *Electron micrograph of cleavage area in vivo. Note stretching and tearing of pigment epithelium villous processes of albino rabbit. (×1,740.)*

hours. Although the mechanism of fluid removal from the subretinal compartment is not well understood, it may be related to metabolic pump mechanisms in the pigment epithelium. Lasansky and de Fisch (1966) were the first to demonstrate a chloride flux in the toad pigment epithelium in the direction of the choriocapillaries. The fact that it took place against an electrochemical gradient suggested that it was related to the short circuit current existing in this retinal area.

In the rabbit, differences in ion concentrations across the blood-retinal barrier have been demonstrated in experiments showing active transport of iodopyracet (Forbes and Becker 1960) as well as of ions such as calcium, magnesium, and potassium (Bito 1970; Noell, Crapper, and Paganelli 1965). Using fluorescein as a tracer, it has been shown that in rabbits, in vivo, the dye is transported against a gradient from the vitreous across the pigment epithelium to the choroid (Cunha-Vaz and Maurice 1967). The latter experiments also showed that inhibitors of anion transport such as benzylpenicillin and Benemid not only interfere with the efflux of fluorescein, but also produce an acute detachment of the retina. Similarly, injection of 10^{-4}M ouabain into the vitreous causes a deterioration of retinal adhesion, as shown by in vivo traction studies (Zauberman 1972b). It is not clear whether ouabain acts directly, on the ion pump, or indirectly, on the membranes of the pigment cells, damaging their structure and thus increasing their permeability. Steinberg and Miller (1973) found that a sodium-potassium pump sensitive to ouabain exists on the apical side of the pigment epithelial cells of the frog. If ouabain, in these experiments, is inhibiting the transport of ions and water out of the subretinal space, then one could speculate that in the absence of inhibitors—that is, in the normal living eye—there could be mechanisms at work that are constantly removing small amounts of fluid from the subretinal space. This would maintain the interphotoreceptor matrix at levels of optimum thinness, which theory suggests promotes the maximal retina–pigment-epithelium apposition.

Whether a "metabolic pump" in the pigment

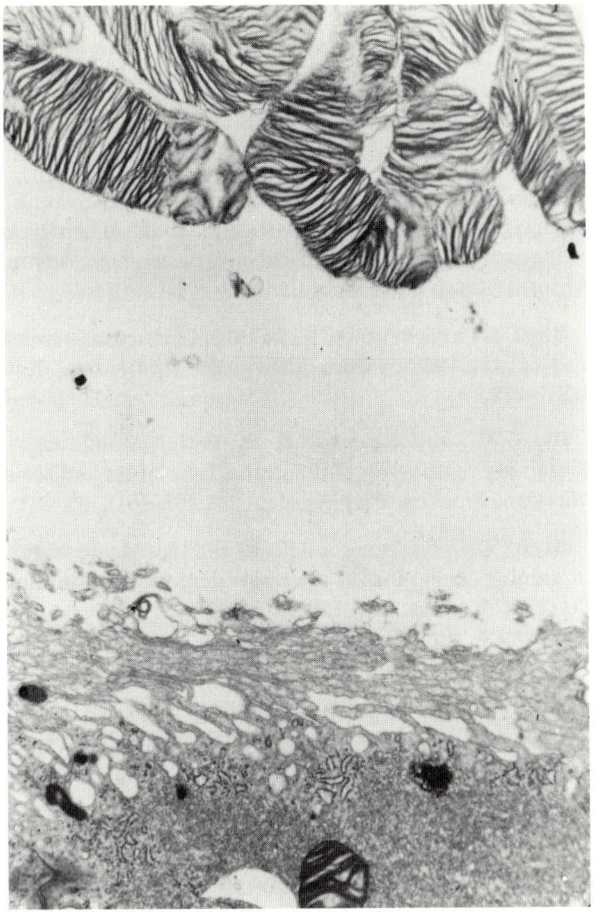

Figure 11.15 *Electron micrograph of cleavage area after death. There is no evidence of obvious stretching of pigment epithelial processes.*

epithelial cells could in addition give rise to pressure differences between the subretinal and the vitreous compartments, and thus play a role in retinal adhesion, has been the object of experimental study (Maurice, Salmon, and Zauberman 1971). This hypothesis is based on the required presence of a hole before rhegmatogenous retinal detachment can develop, and on the observation that intraocular pressure is lower in eyes with detached retina, indicating a drop in the pressure value in the subretinal compartment. Such a pressure gradient could be provided by the presence of a pumping mechanism in the pigment epithelium that removes fluid from the subretinal compartment. Furthermore, the effect of closing the retinal hole in reestablishing retinal adhesion is well known, and is usually followed by a rise in intraocular pressure to normal values. However, experiments designed to measure simultaneously the pressure in the two compartments at different intraocular pressure levels have shown no pressure differences (Fig. 11.16) (Maurice et al. 1971).

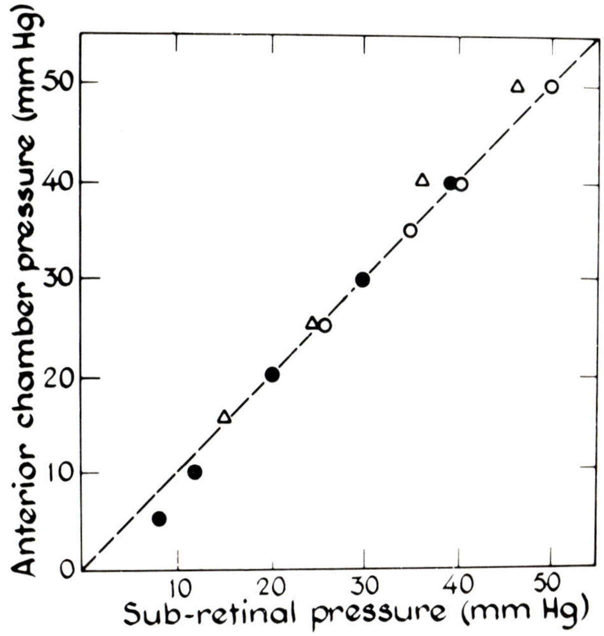

Figure 11.16 *Simultaneous measurements of pressure in anterior chamber and subretinal compartment. No difference was detected.*

HYDROSTATIC FORCES THAT MAINTAIN RETINAL ADHESION. Fatt and Shantinath (1971) hypothesized that hydrostatic forces could produce a differential pressure across the intact retina, thus providing a factor promoting retinal adhesion. Theoretically, factors such as intraocular pressure and the difference between retinal and scleral flow conductivity would tend to support this theory. However, this model must be "leak proof"—that is, it does not allow for tears in the retina. The theory is therefore inconsistent with the fact that in most instances in which one or more holes are present, other areas of the retina still remain adherent to the pigment epithelium. Indeed, it has been shown by retinal traction experiments in vivo that the normal adhesion of the retina adjacent to one or more breaks is not altered (Zauberman and de Guillebon 1972). Also, clinically a very small proportion of eyes with retinal tears develop retinal detachment (Byer 1972; Neumann and Hyams 1972). Of course, it could be hypothesized that the vitreous cortex plugs the retinal tears, and prevents fluid leakage into the subretinal space. Yet, in many in-

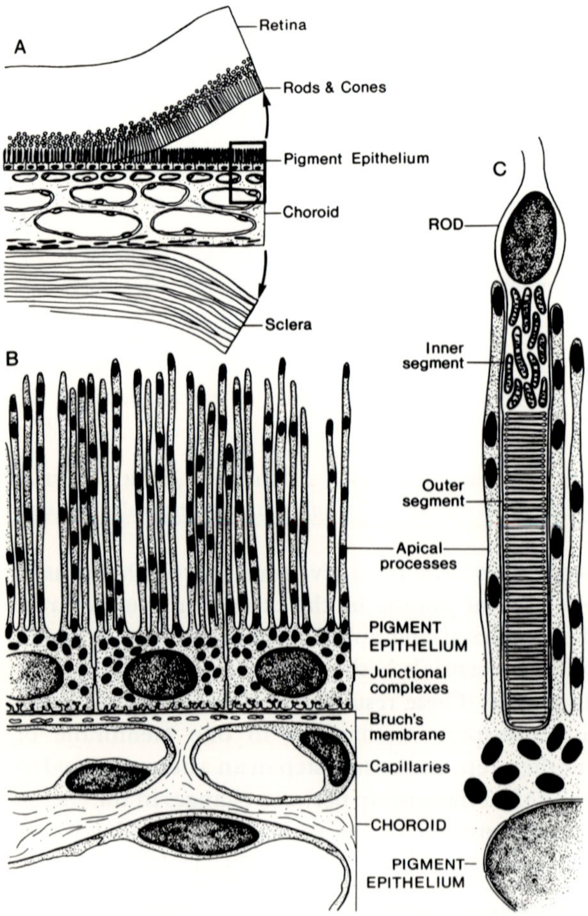

Figure 12.1 *Diagram of the tissue.* A, *removal of sensory retina and sclera from RPE-choroid.* B, *detail of RPE-choroid.* C, *detail of association between pigment epithelial apical processes and red rods. (From Miller and Steinberg 1977a, after Nilsson, 1964.)*

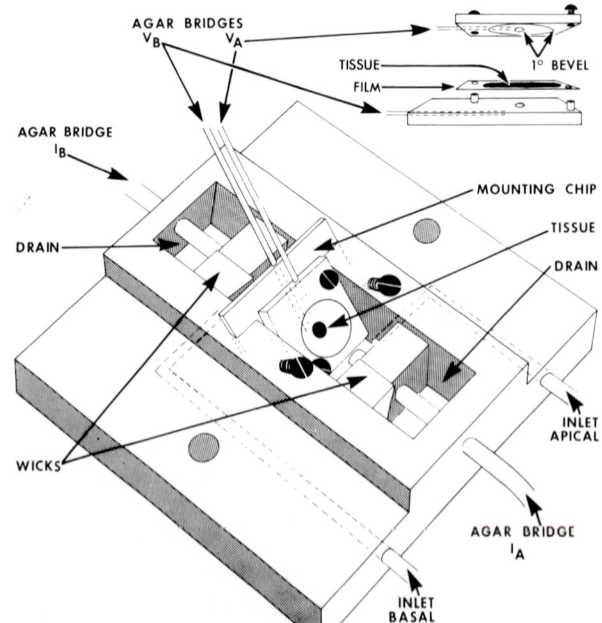

Figure 12.2 *Diagram of the experimental chamber. The insert shows the three parts of the mounting chip: top and bottom plates and x-ray film. (From Miller and Steinberg 1977a.)*

mounted in a plastic chip, which consisted of two lucite plates that compressed the tissue at its edges (Fig. 12.2). These plates had .07 cm^2 centered holes that exposed the choroid on one side and the apical RPE surface on the other.

For microelectrode studies the chip was mounted in a perfusion chamber at a 45° angle so that the apical RPE surface could be penetrated with microelectrodes (Fig. 12.2). The two sides of the pigment epithelium were immersed in 3.3 ml baths separately perfused with a modified amphibian Ringers. The potential generated by the epithelium—that is, the transepithelial potential (TEP), was recorded through Agar-Ringer bridges symmetrically placed on each side of the tissue (Fig. 12.2, V_A, V_B). The microelectrode penetrated the apical membrane of the cell and recorded either the apical membrane potential, when it was referred to the apical bath, or the basal membrane potential, when it was referred to the basal bath.

For flux studies the tissues were mounted perpendicularly, and the baths were not perfused; instead, the fluid in each chamber was stirred and oxygenated by a stream of water-saturated gas bubbles. To determine transepithelial unidirectional fluxes, a tracer amount of the isotope was added either to the basal solution (the solution facing the choroid) or to the apical solution. The fluxes—apical to basal (retina to choroid) and basal to apical (choroid to retina)—were then measured by sampling the "cold" solution on the opposite side of the tissue every 20 minutes. We often ran unidirectional fluxes in opposite directions on paired tissues from the same eye. The difference between these unidirectional fluxes is the net flux, which can be used to assess the direction of active transport across the tissue. Fluxes in both directions can also be measured on the same piece of tissue. For example, after a flux experiment was performed in one direction, the chip containing the tissue was removed from the flux chamber and placed in a beaker where it was washed for 90 minutes. It was then returned to a clean flux chamber, and when the background counts stabilized, a second unidirectional flux experiment was performed in the opposite direction.

Electrical Characteristics of the Isolated RPE-Choroid

The transepithelial potential developed across this tissue is a consequence of the asymmetric character of the apical and basal cell membranes. These membranes can differ in their resistances, relative ionic conductances, and in the presence or absence of certain pumps or carrier mechanisms. The pumps may be located at one or both membranes, and they can be electrogenic, thereby contributing directly to the membrane potential. All of these factors determine the cell membrane potentials and the transport characteristics of the pigment epithelial cells.

VOLTAGES. When set up in the experimental chamber and perfused on both sides with 27.5 mM HCO_3^- Ringer's, the bullfrog RPE-choroid develops an average transepithelial potential of 10 mV (range 8–20). The potential is oriented so that the apical surface of the RPE is electrically positive relative to the basal and, since the potential disappears when the RPE cells are scraped away from the choroid, it must depend on an intact epithelial cell layer. The transepithelial potential is the difference between the membrane potentials of the apical and basal cell membranes (Fig. 12.3).

The internal potential of the pigment epithelial cell is large and negative with respect to both bathing solutions. In a large sample of cells (345, 67 tissues), for example, over 85 percent of the membrane potentials recorded across the apical membrane were between −75 and −95 mV; the mean was −88 mV. In each preparation, the potential across the basal membrane is depolarized with respect to the apical membrane by the amount of the transepithelial potential. In any one tissue, all of the undamaged cells show, essentially, identical membrane potentials.

The size of the transepithelial potential is also affected by the presence of electrical shunting between the apical and basal cell membranes. If it were not for shunting, the transepithelial potential would be larger, and the apical membrane potential would be more hyperpolarized with respect to the basal. It can be calculated that if the shunt resistance were infinite, the maximum theoretical transepithelial potential for the frog pigment epithelium would be 28 mV (Miller and Steinberg 1977a). The shunting can be greatly reduced by placing the tissue in a low Na+ Ringer's solution (11 mM Na+); transepithelial potentials as high as 25 mV have then been recorded (unpublished observations). Also, in two species of toads, where the average transepithelial re-

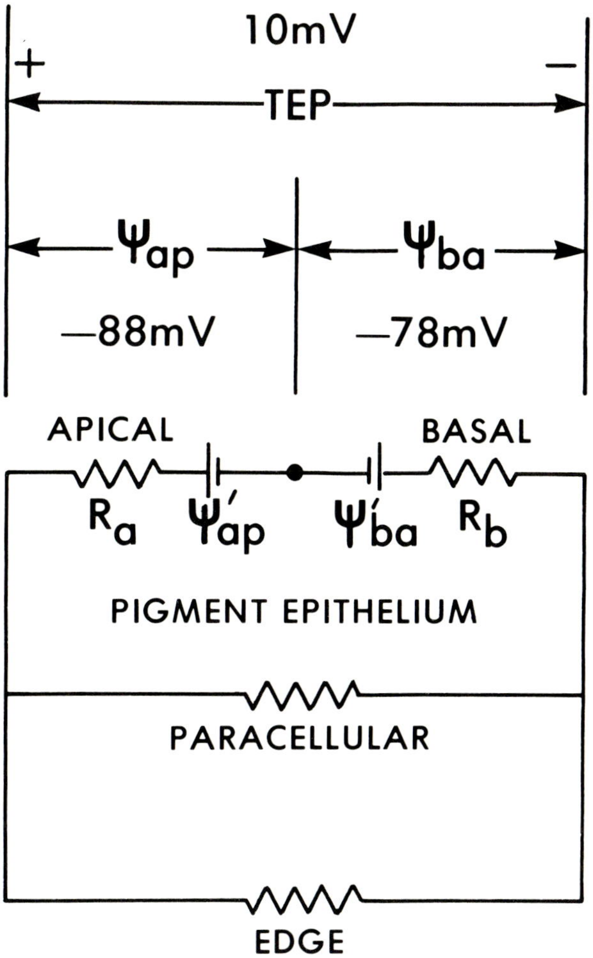

Figure 12.3 *Equivalent circuit and average voltages for the RPE-choroid.* ψ_{ap} *and* ψ_{ba} *refer to the membrane potentials recorded, respectively, between the inside of an RPE cell and the apical and basal solutions. The junctional complexes connecting the pigment epithelial cells in frogs differ from those of some other species in being absent at the cellular apices; they begin about halfway down the lateral surface of each cell (Porter and Yamada 1960; Hudspeth and Yee 1973). Both the apical and basal membranes, therefore, include portions of the lateral membrane (Porter and Yamada 1960). The apical and basal cell membranes are each represented by a battery* ψ'_{ap}, ψ'_{ba}, *and a resistor,* R_a, R_b. *The shunt resistance is represented by separate resistors for its two components, paracellular and edge. The choroidal resistance is negligibly small and is not represented.*

sistances were considerably higher than in frogs, the usual transepithelial potentials were 20–30 mV (Lasansky and DeFisch 1966).

TRANSEPITHELIAL RESISTANCES. It has been known for a long time that the retina contains a high resistance barrier, and it has been suspected from electrophysiological studies that the barrier is the pigment epithelium (Brindley and Hamasaki 1963). Cohen (1965) suggested that the pigment epithelial "tight" junctions could be the locus of the resistance barrier, or "R-membrane," as it was earlier referred to by Brindley (1956). The resistance of the pigment epithelium depends on two major factors, the resistance through the cells themselves and a shunt resistance. The shunt resistance consists of two distinct components. One is the pathway through the junctional complexes between the cells, and the other is a pathway around the edge of the tissue (Fig. 12.3) whether from damage in an in vitro experiment or via adjacent ocular tissues (Rodieck 1973). The edge pathway can vary considerably between in vitro preparations, while it is probably quite constant in situ.

In looking for the origin of the R-membrane at the pigment epithelium, the major consideration must be the properties of the intercellular pathway, since in other epithelia differences in transepithelial resistance originate in differences in the conductance of this pathway (Frömter 1972; Lewis and Diamond 1976; Lewis, Eaton, and Diamond 1976). Depending on the "leakiness" of the epithelia to salt and water movement, they can be classified as "leaky" or "tight." Leaky epithelia are physiologically characterized, in part, by their low transepithelial resistances ($<$200 ohm-cm^2), high hydraulic conductivities, and absent or low transepithelial potentials. Tight epithelia have higher transepithelial resistances ($>$400 ohm-cm^2) and potentials, and relatively lower hydraulic conductivities (Rose and Schultz 1971; Frömter and Diamond 1972). Anatomically, there is evidence that the intercellular pathway is identical with the "tight" junctions or zonulae occludentes. Ionic lanthanum traverses the "tight" junctions of leaky epithelia (Machen, Erlij, and Wooding 1972; Castel, Sahar, and Erlij 1974). Freeze-fracture studies show that tight and leaky epithelia can be distinguished by the substructure of the zonula occludens. Tight epithelia have relatively deep zonula occludens that contain at least five junctional strands interconnected in a complex mosaic (Claude and Goodenough 1973).

The leakiness of the RPE-choroid preparation can be assessed in the most well-sealed tissues, where the edge resistance is highest. In that case, the tissue usually develops a TEP of 15 mV or higher, and has a transepithelial resistance of 350–600 Ωcm^2 (Brindley and Hamasaki [1963] found lower values for the transepithelial resistance, possibly because of shunting). These values indicate that the pigment epithelium is probably tight, and this is confirmed by calculating that in well-sealed tissues the shunt resistance is ten times larger than the total cellular resistance (Miller and Steinberg 1977a). This evidence is directly supported by Hudspeth and Yee's (1973) freeze-fracture study in the frog (*R. pipiens*), which showed the RPE zonula occludens to contain numerous anastamosing strands. By contrast, in a leaky epithelium, such as *Necturus* gall bladder, the cellular resistance is twenty-four times higher than the intercellular shunt resistance (Frömter 1972).

CELL MEMBRANE RESISTANCE. The resistance of the apical membrane is approximately 3500 Ω. From transmission electron micrographs (Nilsson 1964) and scanning electron micrographs (Steinberg 1973) we estimate that the apical processes enlarge the geometrical area (.07 cm^2) by a factor of about 30. This means that the specific resistance is about 6,000–7,000 Ωcm^2, about 50 percent larger than the values reported for the apical membrane of several other epithelia (Frömter 1972; Reuss and Finn 1974; Spenny, Flemstrom, Shoemaker, and Sachs 1975). The calculated basal membrane resistance is about 6400 Ω, and we estimate that the geometrical area (.07 cm^2) is, at most, increased by a factor of three by the sparse basal infoldings (see Chapter 5). This yields a specific resistance of 450–1200 Ωcm^2, considerably lower than the apical membrane.

Relative Ionic Conductances

The relative ionic conductances are determined by changing the extracellular concentration of one permeant ion at a time and then recording the resultant change in membrane potential. In our preparation, the apical and basal membranes could be studied separately by changing the concentration on only one side, and recording the voltage response of that membrane (apical or basal). Each change was made from a solution of fixed composition—the reference solution. This experiment was repeated over a fairly wide range of concentration changes and the data were used to generate a curve relating log (ion concentration) to membrane potential change. The slope of this curve, evaluated at the concentration in the reference solution, was used to calculate the relative ionic conductance, T_i. If the voltage responses are due only to diffusion potentials, and if the extracellular changes do not produce intracellular ion concentration changes, then the relative ionic conductances, at each membrane, should sum to one (Hodgkin and Horowicz 1959). There are a number of problems that complicate what appears to be a relatively simple experiment. The ion concentration changes must be made fast enough so that the intracellular concentrations do not change. There are general problems in interpreting the shifts of membrane potential when electrogenic pumps are present or membrane conductance is altered. In epithelia, transepithelial electrical shunting creates additional problems. In this preparation, for example, the *measured* change in membrane potential ($\Delta\psi$) cannot be directly used to calculate the relative ionic conductance, for it has been reduced by electrical shunting across the paracellular pathway. This also causes the opposite membrane potential to change (Fig. 12.4). In addition, a voltage can develop across the paracellular pathway when ion concentrations on each side are different from each other, and this shunt voltage will also change both membrane potentials. Finally, there is some evidence that a change in membrane potential initiated at one membrane of an epithelium can, by an unknown mechanism, lead to changes in the potential of the opposite membrane (Reuss and Finn 1975; Finn 1976).

These problems can be overcome (Miller and Steinberg 1977a), and we have been able to make fairly good estimates of the relative ionic conductances. It should be pointed out that the potential changes plotted in Figures 12.5 and 12.7, which are used to determine the relative ionic conductances, are *shunt* corrected potentials ($\Delta\psi'$).

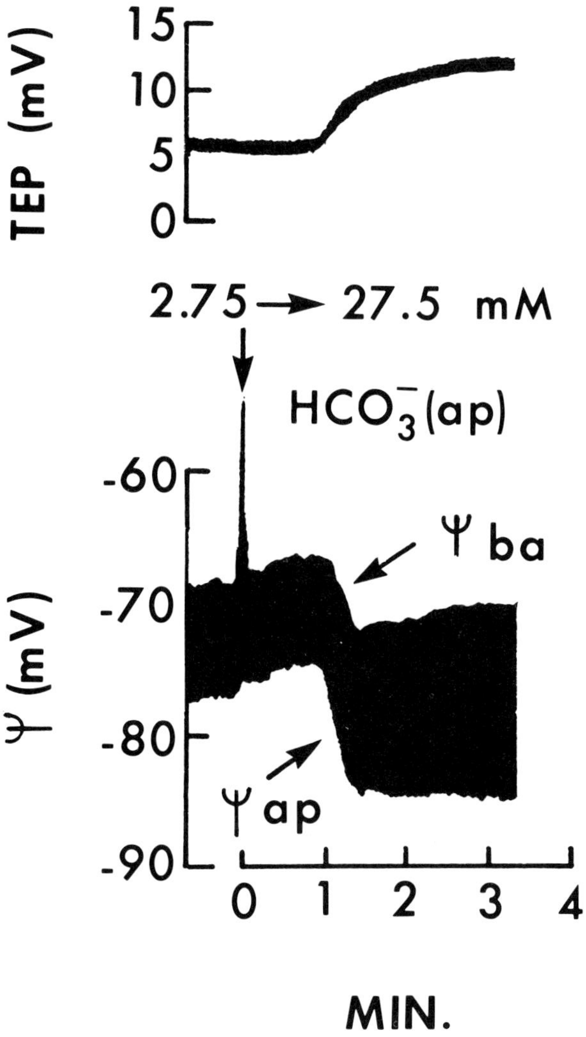

Figure 12.4 *Changes of both ψ_{ab} and ψ_{ba} produced by an ion concentration change on only one side. The arrows in the membrane potential trace (bottom) indicate that ψ_{ba} and ψ_{ap} are, respectively, the upper and lower borders of the trace. The width of the trace equals TEP plus 0.8 mV noise. The increase in apical $[HCO_3^-]_o$ (first arrow) rapidly and simultaneously hyperpolarizes both membrane potentials, $\Delta\psi_{ap} = 10$ mV, and $\Delta\psi_{ba} = 4$ mV. This increases TEP (upper record) from its low level of 6.6 mV in 2.75 mM $[HCO_3^-]$ to 12.6 mV; the increase in TEP appears also in the lower record as the widened trace. (From Miller and Steinberg 1977a.)*

BASAL MEMBRANE. Of all the ions tested (Na^+, HCO_3^-, Cl^-, K^+, Ca^{+2}), potassium was the only one that appreciably altered the basal membrane potential during the observation period (3 min). An elevation of potassium rapidly depolarized the basal membrane, as expected for a potassium diffusion potential. The relative potassium conductance of the basal membrane was estimated to be 0.90 from the slope of the curve *above* 20 mM K^+ in Fig. 12.5. This slope is 52 mV for a tenfold change in potassium concentration and, therefore, the basal membrane acts almost like a potassium electrode.

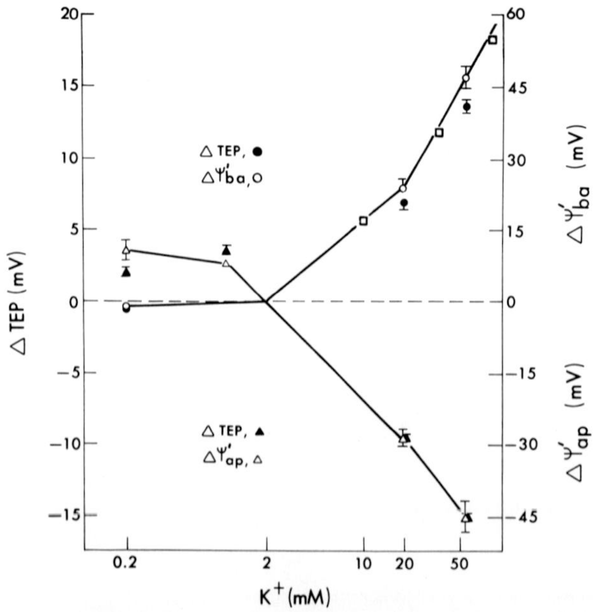

Figure 12.5 *$\Delta\psi'_{ba}$ (top) and $\Delta\psi'_{ap}$ (bottom) as a function of $[K+]_o$. The points are means and SEM's. The solid lines connect the intracellular data points. The value of $\Delta\psi'_{ap}$ at 1 mM was calculated from a series of TEP measurements between 1 and 2 mM that gave TEP = 3.0 ± .3 mV (7 tissues). The open squares were also calculated from a series of TEP measurements. (From Miller and Steinberg, 1977a.)*

The membrane potential only hyperpolarized by 1–2 mV on changing extracellular potassium from 2 mM to 0.2 mM, and, therefore, there was a sharp change in slope near 2 mM (and to a lesser degree near 20 mM) that presented a problem in interpretation. These changes in slope could not be explained by changes in basal membrane potassium conductance. A more likely explanation was that the choroid prevented the basal membrane from *seeing* the full change in concentration. Since these experiments were completed, Dr. Burks Oakley has placed a K-sensitive microelectrode in the choroid near the basal membrane and has shown this to be true. The choroid acts, probably, in two ways to produce this effect. First, it represents a large unstirred layer that significantly slows the rate of diffusion of molecules between the bathing solution and the basal membrane. This alone cannot explain the slope changes near 2 mM and 20 mM, however, and it is possible that potassium release from intracellular compartments also contributes.

The intracellular potassium concentration, 110 mM to 150 mM, was estimated from the basal membrane voltage responses to changes in basal potassium (Miller and Steinberg 1977a). The lower figure is probably the more valid estimate because it is also predicted from the size of changes in the basal membrane potential in experiments where the conditions were optimal for recording the full magnitude of the change.

APICAL MEMBRANE. Of all the ions tested, only potassium and bicarbonate appreciably altered the apical membrane potential. The effects of bicarbonate were discovered in the earliest experiments, which were performed in phosphate-buffered Ringer's. It became clear that tissues exposed to HCO_3^-–CO_2 solutions consistently showed higher transepithelial potentials (Fig. 12.6). Finally, it could be shown that the bicarbonate effect was limited to the apical side, since changes in basal bicarbonate concentration did not appreciably alter the transepithelial potential or the basal membrane potential.

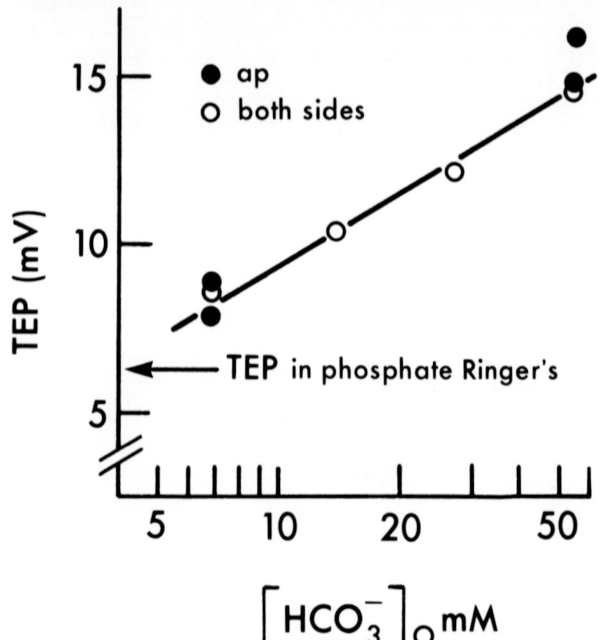

Figure 12.6 *TEP increase due to the addition of HCO_3^-. Tissues were maintained in a phosphate-buffered solution, pH = 7.2, which was then exchanged for solutions containing HCO_3^- at four concentrations, Cl− substitution. Exchanges were made on the apical side only, or on both sides simultaneously. The HCO_3^- solutions also were gassed with 100 percent O_2, and the pH's of the 6.8 mM, 13.8 mM, 27.5 mM, and 55 mM HCO_3^- were, respectively, 7.8, 8.1, 8.2, and 8.3. Data combined from two tissues.*

That the effects attributable to HCO_3^- were not due to pH, nor, perhaps, to P_{CO_2}, was supported by experiments in which the P_{CO_2} was varied while $(HCO_3^-)_o$ was kept constant. For example, equilibrating the 27.5 mM HCO_3^- solution with 70 percent O_2/30 percent CO_2, instead of 95 percent O_2/5 percent CO_2, produced a more than fivefold calculated increase in P_{CO_2} and a measured reduction of pH to 6.6, but did not change the TEP. Similarly, changing to 100 percent O_2 elevated the pH to 8.3 without affecting TEP. The removal of HCO_3^- from the 100 percent O_2 solution, however, rapidly decreased the TEP.

Elevating apical bicarbonate concentration raises the TEP by hyperpolarizing the apical membrane. At 27.5 mM HCO_3^- the slope of the curve that relates apical membrane potential (ψ'_{ap}) and log $(HCO_3)_o$ concentration is about 22 mV for a tenfold change in apical bicarbonate (Fig. 12.7). This curve was actually generated by the transepithelial potential data, since we could show, in the case of HCO_3^-, that the TEP responses and the apical membrane intracellular voltage responses were directly proportional. The constant of proportionality, a function of all the membrane resistances, was determined in a separate series of experiments. The validity of this curve was directly checked in microelectrode experiments by making bicarbonate concentration changes between 2.75, 27.5, and 55 mM and recording the change in apical membrane potential (open circles, Fig. 12.7). Since there are no known specific inhibitors for putative HCO_3^- pumps, it was not possible to eliminate electrogenic contributions to the apical bicarbonate response. The slope in Figure 12.7, therefore, gives an "apparent" HCO_3^- conductance of 0.39.

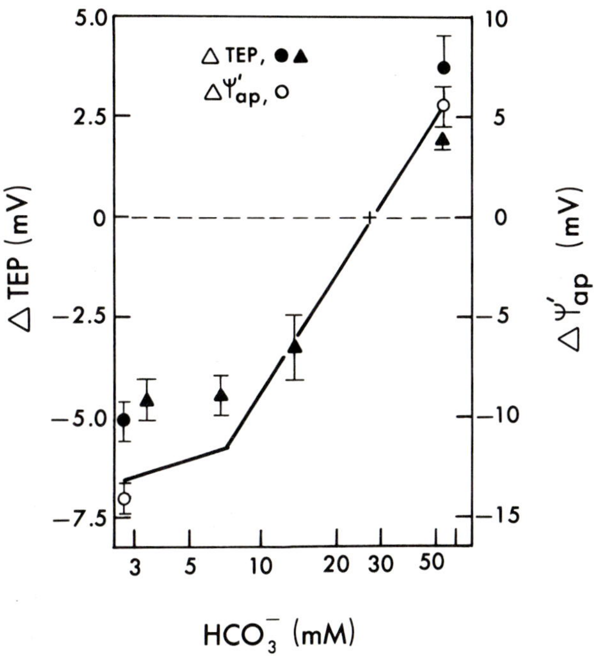

Figure 12.7 $\Delta\psi'_{ap}$ *as a function of apical* $[HCO_3^-]_o$. *The data points are means and SEM's. The solid line was calculated by using the TEP data and the average values of* R_a, R_b, *and* R_s. *(From Miller and Steinberg 1977a.)*

The apical membrane potential is somewhat more sensitive to potassium than it is to bicarbonate. Elevations of apical potassium *depolarize* the apical membrane and, therefore, *decrease* the transepithelial potential. The slope of the apical membrane curve around 2 mM is about 30 mV per tenfold change in concentration (Fig. 12.5), and, therefore, considerably less than the basal membrane slope (52 mV per tenfold change). The relative potassium conductance, T_K, of the apical membrane is therefore 0.52.

EFFECTS OF IONS ON RESISTANCE. Bicarbonate concentration changes on the apical side produce small changes in the membrane resistances that can be neglected, to a first approximation (Miller and Steinberg 1977a). In contrast, potassium concentration changes on either side of the tissue markedly lower the transepithelial resistance, and appear to alter the individual cell membrane resistances. The evidence for the pigment epithelium further suggests that the resistances of the HCO_3^- and K^+ channels are both affected by changing the potassium concentration.

Effects of Metabolic Inhibitors

The transepithelial potential, in general, is readily reduced by the various classes of metabolic inhibitors. These include inhibitors of cell respiration, such as DNP, CN-Iodoacetate; inhibitors of Na, K-ATPases, such as ouabain; and the diuretics—acetazolamide, furosemide, ethacrynic acid. The effects on TEP originate, of course, in changes in membrane potential; a TEP decrease can arise from a depolarization of the apical membrane or a hyperpolarization of the basal membrane, or both of these effects can occur simultaneously. The effect that is easiest to understand is the depolarization of the cell produced by ouabain. Ouabain is known to inhibit Na, K-ATPases and, generally, to act by attaching to a site accessible only from the outside of the cell membrane (Baker and Willis 1972). In some epithelial tissues these sites have been localized by autoradiographic techniques to only the apical or basolateral cell membranes (Quinton, Wright, and Tormey 1973; Stirling 1976). This is probably also true for frog pigment epithelium, since the depolarization only occurs when ouabain, 10^{-4}M, is placed in the apical solution. Perfusion from the basal side, even for one hour, does not reduce the transepithelial potential (Steinberg and Miller 1973).

Other inhibitors also produce a ouabain-like effect—depolarization of the apical membrane relative to the basal. Inhibitors that work intracellularly, such as DNP and CN-Iodoacetate, are effective from either side. The diuretics—ethacrynic acid and furosemide—however, appear to have a greater depolarizing effect when placed in the apical solution. The locus of action of these diuretics is not well understood (Jacobson and Kokko 1976).

There is another effect whose origin is uncertain. Following an initial ouabain-like depolarization, some inhibitors—DNP, for example, but especially furosemide and ethacrynic acid—produce a slow hyperpolarization that originates at the basal membrane. This effect could signal the inhibition of an electrogenic depolarizing basal membrane pump. If that were true, then cold should also produce a basal hyperpolarization as one of its effects, but it only depolarizes the apical membrane, mimicking the effect of ouabain. The late slow hyperpolarization might be caused by a delayed increase in basal potassium conductance due to intracellular (for example, mitochondrial) calcium release (Godfraind, Kawamura, Krnjevic, and Pumain 1971; Meech 1976). This hypothesis was tested by elevating the basal potassium concentration and recording the basal membrane voltage response before and after the application of furosemide or ethacrynic acid, 10^{-3}M. We found that the responses were larger by 30 percent after the application of these inhibitors (unpublished observations), thus supporting the conductance mechanism.

OUABAIN EFFECT: EVIDENCE FOR AN ELECTROGENIC CONTRIBUTION TO MEMBRANE POTENTIAL. When presented to the apical surface of the epithelium ouabain depolarizes the apical membrane in two distinct phases—rapid and slow. The rapid phase is usually 90 percent complete 3 minutes after the start of depolarization and averages 9.6 ± .5 mV (10 tissues). In the initial minute from the start of depolarization, the potential falls at a rate of 4.4 mV/min. This is followed by the slow phase, lasting hours, where the cell depolarizes at the much reduced rate of 0.4 mV/min. An effective "sealing" of the transepithelial shunt by lowered external Na concentrations shows that the rapid phase originates solely at the apical membrane, while the slow phase is seen at both membranes (unpublished observations).

The slow depolarization of membrane potential following ouabain is most probably due to the Na^+, K^+ concentration gradients running down with time. The evidence, so far, strongly indicates that the *rapid phase* originates from the decline of a *direct* contribution to apical membrane voltage by an electrogenic pump (see Miller, Steinberg, and Oakley 1978). For example, at the fastest flow rates, it can be demonstrated that the latency and time course of the rapid phase coincides with the build-up of ouabain concentration outside the apical membrane. Therefore, the rapid phase depolarization appears to occur "instantaneously" with the attachment of ouabain molecules to their receptors. The transepithelial resistance is unaltered during the time course of this rapid phase, and although the ratio of apical to basal cell membrane resistance does rise slightly, it cannot explain the magnitude of this response. In addition, lowering the temperature by 10°C also produces a rapid depolarization whose latency and time-course coincide with the decline in temperature (+7 mV, from 21°C to 11°C). Prior exposure of the tissue to ouabain markedly reduces the cold effect and vice versa—the ouabain effect is much smaller in the cold. In sum, the apical membrane seems to contain a ouabain-sensitive electrogenic hyperpolarizing Na pump that contributes about 10 mV to the resting potential and about 5.0 mV to the transepithelial potential (tissue bathed in 2 mM potassium solution). Changes in the rate of this pump may provide the feedback needed to compensate for extracellular ion concentration changes induced by retinal activity.

Mammalian Pigment Epithelium

We have performed a series of experiments on the isolated RPE-choroid of the cat (Steinberg, Miller, and Stern 1978). Albino and pigmented rabbits were tried first, but with very little success, because of difficulty in mounting the tissue without damaging it. The cat preparation was somewhat more successful than rabbit, but still much more difficult than bullfrog. One problem is that the choroid is more adherent to the sclera, so that a special suction stage had to be developed to hold the tissue for dissection. The RPE-choroid was then mounted in the same chamber used for frog, except that it was perfused at 37°C with a mammalian salt-solution–tissue-culture medium.

When mounted in this way the best preparations showed electrical characteristics that were similar to bullfrog; TEP's of 10–15 mV, apical positive, but R_t's no higher than 3000 Ω. Intracellular recordings were obtained from RPE cells by penetrating the apical surface, and the cleanest pentrations provided apical resting potentials that averaged −70 mV, −80 mV being the highest. The ratio of apical to basal cell membrane resistance, R_a/R_b, obtained in response to transepithelial 6 μA current pulses, also resembled the values obtained in frog, 0.3–0.4 (Miller and Steinberg 1977a). The transepithelial potential also was rapidly depressed by exposure of the tissue to metabolic inhibitors such as DNP, 0.5 mM. Ouabain, 10^{-4}M apical, *rapidly* reduced the TEP and the effect was as fast or faster than in frog. These results suggest that the transepithelial potential, in cat, is also dependent on active ion transport, and that the apical membrane contains a ouabain-sensitive mechanism that directly contributes to the apical resting potential.

Experiments were performed to assess the passive ionic properties of the cell membranes by recording TEP and membrane potential changes brought about by tenfold changes in the extracellular concentrations of Na^+, K^+, Cl^-, HCO_3^-. Ion concentration changes in the solution facing the choroid were totally ineffective, within a 5-minute observation period, in modifying the TEP. Apparently, the cat choroid, in vitro, acts as a much more effective diffusion barrier than in frog. Our preliminary results indicate that the apical membrane of cat, like the frog, also has conductances to K^+ and HCO_3^-. The differences are that K^+ conductance is higher, and the HCO_3^- conductance lower, as compared with frog. The cat apical membrane also has a larger Na^+ response than the frog. It is not yet possible to quantify these apparent conductances.

Active Transport

Demonstrating that a given ion or metabolite is actively transported across the RPE is a step toward understanding the relationships between the choroidal blood supply, pigment epithelium activity, and the functional needs of the neural retina. These active transport mechanisms should respond to changes in the chemical composition of choriocapillaris blood and the milieu surrounding the photoreceptors; and they also should be coupled to the movements of other solutes and of water.

During most of the flux experiments the compositions of the fluids bathing the apical and basal membranes were kept identical, and the transepithelial potential was short-circuited to zero. The current flowing through the external circuit, which kept the transepithelial potential at zero, is called the short-circuit current (SCC). In this circumstance—identical solutions and zero transepithelial potential—the finding of a net flux for a labeled substance (Ussing 1949; Ussing and Zerhan 1951) implies that it is actively transported. In principle, the sum of the net fluxes of all actively transported ions should add up to the short-circuit current.[1]

Let us summarize the commonly accepted criteria for active transport (Wright 1974), namely: (1) a net flux (or transport) against or in the absence of an external electrochemical potential, (2) a *net* flux that exhibits saturation kinetics, and (3) a net transport that is blocked by metabolic inhibitors or other molecules. In all of the experiments to be described at least two of these criteria are met. Criterion (1) is usually accepted as evidence for active transport, and in most cases this conclusion has proved correct. From a rigorous point of view, however, this conclusion is contingent upon the exclusion of solute-solvent coupling (solvent drag), solute-solute coupling, or both (Kedem and Essig 1965; Rehm 1975).

TOAD RETINAL PIGMENT EPITHELIUM. Work by Lasansky and DeFisch (1966) on *Bufo arenarum* and *Bufo marinus* showed that there was a net apical to basal ^{36}Cl flux that could explain about two-thirds of the short-circuit current. Manometric measurements indicated HCO_3^- absorption at

1. The relationships between equivalent circuits, ion flux, short-circuit current and active transport, as they apply to epithelia, have their origin in the work of Ussing (1949) and Ussing and Zerhan (1951). This material has been carefully and extensively reviewed in recent times (Finkelstein and Mauro, 1963; Schwartz, 1971; Schultz, 1975; Rehm, 1975), and the reader is referred to these papers for a detailed theoretical background. A particularly short and elegant paper about what the short-circuit current measures has been written by Ginzburg and Hogg (1967).

the apical membrane, but this could not be related to the short-circuit current. For comparison to the frog (Fig. 12.11), it should be noted that the toad unidirectional chloride fluxes were not affected by acetazolamide although the short-circuit current was reduced by 20 percent. In one species of toad, *Bufo marinus,* there was a small net flux of ^{22}Na in the choroid to retina direction (0.13 μEq/cm^{2}hr) but in contrast to the frog, ouabain, 10^{-5}M, had no effect on the transepithelial potential or current. The average transepithelial resistance was 450 Ωcm^2 for *Bufo arenarum* and 600 Ωcm^2 for *Bufo marinus.*

FROG RETINAL PIGMENT EPITHELIUM. *Sodium.* The effect of ouabain on the apical membrane suggests that this membrane contains Na, K-ATPases that are responsible for moving sodium and potassium across the cell membrane. It was anticipated, therefore, that this would create a net sodium flux in the choroid→retina direction. In bullfrog it was only possible to measure a net flux of ^{22}Na in tissues whose transepithelial resistance was greater than 3500 Ω. We assumed that, at lower resistances, the edge damage was large enough so that most of the flux in both directions took place via the edge-pathway, thus obscuring the net flux. Figure 12.8 is a plot of transepithelial resistance as a function of unidirectional flux. Below 3500 Ω the fluxes increased precipitously and were much more variable between tissues, and from one steady-state measurement to another, in a given tissue. Above 4500 Ω the fluxes leveled off, the variability in and among tissues was reduced, and the net flux could be detected.

In a series of 49 experiments, in which the mean transepithelial resistance was 4,700 Ω, there was a highly significant net ^{22}Na flux of 0.23 μEq/cm^{2}hr (Table 12.1). Desiring to corroborate this result on single tissues, reversal experiments were performed in which the unidirectional fluxes were measured successively on the same tissue. In this way we hoped to obtain a more precise estimate of the net flux. An example of a typical reversal experiment appears in Fig. 12.9, in which there was a net flux of 0.40 μEq/cm^{2}hr in the choroid→retina direction. In every one of twenty such experiments, there was a net flux from choroid to retina, and the mean difference for all experiments was 0.47 μEq/cm^{2}hr, or about twice the value of the unidirectional experiments. The net fluxes of Na+ did not show a seasonal variation (June-September versus November-April).

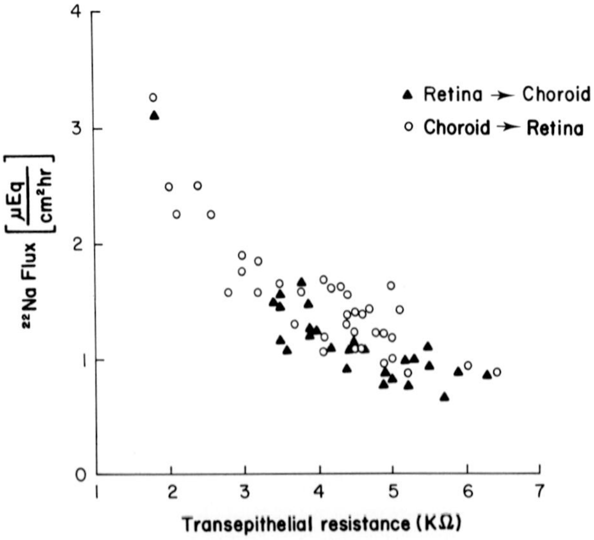

Figure 12.8 *Unidirectional ^{22}Na fluxes as a function of transepithelial resistance. Each point is the mean of 6–8 steady state measurements taken over a period of 2–3 hours on a single tissue; represented are thirty-seven experiments in the choroid → retina direction and twenty-seven in the retina → choroid direction. While the average short-circuit current was not significantly different in these two sets of experiments, the unidirectional fluxes decreased precipitously as R_t increased from less than 2.0 KΩ to greater than 3.5 KΩ. Above 3.5 KΩ, the flux continued to fall as resistance increased, but the slope of the relationship was much reduced and tended towards zero above 4.5 KΩ. For these reasons it was impossible to accept data from any experiments with R_t's lower than 3.5 KΩ. (From Miller and Steinberg 1977b.)*

As one might expect for active transport, the *net* ^{22}Na flux was inhibited by ouabain. The effect appears, however, as an increase in the retina→choroid flux (Fig. 12.10) of 0.3 ± 0.09 μEq/cm^{2}hr (6 experiments), while there was little or no change in the choroid→retina flux. This pattern of flux change cannot be explained by effects of ouabain on membrane resistance, since the transepithelial resistance, and the ratio of apical to basal cell membrane resistance, remained relatively constant during the initial rise in flux, and for 20 minutes *after* the addition of ouabain to the apical solution. After that period, both of these resistances rose (Fig. 12.10), a change which is in the wrong direction to explain the increase in retina to choroid flux. Qualitatively similar effects of ouabain on net sodium transport have been described in frog gastric mucosa, cornea, and turtle bladder (Cooperstein 1959; Candia 1972; Solinger et al. 1968); arguments based on irreversi-

Table 12.1 ^{22}Na experiments.

Values	*Choroid → Retina*	*Retina → Choroid*	*Net flux (Choroid → Retina)*	*P (t-test)*
Unidirectional experiments				
Flux $\left[\frac{\mu Eq}{cm^2\,hr}\right]$	1.28 ± .05 (26)	1.05 ± .05 (23)	.23	= .0013 (unpaired t)
R_t KΩ	4.7 ± .12	4.7 ± .16		
SCC $\left[\frac{\mu Eq}{cm^2\,hr}\right]$	1.34 ± .23	1.40 ± .06		
Reversal experiments				
Flux $\left[\frac{\mu Eq}{cm^2\,hr}\right]$	1.56 ± .06 (20)	1.09 ± .05 (20)	.47	$<$ .001 (paired t)
R_t KΩ	4.7 ± .04	4.6 ± .15		
SCC $\left[\frac{\mu Eq}{cm^2\,hr}\right]$	1.18 ± .05			

Note: The values in each column are mean ± SEM. The numbers in parentheses in columns 2 and 3 refer to the number of experiments. In the unidirectional experiments the criterion for inclusion was a transepithelial resistance $R_t \geq 3.5$ KΩ. In the reversal experiments an additional criterion was that the resistance (R_t) and transepithelial potential (TEP) remained essentially unchanged after the wash.
SCC = short-circuit current.

ble thermodynamics have been advanced to explain this kind of result (Candia 1972; Essig 1968), but the mechanism is not understood.

Chloride. We found a relatively large net flux of ^{36}Cl in the retina-to-choroid direction, which is opposite in direction to the net sodium flux. This confirmed the finding of Lasansky and DeFisch (1966) in the toad. Table 12.2 shows, moreover, that there was a distinct seasonal variation in the ^{36}Cl flux. We followed these fluxes over two consecutive years, finding that the net flux is highest in the spring, drops sharply to its lowest level in the summer, and then gradually returns to a middle level in the winter (although there was some variability in the absolute size of the ^{36}Cl flux from year to year).

The chloride transport system, therefore, cycles during the year, no doubt in response to factors in

Table 12.2 ^{36}Cl experiments.

	Flux or SCC $\left[\frac{\mu Eq}{cm^2\,hr}\right]$				
	Year 1		*Year 2*		
Experiment	*May-July*	*Feb-April*	*May-July*	*Nov-Dec*	*Feb-March*
Retina → Choroid	.93 ± .05 (9)	1.18 ± .06 (13)	.71 ± .05 (10)	1.06 ± .15 (4)	1.18 ± .18 (6)
SCC	1.28 ± .05	1.23 ± .03	1.46 ± .09	1.73 ± .09	1.60 ± .07
Choroid → Retina	.46 ± .04 (12)	.44 ± .06 (9)	.52 ± .05	.70 ± .11 (6)	.68 ± .10 (9)
SCC	1.36 ± .06	1.40 ± .7	1.40 ± .17	1.75 ± .10	1.62 ± .7
Net Flux (Retina → Choroid)	.47	.74	.19	.36	.50
P (t-test)	$<$ .001	$<$ .001	.02	.08	$<$ .02

Note: The values in each column are mean ± SEM and the numbers in parentheses refer to the number of experiments. In both years the experiments were unidirectional. In Year 1, the average transepithelial resistances in the four sets of experiments were not significantly different (range: 4.2 KΩ ± .19 to 4.4 KΩ ± .08). The same was also true for Year 2 (range: 4.2 KΩ ± .27 to 4.7 KΩ ± .26).
SCC = short-circuit current.

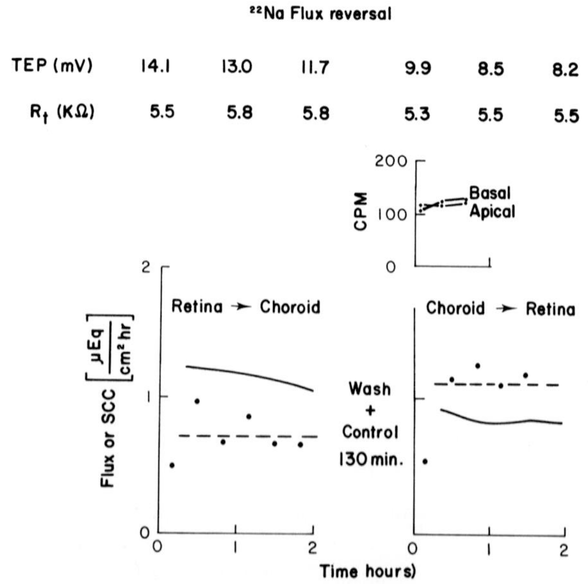

Figure 12.9 ^{22}*Na flux* reversal *experiment. The transepithelial potential (TEP) was short-circuited to zero, and the short-circuit current (SCC, in μEq/cm²hr) was monitored continuously (solid line). The TEP and the transepithelial resistance* (R_t) *were monitored hourly by briefly interrupting the short-circuit current and passing a 1μA current pulse. A tracer,* ^{22}*Na, was first added to the* apical *solution at* t = *0 and the retina → choroid (apical to basal) flux was sampled at 20-minute intervals (closed circles). The sodium flux reached a steady state in about 30 minutes and was monitored for 2 hours (interrupted horizontal line). The tissue was then washed until the background level of radioactivity reached a low steady-state level on each side. This took 130 minutes (wash and control). Tracer was then added to the* basal *solution and the experiment repeated in the choroid → retina (basal to apical) direction. (From Miller and Steinberg 1977b.)*

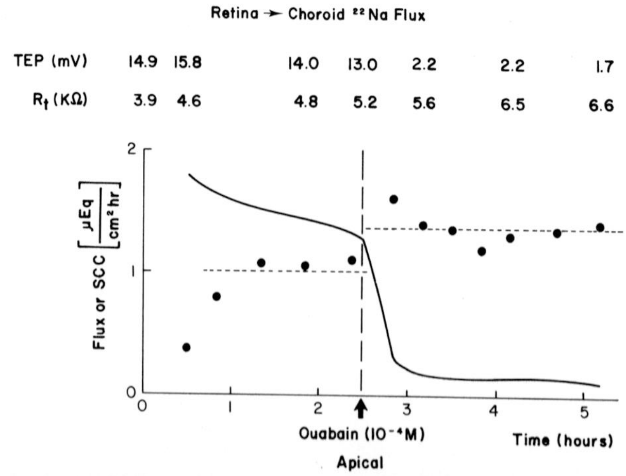

Figure 12.10 ^{22}*Na flux in the retina → choroid direction before and after ouabain,* 10^{-4} *M, was added to the apical solution. The transepithelial potential (TEP) was short-circuited to zero, and the short-circuit current (SCC, in μEq/cm²hr) was monitored continuously (solid line). The TEP and the transepithelial resistance* (R_t) *were monitored hourly by briefly interrupting the short-circuit current and passing a 1 μA current pulse. A tracer,* ^{22}*Na, was added to the apical solution at* t = *0 and the retina→ choroid (apical to basal) flux was sampled at 20-minute intervals (closed circles). The sodium flux reached a steady state at 1 hour (first interrupted horizontal line) and at 2 hours 40 minutes ouabain* (10^{-4} *M) was added to the apical solution alone. (From Miller and Steinberg 1977b.)*

the environment. This variation in net chloride transport could be due to an increase in the rate of active Cl^- transport or to a change in membrane conductance/driving force in the passive Cl channel(s) of one or both membranes. Since it is the unidirectional Cl flux in the direction of net transport that increases from summer to spring, and because seasonal variations in membrane resistance are not obvious, it is most likely the rate of Cl pumping that changes.

Chloride transport systems frequently can be inhibited by acetazolamide or a lowering of external HCO_3^- (Keynes 1969; Emanović, Harrison, Keynes, and Rankin 1976; Leslie, Schwartz, and Steinmetz 1973), and this is also true for the pigment epithelium. Acetazolamide significantly reduced the net chloride flux and, equally, the short-circuit current (Fig. 12.11); and lowering HCO_3^- to 10 percent of its usual concentration was equivalent to 10^{-3}M acetazolamide in its effect on the unidirectional Cl

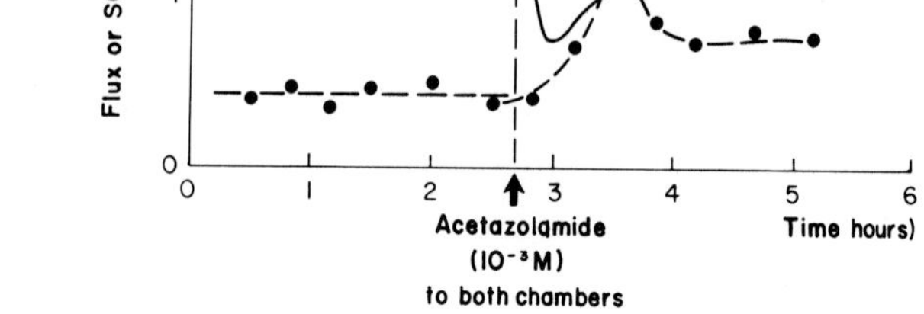

Figure 12.11 *Effect of* 10^{-3} *M acetazolamide on choroid → retina* ^{36}*Cl flux. Experimental protocol as in Figure 12.10. (From Miller and Steinberg 1977b.)*

fluxes and the short-circuit current. The reduction of net flux came about by an *increase* in the choroid→retina flux and a smaller *decrease* in the direction of net transport (retina→choroid); a pattern of changes also seen in the net ^{22}Na inhibition by ouabain. It should be also mentioned that ouabain had no effect on the unidirectional chloride fluxes, and acetazolamide had no effect on the sodium fluxes. We do not yet know whether acetazolamide works intracellularly (Maren and Wiley 1970) or at the cell membrane (Nellans, Frizzell, and Schultz 1975; Cousin, Motais, and Sola 1975).

Potassium and calcium. The unidirectional fluxes of K and Ca are less than 10 percent of the Na and Cl fluxes. In contrast, Noell et al. (1965) found in a rabbit posterior eye segment preparation, in situ, that there was a large net flux of ^{42}K in the opposite direction (choroid to retina), which accounted for at least 65 percent of the short-circuit current. It is difficult to reconcile this finding with our observations in the frog, but there may be very large species differences, of course, and detailed studies of ion transport are needed in mammals.

Calcium fluxes have been more thoroughly studied because of the importance of calcium for photoreceptor function. The results are very clear. The frog pigment epithelium actively transports calcium from the choroidal space into the retinal space. This average net flux, 5.9 nM/cm²hr, is small relative to the ^{22}Na and ^{36}Cl fluxes, but is comparable to the net flux across other epithelial tissues that are presumably specialized for calcium transport (Moriarty and Terepka 1969; Walling and Rothman 1969; Ehrenspeck, Schraer, and Schraer 1970; Pearson and Goldner 1973, 1974; Terepka, Coleman, Armbrecht, and Gunter 1976). The net ^{45}Ca flux responds strikingly to ouabain. In the initial two hours following treatment with 10^{-4}M ouabain it essentially doubles because of an increase in the choroid→retina flux (Fig. 12.12), with no change in the opposite direction.

Bicarbonate-inferred. Because of seasonal variations, the net chloride and sodium flux together only account for about 55–90 percent of the short circuit current (Tables 12.2, 12.3); and since the ^{45}Ca and ^{42}K fluxes represent, at most, 10 percent of the short-circuit current, this leads us to infer that the remaining current is probably carried by a HCO_3^-/H^+ transport system that is also seasonally dependent. In support of this hypothesis, Lasansky and DeFisch

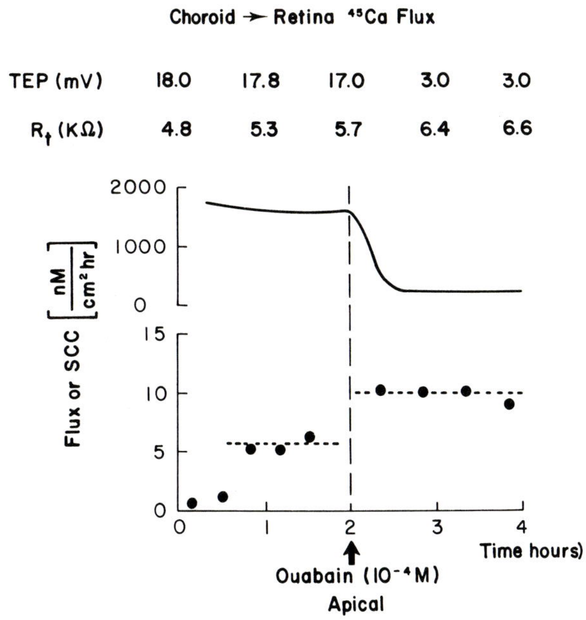

Figure 12.12 *^{45}Ca flux in the choroid → retina direction before and after 10^{-4} M ouabain was added to the apical solution. Experimental protocol as in Figure 12.10. (From Miller and Steinberg 1977b.)*

(1966) showed by manometric measurements that the apical membrane of the toad pigment epithelium absorbs HCO_3^-, but it remains to be determined whether this is by an active process. The large effect of ouabain on short-circuit current also points to HCO_3^-. Ouabain reduced the short-circuit current by 0.6–1 μEq/cm²hr, for example (Fig. 12.11), but only reduced the net ^{22}Na flux by about 0.3 μEq/cm²hr. Ouabain does not alter ^{36}Cl fluxes within this observation time and, again, the ^{45}Ca and ^{42}K fluxes are too small to matter. The only ion available to carry this current is HCO_3^-.

Table 12.3 Electrical properties of cell membranes of frog pigment epithelium

Apical Membrane	*Basal Membrane*
Conductances	
$G_K \approx G_{HCO_3} >> G_{Na}, G_{Cl}$	$G_K >> G_{Na}, G_{Cl}, G_{HCO_3}$[a]
Specific Resistances	
~ 7000 Ωcm²	450 — 1400 Ωcm²
Potential	
≈ −88 mV[b]	≈ −77mV

a. $[K^+]_i \cong 110$ mM.

b. There is a contribution to the membrane potential from a ouabain-sensitive mechanism.

Location of sites. Some inferences concerning the localization of the active transport mechanisms can be made by combining these results with those obtained with microelectrodes. The demonstration of a net choroid→retina flux of ^{22}Na, which accounts for 30–40 percent of the short-circuit current, and its inhibition (50–60 percent) by apical ouabain indicate that active transport takes place across the apical membrane. Since ouabain inhibits an electrogenic pump at the apical membrane, it is probably these apical Na, K-ATPase sites that transport Na^+. The existence of an active Ca^{+2} transport, also at this membrane, is suggested by the net choroid→retina flux of Ca^{+2} in the short-circuited state, by the likelihood of a very unfavorable electrochemical gradient for Ca^{+2} across this membrane (Blaustein 1974), and by the sensitivity of the net flux to apical ouabain.

It is very likely that the basal membrane contains one or more active transport mechanisms, as well. Since the basal membrane acts very nearly like a K^+ electrode, the net current moving across it must be very close to zero (Ginzburg and Hogg 1967). This means that the SCC would have to be very close to zero if there were no active transport mechanisms at the basal membrane, but this is not the case. We have no direct evidence, however, that either the Cl^- or putative HCO_3^- transport systems are located at the basal membrane. We showed, however, that the basal membrane has a very small HCO_3^- conductance relative to the apical membrane, so that if HCO_3^- is to cross the pigment epithelium an active transport system is required at the basal membrane.

Sugar: 3-O-methyl-D-glucose. Fluxes of this nonmetabolizable sugar were measured because of the importance of glucose for retinal function (Miller and Steinberg 1976). We were unable to confirm a report by Zadunaisky and Degnan (1976) of a large net choroid-to-retina flux. Their measurements, also in bullfrog, were based on unidirectional flux measurements, but, as shown in Fig. 12.13, we found no significant flux of this molecule across a four log unit concentration range in all three different types of experiments—unidirectional flux measurements, reversals, and double-label experiments.

Amino acids: taurine and L-methionine. Taurine was chosen to probe, initially, amino acid transport, because of the increasing evidence for its important role in retinal function (Starr and Voaden 1972; Pasantes-Morales, Urban, Klethi, and Mandel 1973; Pasantes-Morales, Klethi, Urban, and Mandel 1974; Hayes, Carey, and Schmidt 1975; Orr, Cohen, and Lowry 1976).

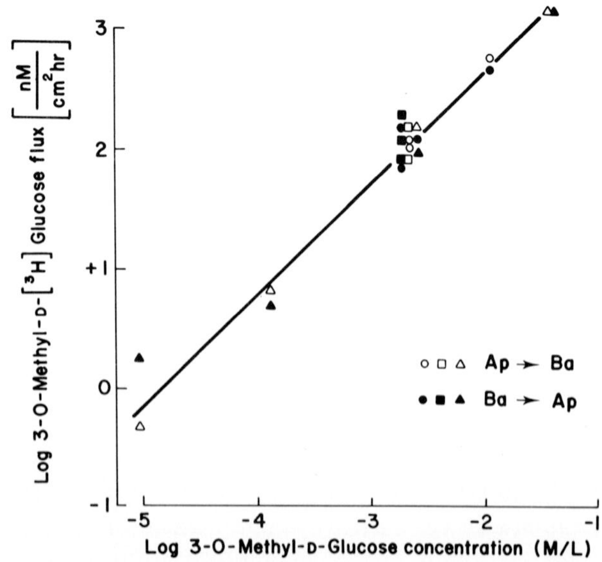

Figure 12.13 *Unidirectional 3-O-methyl-D-glucose fluxes as a function of concentration. Each point is the mean steady-state unidirectional flux from one experiment, and there are three different types of experiments represented: unidirectional flux (6 circles), reversal (6 squares), and double-label (8 triangles). The SEM's were $\leq \pm .03$ log units. (From Miller and Steinberg 1976.)*

Taurine unidirectional fluxes were measured in both directions over a wide range of taurine concentrations, $1.7 \times 10^{-4} - 3 \times 10^{-2}$M (Miller and Steinberg 1976). In both directions, the unidirectional flux increased linearly with concentration and then saturated (Fig. 12.14). At all concentrations, as can be seen, there was a significant net flux of taurine from the retina to choroid. The net taurine flux was highly sensitive to apical ouabain. After about 1 hour in ouabain, 10^{-4}M, the two unidirectional fluxes were about equal, because of a decrease in the retina→choroid, and increase in the choroid→retina fluxes.

These results indicated that there is active transport of taurine across the pigment epithelium. Lake et al. (1975) also showed, in uptake studies on frog pigment epithelium, that the cells accumulate taurine to a high tissue/medium ratio. The ouabain inhibition suggested that the site of active transport is at the apical membrane. These data did not, however, allow us to determine if there is a direct link between metabolic energy and taurine transport, or if the energy necessary for active transport is supplied by Na^+ or K^+ concentration gradients be-

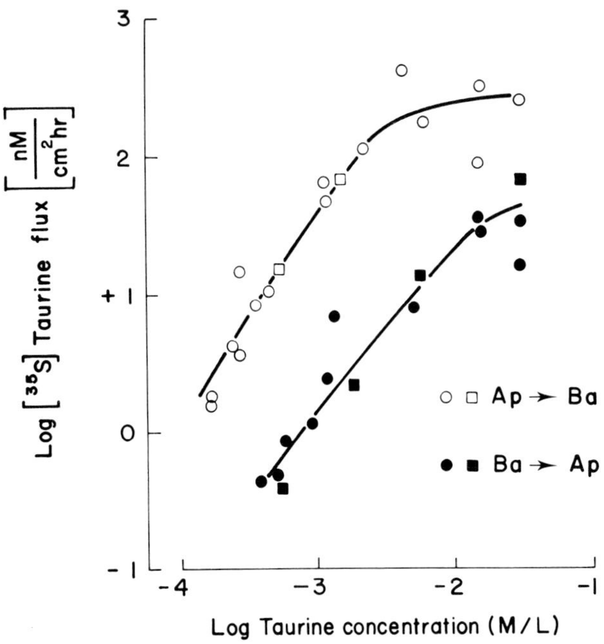

Figure 12.14 *Unidirectional taurine fluxes as a function of taurine concentration. Each point is the mean steady-state unidirectional flux from one experiment. The SEM's were usually ≤ ± .05 log units and the maximum was ± 0.1 log units. In eight experiments (16 points) the apical-to-basal flux and the basal-to-apical flux were measured simultaneously on paired tissues from the same eye. The circles indicate experiments in which the transepithelial potential had been short-circuited to zero voltage. The open-circuit experiments are indicated by squares; in these six experiments, the mean transepithelial potential was 9.5 mV, Rg. 7–13 mV. The best-fit lines were estimated by eye. (From Miller and Steinberg 1976.)*

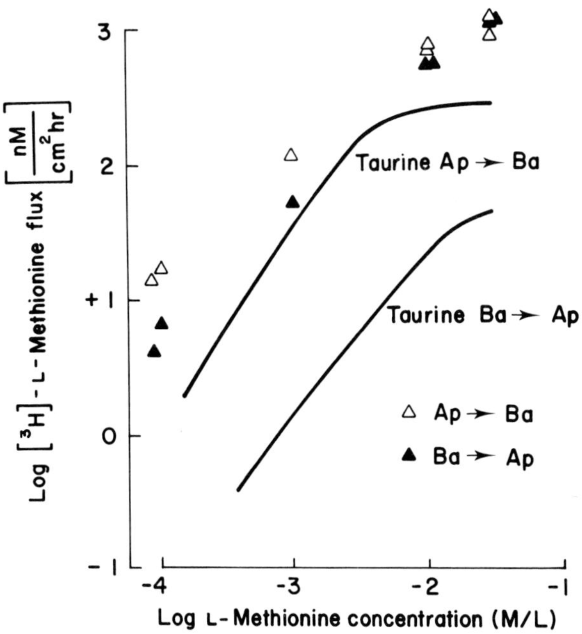

Figure 12.15 *Unidirectional* L-*methionine fluxes as a function of* L-*methionine concentration. Each point is the mean steady-state unidirectional flux from one experiment. The SEM's were ≤ ± .03 log units. The transepithelial potential had been short-circuited to zero in each experiment. The taurine unidirectional flux curves from Figure 12.3 are shown for comparison (solid lines). (From Miller and Steinberg 1976.)*

tween the external solution and the cell cytoplasm—the "Na gradient hypothesis" (Riggs, Walker, and Christensen 1958; Crane 1962, 1965; Schultz and Curran 1970).

To begin to assess the specificity of the taurine results we compared taurine with one of its precursors, L-methionine, another sulfur-containing amino acid. Taurine and L-methionine are transported in other systems and in rabbit ciliary body and retina by separate mechanisms (Oxender and Christensen 1963; Kromphardt 1963; Reddy 1967; Starr and Voaden 1972). The most important difference between taurine and L-methionine fluxes is that the net retina-to-choroid flux is much larger with taurine (Fig. 12.15). Also, the comparison shows that the large net taurine flux is created by a relatively small choroid-to-retina flux. The apparent permeability for taurine in the choroid to retina direction is approximately 19 times less than in the retina to choroid direction, 4.9×10^{-7} cm/sec versus 9.2×10^{-6} cm/sec.

These experiments raise the question of why the mechanism for clearing taurine from the retina functions at a much greater rate than the mechanism for bringing it into the retina. Although the physiological role of taurine in the retina is not known, it appears to be released by light and, therefore, there may be a need to remove it from the extracellular space (Pasantes-Morales et al. 1973; Pasantes-Morales et al. 1974). If taurine has a significant role in *photoreceptor* function, as indicated by its high concentration in these cells (Orr et al. 1976) and by the apparent coupling of the light-induced taurine efflux and calcium movement (Salceda and Pasantes-Morales 1975), then there may be a need to clear it from the extracellular space around the photoreceptors. As pointed out by others, taurine is only very slowly metabolized in the retina and a transport system may be needed to terminate its physiological action (Starr and Voaden 1972; Pasantes-Morales, Klethi, Ledig, and Mandel 1972).

Effects of K_o and Na_o. It is well known that active transport rates for such metabolites as sugars and amino acids are influenced by ionic gradients across the cell membrane, particularly those of Na^+ and K^+ (Riggs et al. 1958; Crane 1962, 1965; Schultz and Curran 1970). Our experiments showed clearly that the taurine flux in the direction of active transport (retina→choroid) depends upon extracellular Na and K concentrations (data not shown). Furthermore, only the apical concentration of these ions exerts an effect. The retina→choroid taurine flux is a logarithmic function of apical sodium concentration. At one-half the normal extracellular concentration of sodium, for example, the taurine flux is 20 percent of normal. The potassium effects are remarkable in that the taurine flux is maximum at the normal concentration of 2.0 mM potassium and falls off as the extracellular concentration of apical potassium is increased or decreased. At 1.0 mM and 3.0 mM potassium, for example, the retina→choroid taurine flux has decreased to about 80 percent of the flux at 2.0 mM. Below 1.0 mM potassium, the taurine flux continues to fall and it is less than 10 percent of normal at 0.2 mM.

The Origin of the Transepithelial Potential

The transepithelial potential is the difference between the apical (ψ_{ap}) and basal (ψ_{ba}) membrane potentials. These potentials depend, in part, on the relative ionic conductances for each membrane, the shunt corrected values of the apical and basal membrane potentials, ψ'_{ap} and ψ'_{ba}, and the potassium and bicarbonate equilibrium potentials, E_K and E_{HCO_3}.[2] In addition to these passive components the apical or basal membrane may contain electrogenic pumps that directly contribute to the membrane potential.

The relative ionic conductances for each membrane and the potassium equilibrium potential can be determined (Miller and Steinberg 1977a), and these data suggest that the apical and basal membrane both contain electrogenic pumps. The existence of the apical pump was confirmed by experiments demonstrating a ouabain sensitive hyperpolarizing sodium pump contributing about 10 mV to the apical membrane potential and, therefore, about 13 to 15 mV to ψ'_{ap} (Miller, Steinberg, and Oakley 1978). The basal membrane *may* contain a depolarizing pump, but the existence of this pump awaits more definitive measurements. Therefore, the apical membrane potential has two components, one that is electrogenic (10 mV) and another that accounts for most of the potential and is due to the passive diffusion of potassium and bicarbonate down their electrochemical gradients. The basal membrane potential is mostly due to the passive diffusion of potassium down its electrochemical gradient.

2. The relationship between $\psi_{ap} - \psi_{ba} \equiv$ TEP and $\psi'_{ap} - \psi'_{ba}$ is linear, depending on the constancy of the apical and basal cell membrane resistances, R_a and R_b, and the transepithelial resistance, R_t. That is:

$$\psi'_{ap} - \psi'_{ba} = TEP\,\frac{(R_a + R_b)}{R_t}$$

The relationship between ψ'_{ap} and the relative ionic conductances is:

$$\psi'_{ap} = T_K E_K + T_{HCO_3} E_{HCO_3}$$

where T_K, T_{HCO_3} are the relative ionic conductances of potassium and bicarbonate and E_K, E_{HCO_3} are the equilibrium potentials for these ions; $T_K = 0.52$, $T_{HCO_3} = 0.39$ and $E_K = -100$ mV (Miller and Steinberg 1977a).

Photoreceptor–Retinal-Pigment-Epithelium Interactions and Potassium

It is now known that extracellular potassium concentration varies as a function of retinal exposure to light in the space between the photoreceptor outer segments and the apical membrane of the pigment epithelium (Oakley and Green 1976). In this space the extracellular potassium concentration is maximal during dark adaptation and is decreased by light. In an intact retina-RPE-choroid preparation, prepared and set up in the same manner as the RPE-choroid (Fig. 12.2), the extracellular potassium concentration outside the photoreceptors decreased from a dark-adapted high of 3 mM to 1 mM. In these experiments the concentrations of the bulk solutions were maintained at 2 mM, and the change occurred by the end of a 35-second bright flash (Oakley 1977).

These variations in extracellular potassium probably form an important link between photoreceptor activity and epithelial cell transport. To begin with, we have shown that the apical membrane potential of the pigment epithelium changes as a function of extracellular potassium concentration. As expected, the light-induced decrease in extracellular potassium hyperpolarizes the apical membrane relative to the basal membrane. The resulting increase in transepithelial potential is the pigment epithelial contribution to the c-wave of the electroretinogram (Oakley and Green 1976; Oakley 1977; Oakley, Steinberg, Miller, and Nilsson 1977).

We do not yet know how, if at all, these membrane potential changes might directly influence pigment epithelial functions, such as transport. It is clear, however, that sustained differences in membrane potential, due to light and dark adaptation, will lead, eventually, to changes in intracellular ion concentrations; and that metabolic effects secondary to these changes are likely to occur.

Also, the light-induced changes in transepithelial potential will, transiently, alter the movement of charged species across the epithelium; the choroid-to-retina flux of anions will increase and the retina-to-choroid flux of cations will decrease. The effect of these ion movements on retinal function will depend on concomitant pump-rate changes, the movements of other metabolites, counterion movements, and the coupling of these movements to fluid flow.

Extracellular potassium should also alter rates of active transport. It might affect rates of active sodium transport, as shown in nerve and muscle (Thomas 1972), or sugar and amino acid transport (Riggs et al. 1958). The active transport of sodium across the apical membrane into the retinal space, for example, could be changed by the light and dark effects on extracellular potassium concentration. Similarly, there is a ouabain-sensitive active transport of calcium into the photoreceptor space, and we have shown, in the isolated RPE-choroid, that the back flux of calcium, retina-to-choroid, is voltage dependent. It is possible, therefore, that changes in extracellular potassium and potential also control the movement of calcium into the retinal space, but this remains to be tested. We have shown that the retina→choroid taurine fluxes are very sensitive to changes in apical potassium. It is especially significant that these changes, 1–3 mM, coincide with the range of light-induced extracellular potassium changes shown by Oakley to occur in the frog retina (Oakley 1977).

It is also possible that the pigment epithelium itself may participate in the regulation of extracellular potassium concentration. The need for some type of regulation is suggested by studies demonstrating effects of extracellular potassium on photoreceptor responses. Intracellular recordings from toad rods, for example, showed that relatively small changes in K_o changed inner segment membrane potentials and altered the kinetics of the response to light (Brown and Pinto 1974). This is the same function already postulated for glia in the regulation of extracellular potassium around neurons, and for the choroid plexus in the control of CSF potassium and calcium concentrations. In both cases, the buffering of potassium has been postulated to occur by passive as well as active mechanisms (Orkand, Nicholls, and Kuffler 1966; Henn, Haljamäe, and Hamburger 1972; Kriz, Sykova, and Vyklicky 1975; Husted and Reed 1976). The pigment epithelial cells are equally well endowed to act as spatial buffers of potassium. This is by virtue of their anatomical relationship to the photoreceptors, their electrical connection to other pigment epithelial cells in the plane of the epithelial sheet (Hudspeth and Yee 1973), and because their apical and basal membranes are very sensitive to changes in extracellular potassium. The presence of apical active sodium transport, assuming that it is generated by Na, K-ATPases, also provides a mechanism for an active contribution to extracellular potassium regulation. We hope that the significance of these analogies and hypotheses about pigment epithelial function will be evaluated in future experiments.

Summary

This chapter has reviewed the results of microelectrode and isotope transport experiments on in vitro preparations of frog retinal-pigment-epithelium–choroid. The major findings are summarized in Tables 12.3 and 12.4. We found that the apical membrane has an average resting potential of −88 mV, a specific resistance of 7000 Ωcm², and significant conductances to both potassium and bicarbon-

Table 12.4 Active ion transport across frog RPE choroid

Isotope	*Short-circuit current (%)*	*Direction of net flux*	*Fluxes are changed by*
^{36}Cl	50–70[a]	Retina → Choroid	Acetazolamide, 10^{-3} M Furosemide, 10^{-3} M Ethacrynic acid, 10^{-3} M $\downarrow [HCO_3^-]_o$
^{22}Na	30–50[a]	Choroid → Retina	Ethacrynic acid, 10^{-3} M
^{45}Ca	<1	Choroid → Retina	Ouabain, 10^{-4} M

a. Since there are seasonal variations in the ^{36}Cl flux, the sum of those fluxes accounts for most of the short-circuit current only in the Spring.

ate. There are ouabain-sensitive sites on the apical membrane and not on the basal membrane. The ouabain-sensitive pumps provide an electrogenic contribution, averaging −10 mV, to the apical membrane potential. The basal membrane's potential is more depolarized than the apical's by the amount of the transepithelial potential, an average of 10 mV. The basal membrane's conductance is due almost exclusively to potassium, and from these data we estimate an intracellular potassium concentration of 110 mM. In addition, we showed that the resistance of the paracellular pathway is not small relative to the cellular pathway and that the pigment epithelium can be classified as a "tight" epithelium. The "tightness" of the paracellular pathway is the reason that the pigment epithelium forms a resistance barrier in the retina—the R-membrane. Initial experiments on a mammalian preparation (cat) of RPE-choroid have indicated that the apical membrane also has ouabain-sensitive sites, a high potassium sensitivity, and appreciable conductances to sodium and bicarbonate.

The ion flux measurements showed that there was a large net retina-to-choroid flux of ^{36}Cl. The net flux of ^{22}Na was also significant, and it was in the opposite direction, choroid to retina. The net ^{22}Na flux can explain most of the nonchloride portion of the short-circuit current, but this depends upon time of year because of seasonal variations in the chloride fluxes. The net chloride flux is depressed by acetazolamide and it is similarly depressed by lowering the external bicarbonate concentration. The ^{42}K and ^{45}Ca fluxes were less than a tenth the size of the unidirectional ^{36}Cl or ^{22}Na fluxes and contributed very little, therefore, to the short-circuit current. There is, however, a net calcium flux from choroid to retina, which, although small, may be functionally significant. From these data it is necessary to postulate the presence of an active transport of HCO_3^-/H^+ that is seasonally dependent. The active transport sites for Na^+ and Ca^{+2} are presumably at the apical membrane, while the basal membrane may be the location for either, or both, the Cl^- and HCO_3^- transport systems.

We also measured the fluxes of two amino acids [^{35}S] taurine and [^{3}H] L-methionine, and one sugar, 3-O-methyl-D [^{3}H]glucose. There was a large net flux of taurine from retina to choroid, a smaller net flux of methionine in this direction, and no net flux of sugar. Taurine transport was ouabain-sensitive, suggesting that the active site is located at the apical membrane. The rate of taurine transport was very sensitive to the concentrations of sodium and potassium outside the apical membrane, but not outside the basal membrane.

The light and dark effects upon extracellular potassium concentration, in the space between photoreceptor outer segments and the epithelial apical membrane, suggest that potassium may form one link between photoreceptor and epithelial activity. Changes in extracellular potassium should affect transepithelial transport by altering the cell membrane potentials and rates of active transport. The pigment epithelium also may serve to regulate extracellular potassium, passively, by acting as a spatial buffer, and actively, by the transport of potassium.

REFERENCES

Baker, P. F., and Willis, J. S. 1972. Binding of the cardiac glycoside ouabain to intact cells. *J. Physiol.* 224: 441–462.

Blaustein, M. P. 1974. The interrelationship between sodium and calcium fluxes across cell membranes. *Rev. Physiol. Biochem. Pharmacol.* 70:34–82.

Brindley, G. S. 1956. The passive electrical properties of the frog's retina, choroid and sclera for radial fields and currents. *J. Physiol.* 134:334–352.

Brindley, G. S., and Hamasaki, D. I. 1963. The properties and nature of the R membrane of the frog's eye. *J. Physiol.* 167:599–606.

Brown, J. E., and Pinto, L. H. 1974. Ionic mechanism for the photoreceptor potential of the retina of *Bufo marinus*. *J. Physiol.* 236:575–591.

Candia, O. A. 1972. Ouabain and sodium effects on chloride fluxes across the isolated bullfrog cornea. *Am. J. Physiol.* 223:1053–1057.

Castel, M., Sahar, A., and Erlij, D. 1974. The movement of lanthanum across diffusion barriers in the choroid plexus of the cat. *Brain Res.* 67:178–184.

Claude, P., and Goodenough, D. A. 1973. Fracture faces of zonula occludentes from "tight" and "leaky" epithelia. *J. Cell. Biol.* 58:390–400.

Cohen, A. I. 1965. A possible cytological basis for the 'R' membrane in the vertebrate eye. *Nature* 205:1222–1223.

Cooperstein, I. L. 1959. The inhibitory effect of strophanthidin on secretion by the isolated gastric mucosa. *J. Gen. Physiol.* 42:1233–1239.

COUSIN, J. L., MOTAIS, R., and SOLA, F. 1975. Transmembrane exchange of chloride with bicarbonate ion in mammalian red blood cells: evidence for a sulphonamide-sensitive "carrier." *J. Physiol.* 253:385–399.

CRANE, R. K. 1962. Hypothesis for mechanism of intestinal active transport of sugars. *Fed. Proc.* 21:891–895.

CRANE, R. K. 1965. Na+ dependent transport in the intestine and other animal tissues. *Fed. Proc.* 24:1000–1005.

DIAMOND, J. M. 1977. The epithelial junction: bridge, gate and fence. *Physiologist* 20:10–18.

EHRENSPECK, G. H., SCHRAER, H., and SCHRAER, R. 1970. Calcium transfer across isolated avian shell gland. *Am. J. Physiol.* 220:967–972.

EMANOVIĆ, D., HARRISON, F. A., KEYNES, R. D., and RANKIN, J. C. 1976. The effect of acetazolamide on ion transport across isolated sheep rumen epithelium. *J. Physiol.* 254:803–812.

ESSIG, A. 1968. The "pump-leak" model. *Biophys. J.* 8:53–63.

FINKELSTEIN, A., and MAURO, A. 1963. Equivalent circuits as related to ionic systems. *Biophys. J.* 3:215–237.

FINN, A. L. 1976. Changing concepts of transepithelial sodium transport. *Physiol. Rev.* 56:453–464.

FRÖMTER, E. 1972. The route of passive ion movement through the epithelium of *Necturus* gallbladder. *J. Membr. Biol.* 8:259–301.

GINZBURG, B. F., and HOGG, J. 1967. What does a short circuit current measure in biological systems? *J. Theor. Biol.* 14:316–322.

GODFRAIND, J. M., KAWAMURA, H., KRNJEVIC, K., and PUMAIN, R. 1971. Actions of dinitrophenol and some other metabolic inhibitors on cortical neurones. *J. Physiol.* 215: 199–222.

HAYES, K. C., CAREY, R. C., and SCHMIDT, S. Y. 1975. Retinal degeneration associated with taurine deficiency in the cat. *Science* 188:949–951.

HENN, F. A., HALJAMÄE, H., and HAMBURGER, A. 1972. Glial cell function: active control of extracellular K+ concentration. *Brain Res.* 43:437–444.

HODGKIN, A. L., and HOROWICZ, P. 1959. The influence of potassium and chloride ions on the membrane potential of single muscle fibers. *J. Physiol.* 148:127–160.

HUDSPETH, J. A., and YEE, A. G. 1973. The intracellular junctional complexes of retinal pigment epithelia. *Invest. Ophthalmol.* 12:354–365.

HUSTED, R. F., and REED, D. J. 1976. Regulation of cerebrospinal fluid potassium by the cat choroid plexus. *J. Physiol.* 259:213–221.

JACOBSON, H. R., and KOKKO, J. P. 1976. Diuretics: sites and mechanisms of action. *Annu. Rev. Pharmacol.* 16:201–214.

KEDEM, O., and ESSIG, A. 1965. Isotope flows and flux ratios in biological membranes. *J. Gen. Physiol.* 48:1047–1070.

KEYNES, R. D. 1969. From frog skin to sheep rumen: a survey of transport of salts and water across multicellular structures. *Q. Rev. Biophys.* 2:177–281.

KRIZ, N., SYKOVA, E., and VYKLICKY, L. 1975. Extracellular potassium changes in the spinal cord of the cat and their relation to slow potentials, active transport and impulse transmission. *J. Physiol.* 249:167–182.

KROMPHARDT, H. 1963. Die Aufnahme von: Taurin in Ehrlich-Ascites-Tumorzellen. *Biochem. Z.* 339:233–254.

KUWABARA, T. 1979. Species difference of the pigment epithelium. In *The retinal pigment epithelium,* ed. K. M. Zinn, M. F. Marmor. Boston: Harvard Univ. Press.

LAKE, N., MARSHALL, J., and VOADEN, M. J. 1975. Studies on the uptake of taurine by the isolated neural retina and pigment epithelium of the frog. *Biochem. Soc. Trans.* 3:524–525.

LASANSKY, A., and DEFISCH, F. W. 1966. Potential, current and ionic fluxes across the isolated retinal pigment epithelium and choroid. *J. Gen. Physiol.* 49:913–924.

LESLIE, B. R., SCHWARTZ, J. H., and STEINMETZ, P. R. 1973. Coupling between Cl^- absorption and HCO_3^- secretion in turtle urinary bladder. *Am. J. Physiol.* 225:610–617.

LEWIS, S. A., and DIAMOND, J. M. 1976. Na+ transport by rabbit urinary bladder, a tight epithelium. *J. Membr. Biol.* 28:1–40.

LEWIS, S. A., EATON, D. C., and DIAMOND, J. M. 1976. The mechanism of Na+ transport by rabbit urinary bladder. *J. Membr. Biol.* 28:41–70.

MACHEN, T. E., ERLIJ, D., and WOODING, F. B. P. 1972. Permeable junctional complexes: the movement of lanthanum across rabbit gall bladder and intestine. *J. Cell. Biol.* 54:302–312.

MAREN, T. H., and WILEY, C. E. 1970. Kinetics of carbonic anhydrase in whole red cells as measured by transfer of carbon dioxide and ammonia. *Mol. Pharmacol.* 6:430–440.

MEECH, R. W. 1976. Intracellular calcium and the control of membrane permeability. In *Calcium in biological systems,* Symposia of the Society for Experimental Biology, vol. 30, pp. 161–191. Cambridge: Cambridge Univ. Press.

Miller, S. S., and Steinberg, R. H. 1976. Transport of taurine, L-methionine and 3-O-methyl-D-glucose across frog retinal pigment epithelium. *Exp. Eye Res.* 23:177–189.

Miller, S. S., and Steinberg, R. H. 1977*a*. Passive ionic properties of frog retinal pigment epithelium. *J. Membr. Biol.* 36:337–372.

Miller, S. S., and Steinberg, R. H. 1977*b*. Active transport of ions across frog retinal pigment epithelium. *Exp. Eye Res.* 25:235–248.

Miller, S. S., Steinberg, R. H., and Oakley, B., II. 1978. The electrogenic sodium pump of the frog retinal pigment epithelium. *J. Membr. Biol.* 44: in press.

Moriarty, C. M., and Terepka, A. R. 1969. Calcium transport by the isolated chick chorio-allantoic membrane. *Arch. Biochem. Biophys.* 135:160–165.

Nellans, H. N., Frizzell, R. A., and Schultz, S. G. 1975. Effect of acetazolamide on sodium and chloride transport by in vitro rabbit ileum. *Am. J. Physiol.* 228: 1808–1814.

Nilsson, S. E. G. 1964. An electron microscopic classification of the retinal receptors of the leopard frog (*Rana pipiens*). *J. Ultrastruct. Res.* 10:390–416.

Noell, W. K., Crapper, D. R., and Paganelli, C. V. 1965. Transretinal currents and ion fluxes. In *Transcellular membrane potentials and ionic fluxes,* ed. F. Snell and W. K. Noell. New York: Gordon and Breach.

Oakley, B., II. 1977. Potassium and the photoreceptor-dependent pigment epithelial hyperpolarization. *J. Gen. Physiol.* 70:405–425.

Oakley, B., II, and Green, D. G. 1976. Correlation of light-induced changes in retinal extracellular potassium concentration with the *c*-wave of the electroretinogram. *J. Neurophysiol.* 39:117–133.

Oakley, B., II., Steinberg, R. H., Miller, S. S., and Nilsson, S. E. 1977. The in vitro frog pigment epithelial cell hyperpolarization in response to light. *Invest. Ophthalmol.* 16:771–774.

Orkand, R. K., Nicholl, J. G., and Kuffler, S. W. 1966. Physiological properties of glial cells in the central nervous system of amphibia. *J. Neurophysiol.* 29:788–806.

Orr, H. T., Cohen, A. I., and Lowry, O. H. 1976. The distribution of taurine in the vertebrate retina. *J. Neurochem.* 26:609–611.

Oxender, D. L., and Christensen, H. N. 1963. Distinct mediating systems for the transport of neutral amino acids by the Ehrlich Cell. *J. Biol. Chem.* 238:3686–3699.

Pasantes-Morales, H., Klethi, J., Ledig, M., and Mandel, P. 1972. Free amino acids of chicken and rat retina. *Brain Res.* 41:494–497.

Pasantes-Morales, H., Klethi, J., Urban, P. F., and Mandel, P. 1974. The effect of electrical stimulation, light and amino acids on the efflux of ^{35}S-taurine from the retina of domestic fowl. *Exp. Brain Res.* 19:131–141.

Pasantes-Morales, H., Urban, P. F., Klethi, J., and Mandel, P. 1973. Light-stimulated release of [^{35}S] taurine from chicken retina. *Brain Res.* 51:375–378.

Pearson, T. W., and Goldner, A. M. 1973. Calcium transport across avian uterus. I. Effects of electrolyte substitution. *Am. J. Physiol.* 225:1508–1512.

Pearson, T. W., and Goldner, A. M. 1974. Calcium transport across avian uterus. II. Effects of inhibitors and nitrogen. *Am. J. Physiol.* 227:465–468.

Porter, K. R., and Yamada, E. 1960. Studies on the endoplasmic reticulum. V. Its form and differentiation in pigment epithelial cells of the frog retina. *J. Biophys. Biochem. Cytol.* 8:181–205.

Quinton, P. M., Wright, E. M., and Tormey, J. M. 1973. Localization of sodium pumps in the choroid plexus epithelium. *J. Cell Biol.* 58:724–730.

Reddy, D. V. N. 1967. Distribution of free amino acids and related compounds in ocular fluids, lens, and plasma of various mammalian species. *Invest. Ophthalmol.* 6:478–483.

Rehm, W. S. 1975. Ion transport and short-circuit technique. *Curr. Top. Membr. Trans.* 7:217–270.

Reuss, L., and Finn, A. L. 1974. Passive electrical properties of toad urinary bladder epithelium: intercellular electrical coupling and transepithelial cellular and shunt conductance. *J. Gen. Physiol.* 64:1–25.

Reuss, L., and Finn, A. L. 1975. Dependence of serosal membrane potential on mucosal membrane potential in toad urinary bladder. *Biophys. J.* 15:71–75.

Riggs, T. R., Walker, L. M., and Christensen, H. N. 1958. Potassium migration and amino acid transport. *J. Biol. Chem.* 233:1479–1484.

Rodieck, R. W. 1973. *The vertebrate retina: principles of structure and function.* San Francisco: W. H. Freeman and Co.

Rose, R. C., and Schultz, S. G. 1971. Studies on the electrical potential profile across rabbit ileum. *J. Gen. Physiol.* 57:639–663.

Salceda, R., and Pasantes-Morales, H. 1975. Calcium coupled release of [^{35}S] taurine from retina. *Brain Res.* 96:206–211.

SCHULTZ, S. G. 1975. Principles of electrophysiology and their application to epithelial tissues. In *Gastrointestinal physiology,* ed. E. D. Jacobson, pp. 69–103. London: Butterworth.

SCHULTZ, S. G., and CURRAN, P. F. 1970. Coupled transport of sodium and organic solutes. *Physiol. Rev.* 50:637–718.

SCHWARTZ, T. L. 1971. The thermodynamic foundations of membrane physiology. In *Biophysics and physiology of excitable membranes,* ed. W. J. Adelman, pp. 47–95. New York: Van Nostrand Reinhold Co.

SOLINGER, R. E., GONZALEZ, C. F., SHAMOO, Y. E., HERMAN, R. W., and BRODSKY, W. A. 1968. Effects of ouabain on ion transport mechanisms in the isolated turtle bladder. *Am. J. Physiol.* 215:249–261.

SPENNY, J. G., FLEMSTROM, G., SHOEMAKER, R. L., and SACHS, G. 1975. Quantitation of conductance pathways in antral gastric mucosa. *J. Gen. Physiol.* 65:645–662.

STARR, M. S., and VOADEN, M. J. 1972. The uptake, metabolism and release of [^{14}C] taurine by rat retina in vitro. *Vision Res.* 12:1261–1269.

STEINBERG, R. H. 1973. Scanning electron microscopy of the bullfrog's retina and pigment epithelium. *Z. Zellforsch. Mikrosk. Anat.* 143:451–463.

STEINBERG, R. H., and MILLER, S. S. 1973. Aspects of electrolyte transport in frog pigment epithelium. *Exp. Eye Res.* 16:365–372.

STEINBERG, R. H., MILLER, S. S., and STERN, W. H. 1978. Initial observations on the isolated retinal pigment epithelium-choroid of the cat. *Invest. Ophthalmol.* 17:675–678.

STIRLING, C. E. 1976. High-resolution autoradiography of ^{3}H-ouabain binding in salt transporting epithelia. *J. Microsc.* 106:145–157.

TEREPKA, A. R., COLEMAN, J. R., ARMBRECHT, H. J., and GUNTER, T. E. 1976. Transcellular transport of calcium. In *Calcium in biological systems* (Symposia of the Society for Experimental Biology), vol. 30, pp. 117–140. Cambridge: Cambridge Univ. Press.

THOMAS, R. C. 1972. Electrogenic sodium pump in nerve and muscle cells. *Physiol. Rev.* 52:563–594.

USSING, H. H. 1949. The distinction by means of tracers between active transport and diffusion. *Acta Physiol. Scand.* 19:43–56.

USSING, H. H., and ZERHAN, K. 1951. Active transport of sodium as the source of electric current in the short-circuited isolated frog skin. *Acta Physiol. Scand.* 23:110–127.

WALLING, M. W., and ROTHMAN, S. S. 1969. Phosphate-independent, carrier-mediated active transport of calcium by rat intestine. *Am. J. Physiol.* 217:1144–1148.

WRIGHT, E. M. 1974. Active transport of iodide and other anions across the choroid plexus. *J. Physiol.* 240: 535–566.

ZADUNAISKY, J. A., and DEGNAN, K. J. 1976. Passage of sugars and urea across the isolated retina pigment epithelium of the frog. *Exp. Eye Res.* 23:191–196.

ACKNOWLEDGMENTS

This work was supported by National Eye Institute Research Grants EY-00468 (to Dr. Kenneth T. Brown) and EY-01429 (to R. S.); by Career Development Award EY-18073 (to R. S.) by Postdoctoral Research Fellowship EY-49436 (to S. M.) and by a Fight for Sight Grant (G-525, to S. M.). Figures 12.8–12.15 are used with permission from Experimental Eye Research, *copyright by Academic Press Inc. (London) Ltd.*

13 Light-Induced Electrical Responses of the Retinal Pigment Epithelium

Michael F. Marmor
Mark Lurie

WHEN A CORNEAL recording electrode is placed upon an eye, and the eye is illuminated, one records not just the electroretinogram, but a long series of electrical responses (voltage changes), which can last up to several hours (Fig. 13.1). The first response is the *early receptor potential,* which occurs within a few milliseconds and represents physicochemical changes in the visual pigment molecules after they

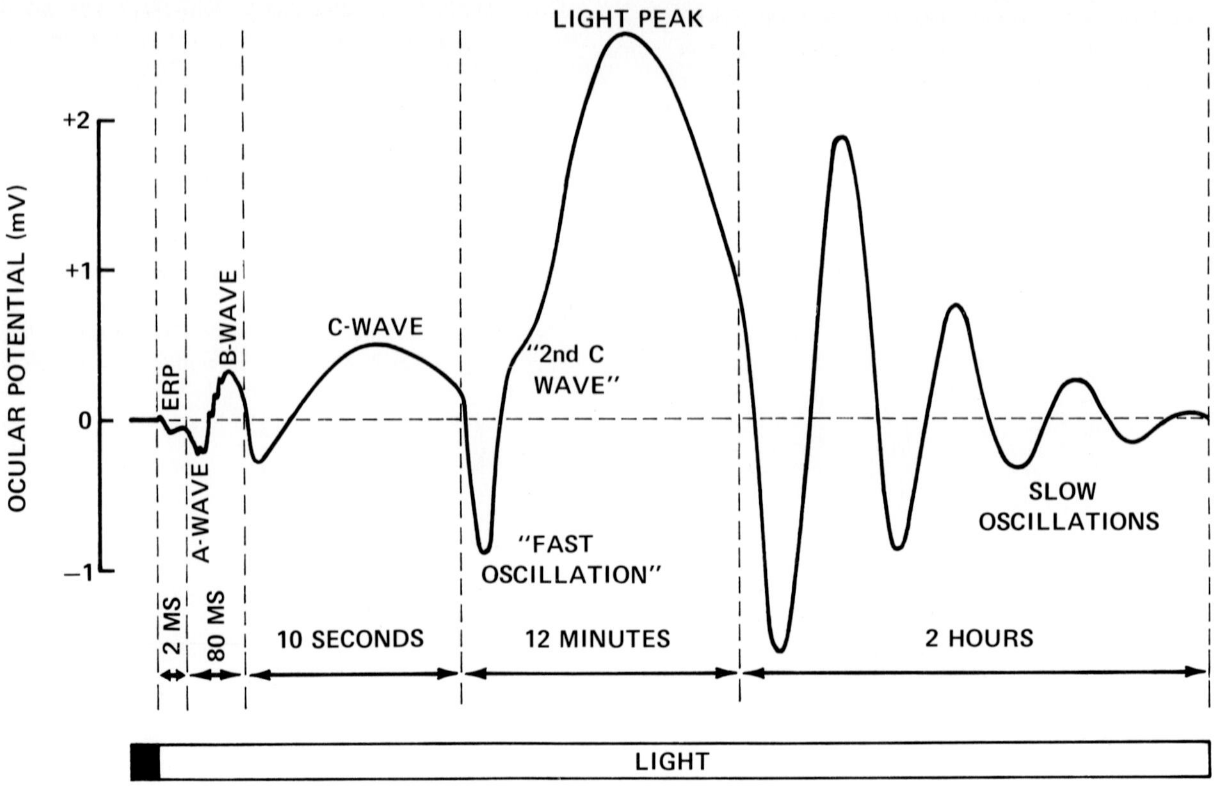

Figure 13.1 *Diagram of the electrical response of the human eye to sustained illumination, as would be recorded at the cornea. The response is drawn as a continuous record that incorporates different time scales, as if a hypothetical recorder were shifted to a slower speed as each component of the response passed its peak. See also Figure 13.7.*

are struck by light. Between 10 and 100 milliseconds after the onset of light, one records the *a* and *b-waves* of the *electroretinogram,* that is, the clinical ERG. The *c-wave* of the ERG is a slower response that appears several seconds after the onset of light. If one continues the recording, the baseline voltage, or *standing potential,* undergoes a slow increase in amplitude, called the *light response,* and this is followed by a series of slow oscillations of the standing potential, which may last more than an hour.

As interest in the pigment epithelium has grown, so has interest in those electrical responses that might reflect or define its functions. Two of the responses just described, the c-wave of the ERG and the light response of the standing potential, are commonly attributed to the pigment epithelium (although neither is in fact produced solely by the pigment epithelium). What is the origin and what is the significance of these so-called pigment epithelial responses? To what extent do they actually reflect pigment epithelial activity? As will be described, the cells of the pigment epithelium are, for all practical purposes, insensitive to light (excepting a tiny signal produced by the melanin granules when exposed to an intense flash), but they can respond passively with an electrical signal to ionic or chemical changes in their environment. Thus, light activation of the photoreceptors, and possibly some inner retinal elements, can lead indirectly to responses of the pigment epithelium. The pigment epithelium is often affected by disease, and a natural question is whether we can apply these responses clinically. A major problem, which we will discuss, is that stable recordings of these responses are much more difficult to obtain than recordings of the conventional ERG, and special techniques must be used.

Our purpose in this chapter is to present the physiologic bases of the light-induced electrical responses of the pigment epithelium, and to define their relevance for present and future clinical applications.

Methodology

The initial deflections of the ERG, the a- and b-waves, are relatively rapid phenomena that may be recorded with alternating current (a.c.) amplification. In a.c. amplification the signal must pass through a capacitor, which damps out any slow changes in voltage. Rapid signals are accurately reproduced, and the advantage of the technique lies in the simplicity of the electrode system and the ready provision of a stable baseline. Direct current (d.c.) amplification can accurately reproduce signals of any duration, but as a result the recordings tend to drift and show artifacts unless the electrode system is very stable and the subject remains still. Figure 13.2 compares d.c. and a.c. amplification of sine waves at different frequencies and demonstrates that pigment epithelial responses such as the c-wave and the light response of the standing potential are too slow to be reproduced by a.c. amplification. Thus, either d.c. amplification has to be used (with the concomitant problems of interference by ocular move-

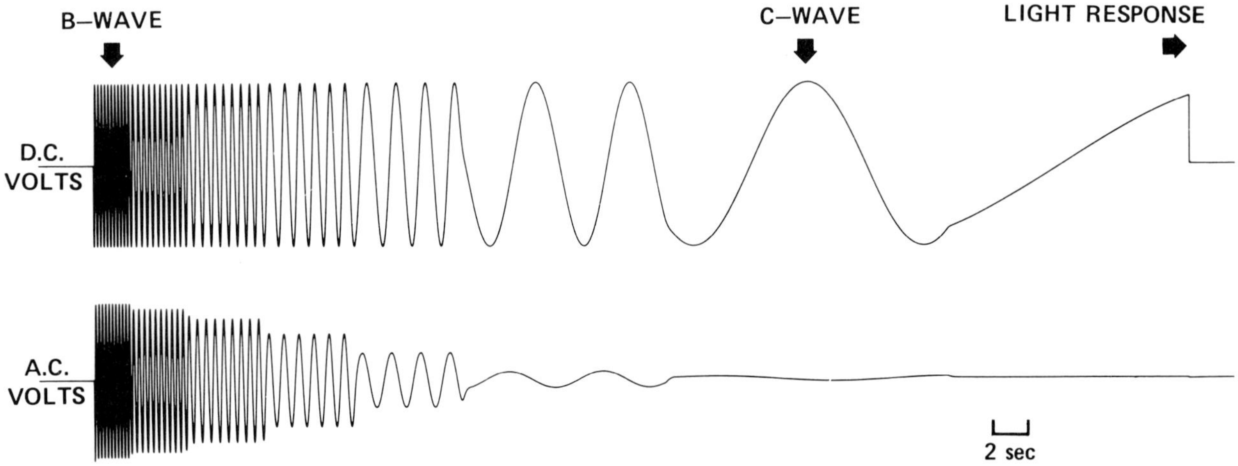

Figure 13.2 *Comparison of d.c. and a.c. amplification. Sine waves at different frequencies were passed through a d.c. (above) and an a.c. (below) amplifier. The a.c. record was filtered at 2–1,000 Hz, typical settings for conventional ERG recording. The arrows show the dominant frequencies of some ocular responses. Note that a.c. amplification is adequate only for recording a- and b-waves, not for slower components.*

ments and the need for specially designed recording electrodes) or one needs to resort to some indirect method of recording such as electrooculography.

D.C. AMPLIFICATION (THE D.C.-ERG). Direct current amplification of the corneal potential requires the use of nonpolarizable electrodes, and the best recordings have used fluid bridges to either silver/silver-chloride amalgam or calomel half-cells for connection with the amplifier. Movement of the eyes, eyelids, and pupils must be controlled, and account must be taken of contact potentials at the electrode/tissue interfaces and of possibly-changing shunt resistances between the electrodes.

Owing to these technical difficulties, there has been a relative paucity of studies on the d.c.-ERG, and few reports have dealt with slow waves in the human ERG. One of the first careful studies using a d.c. recording technique was by Dodt (1951). Hanitzsch, Hommer, and Bornschein (1966) and Heilig, Thaler, and Bornschein (1973) used anesthesia to help stabilize their patients. Ulrich, Bohne, and Ulrich (1974) used a chopper-stabilized amplifier. A new technique for stable recording of the d.c.-ERG in unanesthesized humans was reported by Knave, Nilsson, and Lunt (1973), and later modified by Skoog and Nilsson (1974a), Nilsson and Skoog (1975), and Taumer et al. (1976b). Saline-filled contact lenses under mild suction were used to minimize eye movement artifact and avoid the polarization of metal electrodes. The electrodes were connected by agar bridges to stable matched calomel half-cells, from which the signals were fed to a d.c. amplifier. Although subjects had to be cooperative to hold the fixation target well, satisfactory recordings were possible over long periods of time. Clinical application of these techniques may eventually become feasible, and would be of greatest value for phenomena of intermediate duration (such as the c-wave), which cannot be easily recorded by either electrooculography or the conventional ERG.

ELECTROOCULOGRAPHY. The history of electrooculography has been reviewed in detail by Marg (1951). The technique depends upon alternating eye movements: electrodes are placed at each canthus, and as the ocular dipole (which is cornea-positive) rotates to the side, the electrode nearest the cornea will become positive relative to the other electrode. As the eye rotates back and forth, the polarities reverse and the voltage changes may be recorded by suitable amplifiers. If the amplitudes of the eye movements are kept constant, and d.c.-amplification is used, then the voltage changes will be proportional to the standing potential across the eye. The eyes may be moved voluntarily, or by the experimenter, but in either case involuntary-movement artifacts are minimized and electrical drift is less of a problem than with a corneal electrode. The measurements that are obtained can show relative changes in the standing potential (which may be produced by light, for example) but cannot give absolute values and are highly dependent upon factors such as electrode placement and eye prominence. Thus, recordings are not directly comparable between individuals. The greatest value of electrooculography is that it is easy to use with human subjects, who cannot always be relied upon for the prolonged accurate fixation necessary for more direct recording techniques.

A few further comments about terminology are necessary. The terms "electrooculography" and "electrooculogram" have been in use for many years (Marg 1951). As originally intended, and as used in this chapter, they refer quite simply to the technique just described for recording alternating eye movements and to any record obtained by this technique. Unfortunately, usage in the literature is not consistent. Francois, Verriest, and De Rouck (1955) and Arden, Barrada, and Kelsey (1962) defined protocols for clinical evaluation of the light response, and Arden and colleagues used "electrooculogram" or "EOG" to describe the record from their test. They also defined a numerical value (the light/dark ratio) as a clinical parameter, and this ratio has been referred to as the "EOG" by later authors.

C-Wave of the Electroretinogram

The c-wave was first defined, for the frog, by Einthoven and Jolly (1908) as a slow cornea-positive component of the ERG that developed after the b-wave. In mammals the c-wave (see Figure 13.3) refers to a slow positive wave that follows or rises out of the falling b-wave, and which peaks roughly 2–10 seconds after the onset of light (depending on flash intensity and duration). C-waves have been described in a variety of animals (Kikawada 1968), including mice and rats (Keeler, Sutcliffe, and Chaffee 1928; Pautler and Noell 1976), rabbits (Noell 1953), cats (v. Brücke and Garten 1907; Granit 1933; Tamai 1965; Steinberg, Schmidt, and Brown 1970), sheep (Knave, Møller, and Persson 1972), monkeys (Gouras

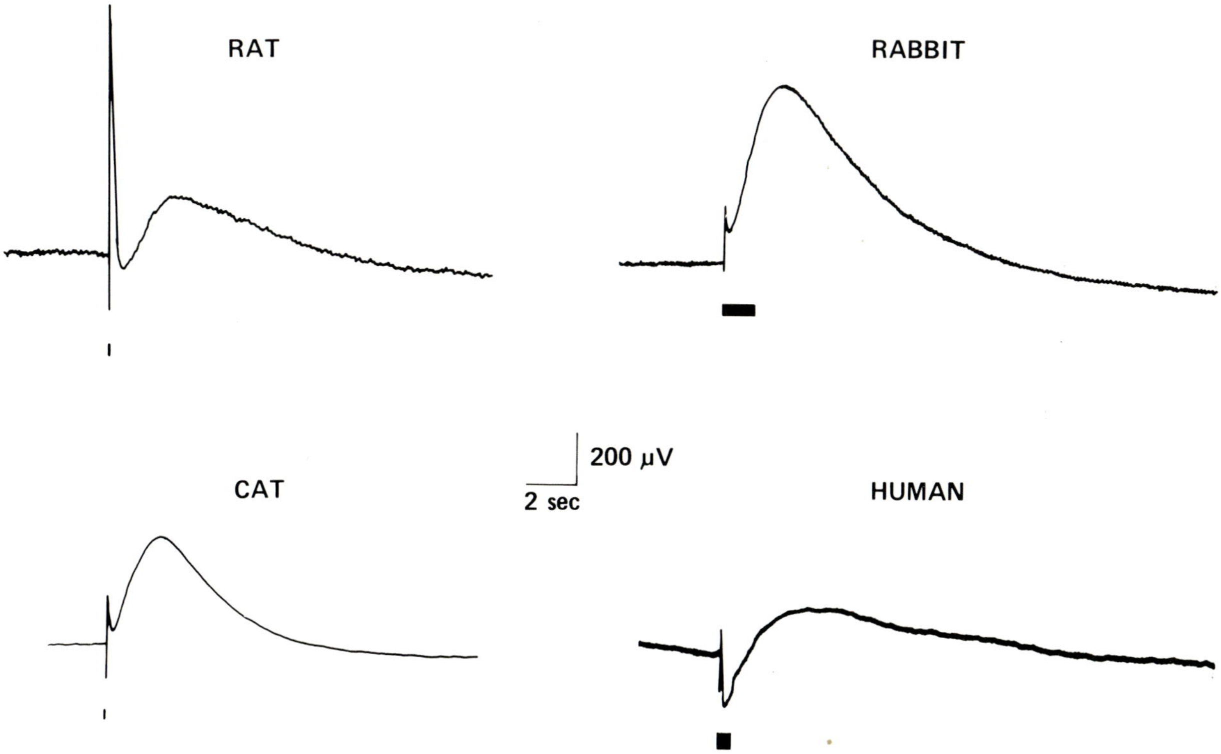

Figure 13.3 *C-waves from rat, rabbit, cat, and man. The rat, rabbit and cat responses were obtained under pentobarbital anesthesia, with the pupils dilated, using flashes of 50 msec, 1 sec, and 10 μsec respectively. The human response was recorded from an awake normal subject with dilated pupils using a saline-filled corneal electrode and a 100 msec flash. This subject had an unusually large a-wave, so that his b-wave barely exceeded the baseline and his c-wave began at a more negative level than in many individuals.*

and Carr 1965a; Fujino and Hamasaki 1965), and humans (to be discussed later).

ORIGIN OF THE C-WAVE. Although the photoreceptors initiate the response, the cornea-positive c-wave voltage is generated primarily by the pigment epithelium. This was demonstrated by Noell (1953, 1954) in a series of experiments on the rabbit eye. First, he documented the "azide reaction," comprising a sudden rise in the standing potential across the eye after an intravenous injection of sodium azide. Then he demonstrated that the pigment epithelium was the site of the azide reaction. Finally, he showed that elimination of the c-wave with a pigment epithelial toxin, sodium iodate (see Fig. 13.4*A*), correlated in rate and time course with the elimination of the azide response, while sparing the a- and b-waves. In confirmation, destruction of the photoreceptors by iodoacetate abolished all components of the ERG (including the c-wave), while sparing the azide reaction.

Conclusive evidence of the involvement of the pigment epithelium in generation of the c-wave was obtained by Brown and Wiesel (1961) and Steinberg, Schmidt, and Brown (1970), who recorded intracellularly from pigment epithelial cells of the cat and found responses that were similar in time course and relative amplitude to the c-waves of the local ERG. These findings have been confirmed in the perfused cat eye by Niemeyer (1976).

As ordinarily recorded, the c-wave has the spectral sensitivity of rhodopsin (Granit and Munsterhjelm 1937; Steinberg, Schmidt, and Brown 1970), but as one would expect, the c-wave cannot be generated by retina alone, separated from the pigment epithelium (Yamashita 1959; Ames and Gurian 1963; Foulds and Ikeda 1966; Dowling and Ripps 1972). Pigment epithelium devoid of photoreceptors also fails to generate a c-wave (Dewar and M'Kendrick 1873; Brown and Crawford 1976b), indicating that the photoreceptors and not visual-pigment-containing phagosomes initiate the response. In the dystrophic rat, the c-wave appears to be altered more than photoreceptor loss would account for, possibly because the accumulation of phagocytic debris affects or blocks the chemical or ionic changes that normally influence the pigment epithelium (Pautler and Noell 1976). Melanin pigment has no known influence on the c-wave. Some authors have failed to record c-waves from albino animals (Dodt and Echte 1961), but more recent studies indicate that albinos do generate c-waves under certain conditions (Reuter 1974; Weidner 1976; Pautler and Noell 1976).

Gouras and Carr (1965a) found that the c-wave of the rhesus monkey was diminished several weeks after interruption of the central retinal artery, and they suggested that the inner retina might play a role in formation of the c-wave. However, Fujino and Hamasaki (1965) found no such effect in acute central retinal artery occlusion. Furthermore, the c-wave still appears in the presence of aspartate (Dowling and Ripps 1972; Oakley and Green 1976; Lurie and Marmor 1979a) or glutamate (Yonemura et al. 1974), which block neural transmission between the photoreceptors and inner retina (Fig. 13.4*B*).

There are no known anatomical connections between rod outer segments and pigment epithelial cells; thus, the generation of the c-wave must be dependent on the diffusion of some "transmitter" substance through the extracellular space between photoreceptors and pigment epithelial microvilli. As described in detail in Chapter 12, the apical membrane of the pigment epithelium is very sensitive to its ionic environment. Oakley and Green (1976) have recently shown in the frog eye that a flash of light causes a decrease in the extracellular potassium concentration around the photoreceptor outer segments, which correlates (in magnitude and time course) with the locally recorded c-wave of the ERG under a variety of stimulus conditions. Intracellular recordings (Oakley 1977) indicate that the apical membrane of the pigment epithelial cell hyperpolarizes in response to this fall in extracellular potassium. The apical hyperpolarization induces a change in the transepithelial potential, which is recorded as the pigment epithelial component of the c-wave. Oakley (1977) has shown that the transepithelial potential and the c-wave amplitude are both proportional to the logarithm of the extracellular potassium concentration at any point in time (once the b-wave process is complete).

A source-sink analysis of the c-wave currents in the rabbit eye by Faber (1969) showed that the c-wave recorded in the vitreous appeared to be an algebraic summation of two processes. One process is vitreal-positive, corresponding to Granit's PI (Granit, 1933), and is generated by the pigment epithelium. The second process, the so-called "slow

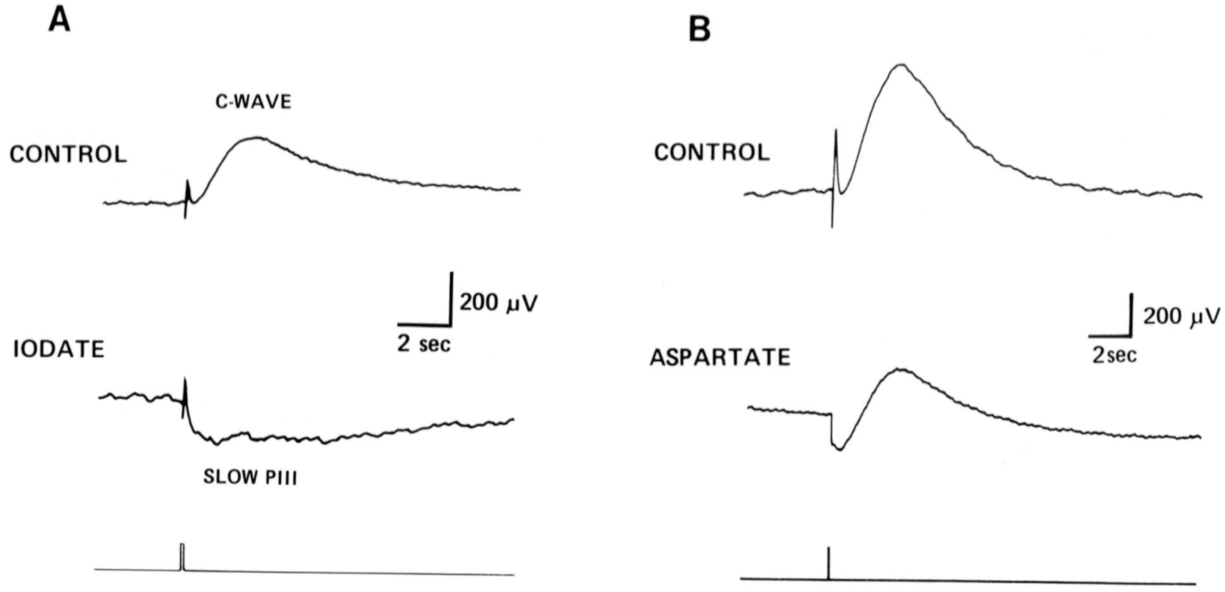

Figure 13.4 *Effects of iodate and aspartate on the d.c.-ERG of the rabbit.* (A) *Intravenous sodium iodate eliminates the c-wave and leaves the negative-going slow PIII.* (B) *Intravitreal sodium aspartate eliminates the b-wave, and deepens the trough that normally follows the b-wave; the c-wave remains intact.*

PIII," is vitreal-negative and is generated in the neuroretina; it becomes visible when the pigment epithelial component is eliminated by iodate (Fig. 13.4*A*). Rodieck (1972) suggested that the c-wave of the cat is a similar summation of positive and negative processes, although he did not define the origin of the negative component. Faber (1969) has suggested the terms *proximal PI* for the neuro-retinal component, and *distal PI* for the pigment epithelial component. Thus, the positive-going c-wave primarily reflects a pigment epithelial response; but its amplitude and physiologic characteristics also reflect the behavior of the proximal PI (slow PIII) component. Pautler and Noell (1976) suggest that the apparent absence of a c-wave in some albino rats, as opposed to pigmented rats, simply results from a predominance of the slow PIII over a diminished distal PI process in albinos.

Current evidence suggests that the c-wave is even more complex, incorporating (under most stimulus conditions) component voltages from at least three different sites in the retina, in addition to the component from the pigment epithelium (Lurie and Marmor 1979a). For example, the slow PIII component of the rabbit eye (isolated by iodate, as in Fig. 13.4*A*) can be further divided, by the addition of aspartate, into a cornea-negative voltage (resistant to aspartate) and a cornea-positive voltage that is generated synaptically in the inner retina. Evidence from our laboratory and others (Faber 1969; Witkovsky, Dudek, and Ripps 1975) indicates that this cornea-negative voltage can, in turn, be considered a summation of signals generated by the photoreceptors and by the Müller's cells. The cornea-positive c-wave may often be a rough measure of the electrical activity of the pigment epithelium, but should *not* be construed simply as a pigment epithelial response.

PROPERTIES AND SIGNIFICANCE OF THE C-WAVE. Recent experiments indicate that c-waves may be generated by both rods and cones, although in most animals, under ordinary conditions, only a rod c-wave can be demonstrated. For example, the c-wave cannot ordinarily be recorded from a light-adapted eye (Granit and Riddell 1934; Brown 1968) and it has the same spectral sensitivity as rhodopsin (Granit and Munsterhjelm 1937; Steinberg, Schmidt, and Brown 1970). However, Schmidt and Steinberg (1971) point out that the cone contribution may simply be too small to detect in the cat, since rods greatly outnumber cones in this animal. C-waves were initially reported to be absent in cone-dominant eyes (Arden and Tansley 1955; Crescitelli 1961), but more recently c-waves have been elicited from the cone-dominant retinas of turtles (Kikawada 1968; Matsuura, Miller, and Tomita 1978) squirrels (Kikawada 1968) and tree shrews (Tigges, Brooks, and Klee 1967). Matsuura, Miller, and Tomita (1978) have shown, furthermore, that the cone c-wave appears to be generated by the same mechanisms as the rod c-wave (although with a faster time course)—that is, by a light-induced decrease in the potassium concentration within the subretinal space, which causes a hyperpolarization of the pigment epithelial cell.

The amplitude of the c-wave is related to both the intensity and the duration of the stimulating light. For example, the rabbit c-wave (Faber 1969; Lurie and Marmor 1979b) varies in a roughly linear relationship with log stimulus intensity, log stimulus duration, or log stimulus energy (intensity multiplied by duration) over a range of several log units, as long as the stimuli end well before the peak of the wave. Schmidt and Steinberg (1971) have shown that similar relationships apply to the intracellularly recorded responses of the cat pigment epithelium.

A striking feature of the c-wave, which distinguishes it from many other electrophysiological signals, is that the time it requires to reach its peak (Fig. 13.5) also increases with the intensity or duration of the stimulus (Faber 1969; Schmidt and Steinberg 1971; Niemeyer 1976). As long as the stimulus duration is not excessive (Lurie and Marmor 1979b), the time-to-peak adheres to Bloch's law (reciprocity law). In other words, stimuli of equal energy (intensity × duration) produce responses that peak at the same time. However, Fig. 13.5 shows that for each stimulus intensity there is a critical duration (typically 3–4 seconds for a strong stimulus in the rabbit) beyond which Bloch's law no longer holds. Below the critical duration the c-wave integrates energy linearly, and reaches its peak *after* the stimulus ends; beyond the critical duration, the c-wave is sustained only *as long* as the stimulus itself. In some cold-blooded vertebrates, such as the frog, the c-wave may remain elevated as long as the light is maintained (Lurie 1976). In contrast, the c-wave in mammals is not sustained by steady illumination, and will fall after 5–10 seconds regardless of the stimulus intensity. However, there is no evidence that the mammalian c-wave generators turn off during sustained illumination; the apparent fall of the mammalian

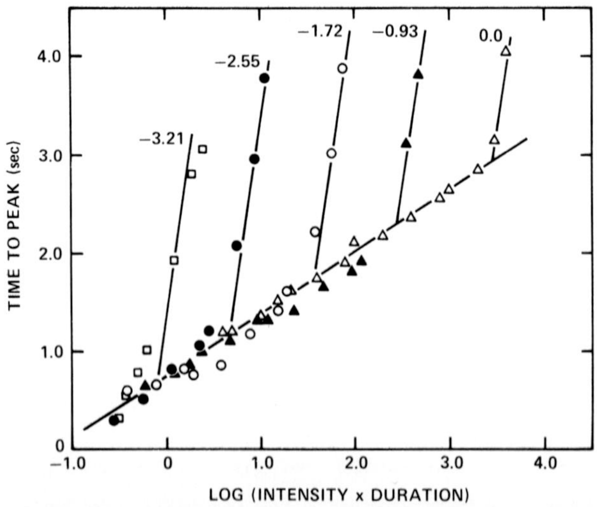

Figure 13.5 *Rabbit c-wave: Relationship between time-to-peak and log energy of the stimulus. For flashes of relatively short duration the points all fall on a straight line. However, at each level of flash intensity (indicated logarithmically by numbers on the graph) a "critical duration" of the stimulus is reached at which the points depart from the primary line. (From Lurie and Marmor, 1979b.)*

c-wave may actually represent the initial negative wave ("fast oscillation") of the light response.

Another interesting property of the c-wave has been described recently by Calissendorff, Knave, and Persson (1974) in sheep. These investigators showed that if a stimulus flash was presented repeatedly, the c-wave amplitude oscillated between 200 and 500 μV roughly twice per hour. Knave, Persson, and Nilsson (1974) pointed out that these c-wave amplitude oscillations are similar in frequency to oscillations of the standing potential. Simultaneous recordings of the c-wave and standing potential in monkeys (Gouras and Carr 1965a) and in man (Nilsson and Skoog 1975) have shown that the oscillations of both are in phase.

The significance of the c-wave for retinal function has not been established. It may be a concomitant of metabolic changes following light-activation of the photoreceptors, and in theory it could influence retinal activity by generating electrical effects that extend through the retina. Early studies found no correlation between the c-wave of the ERG and the discharge in the optic nerve, but Lurie (1976) has shown that in the frog eye the activity of single class IV ganglion cells (dimming fibers) correlates closely with the simultaneously recorded c-wave. The transretinal current associated with the c-wave appears to be sufficient to induce the effect, suggesting (Lurie 1976) that the c-wave can play a role in the transmission of visual information and may be associated with phenomena such as pupillary movement or afterimages.

THE C-WAVE IN MAN. Human c-waves have been reported since the 1920s, but Dodt (1951) showed that the prominent wave in most early records was produced by movement of the pupil and could be eliminated by atropine or a strong miotic. Pearlman (1962) and Tamai (1966) observed pupillary slow waves from undilated eyes and doubted whether a significant human c-wave was recordable. However, later studies using anesthetized patients with dilated eyes showed convincing c-waves (Hanitzsch, Hommer, and Bornschein 1966; Heilig, Thaler, and Bornschein 1973). The current techniques of direct recording from unanesthesized subjects (Nilsson and Skoog 1975; Taumer et al. 1976b; Marmor, Pockrand, and Lurie 1979) allow a characterization of the normal human c-wave in the dilated eye.

The form of the c-wave in human eyes (Fig. 13.3), as in other vertebrate eyes, depends upon conditions of adaptation and stimulation and presumably upon a balance of the various slow potential generators in the eye. C-waves may be elicited in man by brief flashes or by longer steps of light, and both the c-wave amplitude (Fig. 13.6) and time-to-peak in-

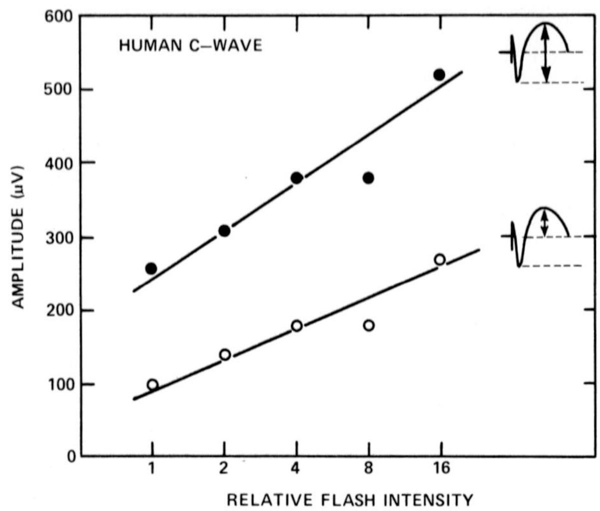

Figure 13.6 *Human c-wave: Relationship between amplitude and log intensity of a 10 μsec flash. The c-waves were measured from both the trough (upper curve) and the baseline (lower curve). From Marmor, Pockrand, and Lurie 1979.)*

crease with the log of stimulus intensity or stimulus duration (Skoog and Nilsson 1974; Taumer et al. 1976c; Textorius 1977; Marmor, Pockrand, and Lurie 1979). The c-wave peaks in about 2 seconds after brief flashes, but not until 5 or 6 seconds after steps of illumination. When shorter stimuli are used, flashes can be repeated after roughly 30 seconds without apparent loss of amplitude.

The eyes must be dilated and dark-adapted in order to record c-waves in man. Taumer et al. (1976c) have confirmed that if no mydriatic is used, a large pupillary wave is recorded that is independent of the c-wave. Unfortunately, there is enormous variability in the amplitude and waveform of the c-wave among normal subjects. Some normals appear to lack c-waves altogether (Taumer et al. 1976d; Marmor, Pockrand, and Lurie 1979), although it is not clear whether these individuals truly lack the c-wave generating process or are showing an atypical balance between the negative slow PIII and positive pigment epithelial responses. Part of the difficulty is in deciding what measurements define the c-wave. For example, is the c-wave properly measured from the starting baseline or from the trough which immediately precedes it? (See Fig. 13.6.) Part of the variability may be a result of variations in the underlying standing potential, which appears to influence the c-wave amplitude. Skoog and Nilsson (1974b) showed that c-waves, recorded at 30-second intervals, showed a two-per-hour cyclic variation in amplitude that correlated in both phase and frequency (Nilsson and Skoog 1975) with simultaneously recorded oscillations of the standing potential.

Several problems must be resolved before c-wave recording can be used reliably for clinical purposes. First, the recording technique must be improved to accommodate less cooperative subjects. Second, clinical test parameters, such as preadaptation, stimulus length, stimulus intensity, and sequence of stimuli must be defined, but this is a difficult task when the response is so slow and exhibits oscillatory behavior. The task is further complicated by the presence of off-effects that can interfere with the measurement of the c-wave peak (Textorius 1977). Third, normal values must be established. This task is complicated by the variability and occasional lack of c-waves among normal subjects. Finally, the degree to which the cornea-positive wave reflects pigment epithelial activity independent of retinal function must be firmly established. The potential attractiveness of the c-wave as a clinical test depends upon the ability to use it as an index of pigment epithelial involvement by disease.

PATHOPHYSIOLOGY OF THE C-WAVE AND THE EFFECTS OF PHARMACOLOGICAL AGENTS. Various pharmacologic agents can alter the c-wave. Granit (1933) observed that lightly anesthetizing a cat with ether abolished the c-wave, while leaving the more rapid components of the ERG unchanged. However, experiments in the rabbit, using injectable anesthetics such as pentobarbital, ketamine, and urethane, do not show this effect (Lurie and Marmor 1979b). Intravenous sodium iodate can selectively damage the pigment epithelium and will eliminate the c-wave in rabbits (Noell 1953, 1954), cats (Rodieck and Ford 1969), and sheep (Nilsson, Knave, and Persson 1977). One should note, however, that iodate in sufficient dose or for a sufficient time will also affect other retinal elements (Noell 1953; Nilsson, Knave, and Persson 1977); its selectivity is relative and depends upon proper use. Quinine diminishes the c-wave in cats (Hommer, Ulrich, and Wündsch 1968) and rabbits (Jünemann and Schulze 1968). On the other hand, enhancement of the c-wave has been produced by sodium azide in rabbits and albino rats (Noell 1953; Weidner 1976), by L-cysteine in rabbits (Shibata 1975), by elevation of the serum osmolarity in rabbits (Yonemura, Kawasaki, and Ishikawa 1976) and by desipramine in cats (Heiss, Heilig, and Hoyer 1967). The frog c-wave appears to be enhanced by the direct application of epinephrine to the retina (Therman 1938).

Calissendorff (1976a, b) found that a variety of melanotropic drugs, including quinine, chloroquin, phenothiazines, and rifampicin, had an unusual effect on the d.c.-ERG of the sheep. After varying intervals, all of these drugs caused a slow rise in c-wave amplitude, followed by a damped oscillation of the amplitude at about two cycles per hour. Barbiturates have a similar effect on the sheep c-wave (Knave, Persson, and Nilsson 1974). Skoog (1974) found that moderate oral doses of ethanol in man caused a striking enhancement of the c-wave amplitude.

Standing Potential

The existence of a steady potential difference across the eye was first described in 1849 by DuBois-Reymond, who found that the cornea of the fish eye was positive to the optic nerve. (The electrical response of the eye to a *change* in illumination was first discovered by Holmgren in 1865 and later found independently by Dewar and M'Kendrick in 1873). The existence of a steady cornea-positive voltage in the dark, the "standing potential," has since been observed to be a constant feature of all vertebrate eyes. Synonyms in the literature include "dark potential" or "resting potential." In the mammalian eye the standing potential has a value of several millivolts.

ORIGIN OF THE STANDING POTENTIAL. The voltage across the eye represents the sum of contributions from various sources. Potential differences exist across the cornea (Potts and Modrell 1957; Donn, Maurice, and Mills 1959), lens (Brindley 1956), ciliary body (Berggren 1960), retina (Zuckerman 1971) and pigment epithelium (Dewar and M'Kendrick 1873; Steinberg and Miller 1973). An alteration in the function of any one of these elements, therefore, will be reflected in a change in magnitude of the standing potential.

Noell demonstrated in 1952 that the pigment epithelium contributed a significant portion of the standing potential in the rabbit eye. Administering sodium iodate to the animal resulted in a selective degeneration of the pigment epithelium, which correlated with a loss of the azide response (see the section on the c-wave) and a disappearance of the standing potential. Faber (1969) confirmed the importance of the pigment epithelium in the generation of the standing potential by an intraretinal source-sink analysis. The currents generated by the retina–pigment epithelium complex in the dark-adapted rabbit eye in situ were measured by Noell, Crapper, and Paganelli (1964). They found the steady d.c. potential between the vitreous and sclera generally to be between 4 and 5 mV. The potential difference across the isolated pigment epithelium-choroid complex has been determined to be 20–30 mV in the toad (Lasansky and de Fisch 1966) and 5–15 mV in the bullfrog (Steinberg and Miller 1973). The ionic and electrical characteristics of the pigment epithelium are considered more fully in Chapter 12.

THE STANDING POTENTIAL IN MAN. The early experimenters who studied the standing potential of the human eye were unaware that a light response existed and their results consequently showed enormous variability. Eventually, however, a basal standing potential in the dark was recognized as a stable parameter. The amplitude of this potential is probably several millivolts, cornea positive, but an absolute value probably cannot be assigned with present recording techniques, because of variables such as electrode position and shunting within tissue fluids. The literature contains varying assessments of this problem (Skoog 1975; Taumer et al. 1976b).

The contribution of nonretinal structures to the human standing potential has been emphasized by Arden and Fojas (1962), who showed that the standing potential, measured by electrooculography, increased markedly after applying topical anesthetic to the cornea. Presumably, a transcorneal potential of polarity opposite to that from the retina and pigment epithelium was eliminated by the anesthetic. On the other hand, Arden and Barrada (1962) found no difference in the standing potential between the normal and operated eyes of unilateral aphakes, suggesting that the lens is not a major contributor. A puzzling observation has been reported by Pasik, Pasik, and Bender (1965), who found that while evisceration of ocular contents in the monkey (leaving only the scleral shell) abolished the standing potential transiently, a potential reappeared after 3–4 weeks.

PATHOPHYSIOLOGY OF THE STANDING POTENTIAL. There is little information about the dark potential in disease because of the difficulty in obtaining reliable measurements. Many authors have reported low absolute amplitudes in the electrooculographic records of patients with severe retinal dystrophies and similar diseases. Francois, Jonsas, and De Rouck (1969) noted that hypoglycemia caused a decrease in the dark potential. Foulds and Ikeda (1966) produced retinal detachments in the rabbit and reported that both the standing potential and the azide response of the pigment epithelium (see above) were preserved.

Yonemura and co-workers have recently demonstrated that several agents may affect the standing potential independently of light. Most striking is the demonstration that hyperosmotic agents diminish the standing potential (Kawasaki, Yamamoto, and Yonemura 1977). Within 10 minutes of starting an intravenous infusion of 20 percent mannitol the human standing potential (measured directly or by electrooculography) falls by 40–45 percent. Monkey experiments indicate that the effect is independent of the retina since it still occurs after poisoning the retina with aspartate or iodoacetate. The response is diminished in patients with pigmentary retinopathy or with widespread pigment epitheliopathy from diabetes, and Yonemura and Kawasaki (1978) suggest that it may be used clinically to detect diffuse abnormality of the pigment epithelium. Surprisingly, they also note prolonged abnormality of the response after retinal detachment or central artery occlusion. Acetazolamide (500 mg, injected intravenously) causes a similar fall in the standing potential (Yonemura et al. 1977). However, there are significant differences between these two effects. The acetazolamide response was not affected by diffuse disorders such as pigmentary retinopathy, but was diminished in patients with posterior pole damage whose hyperosmolar response was normal. Cysteine augments the standing potential in rabbits (Yonemura and Kawasaki 1978) but has not been used clinically.

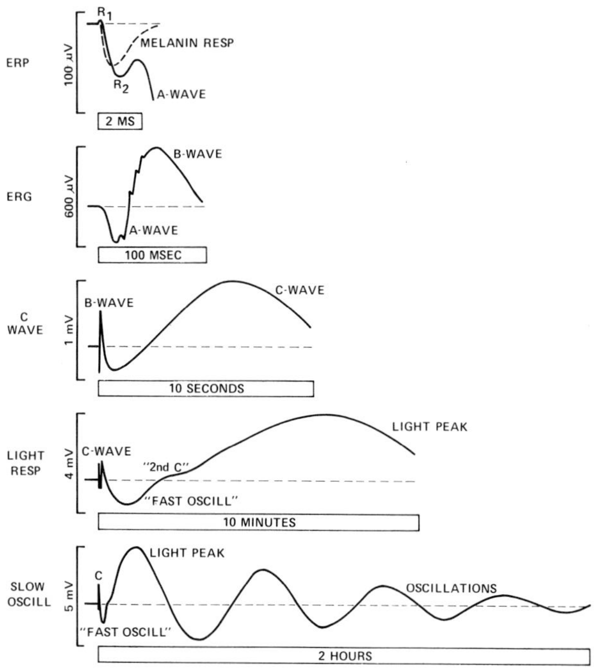

Figure 13.7 *Diagram of the major light-induced electrical responses of the human eye, as would be recorded by a corneal electrode. Note the different amplitude and time scales. See also Figure 13.1.*

Light-Induced Changes in the Standing Potential (EOG)

Exposure of the dark-adapted mammalian eye to light will cause a slow change in the standing potential, which is called the "light response." The initial electrical responses to light (that is, the a- and b-waves of the ERG) are rapid, and they are followed within several seconds by the c-wave. Only *after* the c-wave has occurred do changes in voltage appear that are considered a part of the light response; in fact, the peak of the light response (that is, the "light peak") does not occur until 5–10 minutes after the onset of light (Figures 13.1 and 13.7). The light peak is in general only the first of a set of very slow oscillations in the standing potential. If illumination is suddenly removed from a light-adapted eye, the initial effect on the standing potential will be a fall in voltage to a minimum level termed a "dark trough." The pigment epithelium has been implicated as an important element in the generation of these very slow voltage changes.

ORIGINS OF THE LIGHT RESPONSE. The standing potential represents an algebraic sum of contributions from various structures in the eye, but the structures generating the light response are confined to the retina and pigment epithelium. Noell (1953) showed that the light response in the rabbit was essentially identical in the intact eye and in an eye from which the entire anterior segment had been removed. He and Imaizumi et al. (1968) demonstrated that the light response disappeared upon selective destruction of the pigment epithelium by sodium iodate. The light response also requires contact to be maintained between the photoreceptors and the pigment epithelium, and it is abolished by experimentally induced retinal detachment in the rabbit eye (Foulds and Ikeda 1966). There is also evidence suggesting that inner retinal structures play a part in the generation of the response. Gouras and Carr (1965a) showed that after chronic interruption of the central retinal artery in the monkey eye, the light response disappeared along with the b-wave of the ERG. Gouras and Carr (1965b) also showed that cones as well as rods can initiate a light response in the monkey eye.

The actual mechanism by which the light response is generated, as well as the significance of the voltage changes produced, is unknown. Arden and Kelsey (1962b) have suggested that changes in metabolism of the pigment epithelium might be involved. However, the light response does not correlate with rhodopsin kinetics or any other known biochemical or physiologic process. No measurements have been performed to determine whether local ionic changes coincide with the light response. Although the light response also does not correlate with clinical measures of visual function (see below), shifts in the standing potential voltage are substantial (for example, several times greater than the b- and c-waves) and create electrical effects that can be recorded throughout the retina, and which may have some general influence on neural activity or membrane phenomena. The significance of the slow oscillations remains obscure.

CHARACTERISTICS OF THE LIGHT RESPONSE AND OSCILLATIONS. The sequence of voltage changes after exposure of the dark adapted mammalian eye to a steady light are show diagrammatically in Figure 13.7. The light response of the standing potential begins after the c-wave with a negative deflection, often called the "fast oscillation."

This is followed by a positive wave, peaking at 90–120 seconds, which has only recently been recognized as a discrete entity by Nikara et al. (1976), who called it the "2nd c-wave" (a designation that seems premature since it has not been proven to be related to the c-wave). Finally, the main light peak develops, reaching a maximum at 5–10 minutes and often leading into a set of very slow oscillations of the standing potential, which gradually become damped.

A light response has been observed in a wide variety of animals, including reptiles, birds, and rodents (Kikawada 1968); rabbits and dogs (Kolder and North 1966); sheep (Knave, Persson, and Nilsson 1974); monkeys (Gouras and Carr 1964) and man (see next section). Slow oscillations in the standing potential after the initial light response have been observed in most mammals that have been studied, but not in cold-blooded vertebrates. These oscillations may be induced by constant illumination or by repetitive flashes (Gouras and Carr 1964), and they may be maintained without damping by alternating periods of light and dark (Kolder and North 1966). In amphibians and fish, there is no discrete light peak, but the standing potential rises slowly until it reaches a plateau that is maintained for the duration of illumination (Yosida 1953; Müller-Limmroth 1954; Kikawada 1968).

Extinguishing light after an eye has been light-adapted induces a series of standing potential changes that more or less mirror those induced by light. In mammals, there is a negative off response similar to the c-wave in time course, followed by a positive wave and then descent to a "dark trough" that requires 5–10 minutes to develop (Kikawada 1968; Faber 1969). In amphibia and fish (Kikawada 1968), there is no dark trough, but the standing potential falls slowly to the dark potential level. Lurie (1976) has suggested that this fall in potential may correlate with recovery of sensitivity of the photo receptors.

THE NORMAL LIGHT RESPONSE IN MAN. Miles, in 1940, made the important observation that the amplitude of the human standing potential recorded by electrooculography varied with the state of illumination. Turning on a light caused a slow rise in the standing potential, which continued for at least 5 minutes, and turning off a light caused a fall in amplitude. The Ten Doesschates (1957) discovered that if one waited longer after changing illumination, the electrical response would reach a peak (in the light) or trough (in the dark) and return towards the original level. Then Kris (1958) demonstrated that once a change in the standing potential was initiated, the standing potential amplitude would oscillate through several cycles even though the illumination remained steady.

These observations were confirmed and elaborated upon by Kolder (1959) and Arden and Kelsey (1962a). They showed that turning on a light caused a slow rise in the standing potential, reaching a light peak after 5–10 minutes, whereas turning off a light caused a slow fall in potential to reach a "dark trough" in a similar period of time. Moreover, both responses were shown to be merely the beginning of a complex signal whose amplitude could be set into oscillation by any sudden change (up or down) of illumination (see Figure 13.8, upper record). Once set in oscillation, the amplitude shows a cycle time of about 30 minutes, gradually becoming damped until a steady baseline is again achieved in about two hours. Alternating light and dark can drive the oscillations and maintain their amplitude, but the basic, slow, 30-minute cycle is always present (Figure 13.8, lower record). The amplitude of the initial light peak, recorded by d.c.-electroretinography is in the range of 5 mV (Skoog 1975; Taumer, Hennig, and Pernice 1974).

The early components of the human light response are seen much more clearly with direct d.c. recording than with electrooculography, but both techniques have revealed a waveform similar to that observed in animal studies. For example, both the negative "fast oscillation" and the positive "second c-wave" are visible in published illustrations (Arden

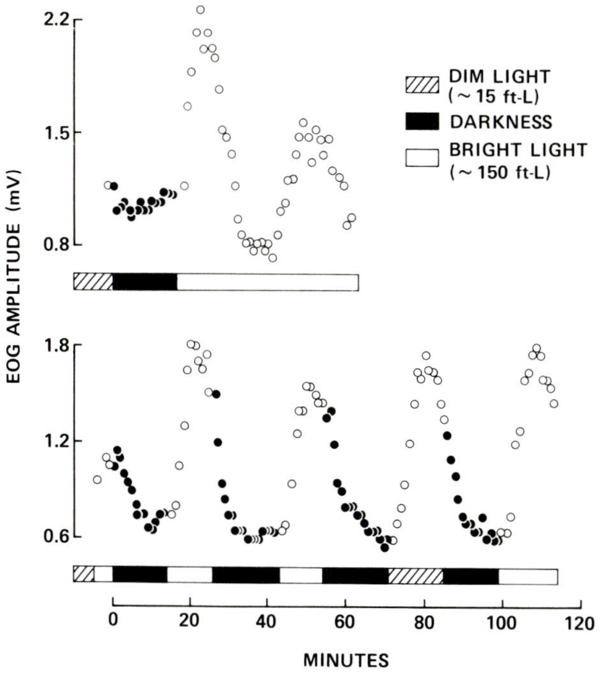

Figure 13.8 *Electrooculographic records of the standing potential from a normal human subject with dilated pupils. The upper record shows that exposure to steady illumination produces a light response followed by a slow oscillation of the standing potential. The lower record shows that alternating cycles of light and dark will maintain the oscillations without a loss of amplitude.*

and Kelsey 1962a; Taumer, Hennig, and Pernice 1974; Taumer et al. 1976b). The off-response, when a light is extinguished, can be similarly analyzed (Skoog, Welinder, and Nilsson 1977).

The amplitude of the light response is proportional to the log intensity of the light step over about three decades, and is graded over an even broader range (Arden and Kelsey 1962b; Francois et al. 1974). Pupillary movement has little effect on the light peak (Arden and Barrada 1962). The action spectrum of the light response is primarily that of the rods (Arden and Kelsey 1962b), but long wavelengths (red light) are more effective than would be predicted from a scotopic curve, so that cones also must contribute (Gouras and Carr 1965b; Elenius and Karo 1966; Kolder 1974). Infrared light is not effective (Anderson and Kolder 1966).

The oscillations of the standing potential correlate with changes in the amplitude of the c-wave recorded at the same time (Gouras and Carr 1965a; Nilsson and Skoog 1975). The a- and b-wave amplitudes have been reported both to correlate with (Gouras and Carr 1964) and to be independent of (Kolder 1974; Francois et al. 1974) the standing potential. In addition to the slow oscillations of the standing potential, there can be superimposed a set of smaller and faster oscillations with a one-minute cycle if the eye is stimulated with lights that alternate on and off (Kolder and Brecher 1966). A number of mathematical models have been proposed (for example, Homer and Kolder 1967; Kolder and Hochgesand 1973; Taumer et al. 1974a) to describe the oscillatory behavior of the standing potential in response to light.

PATHOPHYSIOLOGY OF THE LIGHT RESPONSE: THE CLINICAL EOG. Since absolute potential measurements are unreliable, Arden, Barrada, and Kelsey (1962) and Arden and Barrada (1962) introduced the concept of a ratio between the amplitude of the light peak and the amplitude of the dark trough, to provide a clinical measure of the light response. The normal value of this ratio is above 1.75–2.0 (Arden and Barrada 1962; Adams 1973; Francois et al. 1974). Other clinical measures have been proposed but have not proven to be better (Krogh, 1977). The value of the Arden ratio depends somewhat on conditions of preadaptation and the intensity of light used; the effects of factors such as sex and age are probably not great and are disputed in the literature. Recent authors (Thaler and Heilig 1974; Taumer et al. 1974b; Taumer et al. 1976a) have suggested more stringent criteria of preadaptation to increase the reproducibility of the results.

The major clinical applications and implications of the EOG are discussed in Chapter 21 (see also Taumer 1976). However, a few clinical results need consideration here because they confirm or extend our knowledge of the physiology of the light response. First, the EOG ratio (and hence the light response) does not correlate with the presence or pattern of melanin in the pigment epithelium. For example, the EOG ratio can be normal both in patients without pigment, such as albinos (Reeser et al. 1970; Francois et al. 1974), and in patients with a pigmentary disorder such as rubella retinopathy (Krill 1972) or fundus albipunctatus (Marmor 1977). Conversely, the EOG ratio can be reduced both in patients with a normal peripheral fundus, such as in vitelliform dystrophy, and in patients with pigmentary disorders, such as the pattern dystrophies of the pigment epithelium (Marmor and Byers 1977). Second, the clinical EOG does not necessarily correlate with visual function (and thus the visual significance of the mechanism generating the light response is unknown). Patients with vitelliform dystrophy or pattern dystrophy (Marmor and Byers 1977) may have normal visual acuity, visual fields, color vision, and dark adaptation, along with an abnormal EOG. Third, EOG data from congenital achromats (Elenius and Antaa 1973) support the experimental evidence that cones contribute to the light response. Fourth, clinical data also support the conclusion that inner retinal layers have a role in the generation of the light rise. Central retinal artery occlusion has been well documented (Carr and Siegel 1964; Henkes and Deutman 1973; Thaler and Heilig 1977) to affect not only the b-wave of the ERG but the light rise of the clinical EOG.

Drugs and toxins can have relatively selective effects upon the standing potential and its light response. Both are damaged by iodate (in rabbits) in doses that only mildly affect the photoreceptors and the a-wave (Noell 1959). A moderate oral dose of ethanol (Skoog, Textorius, and Nilsson 1975) induces a large increase in the human standing potential within 10 minutes, regardless of whether the eye is dark-adapted or light-adapted, and the potential will then oscillate slowly for one or two cycles. The peak blood level of ethanol does not occur until about 60 minutes after ingestion, so this effect is triggered by the initial absorption of the drug. Large amplitude slow oscillations can be induced in the sheep by barbiturates (Knave, Persson, and Nilsson 1974). Urethane causes a rise in the standing potential of the rabbit, but reduces its response to light (Kolder and North 1966).

These results point out the potential value—and the limitations of—the EOG as a clinical test. The generation of the response is complex, requiring the photoreceptors, involving the inner retinal layers (although the cell type is uncertain), and finally becoming manifest as changes in voltage across the pigment epithelium. To use the clinical EOG as a monitor of pigment epithelial functions and their involvement by disease (cf. Marmor, 1975a, b) is only valid if other retinal tests are available to judge the integrity of the photoreceptors and inner retinal layers. Even then, "pathology" of the light rise, or the standing potential, must be interpreted with caution since neither are linked to any specific fundus change or to any specific parameter of visual function.

Direct Light Response (Melanin Potential)

PHYSIOLOGIC BASIS OF THE RESPONSE. In 1965, Brown described an electrical response that could be evoked in the isolated pigment epithelium of the toad by very intense flashes of light. A similar response has also been demonstrated in pigmented rats (Arden, Ikeda, and Siegel 1966; Ebrey and Cone 1967; Brown and Crawford 1967b) and in man (Zanen and Debecker 1971). We emphasize that this response is quite different from the conventional ERG or the c-wave, in that it can only be elicited by intense nonphysiological stimuli and it does not require the presence of photoreceptors.

The direct pigment epithelial response has three components (Brown 1965; Brown and Gage 1966; Brown and Crawford 1967a): first, a tiny positive deflection with no detectable latency; second, a brief cornea-negative wave lasting about 1 msec (see Figure 13.7); finally, a slower positive deflection. Melanin was quickly recognized as a probable source of the pigment epithelial response. Brown and Crawford (1967b) and Ebrey and Cone (1967) showed that the spectral sensitivity of the pigment epithelial response is flat throughout the visual range, which correlates with the spectral properties of melanin. They also showed, along with Arden, Ikeda, and Siegel (1966), that the pigment epithelial response was resistant to light adaptation, as is melanin, and that the response was absent in albino rats. Crawford, Gage, and Brown (1967) found that a filter paper impregnated with melanin and stimulated with an intense light would produce an electrical response similar in many respects to the physiologic one.

The slow cornea-positive third phase of the pigment epithelial (melanin) response behaves quite differently from the rapid early phase with respect to physical stress and the influence of chemical agents (Gage, Crawford, and Brown 1968). Membrane damage will eliminate this slow phase, as will the removal of sodium or bicarbonate from the bathing medium. Azide and ouabain poison the slow phase, but act slowly so that active sodium transport does not seem to be directly involved. The slow positive wave of the melanin response probably depends upon ionic conductance changes in the pigment epithelial membrane initiated by the absorption of light.

A natural comparison has been made between the melanin response and the initial response of the photoreceptors to intense flashes of light (Brown and

Murakami 1964). The latter also consists of two rapid phases, called the "early receptor potential," and a slower third phase, called the "late receptor potential" (which contributes the a-wave of the ERG). The early receptor potential is a physicochemical response to light absorption by the photopigment (Pak and Cone 1964; Cone 1967), while the late receptor potential involves membrane changes in ionic permeability. Recordings of the early receptor potential in amphibians may be contaminated by the melanin response, but this interference is generally small in mammals (Brown 1965; Ebrey and Cone 1967; Zanen and Debecker 1971). Arden, Ikeda, and Siegel (1966) and Ebrey and Cone (1967) noted that a variety of pigmented animal or plant tissue could produce electrical responses when strongly illuminated, and they concluded that the melanin response and early receptor potential are examples of a physicochemical reaction common to many oriented light-absorbing molecules.

THE MELANIN RESPONSE IN MAN. Zanen and Debecker (1971) and Debecker and Zanen (1975) showed that a small melanin response could be distinguished from the early receptor potential in man (Figure 13.7), but that it did not significantly interfere with the early receptor potential except at very high flash intensities. The melanin response could also be elicited from the pigment-epithelium–choroid complex of a cadaver eye. The melanin response is unique in being the only purely pigment epithelial response yet discovered, but its physiologic significance is unknown. Clinical use of the response has been considered (Arden 1968) and, theoretically, could aid in an assessment of the pigment epithelial membrane properties or melanin distribution (which may be altered by drugs or disease). It remains a model for a widespread biologic phenomenon, the absorption of light by pigment to initiate an ionic event.

Conclusion

We have described three pigment epithelial responses to light—the melanin response, the c-wave, and the light response of the standing potential. The melanin response is a direct electrical sign of light absorption by the pigment molecules, but requires intense illumination and is probably not of great physiologic significance. The c-wave and light response are both indirect responses, generated in the pigment epithelium following physiologic illumination of the photoreceptors, and both may be influenced by activity of the inner retina as well. The c-wave and light response are remarkable in being very slow (lasting seconds and minutes, respectively) relative to the conventional ERG (cf. Figures 13.1 and 13.7). Current evidence suggests that the major cornea-positive component of the c-wave is a passive response by the pigment epithelial cells to a fall in the extracellular potassium concentration in the vicinity of the photoreceptor outer segments. There are no data regarding the mechanism underlying the light response (or the succeeding slow oscillations), although its extremely slow time course suggests a relationship to metabolic events.

The physiologic significance of the c-wave and light response remains to be established. They are, at the least, indicators of metabolic and ionic events that induce the electrical responses, but they may also have a direct influence on retinal function by creating electrical effects that extend throughout the retina. Neither the c-wave nor the light response has an obvious relationship to clinically measured visual function.

The clinical implications of the c-wave and light response (EOG) will remain tentative until their physiologic significance is better understood. The conditions for reproducing "normal" responses must be better defined before these responses can be used in other than gross distinctions. Nonetheless, the EOG is widely used as a clinical test and practical techniques for recording the c-wave are being developed. The c-wave may ultimately prove more useful than the EOG for studying the pigment epithelium clinically, since it requires less time to record and might be less dependent on the integrity of the inner retina. Neither test can at present be construed as a simple index of pigment epithelial disease, but both may be useful as an adjunct to other tests of retinal function.

We have emphasized the limits of our knowledge about pigment epithelial responses because these limits must be recognized in clinical applications, and because they define the frontiers for further research. There is no question, however, that studies of the melanin response, c-wave, and light response have already provided valuable insight into the physiology of retina and pigment epithelium and into the pathophysiology of ophthalmic disease. As these responses become better understood at a cellular and molecular level, this insight will deepen considerably.

REFERENCES

ADAMS, A. 1973. The normal electro-oculogram (EOG). *Acta Ophthalmol.* 51:551–561.

AMES, A., III, and GURIAN, B. S. 1963. Electrical recordings from isolated mammalian retina mounted as a membrane. *Arch. Ophthalmol.* 70:837–841.

ANDERSON, S. G., and KOLDER, H. 1966. Infrared radiant energy and oscillations of the corneoretinal potential in man. *Invest. Ophthalmol.* 5:242–247.

ARDEN, G. B. 1968. Recent work on the early receptor potential and related rapid responses. In *The clinical value of electroretinography* (ISCERG Symp. Ghent 1966), pp. 51–59. Basel: Karger.

ARDEN, G. B., and BARRADA, A. 1962. Analysis of the electro-oculograms of a series of normal subjects. *Br. J. Ophthalmol.* 46:468–482.

ARDEN, G. B., BARRADA, A., and KELSEY, J. H. 1962. New clinical test of retinal function based upon the standing potential of the eye. *Br. J. Ophthalmol.* 46:449–467.

ARDEN, G. B., and FOJAS, M. R. 1962. Electrophysiological abnormalities in pigmentary degenerations of the retina. *Arch. Ophthalmol.* 68:369–389.

ARDEN, G. B., IKEDA, H., and SIEGEL, I. M. 1966. New components of the mammalian receptor potential and their relation to visual photochemistry. *Vision Res.* 6:373–384.

ARDEN, G. B., and KELSEY, J. H. 1962*a*. Changes produced by light in the standing potential of the human eye. *J. Physiol.* (Lond) 161:189–204.

ARDEN, G. B., and KELSEY, J. H. 1962*b*. Some observations on the relationship between the standing potential of the human eye and the bleaching and regeneration of visual purple. *J. Physiol.* (Lond) 161:205–226.

ARDEN, G. B., and TANSLEY, K. 1955. The spectral sensitivity of the pure-cone retina of the grey squirrel (*sciurus carolinensis leucotis*). *J. Physiol.* (Lond) 127:592–602.

BERGGREN, L. 1960. Intracellular potential measurements from the ciliary processes of the rabbit eye *in vivo* and *in vitro*. *Acta Physiol.* (Scand) 48:461–470.

BRINDLEY, G. S. 1956. Resting potential of the lens. *Br. J. Ophthalmol.* 40:385–391.

BROWN, K. T. 1965. An early potential evoked by light from the pigment epithelium-choroid complex of the eye of the toad. *Nature* 207:1249–1253.

BROWN, K. T. 1968. The electroretinogram: its components and their origins. *Vision Res.* 8:633–677.

BROWN, K. T., and CRAWFORD, J. M. 1967*a*. Intracellular recording of rapid light-evoked responses from pigment epithelium cells of the frog eye. *Vision Res.* 7:149–163.

BROWN, K. T., and CRAWFORD, J. M. 1967*b*. Melanin and the rapid light-evoked responses from pigment epithelium cells of the frog eye. *Vision Res.* 7:165–178.

BROWN, K. T., and GAGE, P. W. 1966. An earlier phase of the light-evoked electrical response from the pigment epithelium-choroid complex of the eye of the toad. *Nature* 211:155–158.

BROWN, K. T., and MURAKAMI, M. 1964. A new receptor potential of the monkey retina with no detectable latency. *Nature* 201:626–628.

BROWN, K. T., and WIESEL, T. N. 1961. Localization of origins of electroretinogram components by intraretinal recording in the intact cat eye. *J. Physiol.* (Lond) 158:257–280.

BRÜCKE, E. TH. V., and GARTEN, S. 1907. Zur vergleichenden Physiologie der Netzhautströme. *Pfluegers Arch.* 120:290–348.

CALISSENDORFF, B. 1976*a*. Melanotropic drugs and retinal functions. I. Effects of quinine and chloroquine on the sheep ERG. *Acta Ophthalmol.* 54:109–117.

CALISSENDORFF, B. 1976*b*. Melanotropic drugs and retinal functions. II. Effects of phenothiazine and rifampicin on the sheep ERG. *Acta Ophthalmol.* 54:118–128.

CALISSENDORFF, B., KNAVE, B., and PERSSON, H. E. 1974. Cyclic variations in the c-wave amplitude of the sheep ERG. *Vision Res.* 14:1141–1145.

CARR, R. E., and SIEGEL, I. M. 1964. Electrophysiologic aspects of several retinal diseases. *Am. J. Ophthalmol.* 58: 95–107.

CONE, R. A. 1967. Early receptor potential: photoreversible charge displacement in rhodopsin. *Science* 155:1128–1131.

CRAWFORD, J. M., GAGE, P. W., and BROWN, K. T. 1967. Rapid light-evoked potentials at extremes of pH from the frog's retina and pigment epithelium, and from a synthetic melanin. *Vision Res.* 7:539–551.

CRESCITELLI, F. 1961. The electroretinogram of the antelope ground squirrel. *Vision Res.* 1:139–153.

DEBECKER, J., and ZANEN, A. 1975. Intensity functions of the early receptor potential and of the melanin fast photovoltage in the human eye. *Vision Res.* 15:101–106.

DEWAR, J., and M'KENDRICK, J. G. 1873. On the physiological action of light. *Trans. R. Soc. Edinburgh* 27:141–166.

DODT, E. 1951. Beitrage zur elektrophysiologie des auges. I. Uber die sekundare Erhebung im Aktionspotential des menschlichen Auges bei Belichtung. *Albrecht von Graefes Arch. Klin. Ophthalmol.* 151:672–692.

DODT, E., and ECHTE, K. 1961. Dark- and light-adaptation in pigmented and white rat as measured by electroretinogram threshold. *J. Neurophysiol.* 24:427–445.

DONN, A., MAURICE, D. M., and MILLS, N. L. 1959. Studies on the living cornea in vitro. I. Method and physiologic measurements. *Arch. Ophthalmol.* 62:741–747.

DOWLING, J. E., and RIPPS, H. 1972. Adaptation in skate photoreceptors. *J. Gen. Physiol.* 60:698–719.

DU BOIS-REYMOND, E. 1849. *Untersuchungen über thierische Elektricität. Bd. II,* pp. 256–257. Berlin: Reimer.

EBREY, T. G., and CONE, R. A. 1967. Melanin, a possible pigment for the photostable electrical responses of the eye. *Nature* 213:360–362.

EINTHOVEN, W., and JOLLY, W. A. 1908. The form and magnitude of the electrical response of the eye to stimulation by light at various intensities. *Q. J. Exp. Physiol.* 1:373–416.

ELENIUS, V., and ANTAA, E. 1973. Light-induced increase in amplitude of electro-oculogram. *Arch. Ophthalmol.* 90:60–63.

ELENIUS, V., and KARO, T. 1966. Cone activity in the light-induced response of the human electro-oculogram. *Pfluegers Arch.* 291:241–248.

FABER, D. S. 1969. Analysis of the slow transretinal potentials in response to light. Ph.D. thesis, State University of New York at Buffalo.

FOULDS, W. S., and IKEDA, H. 1966. The effects of detachment of the retina on the induced and resting ocular potentials in the rabbit. *Invest. Ophthalmol.* 5:93–108.

FRANCOIS, J., DE ROUCK, A., CAMBIE, E., and ZANEN, A. 1974. *L'electro-diagnostic des affections rétiniennes.* Paris: Masson & Cie.

FRANCOIS, J., JONSAS, C., and DE ROUCK, A. 1969. Experimental studies of the effect of hypoglycemia. *Ann. Ophthalmol.* 1:66–71.

FRANCOIS, J., VERRIEST, G., and DE ROUCK, A. 1955. Modification of the amplitude of the human electro-oculogram by light and dark adaptation. *Br. J. Ophthalmol.* 39:398–408.

FUJINO, T., and HAMASAKI, D. I. 1965. The effect of occluding the retinal and choroidal circulations on the electroretinogram of monkeys. *J. Physiol.* (Lond) 180: 837–845.

GAGE, P. W., CRAWFORD, J. M., and BROWN, K. T. 1968. Ionic dependence of the third phase of the light-evoked response from amphibian pigment epithelium cells. *Vision Res.* 8:369–382.

GOURAS, P., and CARR, R. E. 1964. Primate retinal responses: slow changes during repetitive stimulation with light. *Science* 145:413–414.

GOURAS, P., and CARR, R. E. 1965*a*. Light-induced DC responses of monkey retina before and after central retinal artery interruption. *Invest. Ophthalmol.* 4:310–317.

GOURAS, P., and CARR, R. E. 1965*b*. Cone activity in the light-induced DC response of monkey retina. *Invest. Ophthalmol.* 4:318–321.

GRANIT, R. 1933. The components of the retinal action potential in mammals and their relation to the discharge in the optic nerve. *J. Physiol.* (Lond) 77:207–239.

GRANIT, R., and MUNSTERHJELM, A. 1937. The electrical responses of dark-adapted frogs' eyes to monochromatic stimuli. *J. Physiol.* (Lond) 88:436–458.

GRANIT, R., and RIDDELL, L. A. 1934. The electrical responses of light- and dark-adapted frogs' eyes to rhythmic and continuous stimuli. *J. Physiol.* (Lond) 81:1–28.

HANITZSCH, R., HOMMER, K., and BORNSCHEIN, H. 1966. Der Nachweis langsamer Potentiale im menschlichen ERG. *Vision Res.* 6:245–250.

HEILIG, P., THALER, A., and BORNSCHEIN, H. 1973. Slow potentials of ERG in hemeralopia congenita. (Xth ISCERG Symp. Los Angeles 1972.) *Doc. Ophthalmol. Proc. Ser.* 2:219–224. The Hague: Dr. W. Junk.

HEISS, W.-D., HEILIG, P., and HOYER, J. 1967. Retinale Impulsaktivität und Elektroretinogramm unter dem Einfluss von Desipramin. *Experientia* 23:728–729.

HENKES, H. E., and DEUTMAN, A. F. 1973. Electrodiagnostic procedures in drug-induced ocular diseases. *Doc. Ophthalmol.* 34:209–219.

HOLMGREN, F. 1865. Method att objectivera effecten af ljusintryck på retina. *Upsala Läkarförenings Förhandlingar* 1:184–198.

HOMER, L. D., and KOLDER, H. E. J. W. 1967. The oscillation of the human corneoretinal potential at different light intensities. *Pfluegers Arch.* 296:133–142.

HOMMER, K., ULRICH, W. D., and WÜNDSCH, L. 1968. Direkte und indirekte Beeinflussung des ERG durch Antimalariamittel. *Albrecht von Graefes Arch. Klin. Ophthalmol.* 175:121–130.

Imaizumi, K., Atsumi, K., Takahashi, F., and Yoshida, G. 1968. Experimental studies on ERG and EOG of the rabbit's eye. In *The clinical value of electroretinography* (ISCERG Symp. Ghent 1966), pp. 74–82. Basel: Karger.

Jünemann, G., and Schulze, J. 1968. Elektroretinographische Untersuchungen zur Entstehung der Chinin-bzw: Resochin-Retinopathie. *Klin. Monatsbl. Augenheilkd.* 152:562–566.

Kawasaki, K., Yamamoto, S., and Yonemura, D. 1977. Electrophysiological approach to clinical test for the retinal pigment epithelium. *Acta Soc. Ophthalmol. Jap.* 81: 1303–1312.

Keeler, C. E., Sutcliffe, E., and Chaffee, E. L. 1928. A description of the ontogenetic development of retinal action currents in the house mouse. *Proc. Natl. Acad. Sci. U.S.A.* 11:811–815.

Kikawada, N. 1968. Variations in the corneo-retinal standing potential of the vertebrate eye during light and dark adaptations. *Jap. J. Physiol.* 18:687–702.

Knave, B., Møller, A., and Persson, H. E. 1972. A component analysis of the electroretinogram. *Vision Res.* 12:1669–1684.

Knave, B., Nilsson, S. E., and Lunt, T. 1973. The human electroretinogram: DC recordings at low and conventional stimulus intensities. *Acta Ophthalmol.* 51:716–726.

Knave, B., Persson, H. E., and Nilsson, S. E. G. 1974. A comparative study on the effects of barbiturate and ethyl alcohol on retinal functions with special reference to the c-wave of the ERG and the standing potential of the sheep eye. *Acta Ophthalmol.* 52:254–259.

Kolder, H. 1959. Spontane und experimentelle Anderungen des Bestandpotentials des menschlichen Auges. *Pfluegers Arch.* 268:258–272.

Kolder, H. E. 1974. Electro-oculography. *Ophthalmologica* 169:127–140.

Kolder, H., and Brecher, G. A. 1966. Fast oscillations of the corneo-retinal potential in man. *Arch. Ophthalmol.* 75:232–237.

Kolder, H. E., and Hochgesand, P. 1973. Empirical model of electrooculogram. *Doc. Ophthalmol.* 34:229–241.

Kolder, H., and North, A. W. 1966. Oscillations of the corneo-retinal potential in animals. *Ophthalmologica* 152: 149–160.

Krill, A. E. 1972. *Hereditary retinal and choroidal diseases. I. Evaluation.* Hagerstown: Harper & Row.

Kris, C. 1958. Corneo-fundal potential variations during light and dark adaptation. *Nature* 182:1027–1028.

Krogh, E. 1977. Normal values in clinical electro-oculography. III. Numerical evaluation of two dimensionless EOG parameters. *Acta Ophthalmol.* 55:227–236.

Lasansky, A., and de Fisch, F. 1966. Potential, current, and ionic fluxes across the isolated retinal pigment epithelium and choroid. *J. Gen. Physiol.* 49:913–924.

Lurie, M. 1976. Some observations on the c-wave of the electroretinogram in the intact frog eye. *Exp. Eye Res.* 23:197–207.

Lurie, M., and Marmor, M. F. 1979*a*. Light-induced electrical responses of the retinal pigment epithelium. (Proceedings of the XXIII International Congress of Ophthalmology, May 14–20, 1978, Kyoto, Japan.) *Excerpta Medica,* In Press.

Lurie, M., and Marmor, M. F. 1979*b*. Light-integrating capacity of the ERG c-wave under different anesthetics. *Jap. J. Ophthalmol.* In Press.

Marg, E. 1951. Development of electro-oculography. *Arch. Ophthalmol.* 45:169–185.

Marmor, M. F. 1975*a*. Structure and function of the retinal pigment epithelium. *Int. Ophthalmol. Clin.* 15: 115–130.

Marmor, M. F. 1975*b*. The retinal pigment epithelium and ocular disease. *Int. Ophthalmol. Clin.* 15:131–146.

Marmor, M. F. 1977. Fundus albipunctatus: a clinical study of the fundus lesions, the physiologic deficit, and the vitamin A metabolism. *Doc. Ophthalmol.* 43:277–302.

Marmor, M. F., and Byers, B. 1977. Pattern dystrophy of the pigment epithelium. *Am. J. Ophthalmol.* 84:32–44.

Marmor, M. F., Pockrand, P., and Lurie, M. 1979. Experiments toward the development of a clinical c-wave test. *Jap. J. Ophthalmol.* In Press.

Matsuura, T., Miller, W. H., and Tomita, T. 1978. Cone specific c-wave in the turtle retina. *Vision Res.* 18: 767–775.

Miles, W. R. 1940. Modification of the human-eye potential by dark and light adaptation. *Science* 91:456.

Müller-Limmroth, H. W. 1954. Der Einfluss der Belichtung auf das Bestandpotential des Froschauges. *Z. Biol.* 107:275–283.

Niemeyer, G. 1976. c-Waves and intracellular responses from the pigment epithelium in the cat. *Bibl. Ophthalmol.* 85:68–74.

Nikara, T., Sato, S., Takamatsu, T., Sato, R., and Mita, T. 1976. A new wave (2nd c-wave) on corneoretinal potential. *Experientia* 32:594–596.

NILSSON, S. E. G., KNAVE, B., and PERSSON, H. E. 1977. Changes in ultrastructure and function of the sheep pigment epithelium and retina induced by sodium iodate. II. Early effects. *Acta Ophthalmol.* 55:1007–1026.

NILSSON, S. E. G., and SKOOG, K-O. 1975. Covariation of the simultaneously recorded c-wave and standing potential of the human eye. *Acta Ophthalmol.* 53:721–730.

NOELL, W. K. 1952. Azide-sensitive potential difference across the eye-bulb. *Am. J. Physiol.* 170:217–238.

NOELL, W. K. 1953. Studies on the electrophysiology and metabolism of the retina. U.S.A.F. School of Aviation Medicine, Randolph Field, Texas, Project Number 21-1201-0004.

NOELL, W. K. 1954. The origin of the electroretinogram. *Am. J. Ophthalmol.* 38:78–90.

NOELL, W. K. 1959. The visual cell: electric and metabolic manifestations of its life processes. *Am. J. Ophthalmol.* 48:347–370.

NOELL, W. K., CRAPPER, D. R., and PAGANELLI, C. V. 1964. Transretinal currents and ion fluxes. In *Transcellular membrane potentials and ionic fluxes,* ed. F. M. Snell and W. K. Noell, pp. 92–130. New York: Gordon and Breach.

OAKLEY, B., II. 1977. Potassium and the photoreceptor-dependent pigment epithelial hyperpolarization. *J. Gen. Physiol.* 70:405–425.

OAKLEY, B., II, and GREEN, D. G. 1976. Correlation of light-induced changes in retinal extracellular potassium concentration with c-wave of the electroretinogram. *J. Neurophysiol.* 39:1117–1133.

PAK, W. L., and CONE, R. A. 1964. Isolation and identification of the initial peak of the early receptor potential. *Nature* 204:836–838.

PASIK, P., PASIK, T., and BENDER, M. B. 1965. Recovery of the electro-oculogram after total ablation of the retina in monkeys. *Electroencephalogr. Clin. Neurophysiol.* 19: 291–297.

PAUTLER, E. L., and NOELL, W. K. 1976. The slow ERG potentials in hereditary retinal dystrophy of albino and pigmented rats. *Exp. Eye Res.* 22:493–503.

PEARLMAN, J. T. 1962. The c-wave of the human ERG. *Arch. Ophthalmol.* 68:823–830.

POTTS, A. M., and MODRELL, R. W. 1957. The transcorneal potential. *Am. J. Ophthalmol.* 44:284–290.

REESER, F., WEINSTEIN, G. W., FEIOCK, K. B., and OSER, R. S. 1970. Electro-oculography as a test of retinal function. *Am. J. Ophthalmol.* 70:505–514.

REUTER, J. H. 1974. The electroretinogram of albino and pigmented rabbits. (XIth ISCERG Symp. Bad Nauheim 1973.) *Doc. Ophthalmol. Proc. Ser.* 4:483–493. The Hague: Dr. W. Junk.

RODIECK, R. W., and Ford, R. W. 1969. The cat local electroretinogram to incremental stimuli. *Vision Res.* 9:1–24.

RODIECK, R. W. 1972. Components of the electroretinogram: a reappraisal. *Vision Res.* 12:773–780.

SCHMIDT, R., and STEINBERG, R. H. 1971. Rod-dependent intracellular responses to light recorded from the pigment epithelium of the cat retina. *J. Physiol.* (Lond) 217:71–91.

SHIBATA, H. 1975. Effects of L-cysteine on the rabbit ERG. *Acta Soc. Ophth. Jap.* 79:1163–1166.

SKOOG, K.-O. 1974. The c-wave of the human d.c. registered ERG. III. Effects of ethyl alcohol on the c-wave. *Acta Ophthalmol.* 52:913–923.

SKOOG, K.-O. 1975. The directly recorded standing potential of the human eye. *Acta Ophthalmol.* 53:120–132.

SKOOG, K.-O., and NILSSON, S. E. G. 1974*a*. The c-wave of the human d.c. registered ERG. I. A quantitative study of the relationship between c-wave amplitude and stimulus intensity. *Acta Ophthalmol.* 52:759–773.

SKOOG, K.-O., and NILSSON, S. E. G. 1974*b*. The c-wave of the human d.c. registered ERG. II. Cyclic variations of the c-wave amplitude. *Acta Ophthalmol.* 52:904–912.

SKOOG, K.-O., TEXTORIUS, O., and NILSSON, S. E. G. 1975. Effects of ethyl alcohol on the directly recorded standing potential of the human eye. *Acta Ophthalmol.* 53:710–720.

SKOOG, K.-O., WELINDER, E., and NILSSON, S. E. G. 1977. Off-responses in the human d.c. registered electroretinogram. *Vision Res.* 17:409–415.

STEINBERG, R. H., and MILLER, S. 1973. Aspects of electrolyte transport in frog pigment epithelium. *Exp. Eye Res.* 16:365–372.

STEINBERG, R. H., SCHMIDT, R., and BROWN, K. T. 1970. Intracellular responses to light from cat pigment epithelium: origin of the electroretinogram c-wave. *Nature* 227: 728–730.

TAMAI, A. 1965. The c-wave of the cat ERG. *Yonago Acta Med.* 9:199–205.

TAMAI, A. 1966. The c-wave in the ERG and EPG (electropupillogram) of the human eye. *Yonago Acta Med.* 10:153–160.

TAUMER, R., ed. 1976. *Electro-oculography: its clinical importance.* (*Bibl. Ophthal.* vol. 85.) Basel: Karger.

Taumer, R., Hennig, J., and Pernice, D. 1974. The ocular dipole: a damped oscillator stimulated by the speed of change in illumination. *Vision Res.* 14:637–645.

Taumer, R., Kapp, H., Hennig, J., and Rohde, N. 1974*a*. A first electrical analog model of the ODM-oscillations. (XIth ISCERG Symp. Bad Nauheim 1973.) *Doc. Ophthalmol. Proc. Ser.* 4:457–463. The Hague: Dr. W. Junk.

Taumer, R., Mackensen, G., Hartmann, H., Moser, U., Stehle, R., Werner, W., and Wolf, D. 1974*b*. Verhalten des Bestandpotentials des menschlichen Auges nach Belichtungsanderungen. *Albrecht von Graefes Arch. Klin. Ophthalmol.* 189:81–97.

Taumer, R., Rohde, N., Pernice, D., and Kohler, U. 1976*a*. EOG: light test and dark test. *Albrecht von Graefes Arch. Klin. Ophthalmol.* 199:207–213.

Taumer, R., Rohde, N., Wichmann, W., and Rover, J. 1976*b*. A method for DC-ERG recording of alert humans. *Albrecht von Graefes Arch. Klin. Ophthalmol.* 198:45–55.

Taumer, R., Rohde, N., Wichmann, W., and Rover, J. 1976*c*. Experiments concerning the human c-wave. *Albrecht von Graefes Arch. Klin. Ophthalmol.* 198:139–153.

Taumer, R., Wichmann, W., Rohde, N., and Rover, J. 1976*d*. ERG of humans without c-wave. *Albrecht von Graefes Arch. Klin. Ophthalmol.* 198:275–289.

Ten Doesschate, G., and Ten Doesschate, J. 1957. The influence of the state of adaptation on the resting potential of the human eye. *Ophthalmologica* 134:183–193.

Textorius, O. 1977. The influence of stimulus duration on the human d.c. registered c-wave. *Acta. Ophthalmol.* 55:561–572.

Thaler, A., and Heilig, P. 1974. The steady state in EOG. (XIth ISCERG Symp. Bad Nauheim 1973.) *Doc. Ophthalmol. Proc. Ser.* 4:211–216. The Hague: Dr. W. Junk.

Thaler, A., and Heilig, P. 1977. EOG and ERG components in ischemic retinopathy. *Ophthalmol. Res.* 9:38–46.

Therman, P. O. 1938. The neurophysiology of the retina. *Acta Soc. Scient. Fenn. N.S.B.II:* 1–74.

Tigges, J., Brooks, B. A., and Klee, M. R. 1967. ERG recordings of a primate pure cone retina (*Tupaia glis*). *Vision Res.* 7:553–563.

Ulrich, Ch., Bohne, B.-D., and Ulrich, W.-D. 1974. Die langsamen potentiale im menschlichen elektroretinogramm und ihre Ableitungstechnik. In *4th Congress of the European Society of Ophthalmology* (Budapest 1972), part II, pp. 290–294 (530–534). Basel: Karger.

Weidner, C. 1976. The c-wave in the ERG of albino rat. *Vision Res.* 16:753–763.

Witkovsky, P., Dudek, F. E., and Ripps, H. 1975. Slow PIII component of the carp electroretinogram. *J. Gen. Physiol.* 65:119–134.

Yamashita, E. 1959. Some analyses of slow potentials of toad's retina. *Tohoku J. Exp. Med.* 70:221–233.

Yonemura, D., Kawasaki, K., and Ishikawa, C. 1976. Enhancement of the c-wave and abolition of the b-wave by intravenous injection of hypertonic solution. *Acta Soc. Ophthalmol. Jap.* 80:1610–1616.

Yonemura, D., and Kawasaki, K. 1978. Electrophysiological study on activities of neuronal and non-neuronal retinal elements in man with reference to its clinical application. *Jap. J. Ophthalmol.* 22:1–19.

Yonemura, D., Kawasaki, K., Yoshino, Y., and Yanagida, T. 1974. Pharmacological isolation of the P III of the albino rabbit ERG in vivo. *Acta Soc. Ophthalmol. Jap.* 78:858–866.

Yonemura, D., Kawasaki, K., Tanabe, J., and Yamamoto, S. 1978. Susceptibility of the standing potential of the eye to acetazolamide and its clinical applications. *Folia Ophthalmologica Japonica* 29:408–416.

Yosida, Y. 1953. On the nature of steady potential of toad's eye and adaptation. *J. Physiol. Soc. Jap.* 15:133–142.

Zanen, A., and Debecker, J. 1971. Visual pigments and melanin contributions to the fast photovoltage of the human eye. *Vision Res.* 11:169–172.

Zuckerman, R. 1971. Mechanisms of photoreceptor current generation in light and darkness. *Nature* 234:29–31.

ACKNOWLEDGMENTS

Our research is supported by National Eye Institute National Institutes of Health grant EY01678 and by the Medical Research Service of the Veterans Administration.

PART TWO

Pathophysiology and Disease of the Retinal Pigment Epithelium

14 Pathological Responses of the Human Retinal Pigment Epithelium

Donald A. Morris
Paul Henkind

THERE IS a voluminous literature documenting the manifold responses of the retinal pigment epithelium (RPE) to a wide variety of influences. In recent years, with newer methods of study, which include autoradiography, histochemistry, in vitro and in vivo tissue culture techniques, and electron microscopy, much information has been gathered about these cells: what they do, where they go, and what they can become.

We have divided the pathological responses of the pigment epithelium into two broad categories: "Subtraction of Elements" and "Addition of Elements." The former refers to loss of prominent elements of the cells, reduction in size of the cells, or loss of cells. The latter refers to an increase of cellular elements, increase in size and number of cells, and metamorphosis of cells. The categories are subdivided as follows:

Subtraction of Elements

1. Aplasia
2. Hypopigmentation
3. Degeneration and atrophy

Addition of Elements

1. Duplication of RPE
2. Dysplasia
3. Hyperpigmentation and hypertrophy
4. Phagocytosis
5. Simple hyperplasia
6. Hyperplasia with formation of cuticular masses
7. Metaplasia, including osseous formation
8. Migration and membrane formation
9. Pseudoepitheliomatous hyperplasia and neoplasia

We realize that overlap and some ambiguity will occur in the large divisions and within intradivisional boundaries. For example, cells can be both metaplastic and hypopigmented. Also, great difficulty may be encountered in distinguishing a reactive proliferation from a neoplasm, because of the great varieties of histological patterns in each. Furthermore, just as destruction and proliferation can occur side by side in the pigment epithelium, so our categories of subtraction and addition are not mutually exclusive.

Subtraction of Elements

APLASIA. Even though this term denotes a failure of RPE development, we have included it under "Subtraction of Elements," since it represents a loss of RPE cells. There may be a localized absence of retinal and choroidal tissue, including pigment epithelium, in typical colobomata (Frayer 1966). Dysplastic retina may also be found at the point where the pigment epithelium terminates. This and other evidence suggested to Silverstein, Osborn, and Prendergast (1971) that orderly retinal morphogenesis may depend at least in part on the organizing influence of an adjacent pigment epithelium.

HYPOPIGMENTATION. This occurs in ocular and generalized albinism (Fig. 14.1) and in Chediak-Higashi disease (Johnson et al. 1966; Spencer and Hogan 1960). Although the intracellular pigment-bearing apparatus appears to be intact, melanin

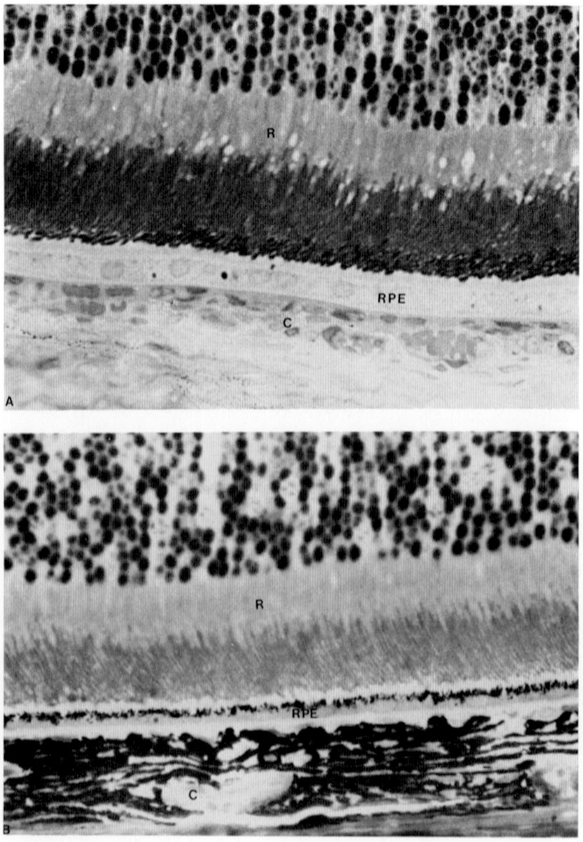

Figure 14.1 *Choroid, pigment epithelium, and outer retina of* (A) *an albino rat and* (B) *a normal rat, showing the lack of pigmentation in the choroid and pigment epithelium in the albino.* C = *Choroid,* RPE = *Pigment epithelium,* R = *Retina. (Montefiore Ophthalmic Path. no. 154B. ×368.)*

cannot be elaborated. In the choroid, pigment granules are virtually absent, while in the RPE, granules of pigment are paler and scantier than usual (Johnson et al. 1966; Spencer and Hogan 1960; Mann 1957; Kurz and Zimmerman 1962). The failure of foveal development may be the result of this defect of the pigment epithelium. The evidence is certainly not conclusive, however, that the absence of melanin causes the foveal defect. Albinism in animals has, in fact, been associated with abnormalities of the higher visual pathways (Creel 1971; Guillery, Avnoin, and Eighmy 1971).

In some atypical colobomata, a lack of pigmentation may be seen in the RPE, corresponding to the involved area. In a case described by Klien (1959), in addition to the failure of the RPE cells to produce pigment, there was a marked folding of the outer neuroectodermal layer, with multiple cysts formed in both outer and inner layers. The choroid was undeveloped, Bruch's membrane absent, and the sclera thinned. Under certain conditions, depigmentation occurs in the RPE cells without loss of their cellular continuity; these cells may become flatter than normal (Klien 1958). This depigmentation is seen ophthalmoscopically as gradually enlarging areas in which the underlying choroidal vasculature may be visualized. This can occur in old age, with or without visible so-called choroidal vascular sclerosis, and in high myopia. Retinal pigment epithelial window defects have been examined by fluorescein angiography and shown to exhibit enhanced choroidal fluorescence without evidence of staining or leakage into the retina. One clinical pathologic study of such a lesion revealed an intact but attenuated retinal pigment epithelium with partial to complete loss of melanin granules. There was also partial atrophy of overlying photoreceptors, and sclerosis of the choriocapillaris (Keno and Green 1978). Depigmentation may also occur in retinitis pigmentosa (Duke-Elder 1967b), and in inflammatory conditions in which the insult is not acute enough to cause the death of the cell. Depigmentation has also been observed in intact RPE cells of the monkey after photic stimulation by the indirect ophthalmoscope (Tso, Fine, and Zimmerman 1972).

When the RPE becomes hyperplastic or neoplastic, hypopigmentation is most easily appreciated in the patternless arrangement of proliferating anaplastic cells (Tso and Albert 1972). The amount of pigmentation may vary from tumor to tumor and in different locations of the same tumor. Pigment granules may be irregularly ovoid, and often aggregate into clumps. The typical cigar-shaped granules of the normal RPE are uncommon in such cases.

In studying the in vitro growth of pure cultures of human RPE, Albert et al. (1972a) noted that, although the pigment epithelium showed growth patterns resembling those seen in pigmented lesions of the RPE in vivo, evidence of pigment production by the growing cells was lacking. Since they found that human ocular and cutaneous melanomas and mouse melanomas, when grown under identical conditions, continued to produce pigment, they concluded that the failure of the RPE to produce pigment was not due to the tissue culture media. Whittaker (1967), working with chick RPE cells, demonstrated that this failure was due to (1) simple dilution of melanin by cell growth, (2) decay of tyrosinase activity, and (3) cessation of tyrosinase synthesis.

Diminished pigment also appeared in the RPE-derived proliferating cells of experimentally produced retinal detachments (Machemer and Laqua 1975). These cells included macrophages, fibrocytes, and other proliferated cells. Because of continuous multiplication, daughter cells lost more and more of their pigment content, until they were finally pigment-free. This was further confirmed by Mandelcorn et al. (1975), studying in vivo growth of transplanted RPE cells within the vitreous, and by Mueller-Jensen and coworkers (Mueller-Jensen, Machemer, and Azarnia 1975) within intravitreal diffusion chambers.

DEGENERATION AND ATROPHY. Under certain conditions, pigment epithelial cells may become flatter than normal, with probable curtailment of their physiological activity (Klien 1958). Many transitional stages may be observed, from simple depigmentation to more marked degeneration of the pigment epithelium with incipient or advanced destruction of the percipient retinal elements. In retinitis pigmentosa, the RPE may be completely absent over large areas in which the rods and cones and the external limiting membrane have disappeared and been replaced by neuroglia (Duke-Elder 1967b) (Fig. 14.2). Choroideremia is another familial condition in which the pigment epithelium degenerates, as do the choroid and outer retina (Duke-Elder 1963b; McCulloch 1950). Central areolar

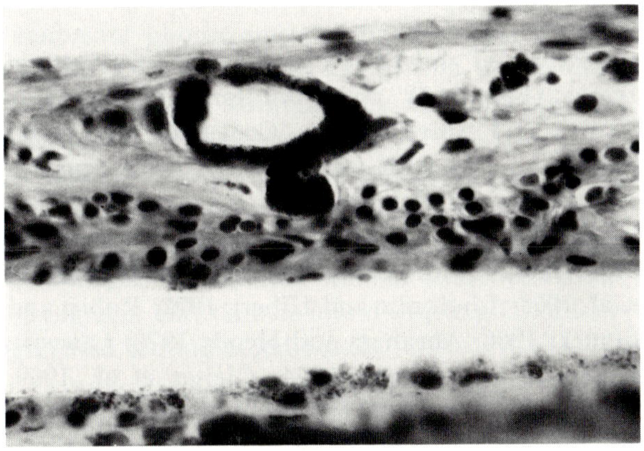

Figure 14.2 *Degeneration and atrophy of the retinal pigment epithelium in a case of retinitis pigmentosa. Note also the migration of epithelial cells into the atrophic retina and their concentration in the wall of a vessel. (Montefiore Ophthalmic Path. no. S-985. ×368.)*

choroidal sclerosis also falls into this category (Ashton 1953; Howard and Wolf 1964; Ferry, Llovera, and Shafer 1972).

In inflammatory conditions, RPE cells may undergo fatty or hyaline degeneration or become necrotic (Wolter 1956). Such conditions may be bacterial (lues), parasitic (toxoplasmosis, onchocerciasis), or viral. In rubella, for example, RPE cells undergo patchy atrophy and depigmentation, most markedly at the posterior pole (Zimmerman 1968; Boniuk and Zimmerman 1967). In subacute sclerosing panencephalitis, RPE destruction has also been reported (Boniuk 1975), with intranuclear inclusion bodies present within the retina. Cytomegalic inclusion disease can result in large areas of RPE dissolution, and intranuclear inclusion bodies are found within the RPE cell (de Venecia et al. 1971). Necrosis is apparently so extensive that the RPE loses its ability to proliferate.

The common condition called paving-stone degeneration (Meyer-Schwickerath 1960) is another age-related degenerative condition of the peripheral retina. Clinically, it consists of whitish, circumscribed, flat lesions with a rim of pigmentation. The whitish areas represent both the disappearance of RPE cells and the outer retinal layers. A focal loss of the RPE also occurs in lattice degeneration. Proliferation and cellular migration into retinal perivascular spaces may also be present (Straatsma et al. 1974). Pigmentary senile degeneration may frequently be seen in the aging eye, particularly anterior to the equator. Histologically, fatty degeneration, vacuolization, and cellular disintegration may occur, in addition to hyperplastic changes (Duke-Elder 1967c).

Hyaline substances may also accumulate in RPE cells in senescence (Wolter 1956). The pigment epithelial cells overlying drusen may appear normal by light microscopy, or they may undergo degeneration, becoming either flattened or absent (Fig. 14.3). Wolter (1956) presented some evidence that indicated that the hyaline degenerative material of drusen may derive from the degenerated RPE cells. Drusen will be discussed in greater detail in the section entitled "Hyperplasia with Cuticular Masses."

In cases of senile macular degeneration, the epithelium shows focal areas of breakdown. Sarks (1976), studying the aging process, presented evidence demonstrating that the earliest microscopic sign of degeneration of the RPE was associated with the

electron microscopy, found diffuse hyperplasia, in which the RPE was organized in multiple sheets, one above the other, in a flattened papillary pattern. In addition, irregular pigmentation with focal hyperpigmentation was observed. There were also plaques of spindle-shaped cells embedded in a fibrillar matrix, about which more will be said under "Metaplasia."

It has been found that in creating retinal detachments experimentally, small areas of pigment epithelium are denuded. These areas again become covered by either a monolayer of RPE cells or by cells lying loosely on top of each other. Such cells contained wedge-shaped pigment granules (Machemer and Norton 1968; Machemer 1968). Machemer and Laqua (1975), studying pigment epithelial proliferation in experimental retinal detachment, noted that whenever a retinal detachment extended beyond the ora and detached the nonpigmented epithelial layer of the pars plana, the underlying layer of pigment epithelium responded by proliferating (Fig. 14.11). There were mounds of these proliferations

Figure 14.11 *Proliferation of the pigment epithelium assuming a wedge-shaped configuration where the retina is detached at the ora. (Montefiore Ophthalmic Path. no. S891. ×145.)*

at the ora, as early as 7 days after the detachment was produced. Some of these mounds consisted of a double or triple layer of cells, while others filled the space under the detached epithelium of the pars plana, as if a wedge were driven between the two epithelial layers. Pigment in these cells was usually heavy at the base and sparse toward the top of the mound. The pigment granules appeared round, and were similar in size to those seen in the pigment cells of the pars plana. Some proliferations continued to extend beyond the pars plana area into the vitreous cavity, forming a membrane that extended toward the inside of the eye along the posterior surface of the vitreous cavity. (The electron microscopic study showed that these cells were epithelial in character. They formed basement membranes and were connected to each other by cell junctions such as desmosomes and zonulae occludentes.) Cells were often either loosely attached to, or freely floating over, the proliferated pigment epithelium of the pars plana. Epithelial cells would often appear as a single layer on the inner retinal surface, in the vitreous cavity, and along the posterior lens surface. Such a layer would not appear before four weeks. The cells were relatively large with convex scalloped apices, and bases intimately joined to the underlying surface. Cells of the same type were also found in multilayered preretinal strands with tubule formation. These cells resembled normal RPE; transitions were found between them and adjacent multilayered, loosely arranged, fibrocyte-like cells.

The Dalen-Fuchs nodule of sympathetic ophthalmia may be due to a swelling and proliferation of the RPE (Duke-Elder 1966). As the nodule enlarges, the cells may become spindle-shaped, eventually undergoing autolysis and depigmentation; the nodule is then invaded by lymphocytes and epithelioid cells. Ishikawa and Ikui (1972), performing an ultrastructural analysis of a Dalen-Fuchs nodule, hypothesized that the densely packed epithelioid cells within the mass were actually derived from the RPE. They believe the RPE cells undergo an alteration that permits them to elongate and form a caplike framework for all the cells contained within the nodule (Fig. 14.12) (Henkind 1973).

HYPERPLASIA WITH FORMATION OF CUTICULAR MASSES. RPE proliferation is often accompanied by homogenous, usually PAS-positive, eosinophilic material, which has generally been regarded as a basement-membrane-like product (Fig. 14.13) (Frayer 1966). Hogan, Alvarado, and Weddell (1971) suggested that the abnormal material in Bruch's membrane arose from pigment epithelial cells; drusen are probably another example of a deposited cuticular product (Fig. 14.3). Farkas and coworkers (Farkas, Sylvester, and Archer 1971; Farkas et al. 1971), in their study of the ultrastructure of drusen, noted progressive alterations in the pigment epithelial cells as drusen developed. Early on, pigment granules became scattered throughout

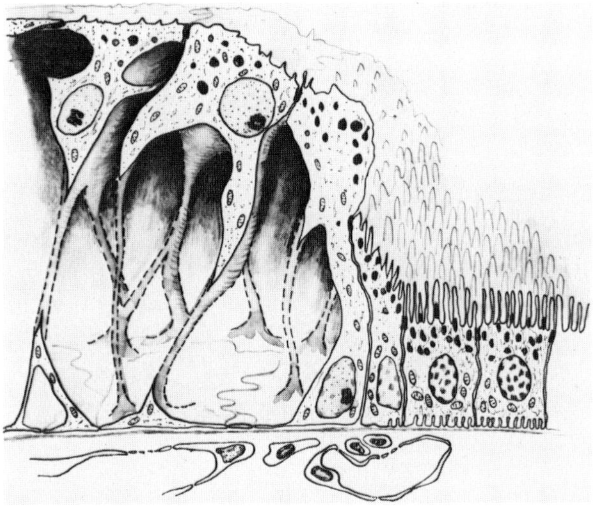

Figure 14.12 *Schematic drawing of pigmented epithelial cells constituting the cage-like framework in Dalen-Fuchs nodule. (From Ishikawa and Ikui 1972.)*

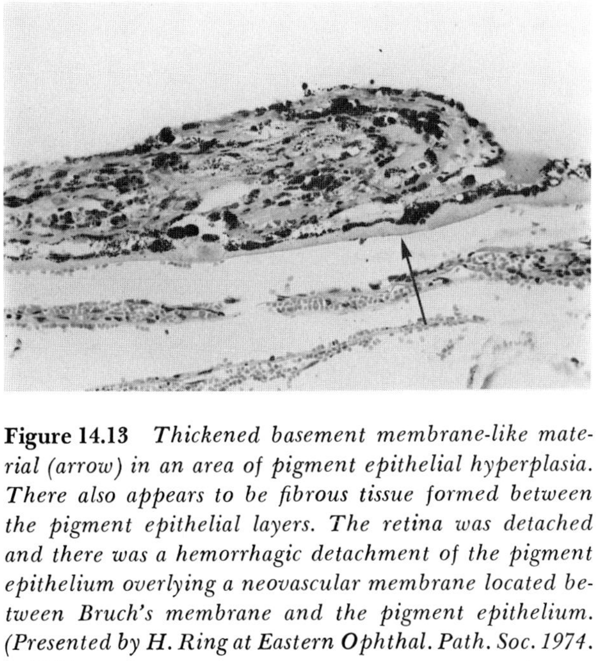

Figure 14.13 *Thickened basement membrane-like material (arrow) in an area of pigment epithelial hyperplasia. There also appears to be fibrous tissue formed between the pigment epithelial layers. The retina was detached and there was a hemorrhagic detachment of the pigment epithelium overlying a neovascular membrane located between Bruch's membrane and the pigment epithelium. (Presented by H. Ring at Eastern Ophthal. Path. Soc. 1974. ×368.)*

the cytoplasm, and the mitochondria showed degenerative changes, then fused with lysosomes. A homogenous fibrillar material was seen within the cytoplasm, and similar material interrupted and obscured the fibers of the inner collagenous zone of basement membrane. Numerous single-membrane-bound vesicles were embedded in this material in the inner collagenous zone. As drusen material accumulated in the inner collagenous zone, it elevated the degenerating pigment epithelial cells; it was felt that these cells were eventually incorporated into the developing drusen. There was also an abundance of photoreceptor remnants present in the degenerated RPE cells. Farkas and coworkers (Farkas, Sylvester, and Archer 1971; Farkas et al. 1971) contended that drusen formation is a pathologic autolysis of the RPE, and that drusen are conglomerates of denatured mitochondria, cytoplasmic debris, pigment granules, and residual bodies, all of which have been extruded from the degenerating RPE cells into the basement membrane.

In addition to aging processes, certain disease entities can lead to the deposition of cuticular material. Klien (1958) has stated that some cases of inflammation, exudative or hemorrhagic stresses, or trauma, particularly in the young, may give rise to large globular lesions consisting of glassy substances alternating with rows and tubules of proliferating RPE cells. Occasionally, such a formation is encountered as an isolated lesion of obscure origin with a predilection for the macular region. Klien feels that these latter lesions may be congenital, but may continue to enlarge. Hyperplasia of the RPE with elaboration of a basement-membrane-like substance may be found in disciform degeneration and in small areas of proliferation accompanying retinal detachment (Frayer 1966). The thickening of Bruch's membrane in the macula might play a role in the development of senile macular degeneration by interfering with the passage of nutritional substances (Reese 1960).

Elongated and sometimes laminated spicules of deeply basophilic calcium salts are frequently found in large areas of basement-membrane-like material (Frayer 1966). In the aging Bruch's membrane a basophilia also occurs, which appears to be associated with the deposition of crystalline calcium salts (Hogan 1965). The deposits found in the basement-membrane-like material apparently represent the same type of degenerative process.

METAPLASIA, INCLUDING OSSEOUS FORMATION. Metaplasia is defined as "abnormal transformation of an adult, fully differentiated tissue of one kind into a differentiated tissue of another kind; metaplasia is an acquired condition, in contrast to heteroplasia" (Stedman 1972). The RPE's ability to become metaplastic has been demonstrated. It is now generally accepted that the RPE is a major source of the fibroblasts forming subretinal scars. The fibroblasts produced by this process appear in most respects similar to those of connective tissue. Differences become more apparent with electron microscopy, and the fibrous tissue that develops from these cells often appears to contain large amounts of basement-membrane-like material.

The classic example of a subretinal scar is found in hemorrhagic disciform retinal degeneration (Frayer 1966). It has been established that this form of macular degeneration is initiated by a hemorrhage beneath the pigment epithelium, secondary to a break in Bruch's membrane. The space beneath the pigment epithelium comes to be occupied by vascular granulation tissue, which originates in the choroid and reaches the subretinal space through Bruch's membrane. A dense avascular layer of connective tissue subsequently develops anterior to the pigment epithelium, probably produced by metaplasia of the RPE cells and marked by the presence of basement-membrane-like material. In some degenerated eyes with complete retinal detachment, the subretinal space can be almost entirely obliterated by a broad membrane of dense, relatively acellular, connective tissue with very little remaining epithelium. Similar scars may be found wherever traction on the retina produces chronic irritation. The largest scars may be found at the ora serrata, completely encircling the globe; they are known as ringschwielen (Fig. 14.14). This process may also be found in cases of angioid streaks and in Coats' disease (Duke-Elder 1967a).

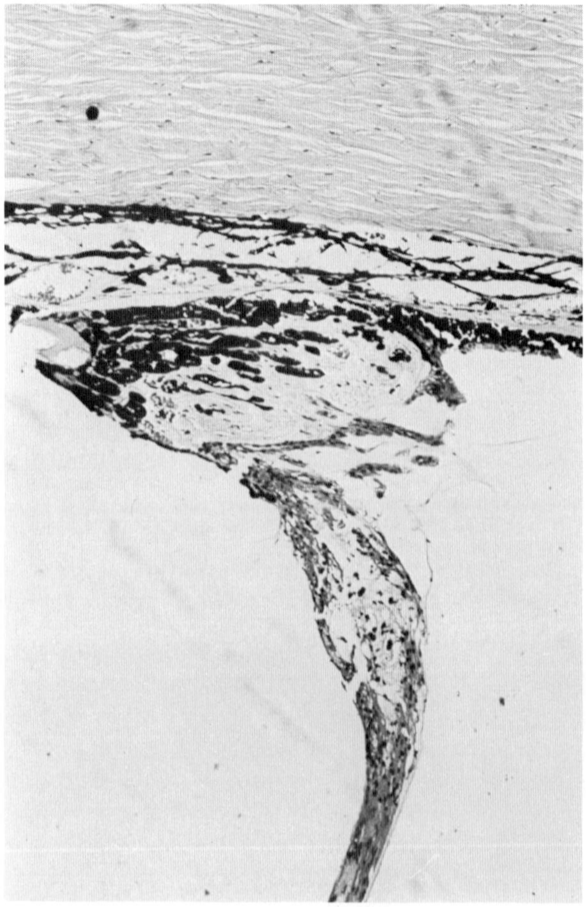

Figure 14.14 *Ringschwiele in a case of retinal detachment. The metaplastic pigment epithelium produces a thickened scar in the region of the ora serrata. (×217.)*

Manschot and deBruijn (1967) demonstrated the transformation of RPE cells into spindle cells capable of laying down fibrous tissue. Kurz and Zimmerman (1962) described a "glial" membrane lining either the inner or outer surface of the detached retina. They showed evidence suggesting that the membrane may be derived from the pigment epithelium, through proliferation of the RPE cells through a retinal hole along the inner retinal surface. Machemer and Laqua (1972), in their retinal detachment experiments, found both preretinal and subretinal membrane formation. They noted varying thicknesses of joined spindle-shaped or elongated cells with oval nuclei and variable amounts of cytoplasmic pigment granules. The retina was either wrinkled or thrown into folds by these membranes. Intravitreal membranes were found either along the posterior vitreous face or upon the lens surface. Electron microscopy revealed epithelial characteristics, consisting of basement membranes and cell junctions. In the cytoplasm, bundles of filaments in parallel arrangement were seen, and the intercellular space was filled with fibrillae that showed a periodicity suggestive of collagen. Zinn (personal communication) reports that the collagen is of the immature type (periodicity 220 Å, diameter 100–200 Å) similar to that found in the vitreous.

Machemer's group (Mandelcorn et al. 1975; Mueller-Jensen, Machemer, and Azarnia 1975) autotransplanted RPE cells from one eye of an owl monkey into the vitreous of the other eye. They concluded that RPE cells are first transformed into macrophages retaining some RPE characteristics; in a second transformation, the round or irregularly shaped cells assume a spindle shape, form coherent membranes, and regain epithelial cell characteristics —in particular, junctional complexes. The cells are surrounded by collagen. It has also been shown that preretinal and subretinal glial proliferation occurs along with RPE proliferation, producing wrinkling and folding of the retina with a clinical picture of massive vitreous retraction (Laqua and Machemer 1975). The work of Wallow and Tso (1972) is in accord with these findings. In addition, Tso (1973) observed similar collagen formation resulting from photic maculopathy in monkeys.

Metaplasia can take other forms. Large plaques of bone lining the inner surface of the choroid can occur in degenerated eyes (Kurz and Zimmerman 1962). Such bone, often associated with fibrous plaques, is interpreted as a further step in the metaplasia of the RPE. Most commonly, ossification is found adjacent to the optic nerve head or near the ora. While this bone is usually embedded in fibrous plaques, hematopoiesis and fat are occasionally noted in the marrow spaces. A patient with chronic myelogenous leukemia, who had sustained trauma in an eye many years previously, was noted to have intraocular bone containing large amounts of leukemic marrow (Fig. 14.15) (Morris 1972).

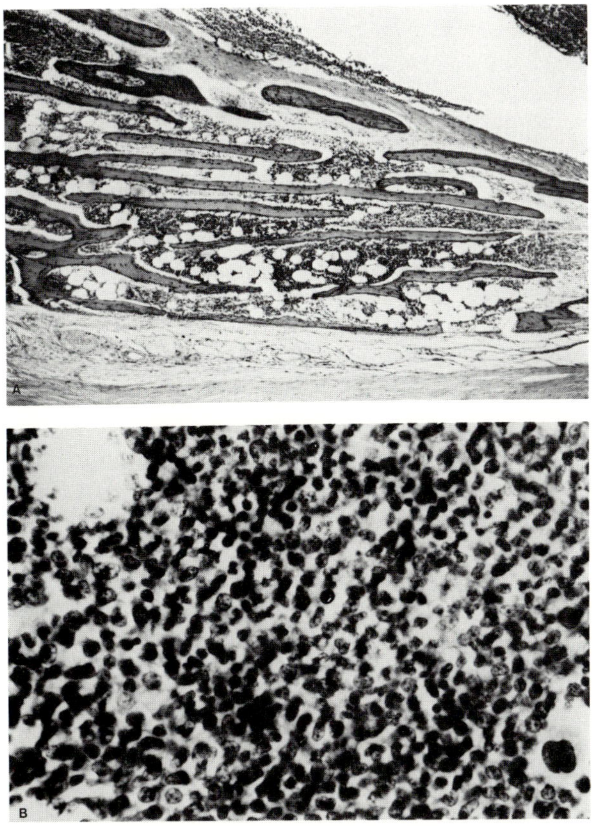

Figure 14.15 (A) *Hematopoiesis present in the marrow spaces between bony plaques within a globe that sustained old trauma.* (B) *The marrow contains leukemic cells. (NYU Ophthalmic Path. no. 2248.* A: *×58.* B: *×368.)*

Bone formation is usually a late change. The source of intraocular ossification has long been controversial. Some authors have felt that it arises from the connective tissue of the choriocapillaris (Knapp 1871; Snowball 1903) while others have implicated the RPE (Brailey and Lobo 1882; Lamb 1935). Frayer (1966) found that whenever Bruch's membrane was intact and recognizable, the bone was anterior to it and usually associated with fibrous metaplasia. Only if Bruch's membrane was destroyed could bone be found beneath it. An exception to this was in the area adjacent to the optic nerve. The pigment epithelium appeared most active in areas surrounding new bone formation.

It is not beyond the realm of possibility that intraocular bone can arise from more than one source, and the choroid may also supply bone-forming cells.

MIGRATION AND MEMBRANE FORMATION. When proliferation of the RPE occurs, it is usually accompanied by migration. The pigment epithelium can migrate in two ways: (1) by passage of free desquamated cells or clusters through the intraocular fluids and tissues; or (2) by proliferation to form continuous sheets along the surface of existing or newly formed surfaces. These surfaces include the vitreous face or strands, and the preretinal or subretinal surface (Frayer 1966). Proliferation of the pigment epithelium can take place through retinal tears (Kurz and Zimmerman 1962). Frayer (1966) found that the pigment epithelium of the ciliary body could form heavy membranes that sometimes covered the entire retina, and that the pigment epithelium might also proliferate along membranes formed from connective tissue, which could also be derived from metaplastic pigment epithelium. In some cyclitic and subretinal membranes he found pigment epithelial growth both on the surface and within the membrane itself. Frayer states: "Although connective tissue membranes appeared to have provided bridges for the growth of epithelium, the epithelium in turn might have an essential ingredient for the formation of still more of it." This was further borne out in his study of the behavior of the pigment epithelium in tissue culture. He observed optimal growth of the RPE cells from explants of the choroid and pigment epithelium of chick embryos when a feeder layer of chick embryo fibroblasts was used. This supportive role of the underlying monolayer was repeatedly demonstrated.

In the eyes of owl monkeys with experimental retinal detachment, Laqua and Machemer (1975b) found proliferation of pigment epithelial and glial cells, forming membranes on all available intraocular structures, such as vitreous structures and inner and outer retinal surfaces.

In retinitis pigmentosa the RPE undergoes atrophy and hyperplasia, and individual cells become hypopigmented or hyperpigmented. The RPE cells migrate forward into the inner retinal layers, becoming arranged around vessel walls. It is usually assumed that the bone corpuscle shape of the pigment is dependent on the disposition of the glial fibers along which the pigment cells migrate forward from the hexagon pigment layer. Ballantyne and Michaelson (1970) give another hypothesis—that the pigment frequently assumes Z, Y, and O shapes because it forms sheaths around parts of the capillary network, the meshes of which thereby become demarcated by the pigment. The cells then migrate farther to the larger veins. Gold et al. (1967) demonstrated the intimate relationship between pigment and vessel walls in a case of retinitis pigmentosa.

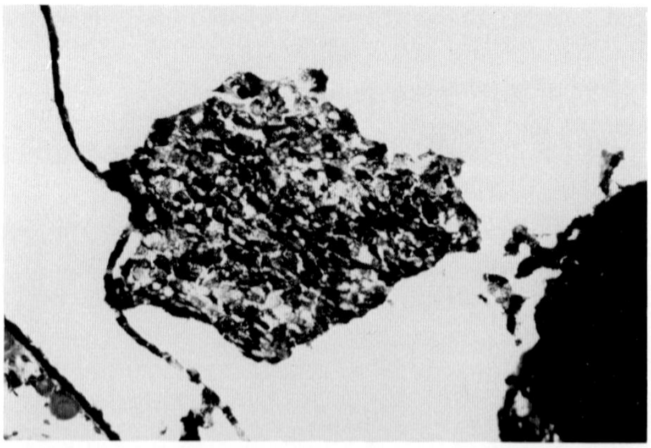

Figure 14.16 *Hyperplastic nodule of retinal pigment epithelium arising in a microphthalmic eye with coloboma, retinal detachment, and retinal dysplasia, in a case of 13 trisomy (Wills Eye Hosp. Path. no. 19233 OD. Presented by C. Steinmetz at Eastern Ophthalmic Pathology Society 1975. ×58.)*

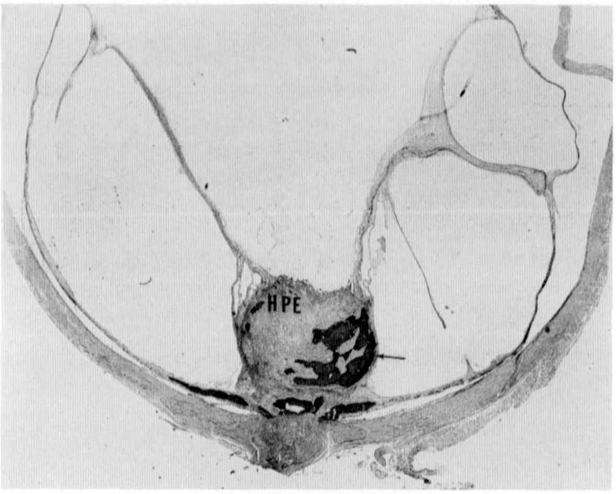

Figure 14.17 *Pseudoepitheliomatous hyperplasia of pigment epithelium* (RPE) *over the optic disc. Osseous metaplasia (arrow) was present within the mass and adjacent to it. The clinical diagnosis was angiomatosis (case reported by Snow). (AFIP Neg. 187762–7. From Frayer 1966.)*

PSEUDOEPITHELIOMATOUS HYPERPLASIA AND NEOPLASIA. Normally there is very little turnover of the pigment epithelium, and this may explain why neoplasms so rarely develop from this tissue. Nodular reactive hyperplasia and a true neoplasm may each have a varied histological pattern. Hereditary pseudotumors of the RPE were first described by Norrie, and were believed to be a result of a defect of the x-chromosome; they may result in retinal detachments (Norrie 1927). Pseudoepitheliomatous hyperplasia also may result from a focus of chronic inflammation. It usually appears in eyes with long-standing disease or is associated with a previous traumatic episode (Fig. 14.16). These eyes are often blind, with uveitis, glaucoma, atrophic retina, neovascularization, and ossification (Fig. 14.17). The proliferative changes can take the form of tubules, with basement-membrane-like material and collagen laid down between tubules and sheets of cells (Fig. 14.18). Infiltration of retina and choroid by hyperplastic cells is common. Occasionally, the hyperplasia is so marked and pleomorphic that it resembles a neoplasm (Tso and Albert 1972); Stow (1949) reported such a case.

In these degenerated eyes, the hyperplastic lesions of the RPE are usually incidental findings, the eyes being removed because of pain and chronic irritation. True neoplasms of the RPE may arise in these eyes, but they may also arise de novo. In the latter case, the eyes are often removed for suspected malignancy.

The primary tumors that most closely resemble the normal RPE have a mosaic pattern, consisting of hexagonal and oval cells with heavily pigmented cytoplasm. Tubular patterns may also be present, with cuboidal cells resting on a thin basement mem-

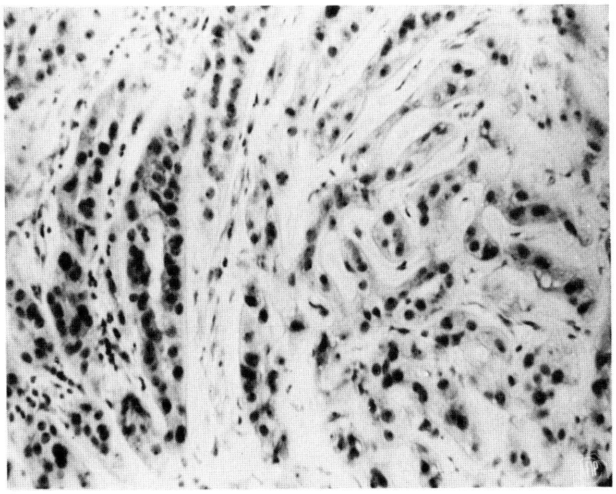

Figure 14.18 *Higher magnification of the lesion shown in Figure 14.17. Well-differentiated pigment epithelial cells are proliferating in orderly columns separated by BML secretions and fibrous tissue. (AFIP Neg. 187762–6. From Frayer 1966.)*

brane; papillary patterns with cells arranged in finger-like projections can be found as well.

Font, Zimmerman, and Fine (1972) reported on a benign tumor arising from the RPE, which was characterized by a prominent vacuolization of many of the pigmented tumor cells. The vacuoles, which were found to contain a sialomucin, appeared to displace the predominantly mature melanin granules to the periphery of the cell. In addition to these cells, there were lightly pigmented cells that contained mostly immature or incompletely developed melanosomes and very few mature melanosomes (Figs. 14.19, 14.20).

At the other end of the spectrum, patternless arrangements of cells may be seen. These cells may be either heavily pigmented signet-ring cells, or anaplastic hypopigmented cells with no definite basement membrane and with large hyperchromic vesicular nuclei with prominent nucleoli. Mitotic figures are common. There is scant extracellular material, and the degree of pigmentation may vary; granules are irregularly ovoid and are often aggregated in clumps. The retina may be extensively infiltrated, with spreading along the nerve fiber layer. Cells can also seed into the vitreous, once the internal limiting membrane has been breached. The choroid and optic nerve anterior to the lamina can also be invaded, but extrascleral extension or metastasis apparently has not been known to occur (Fig. 14.21). Garner (1970) analyzed 19 published reports of benign and malignant tumors of the RPE and cited several cases of extrascleral extension. In many of his cases, however, there does not appear to be sufficient documentation of an RPE origin. Fair (1958), reviewing the literature on RPE tumors, contended that there is no difference between

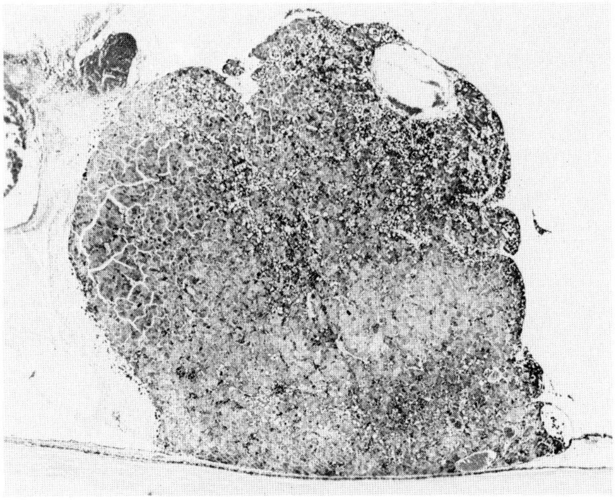

Figure 14.19 *Under low magnification the pigmented tumor has a flat base toward the sclera (bottom of picture). Cells with vacuolization are more numerous toward the apical portion than toward the base. A focal hemorrhage is present in the vitreous at the posterior edge of the tumor. (Stained with hematoxylin and eosin. AFIP Neg. 70–7096. From Font, Zimmerman, and Fine 1972. ×20.)*

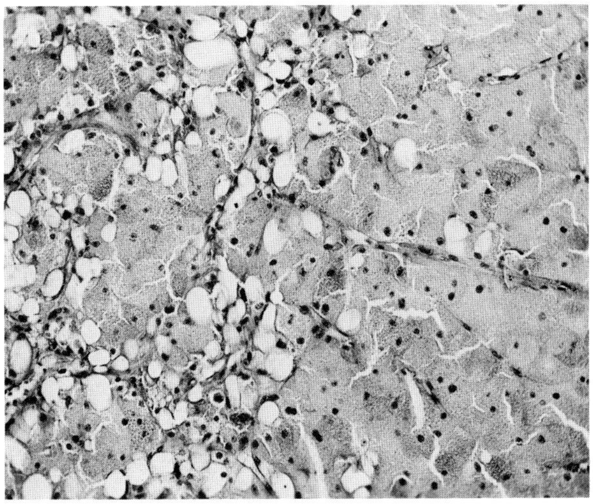

Figure 14.20 *A bleached preparation reveals large polyhedral cells with eccentric, round to oval, pyknotic nuclei and abundant granular cytoplasm. Thin fibrovascular septa are present. Large intracytoplasmic vacuoles are seen in many of the tumor cells. (AFIP Neg. 70–7088. From Font, Zimmerman, and Fine 1972. ×119.)*

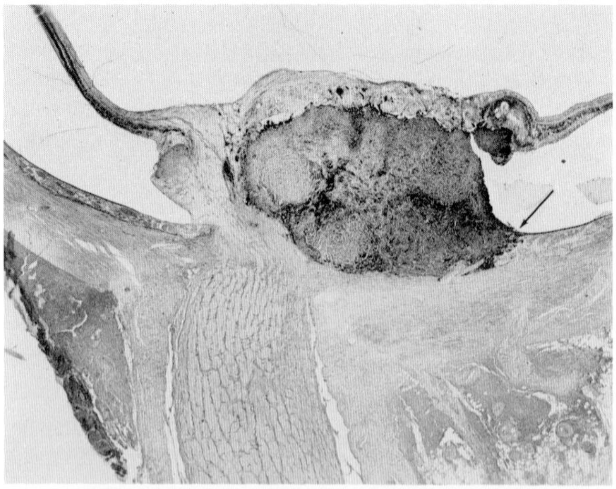

Figure 14.21 *Primary adenocarcinoma of the pigment epithelium. There was clinical evidence to suggest that the tumor might have arisen from an inflammatory focus. Invasion of the choroid (arrow) is one sign suggesting that this is a true neoplasm. Case reported by Fair. (AFIP Neg. 60–6785. From Frayer 1966. ×9.)*

tumors that arise in inflamed eyes and those arising de novo. In each situation, he stated, one finds the same large epithelial cells with oval nuclei, arranged in the characteristic pattern of cords and tubes interspersed with pink-staining homogenous material. Albert et al. (1972b), who produced experimental metastatic RPE tumors by incubating RPE cells with SV40 virus and reinfecting subcutaneously, have suggested that human RPE may also have malignant potential.

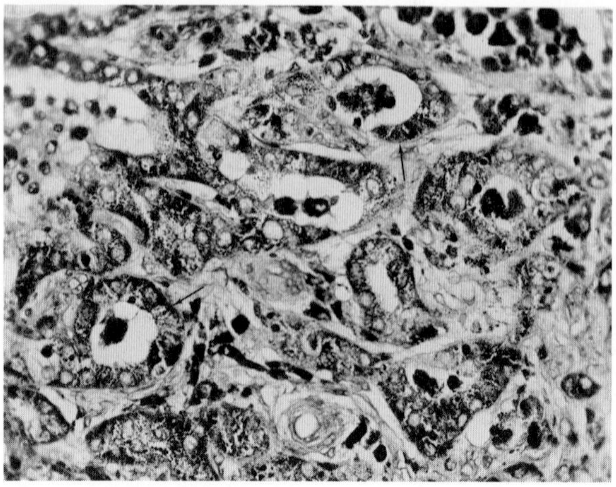

Figure 14.22 *Poorly differentiated cells of the tumor illustrated in Figure 14.21. The glandlike structures formed by these cells (arrows) were found in most areas of the tumor. (AFIP Neg. 6006787. From Frayer 1966. ×220.)*

Two of the usual criteria for malignancy are local invasion and pleomorphism. It is difficult to determine whether a growth is neoplastic—that is, has undergone a change whereby the normal growth-controlling mechanism is permanently impaired, permitting progressive growth—or whether it is a reactive hyperplasia, in that growth will cease if and when the stimulus is removed.

Vogel and coworkers (Vogel, Zimmerman, and Gass 1969) described four cases of an unusual form of proliferation of the RPE, which had the following characteristics: (1) proliferation in and adjacent to the optic nerve head with ill-defined border and a peculiar pattern variously described as mottled, mossy, or filigree; (2) slight elevation of the affected tissues; (3) fine retinal folds or striae radiating from the lesion to the macula; (4) stationary or slowly progressive course; (5) enlargement of the blind spot with deterioration of visual acuity. Two of the cases studied histologically were characterized by a disorganization of the architecture of the nerve head and retina in the area corresponding to the lesion, with severe degeneration and gliosis, proliferation of the RPE into and through the retina, with formation of cords, strands, and sheets of pigment epithelium along the inner surface of the retina and optic disc, absence of real tumefaction, and absence of any apparent causative factor. These tumors have been mistaken for malignant melanomas. Reese (1963) pointed out that these primary tumors have a special predilection for the region of the nerve head.

Conclusion

We have learned that the RPE is a very complex tissue that can respond to outside influences in many ways. Its primary function is to nurture and enfold the percipient retinal elements and act as a light-absorbing screen to initiate the complicated chain of chemical reactions within the rods and cones which leads to the production of an electrical message sent to the brain. However, when destructive forces are unloosed, these versatile cells aid in the repair and mopping-up operations. They become phagocytic, cleaning away debris, which includes the released pigment of other dead or dying RPE cells. They become fibroblastic, helping to bind together damaged elements such as retina and choroid in a healing scar. And once a semblance of order is restored, the RPE may attempt to reconstitute a nurturing layer of cells to receive percipient elements that never arrive. One may compare these changes with those of the lens epithelium, which may metamorphose into fibroblasts or attempt to replace damaged lens fibers. Although the transformations of the RPE can be observed microscopically, at the ultramicroscopic level the RPE cells retain some evidence of their epithelial origins.

The highly specialized RPE is not completely suited for its newly adopted functions, and may not know when to cease and desist. The proliferative and metaplastic stimuli may result in the cells migrating along any available surface, such as vitreous, lens, and retina; subsequent contraction can cause further damage and distortion of tissues.

When the RPE begins to proliferate in response to its new requirements, it may find such joy in reproducing, a function not ordinarily given to it after birth, that a population explosion results. Aggregates and heaped up mounds of the RPE may abound, occasionally achieving neoplastic proportions. After a suitable length of time, these old and tired cells, living in a severely degenerated eye, can actually go about building their own mausoleum, assuming an osteoblastic function and forming bone.

As part of their original function of nurturing the percipient elements, they remove old discs and waste products, breaking them down into soluble substances and sending them into the choroidal circulation. They also maintain and repair their own basement membrane. As the RPE cells become senescent, these functions are impaired. The breakdown products that are processed through the RPE are no longer made completely soluble and instead of entering the bloodstream may languish as deposits within Bruch's membrane. In addition to drusen formation, breaks may occur in Bruch's membrane, leading to subretinal neovascularization and the devastating changes that may be the sequelae of such an event. When the RPE proliferates, its basement-membrane-producing function may go awry, resulting in the formation of cuticular or hyaline membranes or even bony masses. The many pathological responses of the RPE, ranging from the loss of function, depigmentation, degeneration, and atrophy to proliferative and metaplastic alterations, can occur side by side in the same eye in response to similar influences.

REFERENCES

ABRAHAM, R., and HENDY, R. J. 1970. Irreversible lysosomal damage induced by chloroquine in the retinae of pigmented and albino rats. *Exp. Mol. Pathol.* 12:185.

ALBERT, D. M., TSO, M. O. M., and RABSON, A. S. 1972*a*. *In vitro* growth of pure culture of the retinal pigment epithelium. *Arch. Ophthalmol.* 88:63.

———. 1972*b*. Experimental malignant tumors from the retinal pigment epithelium. *Arch. Ophthalmol.* 88:70.

ANDERSON, S. R., and WARBURG, M. 1961. Norrie's disease: congenital bilateral pseudotumor of the retina with recessive x-chromosomal inheritance. *Arch. Ophthalmol.* 66:614.

APPLE, D. J., GOLDBERG, M. F., and WYHINNY, G. 1973. Histopathology and ultrastructure of the argon laser lesion in human retinal and choroidal vasculatures. *Am. J. Ophthalmol.* 75:595.

ASHTON, N. 1953. Central areolar choroidal sclerosis: histopathological study. *Br. J. Ophthalmol.* 37:140.

BALLANTYNE, A. J., and MICHAELSON, I. C. 1970. *Textbook of the fundus of the eye*. Edinburgh and London: E. & S. Livingstone.

BERNSTEIN, H. N., and GINSBERG, J. 1964. The pathology of chloroquine retinopathy. *Arch. Ophthalmol.* 71:238.

BERNSTEIN, H., ZVAIFLER, N., RUBIN, M., and MANSOUR, A. 1963. The ocular deposition of chloroquine. *Invest. Ophthalmol.* 2:384.

BONIUK, V. 1975. Rubeola. *Int. Ophthalmol. Clin.* 15:243.

BONIUK, M., and ZIMMERMAN, L. E. 1967. Ocular pathology in the rubella syndrome. *Arch. Ophthalmol.* 77:455.

BRAILEY, W. A., and LOBO, G. 1882. On choroidal new formations. *R. Lond. Ophthalmol. Hosp. Rep.* 10:405.

BRALEY, A., and SPIVEY, B. 1964. Hereditary vitelline macular degeneration. *Arch. Ophthalmol.* 72:743.

BUETTNER, H. 1975. Congenital hypertrophy of the retinal pigment epithelium. *Am. J. Ophthalmol.* 79:177.

COATS, G. 1908. Forms of retinal disease with invasive exudation. *R. Lond. Ophthalmol. Hosp. Rep.* 17:440.

CREEL, D. J. 1971. Visual system anomaly associated with albinism in the cat. *Nature* 231:465.

DAYAL, Y., and RODGER, F. C. 1959. Mutation of the retinal pigment cells in a case of pseudoglioma. *Arch. Ophthalmol.* 62:785.

DE VENECIA, G., ZURHEIN, G. M., PRATT, M. V., and KISHEN, W. 1971. Cytomegalic inclusion retinopathy in an adult. *Arch. Ophthalmol.* 86:44.

DUKE-ELDER, S. 1963*a*. *System of ophthalmology,* vol. 3, part II, p. 611. London: Kimpton.

DUKE-ELDER, S. 1963*b*. *System of ophthalmology*, vol. 3, part II, p. 619. London: Kimpton.

DUKE-ELDER, S. 1963*c*. *System of ophthalmology*, vol. 3, part II, p. 625. London: Kimpton.

DUKE-ELDER, S. 1963*d*. *System of ophthalmology*, vol. 3, part II, p. 806. London: Kimpton.

DUKE-ELDER, S. 1963*e*. *System of ophthalmology*, vol. 3, part II, p. 626. London: Kimpton.

DUKE-ELDER, S. 1966. *System of ophthalmology*, vol. 9, p. 583. London: Kimpton.

DUKE-ELDER, S. 1967*a*. *System of ophthalmology*, vol. 10, p. 178. London: Kimpton.

DUKE-ELDER, S. 1967*b*. *System of ophthalmology*, vol. 10, p. 601. London: Kimpton.

DUKE-ELDER, S. 1967*c*. *System of ophthalmology*, vol. 10, p. 521. London: Kimpton.

FAIR, J. R. 1958. Tumors of the retinal pigment epithelium. *Am. J. Ophthalmol.* 45:495.

FARKAS, T. G., SYLVESTER, B. A., and ARCHER, D. 1971. The ultrastructure of drusen. *Am. J. Ophthalmol.* 71:1196.

FARKAS, T. G., SYLVESTER, B. A., ARCHER, D., and ALTMAN, M. 1971. The histochemistry of drusen. *Am. J. Ophthalmol.* 71:1206.

FERRY, A. 1964. Lesions mistaken for malignant melanoma of the posterior uvea. *Arch. Ophthalmol.* 72:463.

FERRY, A. P., LLOVERA, I., and SHAFER, D. M. 1972. Central areolar choroidal dystrophy. *Arch. Ophthalmol.* 88:39.

FONT, R. L., ZIMMERMAN, L. E., and ARMALY, M. F. 1974. The nature of the orange pigment over a choroidal melanoma. *Arch. Ophthalmol.* 91:359.

FONT, R. L., ZIMMERMAN, L. E., and FINE, B. S. 1972. Adenoma of the retinal pigment epithelium. *Am. J. Ophthalmol.* 73:544.

FRAYER, W. 1966. Reactivity of the retinal pigment epithelium: an experimental and histopathologic study. *Trans. Am. Ophthalmol. Soc.* 64:586.

GARNER, A. 1970. Tumors of the retinal pigment epithelium. *Br. J. Ophthalmol.* 54:715.

GARTNER, S. 1947. Cyclopia. *Arch. Ophthalmol.* 37:220.

GLEISER, C. A., DUKES, T. W., LAWWILL, T., READ, W. K., BAY, W. W., and BROWN, R. S. 1969. Ocular changes in swine associated with chloroquine toxicity. *Am. J. Ophthalmol.* 67:399.

GOLD, D. H., HENKIND, P., STURNER, W. Q., and BADEN, M. 1967. Oculodermal melanocytosis and retinitis pigmentosa. *Am. J. Ophthalmol.* 63:291.

GUILLERY, R. W., AVNOIN, D. S., and EIGHMY, B. B. 1971. Mutants with abnormal visual pathways: an explanation of anomalous geniculate laminae. *Science* 174:831.

HENKIND, P. 1973. Comment on Ishikawa and Ikui (1972). *Survey of Ophthalmol.* 18:162.

HOGAN, M. J. 1965. Electron microscopy of Bruch's membrane. *Trans. Am. Acad. Ophthalmol. Otolaryngol.* 69:688.

HOGAN, M. J. 1972. Role of the retinal pigment epithelium in macular disease. *Trans. Am. Acad. Ophthalmol. Otolaryngol.* 76:64.

HOGAN, M. J., ALVARADO, J., and WEDDELL, J. E. 1971. *Histology of the human eye: an atlas and textbook*, p. 344. Philadelphia: W. B. Saunders.

HOGAN, M. J., and ZIMMERMAN, L. E. 1962. *Ophthalmic pathology*. Philadelphia: W. B. Saunders.

HOWARD, G. M., and WOLF, E. 1964. Central choroidal sclerosis: a clinical and pathologic study. *Trans. Am. Acad. Ophthalmol. Otolaryngol.* 68:647.

HUNTER, W. S., and ZIMMERMAN, L. E. 1965. Unilateral retinal dysplasia. *Arch. Ophthalmol.* 74:23.

ISHIKAWA, T., and IKUI, H. 1972. The fine structure of the Dalen-Fuchs nodule in sympathetic ophthalmia. *Jap. J. Ophthalmol.* 16:254.

JOHNSON, D. L., JACOBSON, L. W., TOYAMA, R., and MONAHAN, R. H. 1966. Histopathology of eyes in Chediak-Higashi Syndrome. *Arch. Ophthalmol.* 75:84.

KENO, D. D., and GREEN, W. R. 1978. Refined pigment epithelial window defect. *Arch. Ophthalmol.* 96:854.

KLIEN, B. 1958. Diseases of the macula. *Arch. Ophthalmol.* 60:175.

KLIEN, B. 1959. The pathogenesis of some atypical colobomas of the choroid. *Am. J. Ophthalmol.* 48:597.

KNAPP, H. 1871. Formation of bone in the eye. *Arch. Ophthalmol.* 2:1.

KROLL, A. J., and MACHEMER, R. 1968. Experimental retinal detachment in the owl monkey. III. Electron microscopy of retinal and pigment epithelium. *Am. J. Ophthalmol.* 66:410.

KURZ, G., and ZIMMERMAN, L. E. 1962. Vagaries of the retinal pigment epithelium. *Int. Ophthalmol. Clin.* 2:441.

LAMB, H. D. 1935. The pathogenesis of some intraocular osseous tissue: true metaplasia in the eye. *Am. J. Ophthalmol.* 18:409.

LAMB, H. D. 1938. Exudative retinitis: anatomic findings in six early and two late cases. *Am. J. Ophthalmol.* 21:618.

LAQUA, H., and MACHEMER, R. 1975*a*. Glial cell proliferation in retinal detachment (massive periretinal proliferation). *Am. J. Ophthalmol.* 80:602.

LAQUA, H., and MACHEMER, R. 1975*b*. Clinical-pathological correlation in massive periretinal proliferation. *Am. J. Ophthalmol.* 80:913.

LAWWILL, T., APPLETON, B., and ALSTATT, L. 1964. Chloroquine accumulations in human eyes. *Am. J. Ophthalmol.* 71:238.

LINDQUIST, N. G., and ULBERG, S. 1972. The melanin affinity of chloroquine and chlorpromazine studied by whole body autoradiography. *Acta Pharmacol. Toxicol.* (suppl. II):31.

LOWENSTEIN, A., and STEEL, J. 1941. Special case of melanosis fundi: bilateral congenital group pigmentation of the central area. *Br. J. Ophthalmol.* 25:417.

MACHEMER, R. 1968. Experimental retinal detachment in the owl monkey. 2. Histology of the retina and pigment epithelium. *Am. J. Ophthalmol.* 66:398.

MACHEMER, R., and LAQUA, H. 1975. Pigment epithelium proliferation in retinal detachment (massive periretinal proliferation). *Am. J. Ophthalmol.* 80:1.

MACHEMER, R., and NORTON, E. W. D. 1968. Experimental retinal detachment in the owl monkey. I. Methods of production and clinical picture. *Am. J. Ophthalmol.* 66:388.

MANDELCORN, M. S., MACHEMER, R., FINEBERG, E., and HERSCH, S. B. 1975. Proliferation and metaplasia of intravitreal retinal pigment epithelium cell autotransplants. *Am. J. Ophthalmol.* 80:227.

MANN, I. 1957. *Developmental abnormalities in the human eye.* London: British Medical Assoc.

MANSCHOT, W. A., and DE BRUIJN, W. C. 1967. Coats' disease: definitions and pathogenesis. *Br. J. Ophthalmol.* 51:145.

MARMOR, M. F., and ZINN, K. M. 1975. The retinal pigment epithelium. *Int. Ophthalmol. Clin.* 15:111.

MCCULLOCH, C. 1950. The pathologic findings in two cases of choroideremia. *Trans. Am. Acad. Ophthalmol. Otolaryngol.* 54:565.

MEYER-SCHWICKERATH, G. 1960. *Light coagulation.* St. Louis: C. V. Mosby.

MIYAMOTO, M., and FITZPATRICK, T. B. 1957. On the nature of the pigment in retinal pigment epithelium. *Science* 126:449.

MONOHAN, R., and HORNS, R. 1964. The pathology of chloroquine in the eye. *Trans. Am. Acad. Ophthalmol. Otolaryngol.* 68:40.

MORRIS, D. A. 1972. Intraocular leukemic marrow. Presented at Eastern Ophthalmic Pathology Society.

MUELLER-JENSEN, K., MACHEMER, R., and AZARNIA, R. 1975. Autotransplantation of retinal pigment epithelium in intravitreal diffusion chamber. *Am. J. Ophthalmol.* 80:530.

NORRIE, G. 1927. Causes of blindness in children: twenty-five years of experience of Danish Institute for the Blind. *Acta Ophthalmol.* 5:357.

PURCELL, J. J., and SHIELDS, J. A. 1975. Hypertrophy with hyperpigmentation of the retinal pigment epithelium. *Arch. Ophthalmol.* 93:1122.

RAMSEY, M. S., and FINE, B. S. 1972. Chloroquine toxicity in the human eye. *Am. J. Ophthalmol.* 73:229.

REESE, A. B. 1960. The role of the pigment epithelium in ocular pathology. *Am. J. Ophthalmol.* 50:1066.

REESE, A. B. 1963. *Tumors of the eye.* New York: Hoeber.

REESE, A. B., and JONES, I. S. 1956. Benign melanomas of the retinal pigment epithelium. *Am. J. Ophthalmol.* 42:207.

REICH-D'ALMEIDA, F. B., and HOCKLEY, D. J. 1975*a*. In situ reactivity of the retinal pigment epithelium. I. Phagocytosis in the normal rat. *Exp. Eye Res.* 21:333.

REICH-D'ALMEIDA, F. B., and HOCKLEY, D. J. 1975*b*. In situ reactivity of the retinal pigment epithelium. II. Phagocytosis in the dystrophic rat. *Exp. Eye Res.* 21:347.

RUBIN, M., and SIVNICKI, A. 1966. A mechanism for the toxicity of chloroquine. *Arthritis Rheum.* 9:537.

SARKS, S. H. 1976. Ageing and degeneration in the macular region: a clinico-pathological study. *Brit. J. Ophthalmol.* 60:324.

SHIELDS, J. A., and TSO, M. O. M. 1975. Congenital grouped pigmentation of the retina. *Arch. Ophthalmol.* 93:1153.

SILVERSTEIN, A. M., OSBURN, B. I., and PRENDERGAST, R. A. 1971. The pathogenesis of retinal dysplasia. *Am. J. Ophthalmol.* 72:13.

SNOWBALL, T. 1903. Ossification of the choroid. *Trans. Ophthalmol. Soc. U.K.* 23:217.

SPENCER, W. H., and HOGAN, M. J. 1960. Ocular manifestations of Chediak-Higashi Syndrome: report of a case with histopathologic examination of ocular tissue. *Am. J. Ophthalmol.* 50:1197.

Stedman's medical dictionary. 1972. Baltimore: Williams and Wilkins.

Stow, M. W. 1949. Hyperplasia of the pigment epithelium of the retina simulating a neoplasm. *Trans. Am. Acad. Ophthalmol. Otolaryngol.* 54:674.

Straatsma, B. R., Zoegen, P. D., Foos, R. Y., Feman, S. S., and Shabo, A. L. 1974. Lattice degeneration of the retina. *Trans. Am. Acad. Ophthalmol. Otolaryngol.* 78:87.

Streeten, B. W. 1961. The sudanophilic granules of the human pigment epithelium. *Arch. Ophthalmol.* 66:391.

Tso, M. O. M. 1973. Photic maculopathy in rhesus monkeys. *Invest. Ophthalmol.* 12:17.

Tso, M. O. M., and Albert, D. M. 1972. Pathological conditions of the retinal pigment epithelium: neoplasms and nodular non-neoplastic lesions. *Arch. Ophthalmol.* 88:27.

Tso, M. O. M., Fine, B. S., and Zimmerman, L. E. 1972. Photic maculopathy produced by the indirect ophthalmoscope. *Am. J. Ophthalmol.* 73:686.

Vogel, M. H., Zimmerman, L. E., and Gass, J. D. M. 1969. Proliferation of the juxtapapillary retinal pigment epithelium simulating malignant melanoma. *Doc. Ophthamol.* 26:461.

Wallow, I. H. L., and Tso, M. O. M. 1972. Proliferation of the retinal pigment epithelium over malignant choroidal tumor. *Am. J. Ophthalmol.* 73:914.

Wallow, I. H. L., and Tso, M. O. M. 1973. Repair after xenon arc photocoagulation. *Am. J. Ophthalmol.* 75:610.

Wetterholm, D. H., and Winter, F. C. 1964. Histopathology of chloroquine retinal toxicity. *Arch. Ophthalmol.* 71:82.

Whittaker, J. R. 1967. Loss of melanotic phenotype *in vitro* by differentiated retinal pigment cells: demonstration of mechanisms involved. *Dev. Biol.* 15:553.

Wolter, J. R. 1956. Reactions of the elements of the retina and optic nerve: in common morbid entities of the human eye. *Am. J. Ophthalmol.* 42:10.

Zimmerman, L. E. 1968. Histopathological basis for ocular manifestations of congenital rubella syndrome. *Am. J. Ophthalmol.* 65:873.

Zinn, K. Personal communication. 1977.

15 Developmental, Reactive, and Neoplastic Proliferation of the Retinal Pigment Epithelium

Mark O. M. Tso

THE UVEAL MELANOCYTES and the pigment epithelial cells (RPE) of the retina and iris are both melanin-bearing cells of the eye, but they differ remarkably in their response to injury and ability to develop neoplasm. The uveal melanocyte rarely, if ever, undergoes reactive proliferation, whereas the pigment epithelium proliferates readily in response to a variety of stimuli. On the other hand, neoplastic proliferation of the retinal pigment epithelium is rare, but intraocular malignant melanoma arises almost exclusively in the uvea. In this chapter, the normal developmental growth and reactive and neoplastic proliferation of the retinal pigment epithelium will be considered.

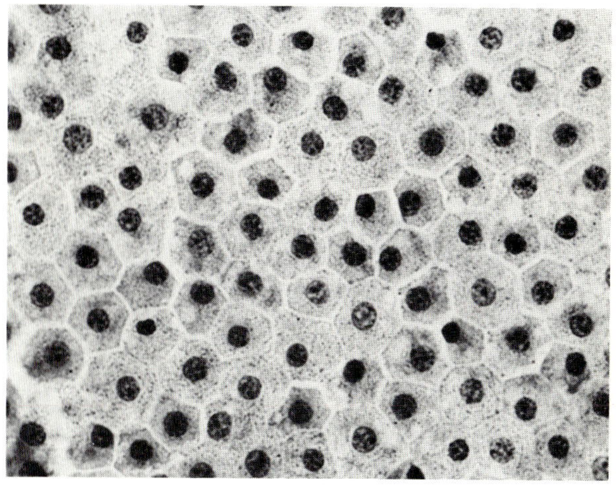

Figure 15.1 *Flat preparation of retinal pigment epithelium in the posterior pole. These highly regular, hexagonal, prismatic cells cover the entire posterior segment of the eye. (Bleached by hydrogen peroxide, stained with H&E. AFIP Neg. 70–730. ×490.)*

Normal Growth and Proliferation

The retinal pigment epithelium is a monolayer of cuboid pigmented cells that covers Bruch's membrane over the entire posterior segment of the eye (Fig. 15.1). As the surface area of the posterior segment of the eye increases from infancy to adulthood by thirty to forty percent, the surface area covered by the retinal pigment epithelium increases correspondingly. The response of the RPE to this growth is twofold.

(1) Even though mitotic figures have been observed in the RPE of normal rats (Tso and Friedman 1967), mitotic figures of adult human pigment epithelium have been noted only in pathologic conditions. The role of cell multiplication in normal development of RPE was examined by quantitative study of ten normal human eyes obtained postmortem from eight individuals between the ages of 4 months of gestation and 96 years (Tso and Friedman 1968). The total number of RPE cells per eye after birth was estimated to be 4.2 to 6.1 million, with greater numbers of cells in the older eyes (Fig. 15.2). This study suggests that in the course of normal growth in the postnatal period, there is mild proliferation among the RPE cells to compensate for the increase in the total surface area of the posterior segment of the eye.

(2) The RPE also responds to the overall increase in area of the posterior segment of the eye by flattening of the cells (Tso and Friedman 1968; Friedman and Tso 1968; Streeten 1969). This process varies with different regions of the eye but appears most

proliferation, may be produced b
depending on the severity of the
Papillary proliferation of the re
thelium is also observed in pa
standing detachments of the retin
posterior pole (Fig. 15.8). This fo

Figure 15.8 *Papillary proliferation lium in a patient with long-standin retina. Note that the proliferation i epithelium cells, without a stromal with H&E. AFIP Neg. 75–14272. ×*

involves only the pigment epit
a stromal component. At times, t
lial cells forming the papillae m
eration. These cells become fl
pigment granules, and their nu
Papillary proliferation of pi
may be modified by environme
pear histologically as diffuse hyp
discussed later in this chapter.

TUBULAR PROLIFERATIO
logic pattern, cuboidal pigment
arranged in an oval fashion aro
and surrounded by a thin layer
brane (Fig. 15.9). Moderate a
connective tissue are present aro
basement membrane of the pig
few pigmented cells may be fou
these acini.
This histologic pattern is ty
patients with longstanding retin
15.10). The detached retina pu
the region of the ora serrata, a

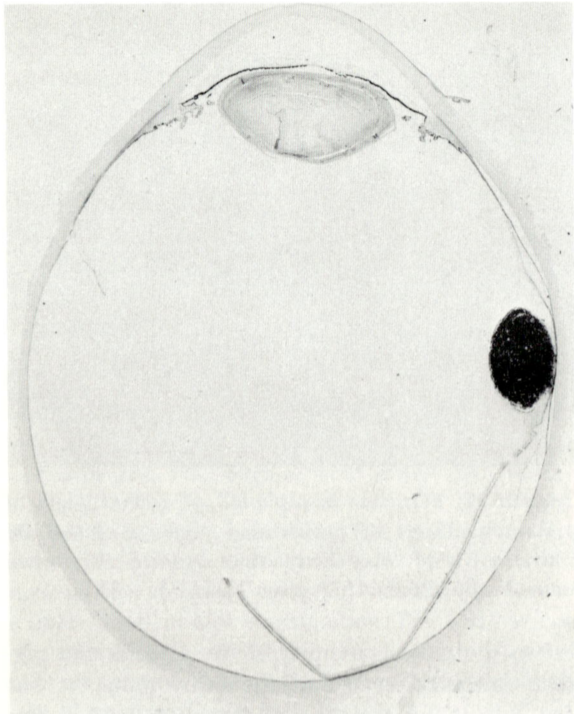

Figure 15.18 *A tumor of retinal pigment epithelium arising in an otherwise normal eye. There is a notable absence of diffuse vitreous-retinal-choroidal disease. (From Tso and Albert 1972. Stained with H&E. AFIP Neg. 70–1033. ×5.2.)*

the pigment epithelium is not noted elsewhere in the eye, nor is there diffuse vitreous, retinal, and choroidal disease. The retina overlying the tumor is atrophic, but the adjacent choroid shows no evidence of inflammatory reactions. A single tumor may exhibit in a variety of histologic patterns, although one histologic pattern usually predominates.

The mosaic pattern consists of hexagonal oval cells with small round nuclei proliferate en masse (Fig. 15.19). Nuclear morphism is not marked and the cytoplasm is heavily pigmented. Very little extracellular material is observed between the tumor cells and there is virtually no fibrovascular stroma.

In the tubular pattern, cuboidal epithelial cells are arranged in the form of tubules and are surrounded by delicate basement membranes and a scanty connective tissue stroma (Fig. 15.20). The cells may appear pleomorphic, and the pigmentation of the cells varies. Groups of pigmented cells are shed into the lumen of these tubular structures. Sometimes the lumen of the tubules is compressed and the tumor cells appear to be arranged in sheets.

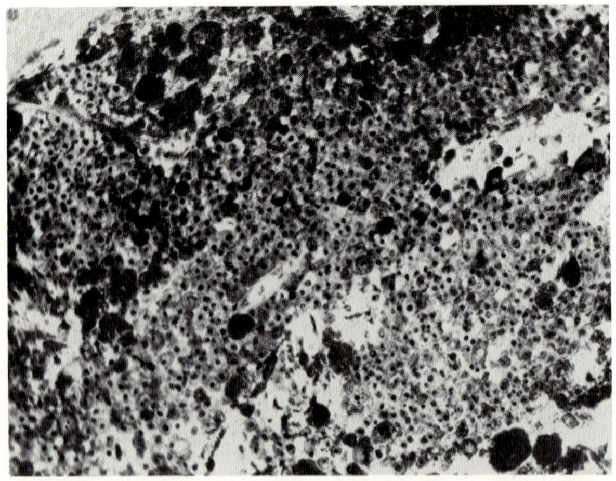

Figure 15.19 *A tumor of retinal pigment epithelium showing a mosaic pattern. Hexagonal and oval tumor cells with small round nuclei are heavily pigmented. There is scanty extracellular material between the cells. (From Tso and Albert 1972. Stained with H&E. AFIP Neg. 70–8798. ×160.)*

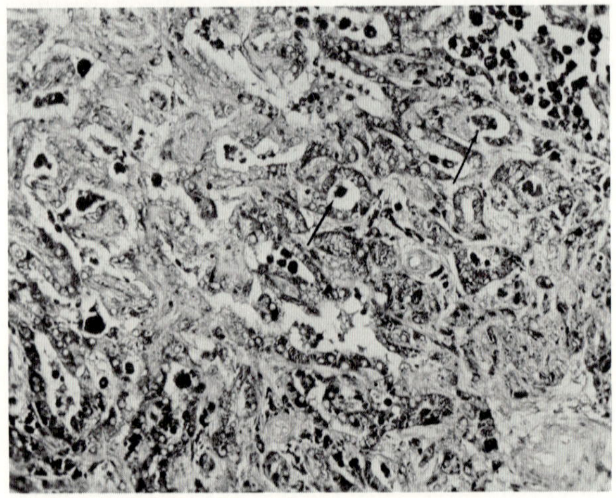

Figure 15.20 *A tumor of pigment epithelium showing a tubular pattern. The cells are arranged in oval acini. Clumps of pigment granules (arrows) are sequestered into lumens of the acini. Note the pigmentation of the cells varies in different locations. (Stained with H&E. AFIP Neg. 70–8471. ×102.)*

In the papillary pattern, a cuboidal epithelium is arranged in finger-like projections. The cells possess a delicate basement membrane, and scanty connective tissue containing blood vessels forms the core of the papillae (Fig. 15.21).

In a patternless arrangement of vacuolated cells, the tumors are composed of masses of heavily pig-

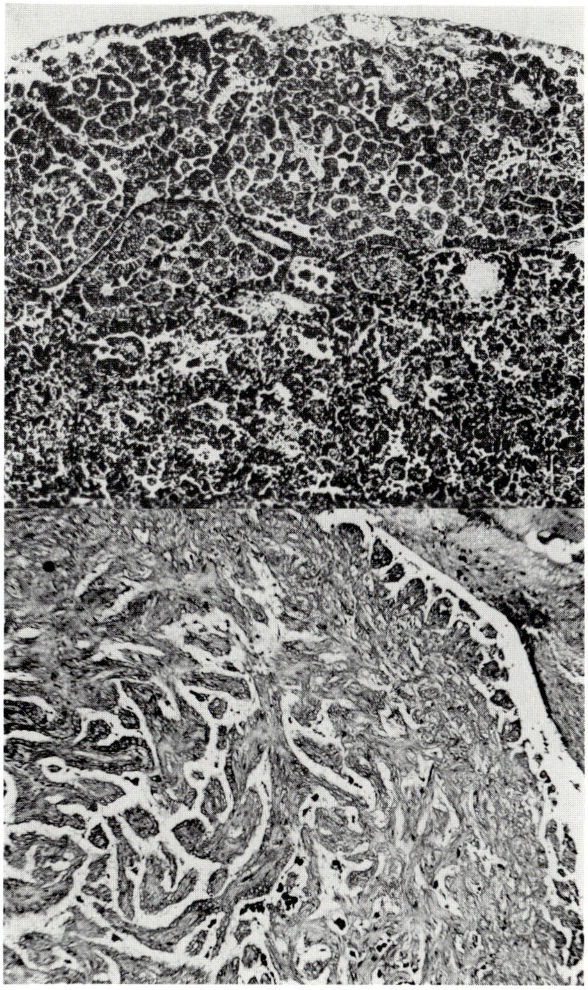

Figure 15.21 Top, *Densely pigmented papillary tumors of pigment epithelium. Papillae are lined by tall columnar pigment epithelium cells. (Stained with H&E. AFIP Neg. 71–7996. ×50.)* Bottom, *Papillary tumor of retinal pigment epithelium. Myxomatous connective tissue forms the core of papillae. (From Tso and Albert 1972. Stained with H&E. AFIP Neg. 70–8472. ×63.)*

mented cells with a delicate connective tissue stroma (Font, Zimmerman, and Fine 1972). Some of the tumor cells characteristically show large intracytoplasmic vacuoles, and the nuclei are pushed to one side and closely mimic signet-ring cells (Fig. 15.22). Polar orientation of pigment epithelial cells into the apical and basal regions cannot be identified.

In a patternless arrangement of anaplastic cells, the tumors consist of polygonal epithelioid cells devoid of definite basement membranes or stroma. The nuclei are pleomorphic and hyperchromatic. Mitotic figures are numerous. Histologically, these tumors appear malignant (Fig. 15.23).

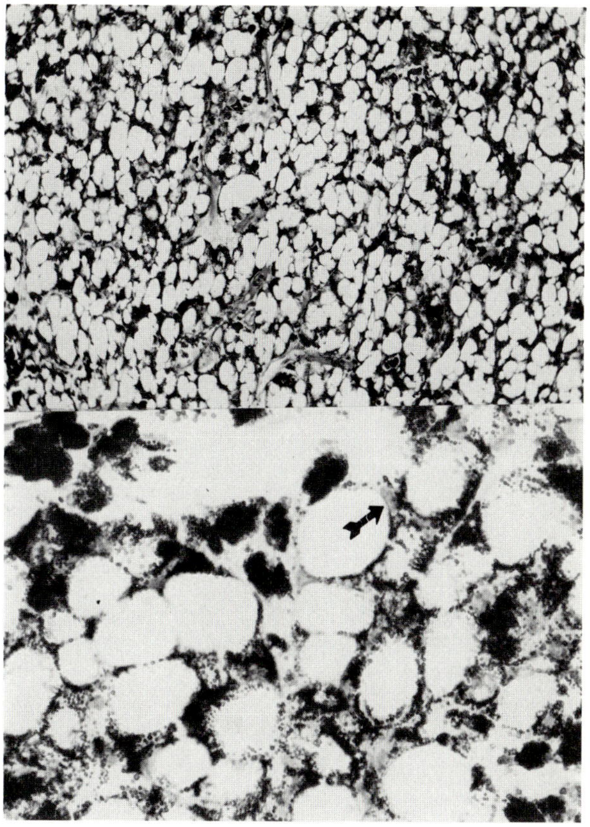

Figure 15.22 Top, *Tumor of pigment epithelium showing vacuolated tumor cells in a delicate connective tissue stroma. (Stained with H&E. AFIP Neg. 60–882. ×87.)* Bottom, *nuclei of tumor cells (arrow) are being pushed to one side by large intercellular vacuoles. Note large oval pigment granules, many of them aggregated in clumps. (Stained with H&E. AFIP Neg. 70–888. ×420.) (From Tso and Albert 1972.)*

BEHAVIOR. Tumors of the RPE may extensively invade into the adjacent retina and choroid. Seeding into the vitreous cavity may be seen, but no extraocular extension, distant metastasis, or tumor death has been convincingly documented. Of the four frequently cited cases of RPE tumors that had supposedly developed extraocular extensions and generalized metastasis, none stands up to close scrutiny (Tso and Albert 1972). Due to the absence of documentation of generalized metastasis, we prefer to refer to these lesions as tumors without attempting to designate benignity or malignancy. It should therefore be pointed out that tumors arising from the pigment epithelium of the iris and ciliary body are not included in this discussion.

COMPARATIVE PATHOLOGY. Tumors of the pigment epithelium differ histologically from malig-

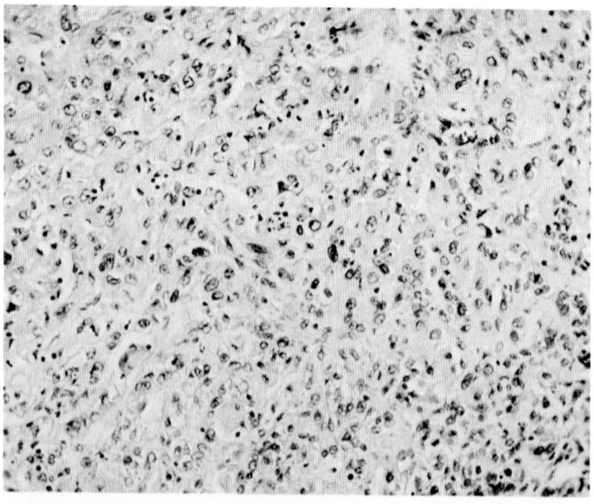

Figure 15.23 *An anaplastic tumor of RPE showing poorly differentiated amelanotic pleomorphic cells. (Stained by H&E. AFIP Neg. 70–8787. ×140.)*

nant melanomas of uveas. They are more comparable to ependymomas. This is not surprising since both ependymomas and the retinal pigment epithelium are derived embryologically from the lining cells of the neural tubes. Both persist into adult life as a monolayer of cuboid cells. The tubular and papillary patterns of RPE tumors are comparable to the epithelial and papillary ependymomas. The anaplastic RPE tumors have their analogues in cellular ependymomas. However, the tumors of vacuolated cells appear to be unique only to RPE tumors, whereas the arrangement of tumor processes around blood vessels observed in some cellular ependymomas is not seen in RPE tumors. Nonetheless, both groups of tumors are relatively benign. Just as ependymomas may seed into the ventrical system, RPE tumors send seedlings into the vitreous cavity.

REFERENCES

Font, R. L., Zimmerman, L. E., and Fine, B. S. 1972. Adenoma of the retinal pigment epithelium. *Am. J. Ophthalmol.* 73:544–554.

Frayer, W. C. 1966. Reactivity of the retinal pigment epithelium: an experimental and histologic study. *Trans. Am. Ophthalmol. Soc.* 64:586–643.

Frayer, W. C. 1972. Neoplasms and related lesions of the retinal pigment epithelium. *Int. Ophthalmol. Clin.* 12:63–69.

Friedman, E., and Tso, M. O. M. 1968. The retinal pigment epithelium. II. Histologic changes associated with age. *Arch. Ophthalmol.* 79:315–320.

Garner, A. 1970. Tumors of the retinal pigment epithelium. *Br. J. Ophthalmol.* 54:715–723.

Hogan, M., and Zimmerman, L. E. 1962. *Ophthalmic pathology: an atlas and textbook.* Philadelphia: W. B. Saunders.

Klien, B. 1958. Diseases of the macula. *Arch. Ophthalmol.* 60:175–186.

Kurz, G. H., and Zimmerman, L. E. 1962. Vagaries of the retinal pigment epithelium. *Int. Ophthalmol. Clin.* 2:441–464.

Reese, A. B. 1960. The role of the pigment epithelium in ocular pathology. *Am. J. Ophthalmol.* 50:1066–1084.

Streeten, B. W. 1969. Development of the human retinal pigment epithelium and the posterior segment. *Arch. Ophthalmol.* 81:383–394.

Tso, M. O. M. 1973. Photic maculopathy in rhesus monkey: a light and electron microscopic study. *Invest. Ophthalmol.* 12:17–34.

Tso, M. O. M., and Albert, D. M. 1972. Pathologic conditions of the retinal pigment epithelium: neoplasm and nodular nonneoplastic lesions. *Arch. Ophthalmol.* 88:27–38.

Tso, M. O. M., and Friedman, E. 1967. The retinal pigment epithelium. I. Comparative histology. *Arch. Ophthalmol.* 78:641–644.

Tso, M. O. M., and Friedman, E. 1968. The retinal pigment epithelium: growth and development. *Arch. Ophthalmol.* 80:214–216.

Tso, M. O. M., Fine, B. S., and Zimmerman, L. E. 1972. Photic maculopathy produced by indirect ophthalmoscope. I. Clinical and histopathologic study. *Am. J. Ophthalmol.* 73:686–699.

Vogel, M. H., Zimmerman, L. E., and Gass, J. D. M. 1969. Proliferation of juxtapapillary retinal pigment epithelium simulating malignant melanoma. *DOC Ophthalmologica* 26:461–481.

Wallow, I. H. L., and Tso, M. O. M. 1972. Proliferation of retinal pigment epithelium over malignant choroidal tumors: a light and electron microscopic study. *Am. J. Ophthalmol.* 73:914–926.

Wallow, I. H. L., and Tso, M. O. M. 1973*a*. Repair after xenon arc photocoagulation. III. An electron microscopic study of the evolution of retinal lesions in rhesus monkey. *Am. J. Ophthalmol.* 75:957–972.

Wallow, I. H. L., and Tso, M. O. M. 1973*b*. Retinal repair after xenon arc photocoagulation. II. Revolution of retinal lesions in rhesus monkey: a clinical and light microscopic study. *Am. J. Ophthalmol.* 75:610–626.

16 Tissue Culture of the Retinal Pigment Epithelium

Daniel M. Albert
Ned Buyukmihci

THE RETINAL PIGMENT EPITHELIUM (RPE) is recognized to be composed of primitive and multipotential cells that are capable of a great variety of activities, including phagocytosis, proliferation, migration, fibrous and osseous metaplasia, and neoplasia (Frayer 1966; Kurz and Zimmerman 1962). Our knowledge that the RPE does have such potentialities is based primarily on the histopathologic examination of enucleated eyes. There is little experimental information elucidating the mechanism involved in these tissue reactions.

By the use of tissue cultures of pure RPE, we can now provide model systems for studying in greater detail the various properties of these cells. By presenting various stimuli to reproducible cell-culture systems, we can better document the reactions that occur. The obvious advantages of the use of tissue culture techniques in the study of the behavior of RPE are:

1. controlled environment
2. ability to study the response to selective insults or abnormal conditions
3. ease of viewing the pathology produced
4. potential for biochemical studies without contamination by other host cells
5. extensive elimination of interference by host protective phenomena

The disadvantages are inherent in the same properties that make the system advantageous:

1. different responses may occur in the artificial environment than in vivo, because of the lack of other ocular tissue with which the RPE cells normally interact
2. the preclusion of immunologic, hormonal, neural, and other responses that may influence the RPE may lead to erroneous conclusions, since these mechanisms may modify the reaction of the RPE in vivo as compared to in vitro

In the past there have been very few tissue culture studies involving pure mammalian RPE. This may have resulted from the difficulty of obtaining and maintaining pure cultures of this tissue. Most of the tissue culture studies have involved pigment epithelial cells of iris, ciliary body, and retina from chick embryos (Adachi 1930; Bisceglie 1932; Dohljanski 1930; Ebeling 1936; Fischer 1922a; Fischer 1922b; Fischer 1924; Fischer and Albert 1924; Fischer 1938; Frayer 1966; Kapel 1927; Luna 1920/21; Meyer 1936; Montalcini and Sacerdote 1938; Smith 1920a; Smith 1920b; Smith 1921; Torö 1931; Wilbur and Chambers 1942). These cultures grew well and in many cases continued to produce pigment after repeated subculturing. However, one worker demonstrated loss of pigment due to (1) simple dilution by cell growth of melanin, tyrosinase activity, and the rate of biosynthesis; (2) decay of tyrosinase activity; and (3) cessation of tyrosinase synthesis (Whittaker 1967). Mammalian pigment epithelial cells have been grown in mixed cultures (Attardi and Moro 1959; Barishak 1960; Pomerat and Littlejohn 1956; Reese and Ehrlich 1958; Tso, Albert, and Zimmerman 1973). Human RPE has also been grown in pure cultures (Albert, Tso, and Rabson 1972; Mannagh, Arya, and Irvine 1973), as has pig RPE (François et al. 1971).

The next section describes the techniques utilized in our laboratory and in our collaboration with others (Albert, Rabson, and Dalton 1968; Albert, Tso, and Rabson 1972; Tso, Albert and Zimmerman 1973).

Methods of RPE Culturing

RPE AND CHOROID AS SEPARATE CULTURES. Globes should be removed as soon after death as possible although those removed several hours after death have been useful. The globes are kept in culture medium RPMI 1640 (Roswell Park Memorial Institute Medium 1640) with 20 percent fetal calf serum added. Under the dissecting microscope, a small window is cut in the sclera of each eye and a portion of the choroid is carefully removed. The specimens of the choroid are minced into 1 mm segments and put into culture. The anterior segment and the lens of each eye are removed and the sensory retina detached and discarded, exposing the RPE. A 5 percent solution of trypsin and Dulbecco's tromethamine (Tris) buffer free of calcium and magnesium is allowed to act on the RPE for 20 minutes at 37° C. Segments of 1 to 3 mm of RPE are dissected from Bruch's membrane and floated in a petri dish containing tissue culture medium and inspected under 30× magnification. Fragments of RPE with no evidence of contamination by other cells should be selected. Acceptable tissue is placed on a 10 mm circle of filter (Millipore), which is then introduced into a Leighton tube and covered with 3 ml of medium 1640. Tissue may be placed directly on a glass cover slip and grown in a similar manner.

Culture fluid should be changed three times weekly and cultures should be incubated at 37° C. After thirty days, cultures are trypsinized with 0.25 percent trypsin in Dulbecco's tromethamine buffer free of calcium and magnesium for 30 minutes at 37° C in preparation for subculturing. The cells are then resuspended in growth medium and passed to other tubes and flasks. Subculturing is done at one to two week intervals thereafter for maximum development.

The histopathologic studies of the cells in cultures may be carried out in several ways: (1) The cells can be examined and photographed as seen in culture flasks. (2) Cells growing on filters (Millipore) and cover slips can be fixed with 4 percent formaldehyde solution and stained with hematoxylin-eosin (HE) and examined after mounting on glass slides. (3) Cells maintained in tissue culture flasks can be scraped from the plastic and centrifuged at 748 *g*'s for five minutes. The resulting pellet is then fixed in 4 percent formaldehyde solution and sections stained with HE.

For electronmicroscopy, we have found initial fixation in 2 percent glutaraldehyde and postfixation in 1–2 percent osmium tetroxide to be adequate, followed by routine embedment in an epoxy resin.

ORGAN CULTURE OF RPE AND CHOROID. Eyes are prepared as in the previous section but with special care to preserve the normal relationship of RPE to choroid. The RPE and uvea are separated from the sclera, placed on a Millipore filter, and immersed in the culture medium for the duration of the experiment. The choroidal surface should appose the Millipore filter. The cultures are maintained in Leighton tubes filled with 2 ml of medium 1640 at 37° C, which is changed twice a week. The organ cultures are studied directly with an inverted microscope. Photographic documentation of their appearance is desirable. Thorotrast (0.1 ml per 2 ml culture medium) can be added to the cultures in order to study phagocytosis.

EXPLORATION OF NEOPLASTIC POTENTIAL OF RPE. Simian vacuolating virus 40 (SV40) strain VAC 777L can be used to infect RPE explants from normal hamster eyes. Cell suspensions of affected cultures are then injected subcutaneously into young hamsters (preferably after compromising their immune system by total body x-radiation). In our studies tissues were fixed in Zenker's formaldehyde solution and sections stained with HE.

Observations of Culture Systems

RPE GROWTH PATTERNS IN CULTURE. In the early stages the cells in the center of the explant remained hexagonal and mononucleate and contained cigar-shaped pigment granules, closely resembling the appearance of the original explants (Fig. 16.1). Migration of explants of RPE occurred 72 hours after the cultures were established. Early migratory cells were usually heavily pigmented and round, with single nuclei.

Mosaic pattern. As growth continued, a compact mosaic of hexagonal- to oval-shaped pigmented cells with large oval nuclei was the most common pattern

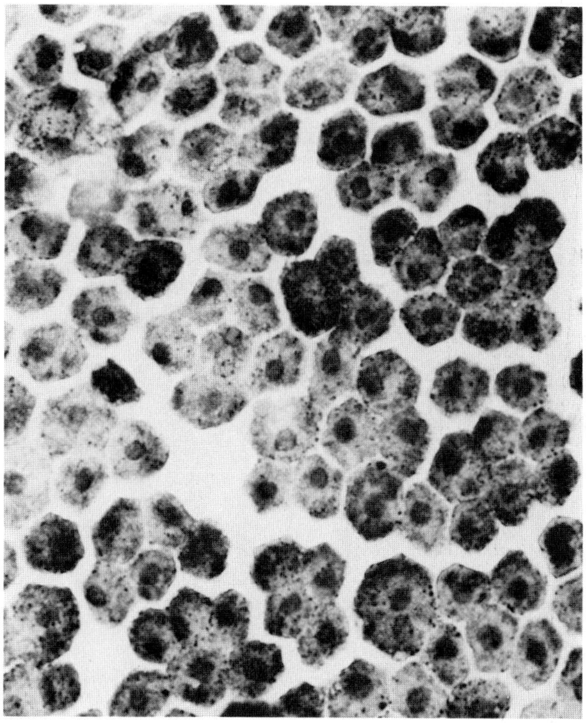

Figure 16.1 *RPE explant prior to placement in tissue-culture flask. Cells are hexagonal and heavily pigmented. (Stained with H&E. ×476.)*

seen (Fig. 16.2). The cells varied in size and in amount of pigment granules. Occasional spindle and dendritic cells were seen in the periphery of these colonies (Fig. 16.3). Long, slender cells were noted that gave the impression that one portion of the cell

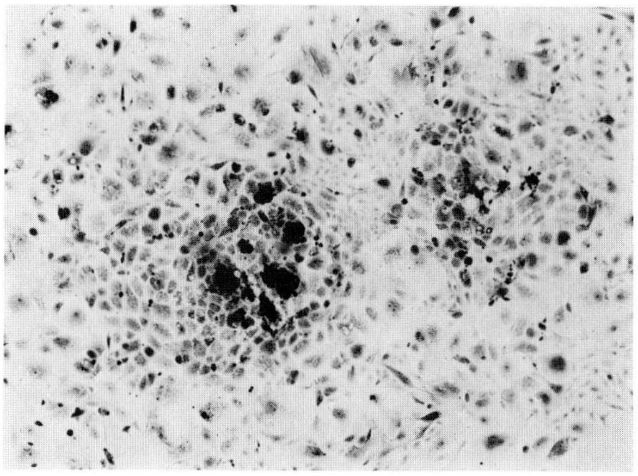

Figure 16.2 *Growth from RPE explant one week after placement in tissue-culture medium. Note epithelioid nature of cells, mitotic figures and variation in size and amount of pigmentation. (Stained with H&E. ×35.)*

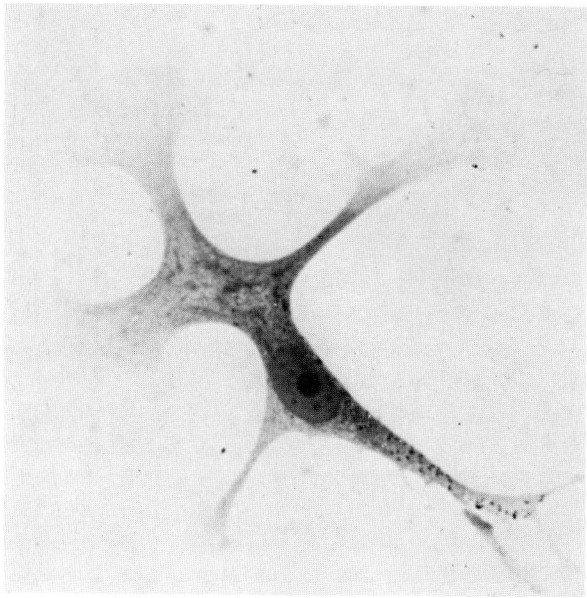

Figure 16.3 *Spindle-shaped cell with dendritic processes. (Stained with H&E. ×476.)*

was fixed to the substrate while the remainder of the cell attempted to migrate (Fig. 16.4).

Tubular pattern. Rosettes of cuboidal or epithelioid cells arranged around a circular lumen were seen most commonly in areas of low cell density. The cells comprising the tubular pattern were sparsely pigmented, and the large oval nuclei were centrally located (Fig. 16.5).

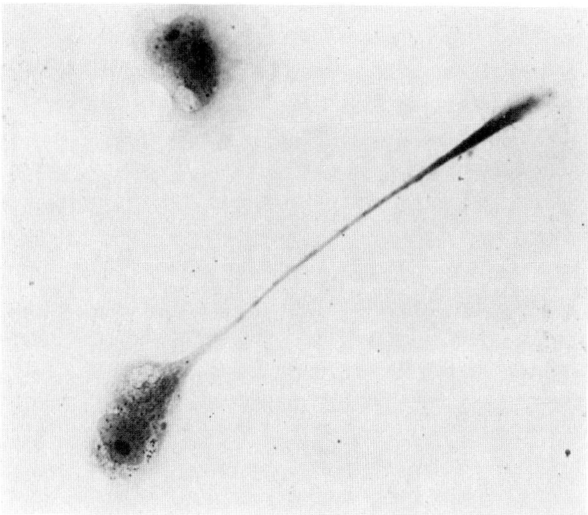

Figure 16.4 *Field showing slender cell with one end apparently fixed to substrate; remainder of cell extends away from perikaryon. (Stained with H&E. ×344.)*

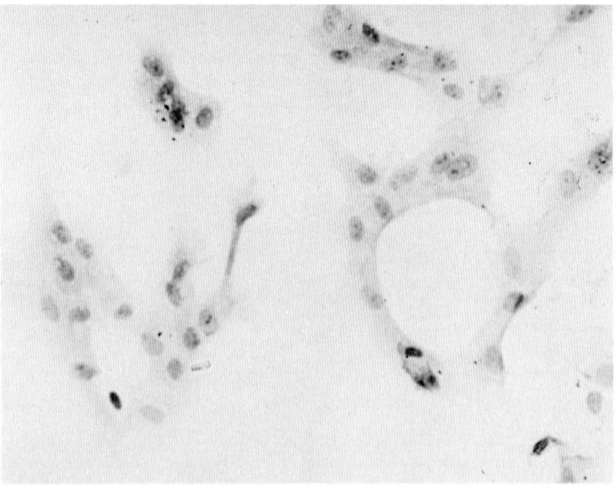

Figure 16.5 *Cuboidal or columnar shaped RPE cells in rosette-like configuration. Note sparsity of pigmentation. (Stained with H&E. ×225.)*

Spindle pattern. Spindle-shaped cells with small, dark, elongated nuclei were arranged in a loosely packed, whorl-like configuration (Fig. 16.6). These cells were mostly devoid of pigment granules. Occasional epithelioid cells were scattered in these areas.

No true anaplasia of the RPE was found in these cultures, nor was a counterpart for the papillary or vacuolated pattern that had been observed in RPE tumors in vivo (Tso and Albert 1972) seen in vitro. Following trypsinization of the original cultures, the predominant pattern of growth was the mosaic arrangement, but all of the patterns seen in the original explant cultures could be identified.

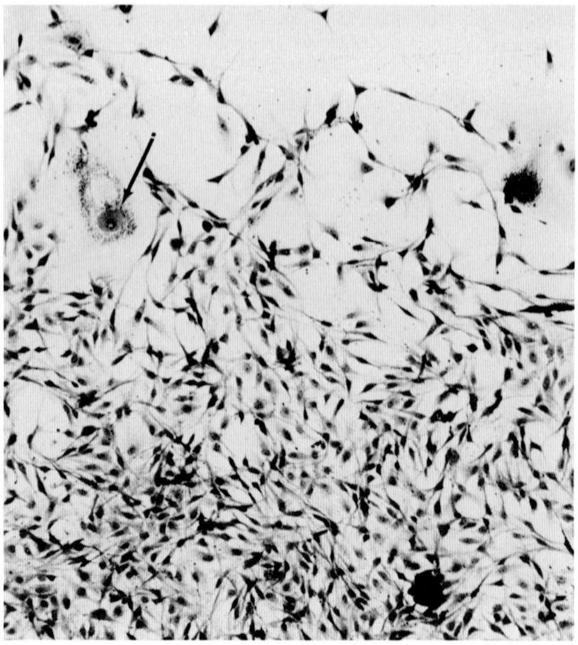

Figure 16.6 *RPE culture showing spindle growth pattern. Note occasional epithelioid cell (arrow). (Stained with H&E. ×70.)*

Histologic features of cells in culture. Multinucleate cells were increasingly common as the duration of culture time increased, with as many as six or eight nuclei being observed. "Giant" cells several times the size of most cells were seen individually and in clusters (Fig. 16.7). Consistent with the rapid rate of growth, mitotic figures were commonly seen. Cells in mitosis had few pigment granules.

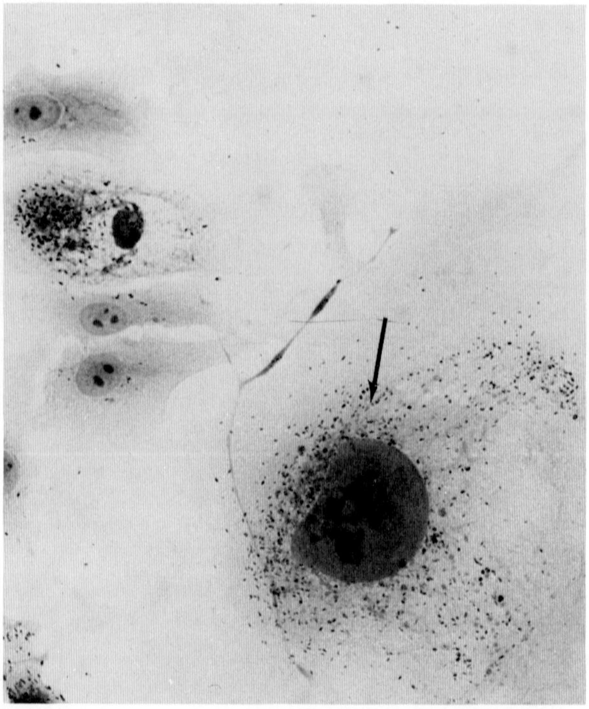

Figure 16.7 *Giant cell with single nucleus. (Stained with H&E. ×476.)*

CHOROID. Explants of choroid grown in a manner identical to that of the RPE showed migration of cells within 72 hours (Fig. 16.8). The outgrowth contained predominantly fibroblastic-appearing cells with little or no pigment, mixed with cells that were heavily pigmented. The heavily pigmented cells were dendritic in shape, some with plump, blunted processes; others contained slender delicate processes (Fig. 16.9). With subsequent sub-

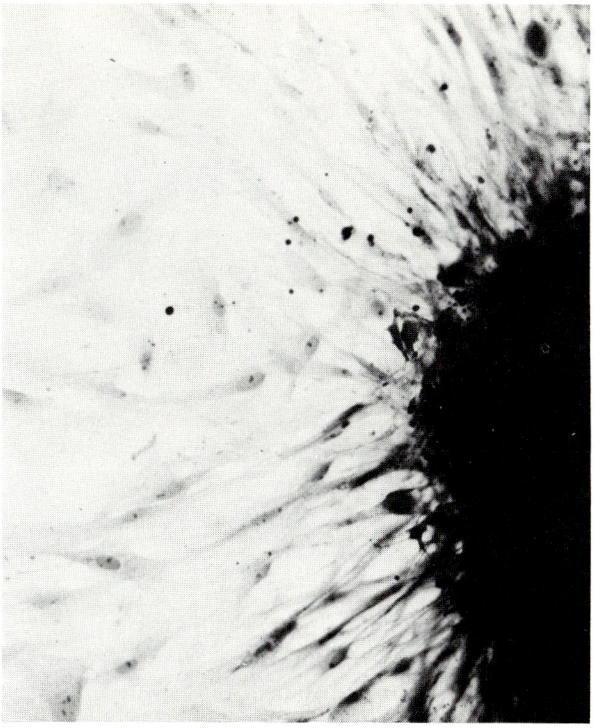

Figure 16.8 *Migration of cells from choroidal explant. Fibroblast-like cells predominate. (Stained with H&E. ×190.)*

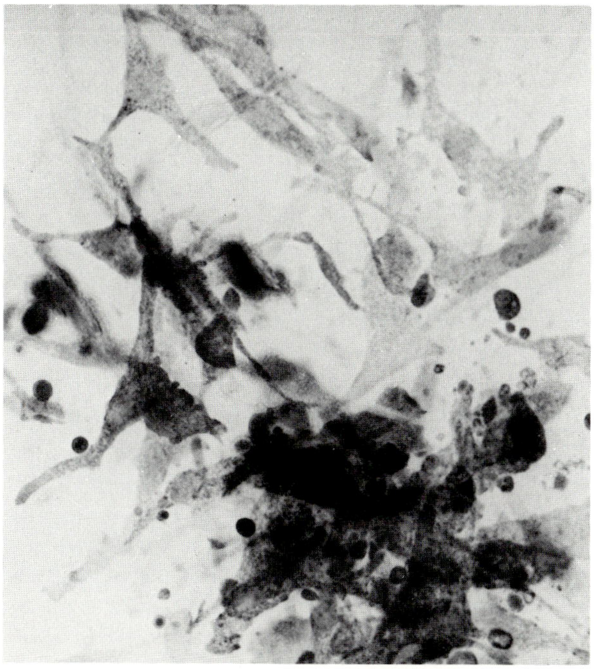

Figure 16.9 *Cell morphology of choroidal explant one week after culture. (Stained with H&E. ×400.)*

culture the fibroblastic cells predominated and the dendritic cells became increasingly infrequent. These dendritic cells did not seem to proliferate and generally appeared isolated. After approximately five passages (generally the twelfth week) these cells could no longer be found.

RPE COMBINED WITH CHOROID. During the first three days in culture the RPE retained its epithelial characteristics, mosaic pattern (Fig. 16.10),

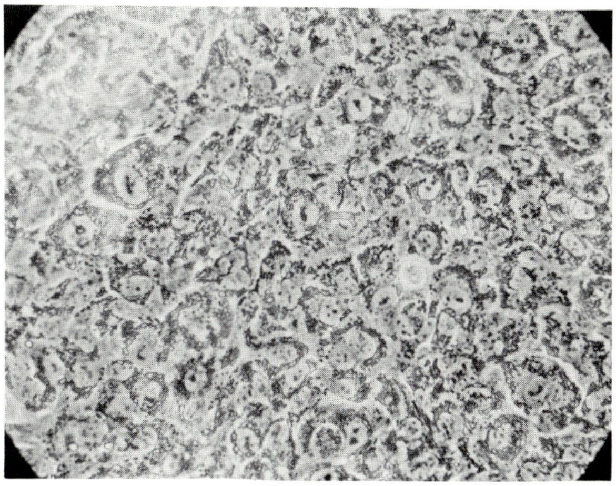

Figure 16.10 *Preparation of RPE and choroid after three days culture, as viewed through inverted microscope. The RPE cells show normal mosaic pattern and regular pigmentation. (AFIP Neg. 72–10803–1. ×378.)*

and relatively uniform pigmentation. The melanin granules remained aggregated in the apical region of the cells, as they are in vivo (Fig. 16.11). Partially depigmented cells were occasionally noted. Initially, Bruch's membrane was flat, but soon it became wrinkled (Fig. 16.11*B*). The choriocapillaris and the uveal stroma appeared unremarkable. Tissues obtained postmortem did not appear to differ from those obtained surgically in either morphologic or growth characteristics. However, different cultures taken from the same patient's eye and handled in similar fashion did show considerable variation in degenerative or proliferative changes. The age of the patients did not seem to affect the culture.

After the first three days the RPE became more irregular. Some of the cells became enlarged, rounded, and columnar. More hypopigmented cells were observed. As Bruch's membrane became wrinkled, it appeared that cells in the RPE were crowded together. No spreading of the RPE over Millipore

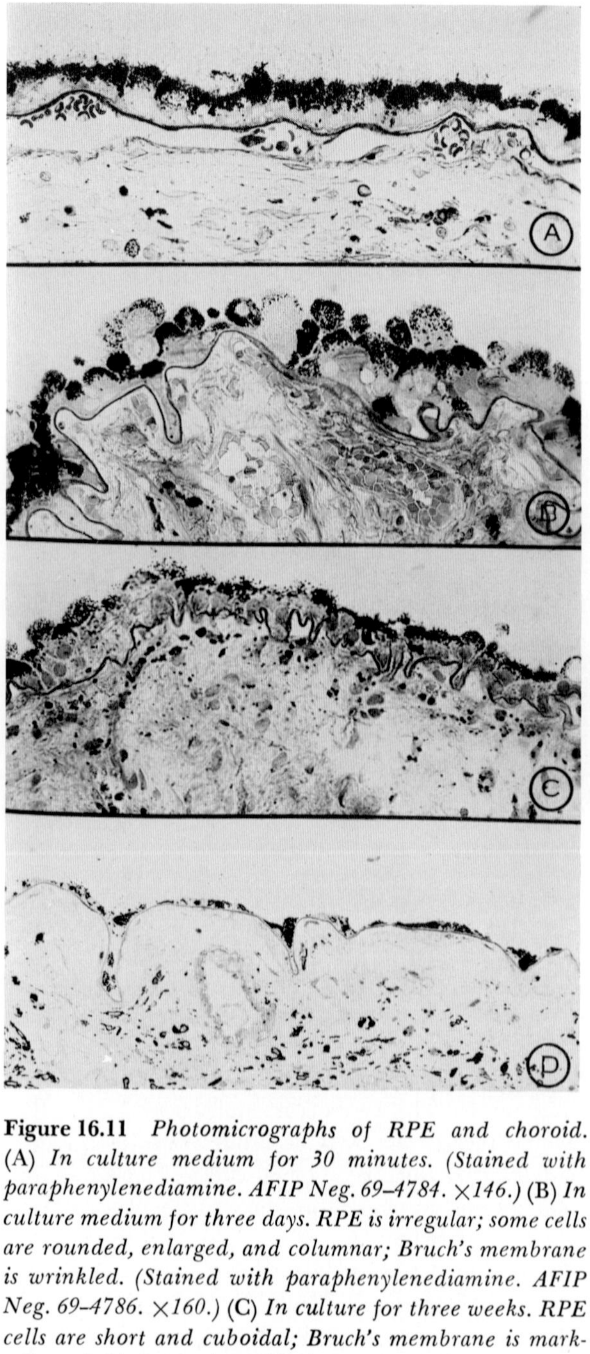

Figure 16.11 *Photomicrographs of RPE and choroid.* (A) *In culture medium for 30 minutes. (Stained with paraphenylenediamine. AFIP Neg. 69–4784. ×146.)* (B) *In culture medium for three days. RPE is irregular; some cells are rounded, enlarged, and columnar; Bruch's membrane is wrinkled. (Stained with paraphenylenediamine. AFIP Neg. 69–4786. ×160.)* (C) *In culture for three weeks. RPE cells are short and cuboidal; Bruch's membrane is markedly wrinkled; choriocapillaris is not distinct; choroidal stroma shows moderate edema. (Stained with paraphenylenediamine. AFIP Neg. 69–4791. ×160.)* (D) *In culture for six weeks. RPE is attenuated and sparsely pigmented; choriocapillaris is absent. (Stained with toluidine blue. AFIP Neg. 70–727. ×131.)*

filters was observed. Definite proliferation of the RPE, forming pigmented nodules, was occasionally observed after 10 days (Fig. 16.12).

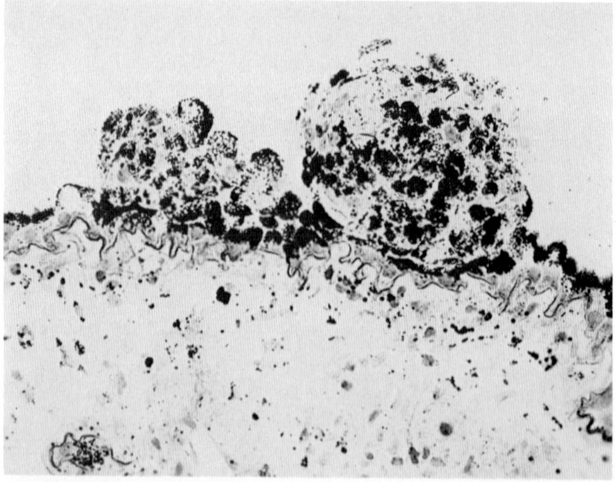

Figure 16.12 *RPE and choroid after three weeks in culture. The RPE has proliferated to form nodules. (Stained with paraphenylenediamine. AFIP Neg. 72–10803–3. ×214.)*

After being in culture for two to three weeks, the RPE flattened and appeared cuboidal (Fig. 16.11*C*). The choriocapillaris was not distinct, and the choroidal stroma showed moderate edema. After six weeks in culture, the RPE was markedly attenuated, and, in some areas, necrotic (Fig. 16.11*D*); it was sparsely pigmented, and the nuclei were flattened. The choriocapillaris at this stage could not be identified.

Electron-microscopic observations. In this organ-culture preparation the apical villi of the RPE were long and sometimes vesicular (Figs. 16.13 and 14). The excellent preservation of the villi was a constant feature even after six weeks in culture. Desmosomes and terminal bars consisting of zonulae adherentes and occludentes joined adjacent epithelial cells (Fig. 16.15). Initially, the typical infoldings of the basal plasma membranes were also preserved (Fig. 16.13), but after the first week, the basal plasma membrane of the retinal pigment epithelium lost its typical infolding and became smooth, lying in close association with the basement membrane (Fig. 16.16).

In the cytoplasm of the RPE, after six days to two weeks in culture, there was an abundance of smooth- and rough-surfaced endoplasmic reticulum, suggesting that these cells were metabolically active (Fig. 16.13). Initially, the cells were heavily pigmented with mature melanin and lipofuscin granules. Laminated bodies were frequently observed as early as three days after culture in the cytoplasm as well as

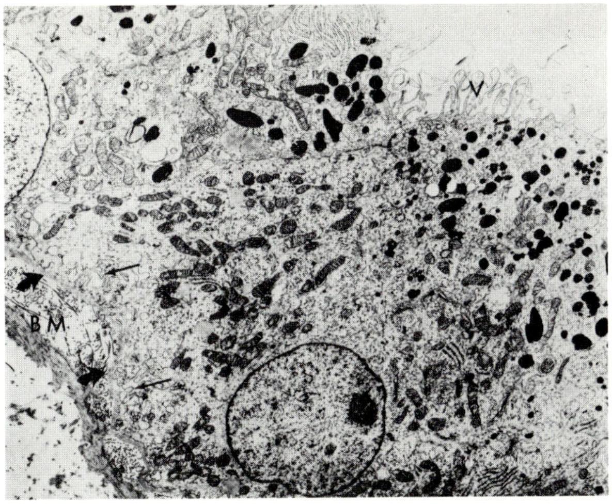

Figure 16.13 *Electron micrograph of RPE in culture for six days. Some apical villi* (V) *appear vesicular. There is abundant smooth- and rough-surfaced endoplasmic reticulum and the pigment granules, mitochondria, and nuclei appear intact. Note presence of basal infolding (thin arrows) of RPE overlying basement membrane (heavy arrows).* BM = *Bruch's membrane. (AFIP Neg. 72–10803–3. ×2,450.)*

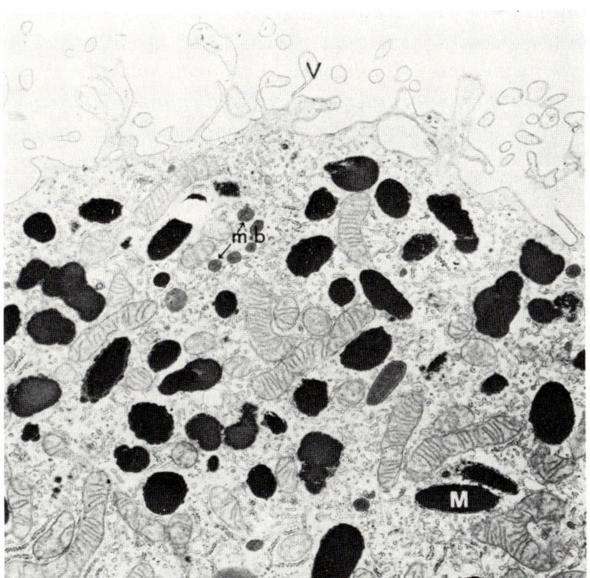

Figure 16.14 *Electron micrograph of RPE in culture for three days. Note vesicular villi* (V), *abundant smooth- and rough-surfaced endoplasmic reticulum, melanin granule* (M), *lipofuscin granule* (L), *and membrane-bound small dense bodies* (mb) *(AFIP Neg. 72–1080–4. ×14,200.)*

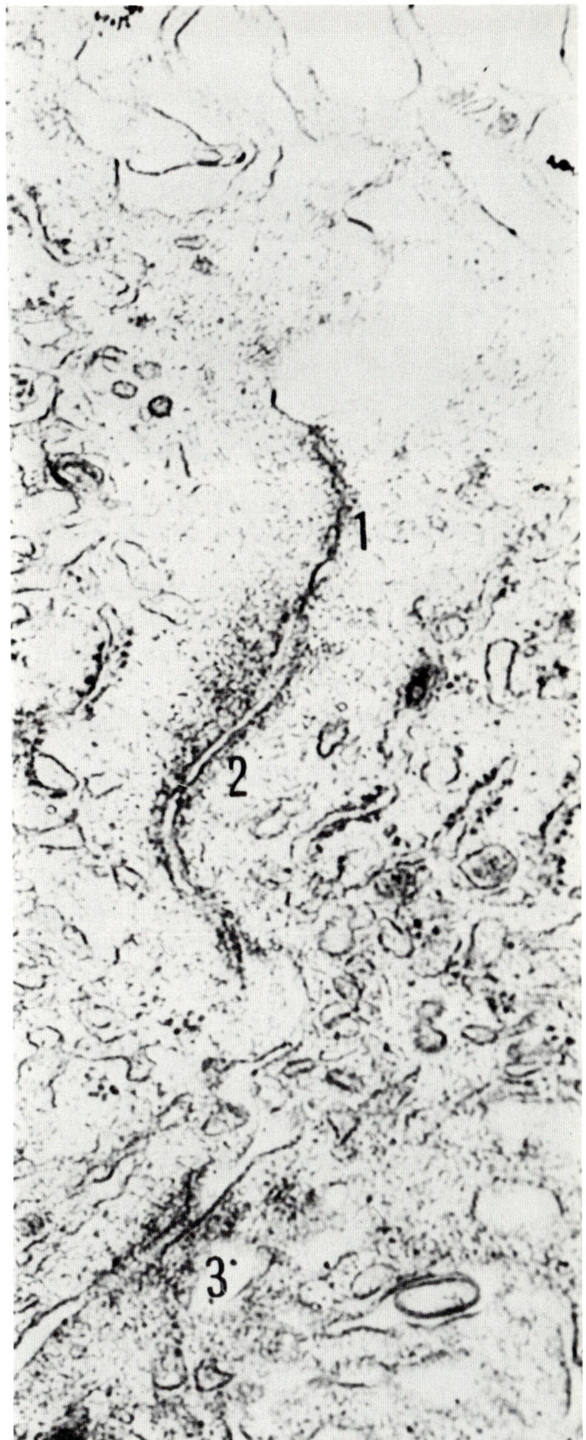

Figure 16.15 *Electron micrograph of RPE in culture for three days. Note intact junctional complex: zonula occludens* (1), *zonula adherens* (2), *desmosomes* (3). *(AFIP Neg. 72–10803–4. ×100,650.)*

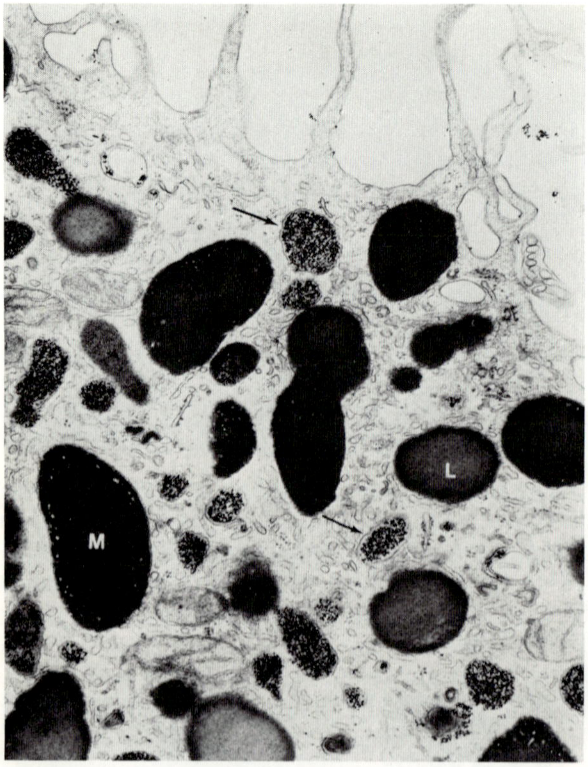

Figure 16.22 *Electron micrograph of apical region of RPE and choroid in culture for three days. Thorotrast added to medium after one day of culture. Membrane-bound dense bodies contain Thorotrast (arrows).* M = *melanin granule;* L = *lipofuscin granule. (AFIP Neg. 72–10803–8. ×21,070.)*

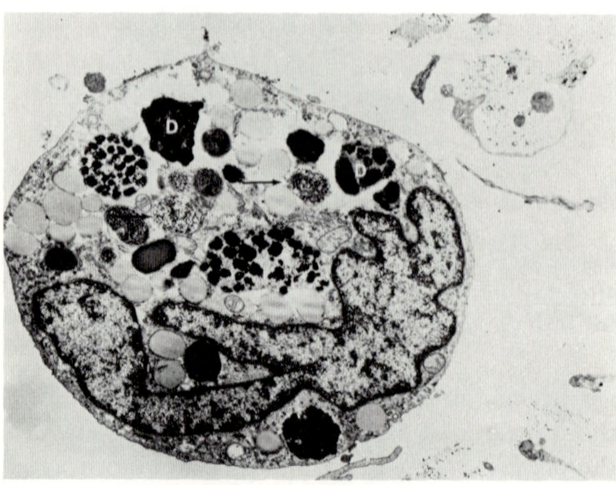

Figure 16.23 *Electron micrograph of macrophage in choroid of cultured RPE and choroid. Note various kinds of cellular debris* (D) *and Thorotrast particles (arrows). (AFIP Neg. 72–10803–8. ×5,770.)*

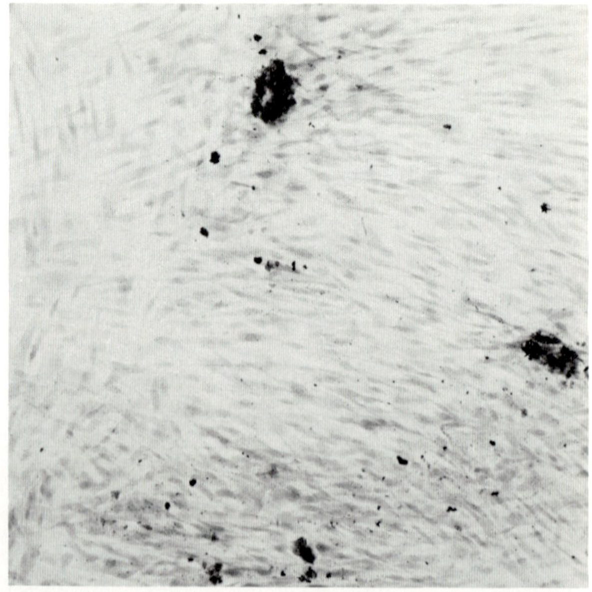

Figure 16.24 *Transformed culture of hamster RPE ten days after infection with SV 40. (Stained with H&E. ×40.)*

weeks after injection of the transformed cells, the hamsters developed palpable neoplasms at the injection sites. Animals that were not killed died within eight weeks after injection, as a result of tumor invasion and metastasis to various sites, including lung, liver, and brain.

On gross examination the tumors were firm and white and most had small areas of central necrosis. Histologically, the tumors consisted predominantly of spindle end epithelioid cells with either round or elongated vesicular nuclei, clumped chromatin, and lightly staining, eosinophilic cytoplasm (Fig. 16.25). Mitotic figures were numerous, and multinucleate giant cells were observed in moderate numbers (Fig. 16.26).

Some tumors contained areas of tubular and glandular structures resembling adenocarcinomas, with various degrees of differentiation (Figs. 16.27–16.30). Some of these were filled with a dense proteinaceous material (Fig. 16.31). No areas of pigment-containing cells were found within the tumors on light-microscopic examination of dopa-stained sections.

Portions of the initial tumors were injected into other hamsters, and transferable tumor lines were easily established. Neoplasms were carried through six generations without alteration in the morphologic features or spontaneous regression of the tumors.

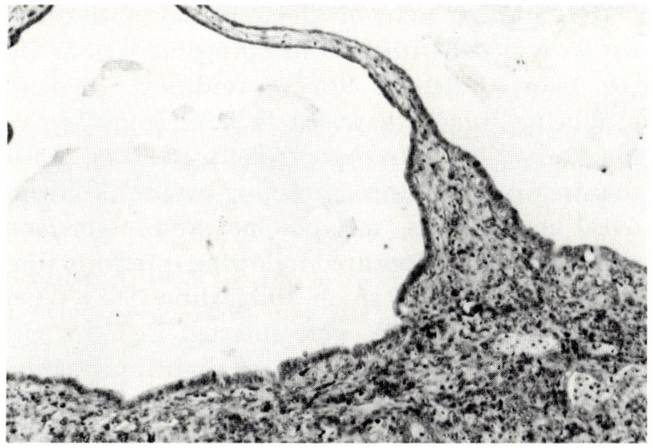

Figure 16.31 *Another tumor showing papillary structure extending into cavity filled with proteinaceous material. (Stained with H&E. ×86.)*

"Authors' Observations" section of this chapter (Barr-Nea and Barishak 1972/73). In their studies, the RPE maintained its pigment while in contact with Bruch's membrane. They also showed transformation of the RPE into stratified squamous epithelium with keratin formation. A similar reaction has not been seen in the human cultures and may represent species variation or pathologic transformation.

The Millipore filter provided distinct advantages in the human culture system. It provided support for the tissues and maintained them in their normal orientation. The excellent preservation of the apical villi and the distribution of the pigment granules along the apical zone of the cells may be evidence of the good mechanical support provided by the Millipore. This was in striking contrast to the observations made by the more prevalent method of obtaining cultured tissue for electron microscopy by making a pellet of trypsinized cells. Examination of such preparations revealed all villi to have disappeared. The Millipore filter was permeable to the culture medium, which could reach the tissue from all sides. The Millipore filter did not provide a surface favorable for spread of cells, as occurs on plastic or glass culture surfaces. Also, from a technical standpoint, the Millipore filter was easily processed for light and electron microscopy and became transparent when put into acetone or propylene oxide. A similar system was employed for culturing other cell types, such as fibroblasts and lymphoblasts, and for harvesting of various viruses, and it showed similar distinct advantages (McCombo, Benyesh-Melnick, and Brunschwig 1968). Synthetic

substrates of other com
similar fashion and m
growth and to be easie
copy than Millipore is.
In the RPE-choroid
RPE in vitro closely re
least during the first tw
villi, terminal bars, a
plasma membranes, as w
branes and melanin an
seen.
The RPE lost infoldi
brane as the culture ag
the basal plasma membr
absorption (Bernstein 19
logically that the disapp
related to the fact that
longer comes via the c
lated bodies in the cyto
thelium are indicative o
tissue culture and differ
photoreceptor cells phag
observed in vivo (Hoga
and Weddell 1971; Spitzn
1967; Young 1971; Young
tive nature of these struct
After two weeks the cells
other cells showed mild
nucleate cells and amelan
dant. After six weeks of c
degenerative changes.
During the period of ti
bled in vivo conditions,
functionally active, inasn
capable of phagocytizing
culture for one to three d
of our organ culture, the
show evidence of phagocyt
An essential morpholog
ognition of phagocytic ac
copy is the presence of a
the phagocytized materia
that are not enveloped by
not be identified as havi
the cell. The Thorotrast t
some of our cultures was,
within membrane-bound in
(phagosomes), clearly sho
tential of the RPE. All Th
cytoplasm were bound by n
It has been shown in re

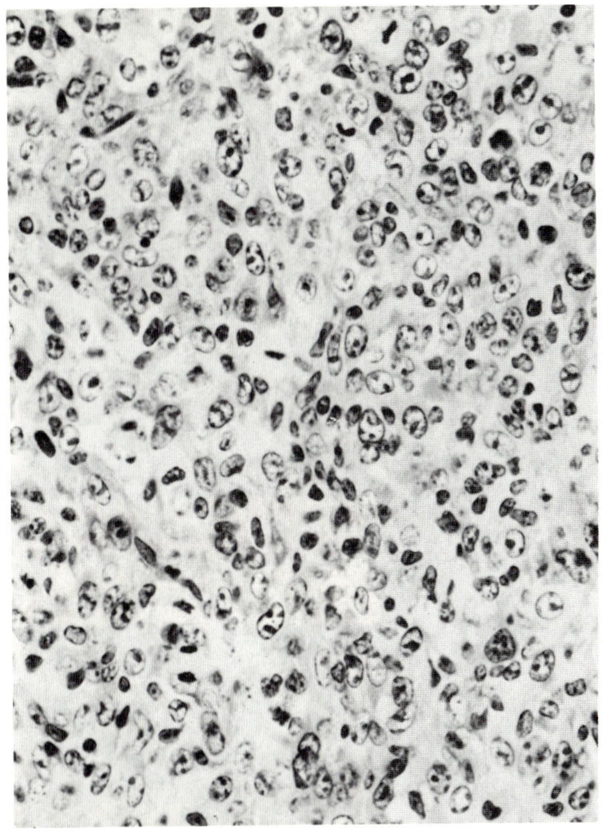

Figure 16.25 *Undifferentiated area of tumor arising in hamster three weeks after injection of transformed RPE cells. Note epithelioid and spindle cell types. (Stained with H&E. ×476.)*

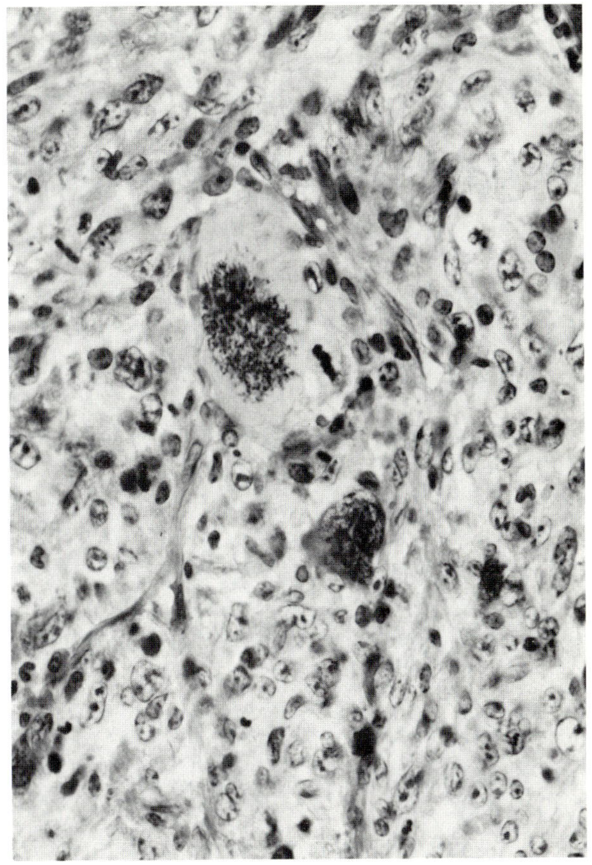

Figure 16.26 *Giant cell in hamster tumor arising three weeks after subcutaneous injection of transformed RPE cells. (Stained with H&E. ×476.)*

Commentary

RPE GROWTH PATTERNS IN CULTURE. In studies done by other workers, the RPE was minced into minute explants or the cells were chemically or physically separated from Bruch's membrane, as in the study reported in the "Authors' Observations" section above (Attardi and Moro 1959; Barishak 1960; Mannagh, Arya, and Irvine 1973; Pomerat and Littlejohn 1956; Reese and Ehrlich 1958). Cells in such cultures also proliferated, their morphologic characteristics became altered, and pigment was gradually lost. Within a short period of time they no longer resembled the original polyhedral cells of the RPE. Autoradiography using L-tyrosine-2-^{14}C indicated that tyrosinase activity was present even in subcultures that appeared not to be pigmented (Mannagh, Arya, and Irvine 1973). The same group of workers also showed a positive dopa reaction in less pigmented or colorless melanin granules in subcultured cells.

RPE cultured from adult pig showed similar morphologic changes to that from humans (François et al. 1971). Pigmentation was progressively lost with subculturing, but "colorless granules" were still produced.

RPE COMBINED WITH CHOROID (ORGAN CULTURE). These experiments have shown that preparations of human RPE and choroid may be maintained as organ cultures for various morphologic studies. Both material obtained postmortem and surgical specimens are suitable. The tissue maintained in culture appears to persist in a relatively active state, without overt proliferation or degeneration, for at least two weeks.

Using chick-embryo RPE organ cultures, others have shown similar results to those described in the

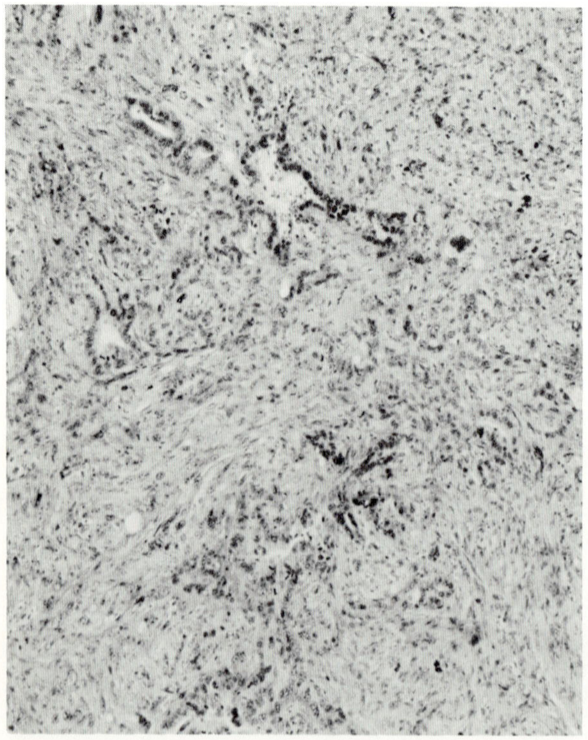

Figure 16.27 *Glandlike proliferation in another area of tumor pictured in Figure 16.26. (Stained with H&E. ×119.)*

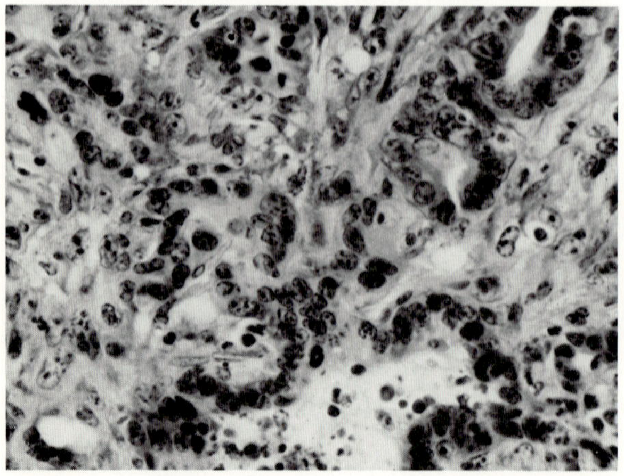

Figure 16.28 *Higher magnification of portion of Figure 16.27. (Stained with H&E. ×312.)*

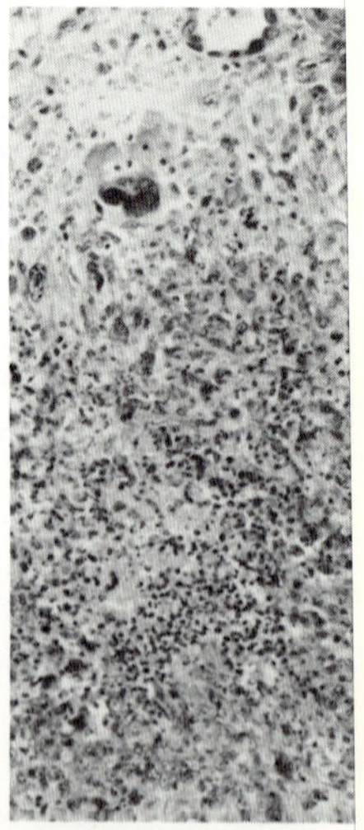

Figure 16.29 *Tubul injected with transfo ×119.)*

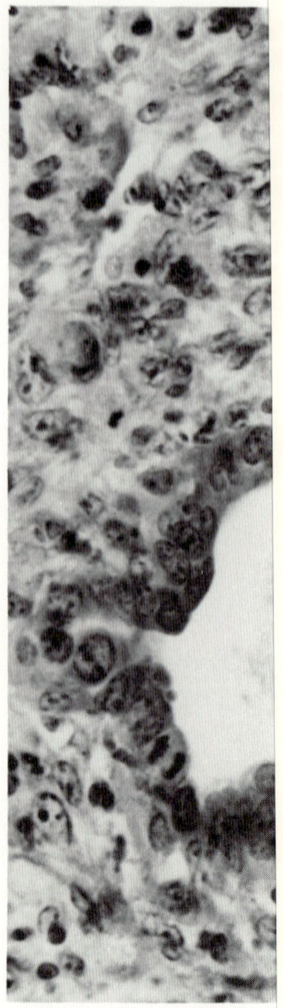

Figure 16.30 *Higher 16.29. Note cuboidal e of differentiation. (Sta*

MANNAGH, J., ARYA, D. V., and IRVINE, A. R., JR. 1973. Tissue culture of human retinal pigment epithelium. *Invest. Ophthalmol.* 12:52–64.

MCCOMBO, R. M., BENYESH-MELNICK, M., and BRUNSCHWIG, M. P. 1968. The use of millipore filters in ultrastructural studies of cell cultures and viruses. *J. Cell Biol.* 36:231.

MCCULLOUGH, B., SCHULLER, J., SHADDUCK, J. A., and YOHNI, D. S. 1972. Induction of malignant melanomas associated with fibrosarcomas in gnotobiotic cats inoculated with Gardner feline fibrosarcoma virus. *J. Nat. Canc. Inst.* 48:1893.

MEYER, H. 1936. Zuchtung der retina des hunnes in vitro. *Z. Mikrosk Anat. Forsch.* 39:151–160.

MONTALCINI, R. L., and SACERDOTE, E. 1938. I caratteri e le modalita di sveiluppo delle fibre collagene e reticolare nelle colture in vitro. *Arch. Ital. Anat. Embriol.* 40:37–121.

POMERAT, C. M., and LITTLEJOHN, L., JR. 1956. Observations on tissue culture of the human eye. *South. Med. J.* 49:230–237.

REESE, A. B., and EHRLICH, G. 1958. The culture of uveal melanomas. *Am. J. Ophthalmol.* 46:163–174.

SMITH, D. T. 1920*a*. Melanin pigment in the pigmented epithelium of the retina of the embryo chick's eyes: studies in vivo and in vitro. *Anat. Rec.* 18:260–261.

SMITH, D. T. 1920*b*. The pigmented epithelium of the embryo chick's eye studied in vivo and in vitro. *Bull. Johns Hopkins Hosp.* 31:239–246.

SMITH, D. T. 1921. The ingestion of melanin pigment granules by tissue cultures. *Bull. Johns Hopkins Hosp.* 32:240–244.

SPITZNAS, M., and HOGAN, M. J. 1970. Outer segments of photoreceptors and the retinal pigment epithelium: interrelationship in the human eye. *Arch. Ophthalmol.* 84:810.

TORÖ, E. 1931. Uber die linsenfaserbildende fahigheit der irizellen bei hubnern und ratten auf grund von implantationversuchen von cornea und iriskulturen ins linsenlose. *Auge. Verh. Anat. Ges.* 40:248.

TSO, M. O. M., and ALBERT, D. M. 1972. Pathological condition of the retinal pigment epithelium: neoplasms and nodular nonneoplastic lesions. *Arch. Ophthalmol.* 88:27–38.

TSO, M. O. M., ALBERT, D., and ZIMMERMAN, L. E. 1973. Organ culture of human retinal pigment epithelium and choroid: a model for the study of cytologic behavior of RPE in vitro. *Invest. Ophthalmol.* 12:554–566.

WALLOW, I. H. L., and TSO, M. O. M. 1972. Proliferation of the retinal pigment epithelium over malignant choroidal tumors: a light and electron microscopic study. *Am. J. Ophthalmol.* 73:914–926.

WHITTAKER, J. R. 1967. Loss of melanotic phenotype in vitro by differentiated retinal pigment cells: demonstration of mechanisms involved. *Dev. Biol.* 15:553–574.

WILBUR, K. M., and CHAMBERS, R. 1942. Cell movements in the healing of microwounds in vitro. *J. Exp. Zool.* 91:287–302.

YOUNG, R. W. 1967. The renewal of photoreceptor cell outer segments. *J. Cell Biol.* 33:61–72.

YOUNG, R. W. 1971. Shedding of discs from rod outer segments in the rhesus monkey. *J. Ultrastruct. Res.* 34: 190–203.

YOUNG, R. W., and BOK, D. 1969. Participation of the retinal pigment epithelium in the rod outer segment renewal process. *J. Cell Biol.* 42:392–403.

17 Photic and Photo-Thermal Effects on the Retinal Pigment Epithelium

Toichiro Kuwabara

ANATOMICAL and functional interactions between the retinal pigment epithelium (RPE) and the photoreceptor cells have been elucidated considerably in recent years by Yamada (1958), Bernstein (1961), Dowling and Gibbons (1962), Bairati and Orzalesi (1963), Hogan et al. (1971), Fine (1961), and Fine and Yanoff (1972). It has become clear that the lamellar membranes of the outer segment are phagocytized by the RPE cells in the course of the normal turnover process (Droz 1963; Young 1967; Young and Droz 1968; Young and Bok 1970; Young 1971). Although the cytologic structure suggesting photosensitivity has not been demonstrated within the RPE cells of mammals, the cells in certain animals appear to be electrophysiologically sensitive to light (Brown 1965). The turnover process of the outer segment appears to be enhanced in certain pathologic conditions, especially mild light exposure. Also, the RPE shows varying degrees of cytologic changes upon stronger or prolonged light stimulations (Noell et al. 1966; Kuwabara and Gorn 1968; Friedman and Kuwabara 1968; Grignolo et al. 1969; Kuwabara and Aiello 1968; Bresnick 1970; O'Steen and Anderson 1971; Tso et al. 1972). Pathologic changes in the photoreceptor cells always accompany damage to the RPE. Thermal energy, especially of xenon light and laser beam, is absorbed primarily by the melanin granules of the pigment epithelium; the severe cellular changes in the pigment epithelium and in the retina are the result of this absorption. The RPE, sensitive to damage, is also active in repair and many of the cytologic changes reflect the repair process. Especially noteworthy is the capacity of the cells to slide over the defective area to proliferate. It has been shown that the regeneration of the pigment epithelium always precedes retinal repair (Kuwabara 1970).

This chapter will describe cytologic findings of the pigment epithelium following light exposure under various conditions. Although the materials described here have been obtained mainly by animal experiments in rats and monkeys, similar changes are regularly found in photocoagulated human eyes. Also, several cytologic changes that are commonly observed in the adult human cell are strikingly similar to those experimentally induced by light exposure. The human retina and RPE may be accidentally or physiologically receiving damaging light exposure as a part of the normal function of the eye. This consideration will be emphasized here.

Our description of animal experiments on light damage will be divided into two categories: photic and thermal damages. Photic damage to the retina and the pigment epithelium, for example, may be produced in the albino rat by exposures of varying durations to diffuse fluorescent light with a brightness of 500–750 foot-candles (Kuwabara and Gorn 1968). Nonthermal damage has been produced in the monkey retina by attaching a stroboflash to the lamphouse of a direct ophthalmoscope (Kuwabara and Okisaka 1976). In these experimental conditions, no significant elevation of temperature in the retina was demonstrated, either by actual measurements or by calculations.

Severe focal thermal damage can be produced in monkeys by exposures to incandescent lamps, xenon arcs, and lasers (Friedman and Kuwabara 1968; Okisaka et al. 1975). In addition, severe diffuse damage of the whole posterior retina may be produced in the albino rat by placing sunlamps around the animal's cage.

Normal Pigment Epithelium

A detailed description of the fine structure of the normal pigment epithelium will be found in Chapter 5. Only fine structures which are specifically related to light damage are emphasized here.

The general structure of the RPE cell is alike in both the rat and monkey. A typical RPE cell is presented diagrammatically in Figure 17.1. The cytoplasm consists mainly of dense, fine smooth endoplasmic reticulum. The cisternae anastomose frequently, and their lumens contain slightly electron-dense substance. Golgi apparatus are present in the apical and parakaryon zones. Rough endoplasmic reticulum is seen sporadically among the smooth endoplasmic reticulum and more numerously in the apical zone, where it forms a lamellar arrangement. Large mitochondria are distributed in the lateral and basal portions. Small primary and larger secondary lysosomes are abundant in the parakaryon zone. Electron-dense inclusion bodies of various sizes are also present in the normal cell. Phagosomes, consisting of fragments of the outer segments and their degrading substances, are seen in the microvilli and in the cytoplasm. The RPE cell of the monkey contains abundant melanin granules, which are mainly spindle shaped, in the apical portion and in the microvilli. In general, melanin granules consist of homogeneous electron-dense substance, but some large round granules have a shredded appearance. Structure suggesting the melanogenetic process is absent in the RPE of adult animals. RPE of albino animals contains nonmelanotic melanosomes only during the developing stage. The microvilli extend inward and surround the photoreceptor outer segments. The striking density and uniformity of the thin microvilli are most clearly demonstrated by scanning electron microscopy (Fig. 17.2).

Although no junctional apparatus is present, attachments between the microvilli and the outer segments appear to be extremely firm. (Sakuragawa and Kuwabara 1976). Separation of the retina from the RPE, even with the gentlest technique, results in breakage of the microvilli and tips of the outer segments. The basal cell membranes are moderately infolded. The microorganelles are absent in the thin microvilli and in the infolded portion of the basal cytoplasm. Beneath the RPE is Bruch's membrane, which consists of the basement membrane of the RPE cell, collagen fibers, elastic fibers, and the basement membrane of the choroidal capillary. The

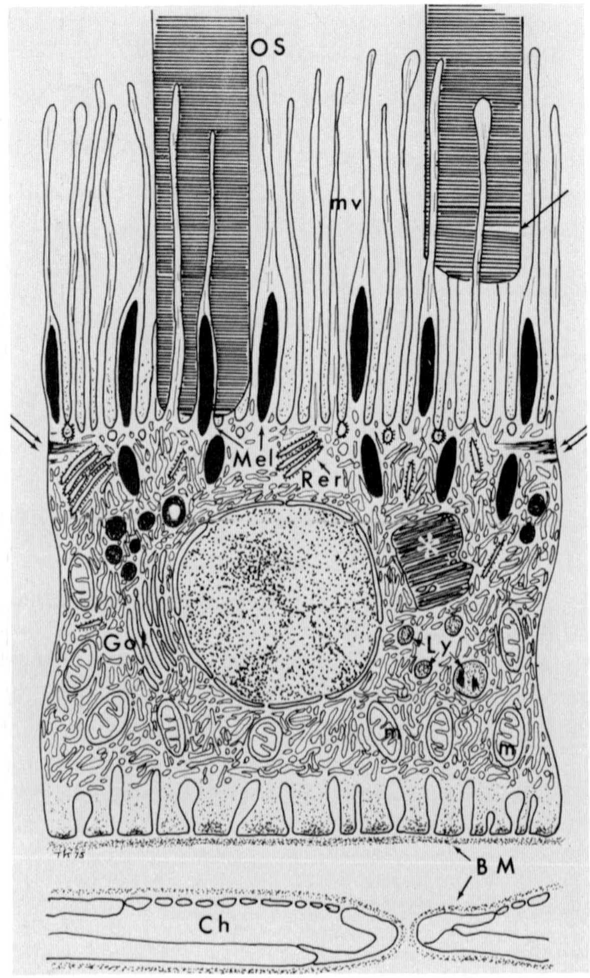

Figure 17.1 *Diagram of a normal pigment-epithelial cell in the posterior part of the monkey eye. Numerous microvilli* (mv) *extend from the apical end and surround the tips of the outer segments* (OS). *The basal cell membrane is moderately infolded and loosely attached to the base. The basement membrane* (BM) *is straight. Cells are joined firmly by apicolateral junctions (double arrows), but the lateral attachment of the rest of the cell is relatively loose. The cytoplasm is packed with smooth endoplasmic reticulum. Other microorganelles are: mitochondria* (m) *in the basal and lateral portions; melanin granules* (Mel) *in the microvilli and in the apical portion; rough endoplasmic reticulum* (Rer) *and small vesicles in the apical area; lysosomes* (Ly), *inclusion bodies, and Golgi apparatus* (Gol) *are mainly in the parakaryon zone. The tips of the outer segments are broken off (arrow) and phagocytized by the RPE cell (white asterisk). The choroidal endothelium* (Ch) *is fenestrated. The basement membrane of the choroid forms the outer limit of Bruch's membrane.*

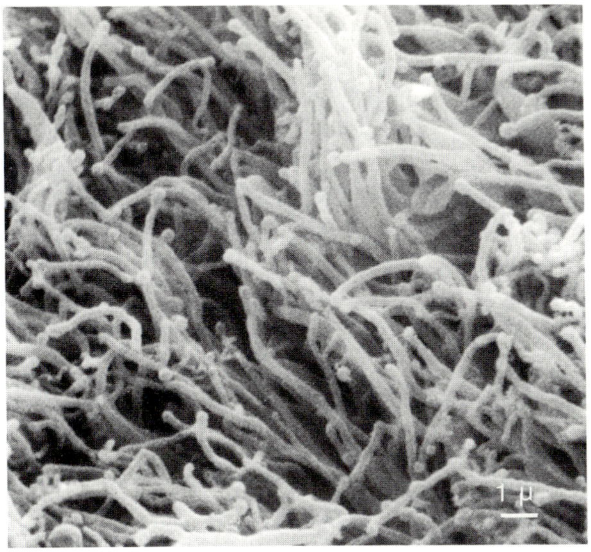

Figure 17.2 *Scanning electron-microscopic view of the apical surface of the RPE cells. The surface is covered with dense microvilli and the cell borders are not visible. (×4,500.)*

endothelium of the choroidal capillary is very thin and fenestrated.

Photic Effects

Although the experimental conditions reviewed in this chapter have been divided into nonthermal and thermal categories, it is difficult to rule out thermal effects completely from the nonthermal category. However, there are certain differences in cytologic appearance between the two conditions. Photic stimulation appears to cause pathologic changes primarily in the photoreceptor outer segments.

The pathologic changes in the photoreceptor elements of the albino rat are considerably different from those of the monkey. When the retinas of albino rats receive fluorescent lamp exposure, the disc membranes of the outer segments become vesiculotubular, starting at the tips (Kuwabara and Gorn 1968). Shedding of the outer segments becomes profoundly active, and lamellar membranes are phagocytized by the pigment epithelial cell (Fig. 17.3).

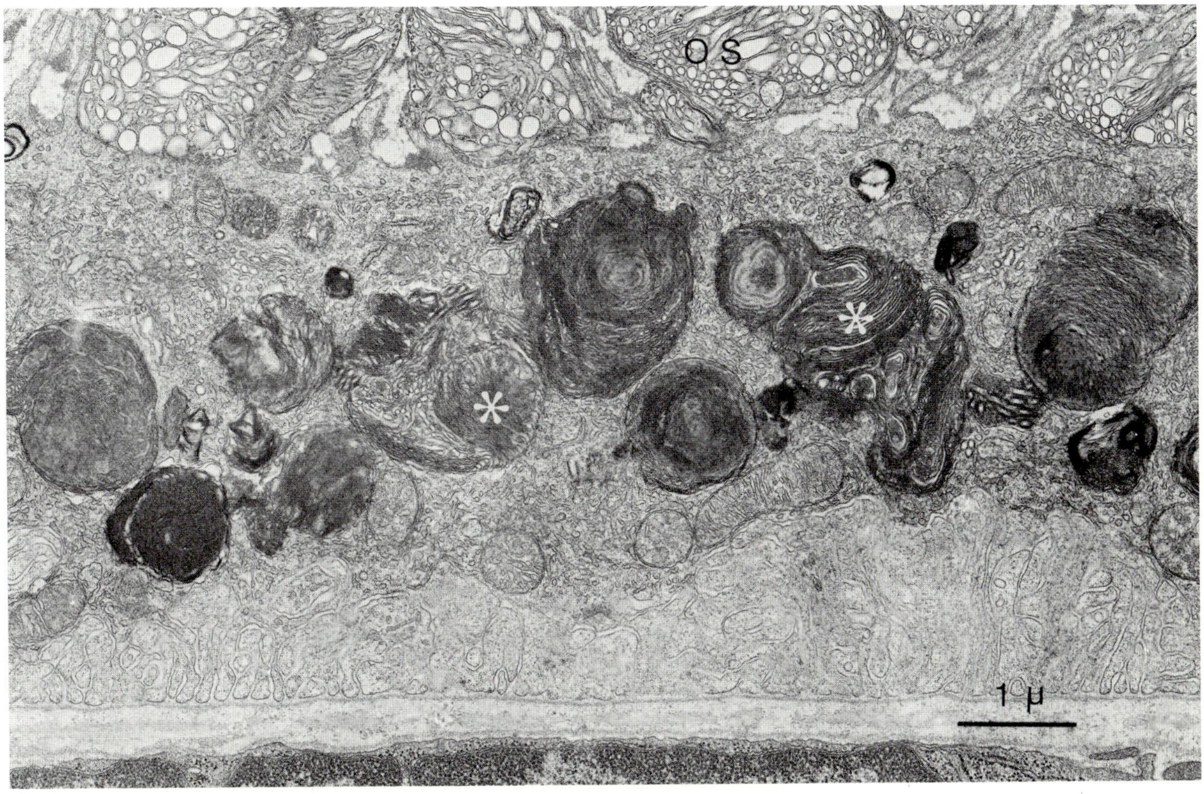

Figure 17.3 *The RPE cell of an albino rat that was exposed to fluorescent light for 10 hours. Phagocytized large lamellar bodies (*) are numerous. Some of them are changing into dark inclusion bodies. The tips of the outer segments (OS) are enlarged and vesiculated. (×16,500.)*

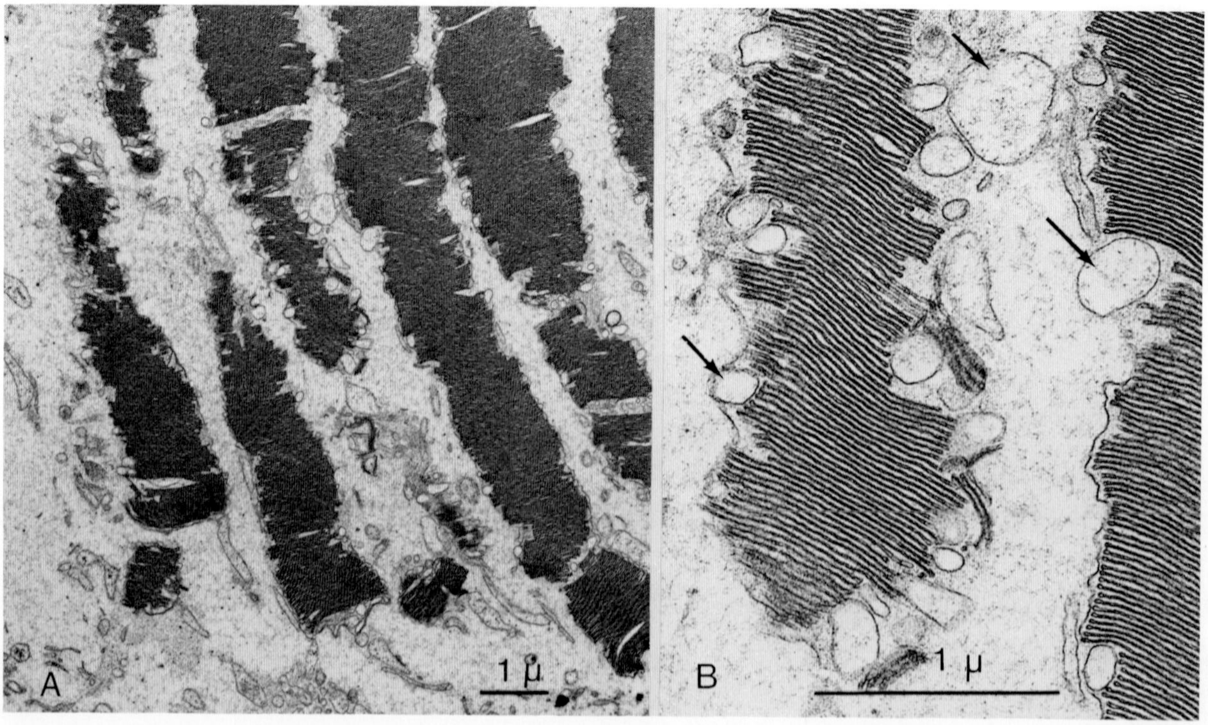

Figure 17.4 *Outer segments of monkey RPE exposed to strobe light (5 flashes per second) for one hour.* (A) *Disarrayed stacks of the lamellar membranes. (×9,440.)* (B) *Higher magnification shows fragmentation and vesicle formation of the plasma membranes of the outer segments (arrows). (×33,040.)*

This enhanced shedding occurs only for a short period, and the continuous exposure results in total degeneration of the photoreceptor cells.

When localized areas of the monkey retina are exposed to a cool strobe flashlight, fragmentation of the plasma cell membrane of the outer segments, instead of disc membranes, occurs as the earliest change. Repeated milder exposures cause the disarray of the stack of discs of the outer segments (Fig. 17.4). Breakage of the outer tips of these damaged rod outer segments is enhanced. Also, disc membranes seem to escape into the subretinal space through the fragmented cell membrane. The vesiculotubular changes of the lamellar membranes are also seen, but these changes occur mainly in the cone outer segments.

The earliest change in the RPE of both albino rats and monkeys following mild light exposure (fluorescent lamp for 1–12 hours in rats and strobe flash for 30–60 minutes in monkeys) appears to be enlargement of the microvilli (Fig. 17.5). The enlarged microvilli often fuse into large cell protrusions. Within them, smooth endoplasmic reticulum and occasionally mitochondria begin to appear. The microvilli also engulf separated outer segments and degenerating substances. Fused microvilli are common findings in the human RPE, especially in old age. Enlarged microvilli of an old human RPE cell have been reported by Spitznas and Hogan (1970).

A thick substance, probably proteinaceous or mucopolysaccharide in nature, accumulates around the microvilli and the outer segments and increases with longer exposure (Fig. 17.5). Since this substance is found constantly in the crypts of the villi and in the vesicles of the apical portion of the cell, it seems to be a secreted substance of the pigment epithelium. Secretion of mucopolysaccharide by the normal RPE cell has been described recently by Feeney (1973). This activity must be enhanced in the light-exposed condition. The photoreceptor inner segment has been identified as a secretor of mucopolysaccharide by Fine and Zimmerman (1963), but no appreciable changes have been demonstrated in the microorganelles or the inner segment during light exposure that correlate with the abnormally accumulating substance in the subretinal space.

The cytoplasmic matrix of the pigment epithelial cells appears to become electron dense at this stage.

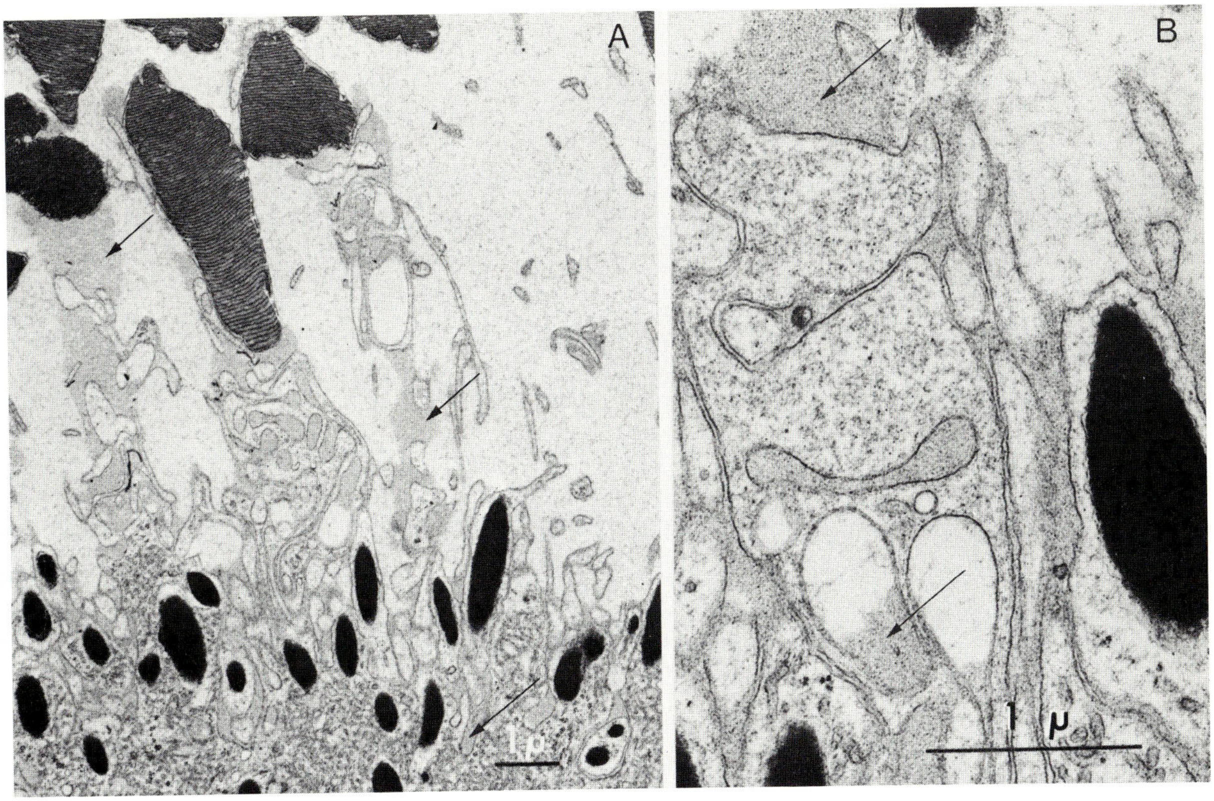

Figure 17.5 *Monkey retina exposed to strobe light for one hour and examined immediately following the experiment. Enlarged microvilli cling to outer tips of the outer segments. Dense granular substance, presumably thicker subretinal fluid, is seen around the outer segment, microvilli, and in crypts of the microvilli (arrows).* (A: ×*8,260.* B: ×*29,500.)*

The smooth endoplasmic reticulum becomes finer and often forms lamellar clusters, the centers of which appear to have been pinched (Fig. 17.6). These clusters are similar to lamellar bodies of the lower animals, though much smaller in size. The rough endoplasmic reticulum increases in size and tends to form lamellar arrangements.

Inclusion bodies and secondary lysosomes in the cytoplasm increase when the retina is exposed to light at a nondamaging dose (Fig. 17.7). Inclusion bodies are of various sizes and shapes. Many are markedly electron dense. The number of the primary lysosomes remains constant. A profound increase in lamellar phagosomes is noted, particularly in rats exposed to a fluorescent lamp for 6–18 hours (Fig. 17.3). The RPE cell is packed with large lamellar membrane bodies for a while, but they disappear within a few days, leaving various degrading products within the cytoplasm (Fig. 17.8). Some become electron-dense amorphous substances, which are surrounded by cell membranes, and others become finer lamellar bodies scattered among the smooth endoplasmic reticulum. These fine lamellar membranes are an average 75Å apart and often form small whorled bodies (Fig. 17.8*C*). The dense inclusion bodies appear to remain in the pigment epithelium for a long period of time even after the removal of the light. The relationship between lysosomal enzymes and the autodigestion process of the lamellar inclusion bodies has been discussed by Ishikawa and Yamada (1970). Some inclusion bodies transform into lipofuscin droplets, which accumulate in the apicolateral portion of the cell and replace the melanin granules. Reduction of melanin granules is apparent in RPE that has been exposed repeatedly for a long period of time. Deposits of large fat droplets are occasionally observed. Similar lipofuscin inclusion bodies are found regularly in the human RPE, though the number is greatly variable. Senescent and pathologic RPE cells are packed with lipofuscin granules (Feeney et al. 1965; Kolb and Gouras 1974).

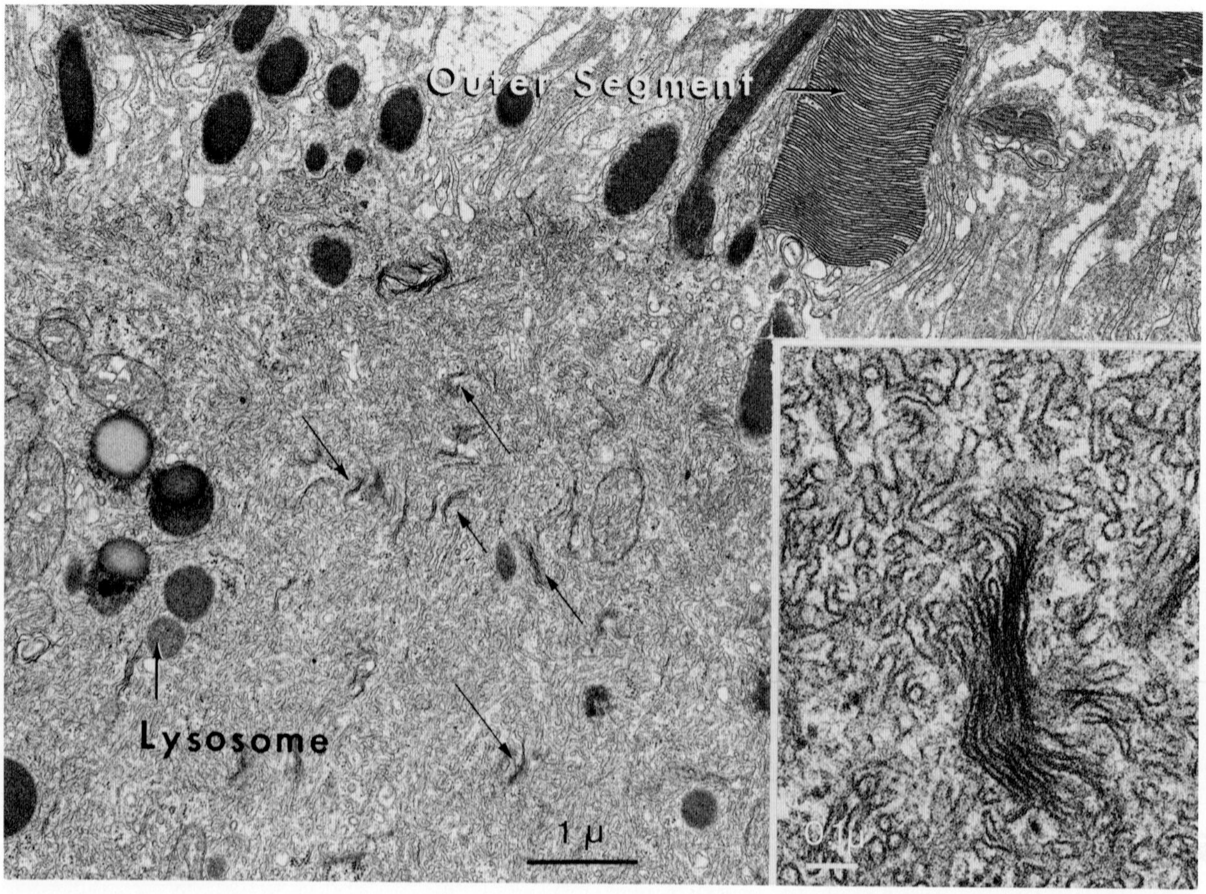

Figure 17.6 *The RPE cell of a monkey exposed to strobe light for one hour and examined 24 hours later. The cytoplasm appears to be darker because of the increased amount of smooth endoplasmic reticulum and the denser content. Smooth endoplasmic reticulum forms small clusters, the centers of which seem to have been pinched (arrows). (×14,160.) Insert shows the cluster of the membrane in higher magnification. (×64,900.)*

A prolonged exposure causes severe damage in the outer segments of the retina. The RPE of the severely damaged retina appears to cease phagocytotic activity. Although numerous fragments of outer segments, many of which are packed with tubular elements, are present in the subretinal space, lamellar inclusion bodies are absent in the RPE. However, electron-dense inclusion bodies are abundant during this stage. Rat retinas that have been exposed to fluorescent lamps for three weeks have no outer segments and the cell bodies of the photoreceptor are degenerated. Despite these severe changes of the retina, the cytologic appearance of the RPE in this stage is near normal (Fig. 17.9). The smooth endoplasmic reticulum and other microorganelles are normal in their distribution and number.

Following an extremely prolonged exposure, the photoreceptor cells of the rat retina often disappear completely. The RPE cells of these retinas remain at the original sites and begin to adhere to the Müller's cells of the retina. The adhesion is formed by interdigitation of the microvilli of both the Müller's and the RPE cells (Fig. 17.10). Similar adhesion occurs in other pathologic conditions in which the photoreceptor cells have disappeared. The cytoplasm of these RPE cells usually contains less dense cisternae and smaller numbers of other microorganelles. The basal infoldings of the RPE are also sparse. These findings indicate that the RPE in this condition is relatively inactive.

The choroidal capillaries maintain their normal structure until the retinal tissues adhere to the RPE, and then the number of the capillaries begins to decrease. Similarly, the number of retinal capillaries decreases after the photoreceptor cells have been

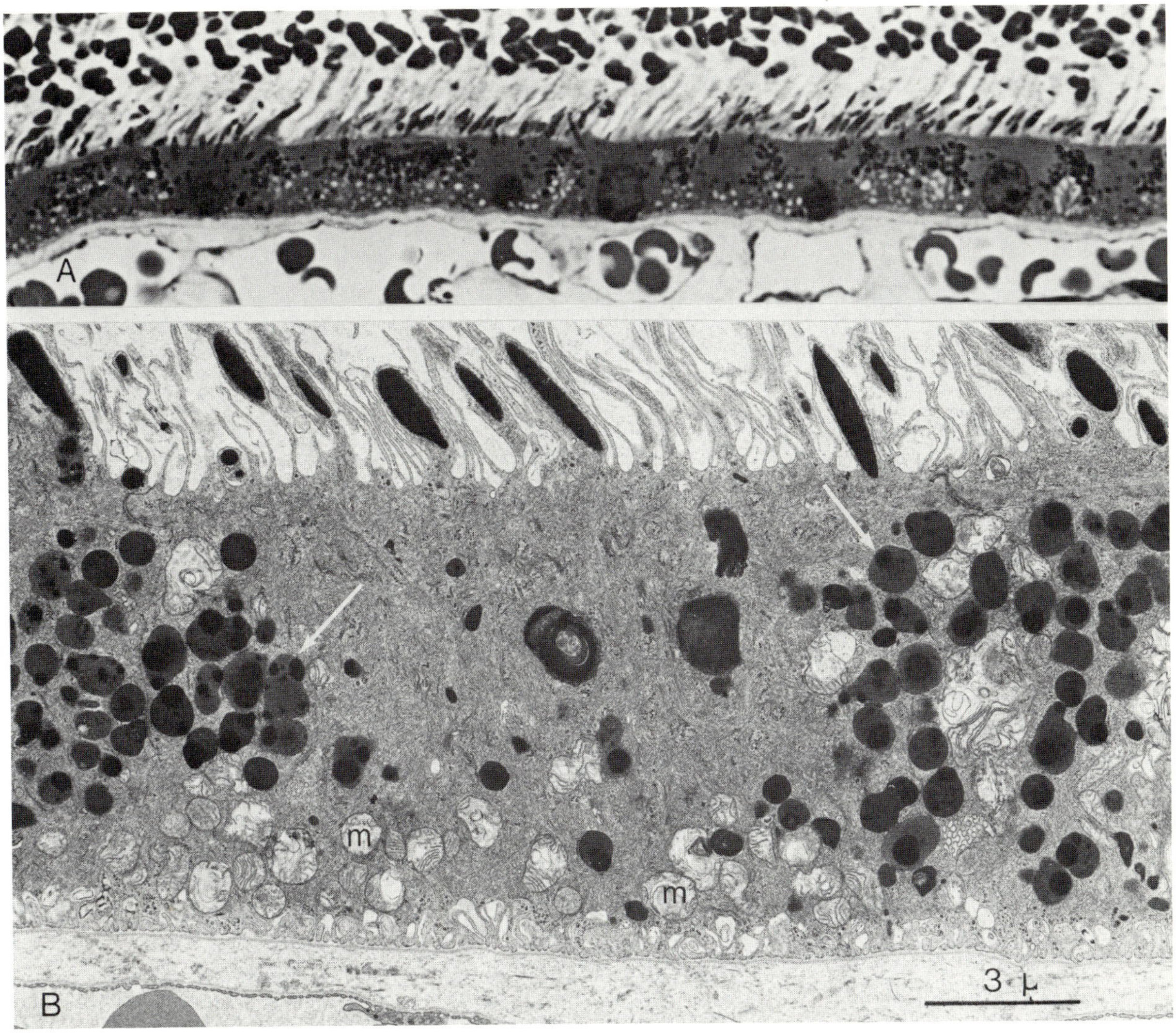

Figure 17.7 *Monkey RPE exposed to strobe light for one hour and examined one week later.* (A) *Epoxy section stained toluidine blue shows small vesicles and dark particles in the cytoplasm.* (×*708.*) (B) *Electron micrograph of the same RPE cell shows swollen mitochondria* (m) *and numerous electron dense inclusion bodies (arrows). Melanin granules are seen only in the microvilli.* (×*7,080.*)

severely damaged by experimental exposures (Okisaka et al. 1975; Dantzker and Gerstein 1969). Marked decrease of patent capillaries in senescent human retinas (Kuwabara and Cogan 1965) seems to be related to gradual damage of the photoreceptor cells, which may have occurred under physiologic light conditions.

Low-Level Thermal Effects

Cytologic changes of varying degrees are produced in the RPE and the photoreceptors when the retina is exposed to laser, xenon, or incandescent light for a short duration or at low energy levels. In most of these experimental conditions, the energy level reaching the retina is less than 1 W/cm^2 sec (Friedman and Kuwabara 1968; Ham et al. 1974). This energy level is not powerful enough to destroy the whole RPE cell directly and the lesion is not always visible clinically. However, cytologic changes are substantial in both the RPE (Ham et al. 1974) and outer segments (Adams et al. 1972). Some cellular changes in the RPE may be specific to injury of this kind, but the pathologic change in the photoreceptor cell is not different from that caused by a mild photic exposure.

The early cytologic changes in the RPE are vacu-

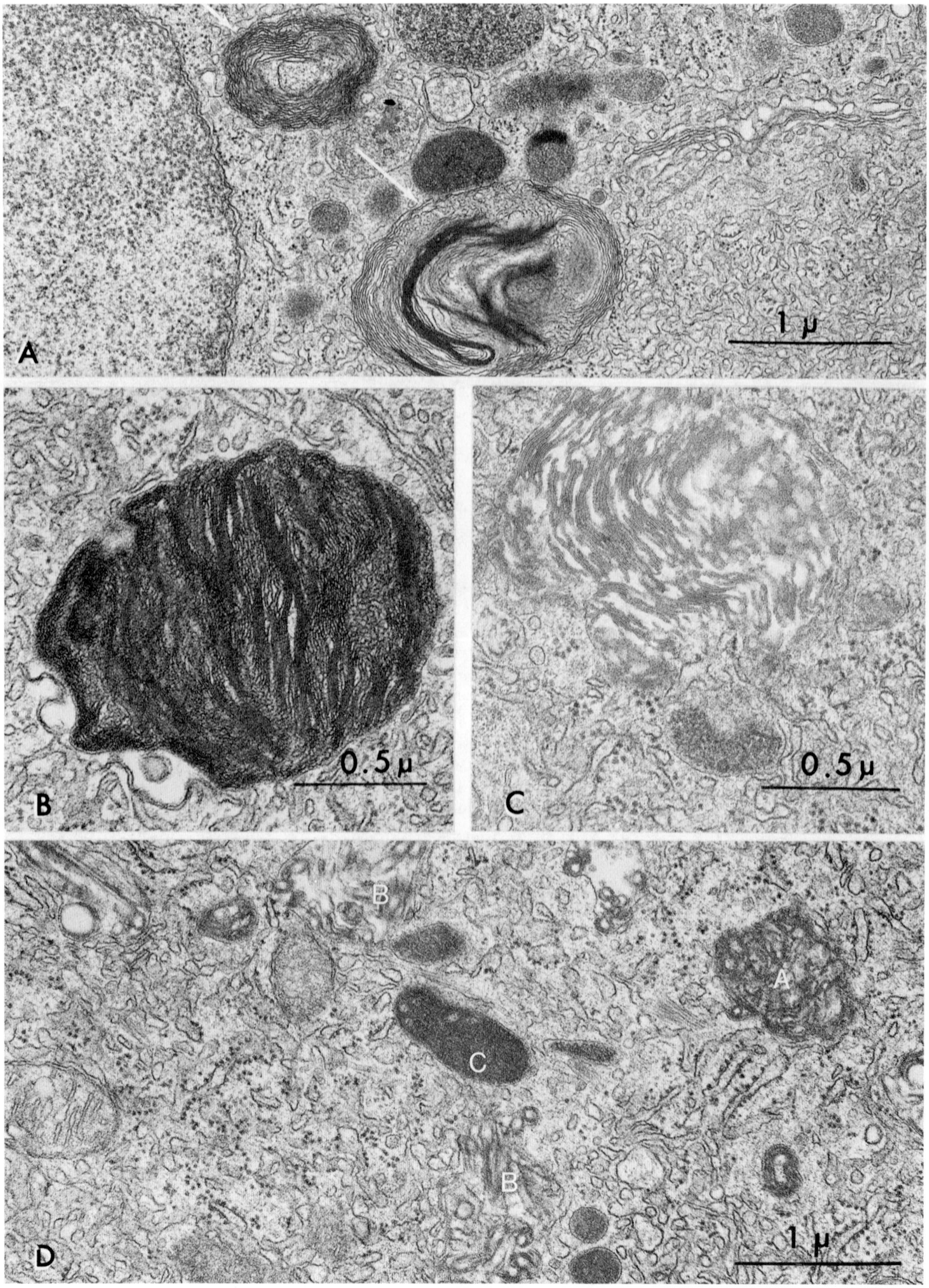
A
1 μ
B
0.5 μ
C
0.5 μ
D
B
A
C
B
1 μ

Figure 17.8 *Different stages of degrading phagosomes that make the cytologic appearance of the RPE of albino rats variable following light exposure.* (A) *Relatively fresh lamellar bodies (arrows) and several lysosomes. (×29,280.)* (B) *Some phagosomes become compact and electron dense membranous bodies. (×48,800.)* (C) *The dense body that is shown in* B *appears to transform into a body consisting of finer membranes (70Å apart). A lysosome is seen nearby. (×48,800.)* (D) *The cytoplasm of the pigment epithelial cell often contains various inclusion bodies following light exposure: partially digested phagosomes* (A), *myelin-like substance* (B), *and enlarged lysosomes* (C). *(×29,280.)*

olation and swelling of mitochondria. Abundant fragmented melanin granules are found immediately following light exposure. These particles are autophagocytized and become large round bodies. Besides these irregular melanin granules, the cytoplasm begins to accumulate numerous electron-dense inclusion bodies. Melanin granules in the microvilli and in the cytoplasm of the exposed cell decrease in number (Fig. 17.11). Lipofuscin particles also accumulate within a short time after exposure (Fig. 17.12).

The smooth endoplasmic reticulum also shows interesting changes. Cells that have received mild thermo-photostimulation develop a peculiar membrane structure within a short period of time. The membranes of enlarged smooth endoplasmic reticulum begin to appose closely to those of the adjacent cisternae and become wavy double membranes. These double membranes are regularly arranged into a lacy pattern and occupy the whole or a portion of the cell (Fig. 17.12). The lumens surrounded by these double membranes are electron-lucent, and the cell matrix contains no other microorganelles except for scanty glycogen particles. These double-membrane structures are often continuous with the double membranes of the abnormally enlarged and

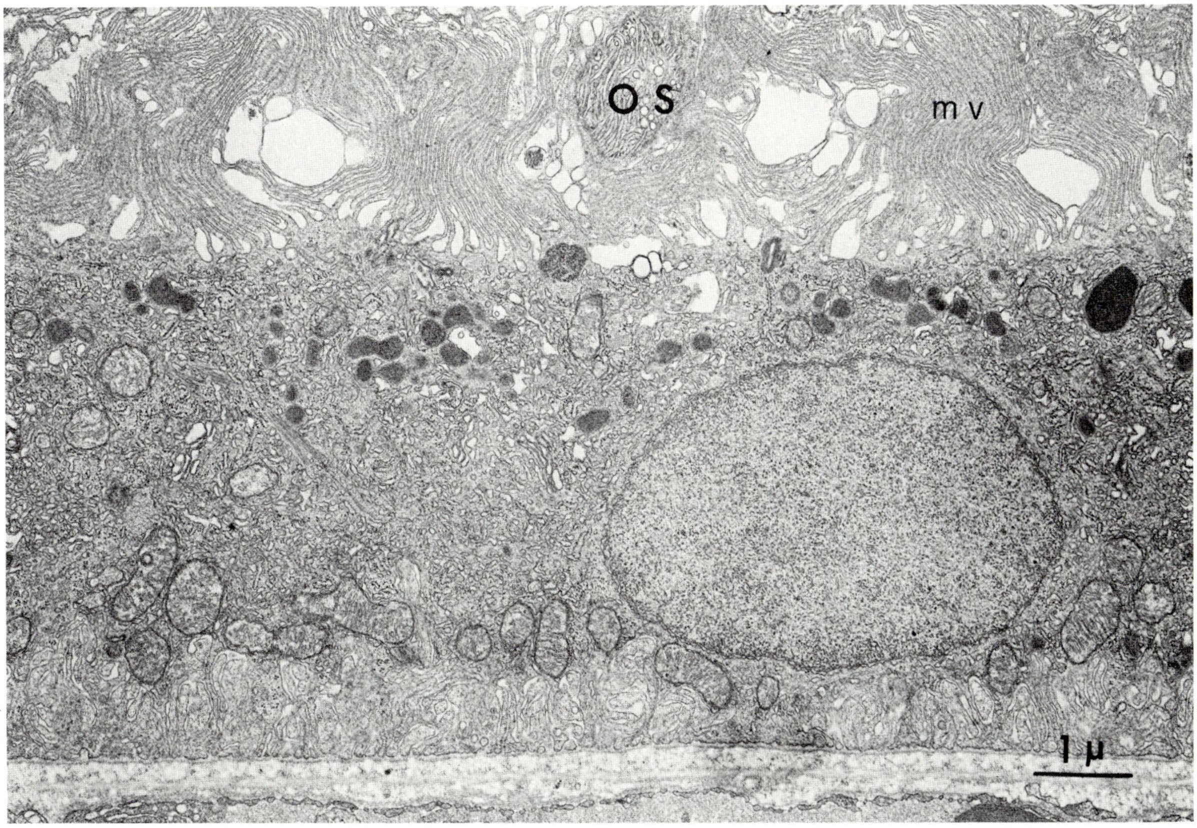

Figure 17.9 *The RPE of an albino rat exposed to bright light for 3 weeks. The outer segments of the retina have degenerated completely. The cytoplasm of the RPE contains numerous lysosomes and electron-dense inclusion bodies, but the general appearance is similar to that of the normal cell. Remnants of outer segments* (OS) *are seen among dense microvilli* (mv). *(×12,980.)*

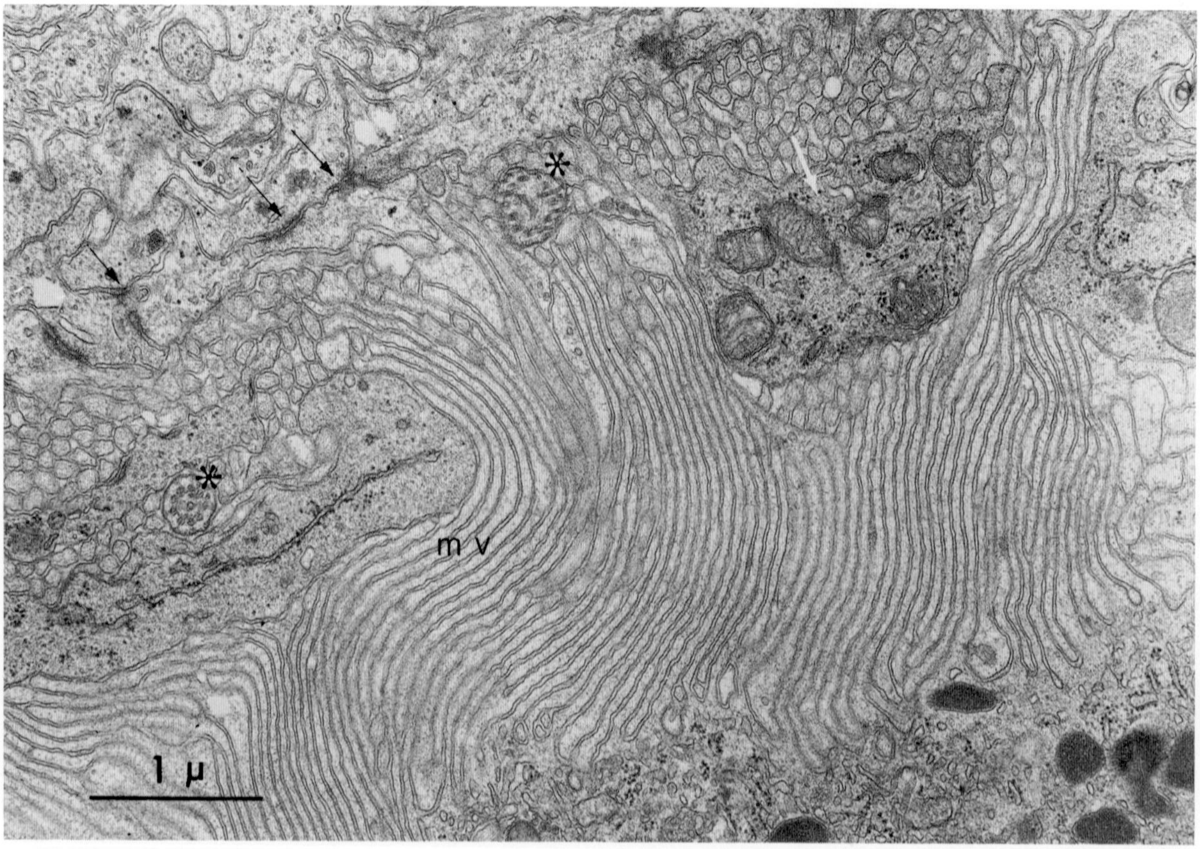

Figure 17.10 *Microvilli of the RPE and Müller's cell have adhered by interdigitation following severe damage to the photoreceptor cells. Remnants of the inner segments (white arrow) and centrioles of the photoreceptor cells (*) are seen among the microvilli. Thin arrows indicate the outer limiting membrane. (×23,600.)*

disrupted mitochondria. Succinic acid dehydrogenase and cytochrome *c* oxidase activities are absent in the peculiar endoplasmic reticulum. Also, the degree of tetrazolium-staining for lactic acid dehydrogenase appears to have been decreased in these cells. This change persists for a long time.

Strong Thermal Effects

The whitish lesions of the fundus that are commonly produced following exposure to focused beams of incandescent lamps, xenon lamps, and lasers are considered to be a result of heat rather than of light itself (Kuwabara and Aiello 1968; Apple et al. 1973; Weingeist 1974). Severe degeneration of cells of the retina and the RPE is observed in the sharply circumscribed focused area (Fig. 17.13). The RPE is most severely damaged, but depending upon the strength of the energy levels, the retina and the choroid are also damaged in various degrees. The damaged RPE cells are markedly vacuolated and often completely disintegrated. Melanin pigment and cell debris are scattered in the subretinal space. These cells eventually disappear from the original site.

The vacuolation of the directly damaged cell occurs by swelling of smooth endoplasmic reticulum and mitochondria. Similar changes occur in the adjacent cells (Figs. 17.14, 17.16). Depending upon the degree of damage, the size and extent of vacuolation varies considerably. In less damaged cells, swollen mitochondria remain in otherwise normal cytoplasm for a long period of time (Fig. 17.14). The damaged mitochondria often change into large bodies consisting of whorled double membranes (Fig. 17.15). The enlarged and deformed mitochondria are often 3–5 μ long and remain in the epithelial cell for several months after the light exposure.

The tissue reactions for repairing the wound produce various cytologic changes in the RPE. The degenerated debris are carried out first by a non-

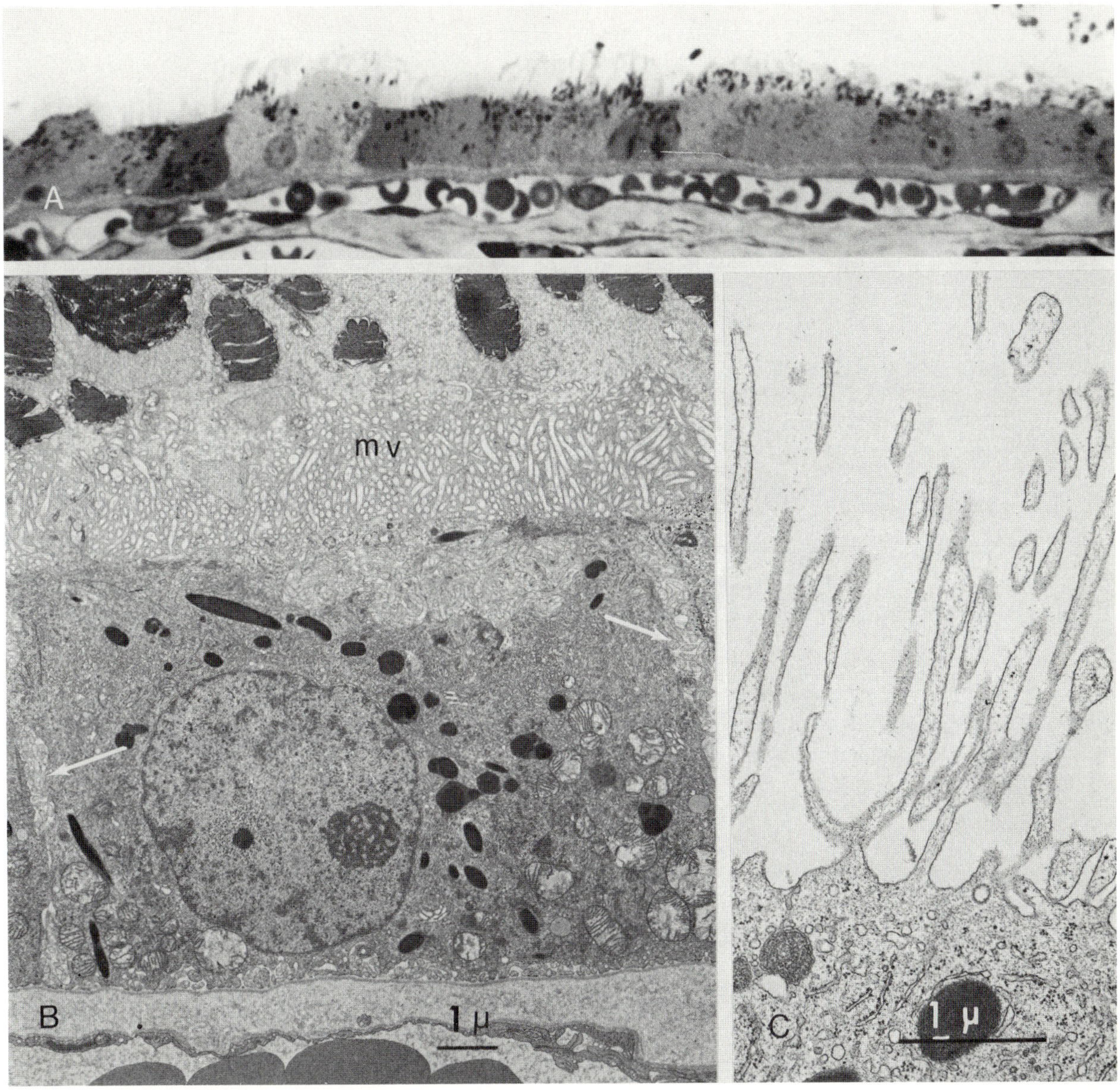

Figure 17.11 (A) *RPE that was exposed to a clinically subthreshold level of xenon lamp 3 days previously. Stainability of the cell is uneven and melanin granules are sparse. (Epoxy section, toluidine blue staining. ×708.)* (B) *RPE that had received a subthreshold argon laser 24 hours previously. Some irregularity of the outer segment is obvious, but the main pathologic change is absence of melanin granules in the microvilli of the RPE* (mv). *A few melanin granules and dark inclusion bodies are seen in the cytoplasm. The appearance of the cytoplasm is normal except for widened lateral intercellular spaces (arrow) and swollen mitochondria. (×8,260.)* (C) *RPE three weeks after the exposure to argon laser at a level similar to that of* B. *Melanin granules are absent in the microvilli. (×21,830.)*

cellular process and moderately by a phagocytotic activity of histiocytes. When the denuded area is smaller than 300 μ in diameter, the neighboring intact cells slide over to cover the damaged area within a few days (Friedman and Kuwabara 1968; Wallow and Tso 1973). The sliding cells are flat in shape and maintain apical junctions. The cytoplasm of these cells is usually electron-dense, due to a higher content of ribosomes and rough endoplasmic reticulum. The mitochondria are irregular in shape but the smooth endoplasmic reticulum is normal in density. Melanin granules are generally absent or

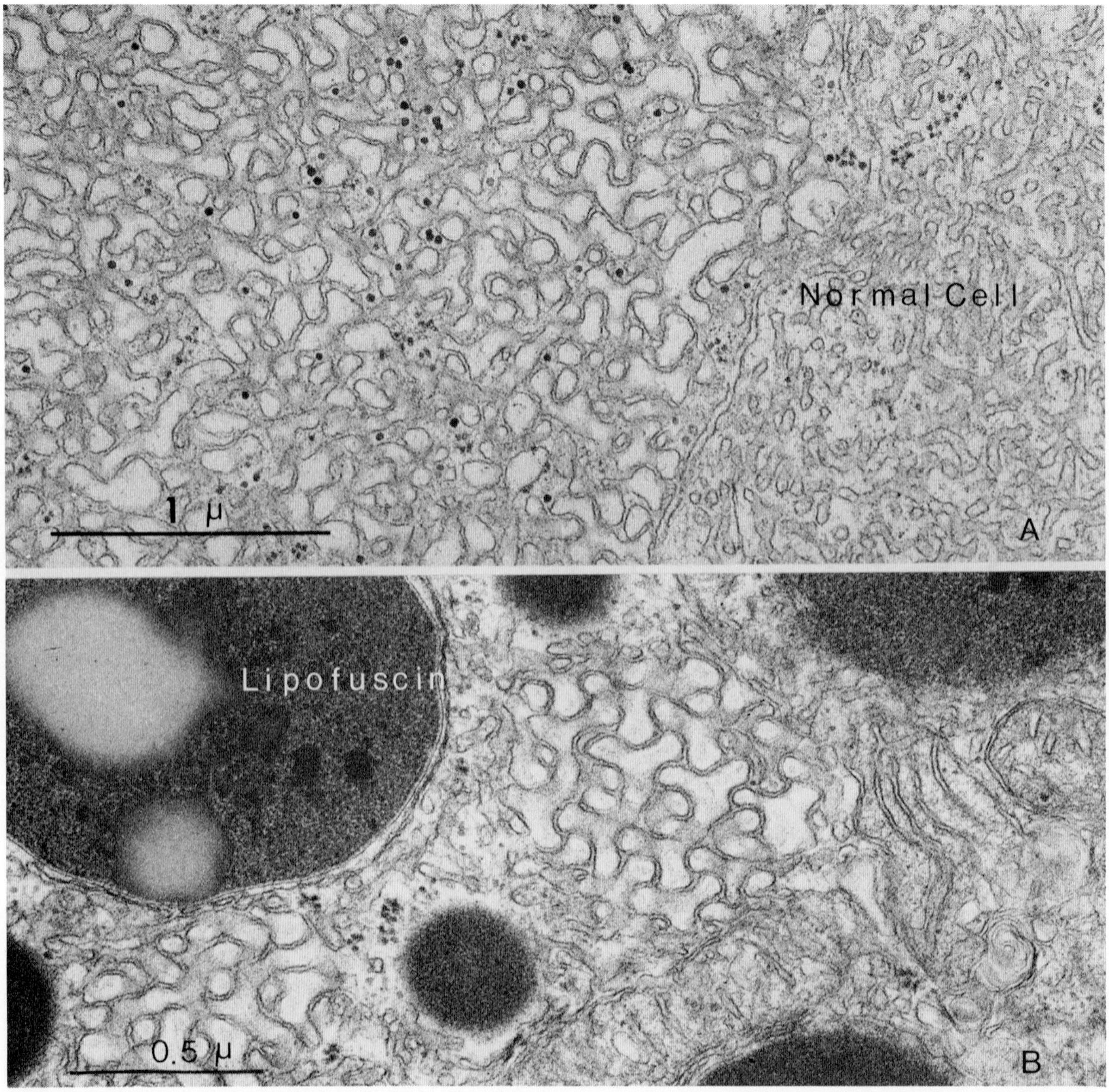

Figure 17.12 (A) *Peculiar smooth endoplasmic reticulum formed following a mild thermal exposure. Convoluted lumens are lined by double membranes. Glycogen particles are seen in the cell matrix. (×41,300.)* (B) *The peculiar smooth endoplasmic reticulum is seen in small localized areas also. A lipofuscin granule is seen in this area. (×56,640.)*

only sparsely found in these cells. Microvilli are sparse in the sliding cells, but after the cells have filled up the defect, microvilli become long and numerous. Although degenerated substances are abundant in the damaged area, the sliding epithelial cells apparently do not take up this debris. The cytoplasm of these cells is usually free from inclusion bodies. Phagocytes that have invaded from the choroidal vessels contain large numbers of melanin granules and other particles. The newly slid cells increase their height within a few days. The cytoplasm remains nonpigmented for a long period of time. The whitish appearance of the damaged area seems to be due to the depigmentation of the newly slid RPE cells.

Larger lesions are repaired considerably more slowly. The epithelial cells often form a loosely packed pile, a few cells thick, at the center of the

Figure 17.13 *Monkey retina exposed to a moderate dose of argon laser 3 days previously. The RPE is most severely damaged. Many photoreceptor cells are necrotic also. (Epoxy section, toluidine blue staining. ×295.)*

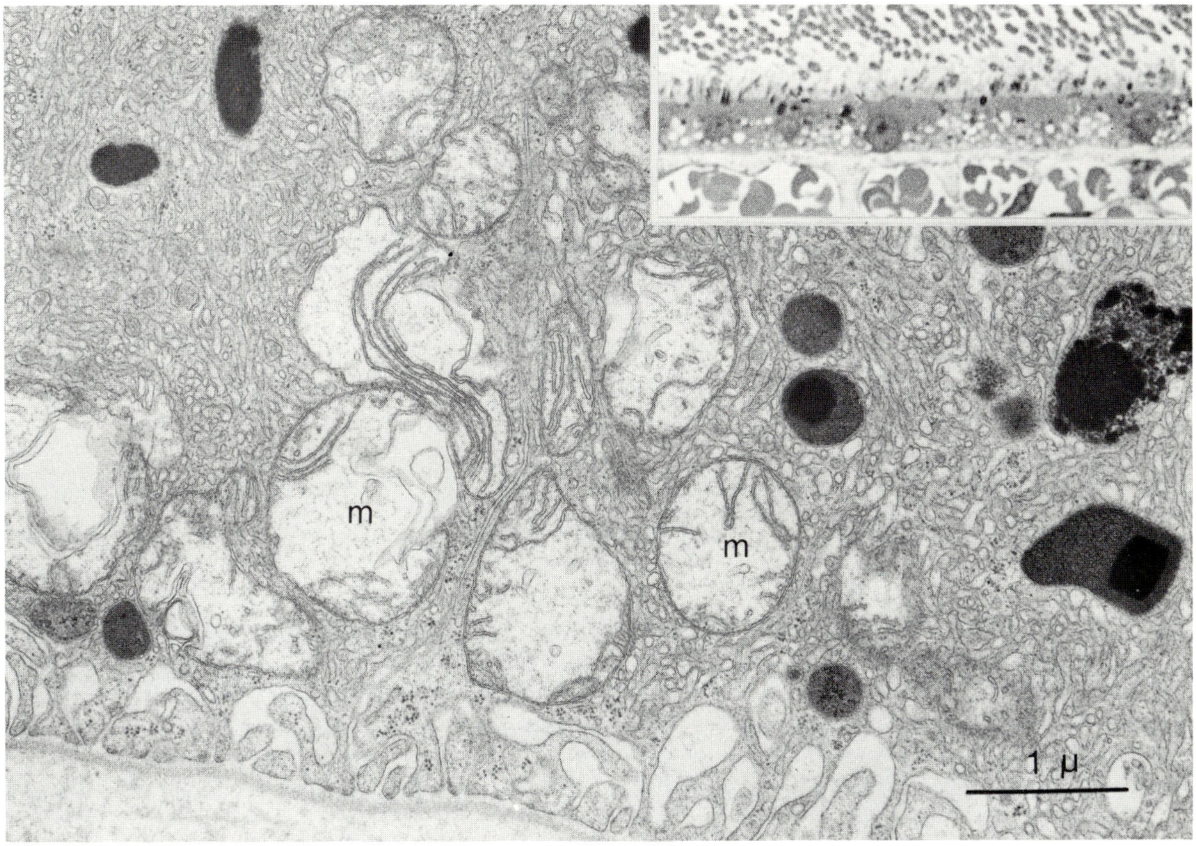

Figure 17.14 *Monkey RPE in the area adjacent to a lesion that had been burned severely by argon laser one week previously. Vacuoles in the cell (insert) correspond to markedly swollen mitochondria* (m). *×22,420.) (Insert: epoxy section, toluidine blue staining. ×295.)*

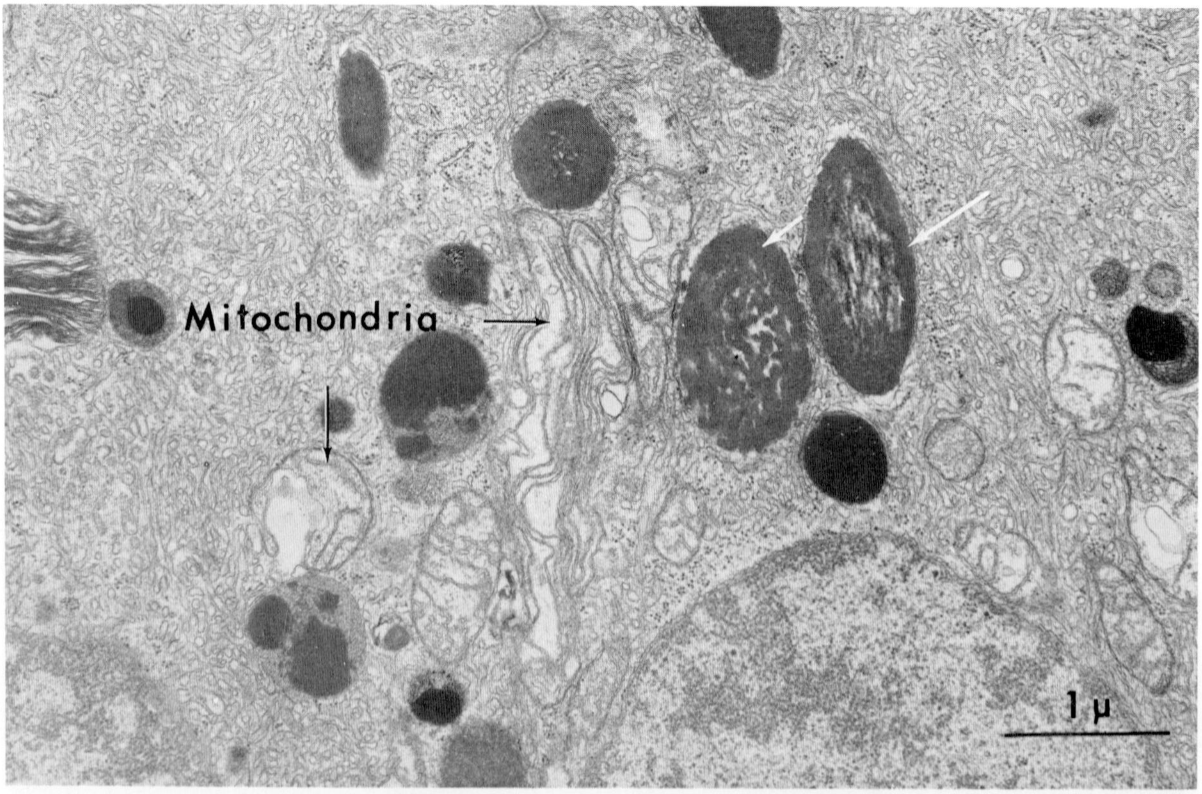
Mitochondria
1 μ

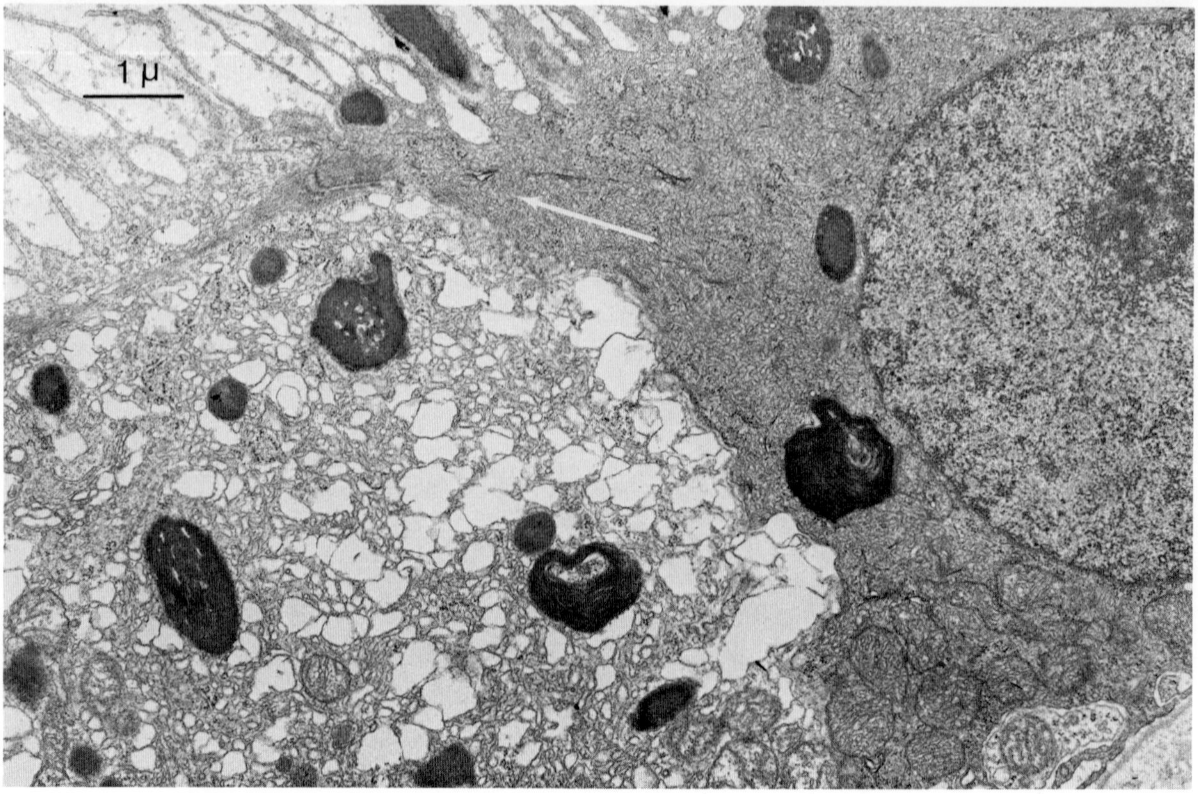
1 μ

Figure 17.15 *Monkey RPE exposed to relatively mild argon laser one week previously. The swollen mitochondria have transformed into whorled double membranes. Melanin granules appear to have been damaged (white arrows). (×22,420.)*

Figure 17.16 *Three days after laser exposure: a survived RPE cell, with dark cytoplasm, is trying to cover a severely damaged cell. The damaged cell is markedly vacuolated. (×13,570.)*

lesion (Fig. 17.17). This phenomenon has been described by Wallow and Tso (1973). These cells maintain the cytologic characteristics of the RPE—rich, smooth endoplasmic reticulum, apicolateral junctions, and numerous microvilli. However, they contain almost no melanin granules or inclusion bodies. The proliferated cells eventually become a monolayer. The nonpigmented cells are usually surrounded by hyperpigmented epithelium. Thus the characteristic appearance of a white lesion with a pigmented ring is formed following photocoagulation treatment. This condition persists for a very long period of time.

A moderate number of pigment-laden cells are present in the subretinal space and in the retinal tissue (Fig. 17.18). These cells remain in the area for a long period after the injury. It is often difficult to identify the origins of these cells, but cytologically they resemble histiocytes. These cells do not attach to each other and each cell has irregular processes. The cytoplasm contains abundant inclusion bodies and scanty smooth endoplasmic reticulum. The phagocytotic nature of these pigment-laden cells has been discussed by several investigators (Wallow and Tso 1973; Tso 1973; Feeney et al. 1975). In general, pigmented cells within retinas with various pathologic conditions exhibit the cytologic characteristics of histiocytes. Although Mandelcorn et al. (1975) have demonstrated an experimental transition of these cells from the RPE cells, true migration of the RPE cell into the retina seems to be rare.

Damaged retinal tissue is repaired in various ways,

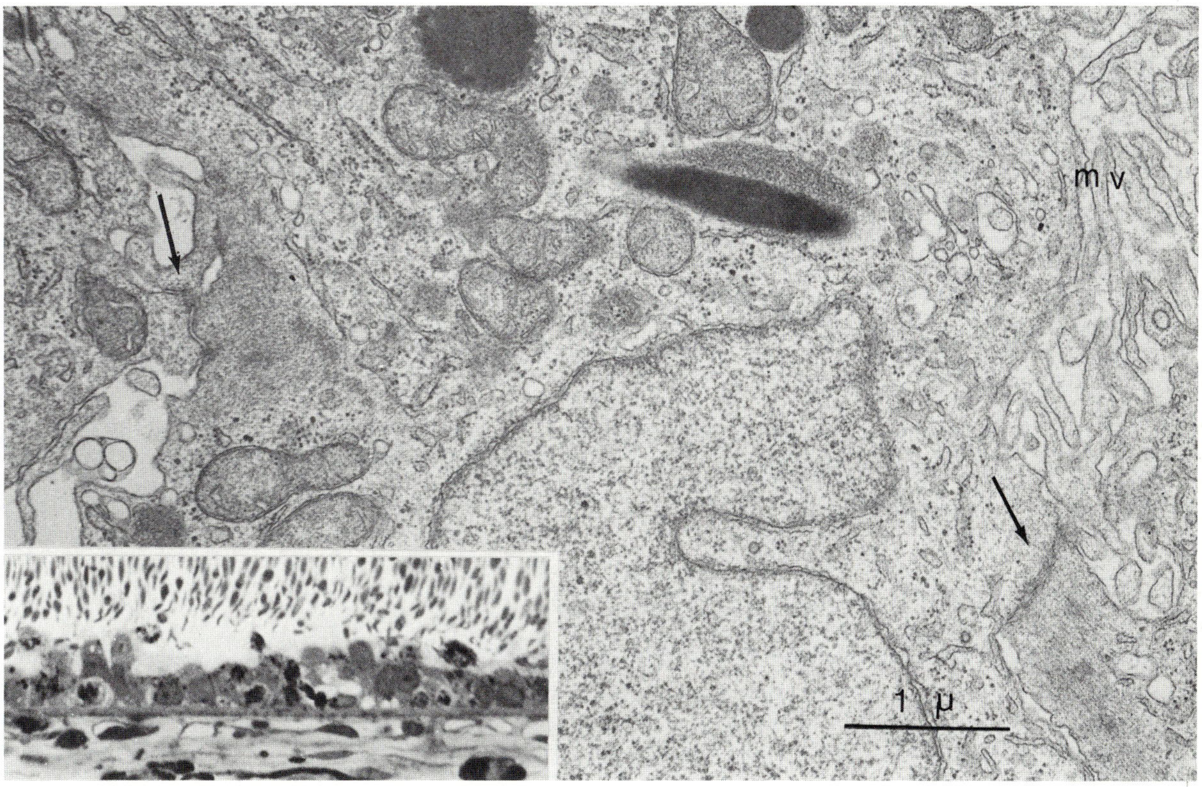

Figure 17.17 *Proliferating RPE cells in a large area that had been injured two weeks previously. The proliferating cells are joined firmly with junctional apparatus (arrows) and have microvilli* (mv). *However, melanin pigment and smooth endoplasmic reticulum are sparse. (×22,420.) (Insert: epoxy section, toluidine blue staining. ×295.)*

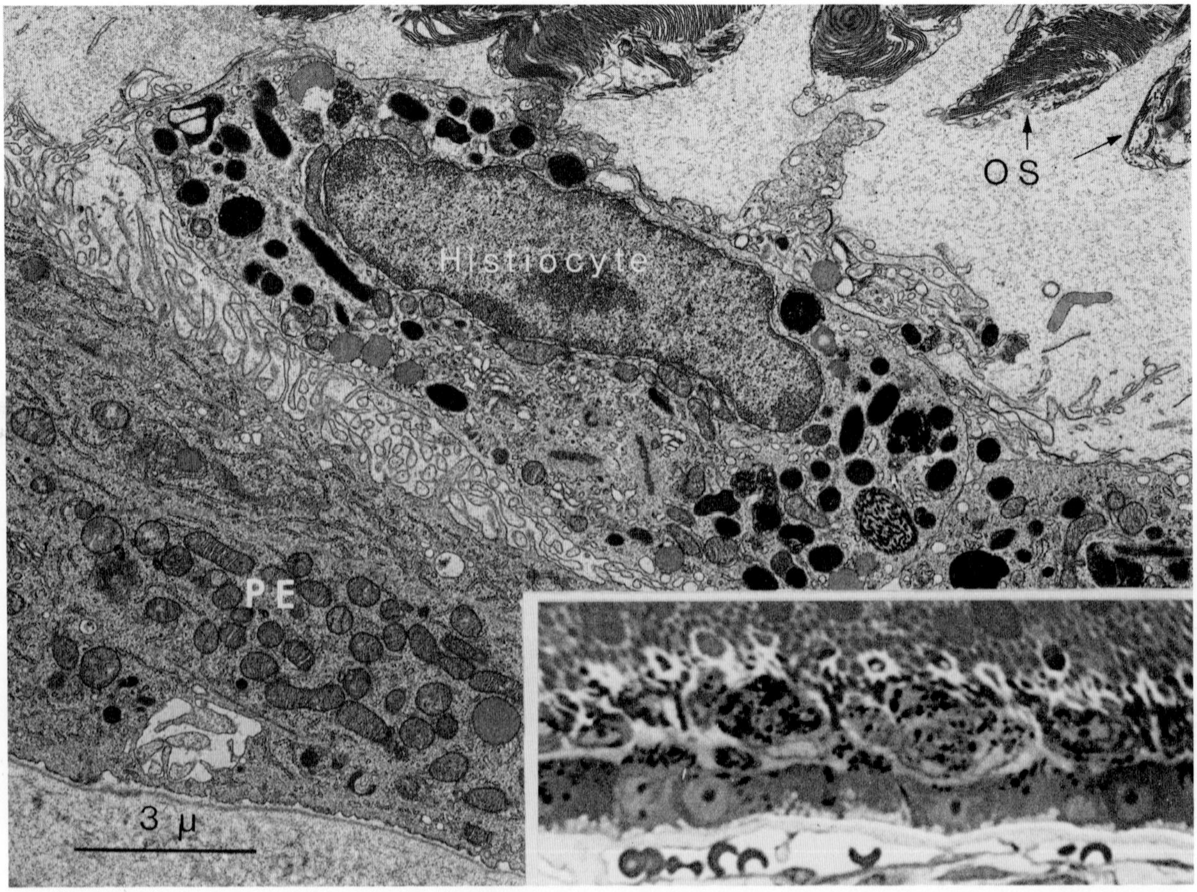

Figure 17.18 *Heavily pigmented cells are found in the subretinal space a few days after xenon lamp exposure. These cells differ from proliferating RPE cells* (PE). *The pigment-laden cells have no junctional apparatus and regular microvilli. The electron-dense particles appear to be fragmented melanin pigment and other cell debris. The pigment-laden cells appear to be histiocytes rather than isolated RPE cells.* **OS** = *Outer segments. (×8,260.) (Insert: epoxy section, toluidine blue staining. ×472.)*

depending upon the degree of the burn. In retinal tissue from which photoreceptor cells have disappeared completely, Müller's and RPE cells adhere to each other at their microvilli. When the damaged area is too large to be replaced by new RPE cells, Müller's cells of the retina adhere to the Bruch's membrane directly. No regeneration of the photoreceptor cells or choroidal capillaries is observed in this area. Clinically, these areas appear extensively whitish.

Choroidal capillaries beneath the damaged RPE are occluded by accumulated platelets at the acute stage (Fig. 17.19). They appear to reopen within a few days if the cellular destruction is not too severe. The choroidal tissue beneath the extensively damaged epithelium becomes an avascular scar tissue. The normal fenestrations of the endothelial cells disappear and the cytoplasm becomes markedly thick when the RPE is moderately damaged either by photic or thermal stimulation (Fig. 17.20). The thickened cytoplasm contains increased numbers of rough endoplasmic reticulum and often contains various inclusion bodies. A proliferation of the endothelial cells and invasion of the pericyte into Bruch's membrane are also observed. These changes may be temporary or chronic. The basement membrane beneath the thickened endothelial cytoplasm becomes thicker and more irregular. The number of electron-dense substances in Bruch's membrane increases also. These changes in Bruch's membranes are similar to those found in senile human eyes. However, no drusen is formed beneath the RPE cell.

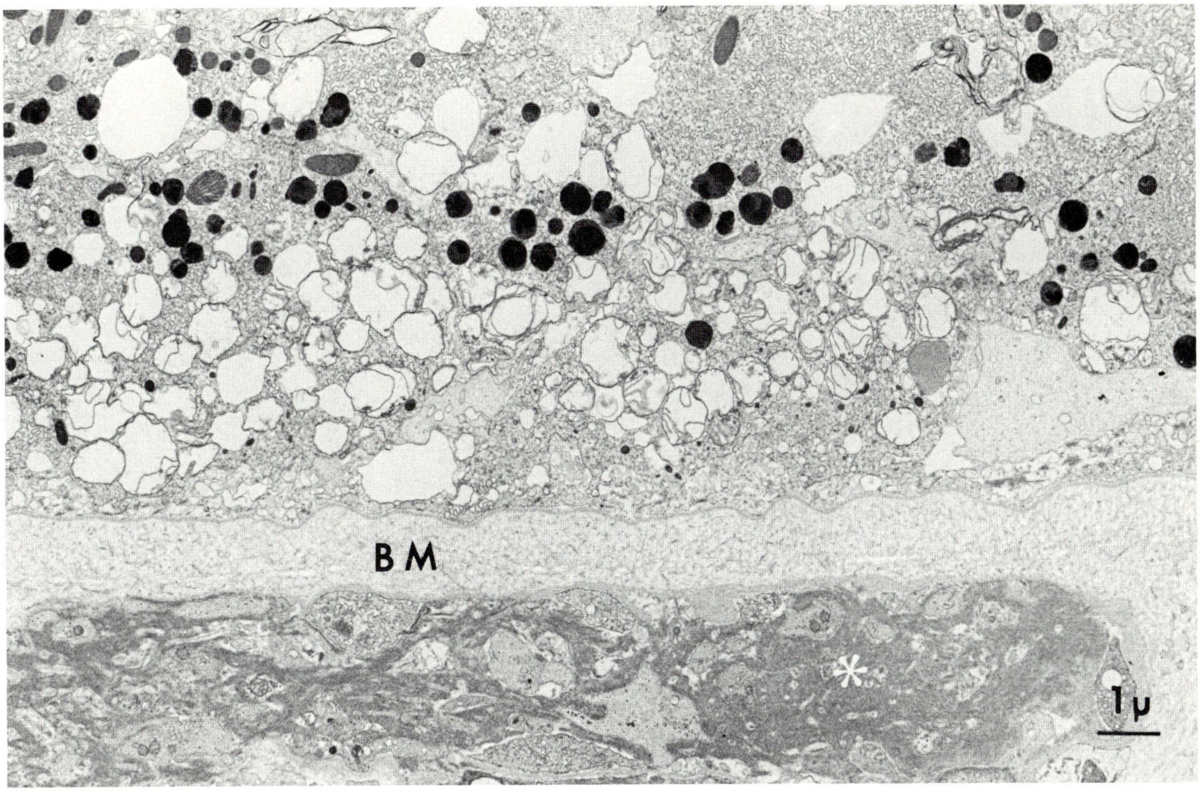

Figure 17.19 *Choroidal capillaries are often occluded immediately after severe thermal light damage. The RPE cells are vacuolated markedly. The choroidal capillaries are filled with fibrin and platelets (white asterisk).* BM = *Bruch's membrane. (×8,260.)*

Conclusion

Although it is difficult to make a strict distinction between photic and thermal changes in the retina and the RPE, in this chapter the changes in the RPE cells have been divided arbitrarily into two categories. Cytologic changes that occur following certain initial changes in the outer segments of the retina are considered to be of photic nature, and acute destructive changes occurring in both RPE and retinal outer segments are considered to be the results of thermal damage. Light damage of the RPE can be either a secondary reaction to the morphological alternation of the photoreceptors of the retina or a repairing process of thermally damaged cells.

Various morphologic changes of the smooth endoplasmic reticulum are observed following photic stimulation. The earliest cytologic change is a localized compacting of the cisternae membranes. With slightly more severe damage, the cisternae form a peculiar double-membrane structure. Also, the amount of rough endoplasmic reticulum increases considerably. Although these changes are most common in photostimulated RPE, similar changes are often demonstrated in other pathologic conditions.

At a certain stage of the photostimulation of the retina, the turnover mechanism of the outer segments appears to be enhanced. The pigment epithelium engulfs a great many lamellar membranes, and profound changes begin to show in the RPE cell. Phagocytosis by the RPE cell is one of the turnover mechanisms of the photoreceptor outer segment (Young and Bok 1969). Experimental phagocytotic activity in the chick embryo (Hollyfield and Ward 1974) and marked decrease of this activity in the RPE of hereditary degenerating rat retina (Bok and Hall 1971; Herron et al. 1969) have been reported. However, the phagocytotic function of the RPE seems to be different from that of phagocytes. In the RPE, profound phagocytosis is seen only in a certain period after photic stimulation. Cell debris are

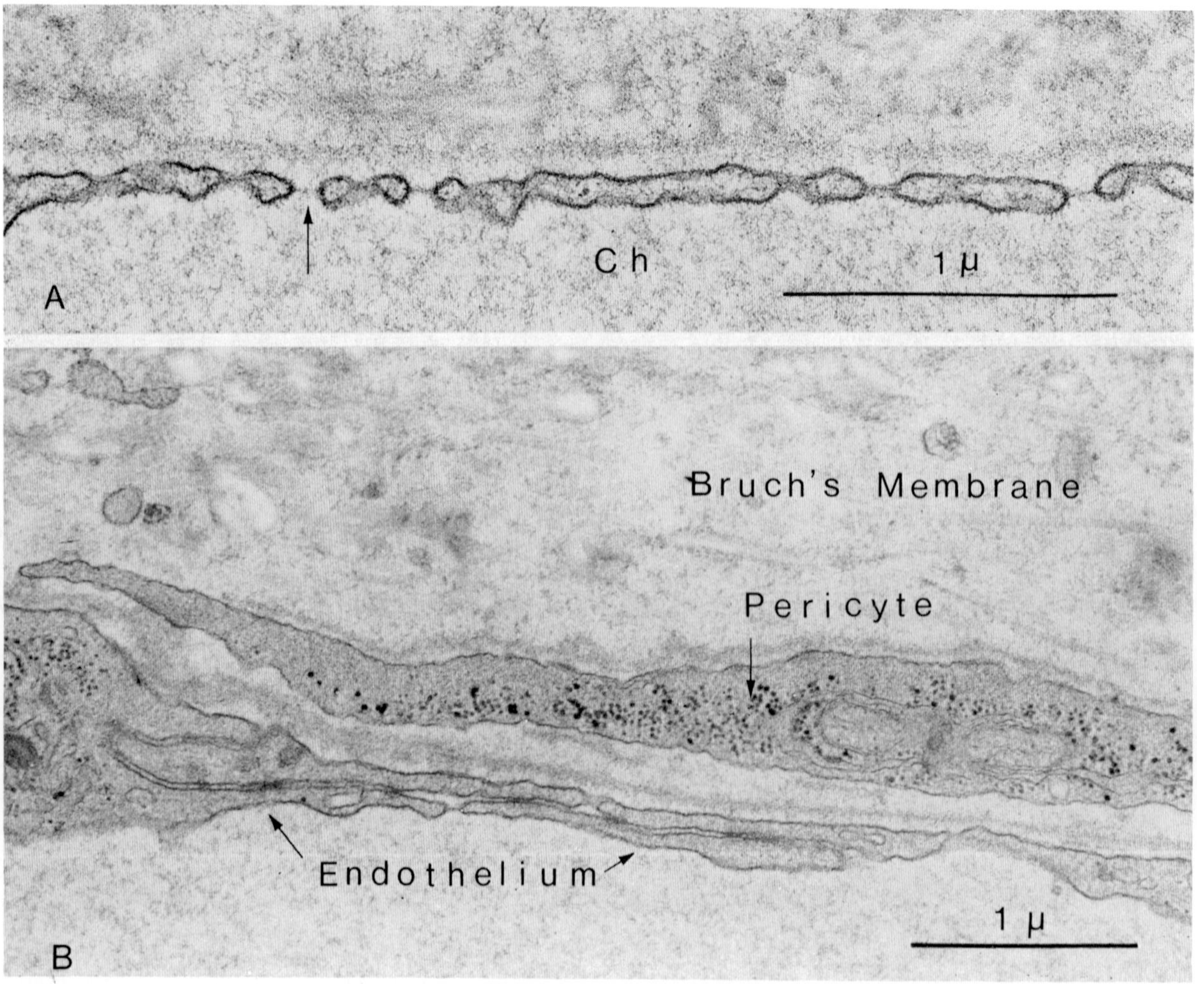

Figure 17.20 (A) *The normal endothelial cell of the inner wall of the choroidal capillary. A thin cytoplasm is fenestrated (arrow).* Ch = *choroidal capillary lumen. (×44,840.)* (B) *After light and thermal injury, the endothelial cell of the choroid becomes thick and often multilayered. The cytoplasm of pericytes often extends into Bruch's membrane. (×30,100.)*

not phagocytized by the RPE cell. The variation in cytologic appearance seems to be due to degradation of the phagocytized membranous substances. Lysosomes play a role in the autodigestion of the phagocytized substances. Lamellar bodies become electron-dense inclusion bodies of various shapes and nature. Lipofuscin appears to be the end product of this process. Accumulation of lipofuscin granules is regularly observed in the late stage of photo damage. It has been reported that lipofuscin granules of the human RPE increase with age (Feeney 1965).

The wide variations in cytologic appearance that are common in the human RPE may be related to the photic conditions of individuals. Active phagocytotic activity seems to last only for a short period of time under continuous light exposure. In addition, autodigestion is active only in healthy cells. Cells in chronically damaged RPE have no phagocytotic activity, and phagocytized membranes are not digested during hibernation of the animal. Cytologic changes of the RPE during hibernation have been described elsewhere (Kuwabara 1975). Similar inactivation is commonly found in RPE cells that have adhered to the damaged retina following various pathologic conditions.

Experimental exposure to focused lights produces discrete burned lesions in the RPE. Because of the heat, RPE cells are damaged severely, but the repairing process begins immediately in the tissue. The first step of the repair is the removal of the damaged cells. The majority of the cell debris in the damaged area is cleaned out without appreciable cellular reaction. Severely damaged cells that have separated from the basal attachment degenerate immediately.

Numerous pigment-laden cells are found commonly in the retina and subretinal space in various pathologic conditions. These pigmented cells are generally considered to originate in the pigment epithelium. However, migration of the true pigment epithelial cells is rather rare following light damage. Pigment-laden cells observed in the light exposure experiments described above have shown cytologic characteristics of the histiocyte or phagocyte. Similar findings have been recently discussed by Wallow and Tso (1973) and Tso (1973). Evidence for the transformation of isolated RPE cells into histiocytes is also mentioned by Feeney et al. (1975) and by Mandelcorn et al. (1975). This chapter has demonstrated that nonpigmented histiocytes pass through Bruch's membrane but fail to show clear evidence of transformation from the RPE cell. After engulfing particles, including melanin, histiocytes migrate into the retina and often gather around venules and near the inner limiting membrane. The basement membrane that is often found around these cells appears to be the product of the adjacent Müller's cells or the blood vessels. These findings are applicable to many other pathologic conditions.

Another prompt reparative process of the RPE is the sliding of the surviving adjacent cells. Mitotic activity is not observed in the vicinity of the injured area, at least in the early stages. The sliding cells cover the defect and become the complete epithelium within a short period of time. Proliferation of RPE cells occurs when the damaged area is large, but these cells soon become a single cell layer. The cytologic appearance of these cells is not quite the same as that of the original cell. Although these cells maintain the fundamental structure of the RPE, melanin pigment granules are sparse or absent. Also, phagocytotic activity diminishes. Despite the abundance of cellular debris in the vicinity, these cells contain very few phagosomes and inclusion bodies.

When the photoreceptor cells have been lost, the RPE cells adhere to Müller's cells with their microvilli, and when the pigment epithelium becomes absent, the retinal glia (Müller's cells) adheres to Bruch's membrane directly. Earlier experiments have revealed that regeneration of the RPE always precedes repair of the retinal tissue (Kuwabara 1970). The whitish appearance of photocoagulated spots is due to either reduction of melanin pigments in the RPE cells or total absence of the epithelial cell.

Decrease of pigmentation in the RPE, similar to that found regularly in senile human eyes, may be one of the changes caused by photic energy. This chapter has demonstrated that RPE cells that are exposed to light become less pigmented. Also, direct photic damage of the melanin granules by laser, especially with an extremely short exposure, has been demonstrated (Ham et al. 1974). Damaged melanin granules appear to be autophagocytized or eliminated from the cell body.

REFERENCES

ADAMS, D. O., BEATRIC, E. S., and BEDELL, R. B. 1972. Retina: ultrastructural alterations produced by extremely low levels of coherent radiation. *Science* 177:58–60.

APPLE, D. J., GOLDBERG, M. F., and WYHINNY, G. 1973. Histopathology and ultrastructure of the argon laser lesion in human retinal and choroidal vasculatures. *Am. J. Ophthalmol.* 75:595–609.

BAIRATI, A., JR., and ORZALESI, N. 1963. The ultrastructure of the pigment epithelium and of the photoreceptor-pigment epithelium junction in the human retina. *J. Ultrastruct. Res.* 9:484–496.

BERNSTEIN, M. H. 1961. Functional architecture of the retinal epithelium. In *The structure of the eye,* ed. G. K. Smelser, pp. 139–150. New York: Academic Press.

BOK, D., and HALL, M. O. 1971. The role of the pigment epithelium in the etiology of inherited retinal dystrophy in the rat. *J. Cell Biol.* 49:664–682.

BRESNICK, G. H. 1970. Ocular effects of argon laser radiation. I. Retinal damage threshold studies. *Invest. Ophthalmol.* 9:901–910.

BROWN, K. T. 1965. An early potential evoked by light from the pigment epithelium choroid complex of the eye of the toad. *Nature* (Lond.) 207:1249–1253.

DANTZKER, D. R., and GERSTEIN, D. D. 1969. Retinal vascular changes following toxic effects on visual cells and pigment epithelium. *Arch. Ophthalmol.* 81:106–114.

DOWLING, J. E., and GIBBONS, I. R. 1962. The fine structure of the pigment epithelium in the albino rat. *J. Cell Biol.* 14:459–474.

DROZ, B. 1963. Dynamic condition of proteins in the visual cells of rats and mice as shown by radioautography with labeled amino acids. *Anat. Rec.* 145:157–167.

FEENEY, L. 1973*a*. Synthesis of interphotoreceptor matrix. I. Autoradiography of ^{3}H-fucose incorporation. *Invest. Ophthalmol.* 12:739–751.

FEENEY, L. 1973*b*. The interphotoreceptor space. II. Histochemistry of the matrix. *Develop. Biol.* 32:115–128.

FEENEY, L., BURNS, R. P., and MIXON, R. M. 1975. Human subretinal fluid: its cellular and subcellular components. *Arch. Ophthalmol.* 93:62–69.

FEENEY, L., GRIESHABER, J., and HOGAN, M. J. 1965. Studies on human ocular pigment. In *The structure of the eye,* ed. J. W. Rohen, vol. 2, pp. 535–548. Stuttgart: Schattauer-Verlag.

FINE, B. S. 1961. Limiting membranes of the sensory retina and pigment epithelium. *Arch. Ophthalmol.* 66: 847–860.

FINE, B. S., and YANOFF, M. 1972. Ocular histology: a text and atlas, pp. 48–107. New York: Harper and Row.

FINE, B. S., and ZIMMERMAN, L. E. 1963. Observation on the rod and cone layer of the human retina. *Invest. Ophthalmol.* 2:446–459.

FRIEDMAN, E., and KUWABARA, T. 1968. The retinal pigment epithelium. IV. The damaging effects of radiant energy. *Arch. Ophthalmol.* 80:265–279.

GRIGNOLO, A., ORZALESI, N., CASTELLAZZO, R., and VITTONE, P. 1969. Retinal damage by visible light in albino rats: an electron microscope study. *Ophthalmologica* 157: 43–59.

HAM, W. T., MUELLER, H. A., GOLDMAN, A. I., NEWMAN, B. E., HOLLAND, L. M., and KUWABARA, T. 1974. Ocular hazard from picosecond pulses of Nd: YAG laser radiation. *Science* 185:362–363.

HERRON, W. L., RIEGEL, B. W., MYERS, O. E., and RUBIN, M. L. 1969. Retinal dystrophy in the rat: a pigment epithelial disease. *Invest. Ophthalmol.* 8:595–604.

HOGAN, M. J., ALVARADO, J. A., and WEDDELL, J. E. 1971. *Histology of the human eye,* pp. 405–420. Philadelphia: W. B. Saunders.

HOLLYFIELD, J. G., and WARD, A. 1974. Phagocytic activity of the pigmented retinal epithelium. III. Interaction between lysosomes and ingested polystyrene spheres. *Invest. Ophthalmol.* 13:1016–1023.

ISHIKAWA, T., and YAMADA, E. 1970. The degradation of the photoreceptor outer segment within the pigment epithelial cell of the rat retina. *J. Electron Microsc.* 19: 85–99.

KOLB, H., and GOURAS, P. 1974. Electron microscopic observations of human retinitis pigmentosa, dominantly inherited. *Invest. Ophthalmol.* 13:487–498.

KUWABARA, T. 1970. Retinal recovery from exposure to light. *Am. J. Ophthalmol.* 70:187–198.

KUWABARA, T. 1975. Cytologic changes of the retina and pigment epithelium during hibernation. *Invest. Ophthalmol.* 14:457–467.

KUWABARA, T., and AIELLO, L. M. 1968. Effect of the ruby laser on the monkey retina: an electron microscopic study. In *Symposium on the treatment of diabetic retinopathy,* ed. M. Goldberg, pp. 665–671. U.S. Department of Health, Education, and Welfare.

KUWABARA, T., and COGAN, D. G. 1965. Retinal vascular pattern. VII. A cellular change. *Invest. Ophthalmol.* 4:1049–1058.

KUWABARA, T., and GORN, R. A. 1968. Retinal damage by visible light: an electron microscopic study. *Arch. Ophthalmol.* 79:69–78.

KUWABARA, T., and OKISAKA, S. 1976. Effect of electronic strobe flashlight on the monkey retina. *Jap. J. Ophthalmol.* 20:9–18.

MANDELCORN, M. S., MACHEMER, R., FINEBERG, E., and HERSCH, S. B. 1975. Proliferation and metaplasia of intravitreal retinal pigment epithelium cell autotransplants. *Am. J. Ophthalmol.* 80:227–237.

NOELL, W. K., WALKER, V. S., KANG, B. S., and BERMAN, S. 1966. Retinal damage by light in rats. *Invest. Ophthalmol.* 5:450–473.

OKISAKA, S., KUWABARA, T., and AIELLO, L. M. 1975. The effect of laser treatment of the retinal capillaries: an experimental study. *Am. J. Ophthalmol.* 80:591–601.

O'STEEN, W. K., and ANDERSON, K. V. 1971. Photically evoked responses in the visual system of rats exposed to continuous light. *Exp. Neurol.* 30:525–534.

SAKURAGAWA, M., and KUWABARA, T. 1976. The pigment epithelium of the monkey: topographic study by scanning and transmitting electron microscopy. *Arch. Ophthalmol.* 94:285–292.

SPITZNAS, M., and HOGAN, M. J. 1970. Outer segments of photoreceptors and the retinal pigment epithelium: interrelationship in the human eye. *Arch. Ophthalmol.* 84:810–819.

TSO, M. O. M. 1973. Photic maculopathy in rhesus monkey: a light and electron microscopic study. *Invest. Ophthalmol.* 12:17–34.

TSO, M. O. M., FINE, B. S., and ZIMMERMAN, L. E. 1972. Photic maculopathy produced by the indirect ophthalmoscope: a clinical and histopathologic study. *Am. J. Ophthalmol.* 73:686–699.

WALLOW, I. H. L., and TSO, M. O. M. 1973*a*. Repair after xenon arc photocoagulation. III. An electron microscopic study of the evolution of retinal lesions in rhesus monkeys. *Am. J. Ophthalmol.* 75:957–972.

WALLOW, I. H. L., and TSO, M. O. M. 1973*b*. Repair after xenon arc photocoagulation. II. A clinical and light microscopic study of the evolution of retinal lesions in the rhesus monkey. *Am. J. Ophthalmol.* 75:610–626.

WEINGEIST, T. A. 1974. Argon laser photocoagulation of the human retina. I. Histopathologic correlation of chorioretinal lesions in the region of the maculopapillar bundle. *Invest. Ophthalmol.* 13:1024–1032.

YAMADA, E., TOKUYASU, K., and IWAKI, S. 1958. The fine structure of retina studied with electron microscope. II. Pigment epithelium and capillary of the choriocapillary layer. *J. Electron Microsc.* 6:42–46.

YOUNG, R. W. 1967. The renewal of photoreceptor cell outer segments. *J. Cell Biol.* 33:61–72.

YOUNG, R. W. 1969. A difference between rods and cones in the renewal of outer segment protein. *Invest. Ophthalmol.* 8:222–232.

YOUNG, R. W. 1971. The renewal of rod and cone outer segments in the rhesus monkey. *J. Cell Biol.* 49:303–318.

YOUNG, R. W., and BOK, D. 1969. Participation of the retinal pigment epithelium in the rod outer segment renewal process. *J. Cell Biol.* 42:392–403.

YOUNG, R. W., and BOK, D. 1970. Autoradiographic studies on the metabolism of the retinal pigment epithelium. *Invest. Ophthalmol.* 9:524–536.

YOUNG, R. W., and DROZ, B. 1968. The renewal of protein in retinal rods and cones. *J. Cell Biol.* 39:169–184.

ACKNOWLEDGMENTS

The author acknowledges Dr. David G. Cogan for his continuous encouragement in this study.

18 Cryogenic and Thermal Effects on the Retinal Pigment Epithelium

Harvey Lincoff
Ingrid Kreissig

WHEN A cryosurgical probe with a temperature of −20° to −30° C is applied to the sclera of the rabbit eye, a snow-like lesion 2 mm in diameter appears in the retina in two to four seconds, depending on the thickness of the wall and intervening vascularity. When the freezing ceases the lesion regresses in seconds, leaving a faint gray patch of edema. Within hours a black inner ring and white outer ring can be perceived at the edges of the patch (Fig. 18.1). In subsequent days the lesion becomes pale and stippled with aggregates of pigment. Depending on the degree of injury, the lesion eventually becomes evenly repigmented with pigment epithelium or becomes pale and atrophic (Lincoff, McLean, and Nano 1964; Lincoff and McLean 1966; Lincoff et al. 1968).

Flat unstained preparations of the pigment epi-

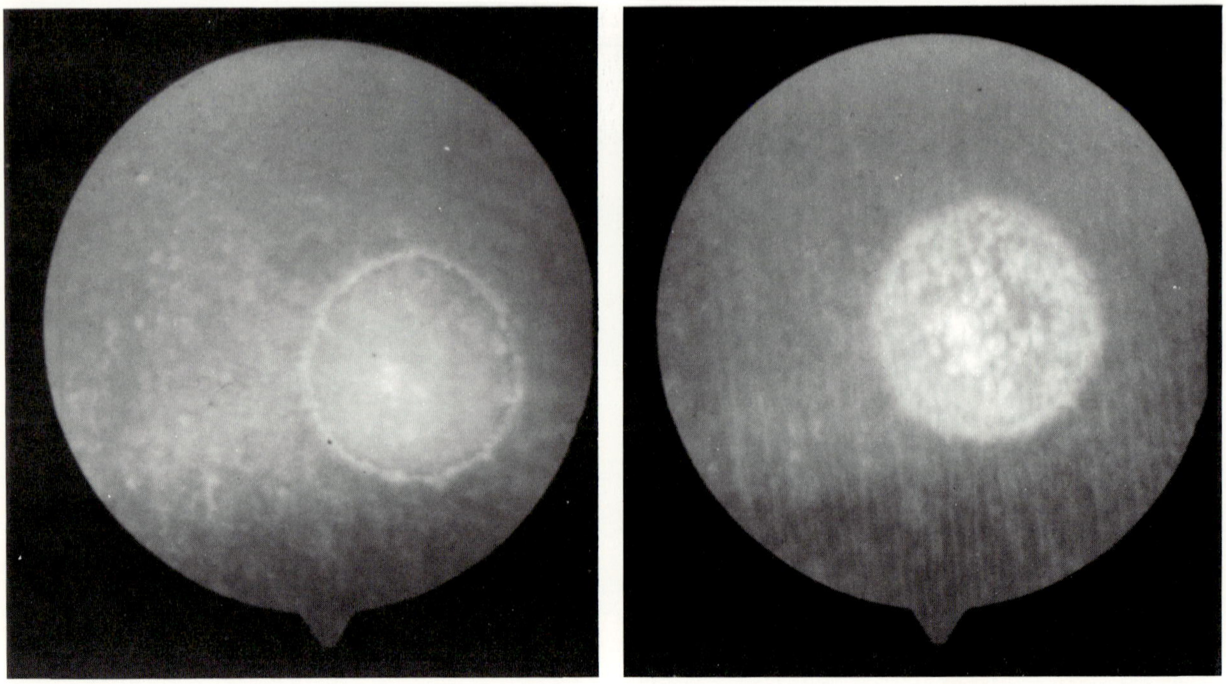

Figure 18.1 Left, *Cryopexy lesion in the rabbit retina after freezing; white outer and black inner rings surround an area of edema.* Right, *Cryopexy lesion on the fifth day; rings have disappeared, the lesion has become pale and stippled with pigment.*

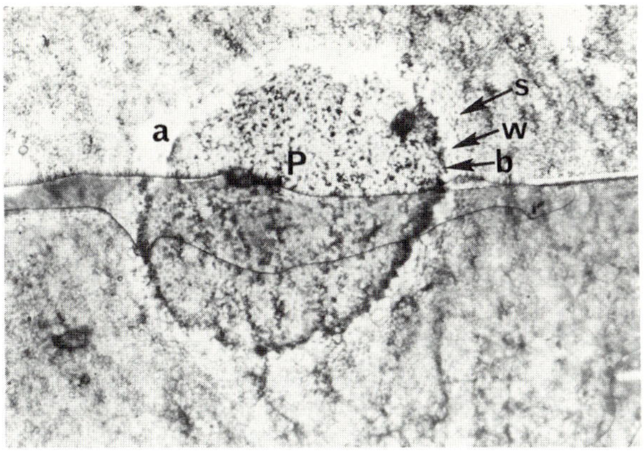

Figure 18.2 *Two-day cryosurgical lesion. The retina has been removed from the upper half. The center contains swollen pigment epithelium* (P). *At the periphery a black ring* (b), *white ring* (w), *and zone of signet cells* (s) *are perceivable. The rings have been artifactually detached in area* (a). *(Flat unstained preparation. ×25.)*

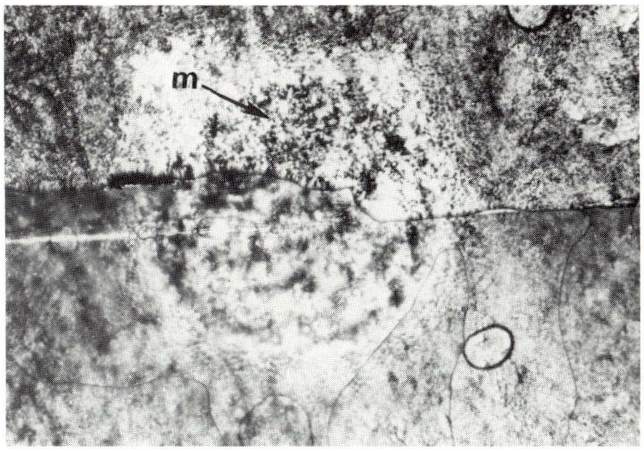

Figure 18.3 *Five-day cryosurgical lesion. The retina has been removed from the upper half. Groups of pigment-laden macrophages* (m) *cover the surface of Bruch's membrane in the center of the lesion. (Flat unstained preparation. ×25.)*

thelium reveal, after freezing, a swelling of the pigment epithelial mosaic (Fig. 18.2). The white ring at the edge of the lesion is the result of ruptured cells that have lost their pigment granules. The dark ring consists of a layer of densely packed pigment epithelial cells, three to six cells across. Whereas the cells in the center of the lesion are swollen and appear lighter, the cells in the dark band are shrunken, with a resultant concentration of pigment. Peripheral to the rings and beyond the edge of the frozen area, the pigment epithelium develops a signet-ring appearance. The granules, which are normally stacked like vertically placed pencils, assume an oblique position and appear to have migrated to the distal side of the cell, as though a strong current emanating from the center of the lesion has driven the pigment granules to the edge of the cell. While there are substantial convection currents around a freezing lesion, signet-ring formation is not present immediately after the application but develops over the next 24 hours. It is more likely, therefore, that the rings mark a critical temperature front or a border of transition between a solid and a fluid state. The signet-ring formation in the periphery may be caused by osmotic flow to the center of the lesion, where pigment epithelial cells have been fatally injured, cell membranes are breaking down, and a pool of protein molecules and electrolytes is coming into existence.

After five days, the peripheral rings and the pigment epithelial mosaic have disappeared (Fig. 18.3). In their place are aggregates of pigment that, on microscopic examination, prove to be collections of large pigment-laden macrophages (Fig. 18.4). A lighter

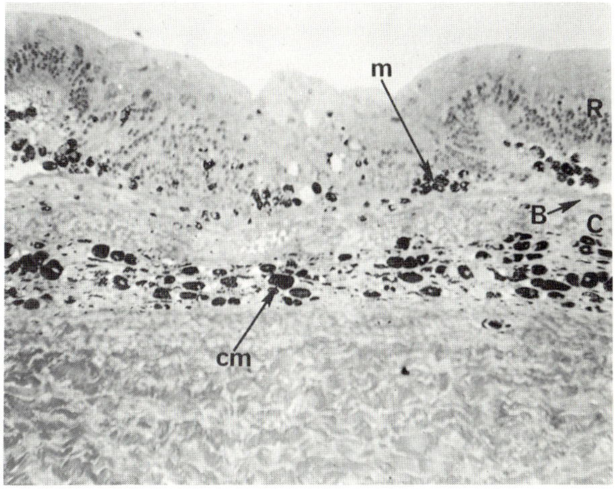

Figure 18.4 *Seven-day cryosurgical lesion.* R, *retina;* B, *Bruch's membrane;* C, *choroid;* m, *an aggregate of retinal macrophages;* cm, *choroidal macrophages. (Stained with toluidine blue. ×60.)*

brown pigment arranged in cords becomes evident at a deeper level (Fig. 18.5). The cords are composed of choroidal macrophages filled with pigment from choroidal melanocytes that have undergone disruption. The linear distribution is in relation to the choroidal blood vessels. The retinal and choroidal macrophages do not mingle but stay on their respective sides of Bruch's membrane. This is con-

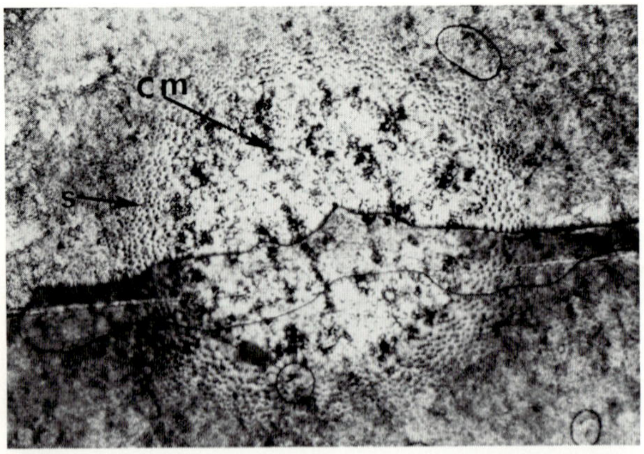

Figure 18.5 *Seven-day medium cryosurgical lesion. The retina has been removed from the upper half. The pigment epithelium has been phagocytized and macrophages in the choroid* (cm), *filled with pigment from melanocytes, show through Bruch's membrane. The choroidal pigment has a linear arrangement. Signet cells* (s) *persist in the periphery. (Flat unstained preparation.* ×25.)

firmed by examination of the granules. The macrophages in the retina contain long granules typical of pigment epithelium, whereas those in the choroid contain the short oval granules of choroidal melanocytes.

While pigment epithelial mosaic cannot be seen in flat preparations after seven days, microscopic sections reveal that a new pigment epithelial layer without granules has covered the lesion (Fig. 18.6). In the 14-day flat preparation, beginning pigmentation of the new epithelium can be seen at the periphery of the lesion (Fig. 18.7). Pigmentation across the lesion becomes complete by seven weeks. The new mosaic is paler; the cells are smaller and less perfect than the original and they remain so, as demonstrated in a 14-month lesion.

Histology of Thermal Injury

An electron micrograph of a three-day lesion reveals that all the layers of Bruch's membrane have survived freezing except for the basement membrane of the endothelium, which is only intermittently present (Figs. 18.8, 18.9). On the retinal side, the pigment epithelium is undergoing dissolution. The cytoplasm of the epithelium shows the granulation effect previously described as following thermal injury (Fine and Geeraets 1966a; Fine and Geeraets 1966b; Orzalesi, Grignolo, Calabria, and Castellazzo 1967). The melanin granules of the cells are in disarray. The villi are fragmented. The mitochondria are degenerated. The outer segments of the visual cells also show a granulation effect. Fibrin lies in the space that has appeared between the pigment epithelium and the outer segments. A new cell with undamaged cytoplasm is sliding over Bruch's membrane beneath the necrotic pigment epithelium.

An electron micrograph of the seven-day lesion shows the persistence of Bruch's membrane, except for the basement membrane of the endothelium. The choriocapillaris is absent. On the inner surface of Bruch's membrane there is a layer of new cells now identifiable as pigment epithelium without granules. The cells have typical basal infoldings and terminal bars. The granules of the former epithelial layer are in the macrophages that lie above. The rods and cones are gone and in their place is retinal glia. An inspection of the junction between the retinal glia and the pigment epithelium in the seven-day lesion reveals desmosomal-like attachments (Figs. 18.10, 18.11).

What is probably the beginning of granule formation in the new pigment epithelium can be seen in the seven-day lesion. By three weeks a definite granule, small and round in shape, and oriented towards the inner edge of the cell, is to be seen (Fig. 18.12). Photographs of four-week lesions reveal multiple layers of granules and more pronounced orientation towards the apex of the cell (Fig. 18.13). The granules eventually become spindle-shaped (Fig. 18.21).

Variations in the Pattern of Regeneration

The character of regeneration of pigment epithelium is dependent upon the degree of injury to the epithelium as well as adjacent structures. It is possible to define ophthalmoscopically three degrees of injury to the rabbit retina: light, medium, and heavy, according to the clinical usage. For light injury, freezing application is stopped the moment the choroid blanches. After warming there is no immediate evidence of freezing, but within an hour edema of pigment epithelium and retina becomes apparent. For the medium injury, freezing is stopped when white first appears in the retina. Upon warming, an area of retinal edema is immediately visible. For the heavy lesion, freezing is sustained for three seconds beyond the first appearance of snow in the retina. Upon warming the retina remains opaque.

The light lesion. Microscopic examination four days after light freezing reveals necrosis in the region of the outer limiting membrane. Bruch's membrane is intact. The pigment epithelium is swollen and irregular. Interior to it are a number of pigment-

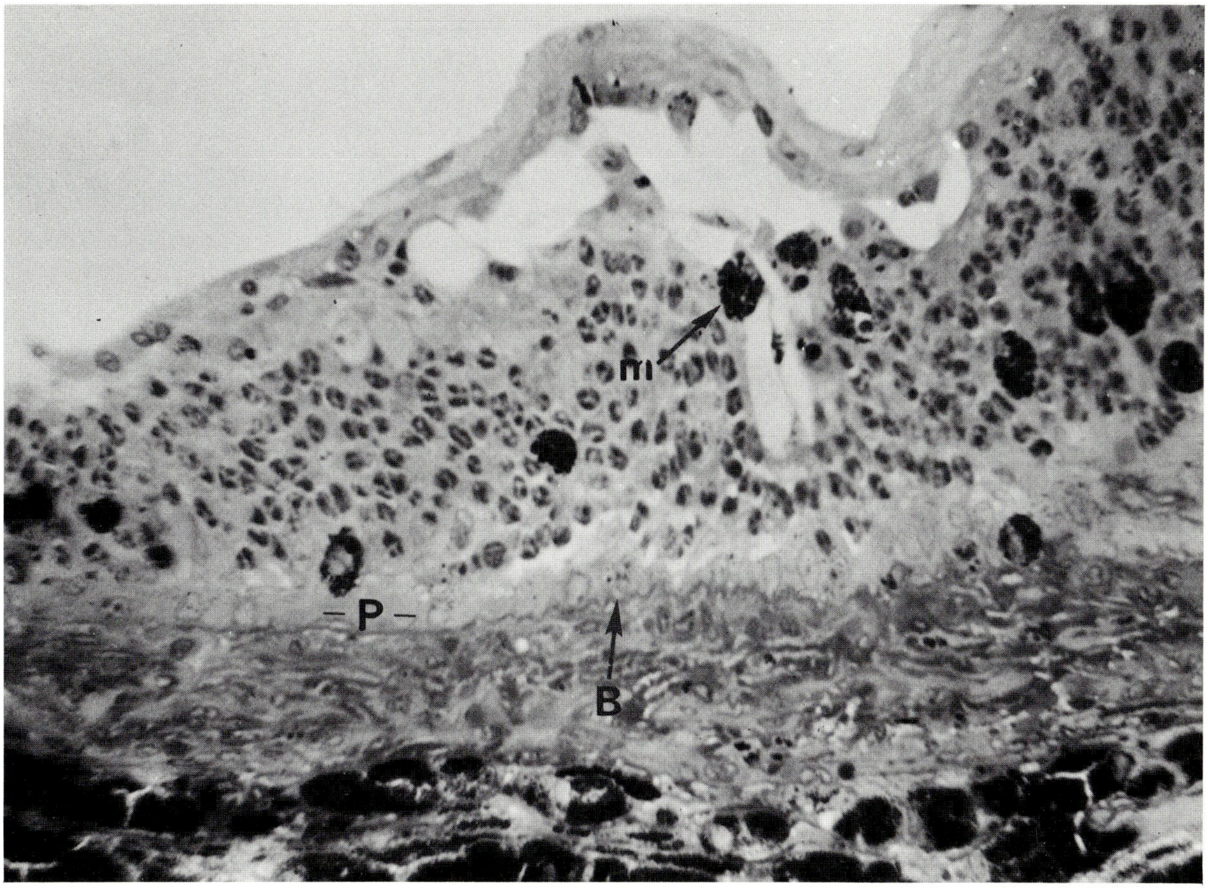

Figure 18.6 *Bruch's membrane* (B) *demonstrated in a seven-day medium cryosurgical lesion. Amelanotic epithelium* (P) *with pale nuclei appears on the surface of Bruch's membrane. Some retinal macrophages* (m) *have migrated into the inner layers of the retina. (Stained with toluidine blue. ×400.)*

laden macrophages. The photoreceptor elements have lost their vertical alignment and appear damaged, but the nuclei internal to the outer limiting membrane usually survive (Fig. 18.14).

Electron micrographs of the four-day lesion confirm that Bruch's membrane survives light freezing intact. Freezing stimulates fibroblastic activity in the membrane. The pigment epithelial cells have been damaged and some are being displaced inward by new cells sliding over Bruch's membrane. The displaced cells appear to be functioning as wandering macrophages in the retina. The outer segments at four days are largely missing and have probably been phagocytized. Damaged inner segments filled with mitochondria are in contact with pigment epithelium. At the level of the outer limiting membrane there are enlarged mitochondria within Müller's cells and processes of Müller's cytoplasm can be seen to extend toward Bruch's membrane.

Microscopic examination of the 14-day lesion shows regeneration of photoreceptor elements (Fig. 18.15). Outer segments are again in parallel columns and interdigitating with pigment epithelial villi. The extensions of Müller's cytoplasm are no longer apparent and seem to have retreated. The electron micrograph reveals that the outer segment discs are again in horizontal order (Fig. 18.16). The pigment epithelium is deformed but active, as indicated by exaggerated basal infoldings and numerous mitochondria. Pigment granules have returned to the internal edge of the epithelium. The granules in some cells are spindle-shaped, indicating that they are surviving rather than new epithelium. The presence of occasional large pigment-bearing macrophages among the outer segments and in the choroid is the most striking evidence of the episode of injury and repair that has occurred.

In summary, the light cryosurgical adhesion results from cryogenic necrosis to the region of the external limiting membrane. Some pigment epithe-

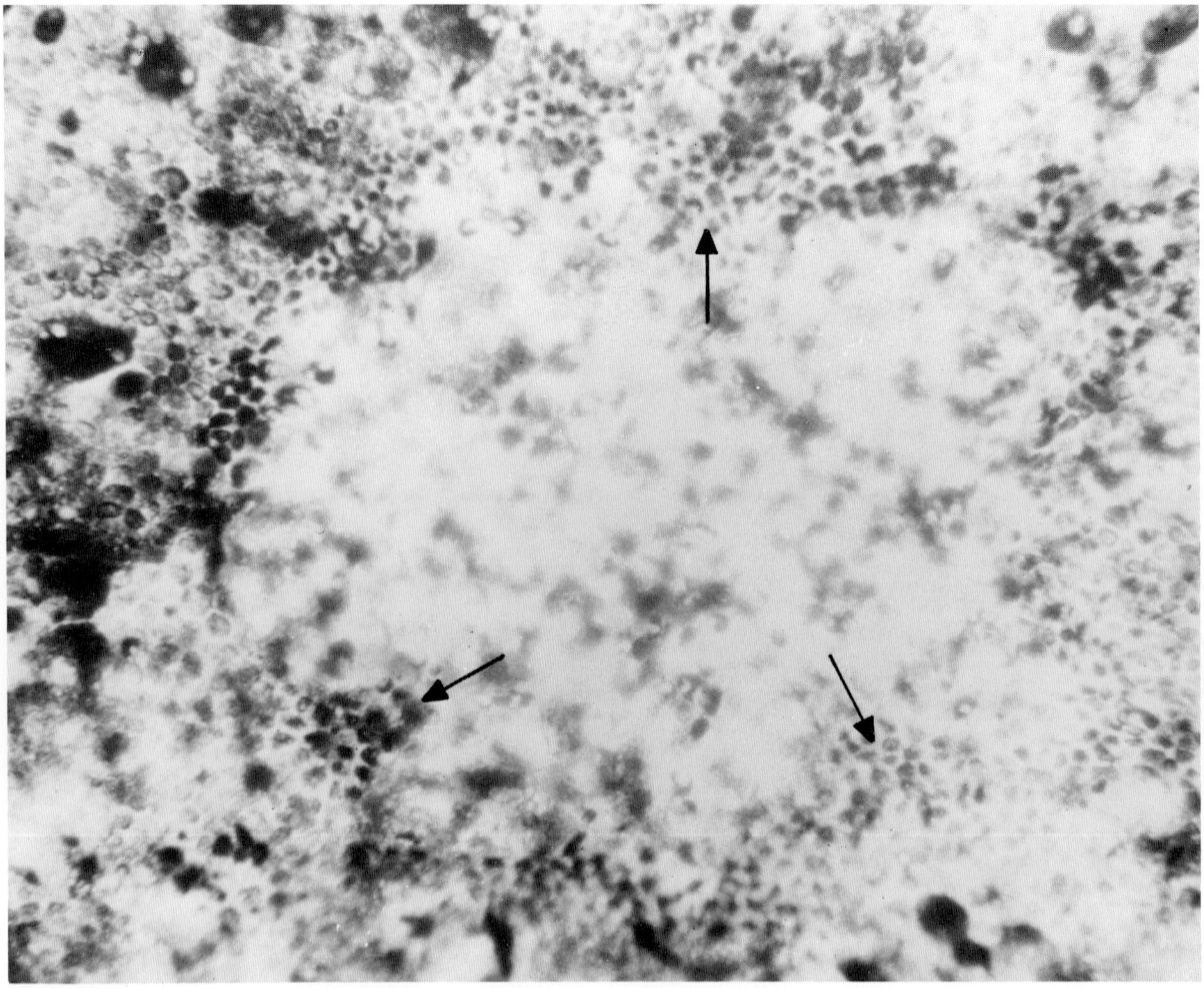

Figure 18.7 *Pigmented epithelium (arrows) reappearing in the periphery of a 14-day cryosurgical lesion. (Flat unstained preparation. ×41.)*

lial cells survive, others are replaced. The controlling aspect of the lesion is the survival of the outer nuclei, which makes for rapid recovery of the inner segments and regeneration of the outer. This is consistent with the observations of Kroll and Machemer (1970) and Kuwabara (1970), who observed regeneration in retina damaged by detachment or by extreme light exposure.

The medium lesion. Microscopic examination reveals that the freezing necrosis has penetrated to the nerve fiber layer and destroyed the pigment epithelium, the photoreceptor elements, the outer nuclei and most of the inner nuclei. The nerve fiber layer and some Müller's cells have survived (Fig. 18.17). The sequel to massive cryogenic necrosis of the retina is not leukocytic invasion, as one might expect, but infiltration by a large mononuclear cell with phagocytic abilities; the cells are swollen with inclusions. New epithelium without pigment slides across Bruch's membrane to replace lost cells. At days four and six, pigment epithelium and macrophages are the predominant cells in the lesion.

The necrotic and phagocytic phase is complete by the ninth day and the macrophages have retreated. It is thought that some enter the vitreous, because they are frequently found internal to the inner limiting membrane. Others move to the margin of the lesion, where they are found in large numbers and are the source of gross pigmentation at the periphery of the lesion. The marginal cells contain, in addition to pigment granules, numerous large inclusions

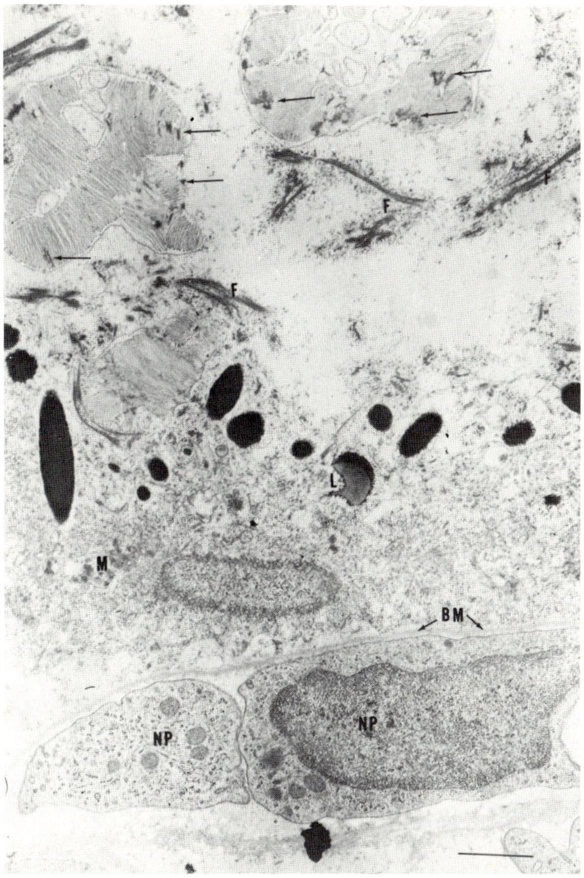

Figure 18.8 *Region of necrotic pigment epithelium three days after a cryopexy application. Necrosis of the pigment cell is evident by the widespread granulation effect throughout its cytoplasm, nucleus, and mitochondria* (M), *as well as within the adjacent photoreceptor outer segments (free arrows). The melanin granules lie in disarray. A single lipofuscin-type granule* (L) *is present. Many strands of fibrin* (F) *are present in the space between the damaged cells. Two normal, nonpigmented cells* (NP) *lie beneath the basement membrane* (BM) *of the pigment epithelial cells. These cells are not normally present and have migrated to this location.* (×9,900.)

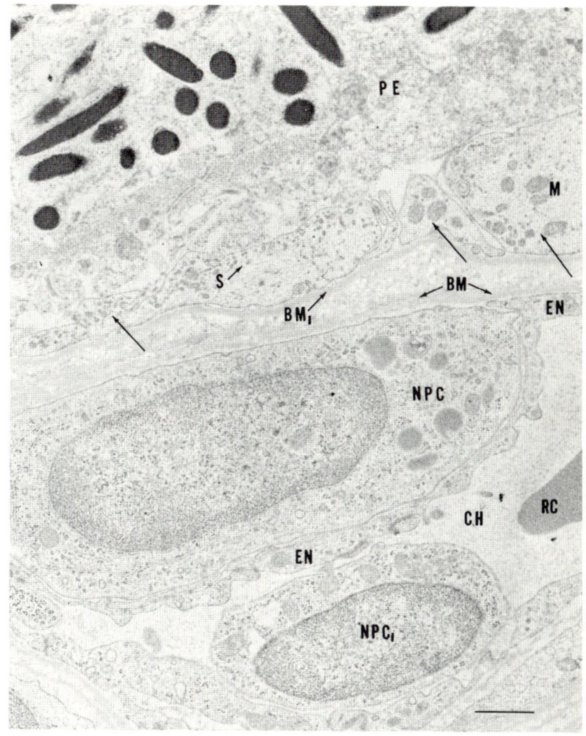

Figure 18.9 *Another region at the base of necrotic pigment epithelial cells* (PE) *three days after freezing. The granulated cytoplasm is evident and is distinct from the lucent cytoplasm of a foreign cell or cells (free arrow), which has interposed itself between the pigment cell and its basement membrane* (BM$_1$). *The new cell has a well-defined plasma membrane surrounding a watery cytoplasm in which many fragments of smooth endoplasmic reticulum* (S) *and mitochondria* (M) *are seen. Deep to the Bruch's membrane lies a large, nonpigmented cell* (NPC) *similar in appearance to the sub-basement-membrane cells in Figure 18.11. The segment of adjacent choriocapillaris* (CH) *appears normal and a fragment of a red blood cell* (RC) *is present in the plasma-filled lumen. The patency of this portion of choriocapillaris indicates that it lies toward the periphery of the lesion. A second, nonpigmented cell* (NPC$_1$) *lies within the lumen and resembles the larger-appearing nonpigmented cell* (NPC) *which lies outside the endothelial lining* (EN), *but still within the endothelial basement membrane* (BM). *These observations suggest that migration of cells from the lumen of the vessels is a source for invading macrophages.* (×10,065.)

filled with a fine granular substance. As they persist and have a single membrane, these inclusions are thought to be phagosomes. The regenerated structure of retina after a medium thermal injury consists of the nerve fiber layer bound to pigment epithelium by layers of undifferentiated neuroepithelium, which appear to be derived from Müller's cells and possibly pigment epithelium (Fig. 18.18). As they mature, the new retinal cells develop a prominent fibrillar structure. Some contain an occasional pigment granule or a myeloid inclusion body (Fig. 18.19). The new retinal cells attach to each other with desmosomes.

The connection of the retinal epithelium to the pigment epithelium can take a number of forms: (1) a smooth inner surface of pigment epithelium may connect to a smooth membrane of Müller's cells

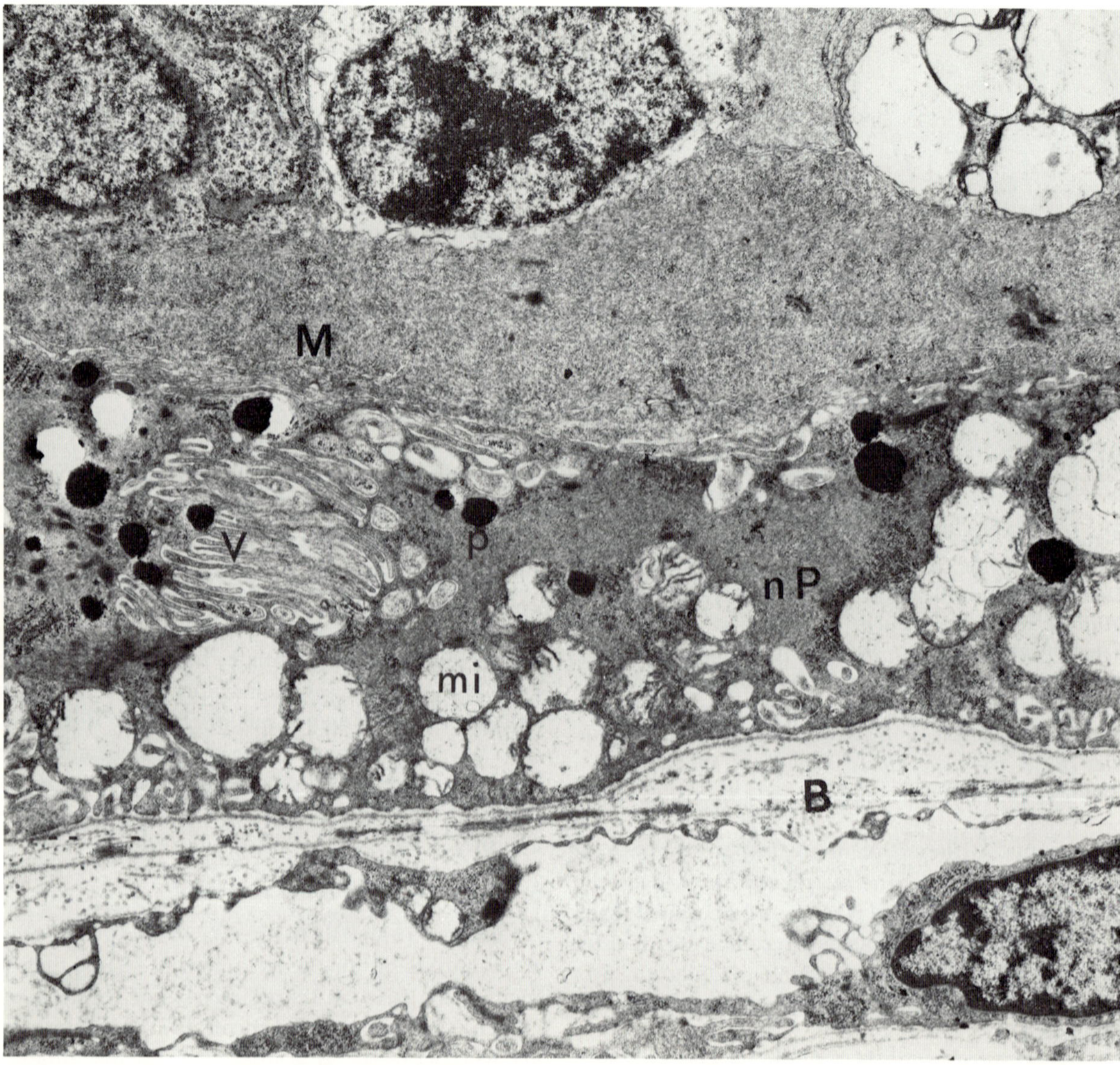

Figure 18.12 *Small round pigment granules* (p) *in new pigment epithelium* (nP) *three weeks after cryopexy. Apical villi* (V) *and mitochondria* (mi) *have evolved.* M, *Müller's cell;* B, *Bruch's membrane. (Courtesy of Ben Fine. ×13,400.)*

and 2 mm for a heavy one. If one considers 2 mm as the average size for a lesion, then lesions caused by light cryopexy have a diameter 20 percent less than average, and those caused by heavy diathermy a diameter 15 percent more than average. While size is a factor in adhesions, the histologic characteristics are far more important. The eyes were removed at intervals of two days to six months. The strength of the adhesion was determined in the unfixed eye by pulling strips of retina through the area of the thermal lesion. A strip of normal rabbit retina 6 mm wide tears when subjected to a force of 300 mg. However, by placing the lesions at the ends of the myelinated fibers, it was possible to utilize the myelinated bundle to pull on the adhesion. A 6 mm strip of retina that includes the myelinated bundle withstands forces as great as 1,850 mg.

From flat preparations of the freshly enucleated eyes, strips 6 mm wide were cut along the course of the myelinated fibers and pinned to a dissection block. The block was clamped under water in a small Petri dish and mounted on a descending stage. The

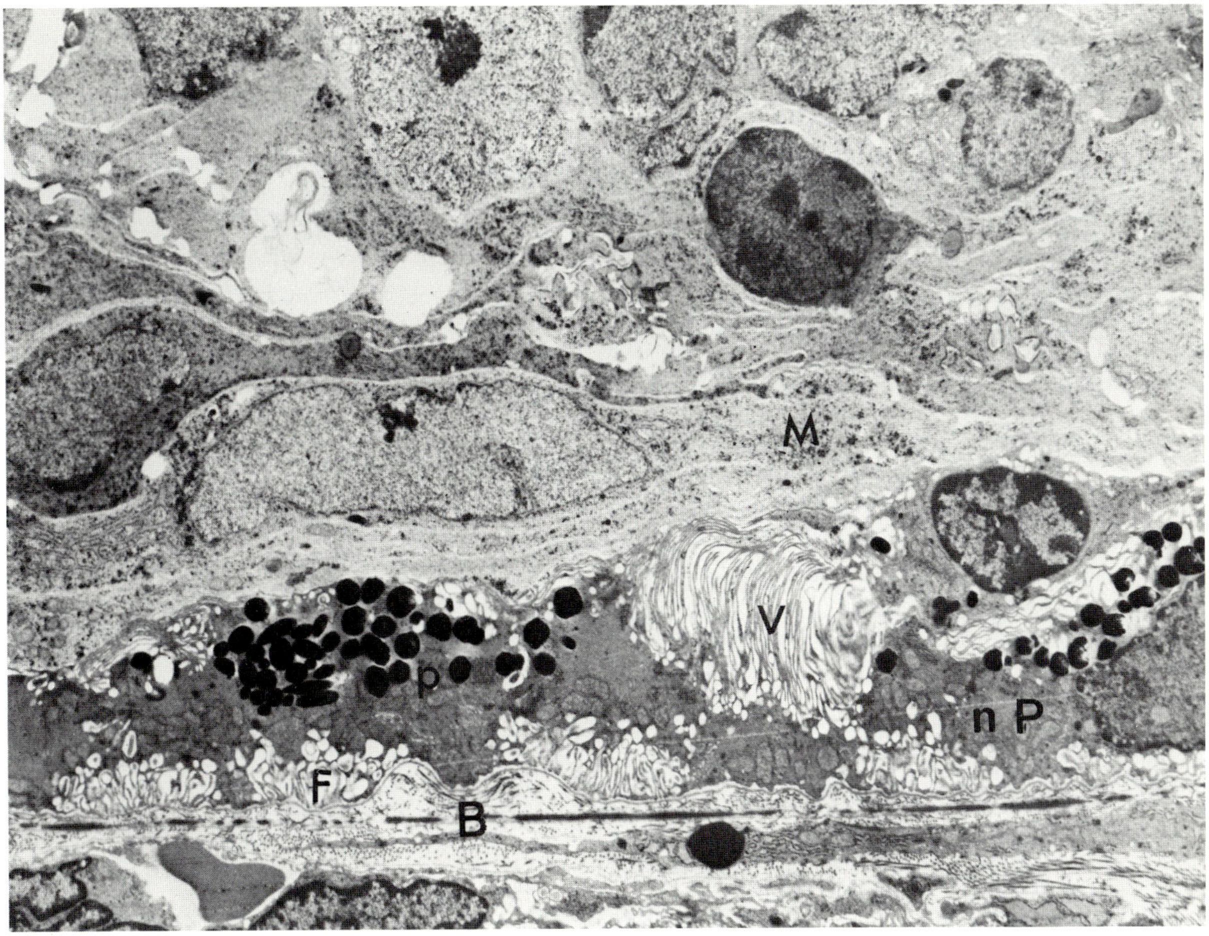

Figure 18.13 *Multiple layers of pigment granules* (p) *oriented towards the apex of new pigment epithelium* (nP) *four weeks after cryopexy. Basal infoldings* (F) *and apical villi* (V) *are more fully developed.* M, *Müller's cell process.* B, *Bruch's membrane. (Courtesy of Ben Fine.* ×*5,400.)*

myelinated fibers and sensory retina were severed at the disc and seized with a 6–0 silk thread. The thread was attached to a strain gauge transducer, which was, in turn, connected to a polygraph. As the stage descended, pigment epithelium was pulled away from the retina and the resistance to the separation was recorded. The rate of descent was controlled by an electric motor and kept constant. The pulling was accomplished within 15 minutes after the eye was enucleated from the anesthetized animal.

The results are reflected in the graphs (Fig. 18.26). The adhesiveness produced by cryopexy was found to vary directly with the intensity of the application. Thus, heavy lesions were significantly stronger than medium lesions, and medium lesions were significantly stronger than light lesions. For lesions produced by diathermy there was no significant increase in the strength of the lesion with increased intensity of application (this may reflect an inability to control the thermal damage initiated by diathermy). When the cryopexy was compared with diathermy at various levels of application, cryopexic lesions were found to be significantly stronger than diathermic lesions at the heavy level of application, equal in strength at the medium level, and occasionally weaker at the light level. All thermal lesions obtained maximum adhesion by 12 days. No increase in strength was observed in lesions tested as late as six months after application.

As the strength of the retinal adhesion varies with the intensity of the cryosurgical application, and the anatomy of the retinal adhesion also varies with the intensity of the application, it is appropriate to correlate the adhesive strength with the anatomy

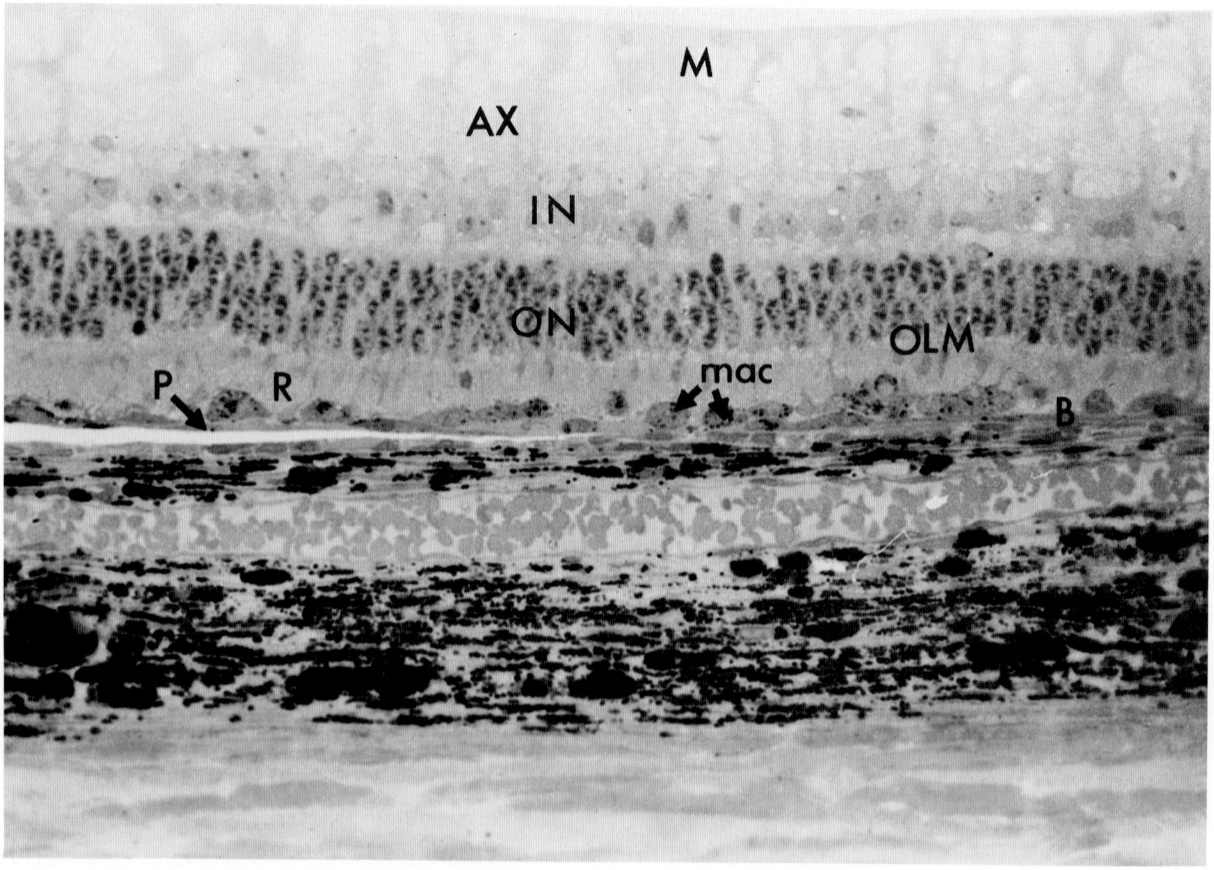
M
AX
IN
ON
OLM
P
R
mac
B

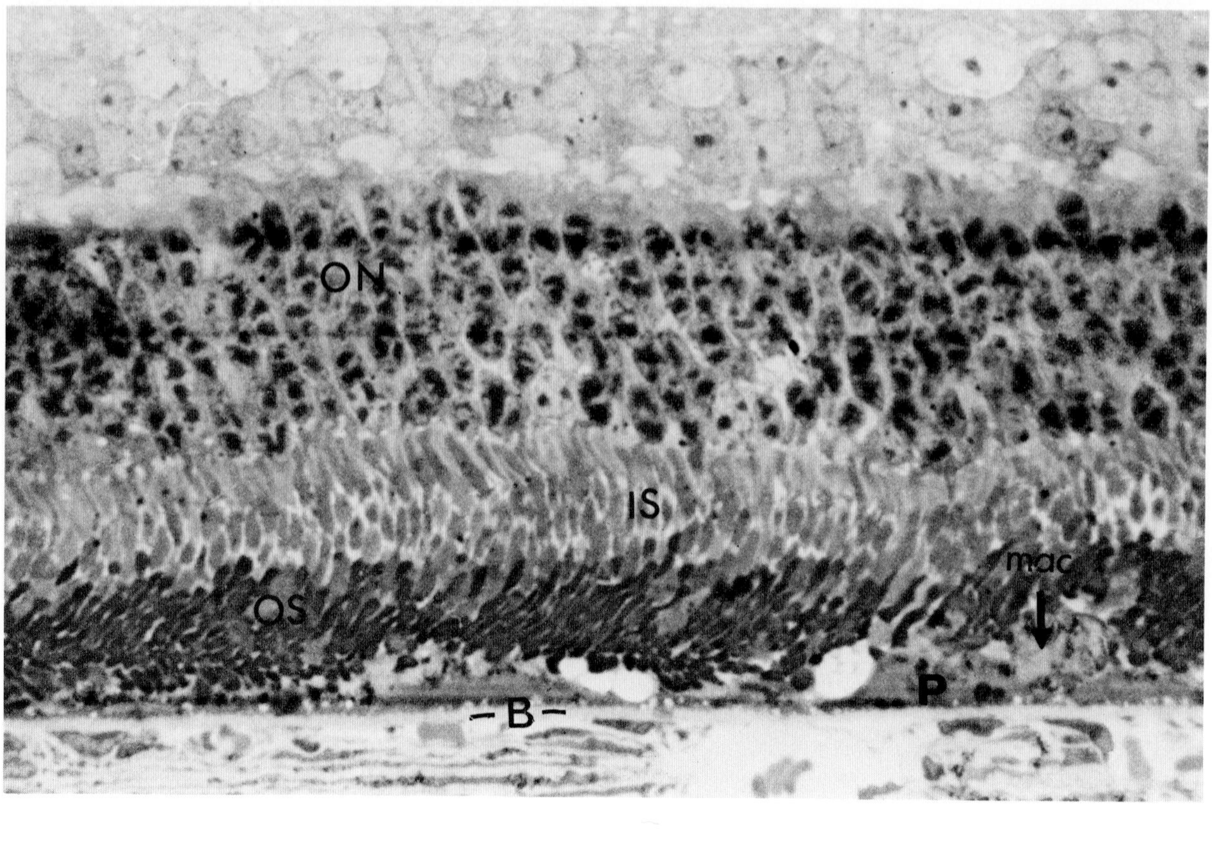
ON
IS
OS
mac
P
-B-

← **Figure 18.14** *Section of retina four days after light cryopexy showing necrosis to the level of the outer limiting membrane* (OLM). *Pigment epithelium* (P) *is flattened. There are macrophages* (mac) *above. The photoreceptor elements* (R) *are necrotic. The outer nuclei* (ON) *are normal, as are the inner nuclei* (IN), *the nerve fiber layer* (AX), *and Müller's cells* (M). B, *Bruch's membrane. (Stained with toluidine blue. Courtesy of Ben Fine. ×76.)*

↙ **Figure 18.15** *Fourteen days after light cryopexy, retina shows recovery of photoreceptor process. Inner segments* (IS) *and outer segment* (OS) *are again in parallel columns and are interdigitating with the pigment epithelium* (P). *A large macrophage* (mac) *sits above the epithelium.* B, *Bruch's membrane. (Stained with toluidine blue. ×190.)*

(Lincoff and Kreissig 1971) and derive some information about the strength of the various anatomical connections. Thus, the adhesion between pigment epithelium and regenerated outer segments, obtained after a light application of cryopexy, is an interdigital connection between villous structures and is probably the weakest form of adhesion, with an average strength of 300 mg. The increased adhesive strength, averaging 600 mg, between retina and pigment epithelium found in the medium lesion is probably mediated by the desmosomal connections that evolve. The strongest adhesion, averaging 1,040 mg, obtained by a heavy application of cryopexy, appears to occur because Bruch's membrane is ruptured and neuroepithelium invades the collagen layers of the choroid, resulting in a chorioretinal scar.

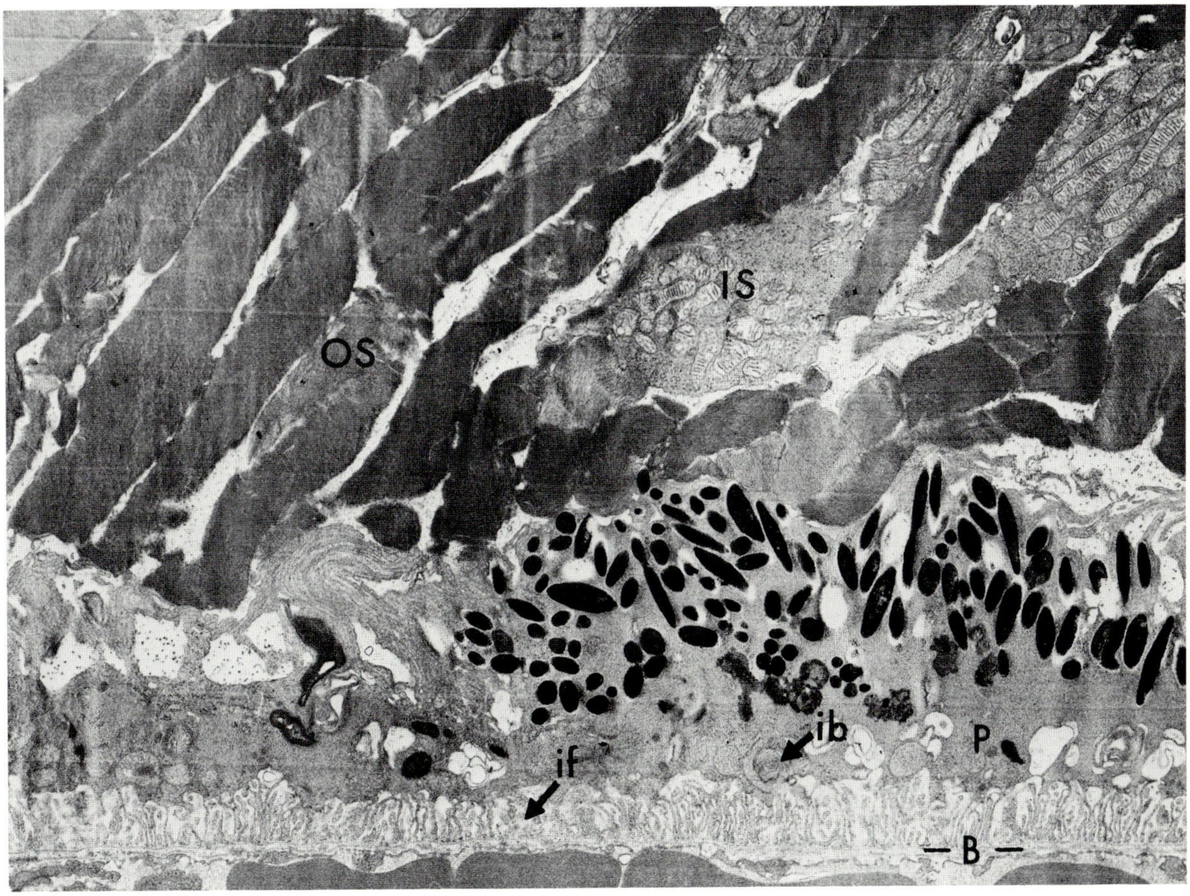

Figure 18.16 *Electron micrograph of retina 14 days after light cryopexy, confirming the regeneration of outer segments* (OS). *Occasional inner segments* (IS) *approach the pigment epithelium* (P). *Granules have returned to the inner surface of the epithelium. Note exaggerated basal infoldings* (if) *and a myeloid inclusion body* (ib). *Bruch's membrane* (B) *is intact. (×6,325.)*

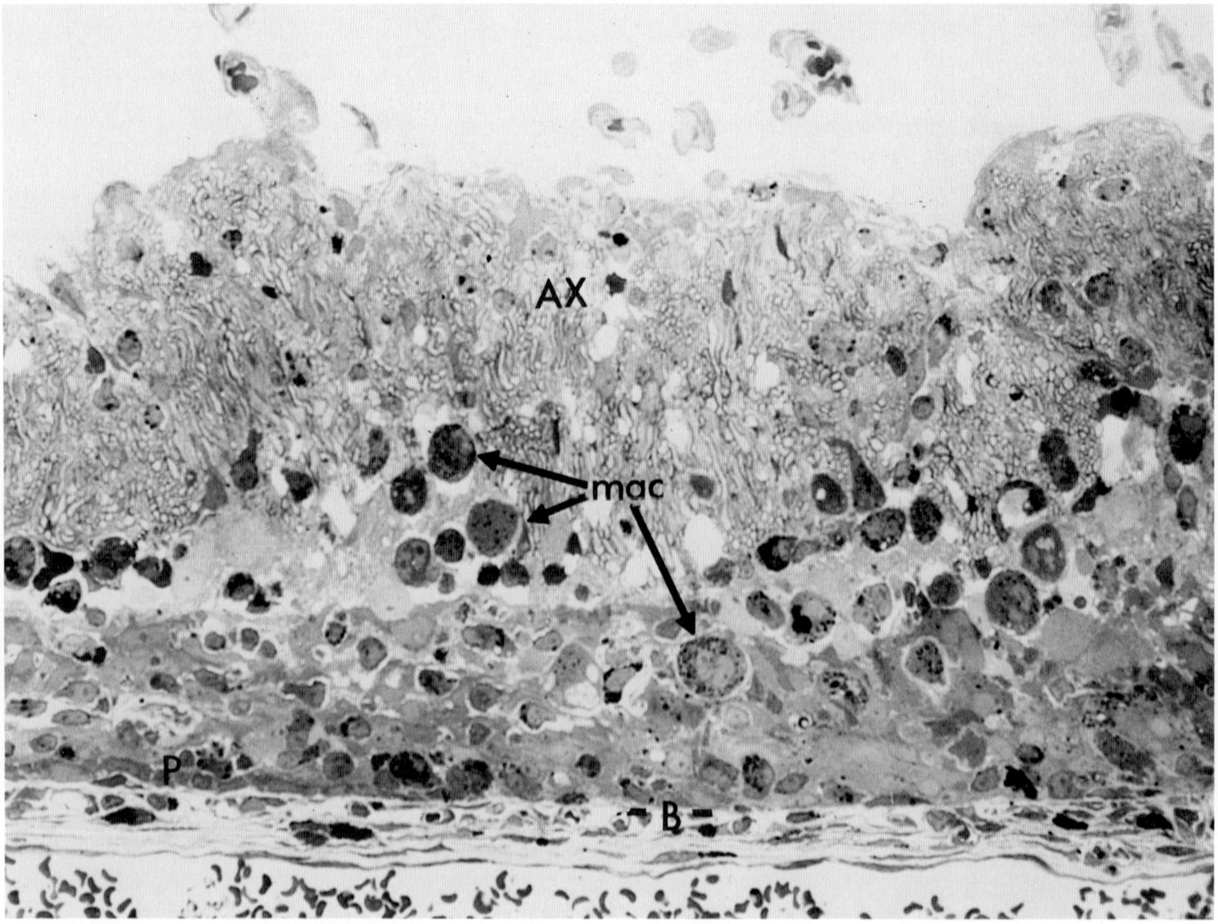

Figure 18.17 *Section of retina six days after medium cryopexy. Freezing necrosis has penetrated to the level of the nerve fiber layer* (AX). *The photoreceptor processes and the nuclear layers are gone. The predominant cell is the macrophage* (mac). B, *Bruch's membrane. (Stained with toluidine blue. ×78.)*

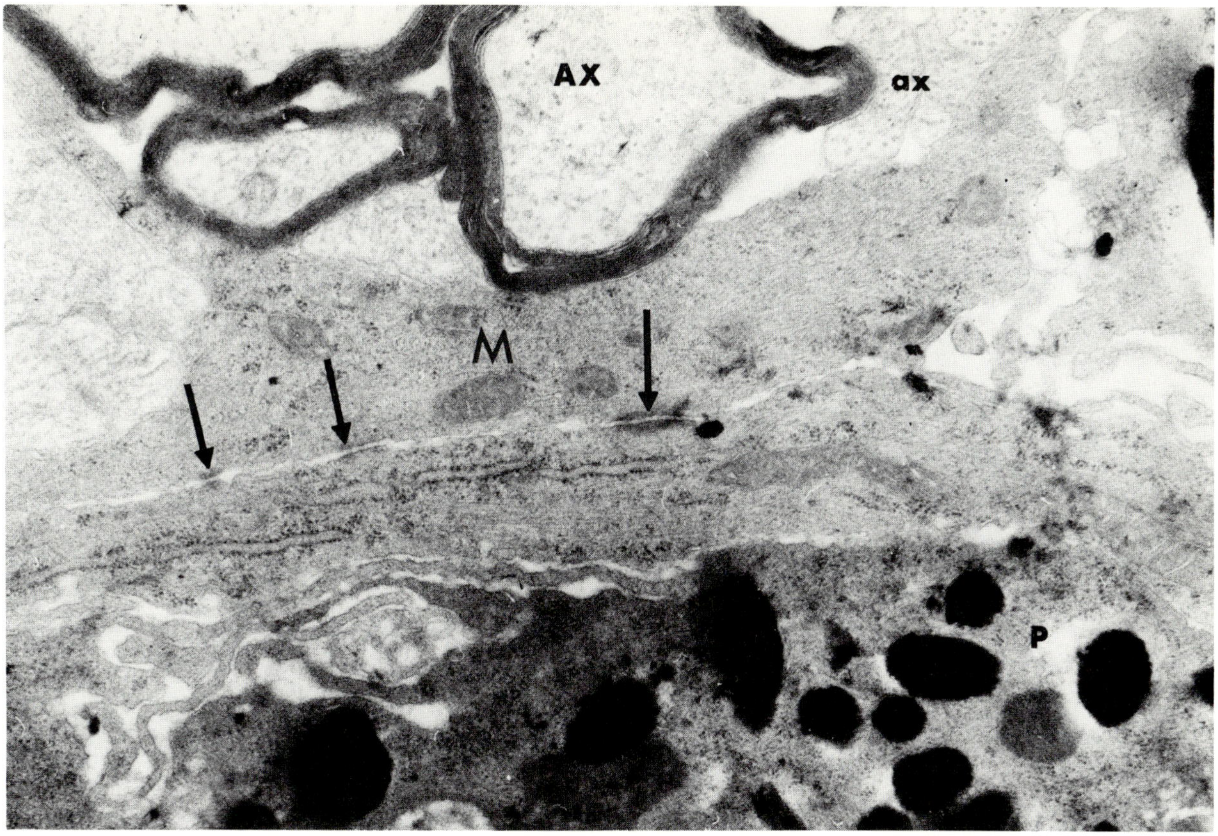

Figure 18.18 *Electron micrograph of retina eight days after medium cryopexy demonstrates glial cells* (M) *connecting nerve fiber bundles to pigment epithelium* (P). *Desmosomal densities (free arrows) connect the cell layers. Note that nonmyelinated nerve fiber bundles* (ax) *as well as myelinated fibers* (AX) *have survived freezing.* (×*18,200.*) B = *Bruch's membrane. Desmosomes are indicated by free arrows.* (×*18,200.*)

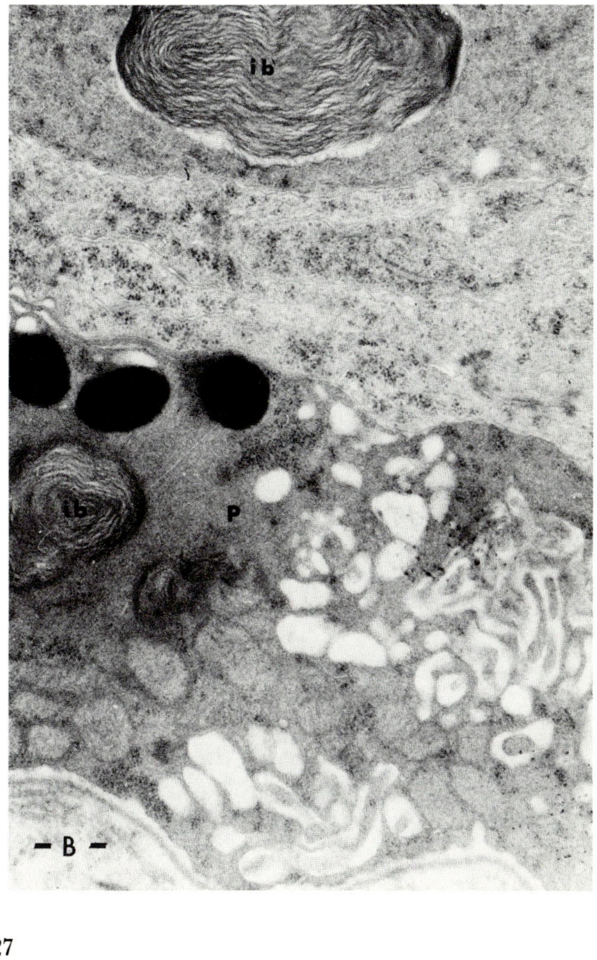

Figure 18.19 *Electron micrograph 28 days after medium cryopexy demonstrates myeloid inclusion bodies* (ib) *in pigment epithelium* (P) *and in the retinal neuroepithelium above.* B, *Bruch's membrane.* (×*13,200.*)

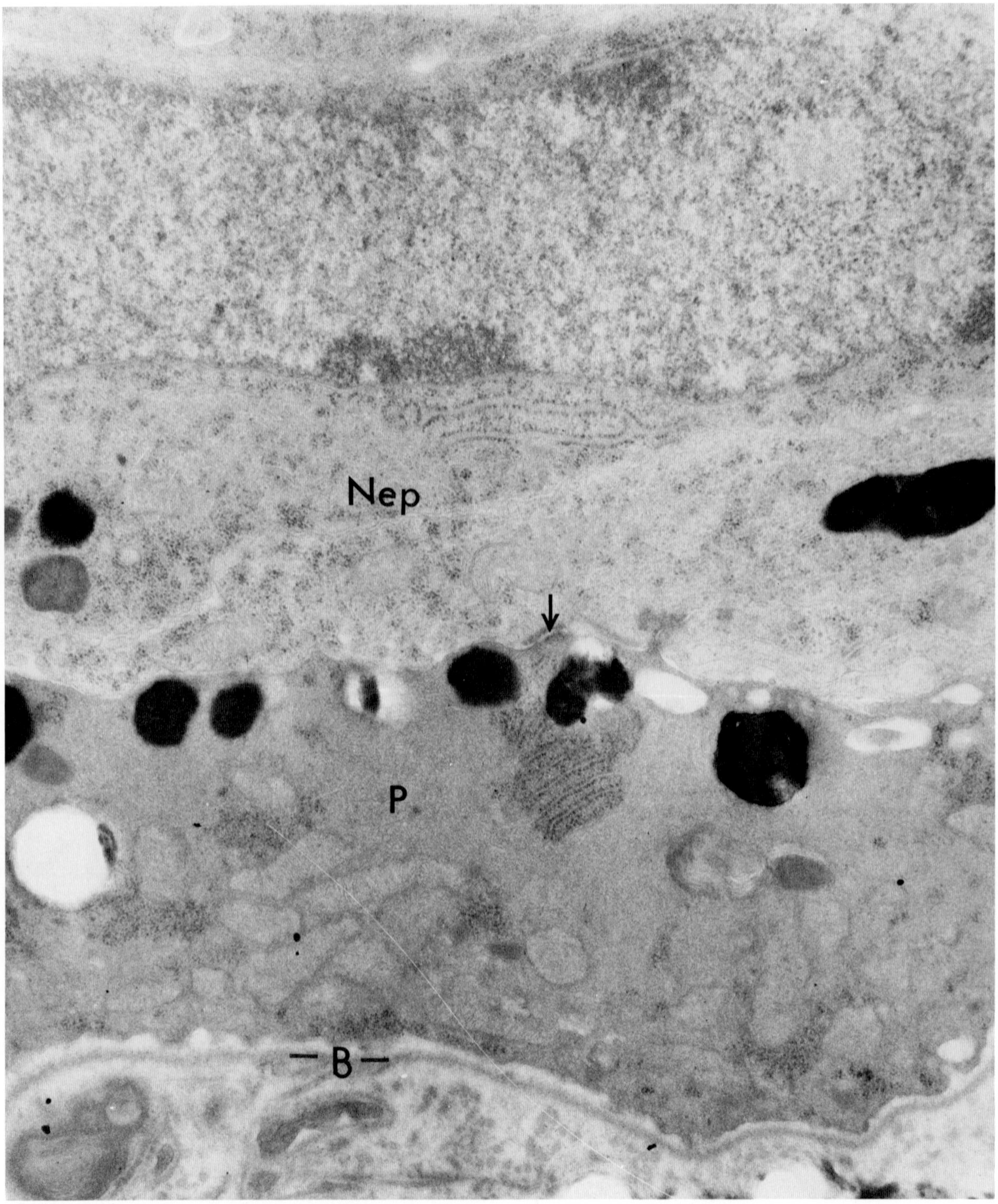

Figure 18.20 *Electron micrograph of the retina 28 days after medium cryopexy shows desmosomal connections (free arrow) between the apical side of pigment epithelium* (P) *and the retinal neuroepithelium* (Nep) *interior to it.* B, *Bruch's membrane.* *(×17,000.)*

Figure 18.21 *Electron micrograph of retina 20 months after medium cryopexy showing microvilli* (V) *of pigment epithelium* (P) *interdigitating with fingerlike projections* (f) *of Müller's cells. There are numerous terminal bars (free arrows) between the cells, and large mitochondria* (m) *internal to them, suggesting a reformation of outer limiting membrane.* B, *Bruch's membrane. (×14,400.)*

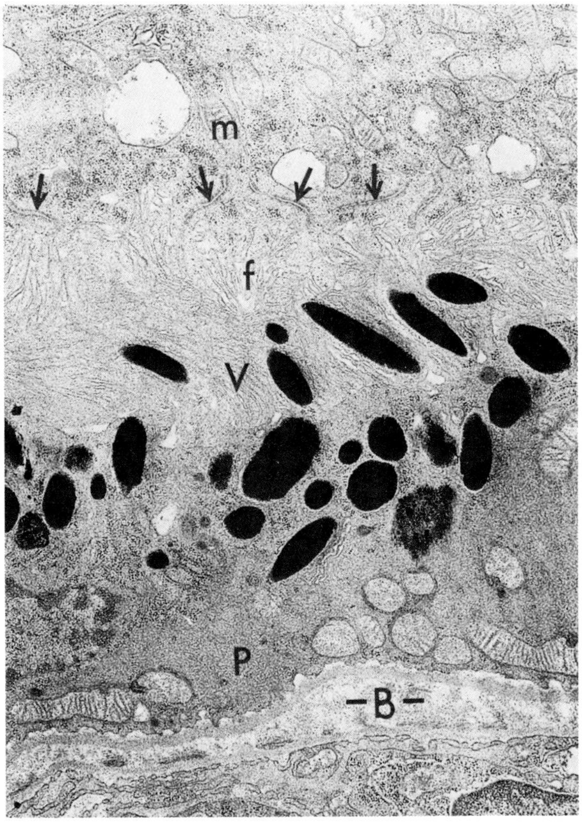

Figure 18.22 *Section of retina four days after a heavy cryopexy application, showing full thickness necrosis. The nerve fiber layer* (AX) *has been destroyed. A few inner nuclei* (IN) *are present and occasional macrophages* (mac). B, *Bruch's membrane. (Stained with toluidine blue. ×195.)*

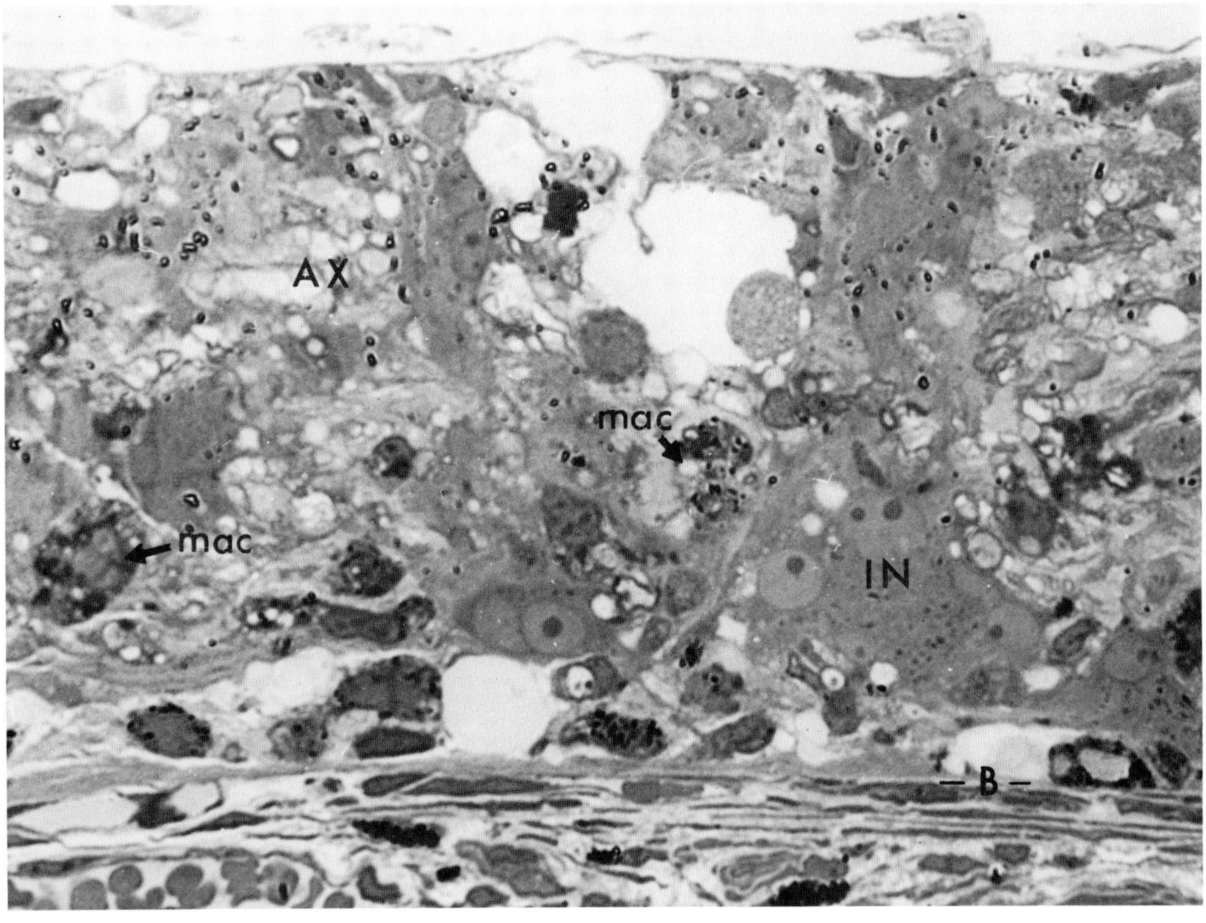

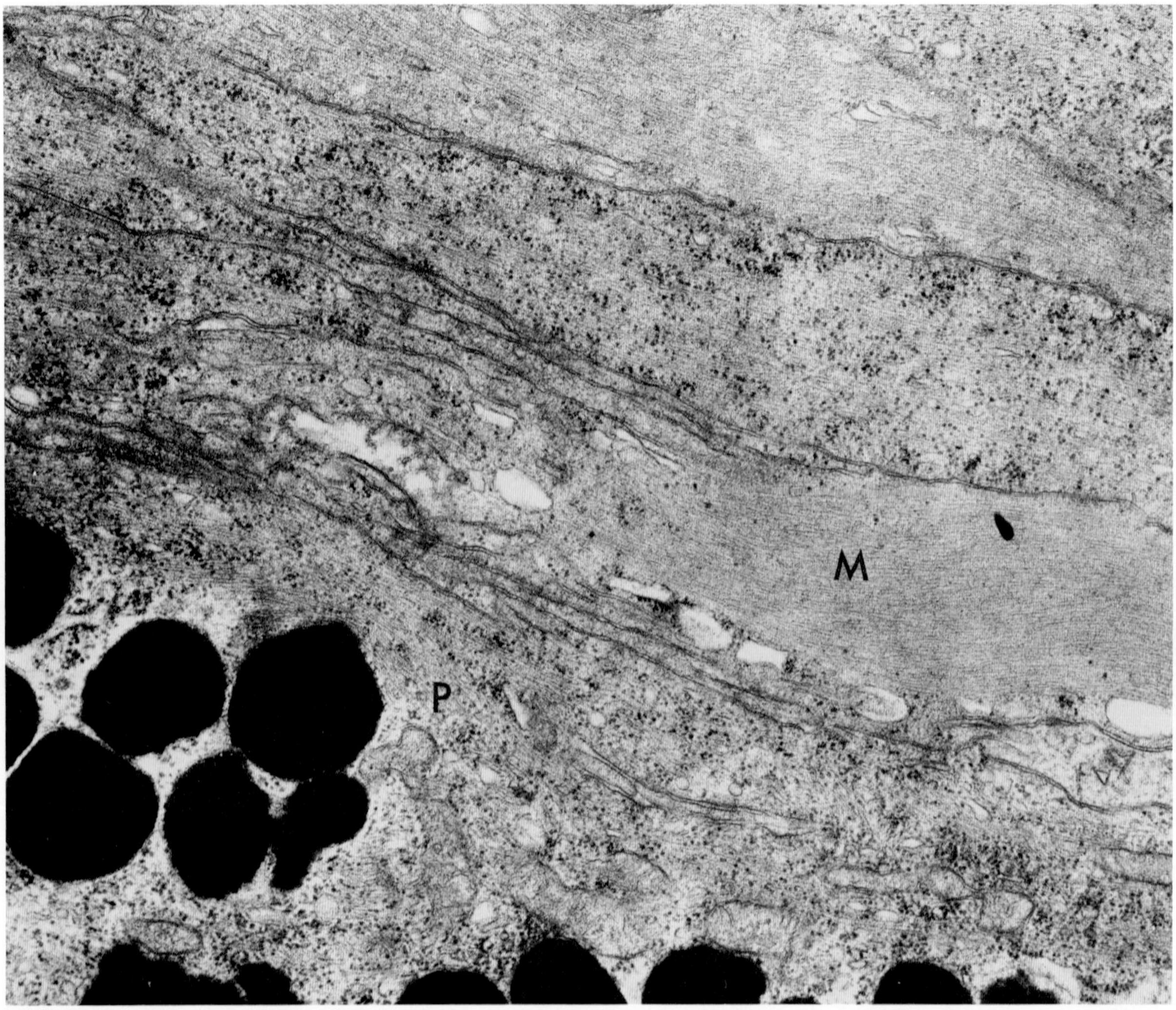

Figure 18.23 *Electron micrograph three years after a medium cryopexy lesion demonstrates pigment epithelium* (P) *in contact with retinal neuroepithelium* (M). *The retinal neuroepithelium has a prominent fibrillar structure. (×42,000.)*

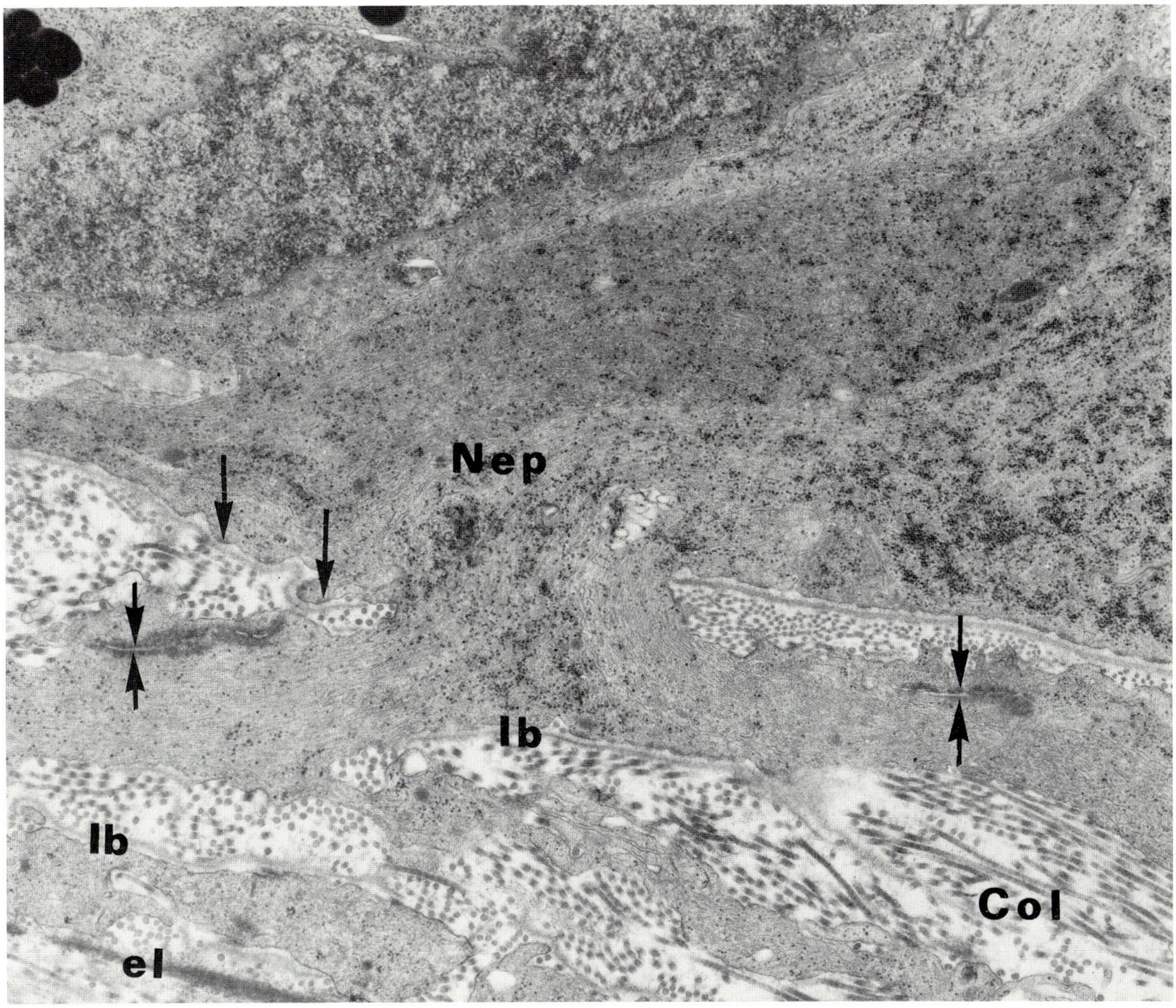

Figure 18.24 *Neuroepithelium* (Nep) *of retina infiltrates the collagen* (Col) *layers of Bruch's membrane through a defect in the membrane, caused by a heavy application of cryopexy. A basal lamina* (lb) *is elaborated on the surface of the neuroepithelial process and interposes between it and collagen. Hemidesmosomes appear between the cell membrane and the basal lamina* (free arrows). *Desmosomal-like densities* (opposing arrows) *occur where the membranes of two invading epithelial processes juxtapose.* el = *elastic lamina of Bruch's membrane.* (×16,200.)

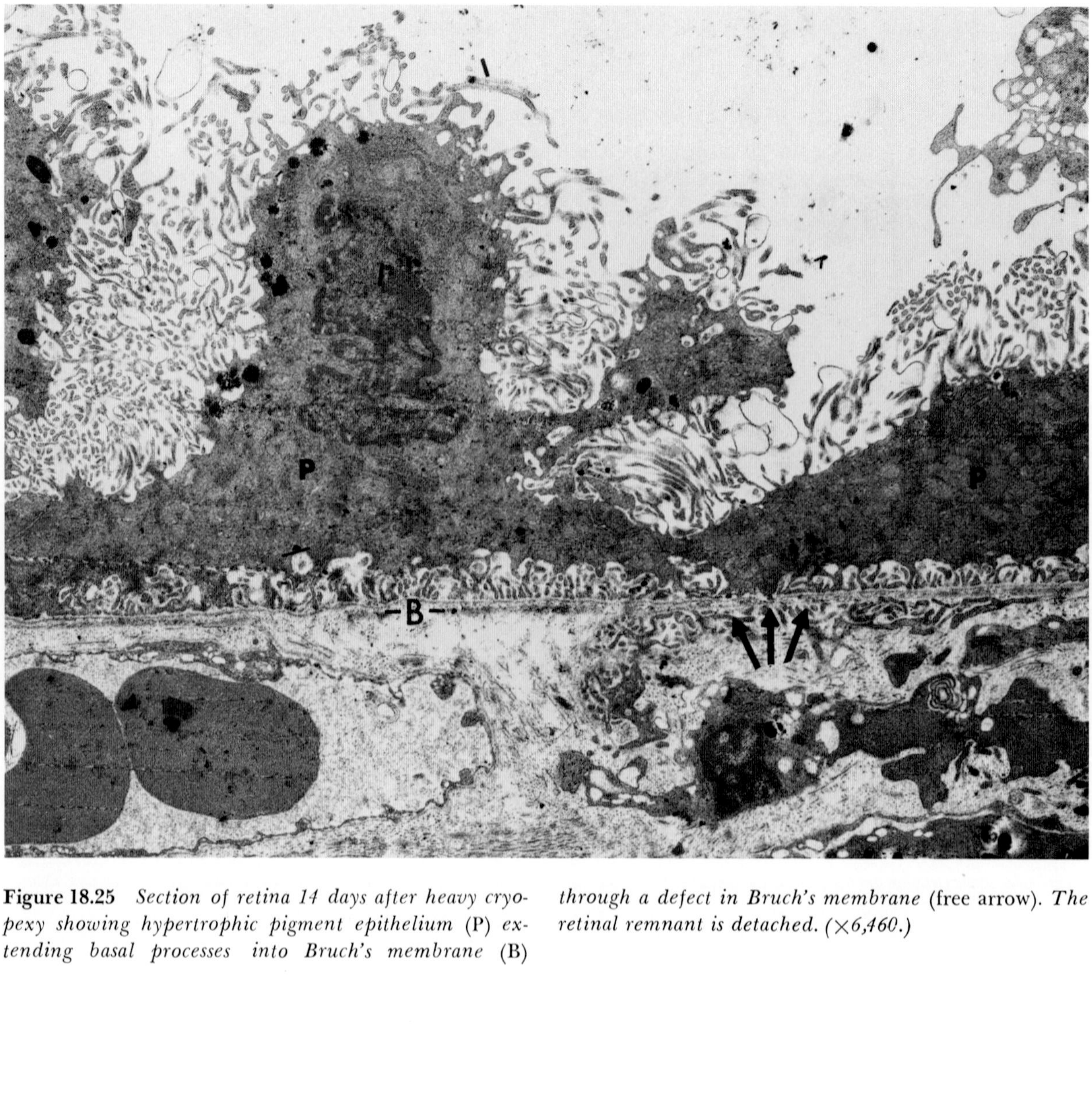

Figure 18.25 *Section of retina 14 days after heavy cryopexy showing hypertrophic pigment epithelium* (P) *extending basal processes into Bruch's membrane* (B) *through a defect in Bruch's membrane* (free arrow). *The retinal remnant is detached. (×6,460.)*

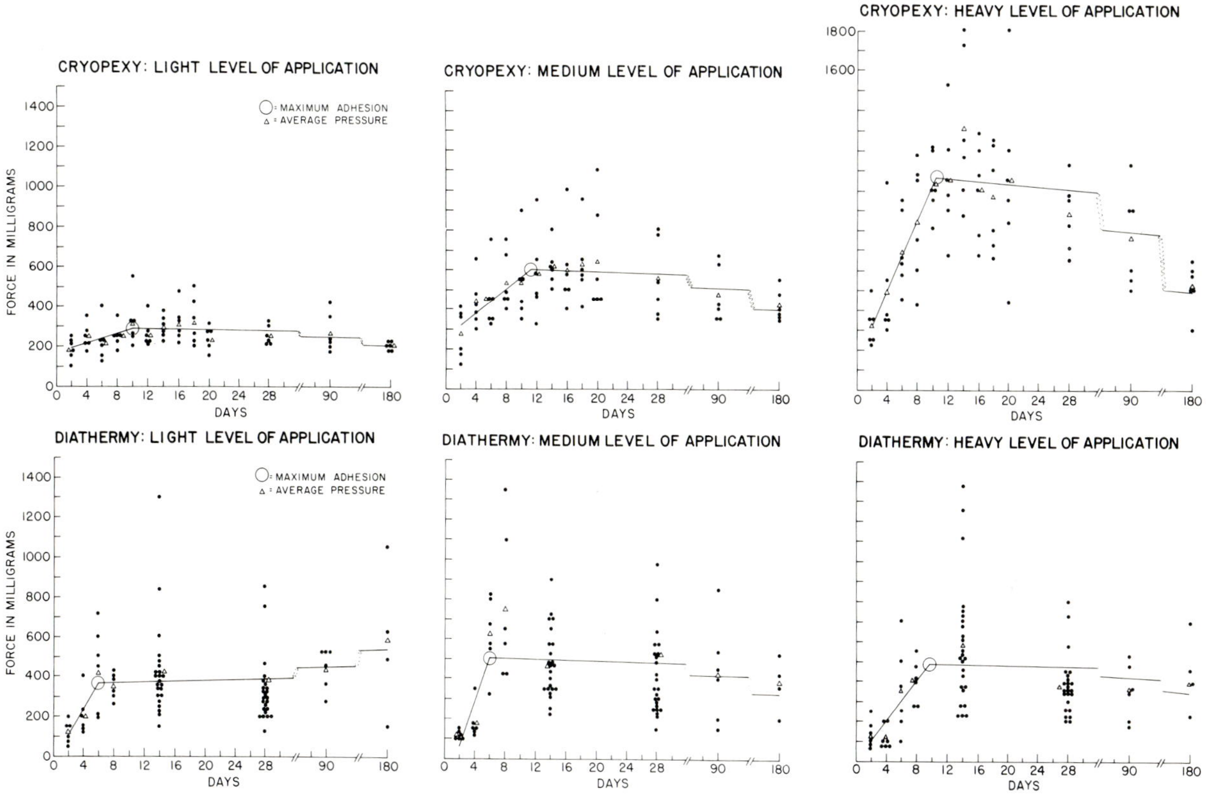

Figure 18.26 *Graphs of the relative adhesiveness of cryopexy and diathermy. A solid dot represents a lesion. A circled point represents the join point (day of maximum adhesion) of the two straight lines fitted to the data. The average adhesiveness of each group is represented by a triangle. The abscissa represents time in days between application and detachment. The ordinate represents milligrams of force required to separate the lesion.*

REFERENCES

Bloch, D., O'Connor, P., and Lincoff, H. 1971. The mechanism of the cryosurgical adhesion. III. Statistical analysis. *Am. J. Ophthalmol.* 71:666–673.

Fine, B. S., and Geeraets, W. J. 1966*a*. Membranes and ground substance in photic injury to the retina. In *Proceedings of the sixth international conference for electron microscopy,* vol. B, p. 503. Tokyo, Japan.

Fine, B. S., and Geeraets, W. J. 1966*b*. Delayed effects of photic injury to the eye. In *Proceedings of the sixth international congress for electron microscopy,* vol. B, p. 505. Tokyo, Japan.

Kreissig, I., and Lincoff, H. 1974. Die Bruchsche Membran und ihr Strukturwandel nach Kryopexie. *Klin. Monatsbl. Augenheilkd.* 164:72–89.

Kroll, A., and Machemer, R. 1970. Experimental retinal detachment and reattachment in the rhesus monkey. *Am. J. Ophthalmol.* 68:58.

Kuwabara, T. 1970. Retinal recovery from light. *Am. J. Ophthalmol.* 70:187.

Lincoff, H., and Kreissig, I. 1971. The mechanism of the cryosurgical adhesion. Part IV. *Am. J. Ophthalmol.* 71:674–689.

Lincoff, H., Long, R., Marquardt, J., and McLean, J. 1968. The cryosurgical adhesion. *Trans. Am. Acad. Ophthalmol. Otolaryngol.* 72:191–202.

Lincoff, H., and McLean, J. 1966. Cryosurgical treatment of retinal detachment. Part II. *Trans. Am. Acad. Ophthalmol. Otolaryngol.* 70:202–209.

Lincoff, H., McLean, J., and Nano, H. 1964. Cryosurgical treatment of retinal detachment. *Trans. Am. Acad. Ophthalmol. Otolaryngol.* 68:412–432.

Orzalesi, N., Grignolo, A., Calabria, G. C., and Castellazzo, R. 1967. A study on the fine structure and the rhodopsin cycle of the rabbit retina in experimental degeneration induced by diaminodiphenoxypentane. *Exp. Eye Res.* 6:376.

19 Diseases of the Retinal Pigment Epithelium in Animals

Gustavo Aguirre
Lionel Rubin

THE DEMONSTRATION that the pigment epithelium is unable to phagocytize rod outer segments in the dystrophic rat (Bok and Hall 1971; Herron et al. 1969) has emphasized the great importance of normal pigment epithelial function in maintaining the integrity of the retina. In most cases, however, there is not a clear-cut correlation between the role of the pigment epithelium and retinal disease. There is also uncertainty about the participation and response of the pigment epithelium in developmental, vascular, inflammatory, and other degenerative diseases of the retina. This chapter presents a brief overview of the various roles played by the pigment epithelium in a number of ocular diseases affecting domestic and laboratory animal species. In many cases it will be evident that the pigment epithelium reaction is not primary, but secondary, to retinal or choroidal disease.

Normal Anatomy

The histology of the retinal pigment epithelium (RPE) of domestic animals is influenced by the RPE's relationship to the tapetum lucidum, a choroidal modification that is located between the large choroidal vessels and the choriocapillaris. In the dog and cat the tapetum (Fig. 19.1) consists of rows of multilayered, flat, often irregularly-shaped cells (tapetum cellulosum). Ultrastructurally, the cytoplasm contains groups of elongated membrane-bound rods that are electron-dense, are oriented parallel to each other, and are uniform in size and spacing (Fig. 19.2). The rods are separated from each other by amorphous ground substance and by very fine fibrils and filaments. Within each cell, there are several groups of rods, each group having its own characteristic and independent orientation (Bernstein and Pease 1959). However, the rods are all arranged in a plane parallel to the pigment epithelial basement membrane.

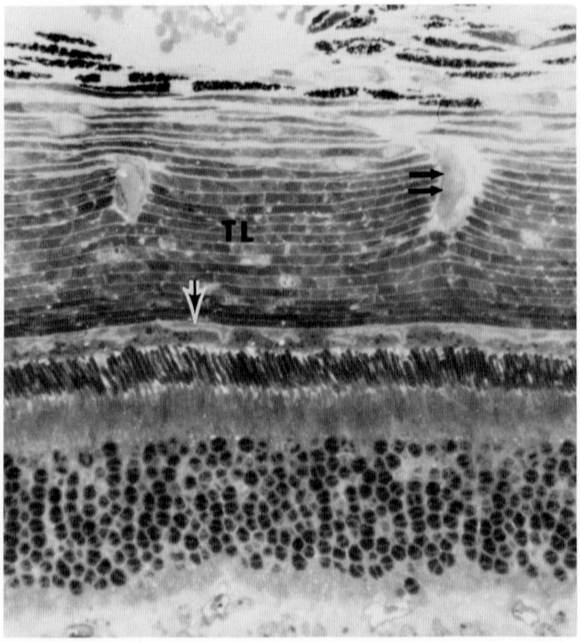

Figure 19.1 *Cat, tapetal retina in area centralis region. The tapetum lucidum* (TL) *consists of layers of tightly packed brick-like cells whose compactness decreases as they merge with the choroid. Large vessels* (double arrows) *penetrate the tapetum. In this section the choriocapillaris* (arrow) *indents the unpigmented basal pigment epithelium.*

In contrast to carnivores, herbivores have a fibrous

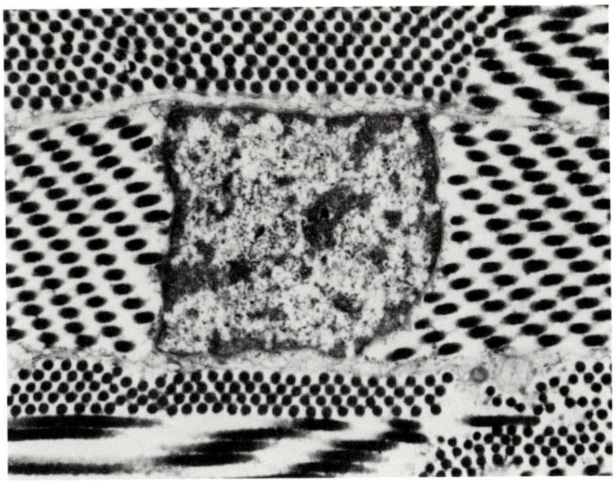

Figure 19.2 *Cat, tapetum cellulosum. The orientation of the rods within three tapetal cells differs. In the upper cell, two groups of rods are observed to be oriented perpendicular and oblique to the plane of section. The tapetal rods are of uniform size and spacing. (×10,350.)*

tapetum (t. fibrosum), consisting of bundles of undulating collagenous fibers arranged somewhat concentrically and with an iridescent surface (Fig. 19.3). In the cow, this layer can be up to 50 μ thick, with the collagenous bundles, which are arranged with various orientations, also in a plane parallel to the basement membrane of the pigment epithelium (Nakaizumi 1964). The pig, rabbit, rat, guinea pig, and most higher primates lack a tapetum.

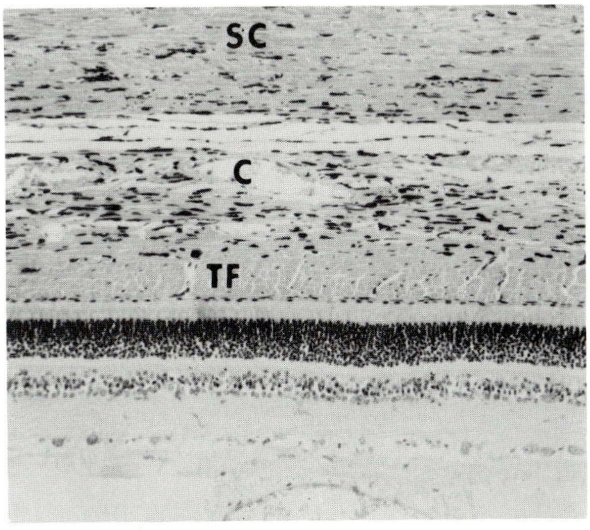

Figure 19.3 *Cow, tapetal retina. The tapetum fibrosum* (TF) *of the herbivore appears morphologically similar to the scleral* (SC) *tissue. A lightly pigmented, wide choroidal layer* (C) *is present between the sclera and tapetum.*

Blood supply to the pigment epithelium and outer retinal layer in the tapetal regions is derived from the large choroidal vessels external to the tapetum. Smaller vessels penetrate this layer and branch extensively under the tapetum, forming an incomplete choriocapillaris layer (Fig. 19.1). In the dog and cat the choriocapillaris forms a thin vascular layer interposed between the tapetal cells and the basal pigment epithelium. Occasionally, however, deep invaginations of the choriocapillaris into the basal pigment epithelium are seen. In the nontapetal regions, the choriocapillaris is adjacent to the larger choroidal vessels. In both tapetal and nontapetal regions, there are multiple fenestrations in the thin endothelial wall of the choriocapillaris adjacent to the pigment epithelium.

Unlike the human, the dog and cat lack a well-defined Bruch's membrane. The pigment epithelial basement membrane is apposed to the basement membrane of the choriocapillaris in the tapetal region, with infrequent collagenous and elastic fibers present between the two basement membranes. In the intercapillary space of the tapetal and nontapetal regions, more collagenous and elastic fibers can be present, and a more classical Bruch's membrane is occasionally recognized (Fig. 19.4*A–C*).

In contrast to carnivores, herbivores have a distinct Bruch's membrane that is similar to that of man, with a very prominent middle elastic layer (Leure-duPree 1968; Mason, Fager, and Abrahamson 1973; Nakaizumi 1964). In areas where the choriocapillaris is absent, the outer collagenous fiber layer blends with the collagenous network of the choroid (Leure-duPree 1968). In the cow, the morphology of Bruch's membrane is uniform throughout the retina, with no variation between the tapetal and nontapetal regions (Nakaizumi 1964). Of all the domestic species, the pig has the most prominent Bruch's membrane (Prince et al. 1960).

The pigment epithelial cells of domestic animal species, in general, are morphologically similar to those of man and of laboratory animal species. The basal zone has numerous infoldings of the cell membrane and has a high mitochondrial density (Fig. 19.4*A–C*). In the dog and cat the nucleus is rounded and occupies a more central location within the cell. The apex of the cells is rich in ribosomes and granular endoplasmic reticulum. From the apex of the pigment epithelium, numerous microvilli extend into the photoreceptor layer and invest the distal tips of the rod outer segments. Larger and more

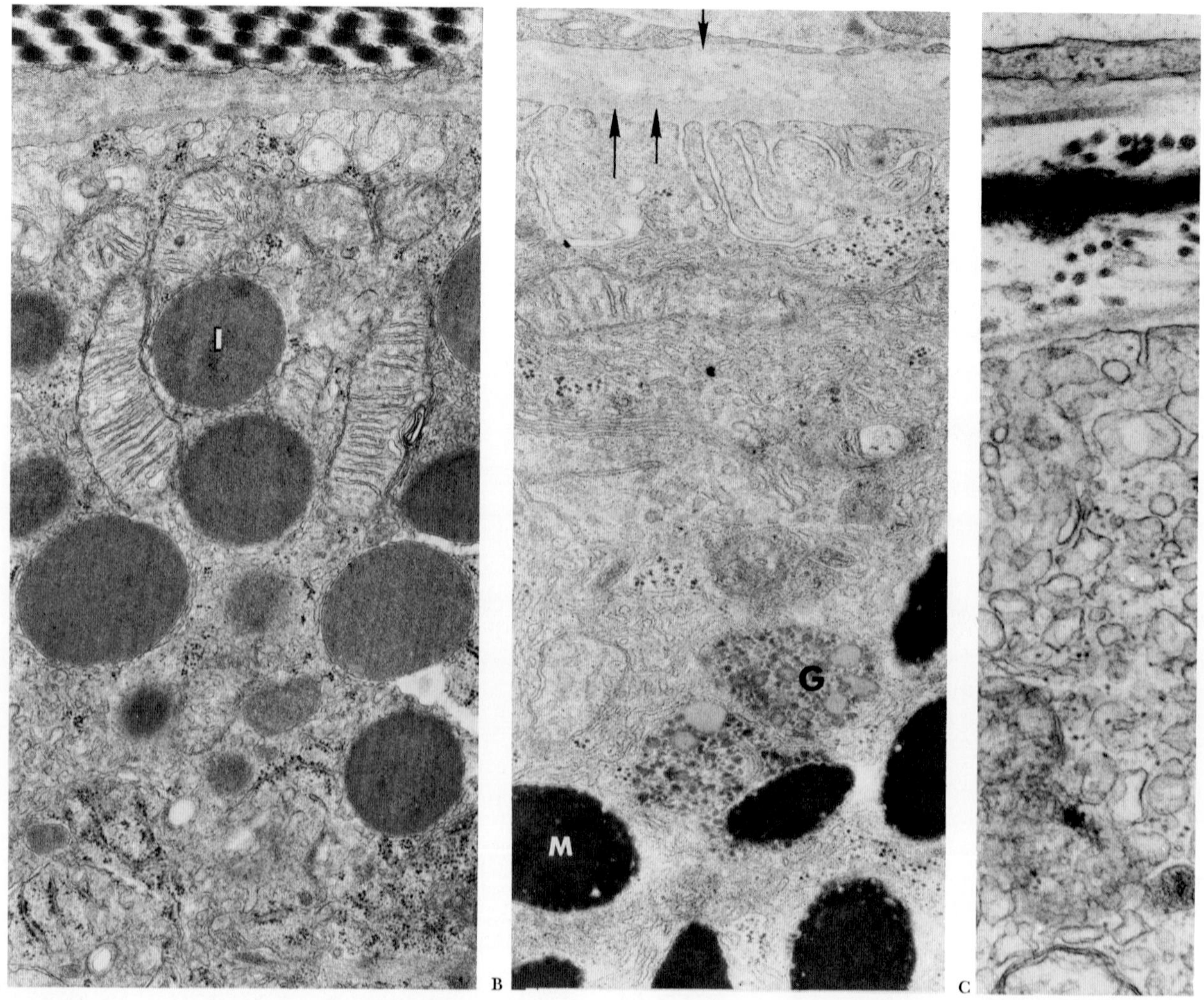

Figure 19.4 A, *Aged cat, tapetal zone. In the intercapillary zone of the tapetal region, Bruch's membrane contains a prominent pigment-epithelial basement membrane. Distinct collagenous and elastic zones are absent. The basal pigment epithelium contains uniformly homogeneous membrane-bound inclusions* (I). B, *dog, nontapetal zone. The basement membranes of the pigment epithelium* (double arrows) *and choriocapillaris endothelium* (single arrow) *are distinct. The collagenous and elastic zones of Bruch's membrane are not present. The pigment epithelium in the nontapetal area has melanin granules* (M). *Granular inclusions* (G) *are probably residual bodies.* C, *dog, nontapetal zone. The pentalaminar organization of Bruch's membrane is evident. The inner and outer collagenous zones contain few collagen fibers, but the elastic layer is prominent. This distinct pentalaminar organization of Bruch's membrane is rarely found in the dog or cat.* (A *and* B, *×21,000.* C, *×42,000.)*

numerous sheetlike processes extend down the distal half of the cone outer segment.

Melanin granules are located in the apical zone of the pigment epithelial cells in the nontapetal region (Fig. 19.5). The pigment epithelium overlying the tapetum lucidum contains no melanin granules. Premelanosomes are rarely found in this region.

Vascular Occlusive Disease

Retinal and choroidal vessels have been experimentally occluded in a number of animals (see Wise, Dollery, and Henkind 1971 for review). In dogs, intraarterial injection of latex spheres at the level of the external ophthalmic artery results in focal vascular occlusion of the retina and choroid. Retinal pathology is absent when a large choroidal vessel or a single choroidal capillary is occluded. Only multiple occlusions of the vessels of the choriocapillaris result in necrosis and atrophy of the outer retinal layer with intraretinal pigment migration (Gay, Goldor, and Smith, 1964).

Naturally occurring retinal or choroidal vascular

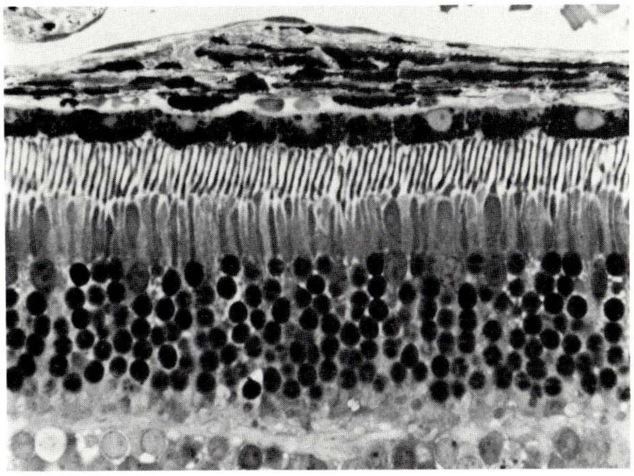

Figure 19.5 *Dog, peripheral retina. The nontapetal pigment epithelium has a high melanin content.*

occlusive diseases are rare in domestic animal species. We have seen, however, atherosclerotic occlusion of multiple choroidal vessels in dogs, which resulted in extensive pigment epithelial damage (Fig. 19.6).

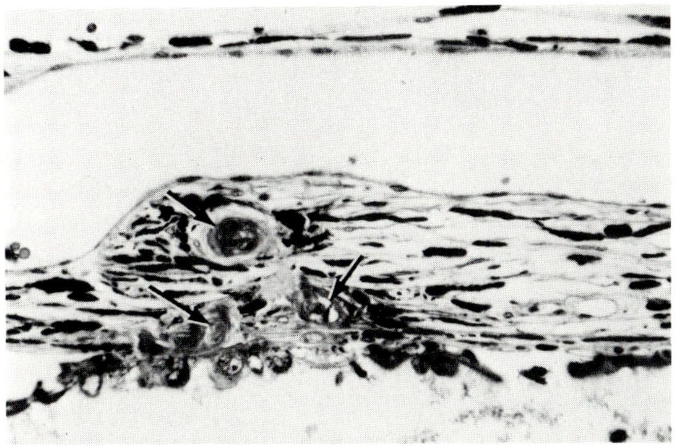

Figure 19.6 *Dog, nontapetal retina. Atheromatous occlusions of vessels* (arrows) *in the choroid and choriocapillaris result in focal damage to the pigment epithelial layer. There is cell loss, as well as ballooning of the remaining RPE cells.*

Inflammatory Disease

ADULT. The response of the RPE of the dog to infectious processes seems to vary within wide limits. This variation may be a function of the disease, although it seems more probable that it is a function of the severity of the inflammation. In canine distemper, a neurotropic virus disease of dogs, the infection of the retina by the virus may cause degeneration of the RPE focally, localized swelling of individual RPE cells, or hypertrophy and hyper-

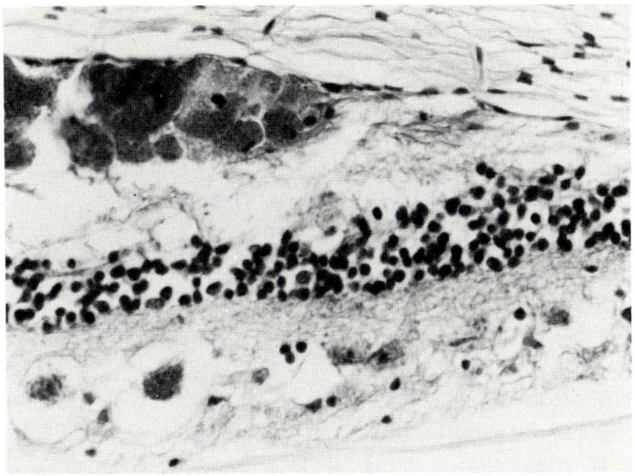

Figure 19.7 *Dog, tapetal retina. Focal inflammation has resulted in pigment epithelial atrophy* (right) *and the formation of a multicellular nest of hypertrophied RPE cells* (left). *The cytoplasm of these altered pigment epithelial cells contains a granular brown material.*

plasia of the RPE, resulting in RPE nests (Fig. 19.7). In addition, pigmented cells may migrate into the damaged zones (Fig. 19.8). These reactions may occur

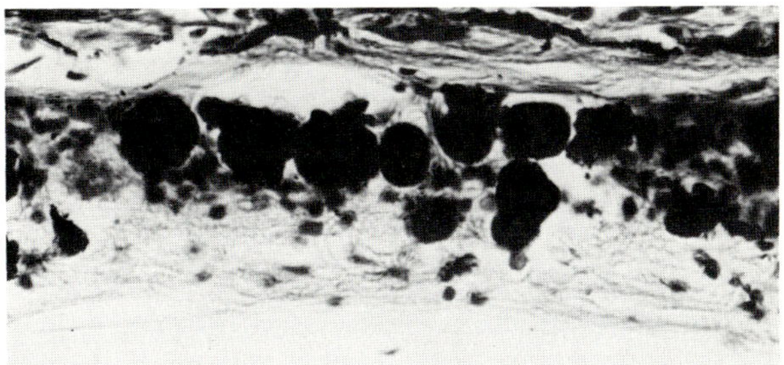

Figure 19.8 *Dog, nontapetal retina. Hypertrophied and heavily pigmented RPE has migrated into the markedly atrophic retina.*

in isolation, or all may be seen in various zones of the previously inflamed eye. In the tapetal zone, the stimulus of the inflammation may also result in the production of pigment by cells not normally pigmented, though the mechanism for this is not known. In rare instances the pigment epithelium of the dog appears to acquire migratory characteristics, and may proliferate and grow along the inner limiting membrane (Fig. 19.9*A–B*).

Abnormalities of the RPE would be anticipated in most types of retinitis, retinochoroiditis, and choroiditis in the dog, although the systemic fungal

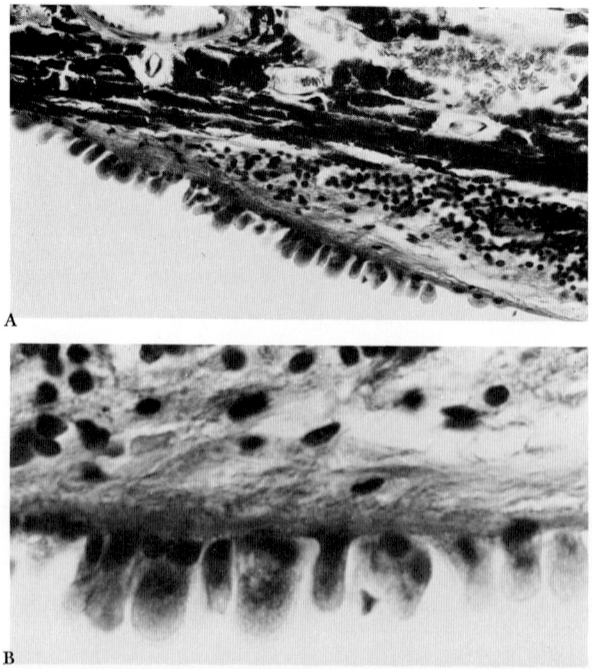

A

B

Figure 19.9 A, *dog, nontapetal retina. Nonpigmented RPE has migrated over the internal limiting membrane of a disorganized and atrophic retina.* B, *higher magnification of* A. *The ectopically located pigment epithelial cells are columnar, and the nucleus is located in the basal region.*

infections probably cause the least pigmentary disturbance. Infections as diverse as herpes virus and toxoplasmosis can cause severe pigmentary disturbances.

In horses, the RPE responds in much the same manner as in dogs. Chronic chorioretinitis, typified by equine recurrent uveitis, is capable of stimulating proliferation of the RPE into large multicellular nests (Figs. 19.10, 19.11). Less marked effects on the

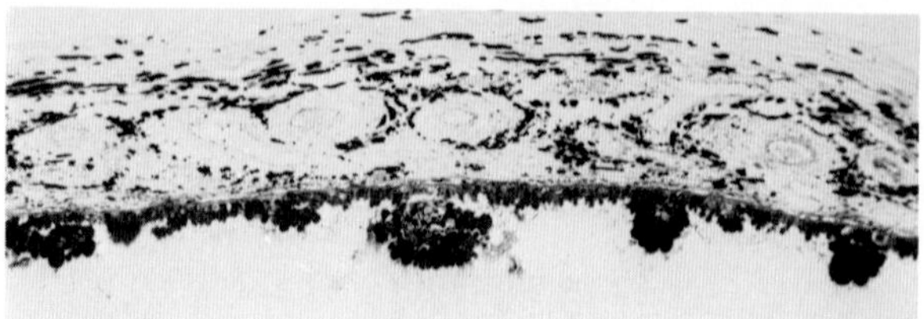

Figure 19.10 *Horse, nontapetal region. Retinal detachment and proliferation of RPE in equine recurrent uveitis.*

RPE are observed with low-grade chorioretinal inflammation, such as that caused by *Streptococcus equi* infection. Unlike the dog, in horses the nonpigmented RPE overlying the tapetal region rarely

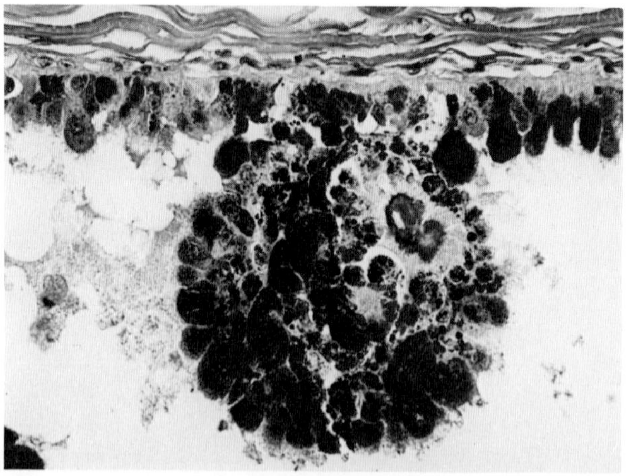

Figure 19.11 *Higher magnification of the same section, different field, as Figure 19.10. The proliferated RPE has formed a large cell cluster in the ventricular space.*

becomes pigmented as a result of inflammatory disease.

DEVELOPING RETINA. Retinal abnormalities are recognized in the differentiating fetal or neonatal retina (Lahav, Albert, and Wyand 1973) in response to infection and radiation. The role of the pigment epithelium in the induction of these developmental abnormalities of the neuroretina has been hypothesized (Silverstein et al. 1971; Silverstein, Osburn, and Prendergast 1971), but never proven.

Ionizing radiation affects the undeveloped or developing retinas of rabbits (Pagenstecher 1961), rats (Tansley, Spear, and Gluckman 1937), mice (Rugh and Wolff 1955), monkeys (Rugh and Skaredoff 1969), and dogs (Shively and Phemister 1967; Shively et al. 1970). In the dog, Shively et al. (1970) have shown that gamma irradiation of the whole body of newborn beagles, at a stage before the outer neuroblastic layer has differentiated into inner and outer layers, results in retinal dysplasia. These dysplastic changes are characterized by agenesis of the photoreceptor outer segments, rosettes, and malformation of the outer retinal layers. The effects are more severe in the peripheral retina, since it is the most primitive portion of the entire retina. The structure of the pigment epithelium, however, appears unaffected by irradiation levels that cause extensive retinal damage (Shively and Phemister 1967).

The effect of ionizing radiation is dependent not only on the dose level, but on the state of retinal differentiation at the time of irradiation. While the developing retina is quite radiosensitive, the differentiated retina is considerably less so and is damaged only by doses of several hundred to a few thousand roentgens (Shively and Phemister 1967). The retinal pigment epithelium appears to be quite insensitive to ionizing radiation.

Dysplastic changes in the retina, often appearing as retinal folds and rosettes, are also caused by fetal and early postnatal viral infections. In the rat, lymphocytic choriomeningitis (LCM) virus or simian virus 40 (SV-40), when inoculated within the first day after birth, results in a widespread retinopathy in the adult animal (Friedman, Bellhorn, and Henkind 1973; Monjan, Silverstein, and Cole 1972). In the case of LCM virus infection, pathologic changes begin in the outer nuclear layer and, with time, progressively affect the inner retinal layers (Monjan, Silverstein, and Cole 1972). These changes are dependent on immunologic competency, for immunosuppression of infected animals prevents the disease, and adoptive transfer of sensitized lymphoid cells into immunosuppressed, chronically infected animals triggers the retinal disease (Silverstein 1974). With SV-40, on the other hand, retinal dysplasia in the form of rosettes, absence of photoreceptors, and retinal gliosis appears to be dependent on the presence of abnormal preretinal neovascularization; in areas without neovascularization, the retina is normal (Friedman, Bellhorn, and Henkind 1973). In both experimental models, although viral antigen is found early in the pigment epithelium, morphologic changes are not present in the pigment epithelium, raising the possibility that infection of the pigment epithelium is inconsequential in the production of these widespread retinal abnormalities. It is possible, however, that a level of pigment epithelial infection that is insufficient to cause morphologic changes is still capable of altering the inductive properties of this cell layer, resulting in neuroretinal abnormalities.

Infection with bluetongue virus in lambs (Silverstein et al. 1971) (Fig. 19.12*A–B*), bovine viral diarrhea–mucosal disease (BVD-MD) virus in calves (Bistner, Rubin, and Saunders 1970) (Fig. 19.12 *C–D*), canine herpes virus in puppies (Percy et al. 1971) (Fig. 19.12*E–F*), and panleukopenia virus in kittens (McMillan 1976) (Fig. 19.12*G*) can also cause extensive neuroretinal abnormalities similar to those produced experimentally in rats. Clinical and experimental studies in dogs (Lahav, Albert and Percy 1974; Lahav et al. 1974; Percy et al. 1971), following neonatal infection with herpes virus have shown that early choroiditis is associated with pigment epithelial damage (Fig. 19.12*E–F*). There is atrophy as well as hypertrophy and folding of pigment epithelium, the latter resulting in outer retinal folds. In some areas, however, there is extensive retinal atrophy with intraretinal pigment migration.

The retinopathy produced by bluetongue virus infection in fetal lambs is probably the best understood model of viral-induced retinopathy (Silverstein et al. 1971). Infection of the fetal lamb near the end of the first trimester of pregnancy results in intraretinal inflammation with focal necrosis, retinal folds, and choroiditis. Pigment-laden cells are present in the nuclear layers. With time, active inflammation subsides and retinal rosettes, atrophy, and gliosis are found. At this stage of the disease the pigment epithelium can be absent from some areas of Bruch's membrane, or can undergo hyperplasia and intraretinal migration (Fig. 19.12*A–B*). In contrast to these changes, infection at mid-gestation produces minimal retinal inflammation and pathology. By the last trimester, infection produces no significant retinal lesions. The type and extent of retinal lesions is dependent on fetal age at the time of infection rather than extent of damage to the pigment epithelium. This suggests that the differentiated, as opposed to the differentiating, retina has a decreased susceptibility to the viral agent, or possibly that the disease is modulated by increasing immunocompetence of the fetal lamb with age (Silverstein 1974). In either case, the fact that the retina in the infected fetal lamb is not separated from the pigment epithelium, and the presence of cytopathic lesions in the retina, argue in favor of an inflammatory cause for the retinopathy. The resultant retinal dysplasia appears to be secondary to this inflammation rather than to a failure or defect in induction of the retina by the pigment epithelium. This would seem to apply also to herpes virus infection in puppies and BVD-MD infection in fetal calves.

In the case of BVD-MD infection in fetal calves, in addition to retinal damage similar to that seen in the other species, thick collagenous and fibrovascular tissue bundles have been observed in the region of the choriocapillaris (Bistner, Rubin, and Saunders 1970) (Fig. 19.12*C–D*). In view of the retinal pigment epithelium's ability to produce collagen (Newsome and Kenyon 1972), this finding suggests yet another possible way in which the pigment epithelium can respond to disease.

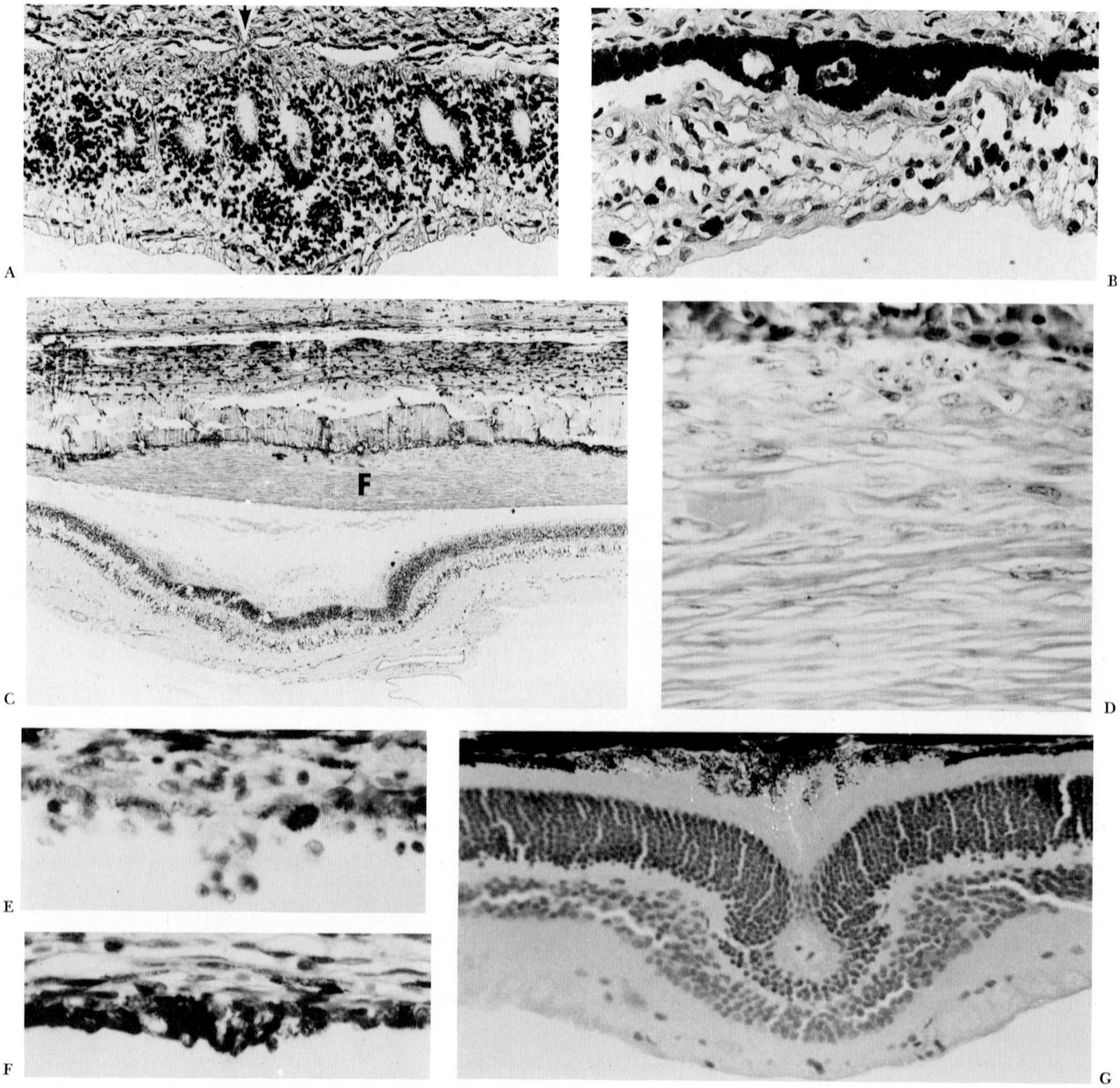

Figure 19.12 *Retinal and pigment epithelial changes following fetal or early postnatal infection with viral agents.* A–B, *bluetongue virus retinopathy in fetal lambs following infection at 50–58 days of gestation. At term the retina shows multiple rosettes and folds. Breaks in the pigment epithelium* (arrow) *allow for focal chorioretinal adhesions* (A). B, *taken from the same eye, shows more severe destruction of the retina as well as hyperplasia and migration of the pigment epithelium. (Reprinted with permission from Silverstein et al. 1971.)* C, *tapetal region of 10-day-old calf infected prenatally with BVD-MD virus. There is a fibrovascular structure* (F) *located between the tapetum fibrosum and retina. The retinal detachment has resulted in degeneration of the photoreceptors. (Reprinted with permission from Bistner 1970.)* D, fibrous tissue *shown in* C *at higher magnification. (Reprinted with permission from Bistner 1970.)* E–F, *following neonatal infection with canine herpes virus, there is round cell infiltration of the choroid with RPE destruction* (E). *Two weeks later* (F), *there is focal proliferation of the RPE. (Courtesy of Drs. M. Lahav, D. M. Albert, and D. H. Percy.)* G, *eighteen-week-old kitten, nontapetal region. Following neonatal inoculation with feline panleukopenia virus, retinal folds are prominent. The pigment epithelium is hypertrophied and there is dispersion of melanin granules in the RPE overlying the retinal fold. (Courtesy of Dr. Alan McMillan.)*

Congenital (Hereditary and Nonhereditary) Neuroectodermal Abnormalities

Retinal dysplasia is recognized in several canine breeds—the Bedlington terrier (Rubin 1968), Labrador retriever (Barnett, Bjorck, and Kock 1970), and Sealyham terrier (Ashton, Barnett, and Sachs 1968), and is inherited as a simple recessive. The dysplastic retina in affected dogs is frequently detached and contains numerous folds and rosettes. Loss of retinal layer organization, with subsequent atrophy or intravitreal hemorrhage originating from proliferative preretinal neovascularization, can be found in the dysplastic retina in later stages (Fig. 19.13*A*).

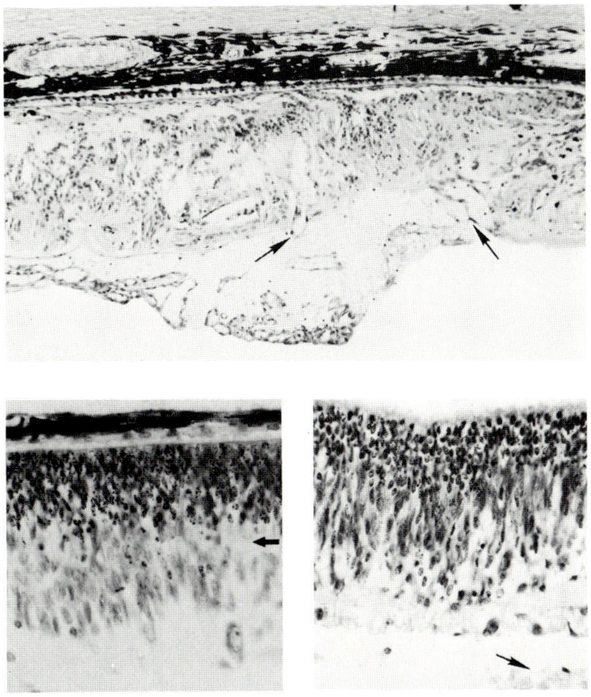

A B C

Figure 19.13 *Retinal dysplasia in Bedlington terriers.* A, *adult affected dog. The retina is gliosed and shows marked loss of layer organization. The pigment epithelial cells are columnar because of loss of photoreceptors and separation of the very thickened external limiting membrane from the apical pigment epithelium. A broad preretinal neovascular tissue originates from two inner retinal vessels* (arrows). B, *the retina of a young affected dog shows an incompletely formed outer plexiform layer* (arrow). *The cells in the inner zone of the outer neuroblastic layer are spindle-shaped and appear less differentiated than the photoreceptor nuclei. The retina is attached and shows incompletely developed photoreceptor inner segments.* C, *detached retina of a young affected dog. The nuclei in the inner zone of the outer neuroblastic layer are spindle-shaped and elongated, indicating lack of retinal differentiation. The preretinal membrane* (arrow) *is part of a proliferative neovascular growth into the vitreous.*

In the Bedlington terrier, abnormal neuroretinal differentiation is first recognized in the early postnatal period when the outer neuroblastic layer fails to separate into inner and outer nuclear layers (Fig. 19.13*B–C*). Retinal folds and rosettes subsequently develop, and the retina detaches before development is completed. The pigment epithelium is morphologically normal in the early postnatal period, prior to the arrest in neuroretinal differentiation and retinal detachment. Following the retinal detachment, the pigment epithelial cells elongate and become columnar in shape, a common finding in dogs with long-standing retinal detachments.

A recessively inherited retinal dysplasia is also recognized in the Australian shepherd dog (Gelatt and McGill 1973). The dysplastic changes appear most severe in areas devoid of retinal pigment epithelium (Fig. 19.14). In this disease, however, there is also a multiplicity of congenital ocular abnormalities.

Collies and Shetland sheepdogs have a recessively inherited disease characterized by hypoplasia of the choroid (Fig. 19.15*A*) and posterior pole colobomas (scleral ectasia syndrome). Retinal detachments occur in this disease, especially in Collies (Roberts, Dellaporta, and Winter 1966) (Fig. 19.15*B*). The area of choroidal hypoplasia is variable in size and occurs in a wedge-shaped configuration temporal to the optic disc at the junction of the tapetal and nontapetal regions (Fig. 19.15*A*). In the affected area, most of the choroidal vessels and the tapetum lucidum are absent, and the retinal pigment epithelium is unpigmented, but appears otherwise normal. Histologically, the retina that lines the scleral ectasia is attenuated, especially with loss of the outer retinal layers. In these areas, pigment epithelial proliferation is commonly found (Fig. 19.15*C–D*).

The ocular abnormality in Collies and Shetland sheepdogs is a defect in mesodermal differentiation that affects the fibrous and vascular tunics of the globe, especially in the posterior pole. Neuroretinal and pigment epithelial abnormalities appear to be secondary changes.

Some Beagle dogs, a breed that normally possesses a tapetum, appear to lack the structure when examined ophthalmoscopically (Heywood 1972; Bellhorn et al. 1975). In this familial anomaly, the area usually occupied by the tapetum appears uniformly

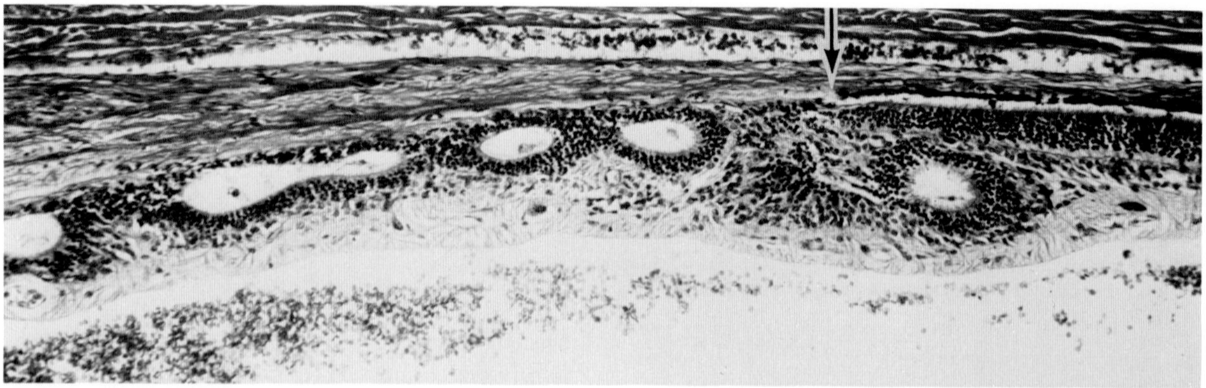

Figure 19.14 *Section of a dysplastic retina from an Australian shepherd dog. The more severe changes are present in areas devoid of pigment epithelium (left of arrow). The retina is more normal in areas where pigment epithelium is present (right of arrow). Free blood is present in the vitreous cavity.*

reddish-pink. Histologic examination of young, apparently tapetum-less, dogs shows tapetal cells in normal position, but their ultrastructure is quite abnormal. The cells are devoid of intracellular rodlets, but have sparse randomly oriented electron-dense structures and extensive membranous profiles (Bellhorn et al. 1975). Because of the absence of cytoplasmic rodlets, light is not reflected and diffracted by the tapetal cells; instead, the uniform reddish-pink is reflected through from the choroidal circulation.

The defect in tapetal-cell structure may progress to complete loss of tapetal cells in older dogs. The pigment epithelium remains unpigmented in the superior retinal quadrants, the areas where the tapetum is defective. Contrary to what is seen in most normal dogs, however, the pigment epithelium in the inferior retinal quadrants—that is, the nontapetal region—has extremely sparse melanin content (Bellhorn et al. 1975).

Degenerative Retinal Diseases

Retinal degenerations, both inherited and sporadic, occur in a number of domestic and laboratory animal species. Except for the dystrophic rat, in which the failure by the pigment epithelium to phagocytize the shed rod outer segments results in retinal degeneration (see Chapter 20), the role of the pigment epithelium in the retinal degenerative process is not well understood.

PROGRESSIVE RETINAL ATROPHY. Generalized retinal atrophies have been recognized in more than fifteen dog breeds and in a number of mixed-breed dogs (Parry 1953; Barnett 1965). These diseases are generally called progressive retinal atrophy (PRA) and are considered the canine counterpart to retinitis pigmentosa. In general, they are characterized by: (1) an autosomal recessive mode of inheritance, in those breeds in which the heredity has been investigated; (2) initial nyctalopia with progression to complete blindness; (3) extinguished ERG; (4) progressive vascular attenuation, retinal thinning, and secondary optic atrophy (Fig. 19.16). In some breeds, posterior subcapsular cataracts that progress to complete cataracts are common.

Recent work has shown that PRA is a general descriptive term encompassing at least three different outer retinal diseases that appear the same clinically, although the age of onset, progression rate, and electrophysiology differ (Aguirre 1976). These diseases are: (1) rod dysplasia with cone degeneration (Aguirre and Rubin 1971), affecting Norwegian elkhounds; (2) rod-cone dysplasia (Aguirre and Rubin 1972; Aguirre 1973), affecting Irish setters; (3) progressive rod-cone degeneration (Aguirre and Rubin 1972; Aguirre 1973) affecting miniature poodles. These findings are similar to those in retinitis pigmentosa, where there are very different electrophysiologic abnormalities in young patients with apparently similar clinical diseases (Berson 1976).

Depending on the type of PRA, the rod or cone photoreceptors, or both, either fail to develop normally, or, once developed, prematurely degenerate (Aguirre 1976). Electrophysiologically this is recognized either by the absence of ERG response or by markedly abnormal ERG responses for the affected photoreceptor group being tested (Fig. 19.17). Mor-

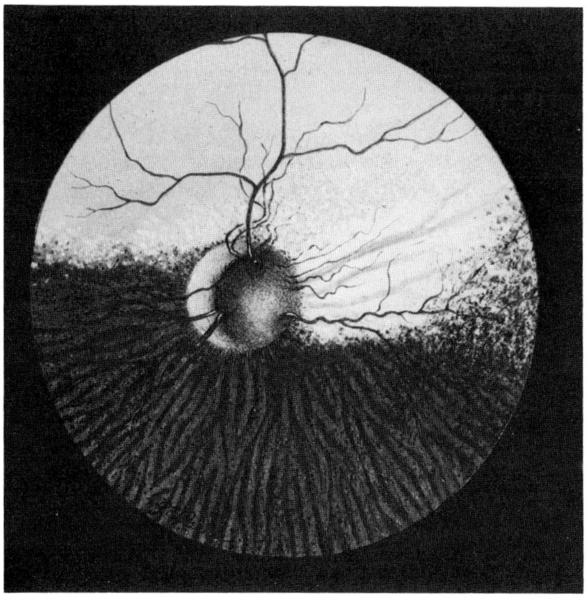
A

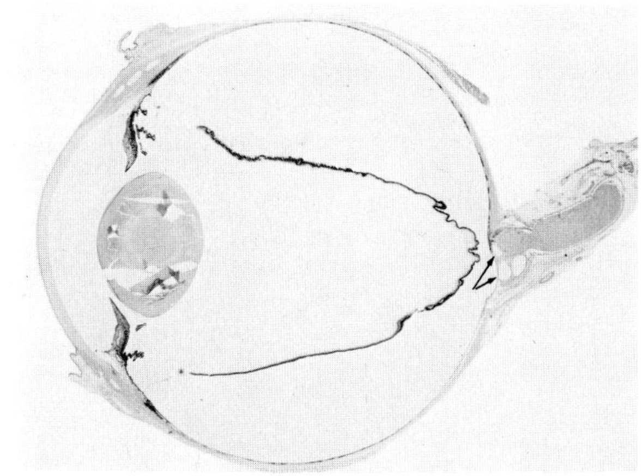
B

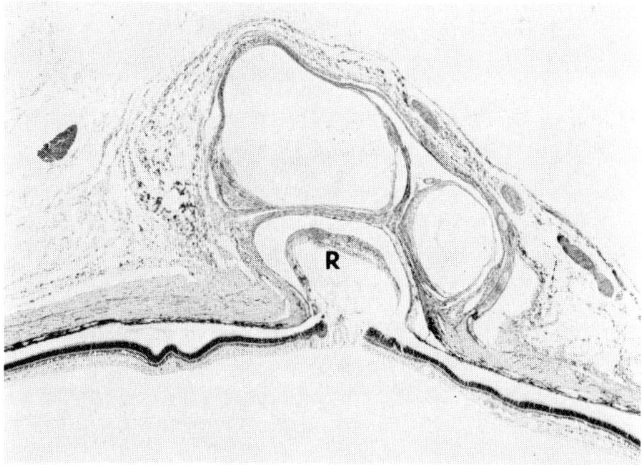

C

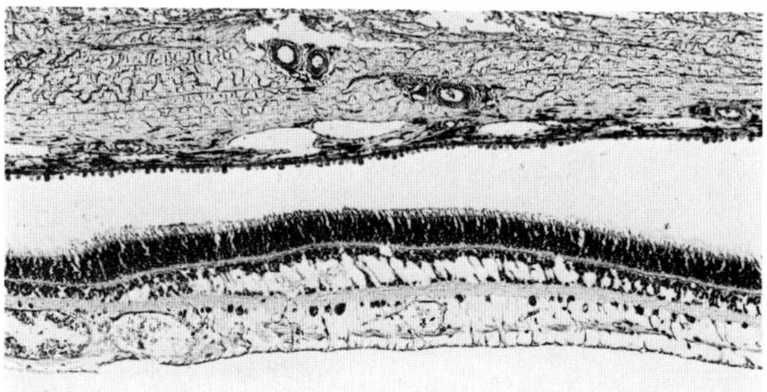
D

Figure 19.15 *Collie ectasia syndrome.* A, *fundus painting of affected dog showing a papillary coloboma and a small wedge-shaped depigmented zone of choroidal hypoplasia. Three large choroidal vessels are seen in this area overlying the white scleral reflection.* B, *affected eye showing giant retinal detachment and papillary and juxtapapillary colobomas* (arrows). C, *the retina* (R) *lining the coloboma is thinned and greatly disorganized.* D, *long-standing retinal detachment in affected eye shows an elongated columnar pigment epithelial layer, loss of the photoreceptor outer segments, and cystoid spaces in the inner nuclear, ganglion cell, and nerve fiber layers.*

phologically, the outer segments of affected photoreceptors undergo extensive lamellar disorganization and loss of orientation before completely degenerating. Loss of the photoreceptor inner segment and nucleus soon follows, and there is marked loss of the outer retinal layers.

In the early stages of these diseases, the pigment epithelium is normal. Phagosomal inclusions, indicative of the uptake of rod outer segments by the pigment epithelium, are present, which suggests that this cell layer functions normally. In areas where there is extensive outer segment degeneration, phagocytic cells, not of pigment epithelial origin, are present in the ventricular space and appear to participate in the removal of cellular debris (Fig. 19.18). Unlike the retinal dystrophy in rats, a buildup of rod outer segment material adjacent to the pigment epithelial apex has not been seen in these canine retinal diseases.

Loss of the outer retinal layers brings the pigment epithelium into apposition with the external limiting membrane (Fig. 19.19). This relationship is maintained until the very late stages of the disease, when there is complete loss of retinal layer organization. At this time the pigment epithelium can undergo atrophy, or can become hypertrophied and migrate intraretinally (Fig. 19.20*A–D*).

The primary abnormalities in the different forms of PRA are in the rod and cone photoreceptors. The pigment epithelial changes are secondary and occur

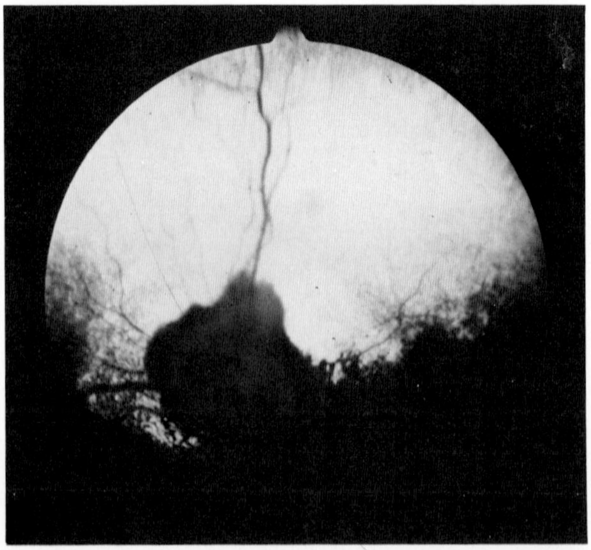

A

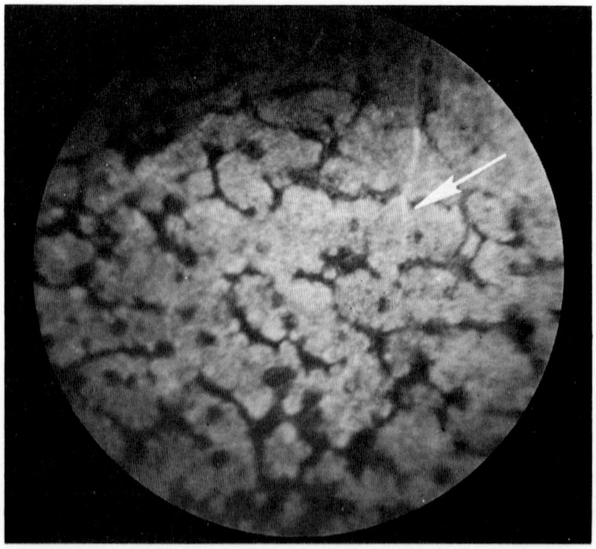

B

Figure 19.16 *Fundus photographs of Norwegian elkhounds affected with rod dysplasia.* A *shows diffuse retinal hyperreflectivity and peripapillary conus, both indicative of advanced retinal atrophy. The retinal vessels are thinned. The nontapetal retina* (B) *shows a markedly narrowed ghost vessel* (arrow) *and darkly pigmented foci that represent areas of pigment epithelial migration into the greatly attenuated retina. (Reprinted with permission from Aguirre and Rubin 1971.)*

in the later stages of the disease. These changes of the RPE are not specific for PRA, as they can occur in other generalized retinal degenerations found in dogs.

Generalized retinal atrophies are also recognized in the cow (Rubin 1974) and cat. In the cat they can be inherited (Rubin and Lipton 1973) or can be nutritionally produced by synthetic diets (Morris 1965; Scott, Greaves, and Scott 1964). In most clinical cases the etiology is unknown (Rubin 1963). Though the conditions have not been studied as extensively

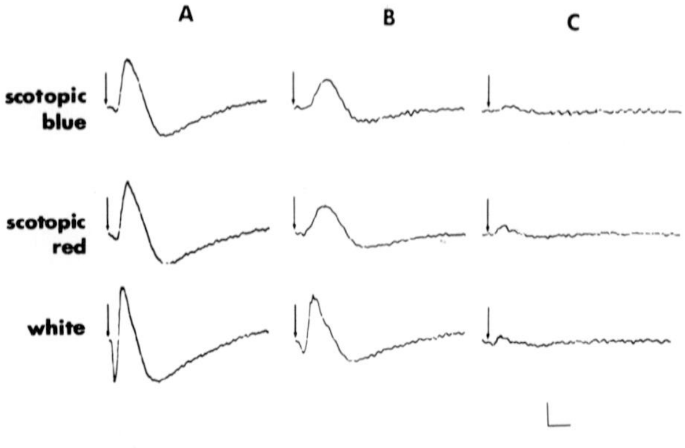

Figure 19.17 *Single-flash dark-adapted electroretinograms recorded from a normal dog* (A), *and from a young* (B) *and an adult* (C) *miniature poodle with progressive rod-cone degeneration. The young affected dog responds to white and to scotopically balanced short (blue) and long (red) wave stimuli, but the response amplitudes are decreased and the peak latencies are prolonged. The adult dog shows a marked decrease in amplitude. Vertical arrows indicate stimulus onset; vertical calibration at lower right, 100 μV; horizontal calibration, 50 msec. (Reprinted with permission from Aguirre and Rubin 1972.)*

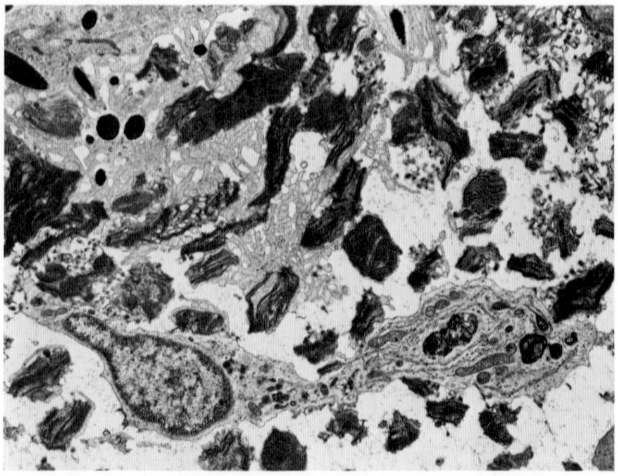

Figure 19.18 *Nontapetal zone of a young miniature poodle affected with progressive rod-cone degeneration. During the stage of the disease characterized by extensive breakdown of photoreceptor outer segments, phagocytic cells invade the ventricular space and aid in the removal of cellular debris.* (×5,075.)

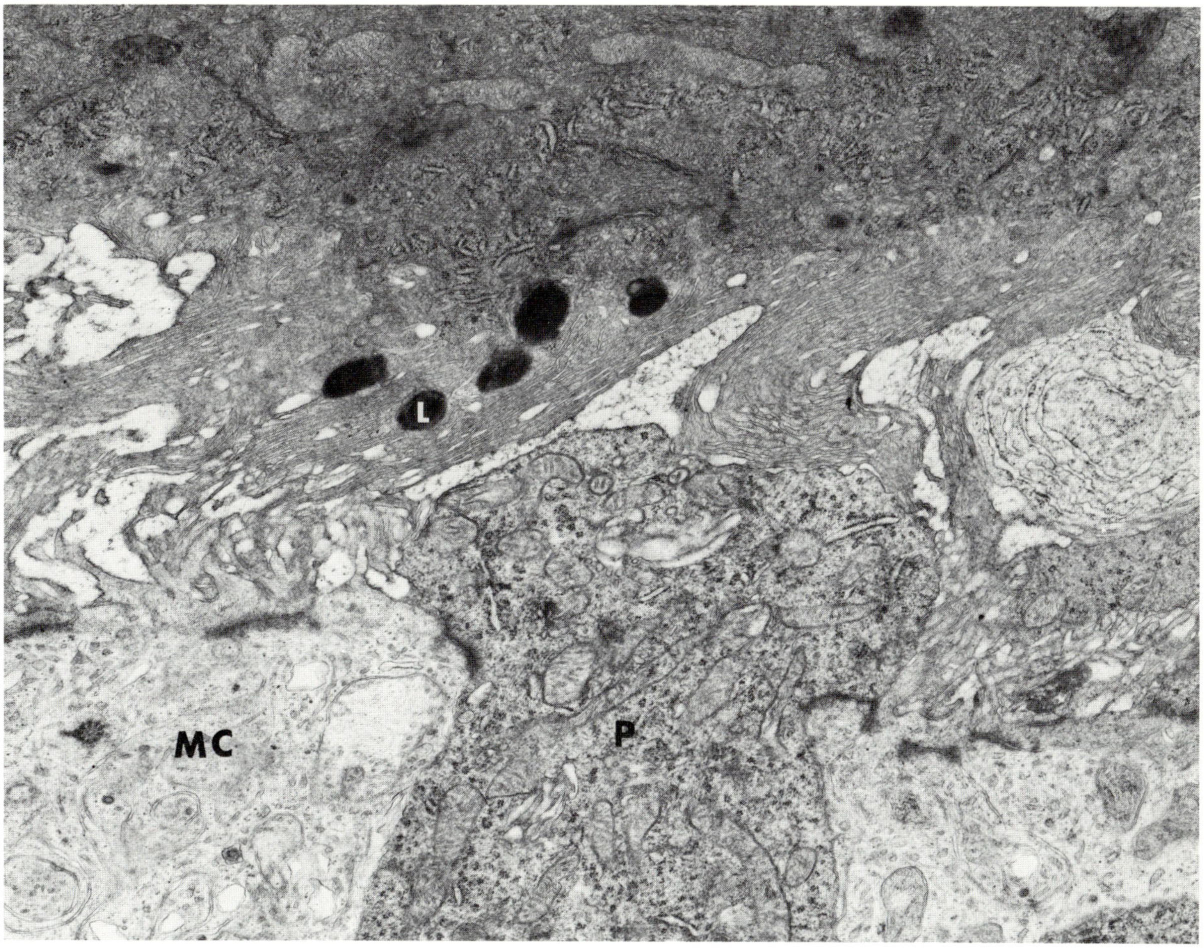

Figure 19.19 *Photoreceptor region of Irish setter with advanced rod-cone dysplasia. Loss of the photoreceptors brings the RPE close to the external limiting membrane. The prominent Müller cell* (MC) *cytoplasm surrounds an unidentified photoreceptor cell* (P). *Granules, apparently lipofuscin* (L), *are in the apex of the pigment epithelium.* (×*14,100.*)

in other animals as in dogs, the pigment epithelium in all cases appears normal and minimally reactive in the early stages of these diseases.

HEMERALOPIA. Hemeralopia is a recessively inherited selective cone degeneration found in Alaskan malamutes (Rubin, Bourns, and Lord 1967; Aguirre and Rubin 1975). Throughout the retina, cone outer segments degenerate and the cone inner segments and perinuclear cytoplasm accumulate large bundles of 100Å filaments (Aguirre and Rubin 1974). Loss of the cone inner and outer segments is often followed by the extrusion of the cone nucleus into the photoreceptor layer. In adult animals, the cone nuclei lie extracellularly in depressions in the pigment epithelial apex (Fig. 19.21). The pigment epithelium remains unreactive and appears unable to phagocytize these large nuclei.

CENTRAL PROGRESSIVE RETINAL ATROPHY. Central progressive retinal atrophy (CPRA) is a specific pigment epithelial dystrophy found in the Labrador retriever, Shetland sheepdog, border collie, and several other dog breeds (Parry 1954; Barnett and Dunn 1969). The genetics of this disorder is not well understood; in one breed, the Labrador retriever, a dominant mode of inheritance with variable gene penetrance is postulated.

Ophthalmoscopically, the disease is characterized initially by small irregular foci of pigment in the tapetal region above the optic nerve head, usually in the area centralis. The tapetal fundus becomes

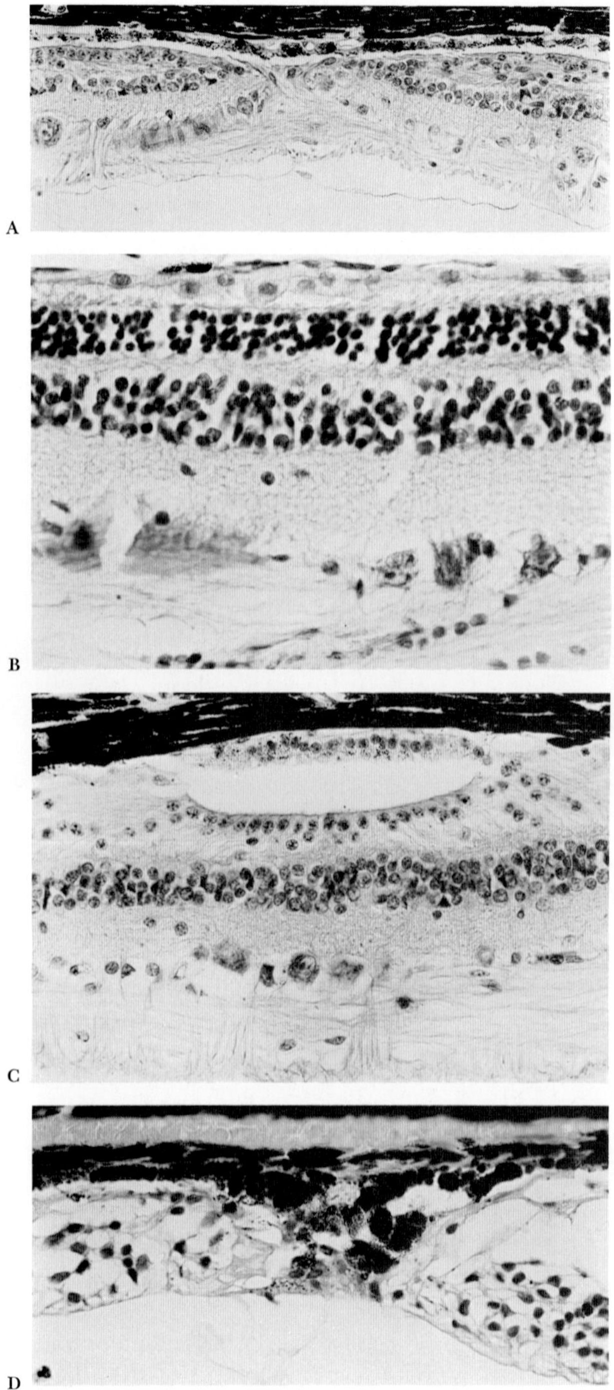

Figure 19.20 *Sections from adult Norwegian elkhounds affected with rod dysplasia.* A, *nontapetal zone. Focal area of retinal gliosis results in loss of retinal layer organization. Retinal pigment epithelial layer is intact.* B, *tapetal zone. There is focal pigment epithelial cell hypertrophy. The receptor and outer nuclear layers are attenuated.* C, *nontapetal zone. An "island" of pigment epithelium and outer nuclear layer is present. Loss of pigment epithelium and outer retinal layers is prominent in adjacent tissue.* D, *nontapetal zone. The focal area of gliosis bridges very attenuated and disorganized retina. There is migration of retinal pigment epithelium into gliosed retina.*

increasingly covered by focal accumulations of irregular pigment clumps, between which there is increased tapetal reflectively, indicative of retinal thinning (Fig. 19.22*A*). In the late stages of the disease, the focal pigment accumulations disappear, and the retina appears atrophic (Parry 1954).

Behavioral abnormalities usually become evident when there is extensive pigmentation of the tapetal zone. There is loss of central vision—that is, a central scotoma, with normal peripheral vision. Peripheral vision is eventually lost, and there is blindness in the late stages of the disease. The age of onset and rate of progression of CPRA is not the same in all affected breeds. In general, the disease is recognized earlier and progresses faster in the border collie (Barnett and Dunn 1969), a breed common in the United Kingdom, than in the Labrador retriever and Shetland sheepdog.

In the early stages of CPRA, the electroretinogram is normal, but later there is a decrease in the b-wave

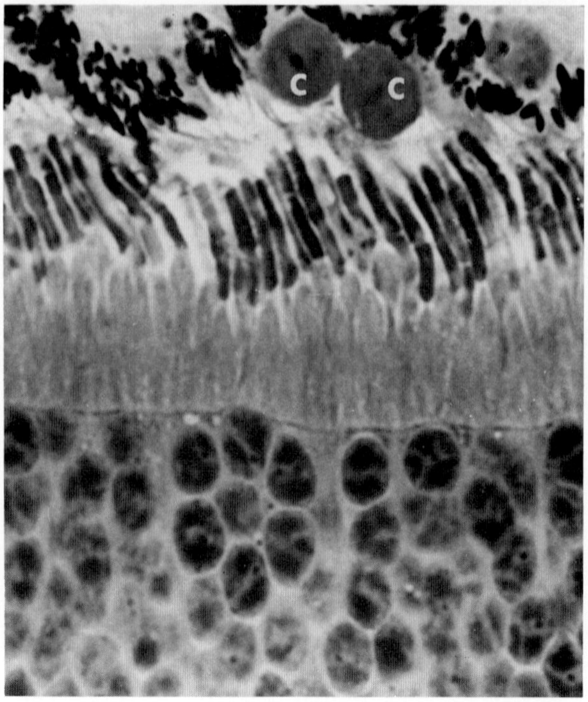

Figure 19.21 *Nontapetal zone of an adult Alaskan malamute with inherited cone degeneration. Selective loss of cone photoreceptors has resulted in a structurally pure rod retina. The two remaining cone nuclei* (C) *are ectopically located at the apex of the pigment epithelium.*

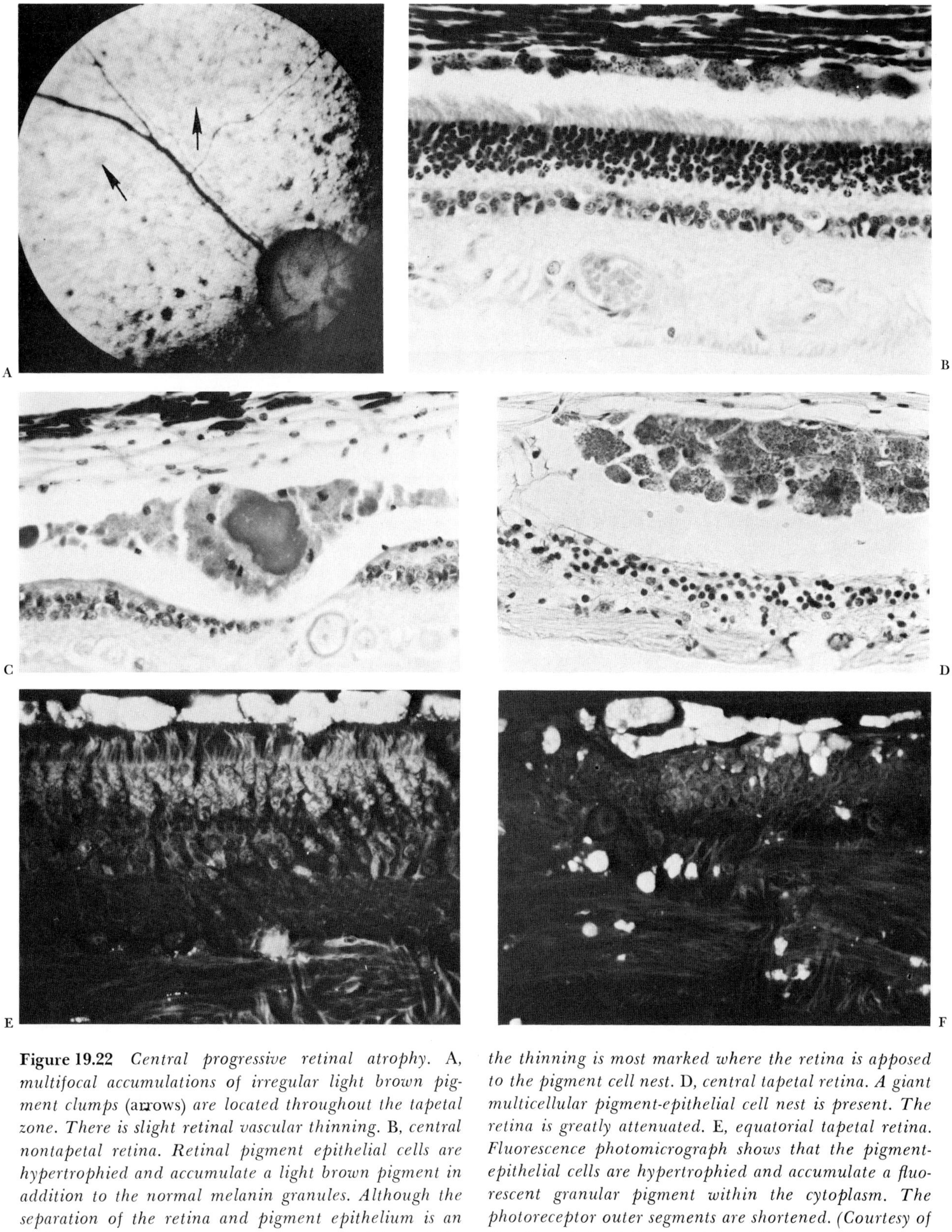

Figure 19.22 *Central progressive retinal atrophy.* A, *multifocal accumulations of irregular light brown pigment clumps* (arrows) *are located throughout the tapetal zone. There is slight retinal vascular thinning.* B, *central nontapetal retina. Retinal pigment epithelial cells are hypertrophied and accumulate a light brown pigment in addition to the normal melanin granules. Although the separation of the retina and pigment epithelium is an artifact, this section demonstrates a marked loss of the photoreceptor outer segments, a finding characteristic of the early stages of this disease.* C, *central tapetal retina. A nest of hypertrophied pigment-epithelial cells surrounds an intracellular lipoidal inclusion within a large pigment epithelial cell. The retina is attenuated throughout, but the thinning is most marked where the retina is apposed to the pigment cell nest.* D, *central tapetal retina. A giant multicellular pigment-epithelial cell nest is present. The retina is greatly attenuated.* E, *equatorial tapetal retina. Fluorescence photomicrograph shows that the pigment-epithelial cells are hypertrophied and accumulate a fluorescent granular pigment within the cytoplasm. The photoreceptor outer segments are shortened. (Courtesy of Dr. Alan Laties.)* F, *central tapetal retina. Fluorescence photomicrograph of a multicellular pigment-epithelial cell nest demonstrating the accumulation of the cytoplasmic fluorescent granular pigment. The outer retina is atrophic and there is intraretinal migration of pigment bearing cells. (Courtesy of Dr. Alan Laties.)*

amplitude. By the time peripheral vision is lost, the ERG is extinguished (Parry 1954). The electrooculogram (EOG) has not been recorded in cases of CPRA.

The histologic changes in CPRA consist of general pigment epithelial cell hypertrophy, which can result in individual giant cells as well as multicellular "nests" of pigment epithelial cells. In the earliest phase of the disease, the neuroretina is normal. Foci of pigment epithelial hypertrophy develop and then spread to involve most of the tapetal zone. Giant pigment epithelial cells are found; their cytoplasm accumulates light-brown refractile pigment granules. There is focal loss of rods, cones, and their nuclei immediately adjacent to these giant pigment cells (Parry 1954). Diffuse loss of receptor outer segments occurs throughout the retinal regions overlaid by hypertrophied pigment epithelial cells (Fig. 19.22*B*).

More advanced changes are seen later in the central retina—the tapetal region and part of the ventral nontapetal region—where there is complete loss of retinal layer organization. In many areas, the pigment epithelium disappears, while in other areas, large multicellular pigment epithelial "nests" enlarge and invade the outer retinal layers. These pigment epithelial "nests" contain small quantities of pigment granules and can measure up to 200 μ in diameter (Fig. 19.22*C–D*). In the end stages of the disease, the loss of retinal layer organization involves the central as well as the peripheral retina. In the central regions, the retina is atrophic and large areas are devoid of pigment epithelium.

In all stages of the disease, the affected pigment epithelial cells accumulate a light-brown granular pigment that is brilliantly autofluorescent under ultraviolet illumination. The density of this pigment within dystrophic pigment epithelial cells increases with the severity and progression of the disease (Fig. 19.22*E–F*). The light and fluorescent microscopic features of this autofluorescent light-brown granular pigment suggest that it is a lipopigment, and that its accumulation may reflect a localized abnormality in the metabolism of lipopigments by the retinal pigment epithelium (Aguirre and Laties 1976). At the moment it has not been possible to further characterize the lipopigment as either ceroid or lipofuscin.

CPRA is the only pigment epithelial dystrophy that has been recognized in dogs. Unlike PRA, where the pigment epithelial changes occur late in the disease and are secondary to an initial photoreceptor abnormality, the pigment epithelial lesions in CPRA result in degenerative changes in the photoreceptor layer. Being a primary pigment epithelial disease, CPRA is similar to retinal dystrophy in rats.

FELINE CENTRAL RETINAL DEGENERATION. Feline central retinal degeneration (FCRD) is a disease found sporadically in domestic and purebred cats (Bellhorn and Fischer 1970). Ophthalmoscopically, the degeneration is bilaterally symmetrical and can range in size from focal lesions limited to the center of the area centralis (Fig. 19.23*A*) to larger ellipsoidal and band-shaped abnormalities.

Histologically, the lesion consists of focal degeneration of the photoreceptor and outer nuclear layers in the area centralis. Throughout the retina, however, ultrastructural examination demonstrates that there are widespread abnormalities affecting only the cone outer segments. This accounts for the absence, or marked abnormality, of cone electroretinographic responses (Bellhorn, Aguirre, and Bellhorn 1974).

The pigment epithelium overlying the area centralis lesion is structurally unaffected by this severe localized destruction of the outer retinal layers. The loss of photoreceptors brings the pigment epithelial microvilli into close apposition to the external limiting membrane and the Müller fiber baskets (Fig. 19.23*B*). However, the pigment epithelium remains nonreactive, and there is no intraretinal migration. Numerous membrane-bound residual bodies and lipofuscin granules in the pigment epithelium may indicate increased phagocytic activity. Fluorescein angiographic and ultrastructural studies of the choroidal vascular supply overlying the area centralis lesion have failed to demonstrate any significant abnormalities (Bellhorn, Aguirre, and Bellhorn 1974).

In the domestic feline population, FCRD is a sporadic, rarely progressive abnormality; it has been observed with increasing frequency, often affecting as many as 4 percent of cats examined ophthalmoscopically. In many cases, the cause of the disease is unknown. However, we have recently found FCRD to occur in many cats whose diet consists almost exclusively of dog food. These foods contain a lower protein concentration than is normally required by cats, and they are deficient in the amino acid taurine (Aguirre 1978).

These findings appear to link the clinically observable disease FCRD with the experimental nutritional retinopathy produced in kittens and cats by

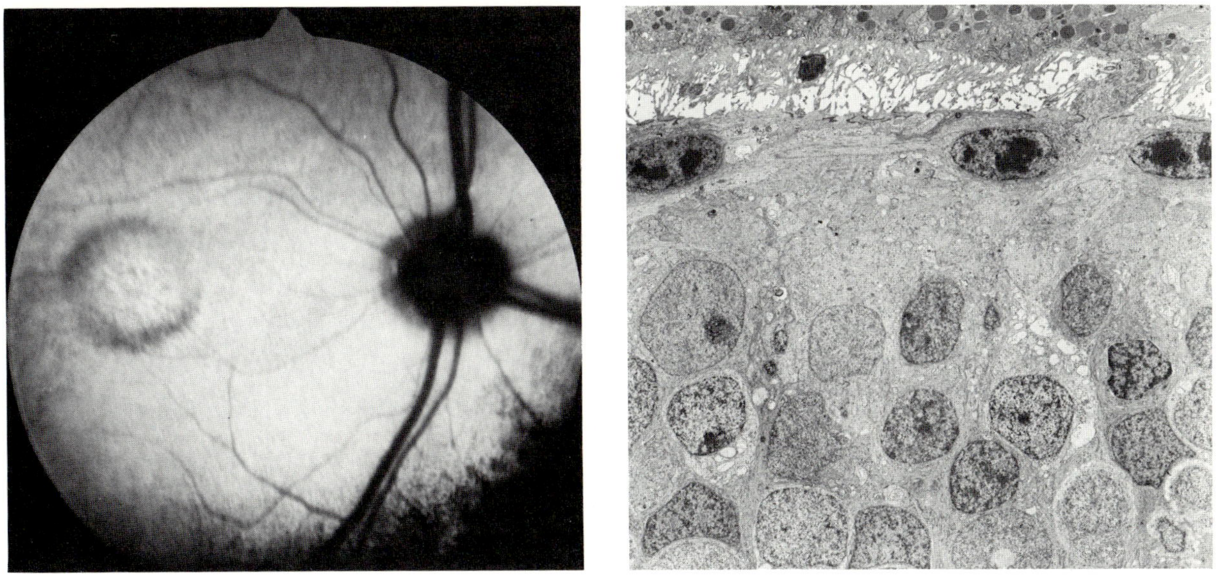

Figure 19.23 *Feline central retinal degeneration.* A, *small, round, focal hyperreflective lesion is situated in the center of the area centralis.* B, *in the center of the area centralis lesion there is almost complete loss of photoreceptors and their nuclei. The apical microvilli of the retinal pigment epithelium interdigitate with the Müller's cell microvilli. The pigment epithelium contains many membrane-bound dense inclusions, presumably lipofuscin. The inner retinal layers are normal. (×2,110.) (Reprinted with permission from Bellhorn, Aguirre, and Bellhorn 1974.)*

feeding them a semipurified diet containing casein as the only source of dietary protein (Rabin, Hayes, and Berson 1973; Hayes, Rabin, and Berson 1975). The experimental casein diet contains no taurine and produces, within 10 weeks, a marked fall in the plasma and retinal taurine concentrations, increases in the cone ERG implicit times (peak latencies), and reduced ERG amplitudes. Later, the retinal DNA concentration decreases, indicating the loss of cells in the retina (Schmidt et al. 1976). Supplementation of the experimental diet with taurine prevents the disease (Berson et al. 1976). Morphologically, FCRD and the taurine-deficient retinal degeneration are similar, although in the latter disease, more severe abnormalities of the rod and cone photoreceptors are observed in those animals chronically maintained on the diet.

MACULAR DISEASE. Focal lesions of the macular area are uncommon in subhuman primates. A unilateral foveal hole with cystoid degeneration of the macula has been described in a baboon. The cystoid changes involved the ganglion cell, inner nuclear, and outer plexiform layers. The pigment epithelium was morphologically normal, even overlying the foveal hole (Ishmael and Bradley 1974).

A bilateral macular degeneration associated with widespread outer retinal disease has also been recently reported in a baboon (Vainisi, Beck, and Apple 1974). The disease may be familial, since several members of a closely bred group of baboons in a public zoological park have developed this macular abnormality (Vainisi, unpublished). In the affected baboons, the maculae were atrophic, with a honeycombed yellow metallic appearance (Fig. 19.24). A prominent "window defect," indicative of pigment epithelial destruction, was observed over the macular region during fluorescein angiography. Histologically, there was complete absence of retina and pigment epithelium in the macula, the internal limiting membrane being apposed to Bruch's membrane. More peripherally, however, the pigment epithelium was intact, although there was marked degeneration of the photoreceptor layer.

In aging rhesus monkeys, the macular area can develop a granular appearance. Clinically, this resembles drusen of the macula, but histologically the only abnormality seems to be slightly swollen, relatively unpigmented RPE cells.

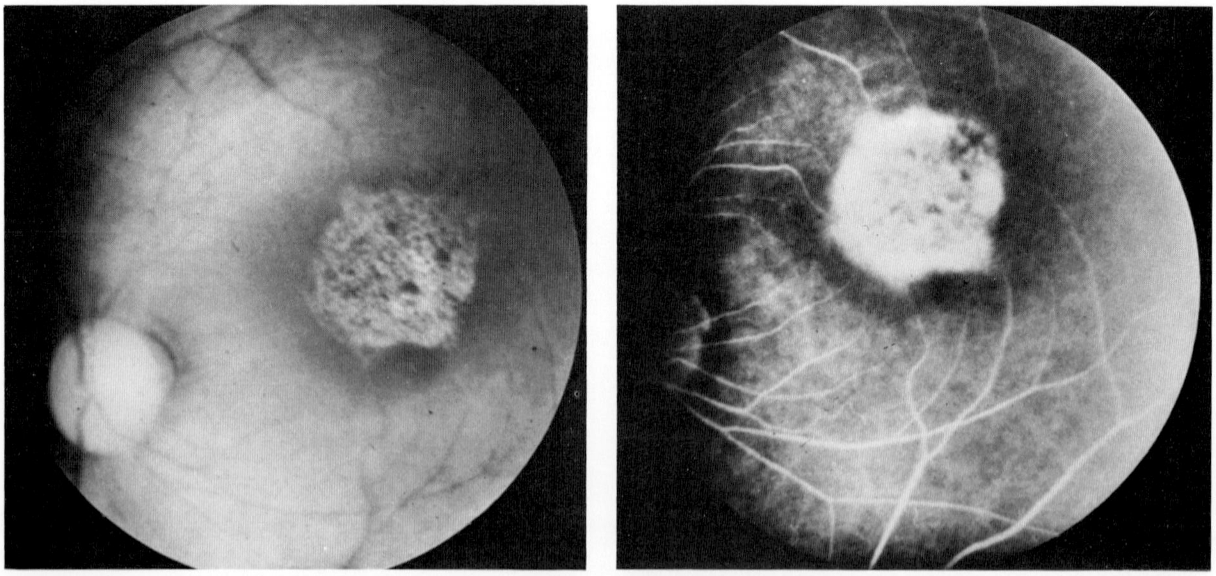

Figure 19.24 *Fundus photographs of a baboon with bilateral macular degeneration. The macula appears honeycombed and atrophic. The fluorescein angiogram shows marked fluorescence of the macular lesion. (Courtesy of Dr. Sam Vainisi.)*

Drusen

Senile or degenerative changes of the Bruch's membrane–pigment epithelial complex have not been recognized in domestic animal species. There has only been one report of drusen in subhuman primates, and this was in a young female baboon (*Papis anubis*). Multiple small, pale, round to ovoid bodies were found scattered throughout the fundus on ophthalmoscopic examination (Fig. 19.25*A*). Histologically, these bodies were round, eosinophilic structures, 10–60 μ in diameter, which were situated in Bruch's membrane. The inner surface of the bodies was bounded by attenuated pigment epithelial cells (Barnett, Heywood, and Hague 1972) (Fig. 19.25*B*).

Nutritional Diseases

VITAMIN A DEFICIENCY. The retinal abnormalities of vitamin A deficiency, characterized chiefly by nyctalopia, are common to most species. In rats, the rod outer segments deteriorate, followed by complete loss of the outer retinal layers as the deficiency continues. In the early stages of the deficiency, the disease process is reversible if vitamin A is provided (Dowling and Wald 1958).

In dogs, the earliest clinically visible ocular change is an alteration of the color of the tapetal region, especially at the junction between it and the nontapetal zone. In addition to the retinal changes, bony overgrowth of the various foramina occurs, causing papilledema followed by optic atrophy, so that both inner and outer layers of the retina eventually degenerate (Mellanby 1943). The role of the pigment epithelium in the degenerative process is not known.

Vitamin A deficiency has been produced experimentally in adult horses (Anderson and Hart 1943). The signs of deficiency take more than one year to develop. Night blindness occurs, as well as signs attributable to epithelial metaplasia. In the retina there is histologically visible widening and vacuolation of the photoreceptor outer segments. In the RPE, there is a decrease in the number of pigment granules, and RPE surfaces are irregular.

Cattle, especially young growing calves, have severe clinical signs associated with vitamin A deficiency. The retinal abnormalities are usually overshadowed by papilledema, which is secondary to constriction of the optic nerve by overgrowth of the surrounding bone (Hayes, Nielsen, and Eaton 1968). The role of the RPE has not been adequately studied in this species, although Barnett et al. have observed local loss of pigment from the RPE (1970).

Vitamin A deficiency in monkeys produces the classic external ocular abnormalities, as well as nyctalopia accompanied by loss of ERG b-wave. The

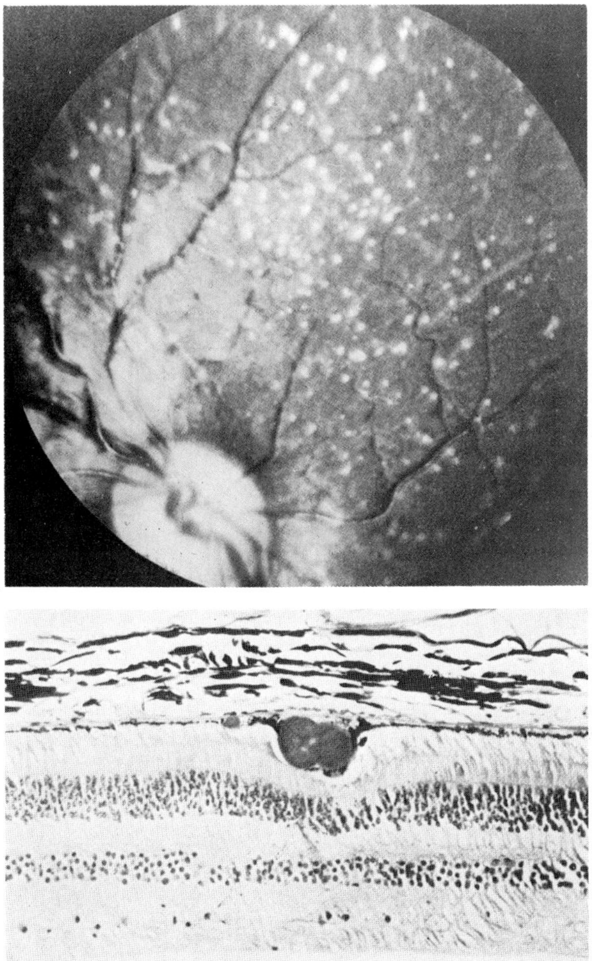

A

B

Figure 19.25 *Drusen. A, fundus photograph showing pale, round to ovoid bodies located deep to the retinal layers. B, histologic section shows that the drusen originate from Bruch's membrane and are covered by attenuated pigment epithelial cells. (Reprinted from Barnett et al. 1972.)*

retinal degeneration is most severe in the macula and paramacular area and less severe in the peripheral retina. The photoreceptor outer segments are shortened and disrupted, and the outer nuclear layer is correspondingly reduced. Curiously, in early degenerative states, the cones seem more severely involved, but in advanced avitaminosis A, rods and cones are equally abnormal. The pigment epithelial cells contain an excess of partially digested lipid droplets, as well as phagosomes (Hayes 1974).

VITAMIN E DEFICIENCY. In addition to anemia, muscle degeneration, testicular degeneration, and accumulation of ceroid material in several tissues, retinal degeneration occurs commonly in vitamin E deficient monkeys. The abnormalities, though not clinically visible, are concentrated in the macular region and in cones in the extramacular zone. The degeneration is confined to the outer segments of the photoreceptors. The pigment epithelium continues its phagocytic activity, the cells becoming distended with lipofuscin-type lysosomes and aggregates of outer segment debris (Hayes 1974). The behavior of the pigment epithelium in vitamin E deficient dogs is similar, but in dogs, the retinal degeneration is extensive, involving both rod and cone outer segments, as well as the outer nuclear and plexiform layers. Many pigment epithelial cells, laden with a granular PAS-positive pigment, migrate into all layers of the degenerate retina (Hayes, Rousseau, and Hegsted 1970). Whether the vitamin E deficiency results in pigment epithelial disease, or whether the photoreceptor degeneration induces pigment epithelial changes, has not been determined.

BRIGHT BLINDNESS. Bright blindness is a progressive degeneration of the outer layers of the retina in sheep. The disease has been recognized only in England, and is caused by chronic ingestion of bracken (*Pteris aquilina*). In advanced cases, only the inner plexiform and ganglion cell layers remain unaffected, all of the outer retinal layers having disappeared. Damage is most severe at the posterior pole. The pigment epithelium does not appear to participate in the degeneration, remaining flat and apparently undisturbed (Barnett and Watson 1970). It has not been determined whether bright blindness is due to a nutritional deficiency or a specific retinotoxic agent in bracken.

Toxicology

ETHAMBUTOL. The antitubercular drug ethambutol, as well as a number of N, N′-alkyl substituted ethylenediamines, has the unusual effect of producing decoloration of the tapetum and disorientation, usually reversible, of the tapetal rods in dogs. The drug acts first where the tapetum is thinnest, but with substantial doses (800 mg/kg body weight), the entire tapetum becomes decolored for as much as 48 hours (Kaiser 1963; Vogel and Kaiser 1963). Although no disturbance of the pigment epithelium or retina has been found in dogs, there appears to be some effect in man, in whom a decrease in visual acuity and loss of the ability to perceive the color green has been observed (Goodman and Gilman

1970). This effect in man may imply some action on the pigment epithelium or photoreceptors.

DIPHENYLTHIOCARBAZONE (DITHIZONE). Intravenous administration of 10 mg dithizone to dogs or rabbits produces severe alterations in the fundus (Budinger 1961). In the dog the initial change is striking; diffuse retinal edema is evident in the periphery in one day. This becomes more intense by 3 days. Pigmentary disruption follows and gives the fundus a mottled appearance. Higher doses of dithizone result in irreversible changes characterized by complete retinal degeneration with pigment cell invasion.

Since dithizone is an efficient chelating agent, and since the canine tapetum is rich in zinc, it has been thought that the blindness in dogs is primarily the result of necrosis of the tapetum. Our investigations have shown that while the tapetum does indeed become necrotic, the overlying retina appears relatively unaffected. By contrast, the retina overlying the nontapetal area becomes severely necrotic. The differing effects of the drug are sharply contrasted in the tapetal-nontapetal junction zone.

The toxic effects of dithizone are not limited to tapetal animals, since Butturini et al. (1953) have demonstrated widespread retinal and pigment epithelial damage in rabbits. Even in those rabbits not having ophthalmoscopically visible signs, electroretinographic abnormalities are present within 4 hours of dithizone administration.

Hydroxypyridinethione also produces blindness in dogs. The blindness is accompanied by retinal edema and detachment, tapetal necrosis, and underlying choroiditis. The changes are limited to the tapetal area (Moe, Kirpan, and Linegar 1960). By contrast, diethyldithiocarbamate produces changes similar to those produced by dithizone in the dog, but not in the rat, cat, rabbit, or monkey. In the dog, the retinal abnormalities are not limited to the tapetal area (Scholler, Brown, and Timmene 1961).

PHENOTHIAZINES. Some chemical compounds, notably the phenothiazines and quinolines, selectively concentrate in the uvea and retinal pigment epithelium. Such accumulation may produce degeneration of the retina, although uveal or pigment epithelial accumulation does not necessarily imply that retinotoxicity must follow. The ability of the retinal pigment epithelium to accumulate drugs varies from species to species (Potts 1962). Some compounds, such as piperidylchlorophenothiazine (NP-207) are limited in their capacity to produce retinotoxicity, affecting only man and the cat, but no other commonly used laboratory species. Administration of 10 to 70 mg NP-207/kg/day to domestic cats results in clinical signs of retinotoxicity in 4 to 6 weeks. Ophthalmoscopically, there appears to be a pigment deposition first in the area centralis. The pigment then spreads to become distributed in a horizontal band near the optic disc. The area affected closely resembles that seen in feline central retinal degeneration. However, as drug administration continues, the entire tapetal area of the fundus becomes pigmented.

In NP-207 toxicity in cats, pigmentary degeneration is thought to be due to lipid deposition in the retinal pigment epithelium. The lipid is believed to contain visual pigment shed from rod outer segments. NP-207 seems to act on the retina partly by activating the shedding process of rod outer segments and partly by totally disrupting these same outer segments (Meier-Ruge and Cerletti 1966).

CHLOROQUINE. In man the chief site of ocular damage by chloroquine is in the rods and cones, primarily in the macular zone. Ganglion cell abnormalities also occur. In animals, the damage produced varies with the species. In dogs and swine, the ganglion cells of the retina accumulate abnormal material to form membranous cytoplasmic bodies, or MCB's (Gleiser et al. 1969). The same phenomenon occurs in both albino and pigmented rats, although in rats the MCB's may occur in all neurons in the retina (Gregory, Rutty, and Wood 1970), and the pigment epithelium is not affected (Hodgkinson and Kolb 1970).

In rabbits, long-term administration of chloroquine results in damage to the pigment epithelium and retina. The fundus contains multiple, often confluent, raised pale white spots, interspersed with pigment deposits. Histologically, selective accumulation of chloroquine occurs in the pigment epithelium, which becomes thickened and wavy and finally migrates into the retina; the migration is coincident with, or causes, outer layer destruction. The abnormalities occur both in pigmented and albino rabbits (Dale, Parkhill, and Layton 1965). Similarly, in cats, swelling and focal proliferation of the pigment epithelium occurs after 4 to 7 weeks of oral administration of chloroquine. There is secondary atrophy of the photoreceptor outer segments (Meier-Ruge 1965).

Amopyroquin, a 4-aminoquinoline similar to chloroquine, causes retinotoxicity in rats and dogs but not in monkeys. Dogs develop pigmentary proliferation with secondary loss of outer segments, while rats primarily lose the outer retinal layers (Kurtz et al. 1967).

When Trenimon (2,3,5-triethylenimino-1,4-benzoquinone), a potent inhibitor of glycolysis, is administered by slow intracarotid injection to rabbits, it produces retinal edema, which is followed by a dystrophic appearance to the retina. The affected retina has extensive accompanying pigment dispersion (Apponi, Tieri, and Rinaldi 1964). Similar effects are produced by the administration of parachloromercurobenzoate (DeBerardinis, Tieri, and Vecchione, 1963).

IODATE AND IODOACETATE. Administration of sodium iodate intravenously primarily and predominantly affects the pigment epithelium. There is a patchy distribution of swollen pigment cells, which later migrate into the retina. The photoreceptor outer segments degenerate secondarily. In some areas pigment epithelial cells become atrophic and disappear (Grignolo, Orzalesi, and Calabria 1966).

Degeneration of pigment epithelial cells occurs following intravenous administration of iodoacetate to experimental animals, but the RPE degeneration occurs only at doses higher than necessary for the death of visual cells (Noell 1958). Bromoacetate produces similar but less severe abnormalities (Lucas, Newhouse, and Davey 1957).

DIAMINODIPHENOXYALKANES. A compound with schistosomicidal activity, M+B 968A is capable of producing impairment of vision in monkeys, dogs and cats after administration of single doses of 100 to 200 mg/kg. In the mouse, rat, rabbit, and guinea pig vision is not affected by single doses. The impairment is due to retinal degeneration, first affecting pigment epithelium, which swells and proliferates into clumps. These clumps distort the overlying retina into arcades. The visual cells then degenerate until the entire outer retina is lost. The tapetal pigment epithelium becomes pigmented, proliferates, and behaves in the same manner as the nontapetal pigment epithelium. Several other diaminodiphenoxyalkanes have similar retinotoxic activity in rabbits (Ashton 1957; Sorsby and Nakajima 1958).

LEAD. Along with other systemic effects, pigment epithelium toxicity occurs in rabbits fed lead compounds (lead subacetate and others). There is accumulation of lipofuscin granules in markedly distended pigment epithelial cells. The photoreceptors atrophy opposite the focal pigment epithelial lesions (Hass et al. 1964; Hughes and Coogan 1974).

N-METHYL-N-NITROSOUREA. When injected into Syrian hamsters, this potent carcinogen produces outer retinal degeneration, accompanied by pigment proliferation and intraretinal migration. The mechanism of action is not known (Herrold 1967).

REFERENCES

Aguirre, G. D. 1973. Progressive rod-cone degeneration in the dog. *Proc. Assoc. Res. Vis. Ophthalmol.*, p. 16.

Aguirre, G. D. 1976. Inherited retinal degeneration in the dog. *Trans. Am. Acad. Ophthalmol. Otolaryngol.* 81: 667–676.

Aguirre, G. D. 1978. Retinal degeneration associated with the feeding of dog foods to cats. *J. Am. Vet. Med. Assoc.* 172:791–796.

Aguirre, G. D., and Laties, A. 1976. Pigment epithelial dystrophy in dogs. *Exp. Eye Res.* 23:247–256.

Aguirre, G. D., and Rubin, L. F. 1971. Progressive retinal atrophy (rod dysplasia) in the Norwegian elkhound. *J. Am. Vet. Med. Assoc.* 158:208–218.

Aguirre, G. D., and Rubin, L. F. 1972. Progressive retinal atrophy in the miniature poodle: an electrophysiologic study. *J. Am. Vet. Med. Assoc.* 160:191–201.

Aguirre, G. D., and Rubin, L. F. 1974. Pathology of hemeralopia in the Alaskan malamute dog. *Invest. Ophthalmol.* 13:231–235.

Aguirre, G. D., and Rubin, L. F. 1975. Rod-cone dysplasia (progressive retinal atrophy) in Irish setters. *J. Am. Vet. Med. Assoc.* 166:157–164.

Aguirre, G. D., and Rubin, L. F. 1975. The electroretinogram in dogs with inherited cone degeneration. *Invest. Ophthalmol.* 14:840–847.

Anderson, A. C., and Hart, G. H. 1943. Histological changes in the retina of the vitamin A deficient horse. *Am. J. Vet. Res.* 4:307–317.

Apponi, G., Tieri, O., and Rinaldi, E. 1964. The retinotoxic action of 2:3:5-triethylenimino-1:4-benzoquinone (Trenimon). *Acta Ophthalmol.* 42:64–67.

Ashton, N. 1957. Degeneration of the retina due to 1:5-di-(p-aminophenoxy) pentane dihydrochloride. *J. Pathol.* (Lond.) 74:103–112.

Ashton, N., Barnett, K. C., and Sachs, D. D. 1968. Retinal dysplasia in the Sealyham terrier. *J. Pathol.* (Lond.) 96:269–272.

Barnett, K. C. 1965. Canine retinopathies. I. History and review of the literature. *J. Small Anim. Pract.* 6:41–55.

Barnett, K. C., Bjorck, G. R., and Kock, E. 1970. Hereditary retinal dysplasia in the Labrador retriever in England and Sweden. *J. Small Anim. Pract.* 10:755–759.

Barnett, K. C., and Dunn, W. I. 1969. The International Sheep Dog Society and progressive retinal atrophy. *J. Small Anim. Pract.* 10:301.

Barnett, K. C., Heywood, R., and Hague, P. H. 1972. Colloid degeneration of the retina in a baboon. *J. Comp. Pathol.* 82:117–118.

Barnett, K. C., Palmer, A. C., Abrams, J. T., Bridge, P. S., Spratling, F. R., and Sharman, I. M. 1970. Ocular changes associated with hypovitaminosis A in cattle. *Br. Vet. J.* 126:561–573.

Barnett, K. C., and Watson, W. A. 1970. Bright blindness in sheep: a primary retinopathy due to feeding bracken (Pteris Aquilina). *Res. Vet. Sci.* 11:289–291.

Bellhorn, R. W., Aguirre, G. D., and Bellhorn, M. B. 1974. Feline central retinal degeneration. *Invest. Ophthalmol.* 13:608–616.

Bellhorn, R. W., Bellhorn, M. B., Swarm, R. L., and Impellizzeri, C. E. 1975. Hereditary tapetal abnormality in the beagle. *Ophth. Res.* 7:250–260.

Bellhorn, R. W., and Fischer, C. A. 1970. Feline central retinal degeneration. *J. Am. Vet. Med. Assoc.* 157: 842–849.

Bernstein, M. H., and Pease, D. C. 1959. Electron microscopy of the tapetum lucidum of the cat. *J. Biophys. Biochem. Cytol.* 5:35–53.

Berson, E. L. 1976. Retinitis pigmentosa and allied retinal diseases: electrophysiologic findings. *Trans. Am. Acad. Ophthalmol. Otolaryngol.* 81:659–666.

Berson, E. L., Hayes, K. C., Rabin, A. R., Schmidt, S. Y., and Watson, G. 1976. Retinal degeneration in cats fed casein. II. Supplementation with methionine, cysteine or taurine. *Invest. Ophthalmol.* 15:52–58.

Bistner, S. I., Rubin, L. F., and Saunders, L. Z. 1970. The ocular lesions of bovine viral diarrhea-mucosal disease. *Pathol. Vet.* 7:275–286.

Bok, D., and Hall, M. O. 1971. The role of the pigment epithelium in the etiology of inherited retinal dystrophy in the rat. *J. Cell Biol.* 49:664–682.

Budinger, J. M. 1961. Diphenylthiocarbazone blindness in dogs. *Arch. Pathol.* 71:304.

Butturini, U., Grignolo, A., and Baronchelli, A. 1953. "Diabete" da ditizone aspetti metabolici oculari ed istologici. *G. Clin. Med.* 34:1253–1347.

Dale, A. J., Parkhill, E. M., and Layton, D. D. 1965. Studies on chloroquine retinopathy in rabbits. *JAMA* 193: 241–243.

DeBerardinis, E., Tieri, O., and Vecchione, L. 1963. Degeneration of the retina caused by the p-chloromercurobenzoate of sodium. *Acta Ophthalmol.* 41:587–588.

Dowling, J. E., and Wald, G. 1958. Vitamin A deficiency and night blindness. *Proc. Nat. Acad. Sci.* 44:648–661.

Friedman, A. H., Bellhorn, R. W., and Henkind, P. 1973. Simian virus 40-induced retinopathy in the rat. *Invest. Ophthalmol.* 12:591–595.

Gay, A. J., Goldor, H., and Smith, M. 1964. Chorioretinal vascular occlusions with latex spheres. *Invest. Ophthalmol.* 3:647–656.

Gelatt, K. N., and McGill, L. D. 1973. Clinical characteristics of microphthalmia with colobomas in the Australian shepherd dog. *J. Am. Vet. Med. Assoc.* 162:393–396.

Gleiser, C. A., Dukes, T. W., Lawwill, T., Read, W. K., Bay, W. W., and Brown, R. S. 1969. Ocular changes in swine associated with chloroquine toxicity. *Am. J. Ophthalmol.* 64:399–405.

Goodman, L. S., and Gilman, A. 1970. *The pharmacological basis of therapeutics,* 4th ed., p. 333. New York: Macmillan Co.

Gregory, M. H., Rutty, D. A., and Wood, R. D. 1970. Differences in the retinotoxic action of chloroquine and phenothiazine derivatives. *J. Pathol.* 102:139–150.

Grignolo, A., Orzalesi, N., and Calabria, G. A. 1966. Studies on the fine structure and the rhodopsin cycle of the rabbit retina in experimental degeneration induced by sodium iodate. Exp. Eye Res. 5:86–97.

Hass, G. M., Brown, D. V. L., Eisenstein, R., and Hemmens, A. 1964. Relation between lead poisoning in rabbit and man. *Am. J. Pathol.* 45:691–727.

Hayes, K. C. 1974. Retinal degeneration in monkeys induced by deficiencies of vitamin E or A. *Invest. Ophthalmol.* 13:499–510.

HAYES, K. C., NIELSEN, S. W., and EATON, H. D. 1968. Pathogenesis of the optic nerve lesion in vitamin A deficient calves. *Arch. Ophthalmol.* 80:777–787.

HAYES, K. C., RABIN, A. R., and BERSON, E. L. 1975. An ultrastructural study of nutritionally induced and reversed retinal degeneration in cats. *Am. J. Pathol.* 78:505–516.

HAYES, K. C., ROUSSEAU, J. E., JR., and HEGSTED, D. M. 1970. Plasma tocopherol concentration and vitamin E deficiency in dogs. *J. Am. Vet. Med. Assoc.* 157:64–71.

HERROLD, K. M. 1967. Pigmentary degeneration of the retina induced by N-methyl-N-nitrosourea: an experimental study in Syrian hamsters. *Arch. Ophthalmol.* 78:650–653.

HERRON, W. L., RIEGEL, B. W., MYERS, O. E., and RUBIN, M. L. 1969. Retinal dystrophy in the rat: a pigment epithelial disease. *Invest. Ophthalmol.* 8:595–604.

HEYWOOD, R. 1972. An anomaly of the ocular fundus of the beagle dog. *J. Small Anim. Pract.* 13:213–215.

HODGKINSON, B. J., and KOLB, H. 1970. A preliminary study on the effect of chloroquine on the rat retina. *Arch. Ophthalmol.* 84:509–515.

HUGHES, W. F., and COOGAN, P. S. 1974. Pathology of the pigment epithelium and retina in rabbits poisoned with lead. *Am. J. Pathol.* 77:237–254.

ISHMAEL, J., and BRADLEY, W. A. 1974. Cystoid degeneration of macula with hole formation in a baboon. *J. Comp. Pathol.* 84:67–71.

KAISER, J. A. 1963. Tapetal depigmentation in dogs produced by ethylenediamines. *Fed. Proc.* 22:369.

KURTZ, S. M., KAUMP, D. H., SCHARDEIN, J. L., ROLL, D. E., REUTNER, T. F., and FISKEN, R. A. 1967. The effect of long-term administration of amopyroquin, a 4-aminoquinoline compound, on the retina of pigmented and nonpigmented laboratory animals. *Invest. Ophthalmol.* 6:420–425.

LAHAV, M., ALBERT, D. M., and PERCY, D. H. 1974. Ocular changes associated with canine herpes virus infection. *Proc. Assoc. Res. Vis. Ophthalmol.* 94.

LAHAV, M., ALBERT, D. M., PERCY, D. H., CARMICHAEL, L. E., and WHITNEY, G. 1974. Unpublished information.

LAHAV, M., ALBERT, D. M., and WYAND, S. 1973. Clinical and histopathologic classification of retinal dysplasia. *Am. J. Ophthalmol.* 75:648–667.

LEURE-DUPREE, A. 1968. Ultrastructure of the pigment epithelium in the domestic sheep. *Am. J. Ophthalmol.* 65:383–398.

LUCAS, D. R., NEWHOUSE, J. P., and DAVEY, J. B. 1957. Experimental degeneration of the retina. II. The lesion produced by bromoacetate: ophthalmoscopic and histologic features. *Br. J. Ophthalmol.* 41:313–316.

MASON, W. T., FAGER, R. S., and ABRAHAMSON, E. W. 1973. Ultrastructure of the receptor epithelial layers of bovine retina. *J. Anat.* 115:289–308.

MCMILLAN, A. 1976. Acquired retinal folds in the cat. *J. Am. Vet. Med. Assoc.* 168:1015–1020.

MEIER-RUGE, W. 1965. Experimental investigation of the morphogenesis of chloroquine retinopathy. *Arch. Ophthalmol.* 73:540–544.

MEIER-RUGE, W., and CERLETTI, A. 1966. Zur experimentelle pathologie der phenothiazine retinopathie. *Ophthalmologica* 151:512–533.

MELLANBY, E. 1943. The effect of bone dysplasia (overgrowth) on cranial nerves in vitamin A deficient animals. *J. Physiol.* 101:408–431.

MOE, R. A., KIRPAN, J., and LINEGAR, C. R. 1960. Toxicology of hydroxypyridinethione. *Toxicol. Appl. Pharmacol.* 2:156–170.

MONJAN, A. A., SILVERSTEIN, A. M., and COLE, G. A. 1972. Lymphocytic choriomeningitis virus-induced retinopathy in newborn rats. *Invest. Ophthalmol.* 11:850–856.

MORRIS, M. L. 1965. Feline degenerative retinopathy. *Cornell Vet.* 55:295.

NAKAIZUMI, Y. 1964. The ultrastructure of Bruch's membrane. II. Eyes with a tapetum. *Arch. Ophthalmol.* 77:388–394.

NEWSOME, D. A., and KENYON, K. R. 1972. Collagen production *in vivo* and *in vitro* by the retinal pigmented epithelium of the chick embryo. A.R.V.O. symposium on the pigment epithelium.

NOELL, W. K. 1958. Studies on visual cell viability and differentiation. *Ann. N.Y. Acad. Sci.* 74:337–361.

PAGENSTECHER, H. E. 1961. *Ber. Dtsch. Ophthalmol. Ges.* 44:447.

PARRY, H. B. 1953. Degeneration of the dog retina. II. Generalized progressive atrophy of hereditary origin. *Br. J. Ophthalmol.* 37:487–502.

PARRY, H. B. 1954. Degeneration of the dog retina. VI. Central progressive atrophy with pigment epithelial dystrophy. *Br. J. Ophthalmol.* 38:653–668.

PERCY, D. H., CARMICHAEL, L. E., ALBERT, D. M., KING, J. M., and JONAS, A. M. 1971. Lesions in puppies surviving infection with canine herpes virus. *Vet. Pathol.* 8:37–53.

POTTS, A. M. 1962. The concentration of phenothiazines in the eye of experimental animals. *Invest. Ophthalmol.* 1:522–530.

PRINCE, J. H., DIESEM, C. D., EGLITIS, I., and RUSKELL, G. L. 1960. *Anatomy and histology of the eye and orbit in domestic animals,* p. 219. Springfield: C. C. Thomas.

RABIN, A. R., HAYES, K. C., and BERSON, E. L. 1973. Cone and rod responses in nutritionally induced retinal degeneration in the cat. *Invest. Ophthalmol.* 12:694–704.

ROBERTS, S. R., DELLAPORTA, A., and WINTER, F. C. 1966. The collie ectasia syndrome: pathologic alterations in the eyes of puppies one to fourteen days of age. *Am. J. Ophthalmol.* 61:1458–1466.

ROBERTS, S. R., DELLAPORTA, A., and WINTER, F. C. 1966. The collie ectasia syndrome: pathology of eyes of young and adult dogs. *Am. J. Ophthalmol.* 62:728–752.

RUBIN, L. F. 1963. Atrophy of rods and cones in the cat retina. *J. Am. Vet. Med. Assoc.* 142:1415–1420.

RUBIN, L. F. 1968. Heredity of retinal dysplasia in Bedlington terriers. *J. Am. Vet. Med. Assoc.* 152:260.

RUBIN, L. F. 1974. *Atlas of veterinary ophthalmoscopy.* Philadelphia: Lea and Febiger.

RUBIN, L. F., BOURNS, T. K., and LORD, L. H. 1967. Hemeralopia in dogs: heredity of hemeralopia in Alaskan malamutes. *Am. J. Vet. Res.* 28:355–357.

RUBIN, L. F., and LIPTON, D. 1973. Retinal degeneration in kittens. *J. Am. Vet. Med. Assoc.* 162:467–469.

RUGH, R., and WOLFF, J. 1955. *Proc. Soc. Exp. Biol. Med.* 89:248.

RUGH, R., and SKAREDOFF, L. 1969. X-rays and the monkey fetal retina. *Invest. Ophthalmol.* 8:31–40.

SCHMIDT, S. Y., BERSON, E. L., and HAYES, K. C. 1976. Retinal degeneration in cats fed casein. I. Taurine deficiency. *Invest. Ophthalmol.* 15:47–52.

SCHOLLER, J., BROWN, D. E., and TIMMENE, E. K. 1961. Toxicological and pathological studies with diethyldithiocarbamate. *Pharmacologist* 3:62.

SCOTT, P. P., GREAVES, J. P., and SCOTT, M. G. 1964. Nutritional blindness in the cat. *Exp. Eye Res.* 3:357–364.

SHIVELY, J. N., and PHEMISTER, R. D. 1967. Alterations in the fine structure of the mature retina of dogs irradiated as neonates. *Exp. Eye Res.* 6:278–282.

SHIVELY, J. N., PHEMISTER, R. D., EPLING, G. P., and JENSEN, R. 1970. Pathogenesis of radiation-induced retinal dysplasia. *Invest. Ophthalmol.* 9:888–900.

SILVERSTEIN, A. M. 1974. The immunologic modulation of infectious disease pathogenesis. *Invest. Ophthalmol.* 13:560–574.

SILVERSTEIN, A. M., PARSHALL, C. J., JR., OSBURN, B. I., and PRENDERGAST, R. A. 1971. An experimental virus-induced retinal dysplasia in the fetal lamb. *Am. J. Ophthalmol.* 72:22–34.

SILVERSTEIN, A. M., OSBURN, B. I., and PRENDERGAST, R. A. 1971. The pathogenesis of retinal dysplasia. *Am. J. Ophthalmol.* 72:13–21.

SORSBY, A., and NAKAJIMA, A. 1958. *Br. J. Ophthalmol.* 42:563–571.

TANSLEY, K., SPEAR, F. G., and GLUCKMAN, A. 1937. Effect of gamma rays on cell division in the developing rat retina. *Br. J. Ophthalmol.* 21:273.

VAINISI, S. J. 1974. Unpublished information.

VAINISI, S. J., BECK, B. B., and APPLE, D. J. 1974. Retinal degeneration in a baboon. *Am. J. Ophthalmol.* 78:279–284.

VOGEL, A. W., and KAISER, J. A. 1963. Ethambutol induced transient change and reconstruction (in vivo) of the tapetum lucidum color in the dog. *Exp. Mol. Pathol.* (suppl.) 2:136–149.

WISE, G. N., DOLLERY, C. T., and HENKIND, P. 1971. *The retinal circulation.* New York: Harper and Row.

20 The Retinal Pigment Epithelium in Mice and Rats with Inherited Retinal Degeneration

Matthew M. LaVail

RODENT DISEASES involving inherited photoreceptor cell degeneration have been known since being described by Keeler in mice (1924) and by Bourne, Campbell, and Tansley (1938) in rats. In both animals, the disorder is caused by an autosomal recessive mutation. Since the early descriptions, considerable interest has been shown in these mutants, because the degenerations share with certain human inherited retinal degenerations the features of heritability, apparent selective destruction of photoreceptor cells, and similar end-stage pathologic features. In particular, the disorder in the rat includes migration of pigment epithelial cells into the neural retina in the late stages of the disease, a feature of retinitis pigmentosa diseases in man.

In the ensuing decades since the original descriptions, morphologic, biochemical, and genetic studies have been carried out in attempts to characterize the disorders and to elucidate the mechanism of the mutations. Recently, a disorder in the pigment epithelial cell—a lack of phagocytosis of rod outer segment discs—was discovered in rats with retinal dystrophy (Herron et al. 1969; Bok and Hall 1969, 1971). This observation has been interpreted to mean that a primary defect in the pigment epithelial cells leads to the subsequent death of photoreceptor cells. While this hypothesis remains to be fully tested, the observed pigment epithelial cell defect has stimulated eye research on several levels. Specifically, the discovery provided researchers with a readily available animal model for studying (1) inherited pigment epithelial cell disorders, (2) control mechanisms for the rod outer segment renewal system, and (3) genetic control of cell differentiation in the pigment epithelial cell. More generally, in focusing attention on the numerous photoreceptor–pigment-epithelial cell interactions, to which much of this book is devoted, and the consequences of defects in any of the interactions, the discovery has raised the possibility that other retinal disorders heretofore considered to be primary photoreceptor cell degenerations might also involve an underlying pigment epithelial cell defect.

In this chapter, the characteristics of the pigment epithelium in mice and rats with inherited retinal degeneration will be described. For each of these species, the following outline will be used: (1) time of onset, pace, and character of the photoreceptor degeneration; (2) early features of the pigment epithelium, either preceding or simultaneous with photoreceptor degeneration; (3) role of the pigment epithelium in the photoreceptor degeneration; and (4) late changes in the pigment epithelium. Needless to say, the separation of early and late changes at the end of the overt photoreceptor degeneration is arbitrary.

Inherited Retinal Degeneration in the Mouse

PHOTORECEPTOR DEGENERATION. The retinal degeneration (*rd*) mutation is present in a number of inbred strains and some noninbred stocks of mice (Sidman and Green 1965).[1] It has been the

1. Notes on genetic terminology. It is conventional to use a capital initial letter for symbols of dominant genes, lower case letters for recessive genes, and a plus sign for the corresponding wild-type allele. An inbred strain is one that has been incrossed (brother mated with sister) for at least 20 generations; this produces homozygosity at almost every genetic locus, so all the animals of a given inbred strain are genetically alike. When a

object of numerous morphologic, genetic, biochemical, electrophysiologic, and behavioral studies (see references in Karli 1952; Sorsby et al. 1954; Noell 1958; Sidman and Green 1965; Karli and Bonaventure 1970; Keeler 1970; Schmidt and Lolley 1973; LaVail and Sidman 1974).

The mutation results in rapid degeneration and disappearance of photoreceptor cells, beginning at late stages of photoreceptor cell differentiation. The appearance of the degenerating (*rd/rd*) retina and normal (+/*rd*) retina at different ages as seen by light microscopy is illustrated in Figures 20.1–6. The mutant and normal retina are almost identical at 10 days of age, with perhaps a slight thinning of the outer nuclear layer in the mutant (Figs. 20.1–2). By 13 days of age the heterozygotes have generated rod outer segments averaging about 9 μm in length (Fig. 20.3), whereas the photoreceptors in the homozygotes already are undergoing severe degeneration and dissolution (Fig. 20.4). By 21 days of age, the rod outer segments in the heterozygote controls (Fig. 20.5) have essentially reached their adult length of 22.2 ± 6.5 μm (LaVail 1973). At the same time, the outer nuclear layer has been reduced to a single row of nuclei in the mice with retinal degeneration (Fig. 20.6). Most of these nuclei slowly and progressively disappear in the following few months. In the course of dissolution of the outer nuclear layer, the remainder of the retina apparently remains intact. Photoreceptor degeneration does not lead to inner nuclear layer cell degeneration, at least up to 41 days of age (Blanks and Bok 1977). By 3–6 months of age, however, subtle changes occur in the ganglion cell population, such as a reduction in cell size by 10–20 percent and a loss of some cells (Grafstein, Murray, and Ingoglia 1972). Gliosis at irregular sites along the eye and disruption of the inner nuclear and inner plexiform layers proceed slowly, beginning at about 6 months of age.

Three recent electron-microscopic studies of *rd* mouse retinas have revealed abnormalities that precede the overt degeneration of photoreceptor cells that begins on about day 10. Sonohara and Shoise

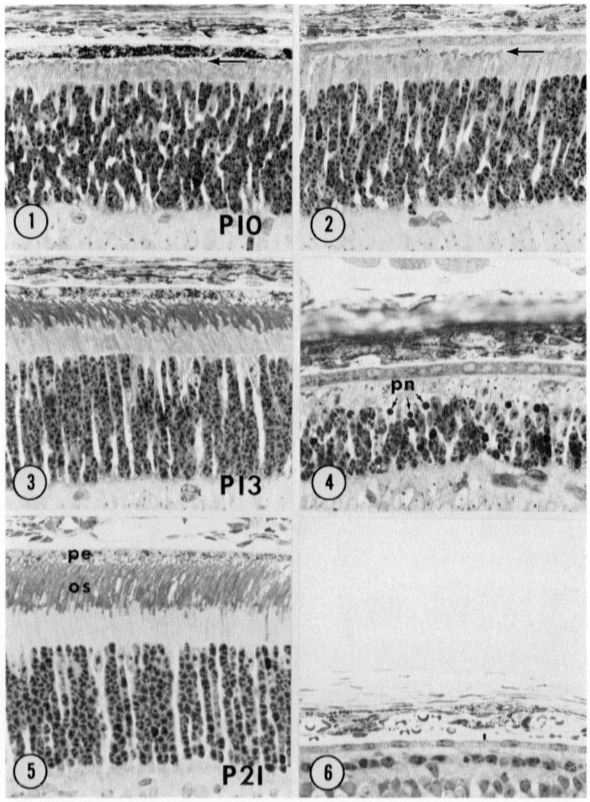

Figures 20.1–20.6 *Retinas from littermate pairs of C57BL/6J* rd le/++ (left) *and* rd le/rd le (right) *mice at different ages. At 10 days of age* (P10;P = *postnatal day), rod outer segments* (arrows) *have just begun to appear in the normal heterozygote (Fig. 20.1) and in the homozygous mutant (Fig. 20.2). Figures 20.3 and 20.4 were taken at 13 days of age* (P13), *and Figures 20.5 and 20.6 at 21 days of age* (P21). *The apparent differences in concentrations of melanosomes in the pigment epithelial cells* (pe) *in the normal retinas (Figs. 20.1, 20.3, and 20.5) are due to slight differences in section thickness. The* le/le *genotype results in a great reduction in number and size of melanosomes in the pigment epithelial cells (Figs. 20.2, 20.4, and 20.6).* os, *outer segments;* pn, *pyknotic nuclei. (From LaVail and Sidman 1974. Reprinted with permission from the American Medical Association, copyright 1974. ×354.) These figures and all other light micrographs in this chapter, unless otherwise indicated, are from 1–2 μm Epon-Araldite sections stained with toluidine blue.*

mutant gene is transferred to an inbred strain by an appropriate number of repeated crosses (Green and Doolittle 1963; Green 1966), that strain is defined as *congenic.* In a congenic strain, the mutant and the normal animals are genetically similar, differing only at the locus of the transferred gene and at closely linked loci. The congenic strains are designated by the inbred strain symbol, followed by a hyphen and italicized gene symbol if the mutation is maintained in homozygous form (for example, C57BL/6J-*rd le* mice are congenic with C57BL/6J mice but are homozygous at the *rd* and *le* loci). If the gene is maintained in heterozygous form, the gene symbol and wild-type symbol are indicated (for example, RCS-*p*/+ rats are congenic with RCS rats but are heterozygous at the *p* locus). Further rules governing the genetic nomenclature of inbred and congenic strains of rodents are found in Festing and Staats (1973).

(1968) have found vacuolization in Müller's cell cytoplasm as early as day 9. Sanyal and Bal (1973), however, have reported differences between 4-day old mutants and littermate controls. In particular, mutants show fewer ribosomes and mitochondria in their inner segments than do controls, and by 6–8 days there is a reduced development of rod outer segment discs and a slower development of inner segments. At the level of the photoreceptor terminals, Blanks, Adinolfi, and Lolley (1974) have noted a virtual absence of triad configurations and suggest that synaptogenesis is arrested in the photoreceptor terminal of the *rd* retina between days 7–10, when the bipolar cell process would normally invaginate the terminal. Many synapses, however, are made between photoreceptors and horizontal cell processes.

Several recent biochemical studies, particularly those of Lolley and his colleagues, have also added considerably to our knowledge of retinal degeneration in the mouse. DNA measurements indicate that cell loss in the mutant begins by about day 5 (Lolley 1973), earlier than the morphology would indicate. Schmidt and Lolley (1973) have demonstrated a deficiency in *rd* mutant retinas, as early as day 7, in the high-K_m phosphodiesterase that is associated with photoreceptors. As a result of the deficiency, cyclic-GMP accumulates in photoreceptors before they begin to degenerate (Farber and Lolley 1974). It seems likely that the accumulation of cyclic-GMP in the cells is responsible in some way for their subsequent degeneration. Lolley et al. (1977) have added further experimental evidence that an abnormality in cyclic-GMP metabolism leads to photoreceptor cell death. When they added the cyclic-GMP phosphodiesterase inhibitor, isobutylmethylxanthine, to developing eye rudiments of Xenopus embryos in culture, the levels of cyclic-GMP rose and the photoreceptors degenerated, thereby simulating the apparent result of the *rd* mutation in the mouse.

The *rd* gene action may prove to be related directly to some aspect of cyclic nucleotide metabolism in photoreceptor cells. In addition to the early abnormal levels in cyclic-GMP phosphodiesterase seen by Schmidt and Lolley (1973), Ferrendelli and Cohen (1976) have recently reported levels of cyclic-GMP in +/*rd* heterozygotes reduced by 40 percent from those seen in +/+ congenic control mice. Although these results are not entirely compatible with those of Lolley and coworkers, such a gene dosage effect is suggestive of being closely related to direct gene action.

EARLY FEATURES OF *rd* MOUSE PIGMENT EPITHELIUM. Very little attention has been paid to the pigment epithelium in the *rd* mutant mouse, probably because some of the roles of the pigment epithelium in retinal physiology have been elucidated only recently. For example, the phagocytic role of the pigment epithelial cell in rod outer segment renewal was not appreciated until 1969 (Young and Bok 1969). We will now consider some of the features of the pigment epithelium that have been described in *rd* mutant mice at ages prior to or during overt photoreceptor degeneration, that is, up to about day 20.

Phagocytosis. It is important to emphasize that pigment epithelial cells in *rd* mice are capable of phagocytizing rod outer segment membranes. Although only a small number of outer segment membranes are synthesized, phagosomes are found within pigment epithelial cells at early stages of the degeneration (Shoise 1968; Sanyal and Bal 1973; Robb 1974).

Pigment epithelial cell processes. Most workers who have examined *rd* mouse retinas of any age by electron microscopy have remarked on the abundant apical pigment epithelial cell processes (Fig. 20.7) (Karli, Stoeckel, and Porte 1965). Numerous apposed processes are conspicuous as early as days 12–15 (Shoise 1968; Caley, Johnson, and Liebelt 1972), and as early as day 8 they may be more abundant than in normal retinas (Sanyal and Bal 1973). There have been no quantitative studies to determine if mutant pigment processes proliferate, that is, increase in cell surface area. The characteristic "lamellated" appearance of the processes may be due simply to the absence or retarded growth of outer segments; thus, this appearance may not be specific to the disease process. A similar appearance is found in retinas damaged as a result of constant light (Kuwabara and Gorn 1968; O'Steen, Shear, and Anderson 1972), in urethane-induced retinopathy (Bellhorn et al. 1973), in late stages of inherited retinal dystrophy in RCS rats (Fig. 20.31), and in lateral surfaces of the pigment epithelial cells of older men (Lerche 1974).

Endoplasmic reticulum. Caley, Johnson, and Liebelt (1972) have described a transition in the development of mouse pigment epithelial cells from cells containing predominantly rough endoplasmic reticulum to those containing predominantly smooth endoplasmic reticulum. These authors report that the transition in the *rd* mouse is delayed to about day 20, approximately one week later than in normal mice, and they have suggested that this delay may play a role in the etiology of the photoreceptor degeneration (discussed below). Sanyal and Bal (1973) found no ultrastructural difference in the appearance of mutant and control pigment epithelial cells at days 10 and 14, although they do not state whether they examined the endoplasmic reticulum carefully at the latter age. This issue warrants classification.

Glycogen. Tamai (see LaVail and Mullen 1976b) has found substantial amounts of glycogen in the pigment epithelial cells of 5-day old normal albino mice (DDK strain), but little, if any, in the same cells of C3H (*rd/rd*) mice. It is not clear whether this difference is due to the action of the *rd* gene, to another gene (C3H mice, for example, are homozygous for *Gus*^{h}, a gene closely linked to *rd* that results in a lower level of β-glucuronidase than in many other strains), or to a polygenic strain difference. The issue might be clarified by the use of congenic mice that have both mutants and controls in the same litter.

Migration. In the late stages of overt photoreceptor degeneration (days 15–20) migrating cells are found within the inner and outer segment zone (Caley, Johnson, and Liebelt 1972). Caley and coworkers suggest that these are pigment epithelial cells that have detached from Bruch's membrane. Alternatively, these may be interstitial cells that are seen normally in mouse retinas (Sidman 1958) or macrophages in the *rd* mouse (Sanyal 1972) that are derived from blood cells or blood vessels.

ROLE OF PIGMENT EPITHELIUM IN PHOTORECEPTOR DEGENERATION. Caley and coworkers (1972) have suggested that a delayed transition from rough to smooth endoplasmic reticulum in developing *rd* mouse pigment epithelial cells leads to photoreceptor cell death; according to this hypothesis, an abnormal vitamin A metabolism results in an inhibition of outer segment production with secondary death of photoreceptor cells. Even if the delayed organelle transition is confirmed, this explanation for the etiology of the disease is not consistent with more recent data. The transition from rough to smooth endoplasmic reticulum begins at about day 12, according to Caley et al. However, abnormal levels of DNA (Lolley 1973), cyclic nucleotide phosphodiesterase (Schmidt and Lolley 1973), and cyclic-GMP (Farber and Lolley 1974) are all present in *rd* mouse retinas before day 10. Furthermore, morphological signs of degeneration have been reported as early as day 4 (Sanyal and Bal 1973). Although these findings show that the proposed hypothesis of Caley et al. (1972) probably is not valid, they do not rule out pigment epithelial cell involvement in the disease.

The role of the pigment epithelium in inherited retinal degeneration has recently been analyzed using experimental chimeras (also called tetraparental mice, mosaic mice, or allophenic mice). Wegmann, LaVail, and Sidman (1971) fused eight-cell embryos of C3H (*rd/rd*) mice with those of C57BL/10 (+/+) mice in culture, allowed the fused embryos to develop to the blastocyst stage and then transplanted them into pseudopregnant female hosts. The C3H ⟷ C57BL/10 chimeric mice were identified soon after birth by the mixture of agouti (C3H) and black (C57BL/10) coat colors. The retinas, after 20 days of age, displayed patches of normal retina interspersed with patches lacking photoreceptor cells. Because both parental strains of mice were pigmented, all of the pigment epithelial cells were fully pigmented. Therefore, the pigment epithelial cells of the *rd/rd* (C3H) genotype could not be distinguished from those of the +/+ (C57BL/10) genotype.

Subsequently, LaVail and Mullen (1976a) used the albino SJL (*rd/rd*) strain instead of the C3H strain, which allowed for the identification of the *rd/rd* mutant pigment epithelial cells on the basis of the absence of melanosomes. In eyes of the SJL ⟷ C57BL/10 chimeras, both normal and mutant pigment epithelial cells were present, overlying both normal (Fig. 20.8) and degenerated retina (Fig. 20.9).[2] Outer segment disc synthesis appeared normal in photoreceptor cells underlying mutant pigment epithelial cells (as determined by autoradiographic analysis), and phagosomes were found in mutant pigment epithelial cells, as well as in normal pigment epithelial cells. It was reasoned that if an abnormality in the pigment epithelial cell caused a secondary degeneration of photoreceptor cells, then all photoreceptors underlying the mutant pigment

2. This lack of correspondence of retinal and pigment epithelial cell types was also seen in comparable mouse chimeras by Mintz and Sanyal (1970) and Sanyal and Zeilmaker (1976).

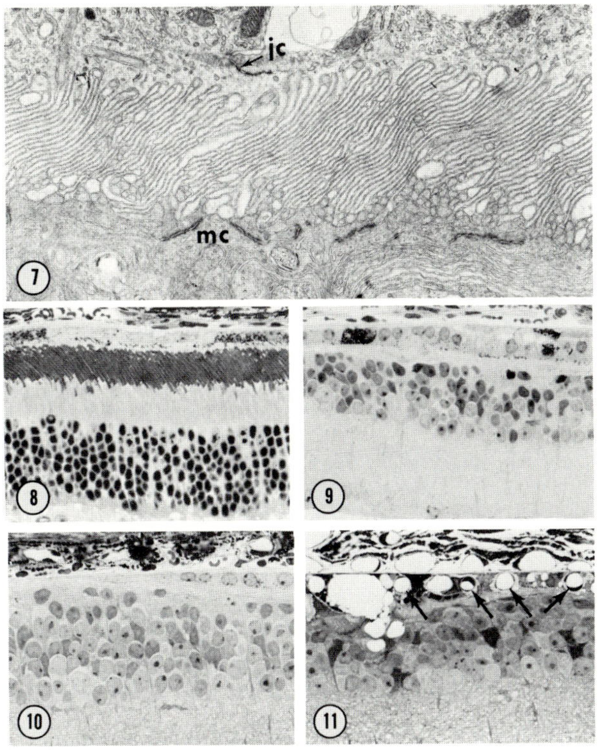

Figure 20.7 *Electron micrograph of the abundant apical pigment epithelial cell processes in a 90-day-old* rd le/rd le *mouse.* jc, *junctional complex between two pigment epithelial cells;* mc = *Müller cell. (×8,925.)*

Figures 20.8–20.9 *Retinas from two one-year-old SJL↔C57BL/10 mouse chimeras. Both albinotic (SJL,* rd/rd *genotype) and pigmented (C57BL/10, +/+ genotype) pigment epithelial cells are found overlying both normal (Fig. 20.8) and degenerated (Fig. 20.9) retina. (From LaVail and Mullen 1976a. ×265.)*

Figure 20.10 *Retina from a 4-month-old* rd le/rd le *mouse. The pigment epithelium is irregularly thinned after the photoreceptors have disappeared. (×345.)*

Figure 20.11 *Retina from an 8-month-old* rd le/rd le *mouse. The pigment epithelium has become vascularized by invading retinal capillaries* (arrows). *The small dilations between pigment epithelial cells are a result of perfusion fixation. (×345.)*

epithelium (regardless of genotype) should degenerate. Since this was *not* the case, and normal photoreceptor cells could be detected adjacent to mutant pigment epithelium, the findings in the chimeric mice indicated that the pigment epithelial cell is not the primary target of the mutant *rd* gene. Further considerations of the use of experimental chimeras to analyze the role of the pigment epithelium in the etiology of inherited retinal degenerations have recently been reviewed by LaVail and Mullen (1976b, c).

LATE CHANGES IN *rd* MOUSE PIGMENT EPITHELIUM.

Thinning. After most of the photoreceptor cells have disappeared (about day 20), some of the pigment epithelial cells thin appreciably (Fig. 20.10) and lose pigment (Sanyal and Bal 1973). Other pigment epithelial cells become abnormally thick (Caley, Johnson, and Liebelt 1972). Why the cells alter and vary in thickness is not apparent. Although Caley et al. (1972) state that the thickness of the epithelium can be related to the extent of debris removal, the data are unconvincing (see figs. 14, 15, 29, and 30 of Caley et al. 1972).

Vascularization. Late in the development of inherited retinal degeneration in the mouse, retinal capillaries vascularize the pigment epithelium. A few capillaries (which are more readily observed after vascular perfusion than after immersion of the retina in fixative) can be found invading the pigment epithelium as early as 4 months of age. As animals become older, the pigment epithelium becomes somewhat more vascularized (LaVail, in preparation) (Fig. 20.11). It should be pointed out, however, that vascularization of the pigment epithelium is not peculiar to the *rd* mouse retina. Retinal capillaries invade the pigment epithelium when photoreceptor cells are missing in rats with (1) inherited retinal dystrophy (Dowling and Sidman 1962; Bok and Hall 1971), (2) slower forms of retinal degeneration such as those found in the Osborne-Mendel (von Sallman and Grimes 1972) and Wag/Rij (Lai et al. 1975) strains, (3) urethane-induced retinopathy (Bellhorn et al. 1973) and (4) constant-light-damaged retinas (O'Steen, Shear, and Anderson 1972). In the constant-light study, it is said that the invading retinal capillaries anastomose with the choriocapillaris; however, this has not been observed in mice or rats with inherited retinal degeneration or in genetically normal albino rats that have a highly vascularized pigment epithelium following long-term exposure to constant light (LaVail 1976a).

The conditions that result in retinal vascularization of the pigment epithelium point to a nonspecific phenomenon that occurs whenever photoreceptor cells are lost. It was surprising, therefore, to learn from discussions with numerous ophthalmologists that retinal vascular invasion of the pigment epithelium is not characteristic of advanced cases of retinitis pigmentosa.

Migration. It has been clear since the original descriptions of inherited retinal dystrophy in the rat (Bourne, Campbell, and Tansley 1938) that pigment epithelial cells migrate into the neural retina. In the mouse disease, however, pigment epithelial cells do not invade the neural retina (Sorsby et al. 1954). Sidman (1961) noticed occasional clusters of melanosomes scattered through the retina of a 2-year-old C3H mouse, but the melanosomes were not localized to a particular cell type. Examination of a series of older (14- to 20-month-old) C57BL/6J-*rd le*[3] mouse retinas has revealed a number of melanosome-containing cells in contact with blood vessels in the inner plexiform and optic fiber layers (Fig. 20.12) (LaVail, in preparation). The pigment-containing cells are identified as pigment epithelial cells, based on their cytological features of (1) abundant smooth endoplasmic reticulum, (2) melanosomes, (3) basal processes apposed to the basal lamina surrounding the blood vessels, and (4) finer processes that face away from the blood vessels and that are identical in appearance to apical pigment epithelial cell processes in the *rd* mouse (Fig. 20.13). Therefore, the *rd* mouse shares the feature of pigment epithelial cell migration into the neural retina with (1) rats with inherited retinal dystrophy, (2) Wag/Rij rats with a slower form of retinal degeneration (Lai et al. 1975), and (3) humans with retinitis pigmentosa.

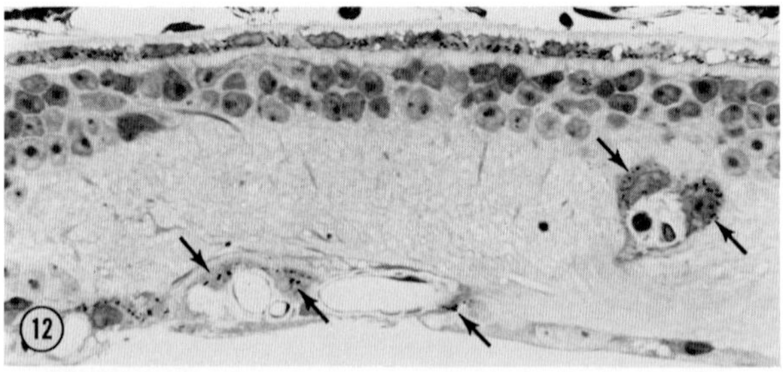

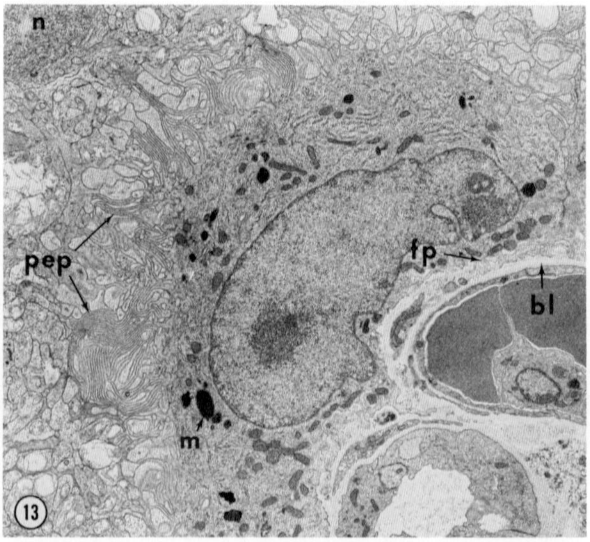

Figure 20.12 *Light micrograph illustrating pigmented cells* (arrows) *adjacent to blood vessels in the inner plexiform layer and optic fiber layer in an 18-month-old* rd le/rd le *mouse. (×408.)*

Figure 20.13 *Electron micrograph of the inner nuclear layer and inner plexiform layer from a section similar to Figure 20.12. A pigment epithelial cell that has migrated into the retina can be identified on the basis of its characteristic apical processes* (pep), *basal foot processes* (fp) *apposed to the capillary basal lamina* (bl), *the abundance of smooth endoplasmic reticulum (seen at higher magnification), and the presence of a melanosome* (m). *Only a few melanosomes are present in the pigment epithelial cells in Figures 20.12 and 20.13 presumably because of the* le/le *pigmentation gene.* n, *nucleus of cell in inner nuclear layer. (×3,700.)*

Inherited Retinal Dystrophy in the Rat

PHOTORECEPTOR DEGENERATION. Cytopathologic, electrophysiologic, biochemical, and behavioral studies on inherited retinal dystrophy (also called inherited retinal degeneration) in rats have been carried out by numerous workers (Bourne, Campbell, and Tansley 1938; Lucas, Attfield, and Davey 1955; Dowling and Sidman 1962; Noell 1963, 1965; Arden and Ikeda 1966; Reading 1970; Herron et al. 1969, 1971; Bok and Hall 1971; Burden et al. 1971; LaVail, Sidman, and O'Neil 1972; Yates et al. 1974; LaVail and Battelle 1975; Custer and Bok 1975; Dewar, Barron, and Richmond 1975; Lolley and Farber 1975, 1976; Dewar, Barron, and Yates 1975; Delmelle, Noell, and Organisciak 1975; Organisciak and Noell 1976; Pautler and Noell 1976; Hall and Nir 1976; Mullen and LaVail 1976; Kaitz 1976). In brief, photoreceptor degeneration proceeds as follows: Outer segment debris (extralamellar material) begins to accumulate at the surface of the pigment epithelium (Fig. 20.14) on about postnatal

3. This congenic strain was described by LaVail and Sidman (1974). Also see footnote 1 on genetic terminology.

day 12, when rod outer segments are just beginning to elongate appreciably. The outer segment zone (intact outer segments and apical debris) becomes thicker than that of normal rats between days 14–20, and it remains thicker until at least 2 months of age. Pyknotic photoreceptor nuclei are first identified about days 16–20. From this age until about days 60–70, the outer segments disintegrate, and the outer segment zone is replaced entirely by a debris zone composed of whorls of outer segment membranes, pigment epithelial cell processes, and invading cells (Figs. 20.15–17). At the same time, photoreceptor

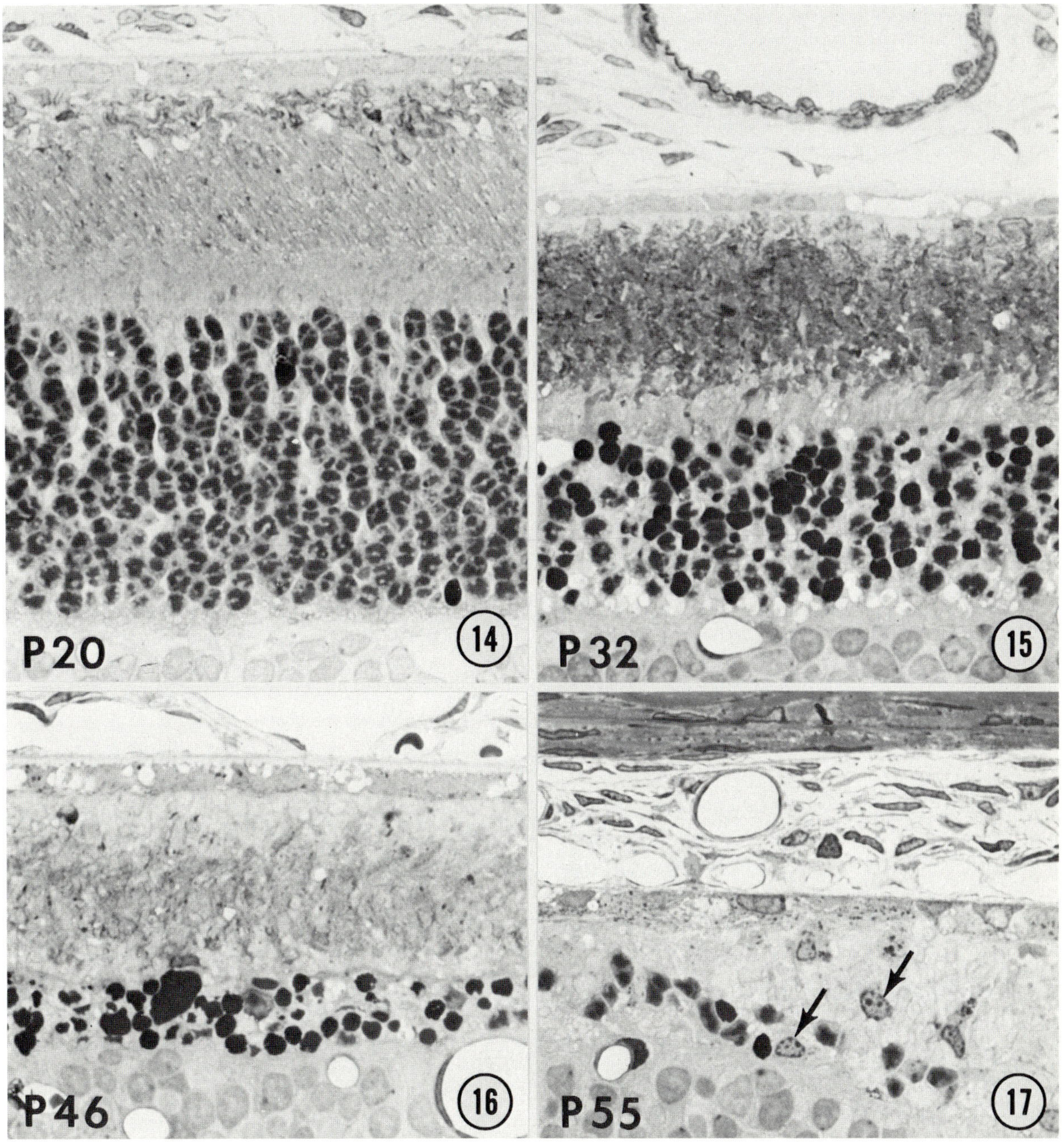

Figures 20.14–20.17 *Light micrographs of retinas from RCS rats of different ages (Fig. 20.14, day 20; Fig. 20.15, day 32; Fig. 20.16, day 46; Fig. 20.17, day 55). In Figure 20.17, the invading cells* (arrows) *have a different nuclear heterochromatin pattern from the pigment epithelial cells. (×875.)*

cells progressively degenerate, and the outer nuclear layer becomes correspondingly thinner until it is virtually eliminated (Figs. 20.15–17). Only a few residual photoreceptor cells remain, and these may persist for the life of the animal (LaVail et al. 1974). The inner retinal layers remain relatively intact until about 5 months of age, when focal gliosis and disruption occurs.

The several stocks and inbred strains of rats with inherited retinal dystrophy that are available today are descendants of the original animals described by Bourne, Campbell, and Tansley (1938). In these animals, the rate of the progression of the disease has apparently remained unchanged by inbreeding programs. Most investigators have studied tan-hooded, pink-eyed animals, of which the RCS (Royal College of Surgeons) inbred strain (Sidman and Pearlstein 1965) is the one most widely used.[4] When the pink-eyed dystrophic rats are reared in the dark, the rate of photoreceptor degeneration is slowed somewhat (Dowling and Sidman 1962). Furthermore, in a congenic strain[5] of RCS rats with black eye pigmentation (strain symbol, RCS-*p*/+), which has recently been developed (LaVail, Sidman, and Gerhardt 1975), the black eye pigmentation slows the progression of the disease to the same extent as does dark-rearing for the pink-eyed RCS rats (LaVail and Battelle 1975; see also Dewar, Barron, and Yates 1975). In the posterior retina, photoreceptor degeneration is slowed by about 10 days. In the far peripheral retina, the disease is slowed by about 30–35 days along the horizontal meridian and in the superior half of the eye, but no slowing of the disease occurs in the inferior half of the eye along the vertical meridian (LaVail and Battelle 1975). Yates et al. (1974) have also described the development of a partially inbred line of pigmented rats with retinal dystrophy (Hunter strain), in which the disease is slowed by about 7 days in the posterior retina, a rate similar to the rate of slowing in the RCS-*p*/+ rats.

EARLY FEATURES OF DYSTROPHIC RAT PIGMENT EPITHELIUM. For the purpose of this review, the early period of the disease has been arbitrarily defined as extending to the time when overt photoreceptor degeneration ends. In pink-eyed RCS rats, most of the photoreceptor nuclei have disappeared by about postnatal day 60–70. It should be noted, however, that small clusters of visual cell nuclei near the optic nerve head and ora serrata may survive at least until day 96 (LaVail and Battelle 1975), and some photoreceptor cell nuclei may persist for the life of the animal (LaVail et al. 1974).

Phagocytosis. The outer segment discs of normal vertebrate rod photoreceptor cells are continually synthesized at the base of the outer segment and subsequently are displaced outward by discs synthesized later (Young 1967; Young and Droz 1968). When the discs reach the apex of the rod outer segment they are either intermittently shed in groups of 10–30 discs and then are ingested by the pigment epithelial cells (Young 1971), or the pigment epithelial cells may play the initial active role in breaking off the apical groups of discs (Spitznas and Hogan 1970). Within the pigment cell cytoplasm, the ingested groups of discs, called phagosomes (Young and Bok 1969), are subsequently degraded (Ishikawa and Yamada 1970; Marshall and Ansell 1971). Further details of outer segment renewal are given by Bok and Young in Chapter 9 of this book.

In the dystrophic rat retina, the inability of the pigment epithelial cell to phagocytize rod outer segment discs (Herron et al. 1969; Bok and Hall 1969, 1971; LaVail, Sidman, and O'Neil 1972) leads to the accumulation of rod outer segment membranes. Herron et al. (1969) has suggested that this accumulation of debris results in the subsequent photoreceptor cell death (discussed below).

The pigment epithelial cells of the dystrophic rat are not incapable of phagocytosis. Custer and Bok (1975) and Bok and Young (see Chapter 9) have found that pigment epithelial cells in both normal and dystrophic rats will phagocytize carbon particles injected via micropipette into the outer segment zone. Reich-d'Almeida and Hockley (1975, 1976) have obtained similar results using the injection of colloidal carbon into the subretinal space after lens extraction and total retinal detachment. These experiments should be credited as the first attempts to study phagocytosis in the mutant rats by invasive methods. Unfortunately, weaknesses in the experiments limited their significance. Each experiment was carried out when the animals were at inappropriate ages, either at postnatal day 4, well before signs of degeneration appear or significant amounts

4. The PETH strain is used by some workers. These animals were obtained from Sidman for the National Institutes of Health colony after the RCS strain reached inbred status, and, therefore, the animals are actually a substrain of RCS. An appropriate substrain designation is RCS/N.

5. See footnote 1 on genetic terminology.

of outer segment membranes have been formed, or at later ages, well after most photoreceptor cells have degenerated and disappeared. It is possible that phagocytic rates or properties are different during the period of active photoreceptor degeneration. In addition, it is possible that phagocytic capability cannot be tested with carbon particles. The particles are so small that they may be taken up by pinocytosis and then sequestered in larger vacuoles. For this reason, some workers believe that a true test of phagocytosis requires particles with a diameter of 1 mm or larger (Allison, in discussion of Tewari, Solotorovsky, and Scherner 1974).

The uptake of carbon particles, nevertheless, suggested to Custer and Bok (1975) that the mutant pigment epithelial cells might be selective in phagocytosis, since the cells usually did not phagocytize outer segment membranes.[6] Such a selectivity raised the possibility that an abnormal constitution of the mutant rod outer segment might preclude its recognition and ingestion by the pigment epithelial cells. In a series of experiments designed to provide more definitive data on this issue, Custer and Bok (1975) incubated retinas from normal rats in a medium containing [^{3}H]amino acids. From these retinas they obtained an outer segment preparation labeled with newly formed outer segment discs; this preparation was injected into the outer segment zone of normal or dystrophic rats. After waiting a given time, the experimenters fixed the eyes and carried out an electron-microscopic autoradiographic analysis of the tissues. Unfortunately, the results did not settle the original issue, because even the control pigment epithelial cells showed little phagocytosis of the labeled outer segment discs. Custer and Bok (1975) suggested that an outer plasma membrane around the shed discs might be necessary for recognition or phagocytosis by the pigment epithelium, and that this membrane may have been disrupted in obtaining the outer segment preparation. Alternatively, since basal discs are the ones labeled in the in vitro incubations, perhaps only apical discs are readily "edible." An additional parameter may have been involved in the experiments. Rod outer segment disc shedding occurs as a cyclic (circadian) event in normal albino rat retinas maintained in cyclic light, with a large burst of shedding during the first 2–3 hours after the onset of light, but very few large phagosomes present later in the day (LaVail 1976b). This temporal factor may have influenced the experiments, although it is not known whether phagocytosis per se (as opposed to shedding by the rod outer segments) follows a circadian rhythm.

Following another strategy, Ansell and Marshall (1976) reported that dystrophic pigment epithelial cells could be induced to phagocytize outer segment debris after the cells had repopulated an area of eye that had been damaged by irradiation with a low-energy pulsed ruby laser. Some of the presumptive pigment epithelial cells may have been blood-derived macrophages. The authors argue otherwise, on the basis of their cytological similarity to pigment epithelial cells, including the presence of melanosomes. Unfortunately, the cytology illustrated is unconvincing on this issue, and the melanosomes may have been scavenged from dead or dying pigment epithelial cells by invading macrophages. Alternatively, the laser irradiation may have stimulated phagocytosis by the pigment epithelial cells. In this case, one must ask what effect the laser had on the cells to stimulate phagocytosis.

Edwards and Szamier (1977) have recently reported a series of experiments in which they tested the phagocytic capability of RCS and normal pigment epithelial cells in vitro. The authors applied polystyrene spheres and rod outer segments obtained from either normal or RCS rats to cultures of both normal and RCS pigment epithelial cells. Cells of both genotypes took up polystyrene spheres, and the cells from normal rats readily phagocytized outer segments from both normal and RCS rats. The RCS pigment epithelial cells, however, only rarely showed phagosomes with application of either the normal or RCS outer segments. These results argued strongly that RCS pigment epithelial cells could phagocytize few, if any, outer segment membranes. These findings have been corroborated by O'Donnell (1977), who carried out similar experiments but used a pigment epithelial cell line that had undergone a number of passages. In addition, Hall (1977 and personal communication) has found very little phagocytosis of outer segments by RCS pigment epithelial cells in primary culture experiments.

With that evidence at hand, it seemed fair to con-

6. Normal pigment epithelial cells apparently are selective phagocytes. For example, Hollyfield and Ward (1974a, b) and Hollyfield (1976) have demonstrated that the pigment epithelial cells in *Rana pipiens* embryos, tadpoles and frogs readily phagocytize spheres but not bacteria (*Sarcina subflava*) of approximately the same size. For further references and discussion of phagocytic specificity in normal pigment epithelial cells see Chapter 9 and Ansell and Marshall (1976).

clude that the phagocytic capability of the retinal dystrophic pigment epithelial cells remained to be demonstrated, at least in a physiological sense. I expressed concern above regarding the demonstration of phagocytosis using colloidal carbon. Feeney and Mixon (1976) have pointed out that latex particles are more negatively charged than most biological surfaces, so one must question the physiological nature of their uptake. Most recently, however, Goldman and O'Brien (1978) have found that a small burst of outer segment disc shedding and phagocytosis by the pigment epithelium occurs in RCS rats following the onset of light in the morning. The number of phagosomes in the pigment epithelium during the burst is small, only about 5 percent of that in control animals, but the presence of any phagosomes indicates that the defect in retinal dystrophy produces a grossly reduced level of phagocytosis of outer segment membranes rather than an absence, as previously thought.

The abnormality in phagocytosis in the RCS pigment epithelial cell may be secondary to another defect in the pigment epithelium. (The primary genetic defect appears to be in the pigment epithelium, as discussed below.) For example, the absence of a substance normally secreted by the pigment epithelial cells (or the presence of an abnormal substance) that must interact with presumptive phagosomes in order to be recognized as "edible" would also lead to the observed defect.

The actual nature of the phagocytic defect obviously remains to be discovered. It is clear that this mutant is a particularly interesting and useful model for the study of genetic control mechanisms of outer segment disposal.

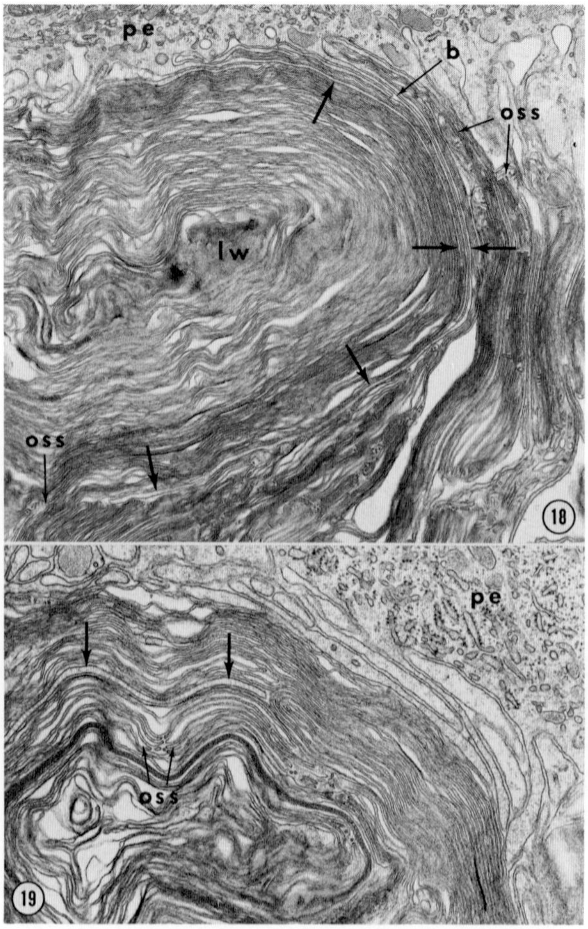

Figures 20.18–20.19 *Electron micrographs of large lamellar whorls* (lw) *adjacent to pigment epithelial cells* (pe) *in 20-day-old rat retinas. Pigment epithelial cell processes can either be traced into the periphery of the whorls* (arrows, *Fig. 20.18*) *or be identified in the central parts of the whorls* (arrows, *Fig. 20.19*). b, *branching of a pigment cell process;* oss, *outer segment saccules (disc membranes). (From LaVail, Sidman, and O'Neil 1972. Fig. 20.18, ×12,030. Fig. 20.19, ×16,270.)*

Pigment epithelial cell processes. The dystrophic rat retina is characterized, in part, by large lamellar whorls in the outer segment zone, which are present almost as early as outer segments are present. The lamellar whorls, which measure several μm in diameter, are found most frequently in the apical third of the outer segment zone, immediately adjacent to the pigment epithelium. Pigment epithelial cell processes can be traced more frequently within the periphery (Fig. 20.18) and less frequently within the central portions of the whorls (Fig. 20.19). Based on observations of (1) morphological configurations of pigment cell processes in relation to outer segment membranes, and (2) the pattern of labeled protein displacement into the lamellar whorls from the pigment epithelium, LaVail, Sidman, and O'Neil (1972) have suggested that the pigment cell processes may be, at least in part, responsible for transforming the degenerating outer segment membranes into circular whorls.

Organisciak and Noell (1976) recently found in 100-day-old retinal dystrophic rats that 10–30 percent of the phospholipids in the debris were derived from a non-outer-segment source. As the authors have pointed out, the pigment epithelial cell processes within the debris may contribute those phospholipids.

The lateral borders of adjacent pigment epithelial cells in normal animals are usually fairly straight and are separated from one another by typical 20–30 nm extracellular spaces, unless distended as a result of perfusion fixation (see Fig. 20.22). In dystrophic rat retinas, however, the lateral borders sometimes form processes that interdigitate with those of the adjacent cells (Fig. 20.20). These interdigitations may, by excluding their cytoplasm, form some of the numerous lamellar whorls that are seen between the pigment epithelial cells (Figs. 20.21–22). Pigment cell processes have been traced into the whorls located within the pigment epithelial cell layer (see fig. 29 in LaVail, Sidman, and O'Neil 1972). The membranes of the whorls between pigment epithelial cells apparently fuse in myelin-like fashion (Fig. 20.21) and continue to do so until the entire whorl is condensed and highly osmiophilic (Figs. 20.22–23). It remains to be determined whether the dense membranes consist of pigment cell membranes, outer segment membranes, or both. The condensation process could represent an attempt to dispose of outer segment membranes. Alternatively, the process could represent an attempt to increase the surface area of the pigment epithelial cells. Whatever the cause, the appearance of the lateral processes of the pigment epithelial cells is clearly abnormal and is consistent with the apparent overabundance of pigment cell processes in general in the mutant rat.

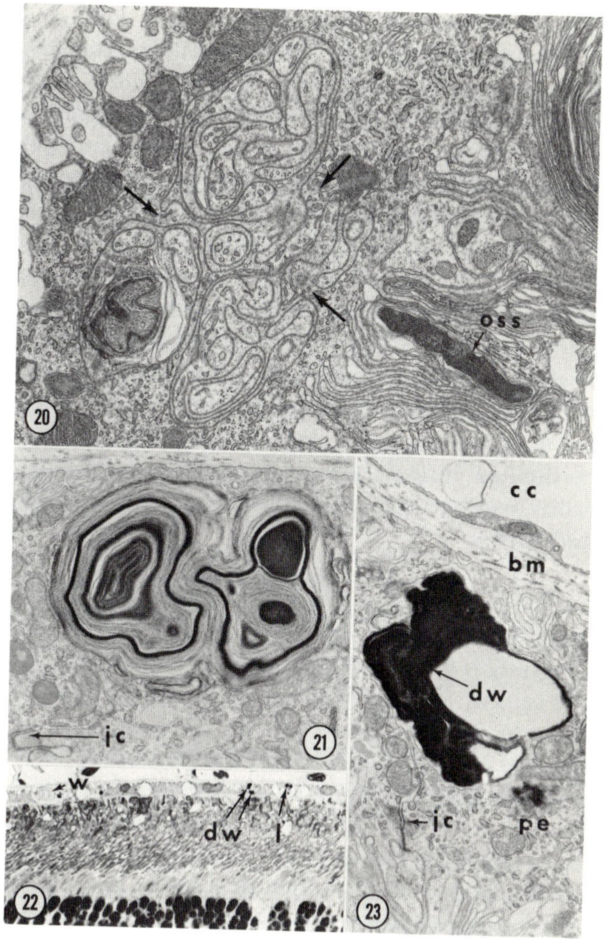

Figures 20.20–20.23 *Micrographs illustrating interdigitating processes of pigment epithelial cells at their lateral borders (Fig. 20.20,* arrows*) and lamellar whorls between pigment epithelial cells (Figs. 20.21–20.23). The lamellar whorl in Figure 20.21 appears to reside within a pigment epithelial cell, but serial sections revealed that it lies between cells. After fixation with osmium tetraoxide, the very dense whorls* (dw) *such as that shown in Fig. 20.23 are strongly basophilic (Fig. 20.22). The partial fusion and condensation of membranes, as shown in Figure 20.21, presumably progresses to the completely condensed configuration shown in Figure 20.23.* bm, *Bruch's membrane;* cc, *choriocapillaris;* jc, *junctional complex between pigment epithelial cells* (pe); l, *lipid droplet;* oss, *outer segment saccules;* w, *lamellar whorl. (From LaVail, Sidman, and O'Neil 1972. Fig. 20.20, × 10,180. Fig. 20.21. ×8,925. Fig. 20.22, ×345. Fig. 20.23, ×8,955.)*

Protein synthesis and displacement. LaVail, Sidman, and O'Neil (1972) demonstrated by light- and electron-microscopic autoradiography following [^{3}H]amino acid injection that protein synthesized in the pigment epithelial cell soma was displaced rapidly into adjacent lamellar whorls, where it apparently remained for 10–14 days (Fig. 20.24). However, the resolution of the autoradiography was not sufficient to permit them to distinguish whether the labeled protein still resided in the pigment epithelial cell processes or had been secreted into the membranes that make up the lamellar whorls. In either case, the labeling was much more conspicuous than that of pigment cell processes in normal rats under identical conditions.

LaVail, Sidman, and O'Neil suggested that the radioactive protein in the lamellar whorls after [^{3}H]amino-acid injection represented a rhodopsin-like protein endogenous to the pigment epithelium or its surroundings, because it was thought that the breakdown of rod outer segments alone could not account for the quantity of rhodopsin accumulated within the eye, as seen in the study of Dowling and Sidman (1962). Recently, however, LaVail and Battelle (1975) reexamined the rhodopsin content and

in the presence of debris (Lolley and Farber 1976). Since the metabolism of the cyclic-GMP becomes abnormal just when the debris begins to accumulate (on about postnatal day 12), Lolley and Farber (1976) have proposed a causal relationship between the accumulation of debris, the induced abnormality in cyclic-GMP metabolism, and the subsequent death of photoreceptor cells. These findings are consistent with a primary pigment epithelial cell abnormality in retinal dystrophy in the rat. If an abnormal substance is secreted from the pigment epithelium (or fails to be secreted and thereby produces some abnormal substance by another reaction), then that substance may be isolated and identified. The studies of Lolley and Farber (1975, 1976), therefore, may prove to be particularly important as first steps in discovering the mechanism that results in photoreceptor cell death.

Vitamin A deficiency has been shown to accelerate the rate of inherited retinal dystrophy in both albino and pigmented dystrophic rats (Delmelle, Noell, and Organisciak 1975). Although the mechanism is unclear, this important observation indicates that the abnormality may lie in vitamin A metabolism, transport, or storage by the mutant pigment epithelial cells.

One additional study on the role of the pigment epithelium in the rat disease is the brief report of Reich-d'Almeida and Rahi (1974) that no autoantibodies to the pigment epithelium could be raised in dystrophic rats. Unfortunately, limited data were presented, so it would seem that the issue of immune factors in retinal dystrophy of the rat remains open.

LATE CHANGES IN DYSTROPHIC RAT PIGMENT EPITHELIUM.

Disruption of diffusion barrier. Pigment epithelial cells in normal vertebrates have a junctional complex between them, which includes a large gap junction and a zonula occludens (tight junction) (Hudspeth and Yee 1973). Since the choroid capillaries are fenestrated, the pigment epithelium serves as a diffusion barrier for the outer retina, just as endothelial cells do in retinal and brain capillaries. For example, intravenously injected tracers such as horseradish peroxidase (HRP) normally do not enter the outer segment zone. Bok (personal communication) has found that in RCS rats, HRP is blocked by the pigment epithelium until 72 days of age, but at 116 days of age, HRP readily enters the outer segment zone. Therefore, the normal barrier function of the PE is lost in late stages of the disease.

Vascularization. Invasion of the dystrophic rat pigment epithelium by retinal capillaries has been described by several investigators (Bourne, Campbell, and Tansley 1938; Dowling and Sidman 1962; Bok and Hall 1971). From my observations (unpublished), it appears that vascularization of the pigment layer occurs only after the outer segment debris has been removed from that area. In pink-eyed RCS rats, some foci of debris removal occur as early as day 55, but large areas are not found until about 3 months of age. At about that time, the first vessels appear to invade the pigment epithelium. In pigmented RCS-*p*/+ rats, the removal of debris is considerably slowed. In the few foci of obliterated debris found at 5 months of age, a few capillaries can be seen invading the pigment epithelium at these sites. Although most of the debris is gone from pink-eyed RCS retinas by one year of age, moderately long stretches of an intact debris zone may still be found in RCS-*p*/+ rats as late as 22 months of age. The ingrowth of blood vessels appears to be one of the factors that disrupts the remaining inner retina. With age, the retinas of pink-eyed RCS rats become much more highly vascularized and much more grossly disrupted than do those of black-eyed RCS-*p*/+ rats.

Migration. It has been suggested that migration of pigment epithelial cells into the retina in pigmented dystrophic rats follows glial fibers (Bourne, Campbell, and Tansley 1938). Many, if not most, of the glial strands are accompanied by retinal capillaries. In every case I have examined thus far, the migration appears to occur along blood vessels rather than along glial fibers in the black-eyed RCS-*p*/+ retinas (Fig. 20.30), as it does in the *rd* mouse retina (described above). In the course of the cells' migration along the outer basal lamina of the vessels, the composition of their cellular organelles appears to remain characteristic of pigment epithelial cells.

The situation is more complex in the RCS rat than in the *rd* mouse because of the persistence of the abundant macrophages in the debris zone, some of which contain melanosomes. However, these macrophages clearly have the nuclear heterochromatin pattern characteristic of this cell type and not of pigment epithelial cells. These cells could be pigment epithelial cells that have changed their struc-

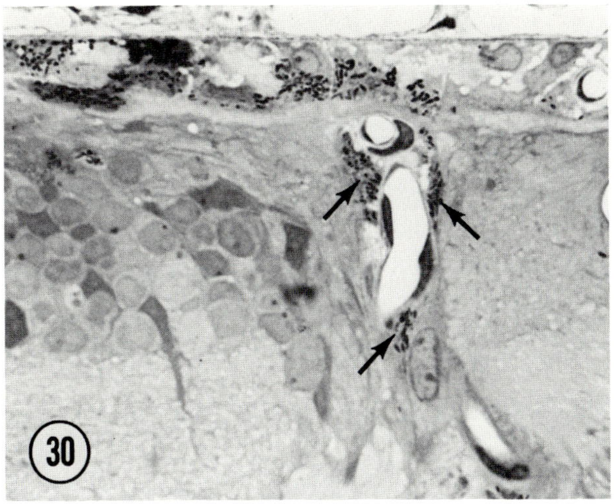

Figure 20.30. *Retina from a 277-day-old pigmented RCS-p/+ rat. Retinal capillaries have invaded the pigment epithelium, and pigment epithelial cells containing melanosomes* (arrows) *appear to be migrating into the neural retina along a retinal blood vessel.* (×*725.*)

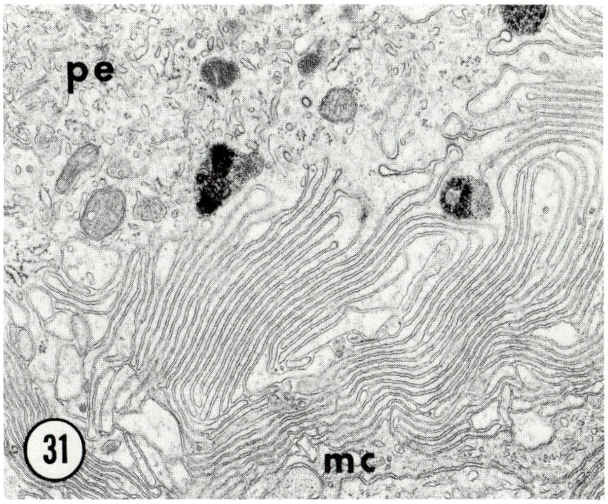

Figure 20.31 *Electron micrograph of the abundant apical pigment epithelial cell processes in a 12-month-old RCS rat.* mc, *Müller's cell;* pe, *pigment epithelial cell.* (×*15,125.*)

ture, but more likely they are macrophages that have engulfed melanosomes lost by pigment epithelial cells.

Pigment epithelial cell processes and surfaces. After most of the photoreceptor cells have disappeared, the pigment cell processes may adopt several configurations. The apical processes may be abundant and interdigitate with one another (Fig. 20.31), such that the tissues are indistinguishable from the *rd* mouse retina at late stages, or the processes may be less abundant and interdigitate more with microvillous processes of contiguous Müller's cell processes (see fig. 7 of LaVail et al. 1974). In other cases, the processes and somas of pigment cells may make a variety of types of junctions with Müller's cell processes (gap junctions, tight junctions, or zonulae adherentes) (Figs. 20.32–34). The significance of the junction formation is unclear, but LaVail et al. (1974) have discussed the possibility that gap junctions may mediate the behavioral responses to light that dystrophic rats show long after most photoreceptors have disappeared.

In some regions in the retinas of older dystrophic rats, the pigment epithelial cell processes become multilayered and envelop packets of outer segment debris membranes (Fig. 20.35). In most cases, the pigment cells appear to have secreted a basal lamina at the surface of at least one portion of the cell facing away from the neural retina (Fig. 20.36). It is worth noting that in one case of human retinitis pigmentosa, cells that make up the bone corpuscle pigmentation (presumed pigment epithelial cells) in the neural retina also have a basal lamina around them (Kolb and Gouras 1974).

Other Hereditary Retinal Degenerations in Mice and Rats

As discussed above, inherited retinal degeneration in the mouse and inherited retinal dystrophy in the rat have been known and studied for many decades. In the past few years, however, several new forms of retinal degeneration have been discovered in both species.

Photoreceptor cell degeneration occurs with a later onset and progresses over a longer period than the RCS condition in a certain proportion of inbred and noninbred Osborne-Mendel (von Sallman and Grimes 1972) and spontaneously hypertensive rats of the SHR/N strain (von Sallman and Grimes 1974), and in all animals of the Wag/Rij strain (Lai et al. 1975). The Osborne-Mendel and Wag/Rij rats show vascularization of the pigment epithelium at late stages of the disease, and pigment epithelial

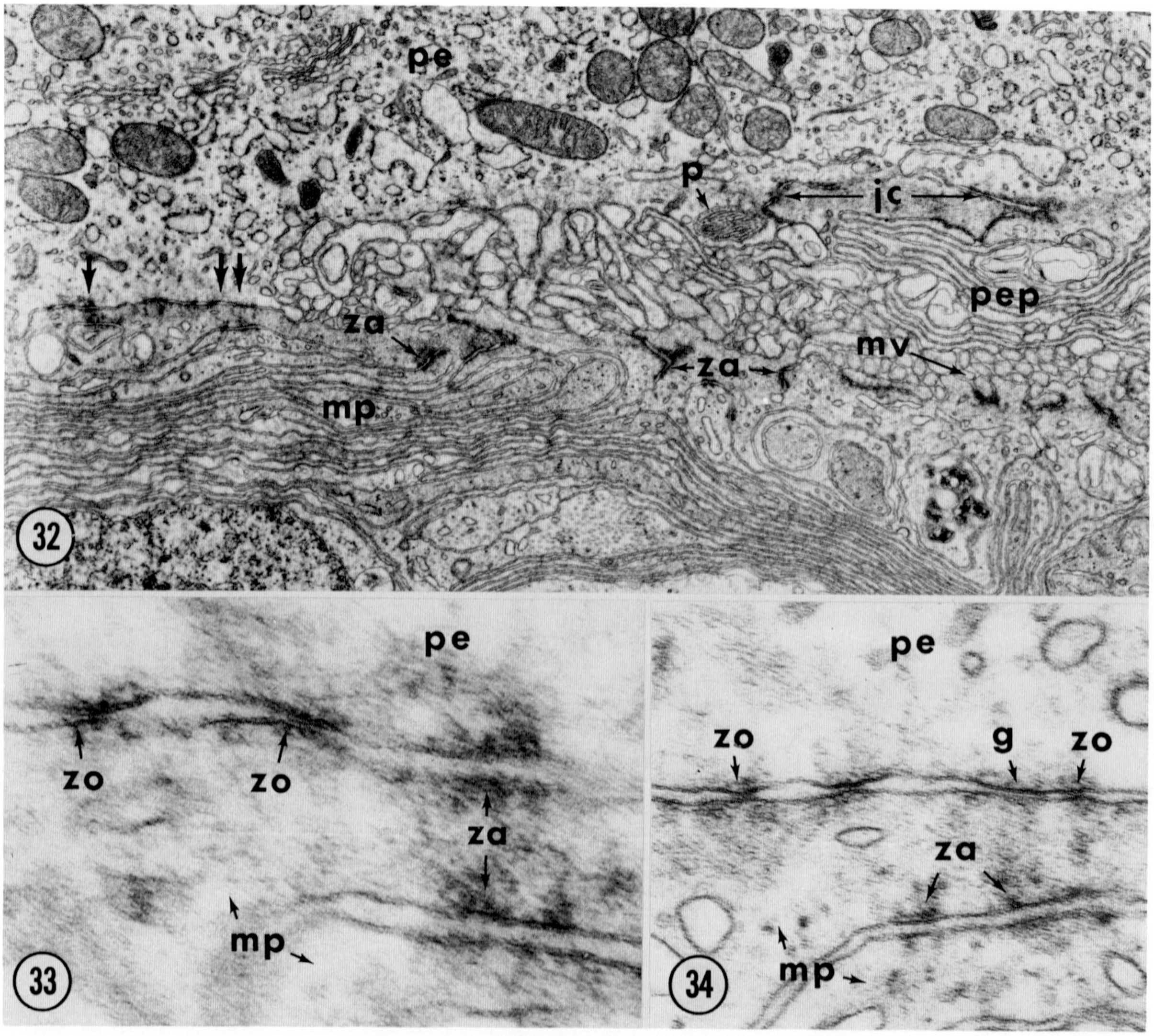

Figures 20.32–20.34 *Electron micrographs from the peripheral retina from a 12-month-old RCS rat, illustrating intercellular contacts between different cell types. Junctional complexes* (jc) *are present between pigment epithelial cells* (pe), *and zonulae adherentes* (za) *are found between adjacent Müller cell processes* (mp). *Zonulae adherentes, zonulae occludentes* (zo), *and gap junctions* (g) *are present between Müller cell processes and pigment epithelial cells. (From LaVail et al. 1974. Fig. 20.32, ×11,830; Fig. 20.33, higher magnification of Fig. 20.32 at region of single arrow, ×169,955; Fig. 20.34, higher magnification of Fig. 20.32 at region of double arrows, ×81,375.)*

cells migrate into the retina in the Wag/Rij animals. The pigment epithelial cells in the Wag/Rij rats show signs of phagocytic activity at all stages of the disease, and no accumulation of outer segment debris is found in any of these disorders. Little else is known yet about the activity and role of the pigment epithelium in these animals.

In about 4 percent of albino Sprague-Dawley rats, a diffuse or focal photoreceptor degeneration occurs (Schardein, Lucas, and Fitzgerald 1975), but evidence for a genetic basis or involvement of the pigment epithelium is lacking.

Two new retinal degenerations have been found in the cerebellar mutant mice: nervous (gene symbol, *nr*) and Purkinje cell degeneration (gene symbol, *pcd*) (Mullen and LaVail 1975). The progressive degeneration of photoreceptor cells in both of these mutants occurs slowly, over the course of about one year. In *pcd*, no accumulation of outer segment debris is found, and phagosomes are present within

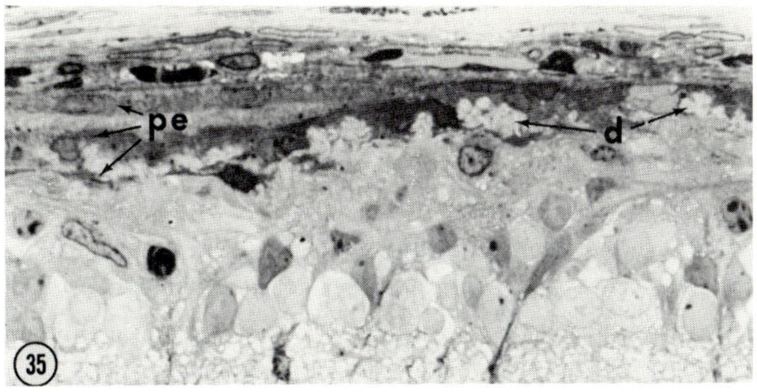

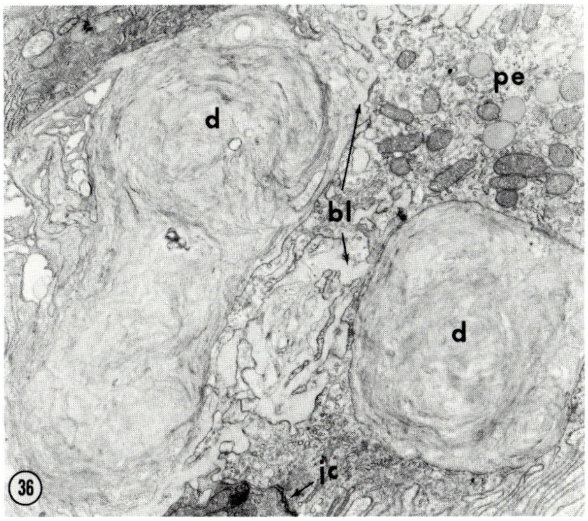

Figure 20.35. *Light micrograph of retina from a one-year-old RCS rat. The pigment epithelial cells* (pe) *are multilayered, and some are enveloping packets of debris membranes* (d). *(From LaVail et al. 1974. ×570.)*

Figure 20.36 *Electron micrograph of an area comparable to that in Figure 20.35. The multilayered pigment epithelial cells* (pe) *still maintain junctional complexes* (jc) *between each other and a basal lamina* (bl) *at their outer surfaces. The enveloped debris membranes* (d) *have lost most of their membrane integrity.* (×*8,140.)*

the pigment epithelial cells. In *nr,* however, some accumulation of debris occurs, and preliminary electron-microscopic examination of the retinas has revealed (1) a paucity, but not a complete absence, of phagosomes, and (2) lamellar whorls quite similar to those found in RCS rat retinas. Further work is in progress to determine if the pigment epithelium has an abnormality in its phagocytic capacity and is involved in the retinal degeneration in the *nr* mouse.

The beige, *bg,* mouse is a mutant in which an abnormality exists in the pigment epithelium. In this mutant, which is currently being studied as an animal model of the Chediak-Higashi syndrome in man, there is an abnormal intracellular digestive capacity and the presence of grossly enlarged lysosomes in leukocytes. In the pigment epithelium, there are abnormally large melanin granules that are thought to fuse with primary lysosomes, resulting in a deficiency in primary lysosomes and an apparent accumulation of phagosomes (Robinson, Kuwabara and Cogan 1975). It is not yet clear, however, whether the apparent abnormality in degradative capacity of the pigment epithelium has an effect on the photoreceptor cells.

Conclusion

The role of the pigment epithelium in the retinal degeneration of the *rd* mouse appears to be negligible. In retinal dystrophy in the rat, the pigment epithelium appears to play a crucial, primary role in the photoreceptor degeneration, and studies are needed to clarify the nature of the genetic defect. Little is yet known about the role of the pigment epithelium in the other forms of retinal degeneration in the mouse and rat.

The *rd* mouse and the RCS rat are currently being studied as animal models for human retinitis pigmentosa. The suitability of the rodents as animal models for the human disorders has been questioned by some, but this would seem to be a moot point. We may not be able to define for many years (or perhaps decades) the early cytopathological stages or the primary involvement of the pigment epithelium in any of the human disorders, yet today we have a number of new cell-biological concepts and tools with which to approach the experimental animals. It would seem that these should be applied to all of the types of animal models for retinal degenerations in an attempt to answer the following questions: (1) What is the cause of photoreceptor cell degeneration? (2)What is the role, if any, of the pigment epithelial cell? (3) Can the degeneration of photoreceptor cells be delayed or prevented? (4) How does the remaining tissue respond to the degeneration of the visual cells? Answers to these questions will require critical examination of all stages of the diseases. Clearly, examination of the early stages of the diseases is required to identify the etiology of the disorders and to attempt therapy. However, since we generally have only older human diseased eyes with which to make comparisons, it also becomes important to define the cell and tissue interactions in the later stages of the animal diseases. The goal should be to identify general principles in retinal cytopathology that are found in mutants with presumed different etiologies. If general principles are found and if the visual cell degeneration can be prevented in the laboratory animals, we then may be in a position to attempt rational therapy on human patients.

REFERENCES

Aguirre, G. D., and Rubin, L. F. 1975. Rod-cone dysplasia (progressive retinal atrophy) in Irish setters. *J. Am. Vet. Med. Assoc.* 166:157–164.

Ansell, P. L., and Marshall, J. 1974. The distribution of extra-cellular acid phosphatase in the retinas of retinitis pigmentosa rats. *Exp. Eye Res.* 19:273–279.

Ansell, P. L., and Marshall, J. 1976. Laser induced phagocytosis in the pigment epithelium of the Hunter dystrophic rat. *Br. J. Ophthalmol.* 60:819–828.

Arden, G. B., and Ikeda, H. 1966. Effects of hereditary degeneration of the retina on the early receptor potential and the corneo-fundal potential of the rat eye. *Vision Res.* 6:171–184.

Bellhorn, R. W., Bellhorn, M., Friedman, A. H., and Henkind, P. 1973. Urethane-induced retinopathy in pigmented rats. *Invest. Ophthalmol.* 12:65–76.

Blanks, J. C., Adinolfi, A. M., and Lolley, R. N. 1974. Photoreceptor degeneration and synaptogenesis in retinal-degenerative (*rd*) mice. *J. Comp. Neurol.* 156:95–106.

Blanks, J. C., and Bok, D. 1977. An autoradiographic analysis of postnatal cell proliferation in the normal and degenerative mouse retina. *J. Comp. Neurol.* 174:317–328.

Bok, D., and Hall, M. O. 1969. The etiology of retinal dystrophy in RCS rats. *Invest. Ophthalmol.* 8:649–650 (Abstract).

Bok, D., and Hall, M. O. 1971. The role of the pigment epithelium in the etiology of inherited retinal dystrophy in the rat. *J. Cell Biol.* 49:664–682.

Bourne, M. C., Campbell, D. A., and Tansley, K. 1938. Hereditary degeneration of the rat retina. *Br. J. Ophthalmol.* 22:613–623.

Burden, E. M., Yates, C. M., Reading, H. W., Bitenskey, L., and Chayen, J. 1971. Investigation into the structural integrity of lysosomes in the normal and dystrophic rat retina. *Exp. Eye Res.* 12:159–165.

Caley, D. W., Johnson, J., and Liebelt, R. A. 1972. The postnatal development of the retina in the normal and rodless CBA mouse: a light and electron microscopic study. *Am. J. Anat.* 133:179–212.

Custer, N. V., and Bok, D. 1975. Pigment epithelium-photoreceptor interactions in the normal and dystrophic rat retina. *Exp. Eye Res.* 21:153–166.

Delmelle, M., Noell, W. K., and Organisciak, D. T. 1975. Hereditary retinal dystrophy in the rat: rhodopsin, retinol, vitamin A deficiency. *Exp. Eye Res.* 21:369–380.

Dewar, A. J., Barron, G., and Reading, H. W. 1975. The effect of retinol and acetylsalicylic acid on the release of lysosomal enzymes from rat retina in vitro. *Exp. Eye Res.* 20:63–72.

Dewar, A. J., Barron, G., and Richmond, J. 1975. Retinal cyclic-AMP phosphodiesterase activity in two strains of dystrophic rat. *Exp. Eye Res.* 21:299–306.

Dewar, A. J., Barron, G., and Yates, C. M. 1975. The effect of light deprivation on the progress of retinal degeneration in two strains of dystrophic rat. *Exp. Eye Res.* 21:507–514.

Dewar, A. J., and Reading, H. W. 1975. The role of retinol in, and the action of anti-inflammatory drugs on, hereditary retinal degeneration. In *Cell impairment in aging and development,* ed. V. J. Cristofalo and E. Holečková, pp. 281–295. New York: Plenum Press.

Dingle, J. T. 1961. Studies on the mode of action of excess of vitamin A.3: release of a bound protease by the action of vitamin A. *Biochem. J.* 79:509–512.

Dowling, J. E., and Sidman, R. L. 1962. Inherited retinal dystrophy in the rat. *J. Cell Biol.* 14:73–109.

Edwards, R. B., and Szamier, R. B. 1977. Defective phagocytosis of isolated rod outer segments by RCS rat retinal pigment epithelium in culture. *Science* 197:1001–1003.

Essner, E., Gorrin, G. M., and Griewski, R. A. 1978. Localization of lysosomal enzymes in retinal pigment epithelium of rats with inherited retinal dystrophy. *Invest. Ophthalmol. and Vis. Sci.* 17:278–288.

Farber, D. B., and Lolley, R. N. 1974. Cyclic guanosine monophosphate: elevation in degenerating photoreceptor cells of the C3H mouse retina. *Science* 186:449–451.

Feeney, L., and Mixon, R. N. 1976. An in vitro model of phagocytosis in bovine and human retinal pigment epithelium. *Exp. Eye Res.* 22:533–548.

Ferrendelli, J. A., and Cohen, A. I. 1976. The effects of light and dark adaptation on the levels of cyclic nucleotides in retinas of mice heterozygous for a gene for photoreceptor dystrophy. *Biochem. Biophys. Res. Comm.* 73:421–427.

Festing, M., and Staats, J. 1973. Standardized nomenclature for inbred strains of rats, 4th listing. *Transplantation* 16:221–245.

GLOOR, B. P. 1974. On the question of the origin of macrophages in the retina and the vitreous following photocoagulation (autoradiographic investigations by means of ^{3}H-thymidine). *Albrecht von Graefes Arch. Klin. Ophthalmol.* 190:183–194.

GOLDMAN, A. I., and O'BRIEN, P. J. 1978. Phagocytosis in the retinal pigment epithelium of the RCS rat. *Science* 201:1023–1025.

GRAFSTEIN, B., MURRAY, M., and INGOGLIA, N. A. 1972. Protein synthesis and axonal transport in retinal ganglion cells of mice lacking visual receptors. *Brain Res.* 44:37–48.

GREEN, E. L. 1966. Breeding system. In *Biology of the laboratory mouse,* ed. E. L. Green, pp. 11–22. New York: McGraw-Hill.

GREEN, E. L., and DOOLITTLE, D. P. 1963. Theoretical consequences of systems of mating used in mammalian genetics. In *Methodology in mammalian genetics,* ed. W. J. Burdette, pp. 3–41. San Francisco: Holden-Day.

HALL, M. O. 1977. Phagocytosis by rat pigment epithelium cells in tissue culture. Presented at the spring meeting of the Association for Research in Vision and Ophthalmology, Sarasota, Florida.

HALL, M. O., and NIR, I. 1976. The binding of concanavalin A to the rod outer segments and pigment epithelium of normal and RCS rats. *Exp. Eye Res.* 22:469–476.

HERRON, W. L., JR., RIEGEL, B. W., MYERS, O. E., and RUBIN, M. L. 1969. Retinal dystrophy in the rat: a pigment epithelial disease. *Invest. Ophthalmol.* 8:595–604.

HERRON, W. L., JR., RIEGEL, B. W., and RUBIN, M. L. 1971. Outer segment production and removal in the degenerating retina of the dystrophic rat. *Invest. Ophthalmol.* 10:54–63.

HOLLYFIELD, J. 1976. Phagocytic capabilities of the pigment epithelium. *Exp. Eye Res.* 22:457–468.

HOLLYFIELD, J. G., and WARD, A. 1974*a*. Phagocytic activity in the retinal pigment epithelium of the frog *Rana pipiens.* I. Uptake of polystyrene spheres. *J. Ultrastruct. Res.* 46:327–338.

HOLLYFIELD, J. G., and WARD, A. 1974*b*. Phagocytic activity in the retinal pigment epithelium of the frog *Rana pipiens.* II. Exclusion of *Sarcina subflava. J. Ultrastruct. Res.* 46:339–350.

HUDSPETH, A. J., and YEE, A. G. 1973. The intercellular junctional complexes of retinal pigment epithelia. *Invest. Ophthalmol.* 12:354–365.

ISHIKAWA, T., and YAMADA, E. 1970. The degradation of the photoreceptor outer segment within the pigment epithelial cell of the rat retina. *J. Electron Microsc.* 19: 85–91.

KAITZ, M. 1976. The effect of light on brightness perception in rats with retinal dystrophy. *Vision Res.* 16:141–148.

KARLI, P. 1952. Rétines sans cellules visuelles: recherches morphologiques, physiologiques et physiopathologiques chez les Rongeurs. *Arch. Anat. Histol. Embryol.* 35:1–76.

KARLI, P., and BONAVENTURE, N. 1970. La mutation *rd* (retinal degeneration) chez la souris: aspects histo- et physio-pathologiques. In *Les mutants pathologiques chez l'animal,* ed. M. Sabourdy, pp. 111–132. Paris: Éditions du Centre National de la Recherche Scientifique.

KARLI, P., STOECKEL, M. E., and PORTE, A. 1965. Dégénérescence des cellules visuelles photoreceptrices et persistance d'une sensibilité de la rétine à la stimulation photique: observations au microscope électronique. *Z. Zellforsch.* 65:238–252.

KEEFE, J. R. 1973. An analysis of urodelian retinal regeneration. II. Ultrastructural features of retinal regeneration in *Notophthalmus viridescens. J. Exp. Zool.* 184: 207–232.

KEELER, C. E. 1924. The inheritance of a retinal abnormality in white mice. *Proc. Nat. Acad. Sci.* 10:329–333.

KEELER, C. E. 1970. Reoccurrence of four-row rodless mice. *Arch. Ophthalmol.* 84:499–504.

KOLB, H., and GOURAS, P. 1974. Electron microscopic observations of human retinitis pigmentosa, dominantly inherited. *Invest. Ophthalmol.* 13:487–498.

KONIGSMARK, B. W., and SIDMAN, R. L. 1963. Origin of brain macrophages in the mouse. *J. Neuropathol. Exp. Neurol.* 22:643–676.

KUWABARA, T., and GORN, R. A. 1968. Retinal damage by visible light. *Arch. Ophthalmol.* 79:69–78.

LAI, Y.-L., JACOBY, R. O., JONAS, A. M., and PAPERMASTER, D. S. 1975. A new form of hereditary retinal degeneration in WAG/Rij rats. *Invest. Ophthalmol.* 14:62–67.

LAVAIL, M. M. 1973. Kinetics of rod outer segment renewal in the developing mouse retina. *J. Cell Biol.* 58:650–661.

LAVAIL, M. M. 1976*a*. Survival of some photoreceptor cells in albino rats following long-term exposure to continuous light. *Invest. Ophthalmol.* 15:64–70.

LAVAIL, M. M. 1976*b*. Rod outer segment disk shedding in rat retina: relationship to cyclic lighting. *Science* 194: 1071–1074.

LAVAIL, M. M., and BATTELLE, B.-A. 1975. Influence of

eye pigmentation and light deprivation on inherited retinal dystrophy in the rat. *Exp. Eye Res.* 21:167–192.

LaVail, M. M., and Mullen, R. J. 1976*a*. Role of the pigment epithelium in inherited retinal degeneration analyzed with experimental mouse chimeras. *Exp. Eye Res.* 23:227–245.

LaVail, M. M., and Mullen, R. J. 1976*b*. Studies on the etiology of inherited retinal degeneration in mice and rats using experimental chimeras. In *Structure of the eye III*, ed. E. Yamada and S. Mishima. *Jap. J. Ophthalmol.*, pp. 363–369.

LaVail, M. M., and Mullen, R. J. 1976*c*. Experimental chimeras: a new approach to the study of inherited retinal degeneration in laboratory animals. In *Retinitis pigmentosa*, ed. M. B. Landers, III, M. L. Wolbarsht, J. E. Dowling, and A. M. Laties, pp. 153–173. New York: Plenum Publishing Corp.

LaVail, M. M., and Sidman, R. L. 1974. C57BL/6J mice with inherited retinal degeneration. *Arch. Ophthalmol.* 91:394–400.

LaVail, M. M., Sidman, R. L., and Gerhardt, C. O. 1975. Congenic strains of RCS rats with inherited retinal dystrophy. *J. Hered.* 66:242–244.

LaVail, M. M., Sidman, R. L., and O'Neil, D. 1972. Photoreceptor-pigment epithelial cell relationships in rats with inherited retinal degeneration: radio-autographic and electron microscope evidence for a dual source of extra lamellar material. *J. Cell Biol.* 53:185–209.

LaVail, M. M., Sidman, M., Rausin, R., and Sidman, R. L. 1974. Discrimination of light intensity by rats with inherited retinal degeneration: a behavioral and cytological study. *Vision Res.* 14:693–702.

Lerche, W. 1974. Elektronenmikroskopische Untersuchungen über strukturveränderun gen im Pigmentepithel der menschlichen Retina. *Albrecht von Graefes Arch. Klin. Ophthalmol.* 189:323–338.

Lolley, R. N. 1973. RNA and DNA in developing retinae: comparison of a normal with the degenerating retinae of C3H mice. *J. Neurochem.* 20:175–182.

Lolley, R. N., and Farber, D. B. 1975. Cyclic nucleotide phosphodiesterases in dystrophic rat retinas: guanosine 3′, 5′ cyclic monophosphate anomalies during photoreceptor cell degeneration. *Exp. Eye Res.* 20:585–597.

Lolley, R. N., and Farber, D. B. 1976. A proposed link between debris accumulation, guanosine 3′, 5′ cyclic monophosphate changes and photoreceptor cell degeneration in the retina of RCS rats. *Exp. Eye Res.* 22:477–486.

Lolley, R. N., Farber, D. B., Rayborn, M. E., and Hollyfield, J. G. 1977. Cyclic GMP accumulation causes degeneration of photoreceptor cells: simulation of an inherited disease. *Science* 196:664–666.

Lolley, R. N., Schmidt, S. Y., and Farber, D. B. 1974. Alterations in cyclic AMP metabolism associated with photoreceptor cell degeneration in the C3H mouse. *J. Neurochem.* 22:701–707.

Lucas, D. R., Attfield, M., and Davey, J. B. 1955. Retinal dystrophy in the rat. *J. Pathol.* 70:469–474.

Marshall, J., and Ansell, P. L. 1971. Membranous inclusions in the retinal pigment epithelium: phagosomes and myeloid bodies. *J. Anat.* (Lond.) 110:91–104.

Mintz, B., and Sanyal, S. 1970. Clonal origin of the mouse visual retina mapped from genetically mosaic eyes. *Genetics* 64:(suppl.) 43–44 (abstr.).

Mullen, R. J., and LaVail, M. M. 1975. Two new types of retinal degeneration in cerebellar mutant mice. *Nature* 258:528–530.

Mullen, R. J., and LaVail, M. M. 1976. Inherited retinal dystrophy: a primary defect in pigment epithelium determined with experimental rat chimeras. *Science* 192: 799–801.

Noell, W. K. 1958. Differentiation, metabolic organization and viability of the visual cell. *Arch. Ophthalmol.* 60:702–733.

Noell, W. K. 1963. Cellular physiology of the retina. *J. Ophthalmol. Soc. Am.* 53:36–48.

Noell, W. K. 1965. Aspects of experimental and hereditary retinal degeneration. In *Biochemistry of the retina*, ed. C. N. Graymore, pp. 51–72. London: Academic Press.

O'Donnell, J. J. 1977. Retinal pigment epithelium phagocytosis in vitro. Presented at the spring meeting of the Association for Research in Vision and Ophthalmology, Sarasota, Florida.

Organisciak, D. T., and Noell, W. K. 1976. Hereditary retinal dystrophy in the rat: lipid composition of debris. *Exp. Eye Res.* 21:101–113.

O'Steen, W. K., and Karcioglu, Z. A. 1974. Phagocytosis in the light-damaged albino rat eye: light and electron microscopic study. *Am. J. Anat.* 139:503–518.

O'Steen, K., Shear, C. R., and Anderson, K. V. 1972. Retinal damage after prolonged exposure to visible light: a light and electron microscopic study. *Am. J. Anat.* 134: 5–22.

Pautler, E. L., and Noell, W. K. 1976. The slow ERG potentials in hereditary retinal dystrophy of albino and pigmented rats. *Exp. Eye Res.* 22:492–503.

Reading, H. W. 1966. Retinal and retinol metabolism

in hereditary degeneration of the retina. *Biochem. J.* 100: 34–35 (abstr.).

Reading, H. W. 1970. Biochemistry of retinal dystrophy. *J. Med. Genet.* 7:277–284.

Reich-d'Almeida, F. B., and Hockley, D. J. 1975. In situ reactivity of the retinal pigment epithelium. II. Phagocytosis in the dystrophic rat. *Exp. Eye Res.* 21:347–357.

Reich-d'Almeida, F. B., and Hockley, D. J. 1976. Primary retinal degeneration: evidence of normal phagocytosis in the retinal pigment epithelium. *Invest. Ophthalmol.* 15:757–760.

Reich-d'Almeida, F., and Rahi, A. H. S. 1974. Antigenic specificity of retinal pigment epithelium and non-immunological involvement in retinal dystrophy. *Nature* 252:307–308.

Robb, R. M. 1974. Electron microscopic histochemical studies of cyclic 3′, 5′ nucleotide phosphodiesterase in the developing retina of normal mice and mice with hereditary retinal degeneration. *Trans. Am. Ophthalmol. Soc.* 72:650–669.

Robison, W. G., Kuwabara, T., and Cogan, D. G. 1975. Lysosomes and melanin granules of the retinal pigment epithelium in a mouse model of the Chediak-Higashi syndrome. *Invest. Ophthalmol.* 14:312–317.

Sanyal, S. 1972. Changes of lysosomal enzymes during hereditary degeneration and histogenesis of retina in mice. II. Localization of N-acetyl-B-glucosaminidase in macrophages. *Histochemie.* 29:28–36.

Sanyal, S., and Bal, A. K. 1973. Comparative light and electron microscopic study of retinal histogenesis in normal and *rd* mutant mice. *Z. Anat. Entwicklungsgesch.* 142:219–238.

Sanyal, S., and Zeilmaker, G. H. 1976. Comparative analysis of cell distribution in the pigment epithelium and the visual cell layer of chimaeric mice. *J. Embryol. Exp. Morphol.* 36:425–430.

Schardein, J. L., Lucas, J. A., and Fitzgerald, J. E. 1975. Retinal dystrophy in Sprague-Dawley rats. *Lab. Anim. Sci.* 25:323–326.

Schmidt, S. Y., and Lolley, R. N. 1973. Cyclic-nucleotide phosphodiesterase: an early defect in inherited retinal degeneration of C3H mice. *J. Cell Biol.* 57:117–123.

Shoise, Y. 1968. Electron microscopic aspects of early changes of inherited dystrophic mouse retina. *Jap. J. Ophthalmol.* 12:181–190.

Sidman, R. L. 1958. Histochemical studies on photoreceptor cells. *Ann N.Y. Acad. Sci.* 74:182–195.

Sidman, R. L. 1961. Histogenesis of mouse retina studied with thymidine-^{3}H. In *Structure of the eye,* ed. G. K. Smelser, pp. 487–505. New York: Academic Press.

Sidman, R. L., and Green, M. C. 1965. Retinal degeneration in the mouse. *J. Hered.* 56:23–29.

Sidman, R. L., and Pearlstein, R. 1965. Pink-eyed dilution (*p*) gene in rodents: increased pigmentation in tissue culture. *Dev. Biol.* 12:93–116.

Sonohara, O., and Shoise, Y. 1968. Electron microscopic study of the visual cell of inherited retinal dystrophic mice. *Folia Ophthalmol. Jap.* 19:77–86.

Sorsby, A., Koller, P. C., Attfield, M., Davey, J. B., and Lucas, D. R. 1954. Retinal dystrophy in the mouse: histological and genetic aspects. *J. Exp. Zool.* 125:171–197.

Spitznas, M., and Hogan, M. J. 1970. Outer segments of photoreceptors and the retinal pigment epithelium. *Arch. Ophthalmol.* 84:810–819.

Tewari, R. P., Solotorovsky, M., and Scherner, N. J. 1974. Macromolecular synthesis and phagocytosis of normal and fused macrophages. In *Activation of macrophages,* ed. W. H. Wagner and H. Hahn, pp. 20–30. Amsterdam: Excerpta Medica.

Tso, M. O. M. 1973. Photic maculopathy in rhesus monkey. *Invest. Ophthalmol.* 12:17–34.

Tso, M. O. M., and Friedman, E. 1970. The retinal pigment epithelium. I. Comparative histology. *Arch. Ophthalmol.* 78:641–649.

Tso, M. O. M., Wallow, I. H. L., and Powell, J. O. 1973. Differential susceptibility of rod and cone cells to argon laser. *Arch. Ophthalmol.* 89:228–234.

von Sallman, L., and Grimes, P. 1972. Spontaneous retinal degeneration in mature Osborne-Mendel rats. *Arch. Ophthalmol.* 88:404–411.

von Sallman, L., and Grimes, P. 1974. Retinal degeneration in mature rats: comparison of the disease in an Osborne-Mendel and a spontaneously hypertensive Wistar strain. *Invest. Ophthalmol.* 13:1010–1015.

Wegmann, T. G., LaVail, M. M., and Sidman, R. L. 1971. Patchy retinal degeneration in tetraparental mice. *Nature* (Lond.) 230:333–334.

Yates, C. M., Dewar, A. J., Wilson, H., Winterburn, A. K., and Reading, H. W. 1974. Histological and biochemical studies on the retina of a new strain of dystrophic rat. *Exp. Eye Res.* 18:119–133.

Young, R. W. 1967. The renewal of photoreceptor cell outer segments. *J. Cell Biol.* 33:61–72.

YOUNG, R. W. 1971. Shedding of discs from rod outer segments in the rhesus monkey. *J. Ultrastruct.* 34:190–203.

YOUNG, R. W., and BOK, D. 1969. Participation of the retinal pigment epithelium in the rod outer segment renewal process. *J. Cell Biol.* 42:392–403.

YOUNG, R. W., and DROZ, B. 1968. The renewal of protein in retinal rods and cones. *J. Cell Biol.* 39:169–184.

ACKNOWLEDGMENTS

The author wishes to thank Patricia Ann Ward, Carolyn O. Gerhardt and Donna M. Colicchio for technical assistance and David Akers for photographic assistance. This investigation was supported in part by USPHS Research Grant EY-01919 and Research Career Development Award EY-70871 from the National Eye Institute.

21 Clinical Evaluation of Retinal Pigment Epithelial Disorders

Ronald E. Carr
Irwin M. Siegel

OPHTHALMOLOGIC SCREENING of patients with suspected disorders of the retinal pigment epithelium (RPE) can be approached with the following questions in mind.

1. What tests are specific for measuring RPE function and can they distinguish between primary and secondary disturbances of the RPE?
2. Do the results of any test procedures provide insight into the underlying disease processes?

Several tests of retinal function are usually required for a complete understanding of the nature and cause of a disorder (Carr and Siegel 1964). Therefore, knowledge of both the capabilities and limitations of psychophysical and electrodiagnostic tests of retinal activity as they relate to the RPE is necessary. Other authors in this volume (see Chapter 9) have already reviewed the RPE contribution to the electroretinogram (ERG) and electrooculogram (EOG). Similarly, the results of these tests as they relate to specific clinical problems of the RPE are discussed in Chapters 24 and 25. What is aimed for here is a guide for the ophthalmologist in the clinical situation. Accordingly, we have selected those tests and procedures that appear to be particularly useful in establishing the presence of RPE disease and in differentiating it from other ocular diseases that may produce similar ophthalmoscopic defects. With few exceptions, the test modalities to be discussed are readily available.

In most instances, one is aware of possible abnormalities in the pigment epithelium because of the funduscopic appearance of a patient's retina. On the other hand, results of certain test procedures may lead to a diagnosis of either a primary or secondary pigment epithelial disorder, the signs of which may be ophthalmoscopically equivocal. Thus, tests of electrical function, as well as those designed to evaluate retinal sensitivity using subjective means, are of value in that they may serve to delineate the extent of an RPE abnormality, besides providing diagnoses of specific forms of RPE disorders. In all the tests to be discussed, the intimate relationship between the pigment epithelium and photoreceptor outer segments (as shown in Fig. 24.4) must be kept in mind, for unlike the inner retinal layers, the receptors are served exclusively by transport through the RPE.

Initial Patient Workup

Just as in general medicine, where a clear and accurate history will provide clues to specific diseases, the visual complaints of a patient and their time of onset are important facts in distinguishing between categories of ocular disease. Thus, a recent loss of central vision in older individuals usually signifies macular or optic nerve disease. Similarly, a slowly progressing, long-standing loss of night vision would be indicative of a generalized retinal disorder. A full family history also can be of value, since the existence of similarly affected relatives or parental consanguinity is evidence favoring a hereditary disorder. Of course, ocular diseases that are congenital or appear in very young patients are most likely to be genetic. It is also important to note any somatic abnormalities in the patient or other family members, since many systemic disorders may occur in association with some of the tapetoretinal degenerations of which the mucopolysaccharide syndromes and certain lipid diseases (for example, Bassen-Kornzweig, Refsum's) are examples. Examination of relatives of the propositus may also reveal a *forme fruste* of the disorder, in which there may be minimal ocular changes but obvious somatic changes. Leber's congenital amaurosis, with its accompanying CNS degenerative changes, is a well-known example of this phenomenon (Dekaban and Carr 1966).

Psychophysical Testing

The extent of psychophysical testing varies from institution to institution, in that a complete battery of tests may be given to all patients regardless of complaint at some clinics, while at others, greater selectivity is used. Often, the individual tests comprising the workup will be predicated on either the patient complaint or on the results of completed procedures. In a busy clinic, where time and personnel may be a factor, test selectivity is a sensible approach.

COLOR VISION TESTING. The theory and administration of the standard tests of color vision can be found in a book by Linksz (1964) as well as in a review article by Fishman (1971). From a practical point of view, it must be kept in mind that only a few color tests can be used. Ideally the tests should measure different aspects of color vision, since this will provide the basis for a broader evaluation of color loss as well as allowing for independent diagnostic assessments.

Table 21.1 lists those tests that have proven useful in many clinics. However, some of the more specialized procedures that are not commonly found in clinics should not be overlooked, since such testing may be required for diseases that affect the color sense in subtle or unique ways. These would include the Nagel anomaloscope, which can be used to measure luminosity losses in the red and green portions of the spectrum, and studies of spectral sensitivity, in which the amount of energy required to elicit a criterion response across the whole range of visible radiation is determined. Some of the tests listed in Table 21.1, such as Ishihara and A.O. H-R-R, re-

Table 21.1 Color vision testing.

Test	*Mode of Operation*	*Advantages*	*Disadvantages*
Ishihara	Color confused pairs	1. Will detect very mild defects 2. Can be used when V.A. < 20/200	1. Does not rigorously classify defects 2. Not easily administered to preschool children 3. Detects only red-green defects
Farnsworth D-15 Panel	Color confused pairs	1. Has good diagnostic scoring 2. Can be used when V.A. is very poor	1. Only detects severe anomalous trichromacies and dichromats 2. Not easily given to preschool children
American Optical Hardy-Rand-Rittler (A.O. H-R-R)	Saturation appearance of colors	1. Quantitative diagnostic scoring 2. Easily given to all ages	1. Not appropriate for patients with V.A. < 20/40
Farnsworth-Munsell 100 Hue Test	Wavelength discrimination	1. Very sensitive indicator of mild color defects 2. Quantitative scoring 3. Can be used with poor V.A.	1. Tedious to administer 2. Not appropriate for preschool children
Sloan Achromatopsia Test	Hue-Brightness	1. Accurately detects complete loss of color vision 2. Easily administered 3. Does not require good V.A.	1. Is nondiagnostic for forms of color loss other than achromatopsia

quire figure discrimination and as such are more acuity dependent than the Farnsworth D-15 panel or 100-Hue, which use chips of colored paper. However, results obtained from patients with very poor acuity (finger counting at 2 feet or worse) should not be heavily relied upon in any case.

The clinician should be aware that color vision tests are designed to test only the foveal and perifoveal area of the retina. Beyond this region, under certain conditions, the retina becomes color anomalous—that is, dichromatic in the midperiphery, and achromatic in the far periphery (Weale 1951, 1953). Therefore, patients who have already learned to fixate eccentrically are not appropriate subjects for such testing. Fundus fixation photography (von Noorden and Mackensen 1962) is one means for quickly establishing whether central fixation exists.

The fact that most of the color tests were originally designed to detect specific forms of *heritable* color defects should not preclude their use in evaluating color defects secondary to ocular pathology. Indeed, it is a diagnostic point when patients give inconsistent or nonclassifiable responses on color tests; such seemingly unreliable results are strong indicators that acquired defects are, in fact, present. Also, unlike congenital color defects, secondary anomalies of color vision may vary considerably from patient to patient. Despite these drawbacks, it is nonetheless important to have some idea of the degree of color loss in patients with ocular disease—and at present no other test procedures with good interindividual reliability are available for the purpose.

In the past it was thought that diseases of the outer retinal layers could be distinguished from lesions in the inner retinal layers by noting which kind of color defect was present. Disorders affecting the optic nerve fibers were said to give rise to red-green defects, while receptor-RPE disease was felt to produce blue-yellow defects (Köllner's law). But recent studies (particularly those of Verriest 1963) have shown that Köllner's relationship does not generally hold. Considering the great degree to which rod receptors and the three types of cone receptors converge on a single nerve fiber (Gouras 1967), it is also difficult to suggest a neurophysiological basis for such a theory. Krill et al. (1970) have even found the inverse relationship—a consistent blue-yellow defect—in several patients with a juvenile form of optic atrophy.

Since all the color vision tests discussed above are designed to evaluate only the color sense of the fovea, other methods are required when this area is normal and a generalized cone dysfunction is suspected. It is useful to recall that only 2 percent of the retinal cones are actually within the fovea; the great majority are located outside it. Color-naming fields can be used to test the crude extrafoveal color sense. They are obtained by having the patient fixate the center of a tangent screen while the examiner brings a colored test object toward fixation until a correct color response is elicited. Color-name isopters for red, green, and blue test objects can thus be obtained. An example of the usefulness of such a procedure is given by Siegel and Smith (1967) who studied a patient having an acquired form of generalized cone dysfunction with sparing of the foveal area.

DARK ADAPTOMETRY. While a history of night blindness should always be regarded as a serious complaint, a certain scepticism should be maintained in examining older patients who profess an "inability" to see at night or feel they cannot see road signs in dim illumination. The pupil dilatation effected by low light levels results in a low degree of myopia (so-called "night myopia") and is easily corrected by an appropriate minus correction. Light sensitivity is not disturbed in such instances. A good question to ask a person suspected of being night blind is how long it takes him to see people nearby after entering a movie theater or dimly lit room. Though dark adaptation depends on the degree of previous exposure to bright lights, most people will become well-adapted within 15 minutes, while those with night blindness will have difficulty even after that time.

The age of onset of night blindness is also important. Patients with a congenital (and often stationary) form of night blindness will have had this disorder all of their lives, while patients with a generalized heredodegeneration may recall having been able to function quite easily at night when they were younger. Individuals who report the onset of night-vision difficulty in middle age usually do not have true night blindness.

The most common method of testing dark adaptation is by means of the Goldmann-Weekers adaptometer, with measurements recorded by the examiner (Carr et al. 1966) or automatically by the patient

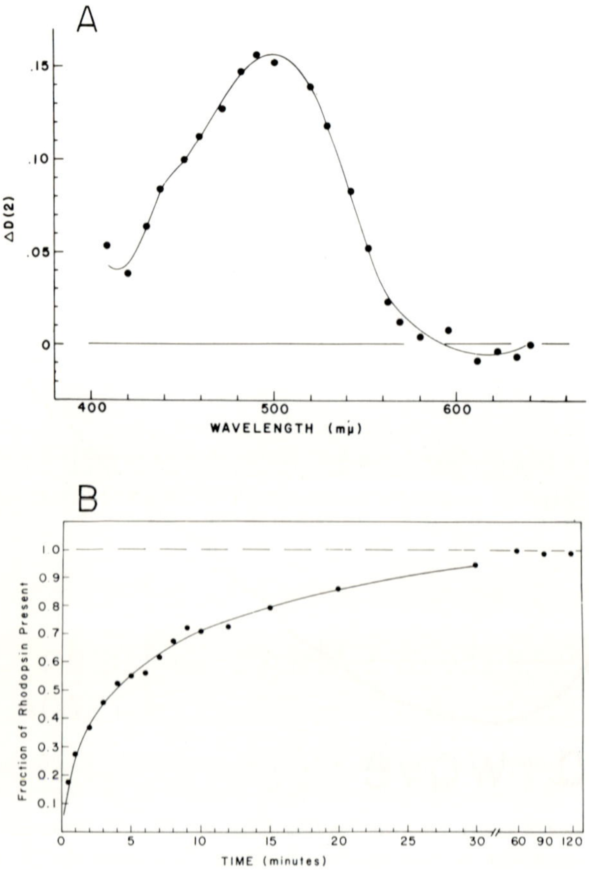

Figure 21.7 *Fundus reflectometry.* A, *density difference* (ΔD_2) *spectrum of rhodopsin measured in the peripheral retina.* B, *temporal sequence of rhodopsin regeneration in the normal human peripheral retina. The half time of regeneration is about 3½ minutes.*

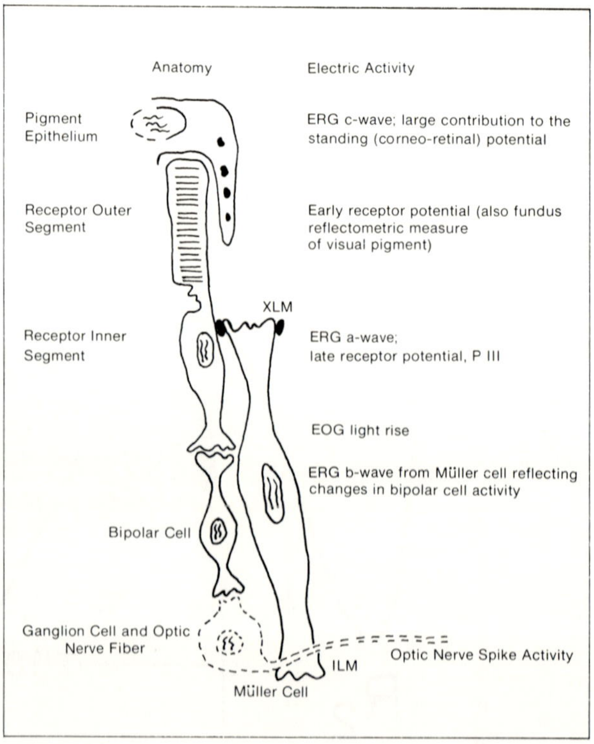

Figure 21.8 *Summary diagram showing, in schematic fashion, the presumed sites of electrical activity in the retina.*

tation of results have improved remarkably, so that the test has substantially increased our diagnostic accuracy and the understanding of many retinal diseases. Several books, including those by Gass (1970) and Archer (1972), provide full descriptions of the procedure. Recent innovations, based on a variety of new photographic materials and vital dyes, have been developed to differentiate between retinal and choroidal vascular anomalies (Brown and Strong 1973; Flower and Hochheimer 1973; Dallow 1974).

The essential steps in performing a fundus angiogram are: (1) sodium fluorescein is injected into an arm vein; (2) light from an electronic flash tube is passed through an excitation filter (usually cobalt blue) that produces dye fluorescence; and (3) the returning light from the fundus passes through a barrier filter placed in front of the film. The filter absorbs most of the light except that being emitted by fluorescein structures, thus allowing these areas to be easily seen against a darkened background. Rapid sequence photographs of the retina can be taken following the dye injection, and the various stages of vessel filling are recorded (Fig. 21.9).

Because the melanin-containing RPE has great light-absorption properties, very little of the choriocapillaris filling is seen (Figs. 21.9*A, B*). If, however, there is a defect of the RPE, the choriocapillaris shows through quite vividly. Thus, not only are abnormalities of the RPE easily detected, but subtle changes in the choriocapillaris become visible, and it is possible to formulate a model of pathogenesis. For example, if the choriocapillaris is abnormal, the primary defect in this tissue *and* the secondary changes induced in the RPE are evident. This is shown in the fluorescein photographs of a patient with peripapillary choroidal dystrophy (Fig. 21.10*A*). On the other hand, a disturbance of the RPE without evidence of abnormal choroidal circulation implies a primary RPE or RPE-photoreceptor defect

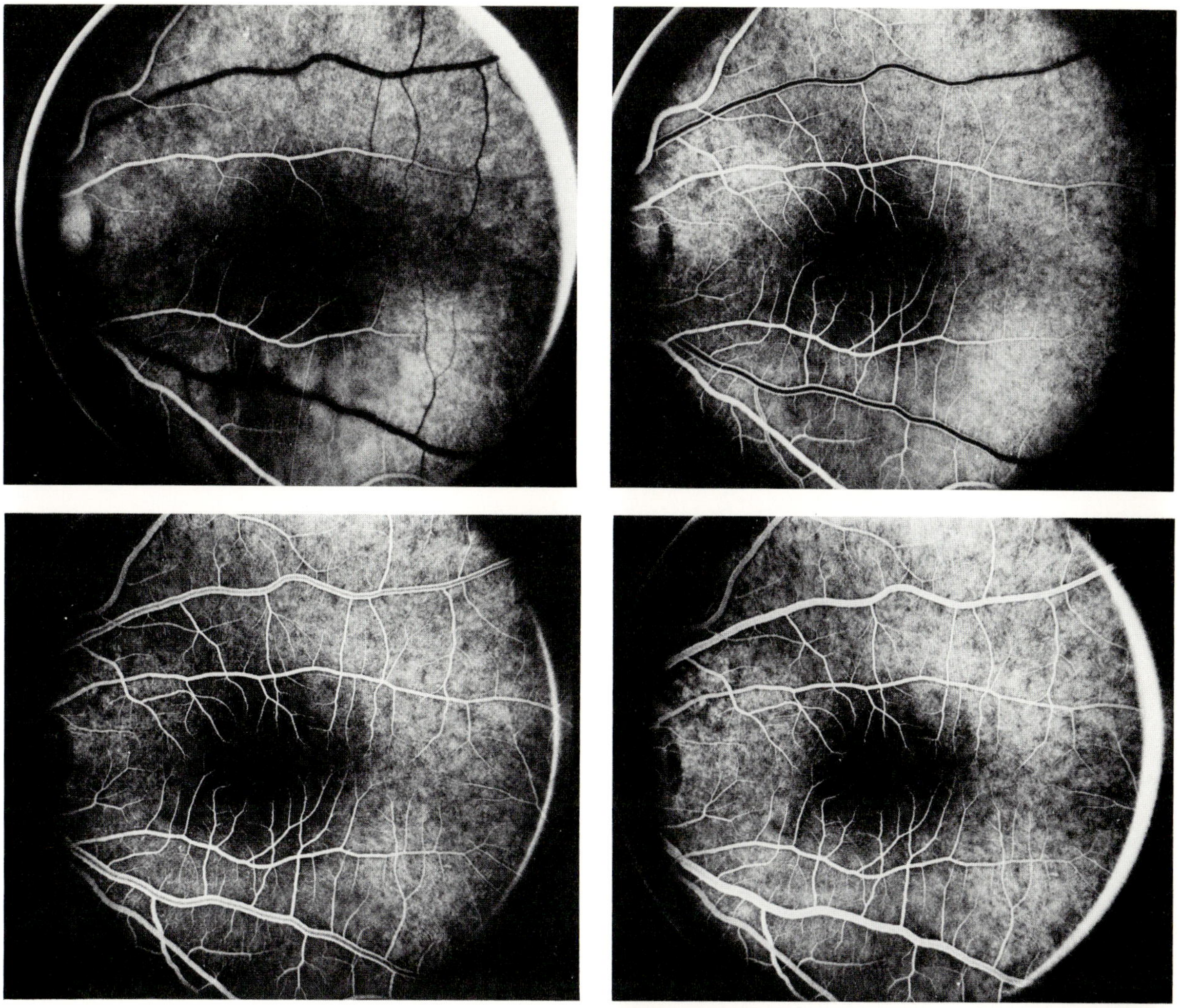

A B C D

Figure 21.9 *Fluorescein angiography showing the normal sequence of vessel filling.* A, *arterial phase with choroidal background fluorescence;* B, *early venous filling with more intense choroidal background fluorescence;* C, *midvenous phase;* D, *late venous phase;* E, *later phase showing less dye in vessels and some coalescence of the choroidal flush.*

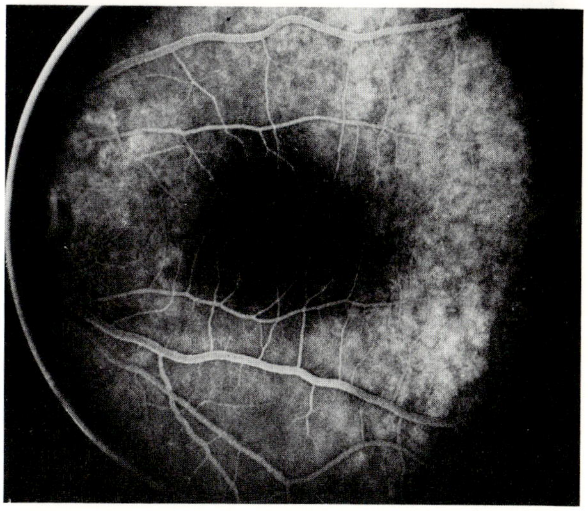

E

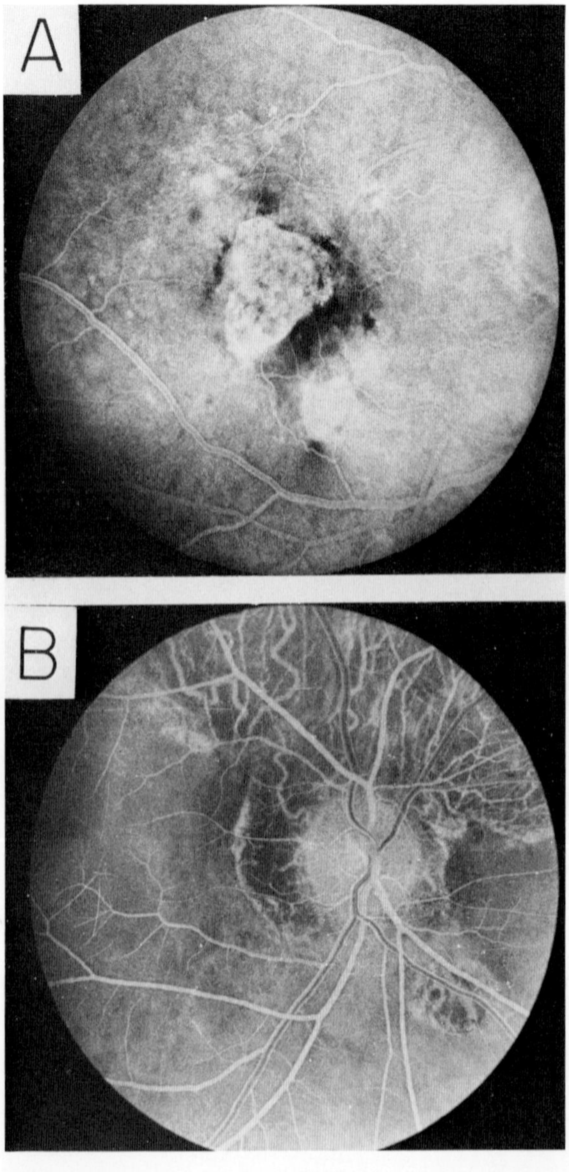

Figure 21.10 *Fluorescein photographs of patients with* (A) *peripapillary choroidal dystrophy with an absent pigment epithelium and choriocapillaris in the affected regions, and* (B) *senile macular degeneration with an absent pigment epithelium and normal choriocapillaris. Both photographs taken 15 seconds after dye injection.*

(Fig. 21.10*B*). Evidence that the photoreceptors are also involved in such diseases can be obtained from some of the electrical tests of retinal function.

REFERENCES

ADAMS, A. 1973. The normal EOG. *Acta Ophthalmol.* 51:551–561.

ARCHER, D. B. 1972. Fluorescein angiography. In *Hereditary retinal and choroidal diseases,* ed. A. E. Krill. I. *Evaluation,* pp. 73–188. New York: Harper and Row.

ARDEN, G. B., and BARRADA, A. 1962. Analysis of the EOG of a series of normal subjects: the role of the lens in the development of the standing potential. *Brit. J. Ophthalmol.* 46:468–482.

ARDEN, G. B., BARRADA, A., and KELSEY, J. H. 1962. New clinical test of retinal function based upon the standing potential of the eye. *Br. J. Ophthalmol.* 46:449–467.

ARMINGTON, J. C. 1968. The electroretinogram, the visual evoked potential and the area luminance relation. *Vision Res.* 8:263–276.

ARMINGTON, J. C., TEPAS, D. I., KROPFL, W. S., and HENGST, W. H. 1961. Summation of retinal potentials. *J. Opt. Soc. Am.* 51:877–886.

BERSON, E. L., and GOLDSTEIN, E. B. 1970*a*. Recovery of the human ERP during dark adaptation in hereditary retinal disease. *Vision Res.* 10:219–226.

BERSON, E. L., and GOLDSTEIN, E. B. 1970*b*. Early receptor potential in dominantly inherited retinitis pigmentosa. *Arch. Ophthalmol.* 83:412–420.

BROWN, N., and STRONG, N. 1973. Infrared fundus angiography. *Br. J. Ophthalmol.* 57:797–802.

BURIAN, H. M., and FLETCHER, M. C. 1958. Visual functions in patients with retinal pigmentary degeneration following the use of NP-207. *Arch. Ophthalmol.* 60:612–629.

CALISSENDORFF, B., KNAVE, B., and PERSSON, H. E. 1974. Cyclic variations in the c-wave amplitude of the sheep ERG. *Vision Res.* 14:1141–1145.

CARR, R. E., and RIPPS, H. 1967. Rhodopsin kinetics and rod adaptation in Oguchi's disease. *Invest. Ophthalmol.* 6:426–436.

CARR, R. E., and SIEGEL, I. M. 1964. Electrophysiologic aspects of several retinal diseases. *Am. J. Ophthalmol.* 58:95–107.

CARR, R. E., and SIEGEL, I. M. 1970. Action spectrum of the human ERP. *Nature* (Lond.) 225:88–89.

CARR, R. E., and SIEGEL, I. M. 1973. Retinal function in patients treated with indomethacin. *Am. J. Ophthalmol.* 75:302–306.

CARR, R. E., GOURAS, P., and GUNKEL, R. D. 1966. Chloroquine retinopathy: early detection by retinal threshold test. *Arch. Ophthalmol.* 75:171–178.

CARR, R. E., MORTON, N. E., and SIEGEL, I. M. 1971. Achromatopsia in Pingelap Islanders: study of a genetic isolate. *Am. J. Ophthalmol.* 72:746–756.

CARR, R. E., RIPPS, H., and SIEGEL, I. M. 1974. Visual pigment kinetics and adaptation in fundus albipunctatus. *Docum. Ophthalmol. Proc. Series* (XI ISCERG Symp.), pp. 193–204.

CARR, R. E., RIPPS, H., SIEGEL, I. M., and WEALE, R. A. 1966. Rhodopsin and the electrical activity of the retina in congenital nightblindness. *Invest. Ophthalmol.* 5:497–507.

CONE, R. A. 1965. The ERP of the vertebrate eye. *Cold Spring Harbor Symp. Quant. Biol.* 30:483–491.

DALLOW, R. L. 1974. Color infrared photography of the ocular fundus. *Arch. Ophthalmol.* 92:254–258.

DEKABAN, A., and CARR, R. E. 1966. Congenital amaurosis of retinal origin. *Arch. Neurol.* 14:294–301.

DEWAR, J., and MCKENDRICK, J. G. 1873. On the physiological action of light. *J. Anat. and Physiol.* 7:273–275.

FISHMAN, G. 1971. Techniques, merits, and limitations of basic tests for color defectiveness. *Surv. Ophthalmol.* 15:370–373.

FLOWER, R. W., and HOCHHEIMER, B. F. 1973. A clinical technique and apparatus for simultaneous angiography of the separate retinal and choroidal circulations. *Invest. Ophthalmol.* 4:248–261.

FRANCOIS, J., DEROUCK, A., and FERNANDEZ-SASSO, D. 1966. L'électro-oculographie dans les dégénérescences vitelliformes de la macula. *Bull. Soc. Belge. Ophthalmol.* 143:547–552.

FRANCOIS, J., VERRIEST, G., and DEROUCK, A. 1955. Modification of the amplitude of the human EOG by light and dark adaptation. *Br. J. Ophthalmol.* 38:398–407.

GALLOWAY, N. R. 1967. Early receptor potential in the human eye. *Br. J. Ophthalmol.* 51:261–264.

GASS, J. D. M. 1970. *Stereoscopic atlas of macular diseases: a funduscopic and angiographic presentation.* St. Louis: C. V. Mosby.

GOURAS, P. 1967. The effects of light-adaptation on rod and cone receptive field organization of monkey ganglion cells. *J. Physiol.* 192:747–760.

GOURAS, P., and CARR, R. E. 1965. Light-induced DC responses of monkey retina before and after central retinal artery interruption. *Invest. Ophthalmol.* 4:310–317.

GOURAS, P., and GUNKEL, R. D. 1963. The EOG in chloroquine and other retinopathies. *Arch. Ophthalmol.* 70:629–639.

GUNKEL, R. D., and BORNSCHEIN, H. 1957. Automatic intensity control in testing dark adaptation. *Arch. Ophthalmol.* 57:681–686.

HENKES, H. E., and LEGEIN, C. P. 1969. Electrodiagnostic procedures in the blind and partially sighted young child. In *Electrical responses of the visual system,* ed. S. J. Fricker, pp. 921–933. International Ophthalmology Clinic Education.

HOLLAND, M. G., and CLARK, F. 1972. An automatic measuring and recording system for clinical EOG. *Ophthalmol. Res.* 3:311–319.

JOHNSON, E. P., RIGGS, L. A., and SCHICK, A. M. L. 1966. Photopic retinal potentials evoked by phase alternation of a barred pattern. In *Clinical electroretinography,* ed. H. M. Burian and J. H. Jacobson, pp. 75–92. New York: Pergamon Press.

KLIEN, B. A., and KRILL, A. E. 1967. Fundus flavimaculatus: clinical, functional and histopathologic observations. *Am. J. Ophthalmol.* 64:3–23.

KNAVE, B., NILSSON, S. E., and LUNT, T. 1973. The human ERG: DC recordings at low and conventional stimulus intensities. *Acta Ophthalmol.* 51:716–726.

KRILL, A. E. 1966. The ERG in fundus flavimaculatus (Proc. IV ISCERG Symp.). *Jap. J. Ophthalmol.* 10:293–300 (suppl.).

KRILL, A. E., MORSE, P. A., POTTS, A. M., and KLIEN, B. A. 1966. Hereditary vitelliruptive macular degeneration. *Am. J. Ophthalmol.* 61:1405–1415.

KRILL, A. E., SMITH, V. C., and POKORNY, J. 1970. Similarities between congenital tritan defects and dominant optic-nerve atrophy: coincidence or identity. *J. Opt. Soc. Am.* 60:1132–1139.

KRIS, C. 1958. Corneo-fundal potential variations during light and dark adaptation. *Nature* 182:1027–1028.

LINKSZ, A. 1964. *An essay on color vision and clinical color-vision tests.* New York/London: Grune and Stratton.

NOVOTNY, H. R., and ALVIS, D. L. 1961. A method of photographing fluorescence in circulating blood in the human retina. *Circulation* 24:82–86.

RABIN, A. R., and BERSON, E. L. 1974. A full field system for clinical ERG. *Arch. Ophthalmol.* 92:59–63.

SIEGEL, I. M., and SMITH, B. F. 1967. Acquired cone dysfunction. *Arch. Ophthalmol.* 77:8–13.

SLOAN, L. L. 1947. Rate of dark adaptation and regional threshold gradient of the dark-adapted eye: physiologic and clinical studies. *Am. J. Ophthalmol.* 30:705–720.

SOKOL, S., and RIGGS, L. A. 1971. Electrical and psychophysical responses of the human visual system to periodic variation of luminance. *Invest. Ophthalmol.* 10:171–180.

VERRIEST, G. 1963. Further studies on acquired deficiency of color discrimination. *J. Opt. Soc. Am.* 53:185–195.

von Noorden, G., and Mackensen, G. 1962. Phenomenology of eccentric fixation. *Am. J. Ophthalmol.* 53:642–661.

Weale, R. A. 1951. The foveal and para-central spectral sensitivities in man. *J. Physiol.* 114:435–446.

Weale, R. A. 1953. Spectral sensitivity and wave-length discrimination of the peripheral retina. *J. Physiol.* 119:170–190.

Weekley, R. D., Potts, A. M., Reboton, J., and May, R. H. 1960. Pigmentary retinopathy in patients receiving high doses of a new phenothiazine. *Arch. Ophthalmol.* 64:65.

22 Toxicology of the Human Retinal Pigment Epithelium

Keith M. Zinn
Michael F. Marmor

THE RETINAL PIGMENT epithelium has a major role in maintaining the health and visual activity of the retina. For example, the pigment epithelium performs phagocytosis of outer segment material and provides substances or enzymes necessary for photoreceptor renewal and for the rhodopsin cycle. Its cell membrane has electrical and transport properties that influence the electrical and photochemical activities of the outer segments, and that allow the pigment epithelium to respond to light. The tight junctions between the cells of the pigment epithelium are critical to these electrical and transport properties, and they serve as the "blood-brain barrier" for the outer surface of the photoreceptors. The pigment epithelium synthesizes an acid mucopolysaccharide material that forms a major part of the intercellular matrix within the subretinal space. The ability of the pigment epithelium to regulate the content and thinness of this subretinal space helps to maintain the normal adhesion between the neurosensory retina and pigment epithelium. Melanin within the pigment epithelium serves as an optical aid to the outer segments by decreasing the scatter of light and serves as a possible site for subcellular molecular reactions.

Various pharmacological agents are known to affect the pigment epithelial cells adversely or to affect the neurosensory retina when they accumulate within the pigment epithelium. When the physiological functions of the pigment epithelium are altered, then the overlying retina is also affected, with a resultant decrease in visual function. This chapter will review the toxicological effects of a variety of substances, especially the quinolines and the phenothiazines, on the human retinal pigment epithelium. Effects that have been primarily observed in animals are covered in Chapter 19. Toxic effects on the electrical responses of the pigment epithelium are discussed in Chapter 13.

The clinical diagnosis that a toxic agent has damaged the pigment epithelium usually depends on a combination of data derived from the patient's history (indicating exposure), the clinical examination (including ophthalmoscopy and fluorescein angiography), and physiologic testing (including color vision, dark adaptation, electroretinography, and electrooculography). These clinical tests by which pigmentary changes and functional changes may be localized to the pigment epithelium are discussed more fully in Chapter 21.

Quinolines

Chloroquine (7-chloro 4-[4-diethylamino-methylbutylamino] quinoline) was first used during World War II as a treatment for malaria. In 1950, chloroquine was used for the treatment of rheumatoid arthritis; in the following year it was employed in the treatment of patients with systemic lupus erythematosus. However, it was not until 1957 that Cambiaggi reported the first case of ocular toxicity in a patient with systemic lupus erythematosus. Corneal epithelial deposits are not uncommon, and rarely cause any visual disturbance. Our concern here will be with the potentially devastating retinal effects.

The total dosages used in treating patients with systemic lupus erythematosus or rheumatoid arthritis are far in excess of the dosages used to treat patients with malaria. In many cases, the daily dosages have averaged 500 mg, with total dosages, in some cases, exceeding 1,000 g. Many studies have been carried out to determine the incidence of retinal toxicity (Arden and Kolb 1966; Henkind and Rothfield 1963; Hobbs, Eadie, and Somerville 1961), which has been found to range from 2.9 percent to 40 percent (the latter incidence was noted when the total dose of chloroquine exceeded 8 g per kg). All these studies indicate that the incidence of retinopathy increases with increasing total dosage, although daily dosage may also be a factor (Scherbel et al. 1965; Elman et al. 1976).

Very few instances of retinopathy have been reported in patients receiving total doses of less than 300 g, and only one case in a patient receiving a total dose of less than 100 g (Nylander 1967). Few cases have been reported in patients receiving less than 250 mg daily (the usual anti-inflammatory dose), even when the total dosage has been several hundred grams (Elman et al. 1976). Also, few if any cases of chloroquine retinopathy have been reported in patients with malaria, because in treating that illness the drug is used only for a brief period of time.

One of the problems with chloroquine is its great affinity for melanin-containing structures in the eye, such as the pigment epithelium (Bernstein et al. 1963; Kinross-Wright 1956; Zvaifler, Bernstein, and Rubin 1962). Experimental studies on rats given chloroquine have shown that the concentration in pigmented tissues of the eye—such as the iris and choroid—is up to eighty times what it is in nonpigmented ocular and body tissues (Bernstein et al. 1963). The toxicity of chloroquine is enhanced because the concentration is cumulative over a long period of time—metabolites of chloroquine have been detected in plasma, red blood cells, and urine as long as five years after the drug has been discontinued (Rubin, Bernstein, and Zvaifler 1963). A single systemic dose of chloroquine has been found to be absorbed immediately by the pigment epithelium, and the drug can remain with that cell for a year or more (Lindquist and Ulberg 1972). Chloroquine can also freely cross the placenta and become heavily concentrated within the melanin-containing tissues of the fetus (Lindquist and Ulberg 1972).

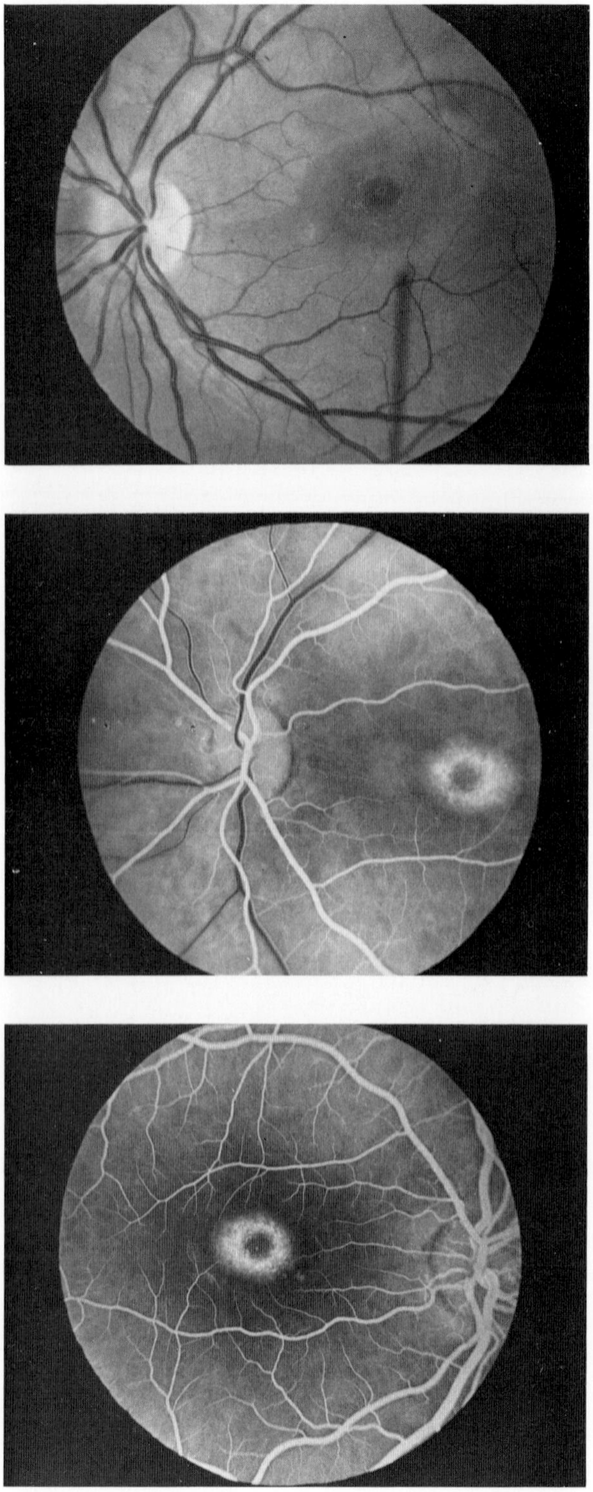

A

B

C

Patients with chloroquine toxicity often present clinically with decreased visual acuity, and visual field examination may show central, paracentral, or peripheral scotomata (Hobbs, Sorsby, and Freedman 1959). The ophthalmoscopic findings in early chloroquine retinopathy consist of a loss of the foveal reflex and increased pigmentation of the macula, but in the later stages, there is a concentric ring of depigmentation surrounded by a ring of increased pigmentation (Figs. 22.1–22.3). This pattern is com-

← **Figure 22.1** *Patient with chloroquine retinopathy.* A, *fundus photograph. Note the altered pigmentation of the macular zone (target lesion).* B, *fluorescein angiogram, early venous phase. Note the depigmented ring within the macula.* C, *fluorescein angiogram, late arteriovenous phase, showing ring of depigmentation within the macular zone. (K. M. Zinn.)*

monly referred to as a "bull's-eye lesion." Arteriolar narrowing, vascular sheathing, and widespread peripheral pigmentation and atrophy may be seen in severe cases. Fluorescein angiography shows the

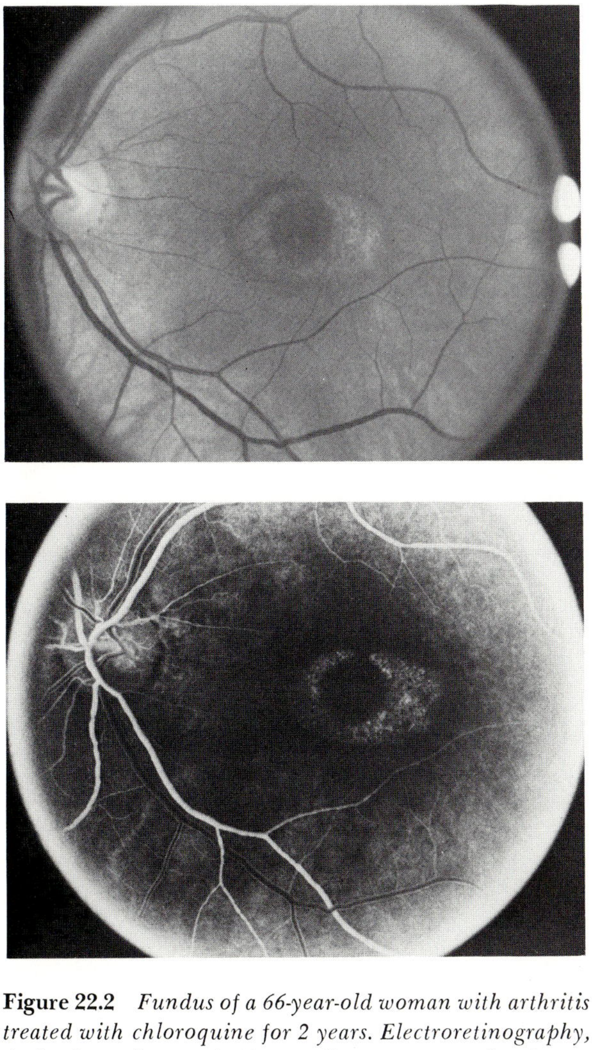

A

B

Figure 22.2 *Fundus of a 66-year-old woman with arthritis treated with chloroquine for 2 years. Electroretinography, dark adaptation, and EOG studies were normal. The patient had a central scotoma with sparing of a small foveal island, allowing a visual acuity of 20/20 O.U. A mild tritan color defect was also noted.* A, *fundus photographs.* B, *fluorescein angiogram, arteriovenous phase, showing the pigment epithelial defect associated with the bull's-eye lesion. (M. F. Marmor.)*

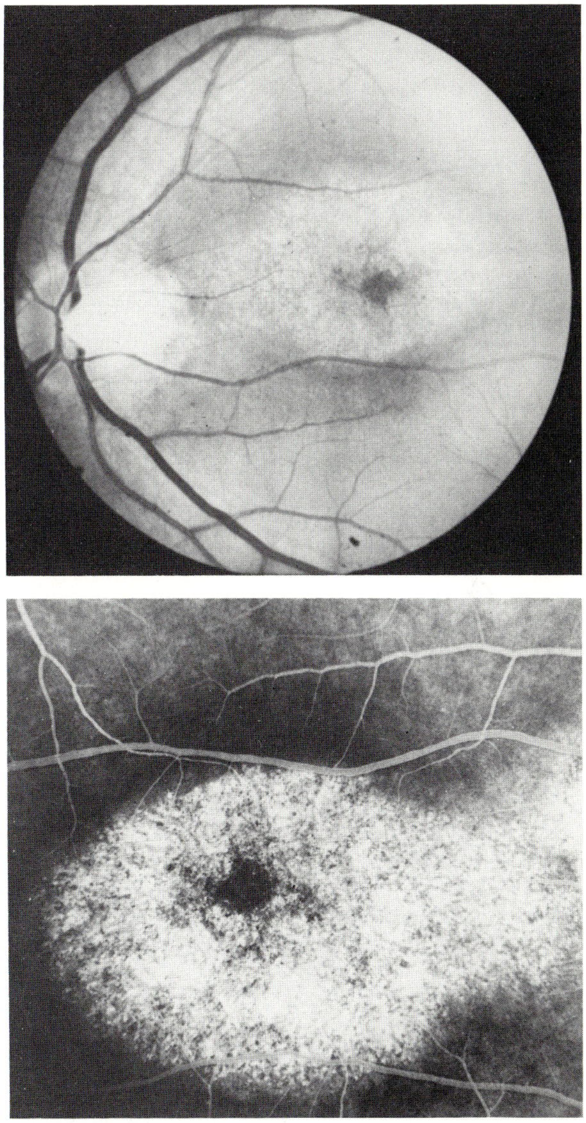

A

B

Figure 22.3 *Fundus of a 46-year-old Caucasian male with lupus erythematosus treated with chloroquine for many years. The patient had scotomas and a visual acuity of 20/60.* A, *fundus photographs.* B, *fluorescein angiogram showing bull's-eye lesion with a ring of depigmentation involving the RPE in the macular zone. (Courtesy of R. F. Carr.)*

pigmentary changes in the macula better than ophthalmoscopy does (Figs. 22.1–22.3).

The functional changes produced by chloroquine retinopathy depend upon the extent and severity of retinal and pigment epithelial damage. In early retinopathy there may be little functional loss except a small subtle ring scotoma, or a mild color deficiency. Severe retinopathy may be indistinguishable from retinitis pigmentosa. Field loss may be variable and may mimic other diseases (Goldhammer and Smith 1974). Color vision may be normal, but is often affected (Earnshaw, Miles, and Stewart 1966). Dark adaptation is sometimes normal even in the late stages of the disease, in contrast to retinitis pigmentosa (Carr, Gouras, and Gunkel 1966; Krill, Potts, and Johanson 1971), but deficient dark adaptation has also been reported (Okun et al. 1963). The electroretinogram (ERG) is normal in mild cases in which the macula is primarily involved, but it reveals progressive reduction of the b-wave, and eventually the a-wave, in more severe cases. Fluorescein angiography may be of value in picking up early retinal pigment epithelium depigmentation (Kearns and Hollenhorst 1966) (Figs. 22.1–22.3).

The EOG is often diminished in severe chloroquine retinopathy, but unless pretreatment baseline studies have been performed, the results of the EOG are of limited value because of the wide variation in normals, and because the EOG is a "mass response" insensitive to limited (for example, macular) lesions. Some studies have suggested that patients taking chloroquine, even though they do not show ophthalmoscopic evidence of retinopathy, may have lowered EOG values that revert to normal upon discontinuation of the drug (Arden, Friedman, and Kolb 1962; Kolb 1965), but other studies give contradictory results (Gouras and Gunkel 1963; Henkind, Carr, and Siegel 1964; van Lith, Mak, and Wijnands 1976). Recent studies on rhesus monkeys (Rosenthal et al. 1978a) have shown histologic damage in early toxicity before any ophthalmoscopic or electrophysiologic changes were evident. Our own experience indicates that no single physiologic test is uniquely diagnostic of chloroquine toxicity.

Once ocular toxicity with chloroquine has developed, it usually does *not* regress even if the drug is withdrawn. In fact, the toxicity may progress despite the discontinuation of chloroquine. Nonetheless, there have been several reports concerning the reversal of depressed electrooculogram (EOG) levels and early visual acuity changes in cases with early macular pigmentation alterations (Arden, Friedman, and Kolb 1962; Crews 1964; Nozik, Weinstock, and Vignos 1964).

The treatment for chloroquine retinopathy is discontinuation of the drug. Administration of ammonium chloride or dimercaprol has been found to enhance the urinary excretion of chloroquine, but has not been shown to be of any significant value in reversing the retinal damage with visual loss. Therefore, prevention appears to be the best form of treatment at the present time (Leopold 1968).

The histopathology of human chloroquine retinopathy, as shown by light microscopy (Bernstein and Ginsberg 1964; Wetterholm and Winter 1964), consists of a loss of melanin material from the retinal pigment epithelium, with subsequent migration of this pigment to the outer nuclear layer of neural retina. There is damage to the photoreceptor cells as well as to the outer nuclear layer, while the ganglion and bipolar cells look normal. However, electron microscopy (Ramsey and Fine 1972) shows that the most striking change is the presence of membranous cytoplasmic bodies in the ganglion cells (see below) and clusters of curvilinear tubules (which may represent early stages in the formation of membranous cytoplasmic bodies) in other neurons, photoreceptor inner segments, pigment epithelium, and striated muscle cells outside the eye. Similar changes can be produced experimentally in the monkey with chronic chloroquine administration (Rosenthal et al. 1978a).

Exposure to chloroquine can stabilize lysosomes (Weissmann 1966), thereby impairing their digestive capacity (Abraham and Hendy 1970). This direct action of chloroquine on lysosomal membranes also involves an alteration in their permeability as well as interference with the transport of enzymes into the lysosomes (Abraham and Hendy 1970; Abraham, Hendy, and Grasso 1968). However there is evidence that chloroquine does not inhibit the synthesis of lysosomal enzymes (Abraham and Hendy 1970).

Chloroquine administered to experimental animals has been reported by some authors to produce membranous cytoplasmic bodies within retinal ganglion cells and the retinal pigment epithelium (Gleiser et al. 1969; Ramsey, Bloodworth, and Engerman 1970). However, in the rat Hodgkinson and Kolb (1970) found these cytoplasmic bodies only in retinal neurons, and not in the pigment epithelium. Gregory, Rutty, and Wood (1970) found similar changes in albino as well as pigmented rats. In monkeys receiving toxic doses of chloroquine (Rosenthal et al. 1978) ganglion cell and photoreceptor degeneration is observed before any degeneration of the pigment epithelium. Most investigators believe that the membranous cytoplasmic bodies are degenerative forms of lysosomes and are identified as such by their acid phosphatase activity (Abraham and Hendy 1970). However, this activity is not very specific and has been detected in melanosomes (Abraham and Hendy 1970) and endoplasmic reticulum (Fishman, Goldman, and Delellis 1967). Chloroquine may also induce hyperplasia of the retinal pigment epithelium and may be responsible for a total increase in the number of melanin granules per cell (Bernstein and Ginsberg 1964; Lindquist and Ulberg 1972; Meier-Ruge 1965; Monohan and Horns 1964).

Several modes of action have been proposed to explain chloroquine-induced retinopathy. Most early investigators felt that the pigment epithelium was primarily involved. Binding of the drug to melanin with damage to the pigment epithelium and subsequent degeneration of the rods and cones has been suggested on the basis of histopathologic changes (Bernstein and Ginsberg 1964; Monohan and Horns 1964; Ramsey, Bloodworth, and Engerman 1970). Rubin and Slonicki (1966) have proposed that chloroquine reduces enzyme activity and protein synthesis in melanosomes, while Gonasun and Potts (1974) have showed that chloroquine (as well as hydroxychloroquine and quinine) inhibits general protein synthesis in pigment epithelial cells.

Recent articles have questioned the role of melanin binding and direct pigment epithelial toxicity in chloroquine toxicity. Abraham and Hendy (1970) have demonstrated that chloroquine can produce irreversible lysosomal damage in retinal bipolar and ganglion cells, and Bednarczyk and Medzihradsky (1977) have showed that an isolated retina can accumulate chloroquine. Gregory, Rutty, and Wood (1970) noted that even albino rats can develop retinopathy. Berson (1970) injected chloroquine directly into the vitreous cavity and was able to produce irreversible damage to the photoreceptor cells in animals with pigmented and nonpigmented RPE. Giuffrida et al. (1976) found that chloroquine inhibited nucleic acid and protein synthesis in isolated rabbit retina, although they (Karlsson et al. 1976) failed to demonstrate inhibition by vitreous injection in vivo. Finally, ultrastructural changes, predominantly in the ganglion cells of a human eye with chloroquine retinopathy (Ramsey and Fine 1972) and of monkeys with early retinopathy (Rosenthal et al. 1978a), also provide evidence that the primary damage is retinal. Nonetheless, a question still remains whether melanin binding can act as a reservoir of toxic material, or can serve to protect adjacent tissues by absorbing the toxin, or whether its relationship to retinal toxicity is coincidental.

Hydroxychloroquine (plaquenil) is an amino quinoline derivative that is related to amodiaquine and chloroquine. It is used in the treatment of systemic lupus erythematosus, scleroderma, and rheumatoid arthritis, as well as in malaria. Several authors (Henkind, Carr, and Siegel 1964; Scherbel et al. 1965; Shearer and Dubois 1967) have noted that hydroxychloroquine can produce pigmentary retinopathy, although its toxicity relative to chloroquine is hard to establish since the incidence of retinopathy with either drug is low in modern series in which the dosages are kept low. High doses of hydroxychloroquine will produce a retinopathy in rats (Legros and Rosner 1971).

Diiodohydroxyquin (diodoquin: 5, 7-diiodo-8-quinolinol), is a quinoline derivative used in amebiasis and trichomoniasis. Etheridge and Steward (1966) reported pigmentary retinopathy in a young boy being treated for "acrodermatitis enteropathica" with a dose of 3.2 grams per day for more than two years. There was an associated nystagmus and pallor of the optic nerve heads.

Quinine is an alkaloid extracted from cinchona bark and is primarily used as an antimalarial compound. For more than 100 years it has been known that quinine has the capacity to cause acute tinnitus, deafness, and blindness (Grant 1974). Visual loss typically develops within a few hours of ingesting a large dose, and at this time the ERG is usually normal. Over the next few days vision may improve, but arterial narrowing becomes evident, the ERG becomes abnormal, and optic atrophy begins. The final acuity and the late changes in the ERG, visual field, or both, are variable.

The mechanism of action of quinine toxicity is still controversial, but has classically centered on either a direct effect on retinal ganglion cells (Francois, Verriest, and DeRouck 1966, 1967), or on the later constriction of the retinal vessels. Potts (1964) has shown that quinine, in vitro, can bind to melanin in the same way that chloroquine and phenothiazine derivatives do, but whether this binding relates to the drug's toxicity is unknown. There is also increasing evidence for a direct toxic effect on the RPE and photoreceptors during the acute amblyopic phase. The EOG may be depressed acutely (Behrman and Mushin 1968) and then recover later when the ERG begins to fail. Experimental work has shown that acute intoxication with quinine can also alter the c-wave (Jünemann and Schulze 1968; Calissendorf 1976a). When the ERG has been recorded within hours of ingestion, the ERG can be markedly reduced, only to recover and then fall again several days later (Hommer 1968). This acute effect on the ERG has been confirmed experimentally (Cibis, Burian, and Blodi 1973) and suggests that the initial toxicity may be to the outer retina with the effects on ganglion cells being secondary or independent.

Ethylhydrocupreine hydrochloride (optochin, numoquin) is a chemical modification of quinine and was formerly used systemically to treat pneumonia. This agent was also used in solutions to treat pneumococcal eye infections topically. The toxic effects are similar to those from quinine (Grant 1974).

Phenothiazines

Phenothiazines have been used widely for many years as tranquilizers and antiemetics, as well as for their anesthetic-potentiating effects. However, the consumption of the phenothiazine derivatives is reaching alarming proportions in the United States. It has been estimated than in 1970 alone, over 1,260,000 pounds of phenothiazines were consumed (Lerner 1972). Furthermore, the affinity of intracellular melanin for phenothiazine derivatives has been well established (Cambiaggi 1957; Potts 1962a, b). These agents—chlorpromazine, for example—can freely cross the placenta (Lindquist and Ulberg 1972) and become heavily concentrated within melanin-containing tissues, such as the pigment epithelium. In fact, a single systemic dose of a phenothiazine can result in almost immediate uptake of the drug by the pigment epithelium, and the drug may remain with the cell up to a year or more (Lindquist and Ulberg 1972). It is not surprising, therefore, that toxic retinopathies involving some of the phenothiazines have been reported clinically (Edler 1966; Kinross-Wright 1956; Siddal 1966). What is surprising is that most of the commonly used phenothiazines are *not* significantly retinotoxic and that retinotoxicity is *not* related to the relative affinity of the drugs for melanin (Weekley et al. 1960).

The first report of toxic retinopathy with a phenothiazine compound was in 1956 when Kinross-Wright (1956) described a retinal pigmentary degeneration with NP-207 (2-chloro-10-[2-(N-methylpiperidyl)ethyl]-phenothiazine). In 1960 (Weekley et al.) a pigmentary retinopathy was described in patients on thioridazine (2 methylmercapto-10-[2-N-methyl-2-piperidyl-ethyl]-phenothiazine). For all of the other commonly used phenothiazines, such as chlorpromazine and trifluoperazine, reports of retinopathy are sparse and questionable. Thus, the action of the phenothiazines in the production of toxic retinopathy cannot be ascribed simply to the drug binding to melanin. Perhaps the piperidyl side chain, present in NP-207 and thioridazine, but not in others, is the critical toxic moiety.

Thioridazine (Mellaril) is a psychotherapeutic piperidyl phenothiazine derivative that has been used widely in the treatment of schizophrenia. The complication of pigmentary retinopathy with visual loss was first reported by Weekley et al. (1960). The toxic effects are clearly related to excessive dosage—for example, a daily dosage exceeding 1000 mg/day (Grant 1974), although Heshe (1961) and Reboton et al. (1962) have reported single cases of retinopathy with dosages as low as 700–750 mg/day. Dosages up to 800 mg/day are generally accepted as being safe (Appelbaum 1963; deMargerie 1962; Forrest and Snow 1968; Hagopian 1966). Higher doses should not be given. Marmor (unpublished data) has followed several patients in whom retinopathy developed within 1–2 months after doubling or tripling the recommended 800 mg dose.

The characteristic pigmentary retinopathy associated with thioridazine toxicity (Davidorf 1973) begins with coarse black pigmentary stippling throughout the posterior fundus, including the macula and sometimes extending (Figs. 22.4*A*, 22.5) to the midperipheral region. The granules of pigment may be larger than the diameter of retinal vessels and they produce a rather unique appearance that cannot be mistaken for retinitis pigmentosa or a bull's-eye maculopathy. With time, even after discontinuation of the drug, these pigment granules coalesce to form large irregular areas of heavy pigmentation or of depigmentation (Fig. 22.4*B*). The end stage resembles in appearance a disseminated chorioretinitis, and it may be accompanied by atrophic changes such as vascular narrowing if the damage is severe.

Pigmentary alterations can precede the disturbance of visual acuity, but are rarely observed this early. Toxicity is usually heralded by a complaint of blurred vision. If the drug is stopped promptly, the vision will often return to normal, possibly leaving some isolated scotomata. However, prolonged exposure to high doses of thioridazine produces irreversible visual loss (although partial recovery may occur). The pigmentary changes are not reversible, even in mild cases, and they progress to a patchy

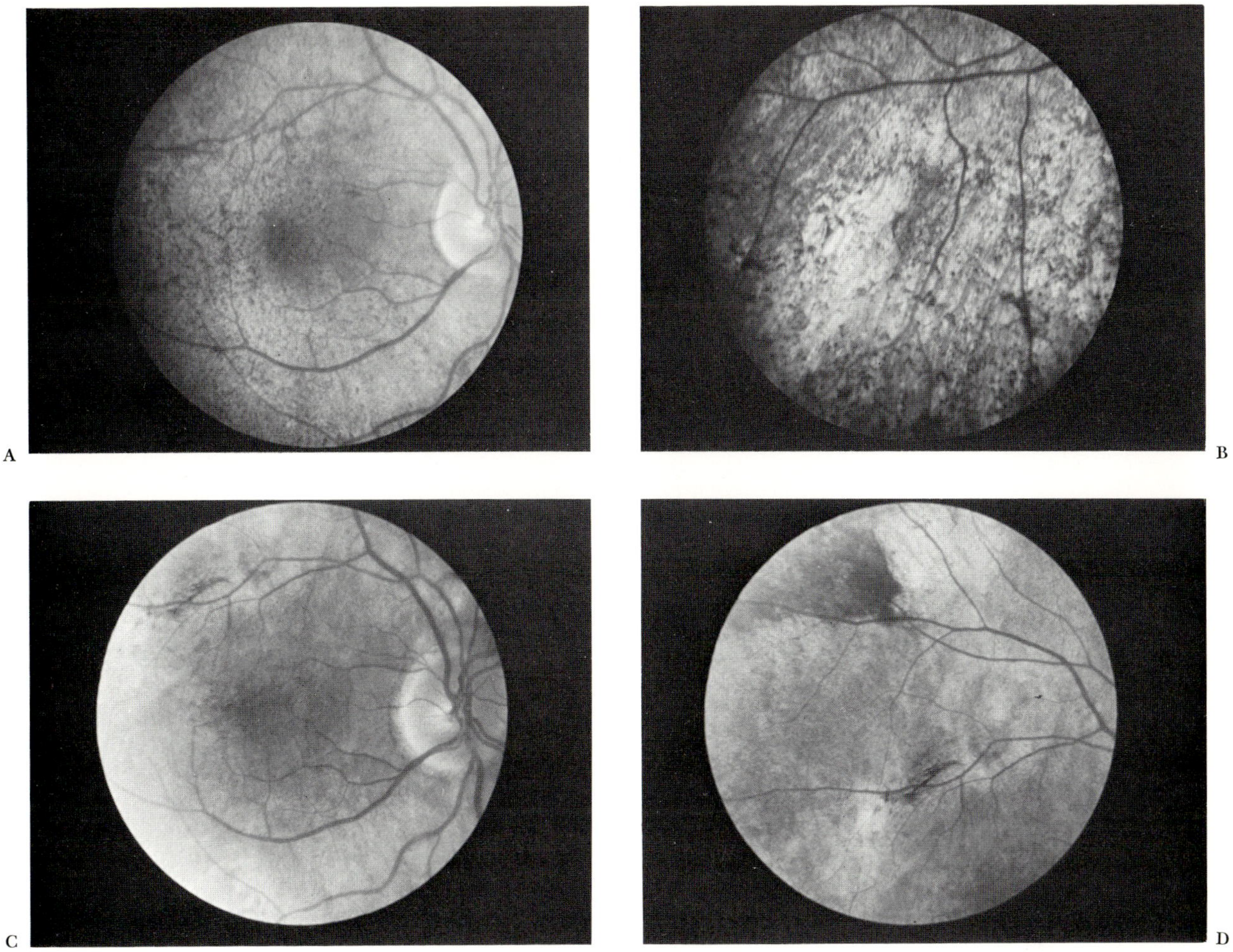

Figure 22.4 *Fundus photographs of a 36-year-old woman who had taken 900–1,200 mg of Mellaril per day for a 2-year period.* A, *right macula; note the coarse granular pigment deposits. Visual acuity was 20/200. The patient had a small 1.5 degree central field remaining and a markedly reduced ERG.* B, *pigmentary changes in the retinal periphery.* C, *right macula, seven years later, showing dispersion of retinal pigmentation. Visual acuity has improved to 20/25 and the central field is 5 degrees.* D, *periphery, seven years later, showing bare atrophic patches and dense black patches. (M. F. Marmor.)*

geographic atrophy of the RPE despite the recovery of visual function (Fig. 22.4).

There is no specific treatment for thioridazine retinopathy other than discontinuation of the drug. Since chloroquine and its derivatives, and the piperydyl-phenothiazines, both produce pigmentary degenerations, many comparisons have been drawn between the two classes of drugs (see Table 22.1). One important difference is that thioridazine retinopathy occurs after a relatively short period of time, during which elevated dosages are used, whereas chloroquine retinopathy occurs after prolonged exposure to the drug. Chloroquine toxicity is also far less reversible than thioridazine retinopathy.

Physiologic tests, including those of color vision, dark adaptation, ERG, and EOG, may all be affected by thioridazine toxicity. Measures of rod function may be relatively more effective than those of cone function, unless damage is severe (Tamai and Holland 1975). In general, however, depression of the ERG and EOG correlate with visible damage to the retina and pigment epithelium, and correlate with psychophysical tests such as those of dark adaptation and of visual fields. All of these parameters may improve if the drug is stopped soon after retinopathy appears, despite evolution of the pigmentary lesions. In severe cases the ERG and EOG can be irreversibly extinguished. None of these tests,

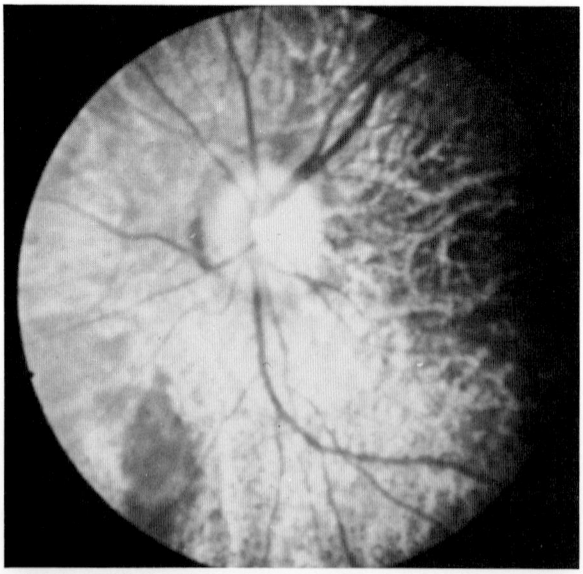

A

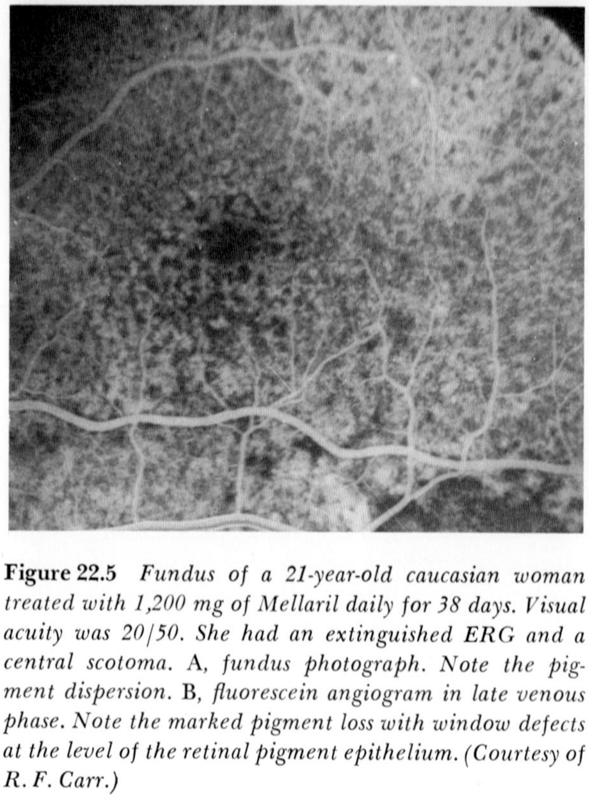

B

Figure 22.5 *Fundus of a 21-year-old caucasian woman treated with 1,200 mg of Mellaril daily for 38 days. Visual acuity was 20/50. She had an extinguished ERG and a central scotoma. A, fundus photograph. Note the pigment dispersion. B, fluorescein angiogram in late venous phase. Note the marked pigment loss with window defects at the level of the retinal pigment epithelium. (Courtesy of R. F. Carr.)*

including the EOG, are selectively sensitive to thioridazine, and most cases will be discovered because of visual complaints or pigmentary changes rather than physiologic tests.

Little is known about the pathophysiology of thioridazine retinopathy, because it has not been satisfactorily produced in animals (Meier-Ruge and Cerletti 1967; Gregory, Rutty, and Wood 1970). However, a similar compound, NP-207, is retinotoxic to animals (as well as humans) and the mode of action is presumed to be similar (see discussion of NP-207 below).

Acute or in vitro studies have shown that thioridazine has a high affinity for melanin (though less than chlorpromazine has) and can abolish the b-wave of the ERG in isolated retina (Meier-Ruge, Kalberer, and Cerletti 1966; Potts 1964; Lutzow and Hommer 1969). In vitro thioridazine inhibits succinic dehydrogenase and activates lactic dehydrogenase, being more effective than chlorpromazine, but less effective than NP-207 (Cerletti and Meier-Ruge 1968). These authors conclude that inhibition of oxidative phosphorylation in the visual cells is the critical event in phenothiazine toxicity. The metabolic products—thioridazine sulfoxide and sulfone—were noted by Meier-Ruge and Werthemann (1967) to have only slight effects on retinal enzymes.

Chlorpromazine (Thorazine, Largactil, Megaphen) was the first widely used psychotropic phenothiazine derivative for the treatment of schizophrenia. When the total dose exceeds roughly 1 kg, pigmentation of the skin, conjunctiva, posterior cornea, and anterior lens may occur (Wetterholm, Snow, and Winter 1965; Siddall 1965). The lens changes in particular may be quite striking, but they very rarely affect vision and may slowly regress if the drug is stopped. Despite the high dose levels of chlorpromazine that are used, there is only marginal evidence that the drug can cause a significant pigmentary retinopathy. Siddall (1966) reported four cases with "fine clumping" of pigment, each of whom had taken about 2400 mg/day. However, there are few cases in the literature, considering the enormous numbers of patients using high doses of chlorpromazine. Boet (1970) reported a high incidence of "retinopathy," but defined the term to include subtle changes that would seem of arguable specificity. Marmor (unpublished data) has examined a number of patients who were reported ten years ago to have had heavy lens and corneal pigmentation from chlorpromazine (Forrest and Snow 1968). All were still using chlorpromazine, or other phenothiazines, and all had cumulative doses of many kilograms, but none had visual loss or significant pigmentary changes in the fundus.

Experimentally, Potts substantiated Green, Meier-Ruge, and Cassano's work indicating that chlorpromazine is bound to melanin in high amounts

Table 22.1 Toxic retinopathies: comparison of findings in *moderately severe* chloroquine and thioridazine toxicity.

	Chloroquine	*Thioridazine*
Dosage	Rare with ⩽300 gm total dose or ⩽250 mg/day	Rare with ⩽800 mg/day
Duration of exposure	Typically years	Typically months
Visual acuity	Decreased	Decreased
EOG light response	Usually decreased	Usually decreased
ERG	Usually decreased	Usually decreased
Dark adaptation	Occasionally remains normal in late cases	Usually abnormal
Ophthalmoscopy	Bull's-eye lesion of the macula with peripheral degenerative changes	Macular and paramacular mottling, with coarse irregular patches of depigmentation and hyperpigmentation
Color vision	Abnormal	Abnormal
Visual field	Central or ring scotomas, with variable peripheral loss	Central and peripheral scotomas
Course	Improves only if drug is discontinued very early	Some, and frequently marked, improvement after discontinuation of medication

(Green et al. 1966; Cassano et al. 1968; Meier-Ruge, Kalberer, and Cerletti 1966; Potts 1962a, b, 1964). However, the relationship between this binding capacity and retinal function remains unclear. Legros, Rosner, and Berger (1973) found that chronic administration of chlorpromazine reduced the ERG in albino but not in pigmented rats. Bornschein et al. (1967) and Calissendorff (1976b) demonstrated an acute reduction of the ERG b-wave in experimental chlorpromazine administration. Chlorpromazine has been shown to inhibit ribosomal RNA synthesis (Breitbart, Mayevsky, and Diamant 1973) as well as to increase the conversion of tyrosine to dopamine (Gey and Burkard 1969; Gey and Pletschner 1968). Other biochemical studies have indicated that chlorpromazine can uncouple oxidative phosphorylation within mitochondria (Abood 1955) as well as inhibit oxidation within the tricarboxylic acid cycle (Berger 1957). The relationship of those effects to possible retinal toxicity is unknown.

Piperidyl chlorophenothiazine (NP-207) was tried as a major tranquilizer for the treatment of various psychotic conditions, but it was soon noted to have the serious side effect of producing a retinal degeneration (Burian and Fletcher 1958; Goar and Fletcher 1956; Kinross-Wright 1956; Rintelen, Hotz, and Wagner 1957; Verrey 1956; Wagner 1956). The degeneration appeared to be dose-related (Goar and Fletcher 1956), and patients receiving 0.4 to 0.8 grams per day of NP-207 for several months had visual symptoms consisting of decreased vision in dim light with eventual decrease in visual acuity and constriction of the visual fields. Abnormal retinal pigmentation developed, similar to that described for thioridazine. Changes on psychophysical and ERG testing ranged from minimal to severe, depending on the extent of toxic damage (as with thioridazine).

The major interest in NP-207 toxicity today is as a model for thioridazine, since their clinical effects and structural similarity suggest a similar mode of action, but only NP-207 retinopathy has been produced in animals. The mechanism of action by which NP-207 affects the retina was studied by Cerletti and Meier-Ruge (1968), Gregory, Rutty, and Wood (1970), and others. These investigators found that NP-207 binds to melanin, as do other phenothiazines. Retinotoxicity could be produced by chronic administration of the drug to cats, but not rabbits, rats, or dogs. Histochemical studies indicated that the main toxic effect of NP-207 was on the ellipsoids of the rods, with inhibition of lactic dehydrogenase, succinic dehydrogenase, and ATPase. Histologic damage to rod outer and inner segments was evident as the retinopathy progressed (largely sparing the cones) and the RPE accumulated lipid debris and lamellar inclusions, presumed to reflect the accelerated shedding of outer segment material. This storage of excess material in the RPE seemed to contribute to the pigmentary retinopathy. The critical mechanism in the toxicology of NP-207 may be an inhibition of oxidative phosphorylation secondary to inhibition of flavine nucleotides (Cerletti and Meier-Ruge 1968).

Trifluoperazine (Stelazine) and other non-piperidyl phenothiazines are widely used in psychiatric practice. One possible case of trifluoperazine retinopathy has been reported (Reboton et al. 1962), but for the most part the nonpiperidyl derivatives are nontoxic to the retina or pigment epithelium by clinical standards.

Antibiotics

Cephaloridine (Loridine) is a broad spectrum antibiotic that has been tested in dogs without showing any apparent retinal toxicity (Atkinson et al. 1966). However, Ballingall and Turpie (1967) reported cephaloridine toxicity in a teenage girl with staphylococcal septicemia, acute renal failure, and blindness. The retinal toxicity was in the form of abnormal pigmentation of the retinal periphery. Papilledema was also noted. The patient had received other drugs, including fusidic acid and cloxacillin; however, none of these agents is known to be retinotoxic (Crosbie 1968).

Sparsomycin is an antibiotic derived from *Streptomyces sparsogenes* and is used as a cancer chemotherapeutic agent. McFarlane, Yanoff, and Scheie (1966) reported two cases of sparsomycin retinal toxicity. In each case, the patient was given sparsomycin intravenously and two weeks later developed pigmentary retinopathy along with ring scotomas. Both patients died two to three months after treatment was begun and on histological examination the primary toxic effect apparent in the eyes was in the retinal pigment epithelium, with secondary damage to the rods and cones.

Aminophenoxyalkanes are a series of compounds that have schistosomicidal activity and toxicologic significance because they seriously affect the metabolism of the RPE (Edge et al. 1956). Most of the work with these compounds has been done on animals; there is a primary degeneration of the retinal pigment epithelium followed by secondary impairment of neurosensory retina, along with destruction of the outer limiting membrane and intraretinal pigment migration in the cat (Ashton 1957) and the rabbit (Sorsby and Nakajima 1958). Histologically, the cellular changes resemble those induced by sodium iodate poisoning rather than those caused by iodoacetic acid.

Diaminodiphenoxypentane, when given to monkeys, dogs, and cats, results in an initial visual loss and optic pallor. In about two months there is a pigmentary retinopathy resembling retinitis pigmentosa. Histopathology studies (Edge et al. 1956) demonstrate degeneration of the retinal pigment epithelium, outer nuclear layer, visual cells, and outer limiting membrane. The bipolar cells, ganglion cells, and optic nerve appear unaffected (Ashton 1957). Studies by Orzalesi et al. (1967) indicate that the RPE is the primary site of damage and rhodopsin synthesis is impaired (Grignolo 1969; Orzalesi et al. 1967). ERG studies have shown alteration of the c-wave, confirming the selective actions of these compounds on the RPE (Arden and Fojas 1962). Moreover, there is a decrease in the rate of dark adaptation in DDA-treated animals, which has been ascribed to an effect on the biosynthesis of retinal photopigments (Arden and Fojas 1962).

Diaminodiphenoxyheptane is another of the aminophenoxy alkanes with schistosomicidal activity that has the toxic effect of reducing the incorporation of radioactive phosphorus into organic phosphates in the retinal pigment epithelium (Glocklin and Potts 1962). This blocking effect is not present in neurosensory retina. Furthermore, it is thought that the toxic action is at the enzyme or cell membrane level.

*1-(*p*-Aminophenoxy-5-phthalimido) pentane* has some retinotoxicity in cats, but Collins has stated that McFadzean reported three patients in which this drug reduced the visual fields when used in the treatment of schistosomiasis (Collins et al. 1967).

Antiinflammatory Agents

Indomethicin (Indocin) is an antiinflammatory agent used to control rheumatoid arthritis and related disorders. There have been reports by Burns (1966, 1968) and by Henkes, van Lith, and Canta (1972) indicating that prolonged usage can cause elevated dark-adaptation thresholds, visual field changes, a reduction in the ERG, a fall in the EOG light/dark ratio, and a diffuse pigmentary disturbance in the fundus. Some improvement was thought to occur on cessation of the drug. These authors infer that indomethacin is toxic to the pigment epithelium. However, other investigators (Carr and Siegel 1973) have not been able to corroborate their findings, and have questioned the existence of indomethacin retinopathy.

Metals

Iron and iron compounds have a variety of toxic effects on the eye. Our discussion will deal primarily with the effects of intraocular iron on the retinal pigment epithelium. Siderosis has been defined as the impregnation of ocular tissues with iron. Intraocular foreign bodies located in the anterior segment rarely affect the retinal pigment epithelium, but iron foreign bodies in the ciliary body or in the vitreous will damage the neurosensory retina and the retinal pigment epithelium.

One of the early signs of retinal toxicity in siderosis bulbi is hemeralopia (Wagenmann 1913), often accompanied by visual field constriction. There is also a progressive loss of central visual acuity, which can end in blindness after several months to years. The fundus may show generalized atrophy or a proliferation of the peripheral retinal pigment epithelium, resembling retinitis pigmentosa (Karpe 1948; Wagenmann 1913). The macula often shows a loss of the normal reflex and an irregular fine pigmentation.

Diagnosis of an intraocular foreign body can be established through soft tissue x-rays and A and B scan ultrasonography. Analysis of aqueous humor for iron can be useful (Baron and Michel 1969; Chechan, Francois, and Hache 1968). Abnormalities in the ERG may precede visual defects in ocular siderosis (Karpe 1948). The ERG may increase transiently in early siderosis (Knave 1969), but later the ERG falls, and as in retinitis pigmentosa, it may be depressed or negative even with good central vision remaining (Cibis, Brown, and Hong 1957; Cleasby 1966). Thus, an abnormal ERG may serve as an important indicator for treatment, with the reservation that the effects of nonspecific trauma and inflammation must be accounted for. Serial testing is more reliable than any single result. If the ERG is severely abnormal, irreversible changes may already have taken place, indicating a poor visual prognosis (Calmettes, Deodati, Bec, and Delpech 1965; Franchino 1967; Gorgone 1966). Knave (1969) has reviewed the effects of intraocular metals on the human ERG.

Cibis et al. (1957, 1959) studied the histopathology of ocular siderosis and demonstrated iron in the retina, especially in the perivascular spaces and in the external limiting membrane, causing narrowing of the retinal vessels. They noted degeneration of the internal and external nuclear layers and proliferation of the retinal pigment epithelium. Later investigators have found the primary histologic damage, after the introduction of iron into the vitreous, to be a degeneration of the photoreceptors, with concomitant changes in the pigment epithelium and the Müller cells (Barber, Catsulis, and Cangelosi 1971; Masciulli, Anderson, and Charles 1972; Yoo 1976). The pigment epithelium may contain large amounts of stainable iron before the overlying retina shows degenerative changes (Declercq, Meredith, and Rosenthal 1977). Metallic and ferrous iron seems more toxic than the ferric form (Grant 1974), possibly because the oxidative process itself is toxic (Declercq, Meredith, and Rosenthal 1977).

Wise (1966) believed that iron ions bind to acid mucopolysaccharides and damage enzymes. There is some recent evidence (Hiramitsu et al. 1976) that iron may induce formation of lipoperoxides, which could damage the visual cells.

Copper has often been described as causing progressive damage to the retina (Grant 1974), but recent data suggest that the effects of copper are very different from those of iron, and that copper may often be tolerated within the eye. Large copper foreign bodies excite a severe inflammatory response that probably accounts for much of their apparent toxicity (Rosenthal, Appleton, and Hopkins 1974). Smaller foreign bodies are variable in their inflammatory effect, and usually cause few or no electrophysiologic changes unless inflammation is also present (Zeimer et al. 1978). Histochemically, copper deposits throughout the retina where it may cause local areas of atrophy, but does not deposit in the pigment epithelium (Rosenthal and Appleton 1975). We have recently found (Rosenthal et al. 1978b) that small intravitreal copper foreign bodies in humans produce little or no change in visual function or electrophysiologic parameters over several years of follow up, unless vitreous inflammation develops. The presence of lenticular and corneal staining has confirmed the dissemination of copper in these eyes.

BERSON, E. L. 1970. Acute toxic effects of chloroquine on the cat retina. *Invest. Ophthalmol.* 9:618–628.

BOET, D. J. 1970. Toxic effects of phenothiazines on the eye. *Doc. Ophthalmol.* 28:1–69.

BORNSCHEIN, H., HEISS, W. D., HEILIG, P., and SANDBACH, G. 1967. Die Wirkung von Chloroquine auf das retinale System. *Wien. Z. Nervenheilkd.* 25:137–142.

BREITBART, H., MAYEVSKY, M. P., and DIAMANT, E. Y. 1973. Brain ribosomal RNA synthesis inhibited by chlorpromazine. *Proc. Soc. Exp. Biol. Med.* 143:204–207.

BROWN, D. V. L. 1975. Reaction of the rabbit retinal pigment epithelium to systemic lead poisoning. *Trans. Am. Ophthalmol. Soc.* 72:404–447.

BULLOCK, J., and ALBERT, D. 1975. Flecked retina: appearance secondary to oxalate crystals from methoxyflurane anesthesia. *Arch. Ophthalmol.* 93:26–31.

BULLOCK, J. D., ALBERT, D. M., SKINNER, C. W., MILLER, W. H., and GALLA, J. H. 1974. Calcium oxalate retinopathy associated with generalized oxalosis: x-ray diffraction and electron microscopic studies of crystal deposits. *Invest. Ophthalmol.* 13:256–265.

BURIAN, H. M., and FLETCHER, M. C. 1958. Visual functions in patients with retinal pigmentary degeneration following the use of NP 207. *Arch. Ophthalmol.* 60:612–629.

BURNS, C. A. 1966. Ocular effects of indomethacin: Slit lamp and electroretinographic (ERG) studies. *Invest. Ophthalmol.* 5:325 (abstr.).

BURNS, C. A. 1968. Indomethacin: reduced retinal sensitivity and corneal deposits. *Am. J. Ophthalmol.* 66 (5): 825–835.

CAINE, R., ALBERT, D. M., LAHAZ, M., and BULLOCK, J. 1975. Oxalate retinopathy: an experimental model of a flecked retina. *Invest. Ophthalmol.* 14:359–363.

CALISSENDORFF, B. 1976*a*. Melanotropic drugs and retinal functions. I. Effects of quinine and chloroquine on the sheep ERG. *Acta Ophthalmol.* 54:109–117.

CALISSENDORFF, B. 1976*b*. Melanotropic drugs and retinal functions. II. Effects of phenothiazine and rifampicin on the sheep ERG. *Acta Ophthalmol.* 54:118–128.

CALMETTES, L., DEODATI, F., BEC, P., DELPECH, J. 1965. Siderose oculaire et ERG. *Bull. Soc. Ophtalmol. Franc.* 65(9):816–819.

CAMBIAGGI, A. 1957. Unusual ocular lesions in a case of systemic lupus erythematosus. *Arch. Ophthalmol.* 57:451–453.

CARR, R. F., GOURAS, P., and GUNKEL, R. D. 1966. Chloroquine retinopathy: early detection by retinal threshold test. *Arch. Ophthalmol.* 75:171–179.

CARR, R. E., and SIEGEL, I. M. 1973. Retinal function in patients treated with indomethacin. *Am. J. Ophthalmol.* 75:302–306.

CASSANO, G. B., SJOSTRAND, S. E., PLACIDI, G. F., and HANSSON, E. 1968. Eye distribution of (^{35}S) chlorpromazine. *Exp. Eye Res.* 7(2):196–199.

CERLETTI, A., and MEIER-RUGE, W. 1968. Toxicological studies on phenothiazine induced retinopathy. *Proc. Europ. Soc. Study Drug Toxicity* 9:170–188.

CHECHAN, C., FRANCOIS, P., and HACHE, J. C. 1968. Application de la spectrophotometrie d'absorption atomique au dosage d'éléments mettalliques dans l'humeur aqueuse. *Bull. Soc. Ophtalmol. Franc.* 68(1):113–120.

CIBIS, G. W., BURIAN, H. M., and BLODI, F. C. 1973. Electroretinogram changes in acute quinine poisoning. *Arch. Ophthalmol.* 90:307–309.

CIBIS, P. A., BROWN, E. B., and HONG, S. M. 1957. Ocular effects of systemic siderosis. *Am. J. Ophthalmol.* 44:158–172.

CIBIS, P. A., YAMASHITA, T., and RODRIGUES, F. 1959. Clinical aspects of ocular siderosis and hemosiderosis. *Arch. Ophthalmol.* 62:180–187.

CLEASBY, G. W. 1966. Siderosis iridis (photograph). *Arch. Ophthalmol.* 75(4):576.

COLLINS, R. F., COX, V. A., DAVIS, M., EDGE, N. D., HILL, J., RIVETT, K. E., and RUST, M. A. 1967. The antischistosomal and retinotoxic effects of some nuclear-substituted aminophenoxyalkanes. *Br. J. Pharmacol.* 29(2):248–258.

CREWS, J. J. 1964. Chloroquine retinopathy with recovery in early stages. *Lancet* 2:436–438.

CROSBIE, R. B. 1968. Cephaloridine toxicity. *Lancet*, Feb. 24, p. 422.

DAVIDORF, F. H. 1973. Thioridazine pigmentary retinopathy. *Arch. Ophthalmol.* 90:251–255.

DECLERCQ, S. S., MEREDITH, C. A., ROSENTHAL, A. R. 1977. Experimental siderosis in the rabbit. *Arch. Ophthalmol.* 95:1051–1058.

DELANEY, W. V., JR., 1975. Presumed ocular chalcosis: a reversible maculopathy. *Ann. Ophthalmol.* 7:278–380.

DE MARGERIE, J. 1962. Ocular changes produced by a phenothiazine drug: thioridazine. *Trans. Can. Ophthalmol. Soc.* 25:160–175.

EARNSHAW, E. R., MILES, D. U., and STEWART, T. W. 1966. Screening for chloroquine retinopathy. *Br. J. Dermatol.* 78:669–674.

EDGE, N. D., MASON, D. F. J., WIEN, R., and ASHTON, N. 1956. The pharmacological effects of certain diaminodiphenoxy alkanes. *Nature* 178:806–807.

EDLER, K. 1966. Ocular changes associated with chlorpromazine therapy: a preliminary report. *Acta Ophthalmol.* (Copenhagen) 44:405–409.

ELMAN, A., GULLBERG, A. E. R., NILSSON, E., RHENDAHL, I., and WACHTMEISTER, L. 1976. Chloroquine retinopathy in patients with rheumatoid arthritis. *Scand. J. Rheumatol.* 5:161–166.

ETHERIDGE, J. E., JR., and STEWART, G. T. 1966. Treating acrodermatitis enteropathica. *Lancet* 1:261–262.

FISHMAN, W. H., GOLDMAN, S. S., and DELELLIS, R. 1967. Dual localization of glucuronidase in endoplasmic reticulum and in lysosomes. *Nature* 213:457–460.

FORREST, F. M., and SNOW, H. L. 1968. Prognosis of eye complications caused by phenothiazines. *Dis. Nerv. Syst.* 29(3):26–28 (suppl.).

FRANCHINO, M. 1967. Evoluzione eletroretinografica della siderosi del bulbo. *Rass. Ital. Ottal.* 35:356–363.

FRANÇOIS, J., VERRIEST, G., and DE ROUCK, A. 1966. Étude des functions visuelles dans deux cas d'intoxication par la quinine. *Bull. Soc. Belg. Ophtalmol.* 143:677–685.

FRANÇOIS, J., VERRIEST, G., and DE ROUCK, A. 1967. Étude des functions visuelles dans deux cas d'intoxication par la quinine. *Ophthalmologica* 153:344–355.

GEY, K. A., and BURKARD, W. P. 1969. Chlorpromazine-induced accumulation of pyridoxal-5′-phosphate and activation of the decarboxylase of acromatic amino acids in rat brain. *Ann. N.Y. Acad. Sci.* 166:213–224.

GEY, K. F., and PLETSCHNER, A. 1968. Acceleration of turnover of ^{14}C-catecholamines in rat brain by chlorpromazine. *Experientia* 24:335–336.

GIUFFRIDA, A. M., SJOSTRAND, J., CAMBRIA, A., SERRA, I., VANELLA, A., AVITABILE, M., JARLSTEDT, J., and KARLSSON, J. O. 1976. In vitro action of chloroquine on nucleic action and protein biosynthesis in the rabbit retina. *Neuropharmacol.* 15:439–441.

GLEISER, C. A., DUKES, T. W., LAWWILL, T., READ, W. K., BAY, W. W., and BROWN, R. S. 1969. Ocular changes in swine associated with chloroquine toxicity. *Am. J. Ophthalmol.* 67:399–405.

GLOCKLIN, V. S., and POTTS, A. M. 1962. The metabolism of retinal pigment cell epithelium. *Invest. Ophthalmol.* 1:111–116.

GOAR, E. F., and FLETCHER, M. C. 1956. Toxic chorioretinopathy following the use of NP 207. *Trans. Am. Ophthalmol. Soc.* 54:129–140. Also *Am. J. Ophthalmol.* 44:603–608.

GOLDHAMMER, Y., and SMITH, J. L. 1974. Bitemporal hemianopia in chloroquine retinopathy. *Neurology* 24: 1135–1138.

GONASUN, L. M., and POTTS, A. M. 1974. In vitro inhibition of protein synthesis in the retinal pigment epithelium by chloroquine. *Invest. Ophthalmol.* 13:107–115.

GORGONE, G. 1966. Importanza clinica dell'erg nella siderosi e calcosi oculare. *Boll. Oculist* 45:638–649.

GOTTLIEB, R. P., and RITTER, J. A. 1977. Flecked retina: an association with primary hyperoxaluria. *J. Pediatr.* 90:939–942.

GOURAS, P., and GUNKEL, R. D. 1963. The EOG in chloroquine and other retinopathies. *Arch. Ophthalmol.* 70:629–639.

GRANT, W. 1974. *Toxicology of the eye,* 2d ed. Springfield, Ill.: C. Thomas.

GREEN, H., and ELLISON, T. 1966. Uptake and distribution of chlorpromazine in animal eyes. *Exp. Eye Res.* 5(3):191–197.

GREGORY, M. H., RUTTY, D. A., and WOOD, R. D. 1970. Differences in the retinotoxic action of chloroquine and phenothiazine derivatives. *J. Path.* 102:139–150.

GRIGNOLO, A. 1969. Rhodopsin cycle and fine structure of rabbit retina in experimental degeneration induced by retinotoxic agents. *Exp. Eye Res.* 8:254.

HAGOPIAN, V. 1966. Pigmentary retinopathy due to thioridazine or Mellaril. *Kresge Eye Inst. Bull.* 10(1): 33–39.

HENKES, H. E., VAN LITH, G. H. M., and CANTA, L. R. 1972. Indomethacin retinopathy. *Am. J. Ophthalmol.* 73: 846–856.

HENKIND, P., CARR, R. E., and SIEGEL, I. M. 1964. Early chloroquine retinopathy: clinical and functional findings. *Arch. Ophthalmol.* 71:157–165.

HENKIND, P., and ROTHFIELD, N. 1963. Ocular abnormalities in patients treated with synthetic antimalarial drugs. *N. Engl. J. Med.* 269:433–439.

HESHE, J., ENGELSTOFT, F. H., and KIRK, L. 1961. Retinal injury developing under thioridazine treatment. *Nord. Psykiat. T.* 15:442–447.

HIRAMITSU, T., MAJIMA, Y., HASEGAWA, Y., HIRATA, K., and YAGI, K. 1976. Lipoperoxide formation in the retina in ocular siderosis. *Experientia* 32:1324–1325.

HOBBS, H. E., EADIE, S. P., and SOMERVILLE, E. 1961. Ocular lesions after treatment with chloroquine. *Br. J. Ophthalmol.* 45:284–297.

HOBBS, H. E., SORSBY, A., and FREEDMAN, A. 1959. Retinopathy following chloroquine therapy. *Lancet* 2:478–480.

Hodgkinson, B. J., and Kolb, H. 1970. A preliminary study of the effect of chloroquine on the rat retina. *Arch. Ophthalmol.* 84:509–515.

Holaday, D. A., Rudofsky, S., and Treuhaft, P. S. 1970. The metabolic degradation of methoxyflurane in man. *Anesthesiol.* 33:579–593.

Hommer, K. 1968. Uber die chininvergiftung der netzhaut. *Klin. Monastsbl. Augenheilkd.* 52:785–805.

Hori, Y., Toda, K., Pathak, M. A., Clark, W. H., and Fitzpatrick, T. B. 1968. A fine-structure study of the human epidermal melanosome complex and its acid phosphatase activity. *J. Ultrastruct. Res.* 25:109–120.

Johnson, N. F., Strang, R., and Wilson, T. M. 1973. Further observations on the retinal fine structure after long term anesthesia. *Exp. Eye Res.* 17:73–85.

Johnson, N. F., Wilson, T. M., and Strang, R. 1973. Retinal fine structure after long term anesthesia. *Exp. Eye Res.* 15:127–139.

Jünemann, G., and Schulze, J. 1968. Elektroretinographische Untersuchungen zur Entstehung der Chinin-bzw. *Resochin-Retinopathie. Klin. Monatsbl. Augenheilkd.* 152:562–566.

Karli, P. 1954. Contribution experimentale a l'étude de la pathogenie des rétinoses pigmentaires. *Ophthalmologica* 128:137–162.

Karlsson, J. O., Stella-Giuffrida, A. M., Jarlstedt, J., McLean, W. G., Serra, I., and Sjostrand, J. 1976. Action of chloroquine on the vivo RNA and protein biosynthesis in the retina and the optic pathway of the rabbit. *J. Neurol. Sci.* 30:237–245.

Karpe, G. 1948. Early diagnosis of siderosis retinae by the use of electroretinography. *Docum. Ophthalmol.* 2:277–296.

Kearns, T. P., and Hollenhorst, R. W. 1966. Chloroquine retinopathy: evaluation by fluorescein angiography. *Arch. Ophthalmol.* 76:378–384.

Kinross-Wright. 1956. Clinical trial of a new phenothiazine compound NP 207. *Psychiat. Res. Rep. Am. Psychiat. Ass.* 4:89–94.

Knave, B. 1969. Electroretinography in eyes with retained intraocular metallic foreign bodies. *Acta Ophthalmol.,* suppl. 100.

Knave, B., Persson, H. E., and Nilsson, S. E. G. 1974. A comparative study on the effects of barbiturate and ethyl alcohol on retinal functions with special reference to the c-wave of the ERG and the standing potential of the sheep eye. *Acta Ophthalmol.* 52:254–259.

Kolb, H. 1965. Electro-oculogram findings in patients treated with antimalarial drugs. *Br. J. Ophthalmol.* 59: 573–590.

Krill, A. E., Potts, A. M., and Johanson, C. E. 1971. Chloroquine retinopathy investigation of discrepancy between dark adaptation and electroretinographic findings in advanced stages. *Am. J. Ophthalmol.* 71:530–543.

Legros, J., and Rosner, I. 1971. Electroretinographic modifications in albino rats after chronic administration of toxic doses of hydroxychloroquine and desethylhydroxychloroquine. *Arch. Ophtalmol.* (Paris) 31:165–180.

Legros, J., Rosner, I., and Berger, C. 1973. Retinal toxicity of chlorpromazine in the rat. *Toxicol. Appl. Pharmacol.* 26:459–465.

Leopold, I. H., 1968. Ocular complications of drugs: visual changes. *J. Am. Med. Assoc.* 205:631–633.

Lerner, W., ed. 1972. *Statistical abstract of the United States,* 93rd annual edition. Washington, D.C.: U.S. Government Printing Office.

Lindquist, N. G., and Ulberg, S. 1972. The melanin affinity of chloroquine and chlorpromazine studied by whole body autoradiography. *Acta Pharmacol. Toxicol.* (suppl. II) 31:3–32.

Lutzow, A. v., and Hommer, K. 1969. Influences of some drugs on the ERG of the isolated rabbit retina. *Ophthalmologica* 158:647–652.

Marmor, M. F. 1977. Fundus albipunctatus: a clinical study of the fundus lesions, the physiologic deficit, and the vitamin A metabolism. *Documenta. Ophthalmol.* 43: 277–302.

Masciulli, L., Anderson, D. R., and Charles, S. 1972. Experimental ocular siderosis in the squirrel monkey. *Am. J. Ophthalmol.* 74:638–661.

McFarlane, J. R., Yanoff, M., and Scheie, H. G. 1966. Toxic retinopathy following sparsomycin therapy. *Arch. Ophthalmol.* 76(4):532–540.

Meier-Ruge, W. 1965. Experimental investigations of the morphogenesis of chloroquine retinopathy. *Arch. Ophthalmol.* (Chicago) 73:540–544.

Meier-Ruge, W., and Cerletti, A. 1967. Experimental pathology of chloroquine and phenothiazine retinopathy. In *Proceedings of the XX International Congress of Ophthalmology,* Internat. Congress Series #146, Excerpta Medica Foundation, pp. 1129–1136.

Meier-Ruge, W., Kalberer, F., and Cerletti, A. 1966. Microhistoautoradiographic investigations of the distribution of tritium-labeled phenothiazine derivatives in the eye. *Experientia* 22(3):153–155. Also *Chem. Abstr.* 64:164–166.

MEIER-RUGE, W., and WERTHEMANN, A. 1967. *Medikamentose Retinopathie*. Stuttgart: George Thieme Verlag.

MONOHAN. R. H., and HORNS, R. C. 1964. The pathology of chloroquine in the eye. *Trans. Am. Acad. Ophthalmol. Otolaryngol.* 68:40–44.

NILSSON, S. E. G., KNAVE, B., and PERSSON, H. E. 1977. Changes in ultrastructure and function of the sheep pigment epithelium and retina induced by sodium iodate. II. Early effects. *Acta Ophthalmol.* 55:1007–1026.

NIWA, T. 1966. Dehydrogenase and retinitis pigmentosa. *Nagoya Shiritsu Daiguku Igakkai* 17(1):45–50.

NOELL, W. K. 1951. The effect of iodoacetate on the vertebrate retina. *J. Cell Comp. Physiol.* 37:283–307.

NOELL, W. K. 1953*a*. Experimentally induced toxic effects on structure and function of visual cells and pigment epithelium. *Am. J. Ophthalmol.* 36:103–116.

NOELL, W. K. 1953*b*. Studies on the electrophysiology and metabolism of the retina. U.S.A.F. School of Aviation Medicine, Randolph Field, Texas, Project Number 21-1201-0004.

NOZIK, R. A., WEINSTOCK, F. J., and VIGNOS, P. J., JR. 1964. Ocular complications of chloroquine: a series and case presentation with a simple method for early detection of retinopathy. *Am. J. Ophthalmol.* 58:774–778.

NYLANDER, V. 1967. Ocular damage in chloroquine therapy. *Acta Ophthalmol.* (suppl.) 92:1–71.

OKUN, E., GOURAS, P., BERNSTEIN, H., and VON SALLMANN, L. 1963. Chloroquine retinopathy: a report of 8 cases with ERG and dark adaptation findings. *Arch. Ophthalmol.* 69:59–71.

ORZALESI, N., GRIGNOLO, A., CALABRIA, G. A., and CASTELLAZZO, R. A. 1967. A study on fine structure and the rhodopsin cycle of the rabbit retina in experimental degeneration induced by diaminodiphenoxypentane. *Exp. Eye Res.* 6 (4):376–382.

POTTS, A. M. 1962*a*. The concentrations of phenothiazines in the eye of experimental animals. *Invest. Ophthalmol.* 1:523–530.

POTTS, A. M. 1962*b*. Uveal pigment and phenothiazine compounds. *Trans. Am. Ophthalmol. Soc.* 60:517–551.

POTTS, A. M. 1964. The reaction of uveal pigment in vitro with polycyclic compounds. *Invest. Ophthalmol.* 3:405–416.

RAMSEY, M. S., BLOODWORTH, J. M. B., and ENGERMAN, R. L. 1970. Chloroquine retinopathy in the rabbit. *Can. J. Ophthalmol.* 5:264–273.

RAMSEY, M. S., and FINE, B. S. 1972. Chloroquine toxicity in the human eye. *Am. J. Ophthalmol.* 73:229–235.

REBOTON, J., JR., WEAKLEY, R. D., BYLENGA, N. D., and MAY, R. H. 1962. Pigmentary retinopathy and iridocycloplegia in psychiatric patients. *J. Neuropsychiatry* 3:311–316.

RINTELEN, F., HOTZ, G., and WAGNER, P. 1957. Zur Klinik und experimentellen Pathologie der Pigmentepithelerkrankung nach Medikation mit einem Piperidinphenothiazin. *Ophthalmologica* 133:277–283.

ROSENTHAL, A. R., and APPLETON, B. 1975. Histochemical localization of intraocular copper foreign bodies. *Am. J. Ophthalmol.* 79:613–625.

ROSENTHAL, A. R., APPLETON, B., and HOPKINS, J. L. 1974. Intraocular copper foreign bodies. *Am. J. Ophthalmol.* 78:671–678.

ROSENTHAL, A. R., KOLB, H., BERGSMA, D., HUXSOLL, D., and HOPKINS, J. L. 1978*a*. Chloroquine retinopathy in the rhesus monkey, *Invest. Ophthalmol.* 17:1158–1178.

ROSENTHAL, A. R., MARMOR, M. F., LEUENBERGER, P., and HOPKINS, J. L. 1978*b*. Chalcosis: A study of natural history. Presented to the American Academy of Ophthalmology, Kansas City, 1978.

RUBIN, M., BERNSTEIN, H. N., and ZVAIFLER, N. J. 1963. Studies on the pharmacology of chloroquine retinopathy. *Arch. Ophthalmol.* 70:474–481.

RUBIN, M., and SLONICKI, A. 1966. A mechanism for the toxicity of chloroquine. *Arthritis Rheum.* 9:537.

SALLMANN, L. 1933. Uber Netzhautschadigung durch Salze der unterjodigen, unterbromigen und unterchlorigen Saure: ein Beitrag zur Frage der Septojodintoxikation. *Z. Augenheilkd.* 80:342–351.

SCHERBEL, A. L., MACKENZIE, A. H., NOUSEK, J. E., HOUSEK, J. E., and ATDJIAN, M. 1965. Ocular lesions in rheumatoid arthritis and related disorders with particular reference to retinopathy. *N. Engl. J. Med.* 273:360–366.

SCHIMMEL, R. 1926. Augenschadigungen nach Septojodinjektionen. *Munch. Med. Wochenschr.* 73:590.

SHEARER, R. V., and DUBOIS, E. L. 1967. Ocular changes induced by long-term hydroxychloroquine (Plaquenil) therapy. *Am. J. Ophthalmol.* 64(2):245–252.

SIDDALL, J. R. 1965. The ocular toxic findings with prolonged and high dosage chlorpromazine intake. *Arch. Ophthalmol.* 74:460–464.

SIDDALL, J. R. 1966. Ocular toxic changes associated with chlorpromazine and thioridazine. *Can. J. Ophthalmol.* 1:190–198.

SKOOG, K. O. 1974. The c-wave of the human d.c. registered ERG. III. Effects of ethyl alcohol on the c-wave. *Acta Ophthalmol.* 52:913–923.

Sorsby, A. 1941. Experimental pigmentary degeneration of the retina by sodium iodate. *Br. J. Ophthalmol.* 25: 58–62.

Sorsby, A., and Nakajima, A. 1958*a*. Experimental degeneration of the retina. III. Inhibitors of glycolysis and of respiration as inducing agents. *Br. J. Ophthalmol.* 42: 558–562.

Sorsby, A., and Nakajima, A. 1958*b*. Experimental degeneration of the retina. IV. Diaminodiphenoxy alkanes as inducing agents. *Br. J. Ophthalmol.* 42:563–570.

Sorsby, A., Newhouse, J. P., and Lucas, D. R. 1957. Experimental degeneration of the retina. I. Thiol. reactors inducing agents. *Br. J. Ophthalmol.* 41:309–312.

Tamai, A., and Holland, M. G. 1975. Electrophysiological studies on a case of thioridazine pigmentary retinopathy. *Yonago. Acta. Med.* 19:188–196.

Toussaint, D., Vereerstraeten, P., Goffin, P., Vanlanduyt, P., Jedwab, J., and Legrand, J. M. 1976. Hyperoxalurie primaire. *Arch. Ophtalmol.* (Paris) 36:97–112.

van Lith, G. H. M., Mak, G. T. M., and Wijnands, H. 1976. Clinical importance of electro-oculogram with special reference to the chloroquine retinopathy. *Bibl. Ophthalmol.* 85:2–9.

van Norren, D., and Padmos, P. 1977. Influence of anesthetics, ethyl alcohol and freon on dark adaptation of monkey cone ERG. *Invest. Ophthalmol. Visual Sci.* 16: 80–83.

Verrey, F. 1956. Dégénérescence pigmentaire de la rétine d'origine médicamenteuse. *Ophthalmologica* 131: 296–303.

Vito, P. 1935. Contributo allo Studio della degenerazione pigmentaria della retina indotta dalla soluzione iodica di Pregl. *Boll. Ocul.* 14:945–959.

Wagenmann, A. 1913. Die Verletzungen des Auges durch chemische Einwirkung. In *Graefe-Saemisch Handbuch der Augenheilkunde,* vol. 9, part 5, chap. 2, pp. 1531–1636. Leipzig: Wilhelm Engelmann.

Wagner, P. 1956. Untersuchungen uber die Wirkung von Phenothiazinderivaten auf den Augenhintergrund des Tieres. *Klin. Monatsbl. Augenheilkd.* 129:772–781.

Webster, S. H., Stohlman, E. F., and Highman, B. 1966. The toxicology of potassium and sodium iodates. *Toxicol. Appl. Pharmacol.* 8(1):185–192.

Weekley, R. D., Potts, A. M., Reboton, J., and May, R. H. 1960. Pigmentary retinopathy in patients receiving high doses of a new phenothiazine. *Arch. Ophthalmol.* 64:65–76.

Weissmann, G. 1966. Lysosomes and joint disease. *Arthritis Rheum.* 9:834–840.

Wetterholm, D. H., Snow, H. L., and Winter, F. C. 1965. A clinical study of pigmentary change in cornea and lens in chronic chlorpromazine therapy. *Arch. Ophthalmol.* 74:55–56.

Wetterholm, D. H., and Winter, F. C. 1964. Histopathology of chloroquine retinal toxicity. *Arch. Ophthalmol.* 71:82–87.

Wise, J. B. 1966. Treatment of siderosis bulbi vitreous hemorrhage and corneal bloodstaining with deferoxamine. *Arch. Ophthalmol.* 75(5):698–707.

Yoo, J. 1976. Responses of Muller cells of rabbit retina in experimental siderosis. *Jap. J. Ophthalmol.* 20:149–158.

Zeimer, R., Gorodetsky, R., Lahav, M., and Belkin, M. 1978. Experimental chalcosis. *Arch. Ophthalmol.* 96:115–119.

Zvaifler, N. J., Bernstein, H., and Rubin, M. 1962. Chloroquine deposition in ocular tissues. *Arthritis Rheum.* 5:667.

23 The Retinal Pigment Epithelium in Ocular Albinism

Ronald E. Carr
Irwin M. Siegel

THE EPITHELIAL LAYER backing the retina, though highly specialized, is composed of cells that are functionally not very different from epithelial cells found in diverse regions of any organism. Such cells usually separate fluids of different compositions and in this capacity help the body maintain a constant internal environment. Regardless of their location, however, all epithelial cells are intricately differentiated to perform specific types of transport activity (Berridge and Oschman 1972). In this larger sense, the retinal pigment epithelium (RPE) possesses many structural features common to other epithelial cells: villous projections, a complex system of endoplastic reticulum, and a rich supply of organelles, such as Golgi bodies and mitochondria. The individual cells making up the RPE, like others of their kind elsewhere, are joined one to the other by an encircling zonula occludens ("tight junction"), the properties of which define several kinds of electrical activity that may be recorded from the eye (Cohen 1965).

But there are some features of the RPE that make it a unique type of epithelial cell. Young has recently found that the RPE is an extraordinarily active phagocytic cell that efficiently lyses large amounts of ingested photoreceptor material (see Chapter 9). Another unusual property of RPE may seem at first too obvious to state, namely, that *it is pigmented.* But the fact is, of all the body tissues, only four contain substantial amounts of melanin, and two of these are in the eye. They are skin, hair, RPE, and uveal tract. One or two small areas of the central nervous system appear pigmented—the substantia nigra and locus caerulus—but these groups of cells probably contain pigment other than melanin, since they are normally pigmented in the completely amelanotic albino (Fitzpatrick and Quevedo 1972).

The importance of a pigment backing to the vertebrate retina (where it has been likened to the antihalation backing of photographic film) has often been commented on, and its image-enhancing qualities have been recognized. It is also well-known that a generalized loss of pigment in the RPE, iris, and choroid usually results in extreme photophobia and deteriorated image formation. Curiously, though, in complete oculocutaneous albinism visual acuity is not significantly improved even in diminished illumination. Clearly, in this disorder there are factors other than the simple absence of melanin that may result in poor vision. This chapter will review the clinical and pathological findings associated with several hypopigmenting disorders. Each of these conditions results in some primary defect of melanin synthesis of uvea and RPE, which in turn apparently retards normal developmental processes of specific portions of the eye and leads to some degree of visual loss.

Melanin Synthesis

Figure 23.1 shows the known steps in the biochemistry of melanin formation from tyrosine (see Chapter 8 for details of this process). Note that early in the schema, the presence of peroxidase is required for the conversion of tyrosine to dopa. Recent investigations have established that tyrosinase, which was initially thought to control the tyrosine-dopa reaction, is actually not required until dopa is already formed. Because of the ubiquitous character of

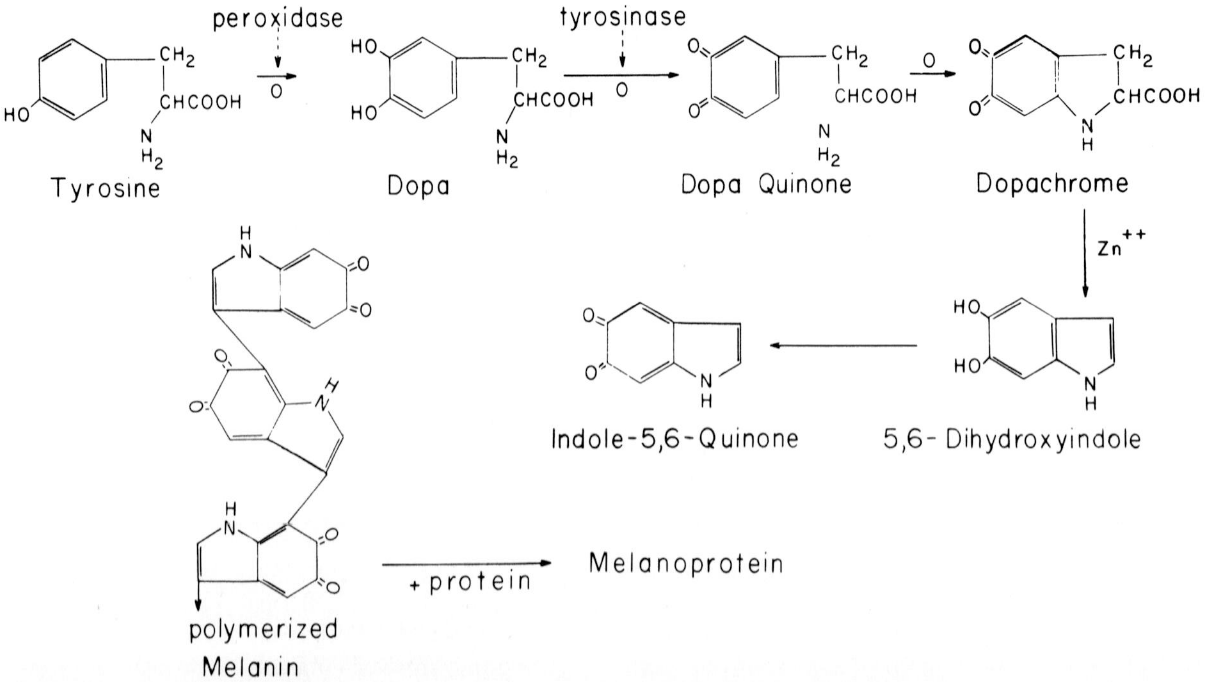

Figure 23.1 *Synthesis of melanin from tyrosine. (After Fitzpatrick and Quevedo 1972.)*

peroxidase in the body, however, "tyrosinase," precisely because it is so specific, is the key catalyst for melanin formation. End-product melanin polymer is attached to membrane protein, presumably through its quinone linkages with the amino or sulfhydryl groups of the protein (Mason 1959). The conjugation of melanin to protein takes place within a unique cellular organelle called a melanin granule or melanosome. It is the ability of a cell to synthesize melanosomes that characterizes it as a melanocyte. While epithelial cells may pursue functions similar to RPE, the RPE has, in addition, the properties of a melanocyte. Finally, though the RPE melanocytes are embryologically distinct from other melanocytes, deriving as they do from the outer layer of the optic cup rather than from the neural crest (Barber 1955), the melanosomes appear identical in both stem lines (Miyamoto and Fitzpatrick 1957).

Melanosome Formation in RPE

There is abundant evidence of a close relation between formation of melanin and the development of the melanosome organelle on whose membranes the melanoprotein is formed (Toda and Fitzpatrick 1972). The developmental sequence for melanosomes is given in Table 23.1. Schematic representations of the granules at each stage are provided to illustrate the standard nomenclature used to describe the different stages. As shown in the table, the ribosome is the subcellular morphological unit for synthesis of the critical enzyme tyrosinase. Primitive melanosomes are first recognized in the Golgi apparatus region, at which time both tyrosinase synthesis and premelanosome formation begins (Ide 1972). Until recently it was not known whether the Golgi apparatus contributed directly to melanosome formation in RPE, as in other melanocytes, since in mouse RPE no such association was noted (Moyer 1963). The sequence has now been clarified (Toda and Fitzpatrick 1972; Hori 1972) as follows: tyrosinase is synthesized in the ribosome, transferred to the Golgi area, and attached to the membrane of this structure; the membrane balloons out from the smooth surface endoplastic reticulum in the Golgi region and forms a Stage I melanosome organelle. As the granule matures, the amorphous, electron-dense content is arranged in a striated or lamellated configuration (Stages I-II), composed of a melanized protein polymer. The mature Stage IV melanosome attains a tightly rolled, spiral, lamellar organization on cross-section; this structure is independent of melanin synthesis, since it is seen in albino melanosomes (Birbeck 1963).

Melanosome development occurs in the RPE

Table 23.1 Developmental sequence of melanin and melanosomes.

Pigment cell organelle	*Function*	*Composition*	*Structure*
Ribosome	Synthesis of tyrosine	RNA + protein	?
Golgi complex and vesicles	Assembly of the early morphological unit	Phospholipid and protein	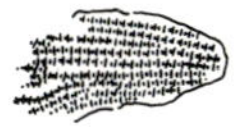 Stage 1
Premelanosome (immature form)	Formation of final structural form and beginning of melanin formation	Phospholipid + protein + tyrosinase and newly formed melanin	 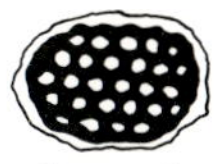 Stage 2 Stage 3
Melanosome (mature form)	Mature granule with fully developed melanin	Relatively inert melano-protein	 Stage 4

Sources: After Moyer 1969; Mund et al. 1972; Riley 1972.

before it is noted elsewhere. Mature granules are already present in the 16 mm (6–7 weeks) human embryo (Hollenberg and Spira 1972; Mund et al. 1972); and in the chick, utilizing electron microscopy, they can even be seen in the rim of the optic cup at Stages 19–20 (Ide 1972). In contrast, pigmented stromal cells are seen in the choroid of rhesus monkey only after 20 weeks gestation. This late appearance, together with the lack of significant amount of enzyme activity throughout their development, suggests that the choroidal pigmented cells originate in the iris and migrate posteriorly as they become melanized (Hu et al. 1973).

Interestingly, the presence of the amino acid precursor tyrosine not only determines pigment synthesis but, at least in ascidian embryos, also influences the extent of granule formation (Whittaker 1973). It has also been noted that of several amino acids used to incubate chick retina, only tyrosine allows normal cellular development (Grau et al. 1965). It is thought that in these circumstances, tyrosinase uses up available tyrosine for melanin production, leaving inadequate amounts for other developmental requirements in the retina. On the other hand, tyrosine concentration may have a more direct (and as yet unknown) role in controlling retinal cell differentiation.

One of the early findings in melanin synthesis is particularly relevant to ophthalmologists. Although tyrosinase activity is detectable very early in RPE preparations from chick, mouse, and human embryos, it quickly reaches a peak and is absent long before gestation ends (Miyamoto and Fitzpatrick 1957). Whatever melanin is present at birth is all the organism ever possesses. This may be related to the fact that malignant melanomas, which occur in any pigmented tissue (and are infrequently seen in nonpigmented tissue) are extremely rare in RPE (Reese 1963).

Disorders of Melanin Synthesis

A block in the synthesis of any precursor in the melanin pathway shown in Figure 23.1 can result in some form of albinism. Acquired albinism may occur if the basic precursor, tyrosine, is not present in sufficient quantity—this commonly results from an amino-acid deficient diet or from protein malnutrition, as in kwashiorkor (Witkop 1971). Albinism may also occur in association with certain systemic disorders, such as Chédiak-Higashi disease (Spencer and Hogan 1960; Stegmaier and Schneider 1965), and Menkes syndrome (Menkes et al. 1962). The most common forms of albinism are, however, hereditary in nature and may result from one or more of the following defects (after Fitzpatrick and Quevedo 1972):

1. lack of tyrosinase synthesis
2. lack of melanin polymerization
3. lack of synthesis of protein matrix of melanosome
4. lack of substrate tyrosine
5. presence of inhibitors of tyrosine-to-melanin pathway

Since albino melanosomes look normal (Birbeck and Barnicot 1959; Dowling and Gibbons 1962), no structural defect in the protein architecture of the granule appears likely. Biochemical studies of albino skin (Foster 1956) indicate that early steps in the melanization process have not taken place, making a defect in the later stages such as polymerization unlikely. The possible circumstances leading to abnormal melanogenesis are therefore reduced to:

1. decrease in synthesis of active tyrosinase
2. unavailability of tyrosine at the site of activity
3. melanogenic inhibition

Regardless of the true nature of the defect, the end result is the same: no melanin is deposited on the membranous lamellae of the melanosome.

Albinism

Hereditary albinism can be separated into two large categories:

1. oculocutaneous albinism, transmitted as an autosomal recessive trait
2. ocular albinism, transmitted as an X-linked recessive trait

Table 23.2 summarizes the major types of albinism and the associated ocular findings (for a complete survey of all the types of albinism, see Witkop 1971 and Siegel 1976). The division of oculocutaneous albinism into two forms is of recent origin and resulted from hair bulb incubation experiments (reviewed by Witkop 1971 and Witkop et al. 1970, 1972), which showed that in the presence of excess tyrosine, hairs from an albino population fell roughly into two types. One group, in which tyrosinase was assumed to be absent, was classified as tyrosinase negative (ty-neg) since no pigment appeared in the hairs after incubation. Such hairs were generally white or very blonde to begin with. The second group of hair bulb preparations did show formation of pigment, even if they originally appeared colorless, and were therefore categorized as tyrosinase-positive (ty-pos). Figure 23.2 shows micrographs of incubated albino hair bulbs.

Ty-neg albinos present the classic picture: white hair, pale skin, and ocular findings such as diaphanous blue irises, nystagmus, photophobia, and very poor vision. Many patients with poor vision but with only mild symptoms of albinism, the so-called "albinoids," were poorly understood until they were shown to have the tyrosinase-positive form

Table 23.2 Forms of albinism with associated skin and ocular findings.

	Oculocutaneous Albinism		*Ocular Albinism (all findings vary according to body pigmentation)*
Characteristics	*Tyrosinase negative*	*Tyrosinase positive*	
Skin color	Pink to red	Pink to normal	Normal
Iris color	Diaphanous blue to grey	Diaphanous blue to brown	Diaphanous blue to brown
Iris trans-illumination	Complete	Complete to partial	Complete to partial
Fundus pigmentation	Depigmented; macula may be pigmented	Variable; periphery generally depigmented	Variable; periphery generally depigmented
Ophthalmoscopic foveal reflex	None	None	None
Nystagmus	Marked	Variable: mild to marked	Variable: mild to marked
Photophobia	Severe	Variable: mild to marked	Variable: mild to marked
Visual acuity	20/200 or less	20/30 to 20/200	20/30 to less than 20/200

Source: After Witkop 1971 and Goodman et al. 1965.

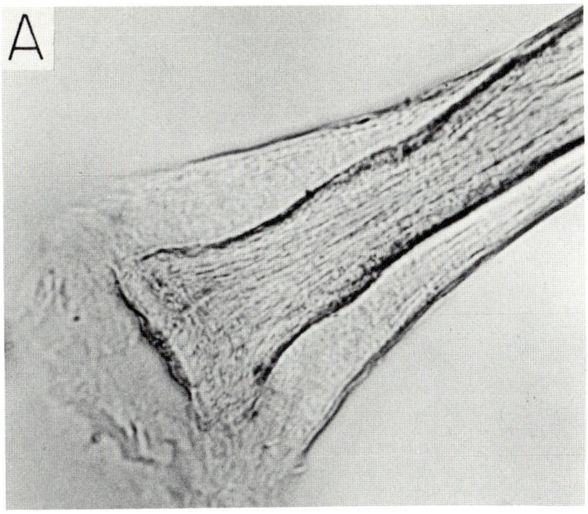

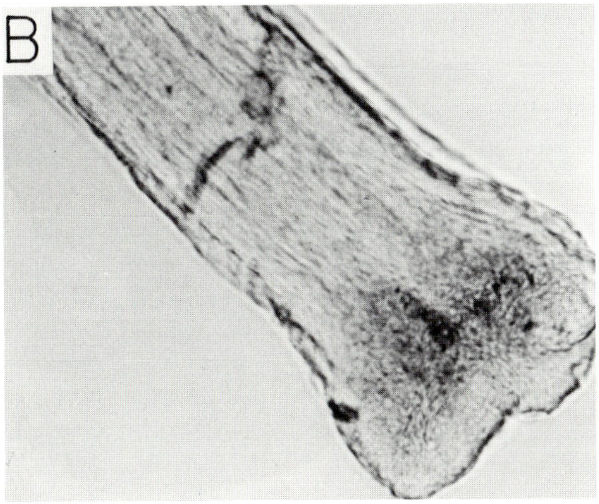

Figure 23.2 A, *the hair bulb from a ty-neg albino after 18 hours incubation in tyrosine. No formation of melanin is deposited at the base of the bulb. Some slight melanization is normally seen in the hair bulb of a ty-pos albino after incubation in a control solution of phosphate buffer* B, *after incubation in tyrosine, however, a hair bulb from the same ty-pos albino shows a dense formation of melanin* (C).

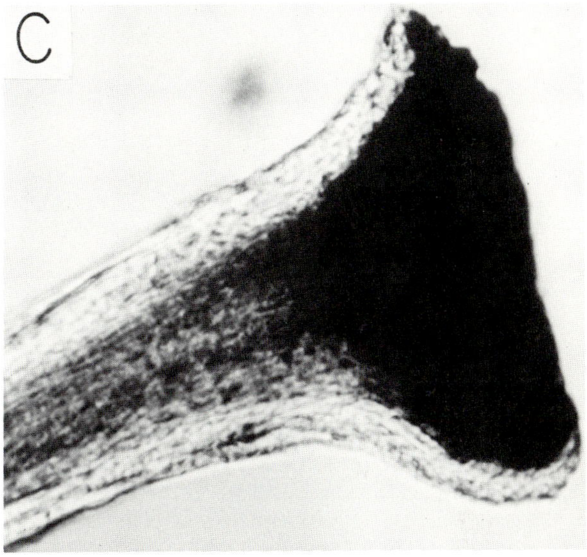

of oculocutaneous albinism. The question then naturally arises: If ty-pos albinos possess tyrosinase, why do they show deficiencies of pigment throughout their bodies? Witkop (1971) has suggested that the ty-pos albino may have a basically different genetic defect than the ty-neg, and what may be involved is either a tyrosinase with reduced activity or one that aggregates poorly into the melanosome. No definitive proof exists for either mechanism. What is certain, however, is that the genes for the two forms, ty-pos and ty-neg, are nonallelic and therefore do not interact. Hence, puzzle cases like Trevor-Roper's (1952) classic family, in which albino parents had normally pigmented offspring, can now be explained. If the mother and father are ty-neg and ty-pos, respectively—a supposition recently proved in Trevor-Roper's case, by performing hair bulb studies on the family (Witkop 1971)—each child is therefore a double heterozygote, that is, carries single recessive genes for each form, balanced by allelic normal pigmentation genes.

Of clinical interest to the ophthalmologist is the observation that ty-pos albinos may display a wide range of pigment variation, associated with an equally broad spectrum of ocular findings. Study of a large group of albinos (Carr et al. 1974 and Siegel 1976) indicated that the more depigmented ty-pos patients showed greater visual problems. Figure 23.3 shows iris transillumination and fundus appearance from ty-neg and ty-pos albino patients.

One other form of oculocutaneous albinism (not shown in the table) closely resembles the ty-pos type and is called yellow mutant. Patients with this condition are usually members of small inbred communities such as the Amish (Nance et al. 1970) and invariably present as fair-skinned, blonde-haired individuals with ocular findings similar to ty-pos albinos. Unlike the ty-pos patients, however, they

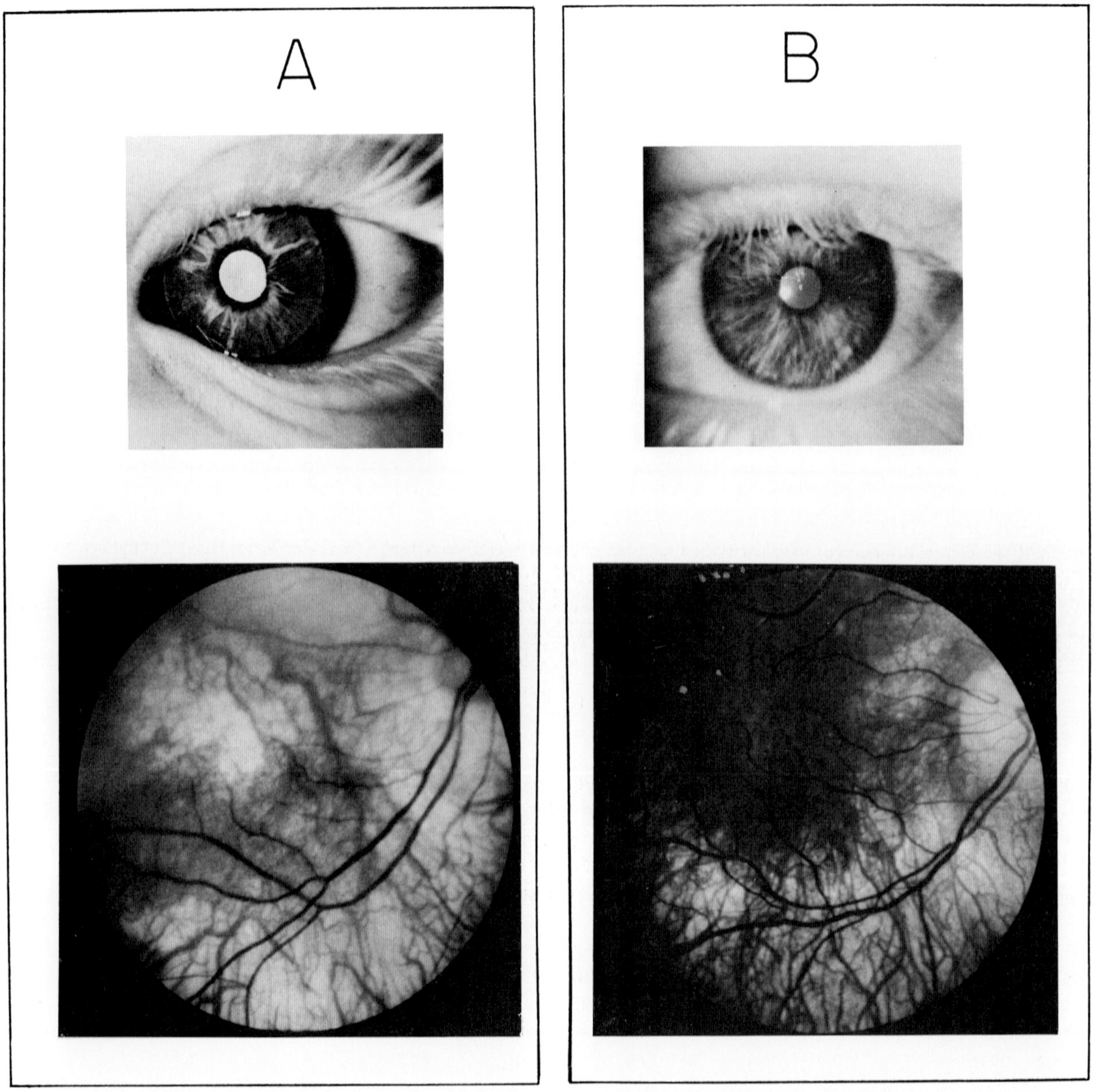

Figure 23.3 *Iris transillumination and fundus photos of a ty-neg* (A) *and ty-pos* (B) *albino. The tip of an optic fiber bundle transmitting high-intensity light was placed against the lid while the anterior segment was photographed with the fundus camera. The rim of the lens can be clearly seen through the diaphanous iris stroma in the ty-neg albino, but is not as evident in the more pigmented ty-pos form. The macular region of the ty-pos albino appears lightly pigmented, while the ty-neg macula contains little or no melanin.*

show an equivocal hair bulb test (Witkop 1971).

Sex-linked ocular albinism occurs less frequently than the oculocutaneous forms and is unique in that only the eye is involved. It must be emphasized that the general body pigmentation against which the eye pigmentation is expressed modifies the amount of pigment produced in the iris and retina and secondarily affects the degree of visual loss. Thus, Goodman et al. (1965) found that all of the visual defects listed in Table 23.2 were much less marked in darker-complexioned patients with ocular albinism than those with very fair skin.

Effect of Melanin on Foveal Development

The major defect associated with albinism is loss of acute vision. However, it has been a matter of conjecture which of the clinical features of albinism —photophobia, nystagmus, or foveal maldevelopment (aplasia)—plays the more prominent role in producing the visual deficit.

Evidence that the structure of the fovea centralis in the albino eye is somehow abnormal has been known to ophthalmologists for a very long time. Early ophthalmoscopic observation by Vogt (1924), using red-free light, clearly demonstrated that the refractile "yellow reflex" of the fovea was absent in albinos. Histological studies of albino eyes also showed, though not convincingly, gross malformation of receptors in the foveal area (Fritsch 1907; Elschnig 1913). No details were given as to thickness of sections or what precautions were taken to assure that the .07 mm^2 area designated as the anatomical fovea was in fact examined. Even as late as 1920, Usher, with six albino eyes at his disposal, could only offer artists' representations of low-power microscopic views. Strangely, no further histological studies on the central area of the albino retina appeared for over fifty years, nor does the largest repository of ocular specimens in the United States, the Armed Forces Institute of Pathology, have any albino sections in their collection (Font 1974). However, Fulton et al. (1978) have recently described a complete lack of foveal differentiation in an albino eye, and O'Donnell and his coworkers (1976) could not find a fovea centralis "despite exhaustive serial sectioning through macular retina." A reading of the literature strongly suggests that the ophthalmoscopic appearance of the albino eye has provided stronger proof of "foveal aplasia" in the albino than actual histology has. Therefore, more pathological material must be examined if we are to decide, among other things, whether albinos have a true receptor aplasia with only the foveal cones involved, or a form of macular hypoplasia in which the entire central area retains features (such as persistence of ganglion cells and nuclear layers) more common to peripheral retina as suggested by Fulton et al. (1978) and Vrabec (1951). The distinction between foveal aplasia and macular hypoplasia, made by Duke-Elder (1963), may be critical to our understanding of the basic visual defect in albinism. It is one thing to theorize on the maldevelopment of certain cells in a small area and quite another to involve a large region of retinal structure. Photophobia, and its attendant symptoms of glare and discomfort, is probably not a major cause of acuity reduction, for foveal aplasia and nystagmus with poor vision are occasionally known to occur in normally pigmented eyes (Waardenburg 1963). The remaining components—foveal structure and nystagmus—are more difficult to separate, since nystagmus invariably accompanies congenital defects of the fovea, such as aniridia (Francois 1961). Nonetheless, a strong functional relation would seem to exist between the presence of a fovea and nystagmus, for even in macular dystrophies that involve most of the retinal cones, no nystagmus is seen as long as the fovea is spared (Deutman 1971).

Since basic retinal function (apart from fine acuity) seems normal in albinos, as judged by at least two measures of massed electrical activity recorded from the eye (Dodt et al. 1959; Gouras and Gunkel 1963), the possibility must be considered that foveal aplasia may in fact be a heritable disorder independent of melanization. For if peripheral receptors develop normally in an amelanotic milieu, why should those few thousand receptors in the fovea be less endowed? If two independent genes are involved they must be located close together on the same chromosome: no instance of albinism with completely normal vision has been reported and, as mentioned, patients with foveal aplasia and no other findings are rare. More to the point, perhaps, is that, with an assumption of a two-gene hypothesis—one gene for albinism and one for aplasia—it becomes necessary to further postulate a series of "foveal aplasia genes," each differing in effectiveness, to account for the varying acuity in the ty-pos albino group. Such complexity is not tenable in view of the simple mode of transmission associated with albinism. A single gene producing two distinct traits (pleiotrophy) such as albinism and aplasia might be a more satisfactory model, but pleiotropic traits (such as those seen in the Laurence-Moon-Biedl syndrome) are variable in expression (Francois 1961) and are thus hard to reconcile with the rather definitive similarities seen among albinos.

Finally, as has already been noted, ty-pos albinos with more pigmentation have better visual function than the lightly pigmented patients of this group (see Table 23.2). This certainly indicates that foveal development, a late structural occurrence, is in some way related to the degree of RPE melanization. The exact nature of such a relation is, however, unclear. Mann (1957) speculated that "the pigment appears just before the beginning differentiation of the retinal layer" and that "the effect of the absence of pigment may be a slowly acting retardation from disturbed metabolism." To this, little can be added, since no data are available that provide a definitive causal relation between the two phenomena.

Whatever proof is lacking for chemical influences, induced by the presence of melanophores, on the foveal anlage, there is dramatic anatomical evidence in normal retina showing the intimate relation that exists between the pigment-filled villous processes of the RPE and the distal ends of the photoreceptors. Figure 23.4 shows three electron micrographs taken through the same area of the cone-dominated ground-squirrel retina (Cohen and Siegel 1973). All the micrographs were taken and printed to the same magnification. The tangential section (A) shows the large well-defined polygonal shapes of the RPE cells, while the transverse section (B) gives some idea how far the pigment-filled villi extend along the length of the outer segment. A second tangential section

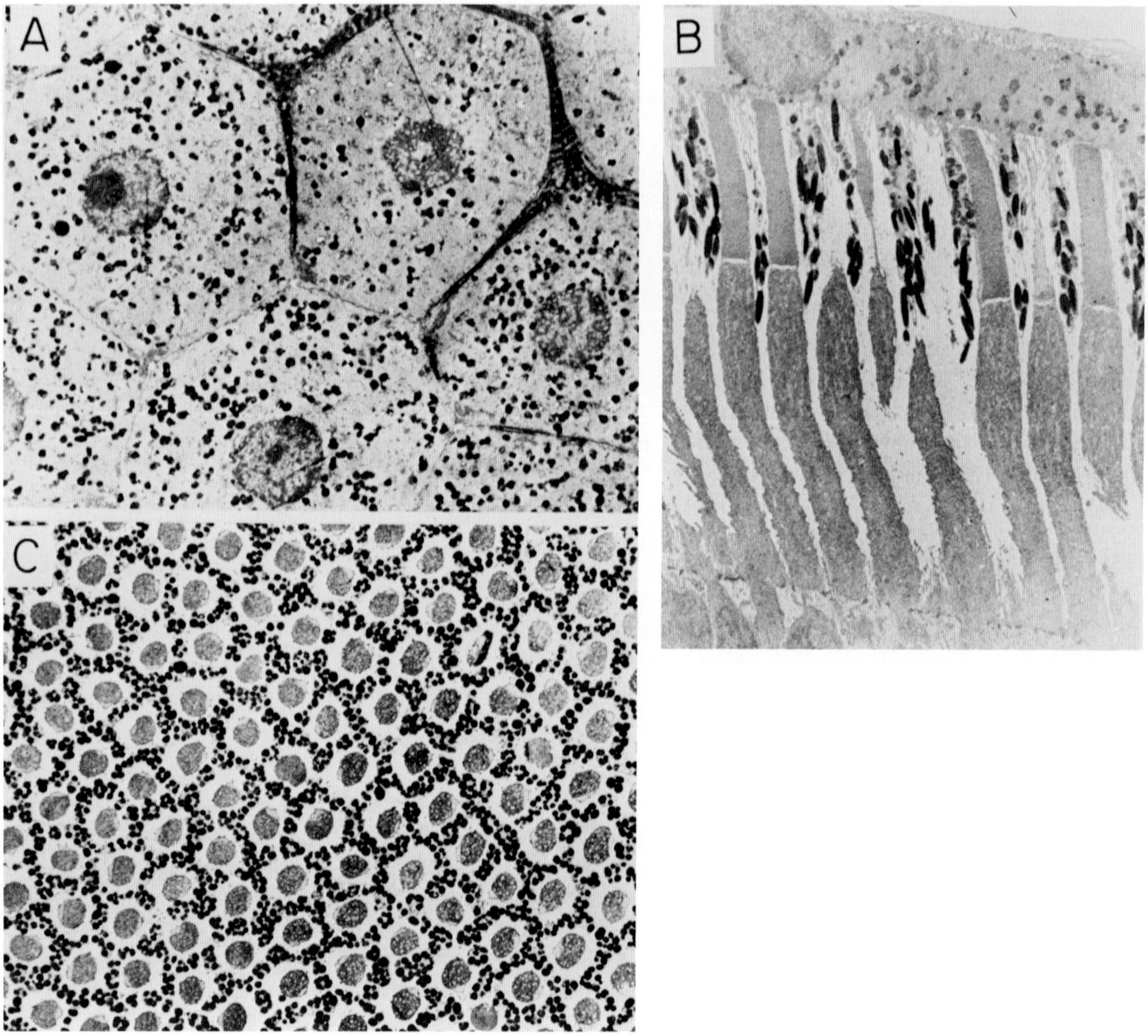

Figure 23.4 *Electron micrographs of the retina and RPE of the Mexican ground squirrel. The same magnification (×5,460) was used throughout.* A, *tangential section through the RPE.* B, *transverse section showing RPE and photoreceptor portion of the receptor.* C, *tangential section at level of receptor outer segments. Each receptor is seen surrounded by a cluster of melanin filled villous projections from the RPE.*

(C) was taken through the middle of the receptor outer segments. When the two tangential sections are compared, a relationship of clinical interest emerges. One RPE cell is seen to serve dozens of receptors, so that if only a few are injured a significant retinal area may be affected. Viewed as a composite picture, the three sections show how each receptor is closely surrounded by several RPE processes. This provides the necessary anatomical proximity for whatever influence the presence of melanin might have on developing foveal receptors.

Albinism and Visual Tract Development

Anatomical studies of albino rats (Lund 1965), rabbits, and guinea pigs (Giolli and Guthrie 1969) have revealed that a substantial proportion of temporal optic fibers cross in the chiasm and have contralateral representation in the higher visual centers. Visually-evoked cortical responses from albino rats (Creel et al. 1970) and Siamese cats (Creel 1971), which are partial albinos, provide functional corroboration to this remarkable finding. The result of the anomalous fiber crossings is a serious disruption of fiber termination locations in the rigorously organized laminar layers of the superior collicular bodies, lateral geniculate nuclei, and occipital cortex (Berman and Cynader 1972; Hubel and Wiesel 1971).

Could the same developmental anomaly be present in the human albino? Creel and his coworkers (1974) have recorded bilaterally asymmetrical light-evoked potentials of monocularly illuminated albino subjects. Their results suggest a disorganization of the nondecussated optic fibers similar to those in other albino mammals.

The fiber-crossing defect in the Siamese cat is probably responsible for the bilateral esotropia displayed by the majority of these animals. Hubel and Wiesel (1971) and Guillery et al. (1974) believe the convergent squint in the Siamese cat may be a result of the animal's attempt to favor the nasal half-retinas over the temporal halves, assuring binocular input to as many cortical cells as possible. Although many animal species may be albinotic, only those with sufficient binocular field overlap may be expected to have a "crossed-eyes" appearance. Now, esotropia is a common finding in human albinos (Witkop 1971), and it is just possible that they too may be attempting to compensate for the topological disruption in the higher centers brought about by the inappropriate incursion of decussating ipsilateral fibers.

No formal causal relationship has yet been established between amelanogenesis and abnormal tract formation. It is difficult to imagine what effect melanin formation could have on the specificity of normally unpigmented fibers. More likely, the genetic factors (or lack of them) responsible for albinism may somehow influence fiber directionality. The relative ease with which cat breeders can separate cross-eyedness from other Siamese traits seems to obviate the presence of a pleiotropic gene, or even two closely linked interacting genes occupying neighboring locations on the same chromosome.

REFERENCES

BARBER, A. N. 1955. *Embryology of the human eye.* St. Louis: C. V. Mosby Co.

BERMAN, N., and CYNADER, M. 1972. Comparison of receptive-field organization of the superior colliculus in Siamese and normal cats. *J. Physiol.* 224:363–389.

BERRIDGE, M. J., and OSCHMAN, J. L. 1972. *Transporting epithelia.* New York: Academic Press.

BIRBECK, M. S. C. 1963. Electron microscopy of melanocytes: the fine structure of hair-bulb melanosomes. *Ann. N.Y. Acad. Sci.* 100:540–547.

BIRBECK, M. S. C., and BARNICOT, N. A. 1959. Electron microscope studies on pigment formation in human hair follicles. In *Pigment cell biology,* ed. M. Gordon, pp. 549–561. New York: Academic Press.

CARR, R. E., SIEGEL, I. M., and SPROTZER, S. 1974. Unpublished observations.

COHEN, A. I. 1965. A possible cytological basis for the 'R' membrane in the vertebrate eye. *Nature* 205:1222–1223.

COHEN, A. I., and SIEGEL, I. M. 1973. Unpublished observations.

CREEL, D. J. 1971. Visual system anomaly associated with albinism in the cat. *Nature* 231:465–466.

CREEL, D. J., DUSTMAN, R. E., and BECK, E. C. 1970. Differences in visually evoked responses in albino versus hooded rats. *Exp. Neurol.* 29:298–309.

CREEL, D. J., WITKOP, C. J., JR., and KING, R. A. 1974. Asymmetric visually evoked potentials in human albinos: evidence for visual system anomalies. *Invest. Ophthalmol.* 13:430–440.

DEUTMAN, A. F. 1971. The hereditary dystrophies of the

posterior pole of the eye (doctoral thesis). Amsterdam: Van Gorcum & Co.

Dodt, E., Copenhaver, R. M., and Gunkel, R. D. 1959. Electroretinographic measurement of the spectral sensitivity in albinos, Caucasians and Negroes. *Arch. Ophthalmol.* 62:795–803.

Dowling, J. E., and Gibbons, I. R. 1962. The fine structure of the pigment epithelium of the albino rat. *J. Cell Biol.* 14:459–474.

Duke-Elder, S. 1963. *System of ophthalmology,* vol. 8, pt. 2. Congenital deformities, pp. 652–653. London: Henry Kimpton.

Elschnig, A. 1913. Zur Anatomie des Menschlichen Albinoauges. *Albrecht von Graefes Arch. Klin. Ophthalmol.* 84:401–419.

Fitzpatrick, T. B., and Quevedo, W. C., Jr. 1972. Albinism. In *The metabolic basis of inherited disease,* ed. J. B. Stanbury, J. B. Wyngaarden, and D. S. Fredrickson, 3rd ed., pp. 326–337. New York: McGraw-Hill Co.

Font, R. 1974. Personal communication.

Foster, M. 1956. Enzymatic studies of the physiological genetics of guinea pig coat coloration. I. Oxygen consumption studies. *Genetics* 41:396–409.

Francois, J. 1961. *Heredity in ophthalmology,* pp. 152–155. St. Louis, Mo.: C. V. Mosby.

Fritsch, K. 1907. Vergleichende Untersuchungen der Fovea Centralis des Menschen. *Anat. Anz.* 30:462–464.

Fulton, A. B., Albert, D. M., and Craft, J. L. 1978. Human albinism. Light and electron microscopic study. *Arch. Ophthalmol.* 96:305–310.

Giolli, R. A., and Guthrie, M. D. 1969. The primary optic projections in the rabbit: an experimental degeneration study. *J. Comp. Neurol.* 136:99–126.

Goodman, G., Ripps, H., and Siegel, I. M. 1965. Sex-linked ocular disorders: trait expressivity in males and carrier females. *Arch. Ophthalmol.* 73:387–398.

Gouras, P., and Gunkel, R. D. 1963. The EOG in chloroquine and other retinopathies. *Arch. Ophthalmol.* 70:629–639.

Grau, C. R., Anstic, R. E., and Matteson, G. C. 1965. Degeneration of the eyes of tyrosine-deficient chick embryos. *Science* 150:1743–1745.

Guillery, R. W., Casagrande, V. A., and Oberdorfer, M. D. 1974. Congenitally abnormal vision in Siamese cats. *Nature* 252:195–199.

Hollenberg, M. J., and Spira, A. W. 1972. Early development of the human retina. *Can. J. Ophthalmol.* 7:472–491.

Hori, Y. 1972. Ultrastructural study of ^{3}H incorporation from 3,4-dopa-2,3-^{3}H into retinal pigment epithelium chick embryo. In *Pigmentation: its genesis and biological control,* ed. V. Riley, pp. 143–154. New York: Appleton-Century-Crofts.

Hu, F., Endo, H., and Alexander, N. J. 1973. Morphological variations of pigment granules in eye of the rhesus monkey. *Am. J. Anat.* 136:167–182.

Hubel, D. H., and Wiesel, T. N. 1971. Aberrant visual projections in the Siamese cat. *J. Physiol.* 218:33–62.

Ide, C. 1972. The development of melanosomes in the pigment epithelium of the chick embryo. *Z. Zellforsch.* 131:171–186.

Lund, R. D. 1965: Uncrossed visual pathways of hooded and albino rats. *Science* 149:1506–1507.

Mann, I. 1957. *Developmental abnormalities of the eye,* 2d ed., pp. 141–143, 287–290. Philadelphia: J. B. Lippincott Co.

Mason, H. S. 1959. Structure of melanins. In *Pigment cell biology,* ed. M. Gordon, p. 563. New York: Academic Press.

Menkes, J. H., Alter, M., Steigleder, G. K., and Sung, J. S. 1962. A sex-linked recessive disorder with retardation of growth, peculiar hair and focal and cerebellar degeneration. *Pediatrics* 29:764–778.

Miyamoto, M., and Fitzpatrick, T. B. 1957. On the nature of the pigment in retinal pigmented epithelium. *Science* 126:449–450.

Moyer, F. H. 1963. Genetic effects of melanosome fine structure and ontogeny in normal and malignant cells. *Ann. N.Y. Acad. Sci.* 100:584–606.

Moyer, F. H. 1969. Development, structure, and function of the retinal pigmented epithelium. In *The retina,* ed. B. R. Straatsma, M. O. Hall, R. A. Allen, and F. Crescitelli, pp. 1–30. Berkeley: Univ. of California Press.

Mund, M. L., Rodrigues, M. M., and Fine, B. S. 1972. Light and electron microscopic observations in the pigmented layers of the developing human eye. *Am. J. Ophthalmol.* 73:167–182.

Nance, W. E., Jackson, C. E., and Witkop, C. J., Jr. 1970. Amish albinism: a distinctive autosomal phenotype. *Am. J. Hum. Genet.* 22:579–590.

O'Donnell, F. E., Hambrick, G. W., Green, W. R., Illie, J., and Stance, D. L. 1976. X-linked ocular albinism. *Arch. Ophthalmol.* 94:1838–1842.

REESE, A. B. 1963. *Tumors of the eye,* 2d ed., pp. 43–47. New York: Harper & Row.

RILEY, V. 1972. Colloquial considerations of the pigment cell. In *Pigmentation: its genesis and biological control,* ed. V. Riley, pp. 639–648. New York: Appleton-Century-Crofts.

SIEGEL, I. M. 1976. The albino as a low-vision patient. In *Clinical low vision,* ed. E. E. Faye, pp. 255–261. Boston: Little, Brown.

SPENCER, W. H., and HOGAN, M. J. 1960. Ocular manifestations of Chediak-Higashi syndrome. *Am. J. Ophthalmol.* 50:1197–1203.

STEGMAIER, O. C., and SCHNEIDER, L. A. 1965. Chediak-Higashi syndrome. *Arch. Dermatol.* 91:1–8.

TODA, K., and FITZPATRICK, T. B. 1972. Ultrastructural and biochemical studies of the formation of melanosomes in the embryonic chick retinal pigment epithelium. In *Pigmentation: its genesis and biological control,* ed. V. Riley, pp. 125–141. New York: Appleton-Century-Crofts.

TREVOR-ROPER, P. D. 1952. Marriage of two complete albinos with normally pigmented offspring. *Br. J. Ophthalmol.* 36:107–108.

USHER, C. H. 1920. Histological examination of an adult human albino's eyeball, with a note on mesoblastic pigmentation on foetal eyes. *Biometrika* 13:46–56.

VOGT, A. 1924. Uber Angeborenes und Vererbtes Fehlen der Macula Lutea. *Klin. Monatsbl. Augenheilkd.* 72:806–808.

VRABEC, F. 1951. Un Cas de la Dégénérescence Pigmentaire Congenitale de la Rétine Examine Histologiquement. *Ophthalmologica* 122:65–75.

WAARDENBURG, P. J. 1963. *Genetics and ophthalmology,* vol. 2, pp. 1722–1723. Oxford: Blackwell Science Publishing.

WHITTAKER, J. R. 1973. Tyrosinase in the presumptive pigment cells of ascidian embryos: tyrosine accessibility may initiate melanin synthesis. *Dev. Biol.* 30:441–454.

WITKOP, C. J., JR. 1971. Albinism. In *Advances in human genetics,* ed. H. Harris and K. Hirschhorn. New York: Plenum Press.

WITKOP, C. J., JR., NANCE, W. E., RAWLS, F. R., and WHITE, J. G. 1970. Autosomal recessive oculocutaneous albinism in man: evidence for genetic heterogeneity. *Am. J. Hum. Genet.* 22:55–74.

WITKOP, C. J., JR., WHITE, J. G., NANCE, W. E., and UMBER, R. E. 1972. Mutations in the melanin pigment system resulting in features of oculocutaneous albinism. In *Pigmentation: its genesis and biological control,* ed. V. Riley, pp. 359–377. New York: Appleton-Century-Crofts.

24 Dystrophies of the Retinal Pigment Epithelium

Michael F. Marmor

CONSIDERING the diverse functions of the pigment epithelium, one might expect to find a large group of dystrophic or congenital disorders that are characterized by a specific abnormality of function such as defective phagocytosis or vitamin A binding. However, such disorders have not been recognized clinically, probably because our ability to localize disease and analyze cellular dysfunction within the human eye is rather crude. At present, only a few human dystrophies have been localized to the pigment epithelium and the functional consequences of these are for the most part unknown. The pigment epithelium may, of course, be a primary site of dysfunction in some of the so-called retinal or choroidal dystrophies, but we can only speculate about this until more specific clinical and biochemical information becomes available.

Much of the discussion in this chapter will necessarily be descriptive, but an attempt will be made to define how and why the pigment epithelium is involved in various dystrophies. In some cases, the pigment epithelium seems to be the primary site of dysfunction; in others it is damaged secondarily. Proof of one or the other of these pathophysiological alternatives is very difficult within the eye, and ocular tissue for histology or histochemistry has usually become available only at the end stages of disease. Our understanding and classification of these dystrophies relies largely upon clinical criteria such as electrophysiologic findings and funduscopic or angiographic changes (see Chapters 13 and 21), and these limitations should be kept in mind in reading this chapter. A more detailed description of these various disorders can be found in several recent books (Deutman 1971; Franceschetti, François, and Babel 1974; Francois and Bonnet 1976; Krill 1977a).

The first and longest section of this chapter is devoted to dystrophies that are widely believed (though certainly not proven) to derive primarily from the pigment epithelium. Adjacent cell layers are always involved to some degree, and until these conditions can be described in terms of a specific chemical or physical abnormality within the pigment epithelial cell, their designation as pigment epithelial dystrophies must remain provisional. This section will also deal with the differential diagnosis of yellow lesions in the fundus, because of the frequent clinical discussion at meetings and in print regarding flecks, vitelliform lesions, and so forth. The second section deals with retinal and choroidal dystrophies, although this designation, as that of the pigment epithelial dystrophies, is provisional. My main concern will be to define the potential involvement of the pigment epithelium in these diseases. The final section of the chapter deals with congenital pigmentary disturbances of the pigment epithelium.

Pigment Epithelial Dystrophies

There is no disorder that can unequivocally be described as a primary dystrophy of the pigment epithelium. At best, we can describe a group of disorders whose early fundus and angiographic appearances show changes primarily within the pigment epithelium, which initially produce little or no signs of damage to the neurosensory retina, or in which there is a dichotomy between a normal electroretinogram (ERG) and a subnormal light response of the electrooculogram (EOG), suggesting the presence of pigment epithelial damage. Such definition is unsatisfactory because we lack the histologic or biochemical data to prove the localization, and because EOG abnormalities are not always specific for the pigment epithelium (see Chapter 13). Also, we may be excluding some pigment epithelial dystrophies that do not alter pigmentation, or whose functional effects are so closely linked to the visual process that the clinical findings seem to implicate the photoreceptors. An example is retinitis pigmentosa, which may well involve a primary dysfunction of the pigment epithelium. However, its early symptoms and findings are those of photoreceptor dysfunction, so that it is ordinarily classified with the retinal dystrophies.

PATTERN DYSTROPHIES OF THE PIGMENT EPITHELIUM. The pattern dystrophies are a group of hereditary disorders in which slowly progressive granular or reticular pigment changes appear at the level of the pigment epithelium. The lesions are most prominent in the macula, but acuity is only minimally affected and the visual prognosis is good.

The term "pattern dystrophy" was coined independently by Marmor and Byers (1977) and Hsieh, Fine, and Lyons (1977), who used it to describe a group of disorders that had been previously separated into discrete entities. Table 24.1 lists the pattern dystrophies that have been described to date. Some families with reticular dystrophy (Sjögren 1950; Deutman and Rümke 1969) and butterfly-shaped dystrophy (Deutman et al. 1970) will show a distinctive pattern of fundus pigmentation (Fig. 24.1). However, both Hsieh, Fine, and Lyons (1977) and Marmor and Byers (1977) have shown that different patterns of reticular pigmentation can be present among members of a single family (Fig. 24.2), suggesting that common factors may underlie these varying manifestations of pigment epithelial disturbance. The pathogenesis of these conditions is unknown, and all carry a similar good visual prognosis for the patient. Thus, the proliferation of individual descriptive terms, as noted in Table 24.1, is not very helpful to the clinician who is unlikely to come across an exemplary pedigree. On the other hand, families with less dramatic findings are probably not rare, and the designation "pattern dystrophy" allows these patients to be recognized and the import of their disease to be understood.

Table 24.1 summarizes the clinical findings in the pattern dystrophies. The families that have been described to date have shown either autosomal recessive or autosomal dominant inheritance. The pigmentary changes may be recognized as an incidental finding, or as a result of examination because of the complaint of mildly diminished acuity. The visual acuity generally remains better than 20/30, with the possible exception of elderly individuals. The ophthalmoscopic changes are much less prominent than the pattern visible on fluorescein angiography (Fig. 24.2), and the diagnosis of a dystrophy —as opposed to a mild, nonspecific, macular degeneration—may be missed unless angiography is performed or the clinician is suspicious enough to examine other family members. The fundus changes may range from a coarse granular mottling of the macula to a well-defined reticular pattern. The pattern may change with age, apparently progressing from a granular to a partial reticular pattern in the patients described by Marmor and Byers (1977), and migrating outward with gradual fading in older patients and reticular dystrophy (Sjögren 1950).

The visual fields may occasionally show a slight decrease in central sensitivity, but color vision, dark adaptation, and the ERG are characteristically normal. The one consistent abnormality is the EOG, which typically shows a slightly low light/dark ratio.

No histologic material is available for any of these conditions. They are presumed to be localized to the pigment epithelium, on the basis of the clinical examination and the EOG findings, which, in the presence of a normal ERG, suggest the presence of a diffuse pigment epithelial abnormality. The source of the prominent reticular pattern in these disorders is unknown. Hsieh, Fine and Lyons (1977) have suggested that some of the abnormal pigmentation may result from lipofuscin rather than melanin accumulation, since patterns which block fluorescein are not always pigmented ophthalmoscopically. One cannot help noting a similarity to the lobular architecture of the choriocapillaris (Hayreh 1975), and a vascular factor may influence the genesis or the expression of pigmentation in these disorders. Reticular patterns of pigmentation are also seen in the peripheral retina of older individuals, presumably as a degenerative change (see Chapter 25).

The main concerns in the differential diagnosis of pattern dystrophies are the juvenile and senile macular degenerations. The most common juvenile macular dystrophies, such as vitelliform dystrophy,

Table 24.1 Pattern dystrophies of the retinal pigment epithelium.

Dystrophy	*Probable inheritance (autosomal)*	*Pigment pattern*	*Pigment obscured by red-free light*	*Pigment blocks fluor-escein*	*Typical visual acuity*	*Color vision*	*Dark adapta-tion*	*ERG*	*EOG light response*	*Central visual field*
Dystrophia reticularis[a]	Recessive	Reticular	Yes	Yes	20/20–20/25	Normal	Normal	Normal	Low normal (age 10)	Normal
Dystrophia macroreticularis[b]	Recessive or Weak Dominant	Reticular and granular	Yes	Yes	20/20–20/60	—	Normal	One case "subnormal"	—	Slight decrease
Reticular pigment dystrophy[c]	Dominant	Reticular and granular	Yes	Yes	20/20–20/25	Normal	Normal	Normal	—	Normal
Butterfly dystrophy[d]	Dominant	Partial reticulum	Yes	Yes	20/20–20/25	Normal	Normal	Normal	Low	Decreased sensitivity
Fundus pulverulentus[e]	Dominant	Granular	—	—	"Normal"	—	Normal	Normal	—	Normal
Pattern dystrophy[f]	Dominant	Partial reticulum and granular	Yes	Yes	20/20–20/30	Normal	Normal	Normal	Slightly low	Normal
Patterned dystrophy[g]	Dominant	Partial reticulum	Yes	Yes	20/20	Occ. mild defect	Normal	Normal	Borderline	Occ. decrease

a. Sjögren 1950; Deutman and Rümke 1969; Chopdar 1976.
b. Mesker, Oosterhuis, and Delleman 1970.
c. Benedikt and Werner 1971.
d. Deutman et al. 1970; Starzycka and Bryk 1976.
e. Slezak and Hommer 1969.
f. Marmor and Byers 1977.
g. Hsieh, Fine, and Lyons 1977.

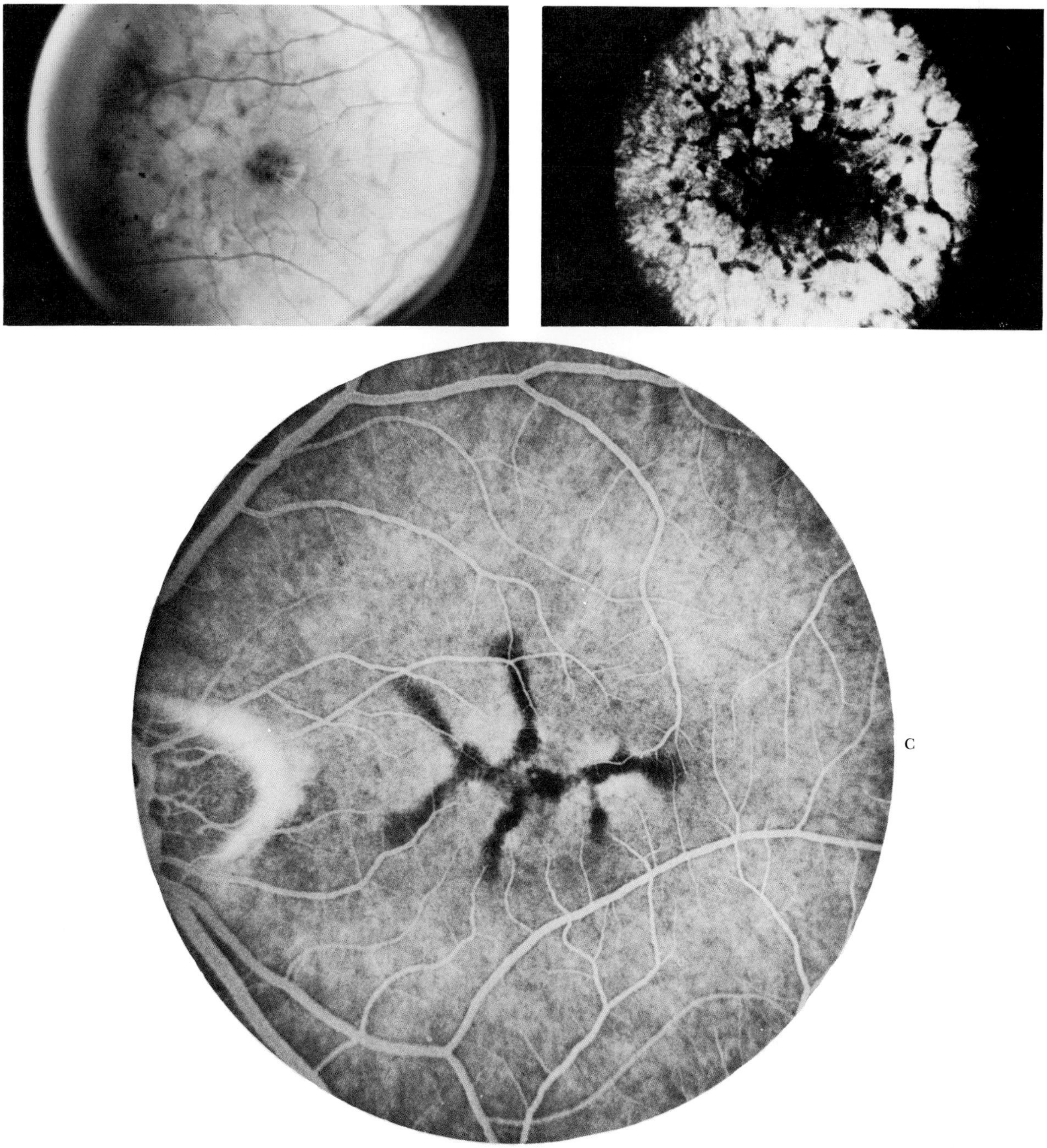

A B C

Figure 24.1 *Pattern dystrophies of the pigment epithelium.* A, *reticular dystrophy.* B, *fluorescein angiogram of the same eye.* C, *butterfly-shaped dystrophy. (Fluorescein angiogram courtesy of A. F. Deutman).*

Stargardt's disease, and x-linked retinoschisis, should ordinarily be easy to distinguish because of their characteristic appearances and their severe early effects upon visual acuity. The less specific changes of early senile degeneration, such as foveal drusen, may pose more of a problem, and the diagnosis of a pattern dystrophy will depend upon angiographic demonstration of a characteristic pigmentary pattern and evidence for a familial incidence. The EOG is not sufficiently impaired in these conditions to be

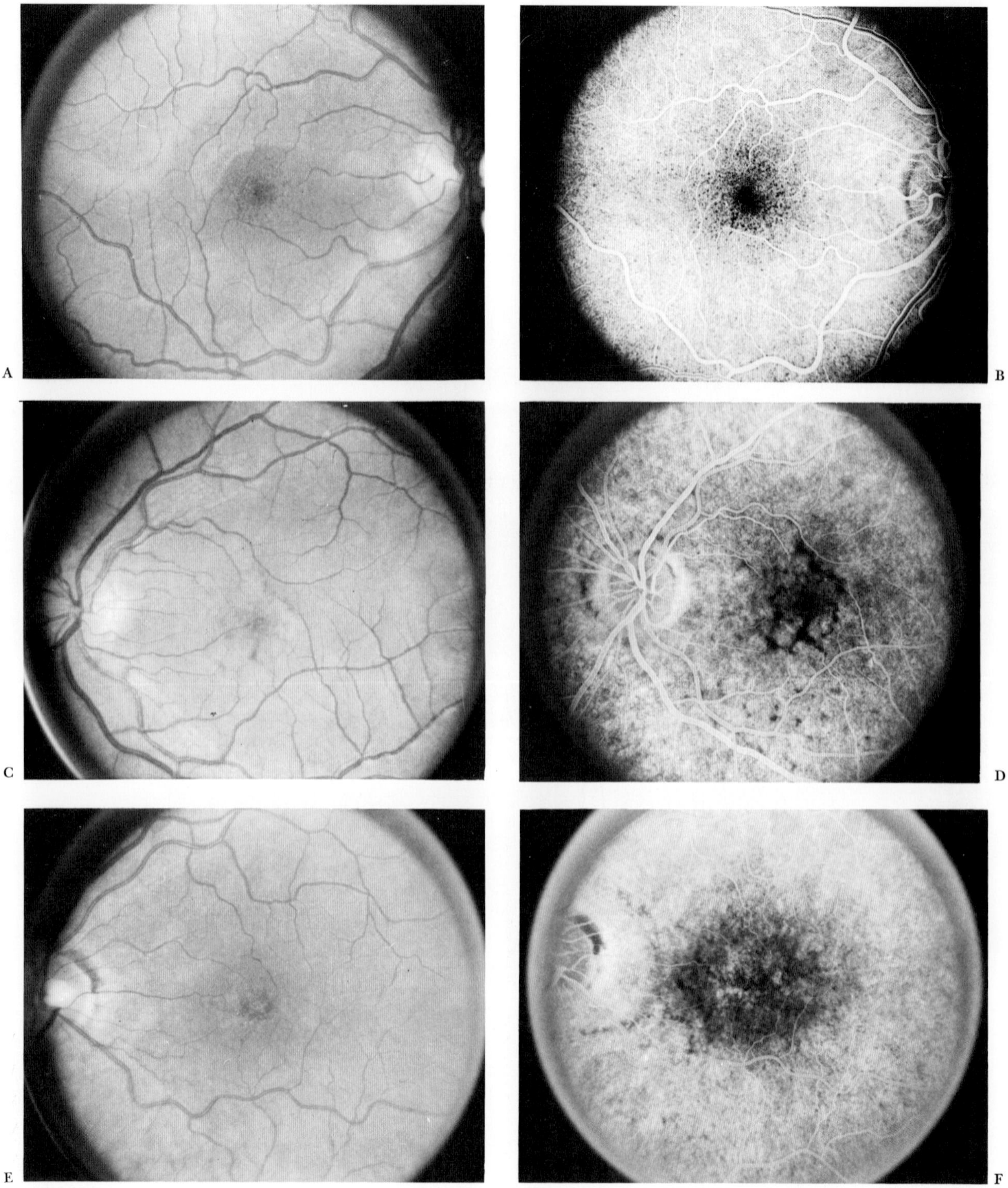

Figure 24.2 *Pattern dystrophy of the pigment epithelium. Fundus photographs and fluorescein angiograms showing the variation in pigmentation among 3 members of the same family (described in Marmor and Byers 1977).* A–B, *fifteen-year-old daughter; visual acuity was 20/20. Her fundus pigmentation was granular.* C–D, *twenty-year-old son; visual acuity was 20/25. His fundus pigmentation was reticular.* E–F, *forty-three-year-old mother; visual acuity was 20/25 to 20/30.*

diagnostic, since it can be borderline in individuals who have widespread degenerative changes in the pigment epithelium from other causes.

VITELLIFORM MACULAR DYSTROPHY. Vitelliform dystrophy, also called Best's disease (Best 1905) or vitelliruptive macular dystrophy (Krill et al. 1966) is a dominantly inherited disorder characterized by bilateral yellow subfoveal cysts (in the early stages) and a strikingly depressed EOG light response. The yellow cysts (Figs. 24.3*A*, 24.4*A*) resemble egg yolks and give the disease its name (from the Latin *vitellus*).

The lesions in vitelliform dystrophy may be present at birth (Barkman 1961), but are typically discovered between ages 3 and 15 (Deutman 1971). They have developed in patients who previously had a normal-appearing fundus (Deutman 1971; Barricks 1977). The typical cyst is 1–2 disc diameters across, mildly elevated, but clearly subretinal by contact lens examination. At this stage of the disease, the fovea overlying the cyst usually appears normal, and the visual acuity is minimally, if at all, affected. Using the Stiles-Crawford effect, Benson et al. (1975) showed that receptor orientation in the fovea is remarkably well maintained despite the elevation by cystic material.

A vitelliform cyst can persist for a number of years, but characteristically begins to break up in adolescence or early adulthood. The lesions may take on a "pseudohypopyon" appearance, or may resolve into an irregular complex of yellowish pigment or of gliotic and pigmented scar tissue (Figs. 24.3*B–C*). As the "egg yolk" becomes "scrambled," the visual acuity usually falls to between 20/50 and 20/200, depending upon the extent of the resulting scar. Rarely, a neovascular membrane may develop under a vitelliform scar, causing hemorrhage and more severe loss of acuity. On the other hand, some individuals resorb their cysts with minimal scarring and retain surprisingly good vision throughout their lives. Bard and Cross (1975) estimate that about 40 percent of people who have macular lesions maintain visual acuity in the best eye of 20/25 or better, and fewer than 10 percent have acuity worse than 20/200. However, the visual acuity can fluctuate and is rather unpredictable for any given individual.

The angiographic appearance of vitelliform dystrophy depends upon the stage of the disease. The

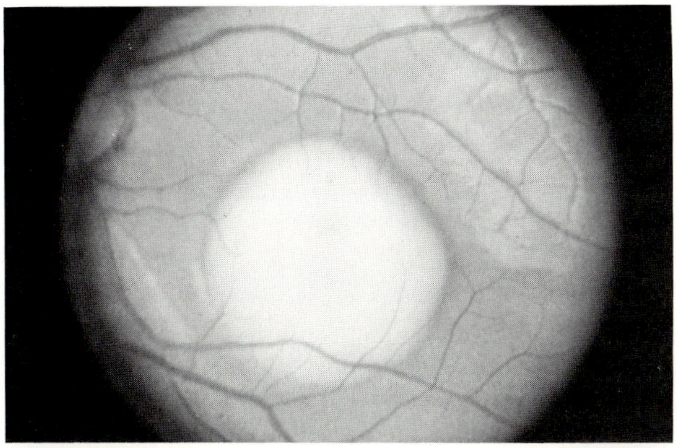

A

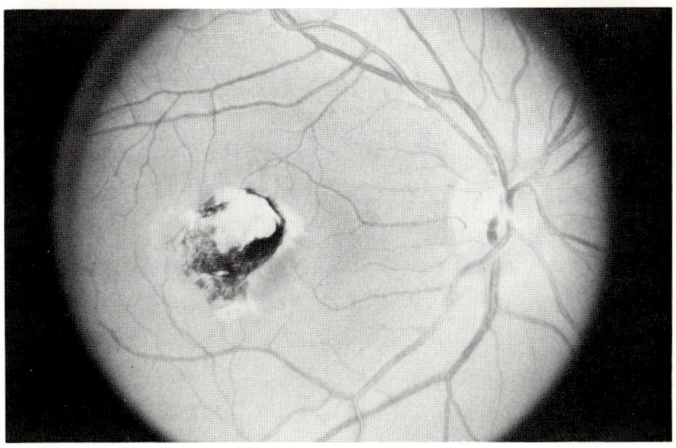

B

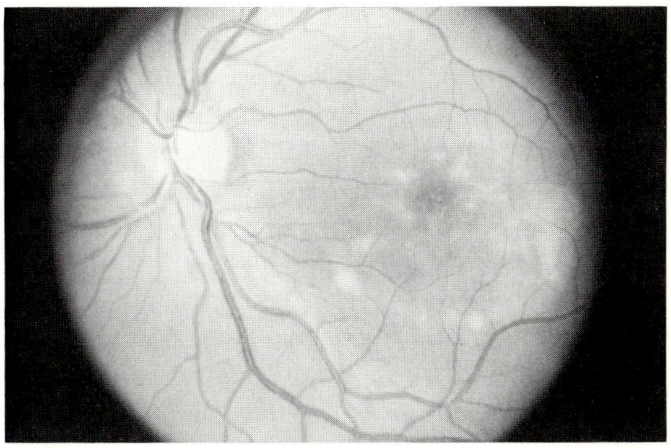

C

Figure 24.3 *Vitelliform macular dystrophy. Fundus photographs from two members of the same family.* A, *five-year-old son with 20/30 acuity.* B–C, *mother.*

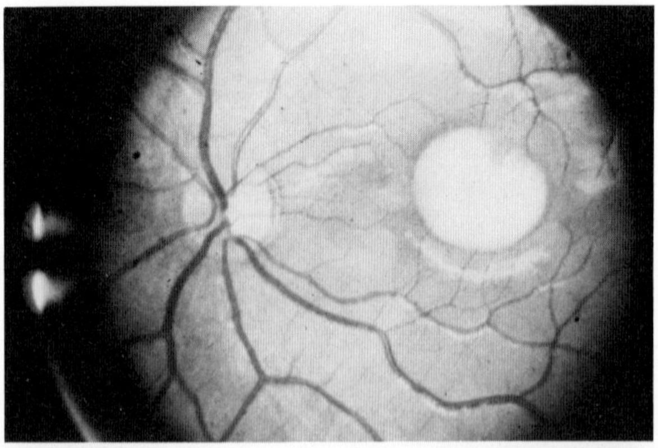

A

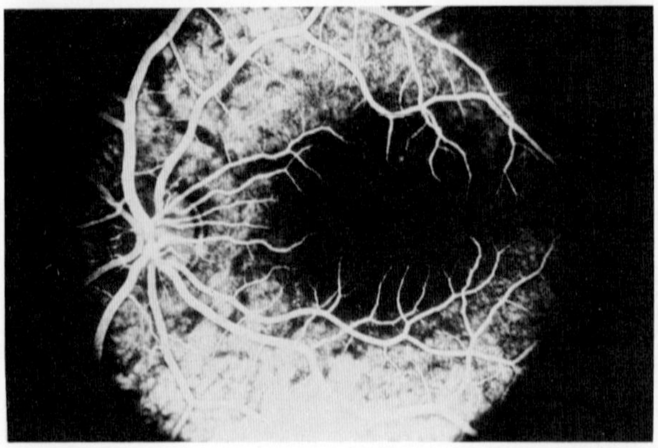

B

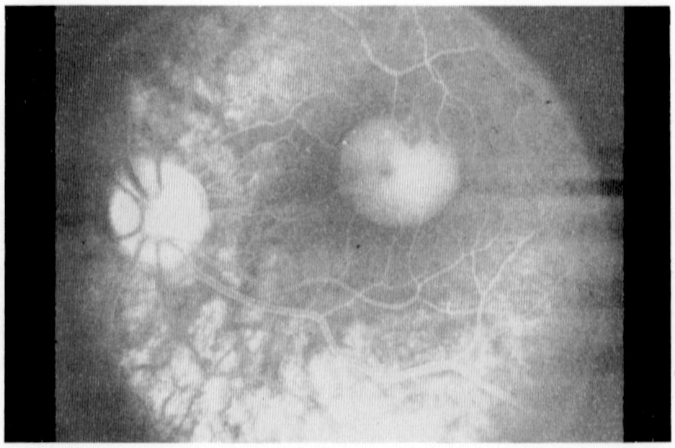

C

Figure 24.4 *Vitelliform macular dystrophy.* A, *fundus photograph.* B, *early phase of the fluorescein angiogram showing blockage of dye by the vitelliform cyst.* C, *late phase of the angiogram showing faint staining of pseudofluorescence.*

vitelliform cyst blocks fluorescein in the early phases of angiography, but may show slight staining or pseudofluorescence later on (Fig. 24.4). As the cyst breaks down, areas of damaged pigment epithelium transmit fluorescence and areas of scar tissue may show staining, an appearance that is indistinguishable from other end-stage macular lesions.

The peripheral visual field in vitelliform dystrophy is normal, but a central scotoma is present when vision is decreased. Patients may show varying degrees of acquired dyschromatopsia. Dark adaptation is normal, and the ERG is normal in the vast majority of cases. The most consistent abnormality is a severe depression of the EOG light response. Affected individuals almost always show a light/dark ratio less than 1.55 (Cross and Bard 1974), and the ratio is often close to 1.0.

The inheritance pattern of vitelliform dystrophy is autosomal dominant, but expression of the disease varies considerably. Bard and Cross (1975) estimate that only about two-thirds of genetically affected individuals have visible macular pathology. However, there is virtually complete penetrance of the EOG abnormality (Cross and Bard 1974; Krill 1977b). The low EOG is thus one of the most characteristic and diagnostic facets of this disease, and one which is all the more remarkable in the absence of any visible diffuse pathology in the fundus.

Typical cases of the dystrophy, with vitelliform lesions in childhood, a positive family history, and a severely depressed EOG, offer no diagnostic difficulty. The main problem in diagnosing vitelliform dystrophy is to differentiate it from nonspecific macular scars, and from bilateral yellowish macular lesions in older individuals (see the following section). The fundus findings can be extremely variable within a single pedigree of the dystrophy (Maloney, Robertson, and Duboff 1977), and once a vitelliform cyst has scarred down there is nothing at all specific about the lesion. Thus, the diagnosis of vitelliform dystrophy should be considered in any child or young adult with unexplained macular lesions. Some difficulty in diagnosis may also be posed by the occasional individual with multifocal or extramacular cysts (Deutman 1971; Miller 1977). These lesions may mimic pigment epithelial detachments or even an inflammatory condition such as placoid pigment epitheliopathy. In all of these cases, the usual criteria of a positive family history and a severely depressed

EOG should prove the diagnosis. The question of false negatives—that is, of whether the EOG can ever be normal in true vitelliform dystrophy—seems of dubious interest: the occurrence must be so rare that one should severely question the diagnosis unless all of the other circumstances (especially a strong family history) are present.

The precise location of the cystic material in vitelliform dystrophy is unknown (see Deutman 1971). The only available histologic material (McFarland 1955; Anderson, presented in Krill 1977b) is from older patients who had macular scars and showed only nonspecific pigment epithelial pathology. Krill (1977b) has speculated that the cystic material is within the cells of the pigment epithelium, whereas François (in François and Bonnet 1976) believes the material to be between the pigment epithelium and sensory retina. The severe abnormality of the EOG, despite a normal ERG and a normal appearing peripheral fundus, is generally cited (for example, Deutman 1971) as evidence for diffuse functional (if not visible) involvement of the pigment epithelium in this disease. At present we can only speculate on how a cellular abnormality can destroy the light response of the standing potential without having any obvious effect on the visual process.

VITELLIFORM LESIONS IN ADULTS. Adult patients will occasionally present with symmetric yellowish macular cysts of uncertain significance. These lesions may resemble those of vitelliform dystrophy (Fig. 24.5*A*), but they have a late onset and they occur in the absence of a dominant family history and, in general, in the presence of a normal EOG. Unfortunately, a problem in terminology is developing. The word *vitelliform* ("yolk-like") is by itself a purely descriptive term. However, the combined terms "vitelliform dystrophy" or "vitelliruptive dystrophy" have been widely accepted as referring only to Best's disease, a dominantly inherited condition characterized by macular lesions and a severely depressed EOG. The situation becomes obfuscated when patients without a family history of disease and with a normal EOG are described as having vitelliform *dystrophy* (Birndorf and Dawson 1973) or when lesions that do indeed look like egg yolks are called "*pseudo*vitelliform" (Fishman et al. 1977). These latter are true vitelliform *lesions,* but the patients may have a "pseudo-dystrophy."

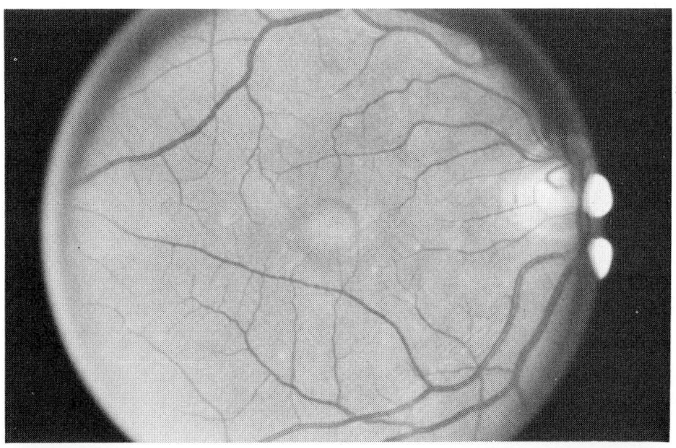
A

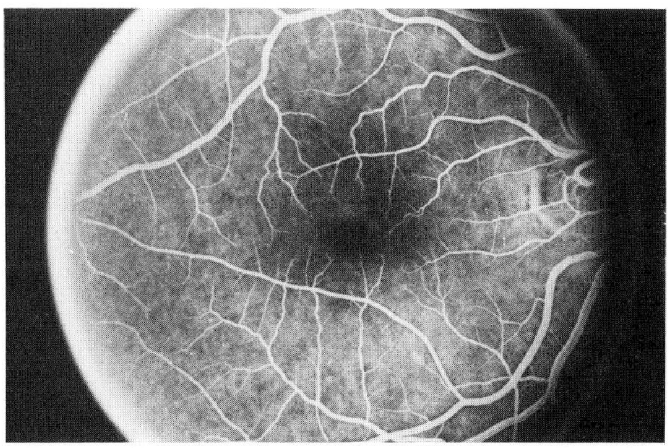
B

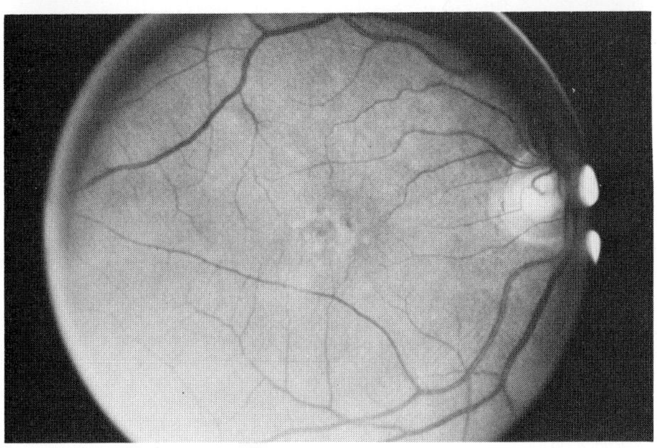
C

Figure 24.5 *Adult vitelliform lesion. Fifty-six-year-old male with symmetric foveal lesions; visual acuity was 20/40 on the right and 20/25 on the left. The EOG showed a light-to-dark ratio of 1.5 on the right and 1.7 on the left.* A, *lesion in the right fovea.* B, *fluorescein angiogram showing the absence of early fluorescence or dye leakage.* C, the *right macula two years later, showing a nonspecific scar; visual acuity was 20/50. (See Marmor 1979a.)*

Gass (1974) described a group of such patients as having a "peculiar foveo-macular dystrophy," and similar cases have been described by Rosenberg, Faurschou, and Nielsen (1977), Kingham and Lochen (1977), Fishman et al. (1977), and Marmor (1979a). These patients are typically 30–50 years old and have mild symptoms of visual loss or metamorphopsia. The foveal lesions may be small enough to raise questions about solar retinopathy, or large enough to mimic the cysts of vitelliform dystrophy. The yellowish material is clearly subretinal and is generally quite thick, so that fluorescein is blocked in the early stages of angiography (Fig. 24.5*B*). The peripheral field, dark adaptation, and ERG are normal, but the EOG may be either normal or mildly subnormal. Gass (1974) has noted that many of these vitelliform lesions in adults have pigment spots in the central portion. These lesions may gradually resolve into scars (Fig. 24.5*C*), which appear similar to those from old pigment epithelial detachments (Marmor 1979a). The cases described by Fishman et al. (1977) showed retinal capillary leakage in the fovea, but my own cases and the cases described by others have not. Some of Gass's patients were related, suggesting dominant inheritance, but most patients do not give a positive family history.

My own belief (Marmor 1979a), consistent with the views of Gass (1974) and of Kingham and Lochen (1977), is that these lesions represent a process similar to detachment of the pigment epithelium, and that they are related to drusen or presenile degenerative changes. It is not rare for patients to develop vitelliform lesions as a "complication" of preexisting drusen, pigment epithelial detachment, or exudative senile macular degeneration (Fig. 24.6). Only when such lesions are bilateral does the question of a new dystrophy arise. Some individuals with bilateral lesions may represent a different process, and possibly a unique dystrophy, but at present there is insufficient data to separate them.

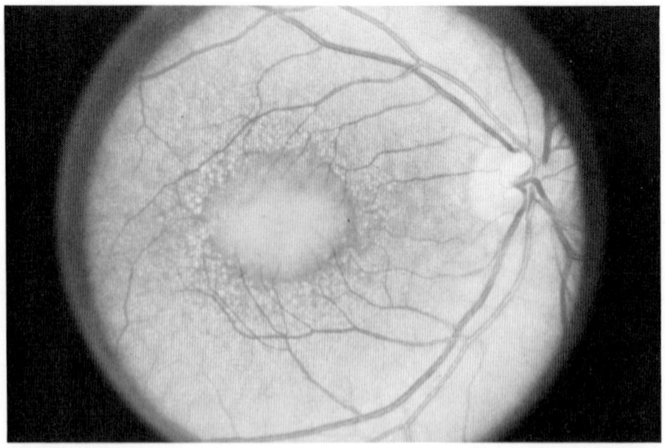

Figure 24.6 *Adult vitelliform lesion. Thirty-four-year-old female with drusen and symmetric foveal lesions; visual acuity was 20/25. The lesions blocked fluorescein during the early phases of angiography, but filled very slowly with dye later on. (See Marmor 1979a.)*

DRUSEN. The term drusen (from the German *Drüse* = gland) or colloid bodies refers to discrete rounded yellowish spots, of varying size, characteristically grouped in the macular region of both eyes, but sometimes found scattered more widely in the fundus (Fig. 24.7). On clinical examination, they appear to reside at the level of the pigment epithelium. Drusen are common in older individuals, and may be associated with varying degrees of pigment epithelial damage and macular degeneration. There is an argument in the literature about whether all drusen represent a dominant hereditary dystrophy, or whether some represent an acquired aging process.

Most elderly people with drusen do not give a positive family history, and family members are often unavailable for examination. However, where family members have been studied, autosomal dominant inheritance has generally been found (Deutman 1971; Krill 1977c). Gass (1973) doubts that any valid distinction can be made between familial and senile drusen. Because the pattern of drusen in the eye is so variable, over the years pedigrees have been given discrete names (for example, Doyne's honeycomb dystrophy, malattia léventinese, Tay's guttate choroiditis), which it serves no purpose to perpetuate (Deutman 1971).

The clinical course and significance of drusen are extremely varied. In younger individuals they are often an incidental finding, but they may be associated with serous exudation, pigment epithelial detachments (see Fig. 24.6), or other manifestations of senile macular degeneration. The degree of visual loss with drusen is usually related to the secondary changes that occur (a more complete discussion of these changes will be found in Chapter 26). Individuals with extensive drusen are probably at greater risk than others for the early development of macular degeneration, but by no means are degenerative changes certain to appear in any given individual, no matter how dense the drusen may be.

The angiographic appearance of drusen is quite characteristic (Fig. 24.7*D*). They light up early in

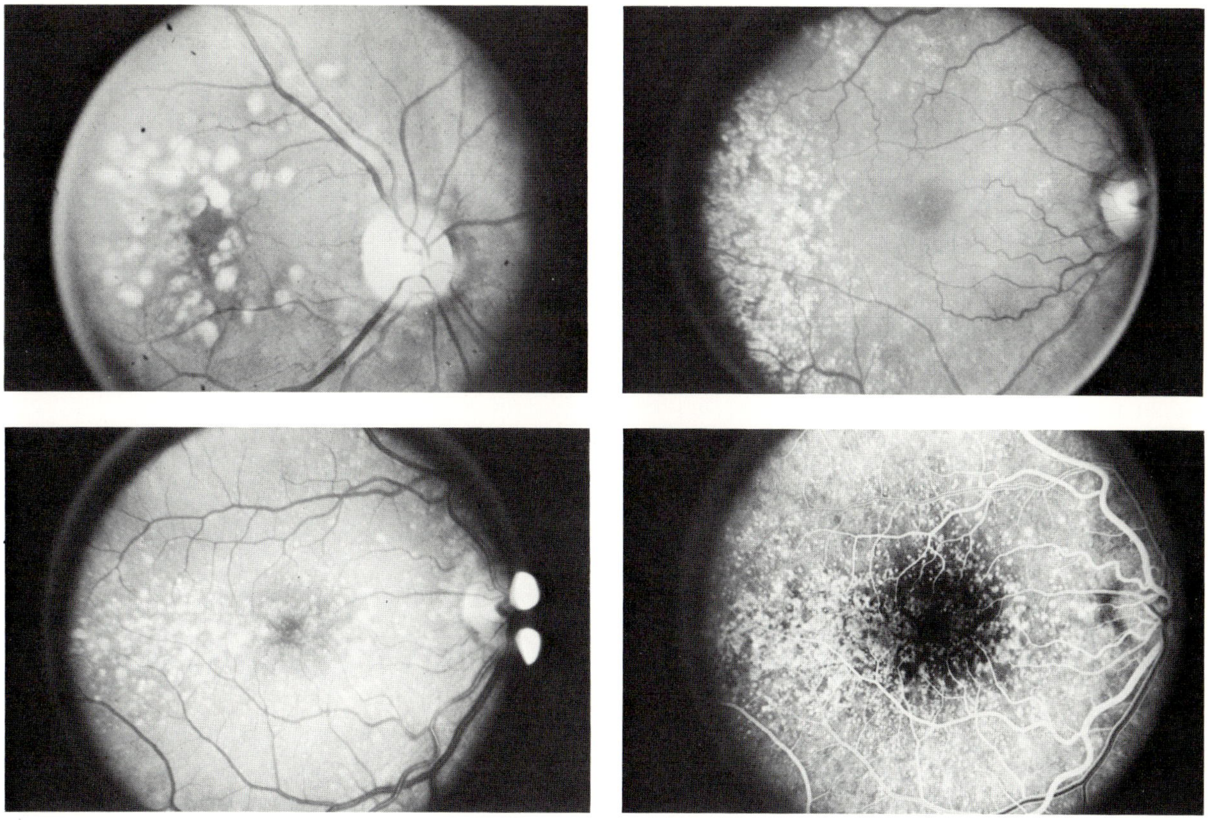

A B C D

Figure 24.7 *Drusen.* A, *large central drusen.* B, *finer, more peripheral drusen.* C, *macular drusen.* D, *fluorescein angiogram of the same eye showing the drusen in sharp relief. There was no leakage in later pictures.*

the angiogram, because the thinned pigment epithelium overlying the drusen allows the choroidal fluorescence to be seen. The drusen continue to fluoresce late in the angiogram, probably because of a combination of scleral fluorescence and staining of the lesions themselves, but there is no leakage of dye.

The visual field and color vision are generally normal unless macular degeneration has begun. Dark adaptation is usually normal, although Krill and Klien (1965) found an occasional very mild defect. The ERG is normal, except for some elderly individuals who show a mild diminution of signals, which in my experience is not uncommon in the general elderly population. The EOG findings in drusen have been of considerable interest because of arguments about whether the disease is fundamentally focal or diffuse. In individuals whose drusen are strictly limited to the macula, the EOG light response is generally normal; the EOG light response may be subnormal in individuals with very widespread drusen, or in the elderly population (Deutman 1971; Gass 1973).

The histology and histochemistry is better known for drusen than for any of the other dystrophic lesions we are considering (see also Chapter 26). Drusen are characterized by a hyaline thickening of Bruch's membrane, which pushes up into the pigment epithelial layer, compressing the pigmented cells and leading ultimately to pigment epithelial degeneration. The hyaline material is composed predominantly of mucopolysaccharides (sialomucins) and cerebroside lipids (Farkas et al. 1971b). The histochemical characteristics seem the same in hereditary forms and sporadic lesions (Farkas et al. 1971a).

There is no unanimity of opinion on the origin of drusen, and an extensive literature exists to support each of several alternatives (for example, see discussions by Krill 1977c; Gass 1973; Hogan 1972). Different authorities contend that drusen derive from vascular changes in the choroid, from a primary degeneration of Bruch's membrane, from the secretion of material by the pigment epithelial cell, or even from retinal changes. In light of our current knowledge about the phagocytic and digestive properties of the pigment epithelium, several authors (for example, Farkas et al. 1971b; Hogan 1972) have proposed that a breakdown of the pigment epithelial mechanisms for digesting and clearing photoreceptor waste products leads to the accumulation of debris in Bruch's membrane.

FUNDUS FLAVIMACULATUS. The term fundus flavimaculatus, proposed by Franceschetti (see Franceschetti and Francois 1965), describes a spattering of irregular ill-defined yellowish flecks, which appear clinically to lie at the level of the pigment epithelium, and which may be concentrated either in the posterior pole or midperiphery of the fundus (Fig. 24.8). The lesions tend to evolve into small pigment epithelial defects. On angiography, the fresh lesions block fluorescein (suggesting the presence of an abnormal substance) and the older lesions transmit fluorescence. In a well-developed case the fundus shows a complex of fresh lesions interspersed with focal regions of depigmentation. The inheritance of the disease is usually autosomal recessive, although an occasional dominant pedigree has been reported.

The clinical appearance and consequences of this disease are quite variable. The lesions may cover much of the fundus or be limited solely to the foveal region. In perhaps 50 percent of cases (Deutman 1971) the lesions are associated with macular degeneration and a loss of central vision during youth or early adulthood. This form of fundus flavimaculatus, associated with macular degeneration, is indistinguishable from Stargardt's disease, and most authors nowadays consider the two terms to be synonymous. There seems to be a continuum of disorders, ranging from purely peripheral fundus flavimaculatus without macular changes, to the more typical forms of Stargardt's disease, which are characterized by a ring of flecks about the fovea and a degeneration of the macular photoreceptors. Our primary concern in this section on pigment epithelial dystrophies is with

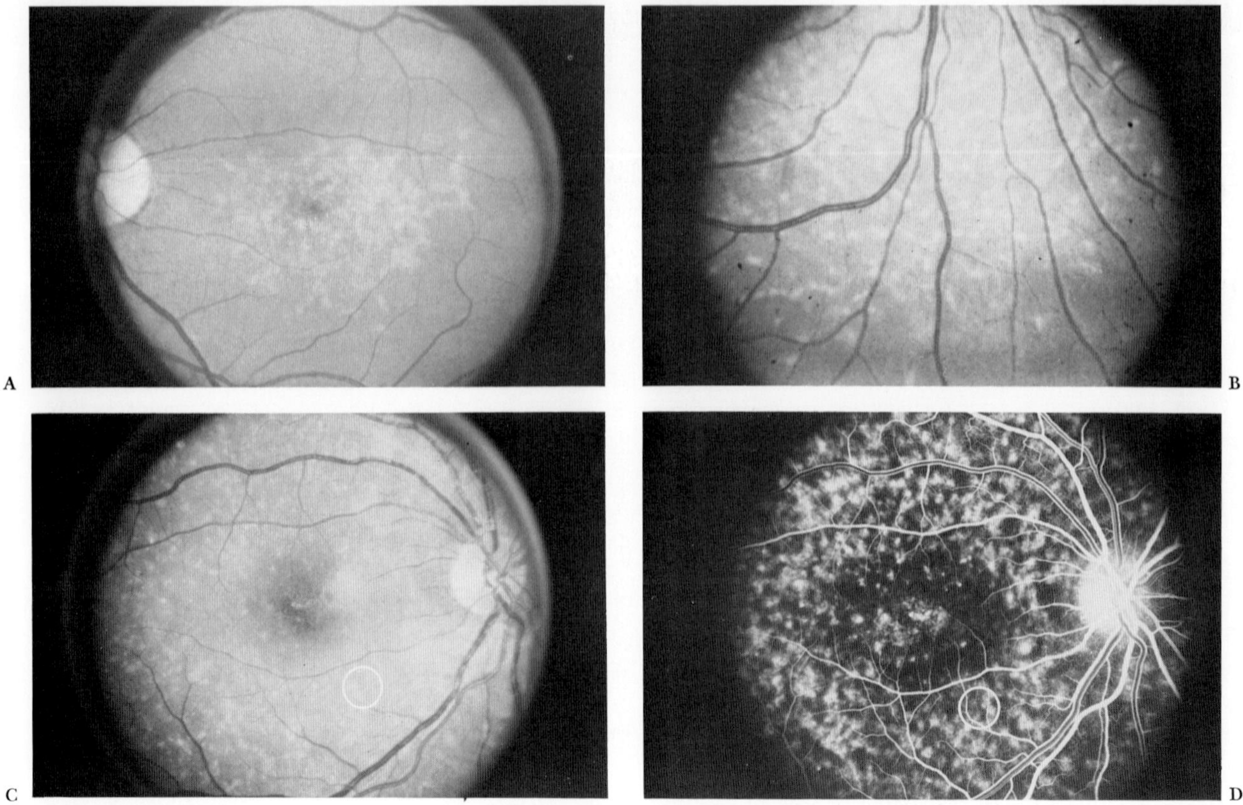

Figure 24.8 *Fundus flavimaculatus.* A, *foveal flavimaculatus with macular degeneration and 20/200 acuity.* B, *peripheral flavimaculatus in a different patient.* C, *macular and peripheral flavimaculatus in a 13-year-old, with 20/200 acuity. The circle encloses both early (toward the left) and late lesions.* D, *fluorescein angiogram from the same patient. Note that the early lesion in the circle does not transmit fluorescence readily as does the older lesion.*

the flavimaculatus lesions themselves, which are not necessarily associated with any visual symptoms even when they are quite extensive.

In routine testing, dark adaptation has been reported as normal in many cases of fundus flavimaculatus, but Krill (Klien and Krill 1967; Krill, 1977c) has argued that a mild prolongation of dark adaptation is characteristic of fundus flavimaculatus. This is of particular interest because dark adaptation is severely abnormal in another condition with flecks and dots in the fundus, namely fundus albipunctatus. The ERG and the EOG in fundus flavimaculatus may be either normal or subnormal, depending on the degree of diffuse pigment epithelial and retinal involvement. The EOG light response is not as severely depressed as in vitelliform dystrophy.

The judgment that the flavimaculatus lesions are localized in the pigment epithelium rests partly on their clinical appearance and partly on the findings in one case that was studied histologically by Klien and Krill (1967). The retina showed extensive autolytic changes, but pigment epithelial cells within the lesions showed an accumulation of material that stained as if it were a mucopolysaccharide. Later electron-microscopic examination of the same tissue (Newell, Krill, and Farkas 1972) showed excrescences of foamy material from the apex of the pigment epithelial cells. Confirmation from a younger and better preserved case is needed.

The differential diagnosis of fundus flavimaculatus poses no problem when the typical lesions are visible, since no other fundus disorder has quite the same appearance. Peripheral lesions in some fundus albipunctatus cases may be similar, but the functional disorder is easily discriminated. Flecks that appear after choroidal embolization or crystalline deposition (see below) should rarely, if ever, cause confusion. The main concern in evaluating fundus flavimaculatus is to determine the degree of macular degeneration (which defines the visual prognosis).

FUNDUS ALBIPUNCTATUS. Fundus albipunctatus is a rare stationary disorder characterized by myriads of white dots in the fundus and an abnormally slow rate of dark adaptation. Lauber (1910) distinguished fundus albipunctatus from retinitis punctata albescens, which can have a similar fundus appearance but is a severe progressive dystrophy akin to retinitis pigmentosa. Some authors feel these disorders form a continuum (Franceschetti, Francois, and Babel 1974), but there is convincing evidence (Carr 1974; Marmor 1977a) that they are discrete and separate entities. Carr, Ripps, and Siegel (1974) studied two brothers with fundus albipunctatus and showed, by fundus reflectometry, that the slow dark adaptation that characterized the disorder clinically corresponded to an abnormally slow rate of visual pigment regeneration.

The inheritance of fundus albipunctatus is autosomal recessive. The disease is probably present from birth or early childhood, with little or no progression thereafter (Marmor 1977b). Affected individuals complain not simply of "night blindness," but of impaired adaptation upon any change from bright to dimmer illumination. One of my patients first noted difficulty on going into a dimly lit dressing room from a sunny beach. Another stated that he could not see the seats in a darkened movie house until near the end of a double feature.

The characteristic fundus lesions (Fig. 24.9) are small yellow-white dots, about the size of retinal arterioles, which appear to most observers, including myself, to be situated at the level of the pigment epithelium (although Carr, Margolis, and Siegel [1976] feel that some lesions are more superficial). The fundus lesions may be scattered irregularly about the posterior pole (normally sparing the fovea) or they may show a striking pattern of radiation out from the macula (Fig. 24.9*A–C*). As one moves peripherally from the macula, the lesions tend to become less punctate and more diffuse or fleck-like, taking on an appearance similar to fundus flavimaculatus lesions. The lesions are basically stable, unlike the continually evolving ones in fundus flavimaculatus, but they are not absolutely constant. Close monitoring by fundus photography (Marmor 1977b) has showed that isolated lesions can come or go at the same location over a period of several months. With fluorescein angiography (Fig. 24.10), the albipunctate lesions are either invisible or block fluorescein, and a mild diffuse pigment epithelial abnormality may be present, which does not correlate with the punctate lesions.

Visual acuity in fundus albipunctatus is normal or only minimally affected, as is the visual field. Careful color testing shows a very mild dyschromatopsia (Marmor 1977b). The most characteristic psychophysical finding in this disorder is the delayed dark adaptation (Fig. 24.11). Whereas a normal subject (after a bleaching exposure to light) will reach a cone threshold in about 7 minutes and a final rod threshold in about 30 minutes, patients with fundus

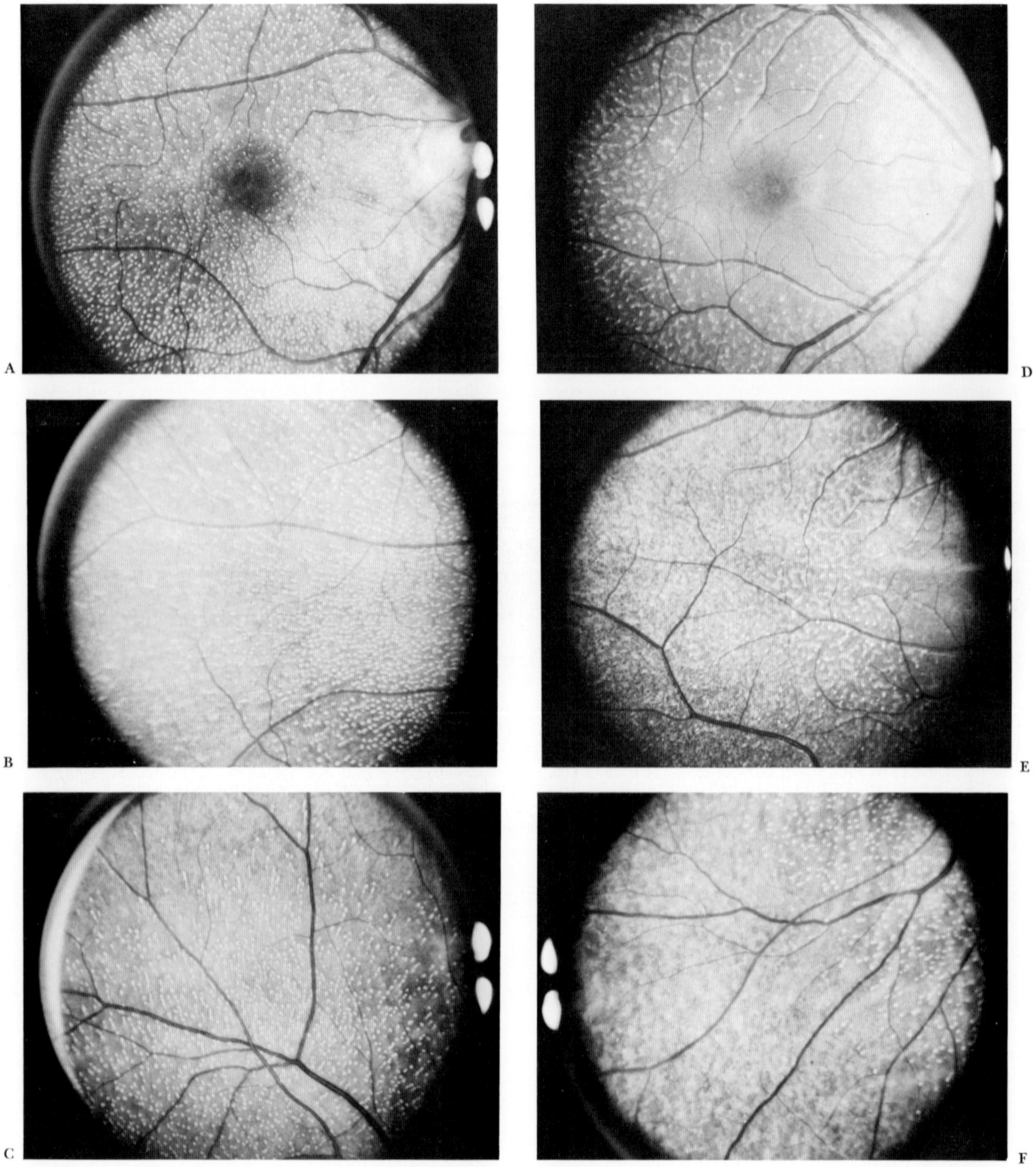

Figure 24.9 *Fundus albipunctatus. Fundus photographs from two unrelated patients (described in Marmor 1977b).* A–C, *thirty-three-year-old male with 20/20 acuity. Note the striking radial orientation of the punctate lesions.* D–F, *eight-year-old female with 20/20 acuity. The lesions in this patient are more irregular and become flecklike in the periphery.*

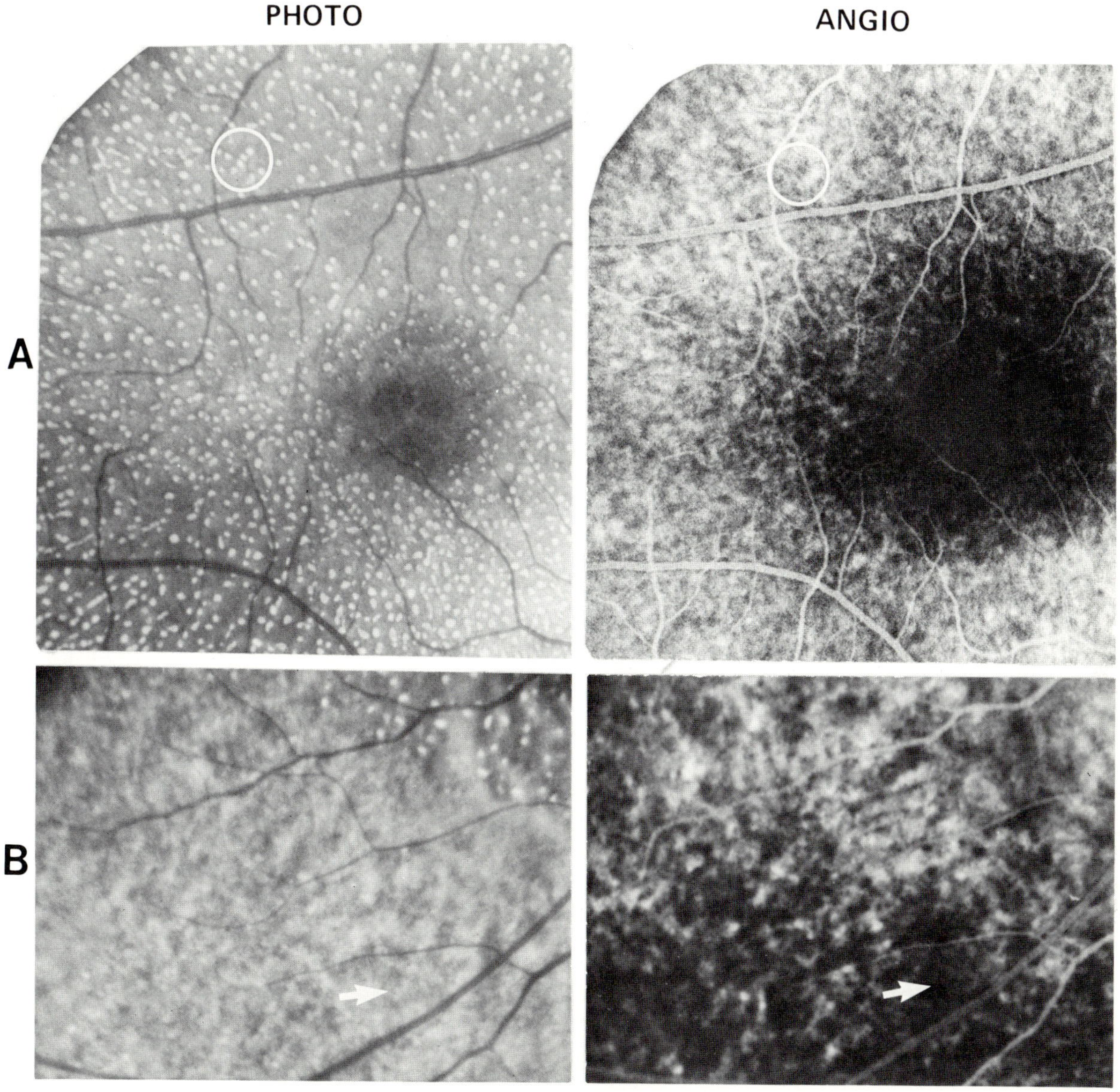

Figure 24.10 *Fundus albipunctatus. Comparison of fundus photographs with fluorescein angiograms from the cases shown in Figure 24.9.* A, *macula of the 33-year-old male. The circle encloses a row of four punctate lesions that block, rather than transmit, fluorescence.* B, *more peripheral view of the 8-year-old female. The arrow shows flecks that are difficult to identify on the angiogram.*

albipunctatus require 1–2 hours to reach normal cone threshold and 3–4 hours to reach normal rod threshold. Electrophysiologic findings correlate exactly with the state of dark adaptation. The scotopic ERG shows no rod function after 15 minutes of dark adaptation, but reaches normal amplitude after 3–4 hours of dark adaptation. Similarly, the EOG shows no light rise if performed in the usual manner (with 10–15 minutes dark adaptation), but will show a normal light rise if an eye has been fully dark adapted (more than 4 hours) before exposure to light.

White dots in the fundus have also been observed (Fig. 24.12) in vitamin A deficiency (Uyemura 1928; Teng 1965; Levy and Toskes 1974; Fells and Bors 1969; Sommer et al. 1978), and in abetalipoproteinemia (see below), in which there is a relative vitamin A deficiency (Gouras et al. 1971). However, recent studies have shown that vitamin A and retinol binding protein levels are normal in fundus albi-

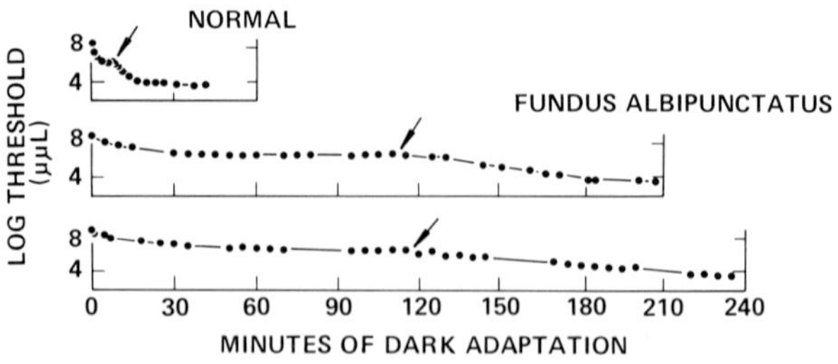

Figure 24.11 *Fundus albipunctatus. Dark-adaptation curves from a normal subject and from the cases of fundus albipunctatus shown in Figure 24.9. An arrow shows the cone-rod break on each curve. (After Marmor 1977b.)*

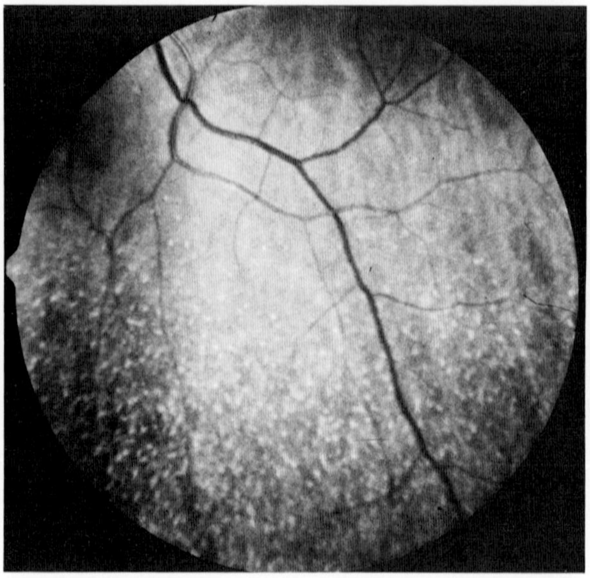

Figure 24.12 *Vitamin A deficiency. Note the punctate white lesions. (From Levy and Toskes 1974.)*

punctatus (Carr, Margolis, and Siegel 1976; Marmor 1977b). Furthermore, the administration of vitamin A in high doses, which may reverse the lesions of vitamin A deficiency (Bors and Fells 1971; Sommer et al. 1978), has no effect on either the fundus lesions or the functional deficit of fundus albipunctatus. The defect in this disorder seems likely to reside within ocular systems for transporting or regenerating vitamin A, but the precise abnormality and its relationship to the white lesions remains obscure. The most likely sites for the metabolic dysfunction are the pigment epithelium (especially if the albipunctate dots are in fact located within the pigment epithelium) or the outer segments of the photoreceptors.

The major concern in the differential diagnosis of fundus albipunctatus is to distinguish it from the progressive dystrophy, retinitis punctata albescens (see below). The lesions of fundus flavimaculatus may occasionally resemble those of fundus albipunctatus, but the more severe dark-adaptation defect of the latter should allow differentiation. The vitamin A deficiency disorders noted above should pose no problem in differential diagnosis, nor should disorders such as systemic oxalosis or cystinosis in which crystalline material is deposited in the fundus (see below). Because the term "fundus albipunctatus" describes a well-defined disorder, it should not be used as a descriptive term for these other conditions.

RETINITIS PUNCTATA ALBESCENS. Retinitis punctata albescens is a hereditary dystrophy with white dots in the fundus, which shows the clinical and electrophysiologic characteristics of retinitis pigmentosa. The punctate lesions may appear similar to those in fundus albipunctatus, but the fundus also shows diffuse retinal and pigment epithelial atrophy, arterial narrowing, disc pallor, and often some black pigment spicules intermingled with the white dots (Fig. 24.13). The symptoms include permanent night blindness, peripheral field loss, and sometimes a loss of central vision.

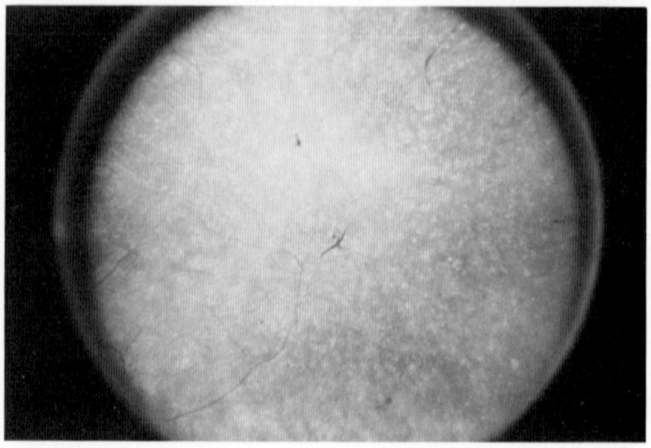

Figure 24.13 *Retinitis punctata albescens. Twenty-six-year-old male with an unrecordable ERG.*

In the European literature (see Franceschetti, Francois, and Babel 1974), retinitis punctata albescens is often considered to be a severe form of fundus albipunctatus. Others, including Lauber (who first separated the disorders in 1910), Carr (1974), and myself (Marmor 1977a) feel that they are quite distinct—one is a stationary disorder of visual pigment regeneration, while the other is a progressive dystrophy that permanently damages the

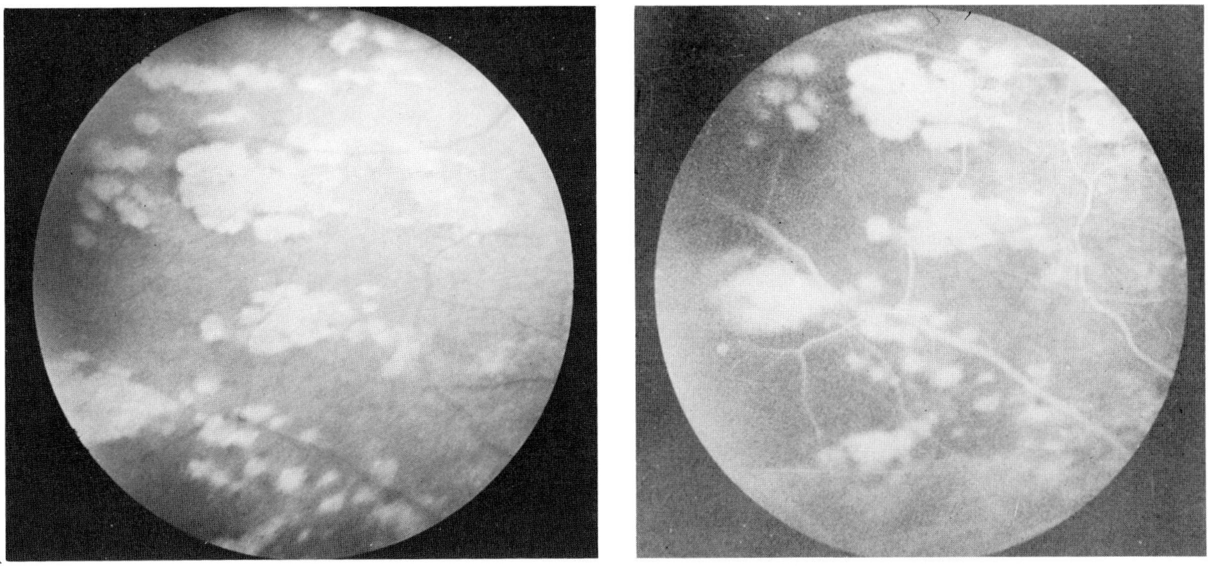

A B

Figure 24.14 *Fleck retina of Kandori.* A, *fundus photograph.* B, *fluorescein angiogram. (From Kandori et al. 1972.)*

photoreceptors and pigment epithelium. Still, it is intriguing to speculate on whether there are any common factors in the genesis of the punctate lesions seen in retinitis punctata albescens and fundus albipunctatus.

FLECK RETINA OF KANDORI. Fleck retina, first described by Kandori (1959), is a rare and stable affliction in which large irregular dirty yellow patches (Fig. 24.14) are present in the fundus periphery; it is associated with a very mild abnormality of dark adaptation.

The lesions appear to be congenital, since they have been observed at a young age and no progression has been observed over 15 years of follow-up (Kandori et al. 1972). The inheritance is probably autosomal recessive, if the condition is hereditary at all. The characteristic fundus lesions are irregularly shaped patches ranging from the size of drusen to as much as a disc diameter across. On fluorescein angiography the lesions transmit fluorescence early in the arterial phase and show sharp borders in a fashion similar to drusen, although they have not been associated with drusen in the fundus. Visual acuity is normal, but most patients complain of decreased night vision all of their lives. Dark adaptometry shows a very mild prolongation of both the photopic and scotopic limbs of the dark adaptation curve, so that the patients' symptoms probably relate to slow rather than deficient adaptation. Color vision, ERG's, and EOG's are normal in the majority of cases. Kandori et al. (1972) speculate that the disease is localized to the pigment epithelium, but is of patchy distribution so that tests of diffuse function such as the ERG and EOG remain normal. Fundus reflectometry has not been done (to determine whether any abnormality exists in the regeneration of visual pigment), and there are no histologic reports to date.

OTHER DISORDERS THAT PRODUCE SCATTERED YELLOW LESIONS IN THE FUNDUS. Many of the disorders described earlier show scattered yellow or white lesions in the fundus, but there should be little difficulty in separating them diagnostically because of their distinctive clinical features or electrophysiologic findings. Lumping these disorders under a term such as "flecked retina syndrome" (Krill and Klien, 1965) serves only to obscure their well-defined differences. The "pattern dystrophies" were grouped together because they share fundus findings, electrophysiologic characteristics, and prognosis. The so-called "flecked retina diseases"—drusen, fundus flavimaculatus, fundus albipunctatus, retinitis punctata albescens, and fleck retina of Kandori—differ in these characteristics, and in my opinion the term "flecked retina syndrome" should be abandoned.

There are also a variety of systemic or nondystrophic conditions that are characterized by scattered yellow lesions in the fundus which may superficially resemble those in the pigment epithelial dystrophies we have described. A few examples will show the spectrum of conditions that can produce multifocal lesions of the pigment epithelium. These conditions are clearly distinct from the dystrophies described earlier, and for reasons outlined in the preceding paragraph should *not* be referred to as flecked retina diseases.

Systemic oxalosis (Bullock and Albert 1975; Gottlieb and Ritter 1977) can cause a diffusely spotted fundus (Fig. 24.15) because of deposition of calcium

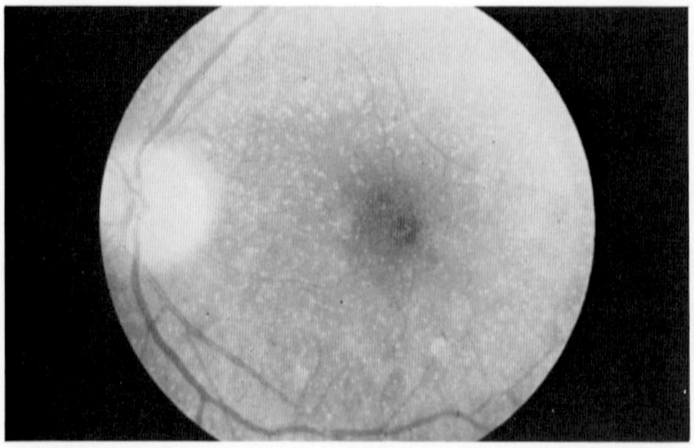

Figure 24.15 *Retinal oxalosis. The punctate lesions are a result of crystal deposition in the pigment epithelium. (From Bullock and Albert 1975; copyright 1975, American Medical Association.)*

oxalate crystals in the pigment epithelium (Bullock and Albert 1975; Timm 1976). The lesions are more punctate than fleck-like and might conceivably be confused with fundus albipunctatus. The oxalosis can result from toxic damage to the kidneys, as may occur after methoxyflurane anesthesia (Bullock and Albert 1975), or from a primary metabolic abnormality (Gottlieb and Ritter 1977). Diffuse pigment epithelial damage can also occur in cystinosis, apparently because of crystal deposition in the pigment epithelium (Francois et al. 1972; Sanderson et al. 1974). These pigmentary changes may be associated with minimal functional abnormality (Francois et al. 1972), with decreased vision if the maculae are involved (Sanderson et al. 1974), or with a diminished ERG in severe cases (Read et al. 1973).

An acquired cause of fleck-like lesions, which has only recently been reported, is inadvertent embolization of the choroidal circulation. This can occur following the injection of steroid particles into the nasal turbinates, via anastomoses between branches

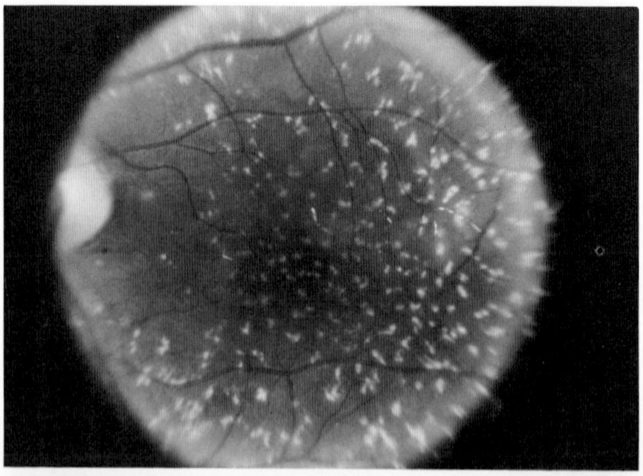

A

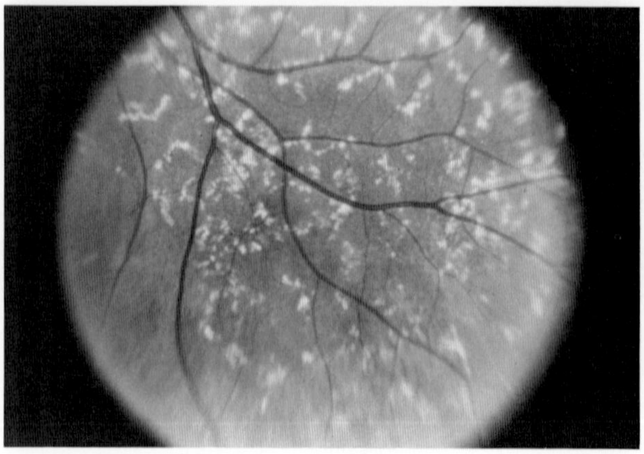

B

Figure 24.16 *Embolization of the choroidal circulation. The patient noted loss of vision following steroid injection into the nasal turbinates.* A, *macula showing diffuse choroidal lesions. Emboli are also visible in the retinal circulation.* B, *inferior and temporal view. (From Byers 1978; copyright 1978, American Medical Association.)*

of the external carotid and ophthalmic arteries (Byers 1978). The fundus appearance (Fig. 24.16) can be quite suggestive of fundus flavimaculatus, but the history of events should make the diagnosis evident. The ability of vascular occlusion to cause this pigmentary pattern is intriguing and suggests the possibility that vascular factors also may contribute to the pathogenesis of the fleck-like lesions in the dystrophies.

We have already considered, in the section on fundus albipunctatus, some of the systemic disorders that may produce white dots in the fundus, namely vitamin A deficiency (Fig. 24.12) and abetalipoproteinemia (Fig. 24.22). Similar lesions have recently been noted in patients with pancreatitis (Toskes, Dawson, and Fitzgerald 1978), in whom the dark adaptation and ERG abnormalities are dependent on zinc as well as vitamin A metabolism. Yellow punctate and fleck-like lesions have also been documented in several patients with hereditary nephropathies (Peterson and Albert 1974), but the functional significance of these lesions is not known.

OTHER PIGMENT EPITHELIAL DYSTROPHIES. New dystrophies are continually being described which involve the pigment epithelium to a greater or lesser degree, but proving them to be primary pigment epithelial dystrophies can prove extremely difficult. A few examples of dominantly inherited dystrophies will demonstrate the spectrum of fundus changes that have been reported.

Frank et al. (1974) described a dominant pedigree in which a progression of foveal changes reminiscent of drusen resulted in poor vision and a central scar. The ERG and EOG were reported to be normal. Fetkenhour et al. (1976) described a central areolar dystrophy in which nonprogressive depigmented lesions lay under the fovea, but were not associated with either visual loss or abnormality of the ERG and EOG. Singerman, Berkow, and Patz (1977) described patients with a complex of pigmentary changes and yellowish flecks in the macula who showed a slow progressive loss of acuity and a diminished ERG. Deutman (1974) reported a family with depigmented annular or bull's-eye lesions in the macula, with minimal loss of acuity, and with variable changes in the ERG and EOG. Noble, Carr, and Siegel (1977) reported an autosomal dominant disorder with widespread and nonprogressive thinning of the pigment epithelium that was present from a young age. The visual acuity and the ERG were often, but not invariably, diminished. A dominant cystoid macular dystrophy was recently described by Pinckers and colleagues (Deutman, Pinckers, and Aan de Kerk 1976; Notting and Pinckers 1977). It was characterized by dilatation and leakage of the macula capillaries (leading to an angiographic appearance of cystoid edema), and by a diffuse disturbance of the pigment epithelium. The ERG was generally normal, but the EOG was reduced.

ANIMAL MODELS. None of the human dystrophies listed above can be described, with certainty, as a purely pigment-epithelial disease. Lest the reader fear that such a condition may not exist, a brief description is provided of two animal disorders that primarily involve the pigment epithelium.

First is pigmentary dystrophy of RCS rats (see Chapter 20). Recent work by LaVail and colleagues (Mullen and LaVail 1976) has demonstrated that the pigment epithelium in these animals is unable to phagocytize the tips of the rod outer segments. Thus the outer segments grow unchecked until the debris accumulates and the photoreceptors finally degenerate. This disorder has been considered a model for retinitis pigmentosa, although there is no evidence that the human disease has a similar pathogenetic mechanism. Nonetheless, the RCS rat provides a superb model of pigment epithelial disease and is teaching us how to approach the study of retinal dystrophies.

Second is a pigment epithelial dystrophy in the dog, called central progressive retinal atrophy (see Chapter 19). Aguirre and Laties (1976) have shown histologically that the earliest abnormality in this disease is hypertrophy of the pigment epithelial cells with an accumulation of autofluorescent light-brown pigment that resembles lipofuscin. Only later, as epithelial damage spreads, does the sensory retina become involved.

Retinal and Choroidal Dystrophies

The pigment epithelium is visibly involved in the majority of chorioretinal dystrophies, even when it is not the primary site of disease. In some of these disorders, the pigment epithelial involvement is purely secondary, but in others it can be an intimate part of the disease process. Retinitis pigmentosa, for example, is characterized by early damage to the photoreceptors, but the causative metabolic defect could reside in either the photoreceptors or pigment epithelium. Choroidal dystrophies cause a loss of vascularity beneath the pigment epithelium, which may affect retinal function directly or by way of damage to the pigment epithelium.

To describe all of the hereditary chorioretinal diseases would be far beyond the scope of this chapter; several comprehensive books on the subject have been published in recent years (for example, Vinken and Bruyn 1972; Franceschetti, Francois, and Babel 1974; Krill 1977a). My purpose in this section is only to outline the major categories of chorioretinal dystrophy in order to demonstrate how the pigment epithelium may be involved in their pathogenesis or clinical appearance.

PHOTORECEPTOR DYSTROPHIES. The prototype diffuse photoreceptor dystrophy (rod-cone dystrophy, tapetoretinal dystrophy) is retinitis pigmentosa. The term "retinitis pigmentosa" (Donders 1857) is a misnomer since the condition is not of inflammatory origin, and as generally used the term does not refer to a single disease but to a group of disorders which, broadly speaking, are characterized by hereditary transmission and a slowly progressive degeneration of the photoreceptors. Autosomal-recessive, autosomal-dominant, dominant with reduced penetrance, and x-linked forms are known. The same disorder, insofar as vision and physiologic abnormalities are concerned, can exist with typical pigment spicules (Fig. 24.17), without spicules (retinitis pigmentosa "sine pigmento"), or with white dots in place of black ones (retinitis punctata albescens). However, there is always equatorial atrophy of the pigment epithelium and retina, and marked narrowing of the vasculature. The disc often has a waxy pallor, the macular reflex is usually dull, and some patients show cystoid edema or frank macular atrophy. Other associations include myopia and posterior subcapsular cataracts. Although "typical" cases are easy to define, the reader should be aware that many patients have atypical combinations of cone and rod, or peripheral and central, involvement. The dividing line between retinitis pigmentosa and other photoreceptor dystrophies, such as the cone dystrophies or Stargardt's disease, is not always clear.

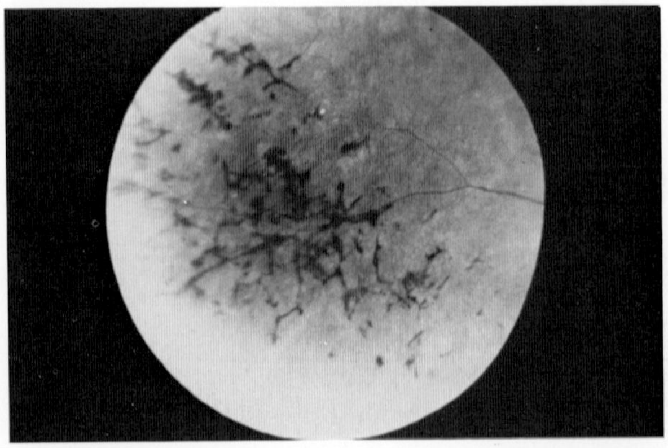

Figure 24.17 *Retinitis pigmentosa.*

Symptoms usually begin in the teens, first with an awareness of night blindness, and later with loss of peripheral field. A small island of central vision is preserved in many patients, but, on the average, visual acuity falls with age (Pearlman 1978). The loss of acuity is relatively abrupt (occurring over 5–6 years) so that most individuals, at any given moment, have either very good or very poor acuity (Marmor, 1979c). Fortunately, it is very rare for individuals to go totally blind. Physiologic testing typically shows an acquired dyschromatopsia, severely impaired dark adaptation, and a ring scotoma. The ERG shows severe diminution in amplitude and usually a delay in the timing of the responses (Berson, Gouras, and Hoff 1969; Marmor 1979b), often before ophthalmoscopic changes are obvious. The ERG and EOG light response are often extinguished by the teens.

The differential diagnosis of retinitis pigmentosa is important because the clinical appearance of the disease is *not* pathognomonic and some similar-appearing conditions may be treatable. An atrophic fundus with scattered spicules of pigment may result from long-standing ischemia, inflammatory disease (such as pars planitis), infection (such as syphilis or measles), trauma (including siderosis), drugs (such as chloroquine), metabolic diseases (such as abetalipoproteinemia or Refsum's syndrome), systemic syndromes (such as spinocerebellar degeneration or progressive external ophthalmoplegia), and other hereditary dystrophies that may have a different prognosis or genetic implication (such as sectorial or delimited retinitis pigmentosa [Marmor 1979b], the vitreoretinal degenerations, or the choroidal dystrophies). The variety of conditions on this list reflects the fact that the pigment epithelium can be involved by many disease processes within the eye, but has only a limited repertoire of responses.

The pathogenesis of retinitis pigmentosa is totally unknown, despite intensive research efforts all over the world. The fact that retinal function and the a-wave of the ERG are affected early in the disease, often before any pigmentary change is evident, has led to the assumption that the primary pathology is in or near the photoreceptors (which generate the a-wave). However, a functional defect in the pigment epithelium, with secondary injury to the photoreceptors, would be equally consistent with this localization. For example, recent electron-microscopic studies (Kolb and Gouras 1974; Szamier and Berson 1977) on patients with advanced disease have shown that the remaining foveal cones had shortened outer segments, and the underlying pigment epithelium had an excessive accumulation of lipofuscin and other debris which suggests that it had been capable of outer segment phagocytosis (Chapter 9). Physiologic studies on patients with retinitis pigmentosa (Ripps, Brin, and Weale, 1978) also give evidence that outer segments become progressively shortened as the disease advances. Thus, in contrast to the RCS rat dystrophy (Chapter 20) in which defective

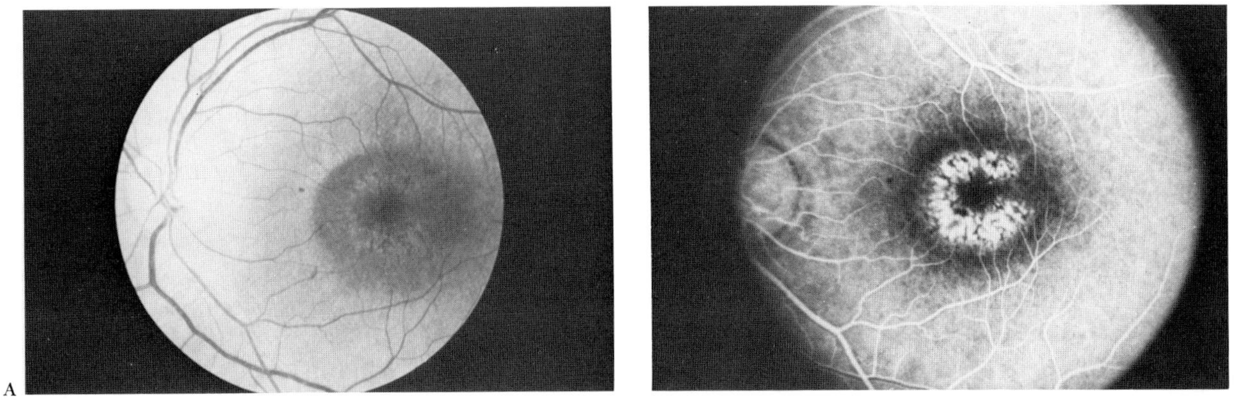

Figure 24.18 *Cone dystrophy. Twenty-five-year-old male with 20/100 acuity and very poor color perception. The ERG showed very low cone signals and mildly reduced rod signals.* A, *fundus photograph.* B, *fluorescein angiogram.*

phagocytosis causes the photoreceptor degeneration, one problem in retinitis pigmentosa may be excessive phagocytosis by the pigment epithelium.

Efforts to find a metabolic cause for retinitis pigmentosa have been unsuccessful (Dewar and Reading 1975). No consistent abnormalities of vitamin A metabolism or transport have been demonstrated (Gouras and Chader 1974). A recent report by Gahlot et al. (1976) that copper metabolism is abnormal in patients from India could not be substantiated in Scandinavia (Ehlers and Bülow 1977) or America (Marmor, Nelson, and Levin 1978). Animal models such as RCS rats (Chapter 20) or dogs with retinal dystrophies (Chapter 19) are very useful, but the models have not been shown to be equivalent to any human form of the disease. The fact that light may damage the retina (Chapter 17), and that light damage may be a function of vitamin A metabolism (Noell and Albrecht 1971), has led Berson (1971) to propose that light deprivation could be a means of slowing down the progression of retinitis pigmentosa. This hypothesis remains to be proven clinically.

The cone degenerations form a subgroup of photoreceptor dystrophies in which the damage is predominantly to the cone rather than the rod system (see Krill, Deutman, and Fishman 1973; Grey et al. 1977). Patients are aware of poor color perception, are usually photophobic, and have poor central vision although the acuity is not usually worse than 20/200 unless there is also significant rod involvement. The inheritance may be either recessive or dominant, and the photoreceptor damage may be localized in the macula or may be diffuse. The most characteristic fundus finding, although it may be absent, is a bull's-eye depigmentation in the fovea, somewhat similar in appearance to the lesions of chloroquine retinopathy (Fig. 24.18). The diagnosis is made on the basis of an extremely deranged color sense, loss of the cone limb on dark-adaptation testing, and an ERG showing depression or extinction of the cone signals with relative preservation of the rod signals. The disease can be distinguished from achromatopsia because the symptoms are progressive and the fundus is abnormal.

The extent of pigment epithelial involvement in cone degeneration is unknown. The bull's-eye lesion involves focal damage to the pigment epithelium, but this may not be a very specific lesion (Weise and Yannuzzi 1974; Bonnet 1976). No histologic material is available from early cases.

Stargardt's disease (or fundus flavimaculatus) is an example of a photoreceptor dystrophy confined primarily to the macula. Although an occasional patient shows peripheral degeneration, the disease in its classic form is characterized by a progressive loss of vision in childhood or adolescence and by fundus pathology limited to the macula. Characteristically, there is an irregular pigmentary dispersion underneath the fovea, often surrounded by an indistinct halo of yellowish flavimaculatus flecks (Fig. 24.19; see also Fig. 24.8). The peripheral retina appears normal and shows normal retinal functions, including visual fields, dark adaptation, ERG, and EOG. The vision will frequently fall to the vicinity of 20/200, but then remain stable for the remainder of life. The inheritance is characteristically recessive, although dominant pedigrees having similar findings have been reported (Deutman 1971). The pathogenesis of Stargardt's disease, like that of retinitis pigmentosa, is unknown and may involve similar

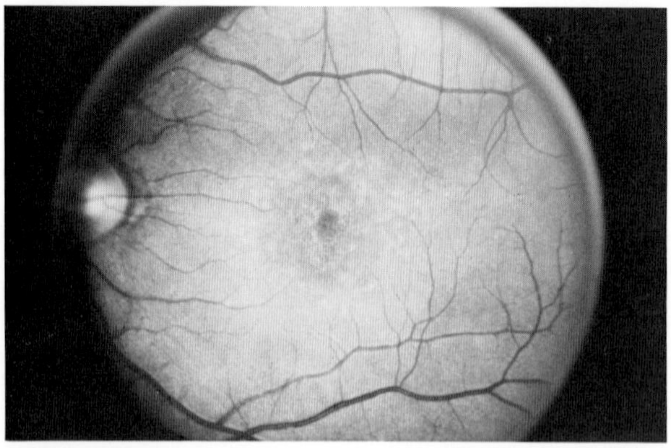

Figure 24.19 *Stargardt's disease. Thirty-nine-year-old woman with 20/70 acuity and a normal ERG.*

aberrations in a focal region. The role of the pigment epithelium has not been established, although the presence of flecks (as in fundus flavimaculatus) is suggestive.

There are dozens, if not hundreds, of published reports on juvenile macular degeneration of various types, and the classification of these diseases falls outside the scope of this book. A few newly described conditions that appear to relate more to the pigment epithelium than to the retina have been mentioned earlier in this chapter.

CHOROIDAL DYSTROPHIES. The choroidal dystrophies are characterized by fundus lesions that show marked atrophy of the choriocapillaris and large choroidal vessels (Fig. 24.20). The lesions may begin in the periphery (as in choroideremia) or be confined to the macula (as in central areolar choroidal dystrophy). In either case, the primary damage occurs to the choroid or choriocapillaris, with ultimate atrophy of the overlying pigment epithelium and retina. Whether the retinal damage is purely ischemic or results from specific functional changes in the pigment epithelium is unknown.

Choroideremia is an x-linked condition whose clinical characteristics are much like retinitis pigmentosa. The carriers are generally identifiable because their fundus is a mosaic showing patchy areas of pigment dispersion. The EOG and ERG are both severely depressed early in the course of the disease. Central areolar choroidal dystrophy is a dominantly inherited condition that only affects central vision.

Gyrate atrophy is a recessively inherited choroidal dystrophy that has been of special interest since Takki (1974) showed that it is associated with a systemic metabolic defect, hyperornithinemia. The disease begins with discrete well-circumscribed patches or gyri, which gradually enlarge and migrate until most of the fundus is involved. Over the long term, the clinical progression and physiologic findings are similar to those of choroideremia. Despite the hyperornithinemia, there are no known systemic effects of the disease. Carriers show normal ornithine levels, but clear ornithine slowly after being given an oral load (Takki and Simell 1974). Recent studies by several groups (O'Donnell, Sandman, and Martin 1977; Valle, Kaiser-Kupfer, and Del Valle 1977) have shown patients with gyrate atrophy to be deficient in the enzyme ornithine aminotransferase. However, the connection between the abnormal ornithine metabolism and the chorioretinal atrophy has not yet been determined.

UNCLASSIFIED DYSTROPHIES. Pigmented paravenous retinochoroidal atrophy (Franceschetti 1962; Pearlman et al. 1975; Takei, Harada, and

A

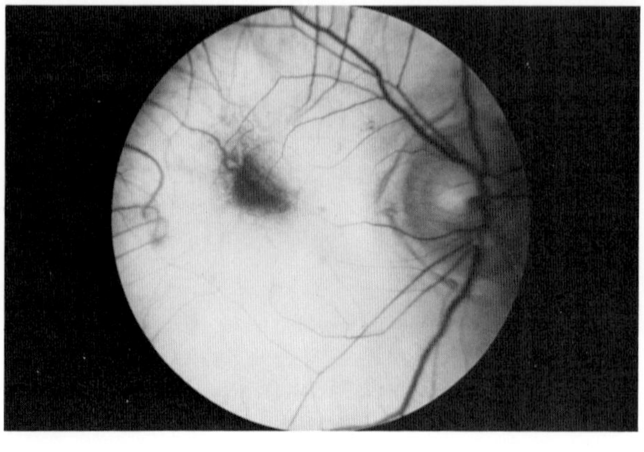

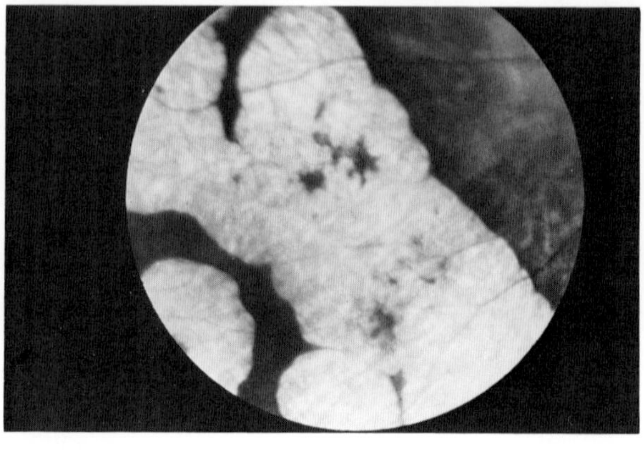

B

Figure 24.20 *Choroidal dystrophies.* A, *choroideremia.* B, *gyrate atrophy.*

Mizuno 1977; Schwab, Hackelbusch, and Gärtner 1978) is a rare disorder of uncertain etiology that causes chorioretinal atrophy beneath the major retinal veins (Fig. 24.21). Depending upon the extent

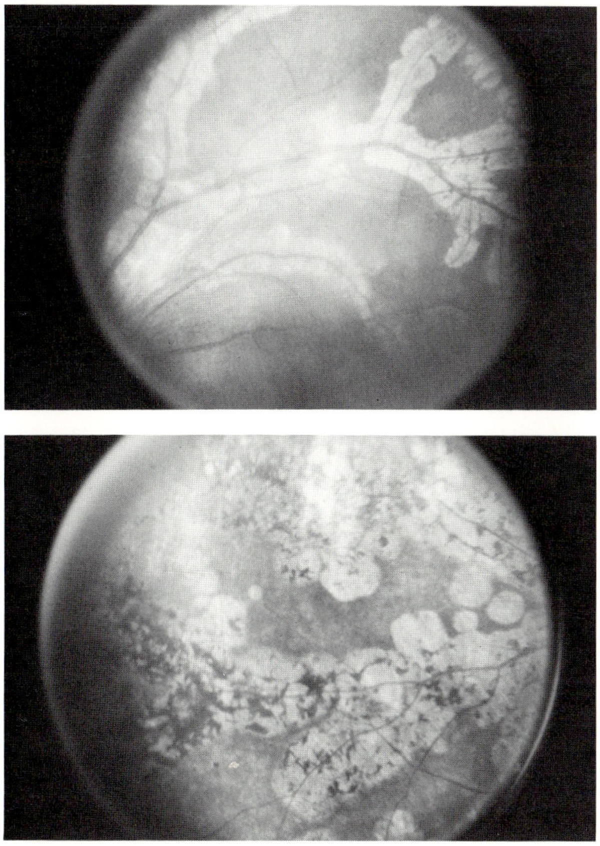

A

B

Figure 24.21 *Pigmented paravenous retinochoroidal atrophy.* A, *early involvement. (From Pearlman et al. 1975.)* B, *more extensive involvement, with bone-spicule formation. (Courtesy of J. T. Pearlman.)*

of fundus involvement, visual acuity and retinal function tests may either be normal or show impairment. The b-wave of the ERG is often reduced. The initial descriptions of this disorder implicated secondary causes such as infection, but Franceschetti (1962) has argued for a primary pathologic process. The most recent descriptions seem to favor a dystrophic or degenerative etiology, although familial occurrence has not yet been documented. Pearlman, Heckenlively, and Bastek (1978) have shown clearly that the disease can progress. The retinal veins obviously influence the development of the lesions, but all layers, including photoreceptors, pigment epithelium, and choriocapillaris, are involved by the atrophic process.

SYSTEMIC HEREDITARY DISEASES. There are a large variety of systemic hereditary diseases that cause pigmentary changes in the fundus, and a complete listing would serve little purpose (see Vinken and Bruyn 1972; Franceschetti, Francois, and Babel 1974; Krill 1977a). Metabolic defects have not yet been found for most of these, such as myotonic dystrophy, progressive external ophthalmoplegia, congenital renal disease (nephronophthisis), or the Laurence-Moon and Bardet-Biedl syndromes. A few examples are given below of disorders in which the metabolic defect is known and may involve the pigment epithelium. The association of hyperornithinemia and gyrate atrophy has already been mentioned.

Abetalipoproteinemia (Bassen-Kornzweig syndrome) is a recessive disorder characterized by reduced or absent β-lipoproteins (Salt et al. 1960) crenated erythrocytes (acanthocytes), ataxia, steatorrhea, and retinal pigmentary degeneration (Bassen and Kornzweig 1950; Kornzweig 1976). The retinal changes (Fig. 24.22) can include the appearance of whitish punctate lesions that resemble those of vitamin A deficiency (Gouras, Carr, and Gunkel 1971). Symptoms include field loss, poor dark adaptation, and severe attenuation of the ERG. Patients are deficient in vitamin A because of the steatorrhea, and administration of parenteral vitamin A improves visual function and the ERG. However, it is not known whether the retinal degeneration or lesions can be totally prevented by vitamin A.

Refsum's disease (Fig. 24.23) is a recessive disorder characterized by an accumulation of phytanic acid (Klenk and Kahlke 1963) and by peripheral neuropathy, ataxia, nerve deafness, and a retinal pigmentary degeneration (Refsum 1946; Steinberg 1971). Treatment with a phytanate-free diet has lowered the serum levels of phytanic acid and improves neuromuscular function, but hearing and vision have not responded (Steinberg et al. 1970). Histologic examination of the eye from one patient (Toussaint and Danis 1971) showed the pigment epithelial cells to be heavily laden with lipid material and the overlying photoreceptors totally degenerated, suggesting that the retinal degeneration in this disease may be a result of pigment epithelial damage by accumulated lipids.

The neuronal storage diseases usually show their most prominent effects on the inner layers of the retina, and the accumulation of abnormal material in the ganglion cells produces the cherry-red spot of

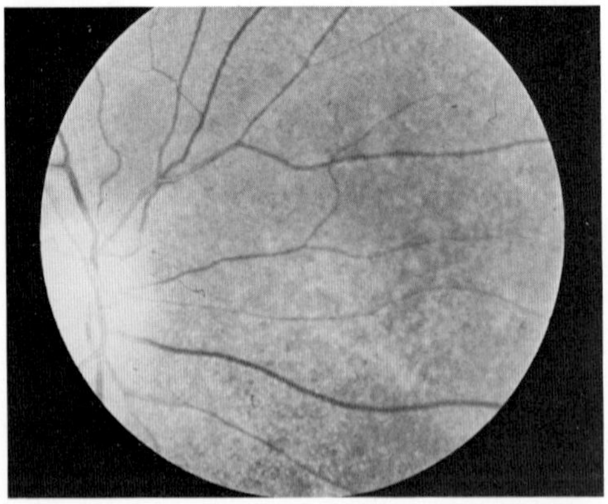

A

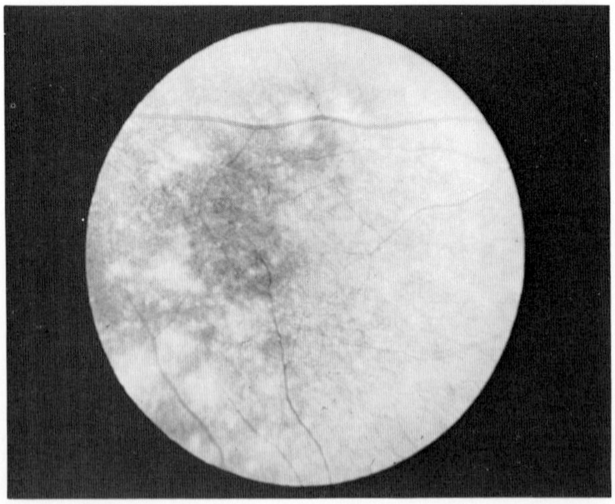

B

Figure 24.22 *Abetalipoproteinemia.* A, *note the whitish lesions.* B, *another case. (From Gouras, Carr, and Gunkel 1971.)*

the gangliosidoses (such as Tay-Sachs disease) or of Niemann-Pick disease. Pigmentary changes develop very late, if at all. An exception to this description is the group of ceroid-lipofuscinoses, including Batten-Vogt syndrome, Batten disease, Jansky-Bielschowsky disease, Spielmeyer-Sjogren disease, and Kuf's disease. In these conditions, there is early involvement of the photoreceptors and pigment epithelium, but no cherry-red spot (Zeman 1976; Hittner and Zeller 1975). The ERG is severely depressed, in contrast to the normal ERG found in early Tay-Sachs disease (Stanescu, Spiritus, and Michiels 1973). Histology (Goebel, Fix, and Zeman 1974) shows extensive loss of photoreceptors, atrophy of the pigment epithelium, and a widespread accumulation of autofluorescent lipopigment that probably is a by-product of uncontrolled peroxidation of unsaturated fatty acids (Zeman 1976). Affected individuals show an absence of leukocyte peroxidase (Armstrong, Dimmitt, and Van Wormer 1974), which is useful as a diagnostic test.

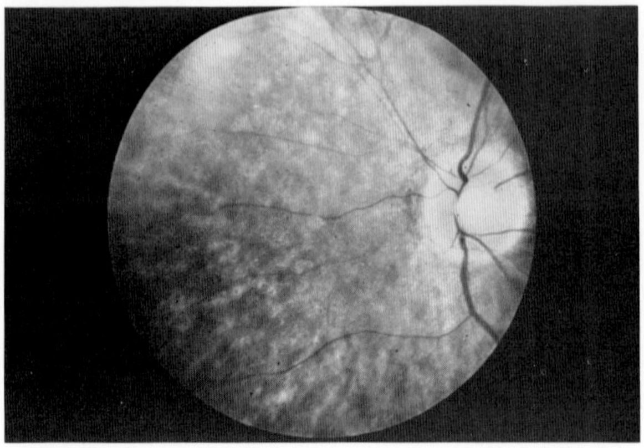

Figure 24.23 *Refsum's syndrome. Male in his mid-twenties with peripheral neuropathy, severe night blindness, and moderate constriction of the visual fields. (Courtesy of D. Steinberg.)*

Congenital Pigmentary Disturbances

The congenital and familial abnormalities of fundus pigmentation do not fit neatly under the earlier headings in this chapter. With the exception of albinism, they are benign insofar as vision is concerned, but recognition of them is important if an incorrect diagnosis of progressive dystrophy, inflammation, or intraocular tumor is to be avoided.

ALBINISM. Both ocular and oculocutaneous albinism are characterized by a lack of melanin in the fundus (Fig. 24.24) and by reduced visual acuity. Since they are discussed in detail in Chapters 8 and 23, only a few words will be added here with regard to how the pigmentary disturbance relates to the clinical symptomatology. Writers in the past have speculated whether the absence of pigmentation allows excess light to damage the eye, but this seems unlikely since many individuals have very albinotic fundi with normal vision. Furthermore, the poor acuity is consistent with clinical impressions that the fovea is undifferentiated (see Fig. 24.24*A*), and with recent reports which confirm histologically that the fovea is absent in albinism (Naumann, Lerche,

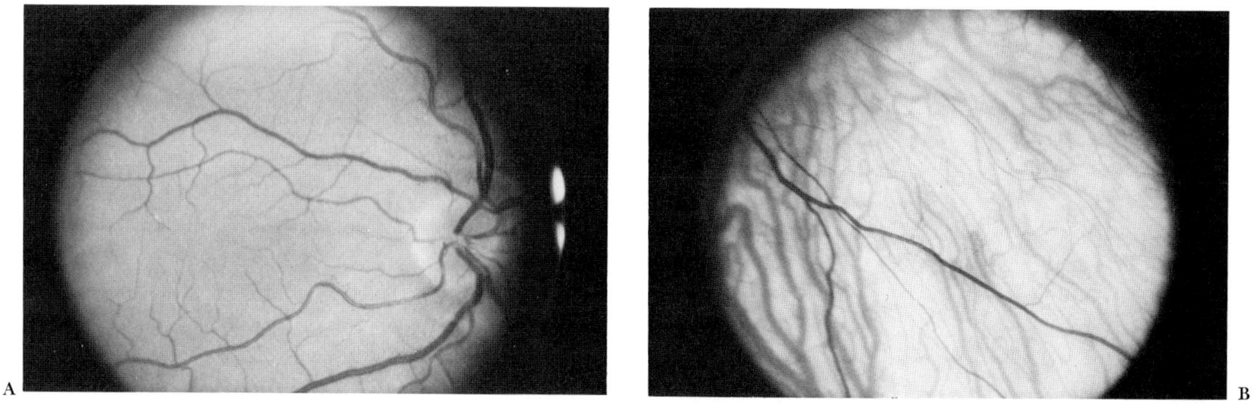

Figure 24.24 *Ocular albinism. Twenty-four-year-old female with 20/200 acuity.* A, *macula. Note the absence of a distinct fovea.* B, *periphery, showing the lack of pigmentation.*

and Schroeder 1976; Fulton, Albert, and Craft 1978). A further complication in determining the cause of the reduced acuity is the fact that albino animals have aberrant visual pathway connections (Guillery and Scott 1973). These presumably exist in the human too, since there is a high incidence of strabismus among albinos and the visually evoked responses from the two eyes are asymmetric (Creel, Witkop, and King 1974). Studies on heterochromic rats (Wise and Lund 1976) suggest that these aberrant pathways correlate with melanin in the pigment epithelium rather than with the albino gene, but the mechanism of this relationship remains to be established.

DIFFUSE HYPERPIGMENTATION. Diffuse unilateral pigmentation of the fundus may occur in melanosis bulbi or in nevus of Ota. The histology of these lesions shows a hyperpigmentation of the uveal tract, but generally no involvement of the retinal pigment epithelium (Duke-Elder 1964).

Large, bilateral, geographic patches of darkened fundus (Fig. 24.25) may occasionally be seen in otherwise normal fundi, particularly in black or darkly pigmented individuals (Nagpal et al. 1975; Condon and Serjeant 1972; Sugar and Wolff 1977). Most patients who have been described showed some degree of hemoglobinopathy, which suggests that the lesions may not be totally benign or incidental, although I know of no adequate survey to confirm this. Fluorescein angiography shows no abnormality in these geographic dark regions, so that the nature of the pigmentation is obscure. The patches may reside at the level of the pigment epithelium, but presumably cannot represent hypertrophy or accumulation of melanin granules, or the angiograms would be abnormal.

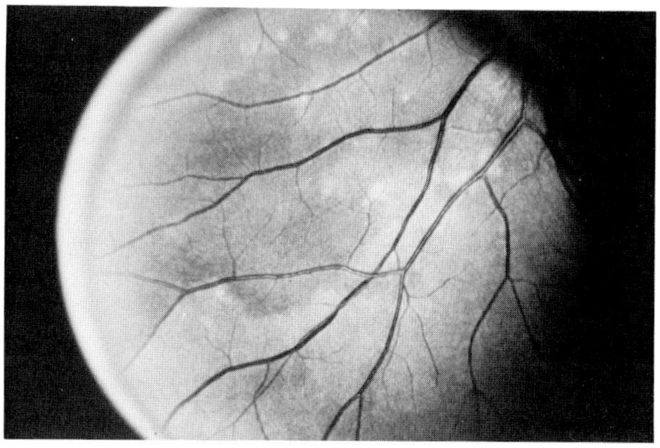

Figure 24.25 *Geographic dark patches in the fundus. Note the difference in coloration between the upper and lower regions of the photograph.*

CONGENITAL GROUPED PIGMENTATION. Congenital grouped pigmentation (Hoeg 1911) of the retina is characterized by collections of well-circumscribed round or ovoid lesions of varying size, usually situated outside the macula and in only one eye (Fig. 24.26). No evidence of hereditary transmission was detected in a series of cases from Scandinavia (Forsius et al. 1971), but there has been one report of two sisters with bilateral grouped pigmentation of the macula (Forgacs and Bozin 1966). The lesions are characteristically flat and of brownish or slate-grey color; because of the clustered appearance they are often called "bear tracks" or "animal tracks." Ordinarily, they are not associated with any visual symptomatology and are most com-

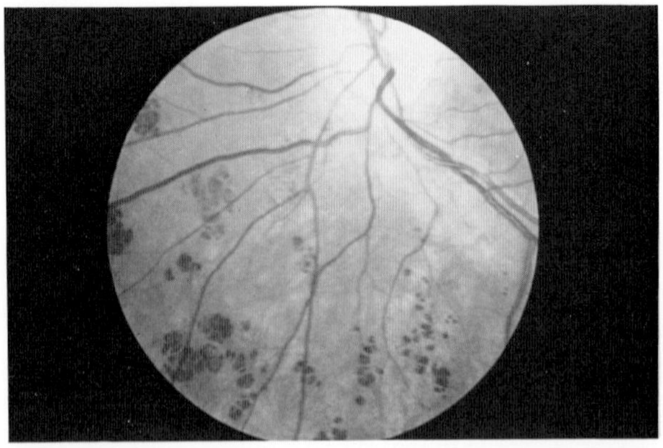

Figure 24.26 *Congenital grouped pigmentation (bear tracks).*

monly found as an incidental finding that need not arouse concern. Rarely, they may be confused with a nevus or degenerative lesion. Perdriel et al. (1973) noted that if the lesions are central, mild functional abnormalities may be present, including a disturbance in color vision and a diminution in the photopic ERG.

There are conflicting reports on the histology of these lesions. Legrand et al. (1964) found the pigment epithelial cells to be normal, but noted layers of pigment in the subretinal space, with overlying damage to the photoreceptors. Shields and Tso (1975) found the pigment epithelial cells to have an increased content of pigment granules, while the overlying photoreceptors appeared normal. Egerer (1976) found the pigment epithelial cells to be hypertrophied and hyperpigmented, and to show layering within the area of the lesions.

CONGENITAL HYPERTROPHY OF THE RETINAL PIGMENT EPITHELIUM. Congenital hypertrophy of the pigment epithelium is characterized by single, flat, greyish-brown or black lesions, well-demarcated, with rounded or scalloped borders (Fig. 24.27). They may be located in any region of the fundus and are characteristically surrounded by a narrow hypopigmented halo. There are frequently irregular depigmented lacunae within the lesion itself, but there are no drusen or orange pigment deposits on the surface as may be seen in nevi or melanomas. Furthermore, the color is not nearly as jet black as in melanocytoma. Curiously, abnormalities of the retinal vasculature may sometimes be present over the lesions (Cleary, Gregor, and Bird 1976).

A

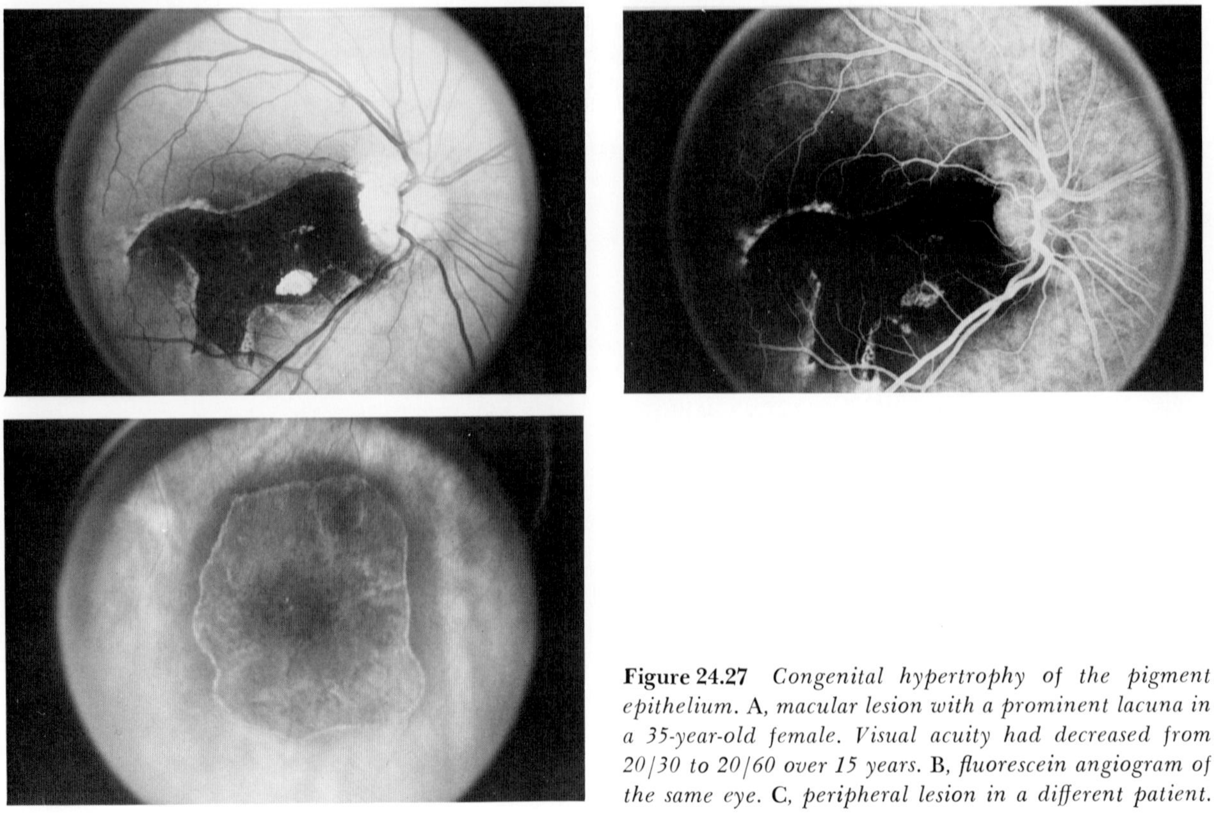

B

C

Figure 24.27 *Congenital hypertrophy of the pigment epithelium.* A, *macular lesion with a prominent lacuna in a 35-year-old female. Visual acuity had decreased from 20/30 to 20/60 over 15 years.* B, *fluorescein angiogram of the same eye.* C, *peripheral lesion in a different patient.*

The clinical characteristics have been clarified by the recent studies of Buettner (1975) and Purcell and Shields (1975). Buettner's cases had no visual symptoms, but showed focal visual field defects corresponding to the lesions. Histology (Reese and Jones 1956; Kurz and Zimmerman 1962) and electron microscopy (Buettner 1975) showed the pigment epithelial cells to be enlarged and packed with round pigment granules that appeared more like lipofuscin than melanin. The normal ovoid melanin granules were absent from hypertrophied regions of pigment epithelium. The photoreceptors overlying the lesions were degenerated, but Buettner speculated that the primary defect was within the pigment epithelium, because the field defects were mildest in young patients, suggesting that the photoreceptor degeneration was secondary and progressive. Purcell and Shields (1975) reviewed the findings in more than fifty cases and confirmed the basic observations of Buettner, although they were not able to demonstrate field defects in most of their patients. The lesions are for the most part stable, but very slow progression can occur (Norris and Cleasby 1976).

Because congenital hypertrophy can resemble congenital grouped pigmentation histologically, both Buettner and Purcell and Shields speculate that these two entities may be variants of the same disease, with different degrees of pigment density. On the other hand, the lesions of grouped pigmentation are inconsistent in producing photoreceptor damage and lack some of the distinguishing features of congenital hypertrophy, such as a surrounding hypopigmented halo or the presence of central lacunae.

The major clinical significance of congenital hypertrophy is the necessity of differentiating it from malignant melanoma and other pigmented lesions of the fundus. In contrast to congenital hypertrophy, malignant melanomas are characteristically elevated, have ragged margins, show overlying orange pigment, show intrinsic vascular changes on angiography, and usually show characteristic findings on ultrasound or ^{32}P testing. Other lesions that may simulate congenital hypertrophy include old chorioretinitic scars, true hyperplasia of the pigment epithelium (which is usually progressive, juxtapapillary, and associated with a history of acquired insult), and, rarely, other isolated dark or pigmented lesions, such as the sunburst sign of sickle-cell disease or focal hemorrhage. Further discussion of pigment epithelial hyperplasia and neoplasia is found in Chapter 15.

REFERENCES

AGUIRRE, G. D., and LATIES, A. 1976. Pigment epithelial dystrophy in the dog. *Exp. Eye Res.* 23:247–256.

ARMSTRONG, D., DIMMITT, S., and VAN WORMER, D. E. 1974. Studies in Batten Disease I. Peroxidase deficiency in granulocytes. *Arch. Neurol.* 30:144–152.

BARD, L. A., and CROSS, H. E. 1975. Genetic counseling of families with Best macular dystrophy. *Trans. Am. Acad. Ophthalmol. Otolaryngol.* 79:865–873.

BARKMAN, Y. 1961. A clinical study of a central tapetoretinal degeneration. *Acta Ophthalmol.* 39:663–671.

BARRICKS, M. E. 1977. Vitelliform lesions developing in normal fundi. *Am. J. Ophthalmol.* 83:324–327.

BASSEN, F. A., and KORNZWEIG, A. L. 1950. Malformation of the erythrocytes in a case of atypical retinitis pigmentosa. *Blood* 5:381–387.

BENEDIKT, O., and WERNER, W. 1971. Retikulare pigmentdystrophie der Netzhaut. *Klin. Monatsbl. Augenheilkd.* 159:794–798.

BENSON, W. E., KOLKER, A. E., ENOCH, J. M., VAN LOO, J. A., JR., and HONDA, Y. 1975. Best's vitelliform macular dystrophy. *Am. J. Ophthalmol.* 79:59–66.

BERSON, E. L. 1971. Light deprivation for early retinitis pigmentosa. *Arch. Ophthalmol.* 85:521–529.

BERSON, E. L., GOURAS, P., and HOFF, M. 1969. Temporal aspects of the electroretinogram. *Arch. Ophthalmol.* 81: 207–214.

BEST, F. 1905. Über eine hereditäre Maculaafektion. *Z. Augenheilkd.* 13:199–212.

BIRNDORF, L. A., and DAWSON, W. W. 1973. A normal electrooculogram in a patient with a typical vitelliform macular lesion. *Invest. Ophthalmol.* 12:830–833.

BONNET, M. 1976. "Schiebscheiben"-Makulopathie (bull's eye disease). *Klin. Monatsbl. Augenheilkd.* 168: 297–302.

BORS, F., and FELLS, P. 1971. Reversal of the complications of self-induced vitamin A deficiency. *Br. J. Ophthalmol.* 55:210–214.

BUETTNER, H. 1975. Congenital hypertrophy of the retinal pigment epithelium. *Am. J. Ophthalmol.* 79:177–189.

BULLOCK, J., and ALBERT, D. 1975. Flecked retina: appearance secondary to oxalate crystals from methoxyflurane anesthesia. *Arch. Ophthalmol.* 93:26–31.

BYERS, B. 1978. Blindness secondary to steroid injections into the nasal turbinates. *Arch. Ophthalmol.*, in press.

Carr, R. E. 1974. Congenital stationary nightblindness. *Trans. Am. Ophthalmol. Soc.* 72:449–487.

Carr, R. E., Margolis, S., and Siegel, I. M. 1976. Fluorescein angiography and vitamin A and oxalate levels in fundus albipunctatus. *Am. J. Ophthalmol.* 82:549–558.

Carr, R. E., Ripps, H., and Siegel, I. M. 1974. Visual pigment kinetics and adaptation in fundus albipunctatus. *Doc. Ophthalmol. Proc. Series* 4:193–204.

Chopdar, A. 1976. Reticular dystrophy of retina. *Br. J. Ophthalmol.* 60:342–344.

Cleary, P. E., Gregor, Z., and Bird, A. C. 1976. Retinal vascular changes in congenital hypertrophy of the retinal pigment epithelium. *Br. J. Ophthalmol.* 60:499–503.

Condon, P. I., and Serjeant, G. R. 1972. Ocular findings in homozygous sickle cell anemia in Jamaica. *Am. J. Ophthalmol.* 73:533–543.

Creel, D., Witkop, C. J., Jr., and King, R. A. 1974. Asymmetric visually evoked potentials in human albinos: evidence for visual system anomalies. *Invest. Ophthalmol.* 13:430–440.

Cross, H. E., and Bard, L. 1974. Electro-oculography in Best's macular dystrophy. *Am. J. Ophthalmol.* 77:46–50.

Deutman, A. F. 1971. *The hereditary dystrophies of the posterior pole of the eye.* Assen: Van Gorcum and Co.

Deutman, A. F. 1974. Benign concentric annular macular dystrophy. *Am. J. Ophthalmol.* 78:384–396.

Deutman, A. F., Pinckers, A. J. L. G., and Aan de Kerk, A. L. 1976. Dominantly inherited cystoid macular edema. *Am. J. Ophthalmol.* 82:540–548.

Deutman, A. F., Rümke, A. M. L. 1969. Reticular dystrophy of the retinal pigment epithelium. *Arch. Ophthalmol.* 82:4–9.

Deutman, A. F., van Blommestein, D. A., Henkes, H. E., Waardenburg, P. J., Soleveld-van Driest, E. 1970. Butterfly-shaped pigment dystrophy of the fovea. *Arch. Ophthalmol.* 83:558–569.

Dewar, A. J., and Reading, H. W. 1975. The biochemical aspects of retinitis pigmentosa. *Int. J. Biochem.* 6:615–641.

Donders, F. C. 1857. Beiträge zur pathologischen Anatomie des Auges. 2. Pigmentbildung in der Netzhaut. *Archiv. f. Ophthalmol.* 3:139–150.

Duke-Elder, S. 1964. *Congenital deformities.* In *System of ophthalmology,* ed. S. Duke-Elder, vol. 3, pt. 2, pp. 794–800. London: Henry Kimpton.

Egerer, I. 1976. Die gruppierte (oder nävoide) Pigmentation des Augenhintergrundes. *Klin. Monatsbl. Augenheilkd.* 168:672–677.

Ehlers, N., and Bülow, N. 1977. Clinical copper metabolism parameters in patients with retinitis pigmentosa and other tapeto-retinal degenerations. *Br. J. Ophthalmol.* 61:595–596.

Farkas, T. G., Krill, A. E., Sylvester, V. M., and Archer, D. 1971*a*. Familial and secondary drusen: histologic and functional correlations. *Trans. Am. Acad. Ophthalmol. Otolaryngol.* 75:333–343.

Farkas, T. G., Sylvester, V., Archer, D., and Altona, M. 1971*b*. The histochemistry of drusen. *Am. J. Ophthalmol.* 71:1206–1215.

Fells, P., and Bors, F. 1969. Ocular complications of self-induced vitamin A deficiency. *Trans. Ophthalmol. Soc. U. K.* 89:221–228.

Fetkenhour, C. L., Gurney, N., Dobbie, J. G., and Choromokos, E. 1976. Central areolar pigment epithelial dystrophy. *Am. J. Ophthalmol.* 81:745–753.

Fishman, G. A., Trimble, S., Rabb, M. F., and Fishman, M. 1977. Pseudo-vitelliform macular degeneration. *Arch. Ophthalmol.* 95:73–76.

Forgacs, J., and Bozin, I. 1966. Manifestation familiale de pigmentations groupées de la région maculaire. *Ophthalmologica* 152:364–368.

Forsius, H., Eriksson, A., Nuutila, A., Vainio-Mattila, B., and Krause, U. 1971. A genetic study of three rare retinal disorders: dystrophia retinae dysacusis syndrome, x-chromosomal retinoschisis and grouped pigments of the retina. *Birth Defects* 7:83–98.

Franceschetti, A. 1962. A curious affection of the fundus oculi: helicoid peripapillar chorioretinal degeneration—its relation to pigmentary paravenous chorioretinal degeneration. *Doc. Ophthalmol.* 16:81–110.

Franceschetti, A., and François, J. 1965. Fundus flavimaculatus. *Arch. Ophtalmol.* (Paris) 25:505–530.

Franceschetti, A., François, J., and Babel, J. 1974. *Chorioretinal heredodegenerations.* Springfield: Charles C. Thomas.

François, P., and Bonnet, M. 1976. *La macula.* Paris: Masson.

François, J., Hanssens, M., Coppieters, R., and Evens, L. 1972. Cystinosis. *Am. J. Ophthalmol.* 73:643–650.

Frank, H. R., Landers, M. B., III, Williams, R. J., and Sidbury, J. B. 1974. A new dominant progressive foveal dystrophy. *Am. J. Ophthalmol.* 78:903–916.

Fulton, A. B., Albert, D. M., and Craft, J. L. 1978.

Human albinism: light and electron microscopy study. *Arch. Ophthalmol.* 96:305–310.

Gahlot, D. K., Khosla, P. K., Makashir, P. D., Vasuki, K., and Basu, N. 1976. Copper metabolism in retinitis pigmentosa. *Br. J. Ophthalmol.* 60:770–774.

Gass, J. D. M. 1973. Drusen and disciform macular detachment and degeneration. *Arch. Ophthalmol.* 90:206–217.

Gass, J. D. M. 1974. A clinicopathologic study of a peculiar foveomacular dystrophy. *Trans. Am. Ophthalmol. Soc.* 72:139–156.

Goebel, H. H., Fix, J. D., and Zeman, W. 1974. The fine structure of the retina in neuronal ceroid-lipofuscinosis. *Am. J. Ophthalmol.* 77:25–39.

Gottlieb, R. P., and Ritter, J. A. 1977. "Flecked retina": an association with primary hyperoxaluria. *J. Pediatr.* 90:939–942.

Gouras, P., Carr, R. E., and Gunkel, R. D. 1971. Retinitis pigmentosa in abetalipoproteinemia: effects of vitamin A. *Invest. Ophthalmol.* 10:784–793.

Gouras, P., and Chader, G. 1974. Retinitis pigmentosa and retinol-binding protein. *Invest. Ophthalmol.* 13:239–242.

Grey, R. H. B., Blach, R. K., and Barnard, W. M. 1977. Bull's eye maculopathy with early cone degeneration. *Br. J. Ophthalmol.* 61:702–718.

Guillery, R. W., and Scott, G. L. 1973. Genetic mechanisms determining the central visual pathways of mice. *Science* 179:1014–1016.

Hayreh, S. S. 1975. Segmental nature of the choroidal vasculature. *Br. J. Ophthalmol.* 59:631–648.

Hittner, H. M., and Zeller, R. S. 1975. Ceroid-Lipofuscinosis (Batten Disease). *Arch. Ophthalmol.* 93:178–183.

Hoeg, H. 1911. Die gruppierte Pigmentation des Augengrundes. *Klin. Monatsbl. Augenheilkd.* 49:49–77.

Hogan, M. J. 1972. Role of the retinal pigment epithelium in macular disease. *Trans. Am. Acad. Ophthalmol. Otolaryngol.* 76:64–80.

Hsieh, R. C., Fine, B. S., and Lyons, J. S. 1977. Patterned dystrophies of the retinal pigment epithelium. *Arch. Ophthalmol.* 95:429–435.

Kandori, F. 1959. Very rare cases of congenital nonprogressive nightblindness with fleck retina. *J. Clin. Ophthalmol.* (Tokyo) 13:384–386.

Kandori, F., Tamai, A., Kurimoto, S., and Fukunaga, K. 1972. Fleck retina. *Am. J. Ophthalmol.* 73:673–685.

Kingham, J. D., and Lochen, G. P. 1977. Vitelliform macular degeneration. *Am. J. Ophthalmol.* 84:526–531.

Klenk, E., and Kahlke, W. 1963. Über das Vorkommen der 3.7.11.15-Tetramethylhexadecansäure (Phytansäure) in den Cholesterinestern und anderen Lipoidfraktionen der Organe bei einem Krankheitsfall unbekannter Genese (Verdacht auf Heredopathia atactica polyneuritiformis [Refsum-Syndrom]) *Hoppe-Seylers Z. physiol. Chem.* 333:133–139.

Klien, B. A., and Krill, A. E. 1967. Fundus flavimaculatus. *Am. J. Ophthalmol.* 64:3–23.

Kolb, H., and Gouras, P. 1974. Electron microscopic observations of human retinitis pigmentosa, dominantly inherited. *Invest. Ophthalmol.* 13:487–498.

Kornzweig, A. L. 1976. Bassen-Kornzweig syndrome: present status. *Metabolic Ophthalmol.* 1:51–53.

Krill, A. E., ed. 1977*a*. *Hereditary retinal and choroidal diseases.* Vol. 2. *Clinical characteristics.* Hagerstown: Harper and Row.

Krill, A. E. 1977*b*. Vitelliruptive macular dystrophy. In *Hereditary retinal and choroidal diseases,* ed. A. E. Krill. Vol. 2. *Clinical characteristics.* Hagerstown: Harper and Row.

Krill, A. E. 1977*c*. Flecked retina disease. In *Hereditary retinal and choroidal diseases,* ed. A. E. Krill. Vol. 2. *Clinical characteristics.* Hagerstown: Harper and Row.

Krill, A. E., Deutman, A. F., and Fishman, M. 1973. The cone degenerations. *Doc. Ophthalmol.* 35:1–80.

Krill, A. E., and Klien, B. A. 1965. Flecked retina syndrome. *Arch. Ophthalmol.* 74:496–508.

Krill, A. E., Morse, P. A., Potts, A. M., and Klien, B. A. 1966. Hereditary vitelliruptive macular degenerations. *Am. J. Ophthalmol.* 61:1405–1415.

Kurz, G. H., and Zimmerman, L. E. 1962. Vagaries of the retinal pigment epithelium. *Int. Ophthalmol. Clin.* 2:441–465.

Lauber, H. 1910. Die sogenannte Retinitis punctata albescens. *Klin. Monatsbl. Augenheilkd.* 48:133–148.

Legrand, M. J., Hervouet, F., Baron, A., Ginguene, A., and Lenoir, A. 1964. Pigmentation groupée de la rétine. *Bull. Soc. Ophtalmol.* 64:313–315.

Levy, N. S., and Toskes, P. P. 1974. Fundus albipunctatus and vitamin A deficiency. *Am. J. Ophthalmol.* 78:926–929.

Maloney, W. F., Robertson, D. M., and Duboff, S. M. 1977. Hereditary vitelliform macular degeneration. *Arch. Ophthalmol.* 95:979–983.

MARMOR, M. F. 1977*a*. Defining fundus albipunctatus. (XIV ISCERG Symposium, Louisville, 1976.) *Doc. Ophthalmol. Proc. Series* 13:277–234.

MARMOR, M. F. 1977*b*. Fundus albipunctatus: a clinical study of the fundus lesions, the physiologic deficit, and the vitamin A metabolism. *Doc. Ophthalmol.* 43:277–302.

MARMOR, M. F. 1979*a*. "Vitelliform" lesions in adults. *Ann. Ophthalmol.*, in press.

MARMOR, M. F. 1979*b*. The electroretinogram in retinitis pigmentosa. *Arch. Ophthalmol.*, in press.

MARMOR, M. F. 1979*c*. The natural history of visual loss in retinitis pigmentosa. In preparation.

MARMOR, M. F., and BYERS, B. 1977. Pattern dystrophy of the pigment epithelium. *Am. J. Ophthalmol.* 84:32–44.

MARMOR, M. F., NELSON, J. W., and LEVIN, A. S. 1978. Copper metabolism in American retinitis pigmentosa patients. *Br. J. Ophthalmol.*, in press.

MCFARLAND, C. B. 1955. Heredodegeneration of the macula lutea. *A.M.A. Arch. Ophthalmol.* 53:224–228.

MESKER, R. P., OOSTERHUIS, J. A., and DELLEMAN, J. W. 1970. A retinal lesion resembling Sjögren's dystrophis reticularis laminae pigmentosae retinae. *Perspect. Ophthalmol. II* (Excerpta Med. Monograph Series) 26: 40–45.

MILLER, S. A. 1977. Multifocal Best's vitelliform dystrophy. *Arch. Ophthalmol.* 95:984–990.

MULLEN, R. J., and LAVAIL, M. M. 1976. Inherited retinal dystrophy: primary defect in pigment epithelium determined with experimental rat chimeras. *Science* 192: 799–801.

NAGPAL, K. C., GOLDBERG, M. F., ASDOURINA, G., GOLDBAUM, M., and HUAMONTE, F. 1975. Dark-without-pressure fundus lesions. *Br. J. Ophthalmol.* 59:476–479.

NAUMANN, G. O. H., LERCHE, W., and SCHROEDER, W. 1976. Foveola-Aplasie bei Tyrosinase-positivem oculocutanen Albinismus. *Albrecht von Graefes Arch. Klin. Ophthalmol.* 200:39–50.

NEWELL, F. W., KRILL, A. E., and FARKAS, T. G. 1972. Drusen and fundus flavimaculatus: clinical, functional and histologic characteristics. *Trans. Am. Acad. Ophthalmol. Otolaryngol.* 76:88–100.

NOBLE, K. G., CARR, R. E., and SIEGEL, I. M. 1977. Pigment epithelial dystrophy. *Am. J. Ophthalmol.* 83:751–757.

NOELL, W. K., and ALBRECHT, R. 1971. Irreversible effects of visible light on the retina: role of vitamin A. *Science* 172:76–80.

NORRIS, J. L., and CLEASBY, G. W. 1976. An unusual case of congenital hypertrophy of the retinal pigment epithelium. *Arch. Ophthalmol.* 94:1910–1911.

NOTTING, J. G. A., and PINCKERS, A. J. L. G. 1977. Dominant cystoid macular dystrophy. *Am. J. Ophthalmol.* 83:234–241.

O'DONNELL, J. J., SANDMAN, R. P., and MARTIN, S. R. 1977. Deficient L-ornithine: 2-oxoacid aminotransferase activity in cultured fibroblasts from a patient with gyrate atrophy of the retina. *Biochem. Biophys. Res. Comm.* 79: 396–399.

PEARLMAN, J. T. 1978. A mathematical model of retinitis pigmentosa. *Jap. J. Ophthalmol.* In press.

PEARLMAN, J. T., HECKENLIVELY, J. R., and BASTEK, J. V. 1978. Progressive nature of pigmented paravenous ret inochoroidal atrophy. *Am. J. Ophthalmol.* 85:215–217.

PEARLMAN, J. T., KAMIN, D. F., KOPELOW, S. M., and SAXTON, J. 1975. Pigmented paravenous retinochoroidal atrophy. *Am. J. Ophthalmol.* 80:630–635.

PERDRIEL, M. G., CHEVALERAUD, J., VIGNAT, J.-P., and DECROIX, G. 1973. Pigmentations groupées de la rétine étude angiofluorographique et électrophysiologique. *Bull. Soc. Ophtalmol.* 73:245–247.

PETERSON, W. S., and ALBERT, D. M. 1974. Fundus changes in the hereditary nephropathies. *Trans. Am. Acad. Ophthalmol. Otolaryngol.* 78:762–771.

PURCELL, J. J., and SHIELDS, J. A. 1975. Hypertrophy with hyperpigmentation of the retinal pigment epithelium. *Arch. Ophthalmol.* 93:1122–1126.

READ, J., GOLDBERG, M. F., FISKMAN, G., and ROSENTHAL, I. 1973. Nephropathic cystinosis. *Am. J. Ophthalmol.* 76: 791–796.

REESE, A. B., and JONES, I. S. 1956. Benign melanomas of the retinal pigment epithelium. *Am. J. Ophthalmol.* 42:207–212.

REFSUM, S. 1946. Heredopathia atactica polyneuritiformis—A familial syndrome not hitherto described. *Acta Psychiatrica et Neurologica*. Suppl. 38:1–303.

RIPPS, H., BRIN, K. P., and WEALE, R. A. 1978. Rhodopsin and visual threshold in retinitis pigmentosa. *Invest. Ophthalmol. Vis. Sci.* 17:735–745.

ROSENBERG, T., FAURSCHOU, S., and NIELSEN, N. 1977. The clinical classification of maculopathies in adults. *Acta Ophthalmol.* [Suppl.] 55:3–23.

SALT, H. B., WOLFF, O. H., LLOYD, J. K., FOSBROOKE, A. S., CAMERON, A. H., and HUBBLE, D. V. 1960. On having no beta-lipoprotein. *Lancet* August 13:325–329.

SANDERSON, P. O., KUWABARA, T., STARK, W. J., WONG, V. G., and COLLINS, E. M. 1974. Cytostinosis: a clinical, histopathologic and ultrastructure study. *Arch. Ophthalmol.* 91:270–274.

SCHWAB, B., HACKELBUSCH, R., and GÄRTNER, J. 1978. Pigmentierte, paravenöse Nez- und Aderhautatrophie. *Klin. Monatsbl. Augenheilkd.* 172:100–105.

SHIELDS, J. A., and TSO, M. O. M. 1975. Congenital grouped pigmentation of the retina. *Arch. Ophthalmol.* 93:1153–1155.

SINGERMAN, L. J., BERKOW, J. W., and PATZ, A. 1977. Dominant slowly progressive macular dystrophy. *Am. J. Ophthalmol.* 83:680–693.

SJÖGREN, H. 1950. Dystrophia reticularis laminae pigmentosae retinae. *Acta Ophthalmol.* 28:279–295.

SLEZAK, H., and HOMMER, K. 1969. Fundus pulverulentus. *Albrecht von Graefes Arch. Klin. Ophthalmol.* 178:177–182.

SOMMER, A., TJAKRASUDJATMA, S., DJUNAEDI, E., and GREEN, W. R. 1978. Vitamin A responsive pan-ocular xerophthalmia in a healthy adult. *Arch. Ophthalmol.* 96: 1630–1634.

STANESCU, B., SPIRITUS, M., and MICHIELS, J. 1973. La rétinopathie dans la maladie de Batten-Mayou-Spielmeyer-Vogt. *Arch. Ophtalmol.* (Paris) 33:827–834.

STARZYCKA, M., and BRYK, E. 1976. Schmetterlingsförmige dystrophie der makula. *Klin. Monatsbl. Augenheilkd.* 169:454–458.

STEINBERG, D. 1971. The metabolic basis of the Refsum syndrome. *Birth Defects* 12:42–52.

STEINBERG, D., MIZE, C. E., FALES, H. M., and VROOM, F. Q. 1970. Phytanic acid in patients with Refsum's syndrome and response to dietary treatment. *Arch. Intern. Med.* 125:75–87.

SUGAR, H. S., and WOLFF, L. 1977. Geographic dark posterior fundus patches. *Am. J. Ophthalmol.* 83:847–852.

SZAMIER, R. B., and BERSON, E. L. 1977. Retinal ultrastructure in advanced retinitis pigmentosa. *Invest. Ophthalmol.* 16:947–962.

TAKEI, Y., HARADA, M., and MIZUNO, K. 1977. Pigmented paravenous retino-choroidal atrophy. *Jap. J. Ophthalmol.* 21:311–317.

TAKKI, K. 1974. Gyrate atrophy of the choroid and retina associated with hyperornithinaemia. *Br. J. Ophthalmol.* 58:3–23.

TAKKI, K., and SIMELL, O. 1974. Genetic aspects in gyrate atrophy of the choroid and retina with hyperornithinaemia. *Br. J. Ophthalmol.* 58:907–916.

TENG, K. H. 1965. Further contributions to the fundus xerophthalmicus. *Ophthalmologica* 150:219–238.

TIMM, G. 1976. Das krankheitsbild der retinopathia oxalogenica. *Klin. Monatsbl. Augenheilkd.* 168:537–543.

TOSKES, P., DAWSON, W., and FITZGERALD, C. 1978. Treatment of the retinopathy associated with chronic pancreatitis. *Clin. Res.* 26:50A.

TOUSSAINT, D., and DANIS, P. 1971. An ocular pathologic study of Refsum's syndrome. *Am. J. Ophthalmol.* 72:342–347.

UYEMURA, M. 1928. Ueber eine merkwurdige Augenhintergrundveränderung bei zwei Fällen von idiopathischer Hemeralopie. *Klin. Monatsbl. Augenheilkd.* 81:471–473.

VALLE, D., KAISER-KUPFER, M. I., and DEL VALLE, L. A. 1977. Gyrate atrophy of the choroid and retina: deficiency of ornithine aminotransferase in transformed lymphocytes. *Proc. Natl. Acad. Sci. U.S.A.* 74:5159–5161.

VINKEN, P. J., and BRUYN, G. W., ed. 1972. *Handbook of clinical neurology.* Vol. 13 *Neuroretinal degenerations.* Amsterdam: North-Holland Publishing Co.

WEISE, E. E., and YANNUZZI, L. A. 1974. Ring maculopathies mimicking chloroquine retinopathy. *Am. J. Ophthalmol.* 78:204–210.

WISE, R. P., and LUND, R. D. 1976. The retina and central projections of heterochromic rats. *Exp. Neurol.* 51: 68–77.

ZEMAN, W. 1976. Batten disease: ocular features, differential diagnosis and diagnosis by enzyme analysis. *Birth Defects* 12:441–453.

ACKNOWLEDGMENTS

My research is supported in part by National Eye Institute grant EY01678.

25 Inflammations and Degenerations of the Retinal Pigment Epithelium

Michael F. Marmor

THE PIGMENT EPITHELIUM, as a neuroepithelial tissue, is probably not the source of inflammatory responses unless actually infected with a microorganism. However, the pigment epithelium has a remarkable ability to react when its neighboring tissues become inflamed or damaged, whether from infection, trauma, or the effects of aging (see Chapters 14, 15, and 26). Mild degrees of local damage or ischemia may be manifest only as swelling or opacification of the pigment epithelial cells. Greater injury causes areas of pigment dispersion and hypo- or hyperpigmentation. After severe damage to adjacent structures, the pigment epithelial cells may migrate or transform into macrophages and fibrocytes. This chapter cannot deal with every fundus disease that stimulates such visible pigmentary responses—to do so would require the inclusion of almost all the toxic, inflammatory, and degenerative conditions in ophthalmology. Instead, it will focus on a selection of disorders in which the pigment epithelium is critically involved in the pathogenesis of fundus lesions or visual dysfunction.

The topics herein include inflammations that originate close to, if not within, the pigment epithelium (such as placoid pigment epitheliopathy), infections involving the pigment epithelium (such as rubella), focal exudative disorders of the pigment epithelium (such as central serous retinopathy), and degenerative conditions in which pigment epithelial changes are a major component (such as angioid streaks). This chapter will provide only a brief description of these diseases, but will attempt to show how damage to the pigment epithelium contributes to their clinical effects. In some of these conditions the pigment epithelium is not the primary site of disease, but secondary involvement of the pigment epithelium is a major cause of the photoreceptor damage and visual disability. For example, histoplasmosis is probably a disease of the choroid, but its ocular significance results from the formation of focal scars involving the barrier functions of Bruch's membrane and pigment epithelium. The question of clinical relevance—with respect to pigment epithelial involvement in a disease—is not whether the involvement is primary or secondary, but whether it significantly influences the course of the disease (Marmor 1975).

Inflammations of the Retinal Pigment Epithelium

There is no firm evidence that the pigment epithelium is the primary target of any inflammatory disorder, although there are several conditions in which an inflammatory process appears to be localized so close to the RPE that the pigment epithelial response has been considered an intimate part of the disease. Examples include placoid pigment epitheliopathy, pigment epitheliitis, and serpiginous choroidopathy; and as the names suggest, there is hardly unanimity about the etiology of these conditions.

ACUTE POSTERIOR MULTIFOCAL PLACOID PIGMENT EPITHELIOPATHY. Acute posterior multifocal placoid pigment epitheliopathy (APMPPE), first described by Gass in 1968, is a syndrome characterized by the development in young adults of multifocal yellowish-white placoid lesions that appear to be at the level of the pigment

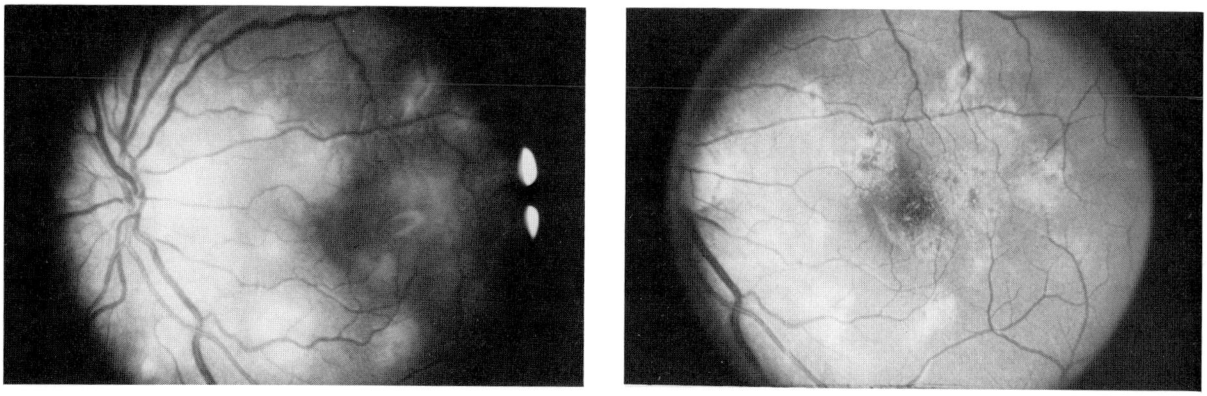

A B

Figure 25.1 *Acute posterior multifocal placoid pigment epitheliopathy. Twelve-year-old boy who noted the onset of blurred vision 10 days after an episode of flu. Mild cells and flare in the anterior chamber were noted.* A, *the left eye on initial examination; the visual acuity was 20/60.* B, *the same eye 10 days later. The visual acuity had fallen to 20/200, but returned to 20/30 during the next week.*

epithelium (Fig. 25.1). The lesions are generally bilateral and characteristically resolve over a period of weeks, leaving pigmentary scars in their place. Vision is significantly decreased during the acute phase, frequently dropping to 20/200 or worse, but virtually always improves as the acute lesions resolve. The ultimate acuity is rarely worse than 20/50 and is frequently normal despite the presence of considerable depigmentation and pigmentary dispersion in the place of the old lesions. The disease afflicts both men and women, and no familial incidence has been demonstrated.

Numerous reports suggest systemic associations that may give a clue to the origin of the condition. Several series (for example, Ryan and Maumenee 1972; Holt, Regan, and Trempe 1976) have noted a high incidence of antecedent illness, most often a mild virus-like syndrome, and adenovirus type 5 infection has been demonstrated in one case (Azar et al. 1975). Other associations include erythema nodosum (Van Buskirk, Lessell, and Friedman 1971; Deutman et al. 1972), mycobacterial infection (Brown, Eberdt, and Ladas 1973), pleocytosis and elevated protein in the cerebrospinal fluid (Bullock and Fletcher 1977; Fishman, Baskin, and Jednock 1977), and acute thyroiditis (Jacklin 1977). Ocular associations include anterior uveitis, which has frequently been described (Holt, Regan, and Trempe 1976; Annesley, Tomer, and Shields 1973; Fitzpatrick and Robertson 1973; Lowes 1977; Jacklin 1977), and the occasional occurrence of retinal vasculitis and papillitis (Kirkham, Ffytche, and Sanders 1972) or serous retinal detachments (Bird and Hamilton 1972). Despite these various associations with inflammatory disease, placoid pigment epitheliopathy does not respond to steroids and no specific therapy is available.

The ERG and EOG light response have been reported as normal in some patients (Ryan and Maumenee 1972; Gass 1977a; Holt, Regan, and Trempe 1976), whereas significant abnormalities have been noted in others (Deutman et al. 1972; Fishman, Rabb, and Kaplan 1974; Smith et al. 1978). In severe cases both the a- and b-waves of the ERG may be moderately reduced, and recovery of the ERG may be incomplete even though the visual recovery is quite good. The EOG light response similarly can fall to a very low level during the acute stage, with varying degrees of recovery. There may also be reversible abnormalities in dark adaptation and in the Stiles-Crawford effect (Smith et al. 1978). These findings indicate that widespread damage to the pigment epithelium and even the photoreceptors can occur, depending on the extent and severity of the disease process.

In the acute stage of the disease, the grey-yellow lesions block fluorescein during the early phases of angiography, suggesting that the pigment epithelium has become edematous and opaque (Fig. 25.2). In the later phases of the angiogram these lesions stain and retain fluorescence for a long time. As the lesions evolve into scars, the angiographic pattern changes, so that the early pictures show a transmission of fluorescence through depigmented areas that do not stain in the later pictures.

Gass (1968) and Ryan and Maumenee (1972)

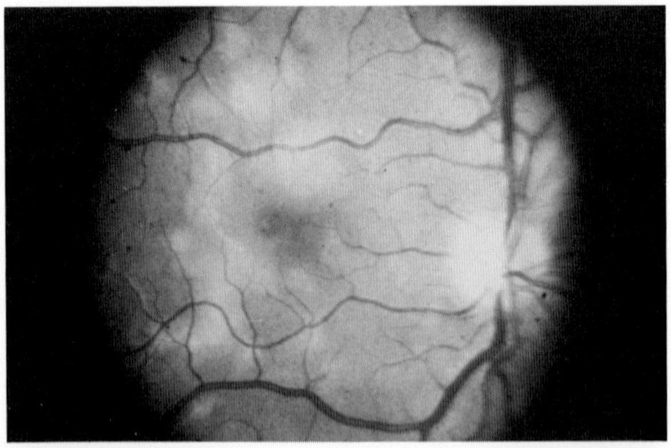

A

Figure 25.2 *Acute posterior multifocal placoid pigment epitheliopathy. Angiographic appearance of the acute lesions.* A, *fundus photograph.* B, *early phase of the fluorescein angiogram, showing blockage of choroidal fluorescence by the lesions.* C, *late phase of the angiogram, showing staining of the lesions.*

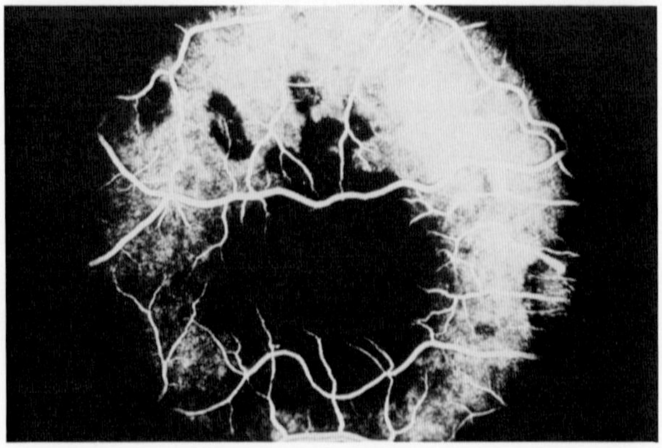

B

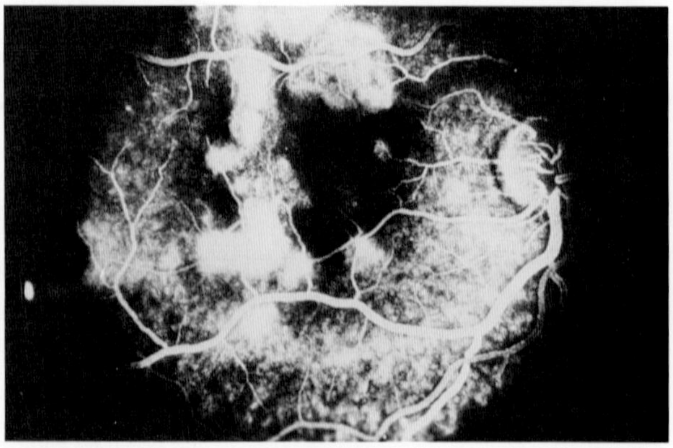

C

originally thought the disease was a pigment epitheliopathy, and that inflammatory edema within the pigment epithelium was the cause of the characteristic lesions. Most other authors (for example, Van Buskirk, Lessell, and Friedman 1971; Deutman et al. 1972; Holt, Regan, and Trempe 1976; Takei and Mizuno 1976; Bird and Hamilton 1972; Savino et al. 1974) believe that the lesions originate in the choroid or choriocapillaris, and that the changes in the overlying pigment epithelium (and photoreceptors) are secondary. Presumably, a systemic viral or inflammatory disease induces a choroidal vasculitis, which in turn causes focal ischemia of the choriocapillaris. Deutman and Lion (1977) recently described a case in which choriocapillary nonperfusion could clearly be seen beneath early lesions, and Hayreh and Baines (1972) have shown that similar lesions can develop after experimental choroidal vessel occlusion. A relationship between placoid lesions and light injury to the retina has also been proposed (Henkes 1977).

Similarities have been noted between placoid pigment epitheliopathy and other forms of choroiditis. The lesions can be quite reminiscent of serpiginous choroidopathy, which may have a related if not common pathogenesis.

SERPIGINOUS CHOROIDOPATHY. Serpiginous choroidopathy is a condition of middle-aged or older individuals in which patchy and confluent grey-yellow lesions develop bilaterally in the posterior pole and gradually extend, leaving pigmentary and choriocapillary atrophy behind (Fig. 25.3). Several patterns of lesions have been observed, so that a variety of terms (serpiginous, geographic, helicoid, peripapillary) describe what is probably the same condition.

The clinical course and prognosis of the disease are variable, because the lesions progress irregularly. Once the lesions cross the fovea, vision drops precipitously, usually to a level of legal blindness. However, patients may be unaware of any visual loss until the fovea is involved, leading to an occasional history of "acute" disease because a long-standing geographic lesion finally impinged upon the fovea. The condition has often been described as progressive or creeping, but Hamilton and Bird (1974) and Gass (1977a) have noted that the progress is better described as stepwise. Each region of the retina that becomes affected evolves from an opaque lesion to a depigmented scar, and extensions are visible as new discrete opaque lesions which are usually, but not necessarily, contiguous. This stepwise progres-

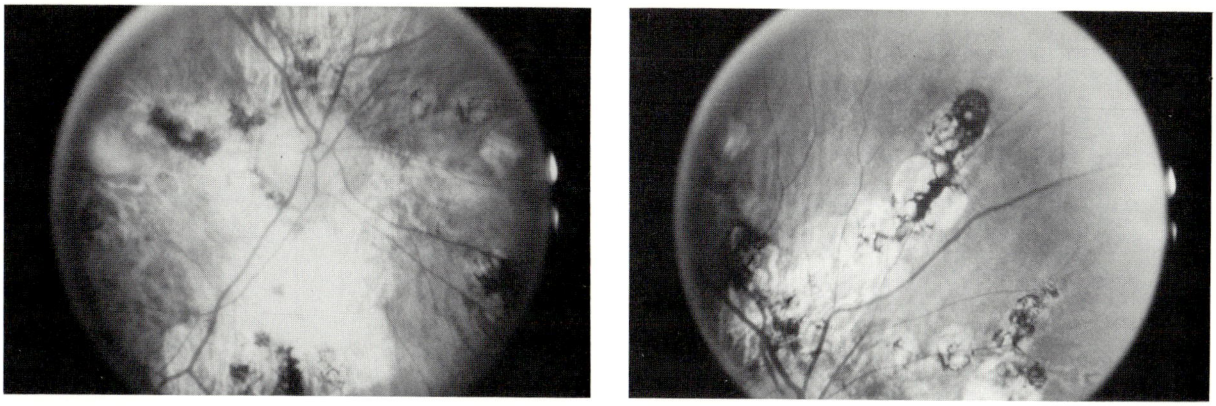

A B

Figure 25.3 *Serpiginous choroidopathy. Sixty-six-year-old woman with failing vision over the past 5 years; present acuity is counting fingers at one foot.* A, *peripapillary lesions.* B, *lesions extending toward the periphery.*

sion suggests that the disease does not spread by "infection" from adjacent cells, but may be based upon segmental structures such as the choriocapillary lobules.

Early lesions characteristically look greyish or yellowish and appear to be at the level of the pigment epithelium. During the early phases of angiography they block fluorescein (Fig. 25.4), although a hyperfluorescent border may be present (Schatz, Maumenee, and Patz 1974); during the later phases of angiography they show staining. As they evolve, they leave areas of depigmentation, pigment clumping, and choriocapillary loss. These late lesions are dark early in the angiogram, because of the absence of choriocapillaris, but transmit fluorescence later in a manner typical of other chorioretinal scars. Electrophysiologic tests have not proven of much value in this condition. The ERG and EOG may both be affected, but only in proportion to the extent of the lesions (Baarsma and Deutman 1976; Chisholm, Gass, and Hutton 1976). Occasionally, the degree of chorioretinal atrophy may be so striking in the late

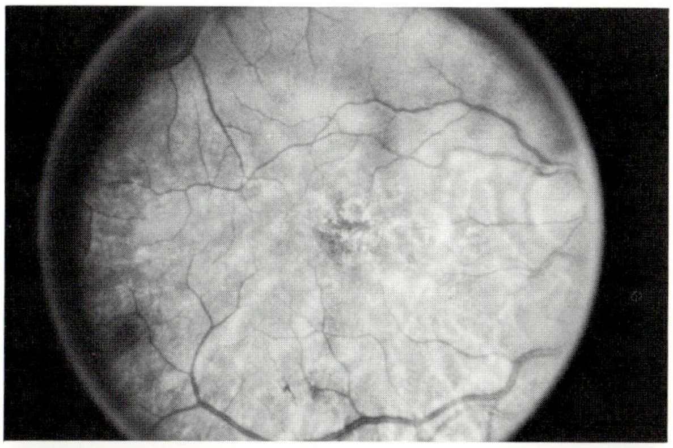

A

Figure 25.4 *Serpiginous choroidopathy. Forty-four-year-old male with 20/400 acuity. The large lesion has destroyed the macula and is still advancing.* A, *fundus photograph. Note the active greyish border zone superior to the fovea.* B, *early phase of the fluorescein angiogram, showing blockage of fluorescence by the active zone.* C, *late phase of the angiogram, showing staining of the active border.*

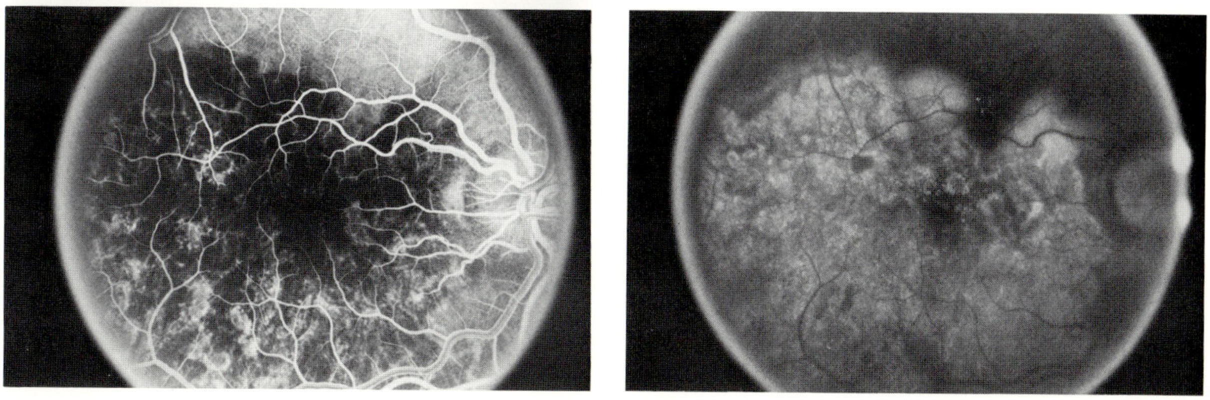

B C

stages that the fundus resembles a diffuse choroidal dystrophy (Chisholm, Gass, and Hutton, 1976).

The angiographic characteristics of serpiginous choroidopathy are very similar to those of placoid pigment epitheliopathy. Even though the two disorders affect different age groups, show a different distribution of lesions, and have a different prognosis, many current reviewers of serpiginous choroidopathy (Hamilton and Bird 1974; Schatz, Maumenee, and Patz 1974; Baarsma and Deutman 1976; Takei and Mizuno 1976; Pallin 1977) feel that the underlying pathology is similar in the two disorders. Presumably, it involves disease of the choriocapillaris, although whether it is primarily occlusive (Hamilton and Bird 1974) or inflammatory (Baarsma and Deutman 1976) is debatable. Serpiginous choroidopathy has not been linked with antecedent illness, as placoid pigment epitheliopathy has, and may represent a deeper and more chronic vasculitic response within the eye. Treatment has been notoriously ineffective. Steroids have been tried without any reported success; antimetabolites have been recently suggested (Laatikainen and Erkkilä 1974), but without any long-term follow-up.

ACUTE RETINAL PIGMENT EPITHELIITIS. Acute retinal pigment epitheliitis, in contrast to the preceding two conditions, is a rather subtle abnormality. It is characterized by discrete clusters of small black spots in the macular region, each surrounded by a tiny depigmented halo (Fig. 25.5*A*). This syndrome was initially described by Krill and Deutman (1972) and Deutman (1974) in adults who noted a relatively acute blurring of vision in the affected eye. One or more clusters of lesions were seen in the posterior pole, and on fluorescein angiography they showed blockage under the pigment

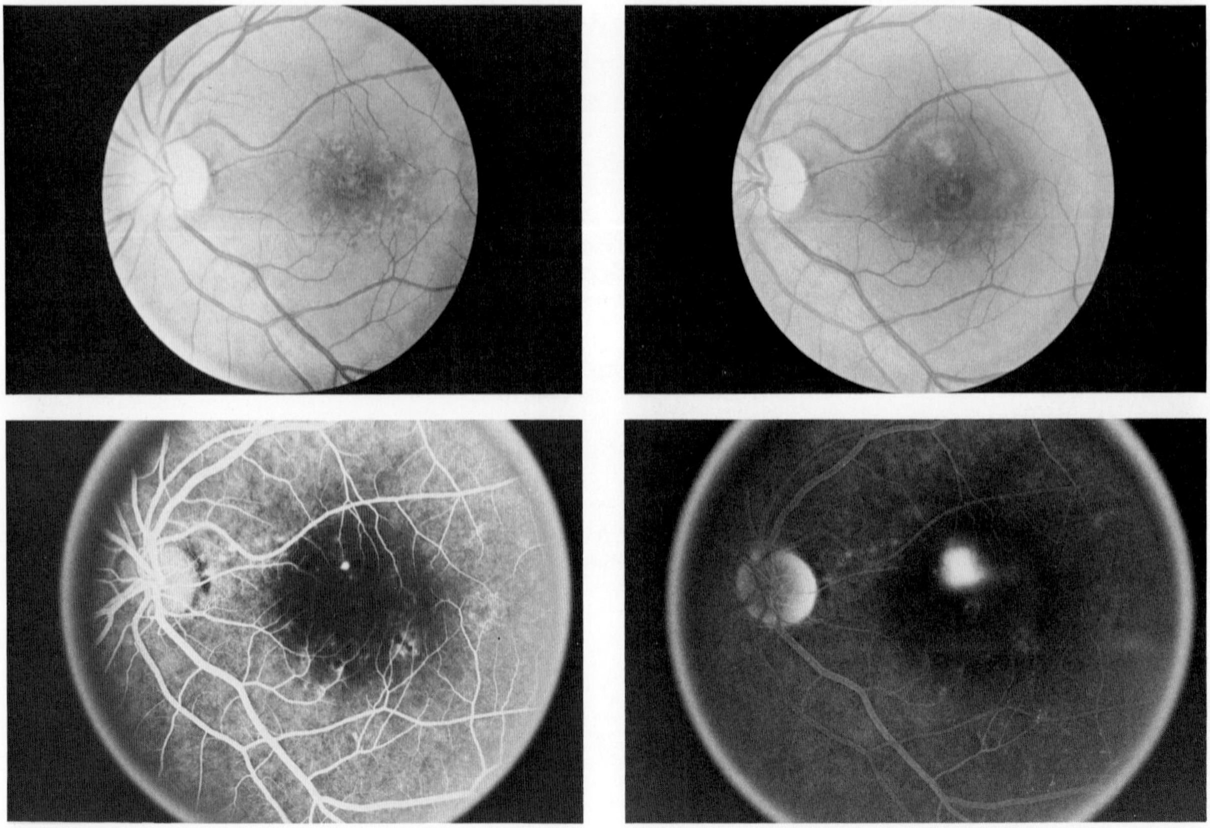

A B C D

Figure 25.5 *Acute retinal pigment epitheliitis. Forty-one-year-old male with 20/25 acuity and mild symptoms of metamorphopsia on the left.* A, *fundus photograph showing multiple characteristic lesions, each containing a spot of pigment.* B, *serous detachment of the retina, which developed one week later. An EOG showed a light-to-dark ratio of 1.8 on the right and 2.2 on the left.* C, *fluorescein angiogram showing the multiple pigment epitheliitis lesions and a focal point of serous leakage.* D, *late phase of the angiogram, showing the slow percolation of dye into the serous detachment.*

spot and transmission through the surrounding halos. There was no leakage of dye, and no choriocapillary abnormalities were noted. Over 6–12 weeks the vision returned to normal, although pigmentary defects remained. The ERG was normal, but the EOG findings were variable, and in two patients a subnormal EOG light response was noted during the acute phase; the EOG returned to normal after the vision had recovered.

Because of the occasional EOG abnormality, the location of pigmentary changes within the pigment epithelium, and the lack of any visible abnormalities in the choriocapillaris, Krill and Deutman (1972) believed the disease was a retinal pigment epitheliitis. They speculated that it might be viral in origin, or in some way related to central serous retinopathy, which occasionally develops in patients with these lesions (Fig. 25.5*B–D*).

Relatively few reports on this condition exist, which is surprising since it has been rapidly accepted as a diagnostic entity in reviews and texts. The initial cases were unilateral, but Friedman (1975) reported a bilateral and recurrent case. Most intriguing, and difficult to interpret, is a series of 61 patients reported by Eifrig, Knobloch, and Moran (1977). They found cases in all age groups, bilateral involvement in almost half, and little change in vision after the initial presentation. However, this series was based on a retrospective examination of angiograms, so that some of the lesions may have represented different stages of the disease or other conditions altogether. The incidence and pathology of retinal pigment epitheliitis seems somewhat less distinctive than those of placoid pigment epitheliopathy and serpiginous choroidopathy. The definition of retinal pigment epitheliitis as a discrete clinical entity may be premature.

GENERALIZED CHORIORETINITIS. The general consideration of uveitis and chorioretinitis is beyond the scope of this book. For the most part, conditions such as Vogt-Koynagi-Harada disease, Behçet's disease, pars planitis, and retinal vasculitis involve the pigment epithelium secondarily through adjacent choroidal or retinal inflammation. Extensive pigmentary changes may occur in uveitis and will sometimes be pathognomonic of a disease; but they are only reactive scars and their significance will depend upon their depth and distribution (Fig. 25.6). Individual uveitic syndromes are described in

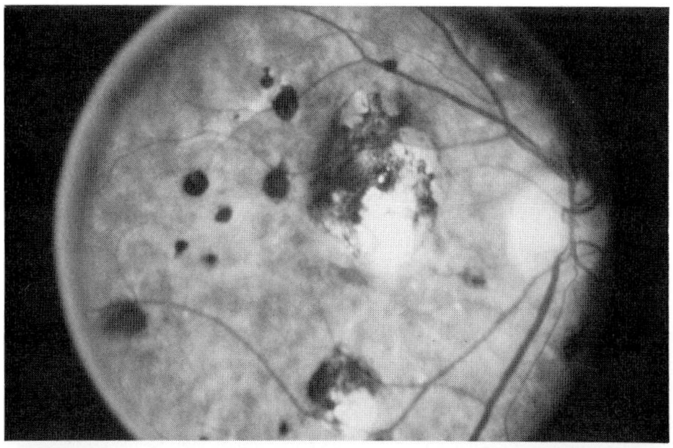
A

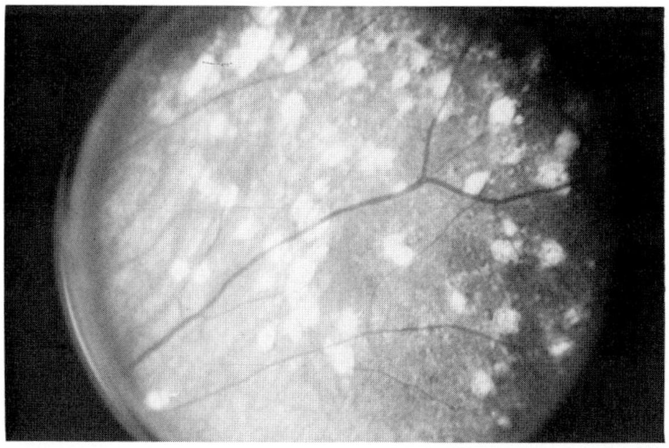
C

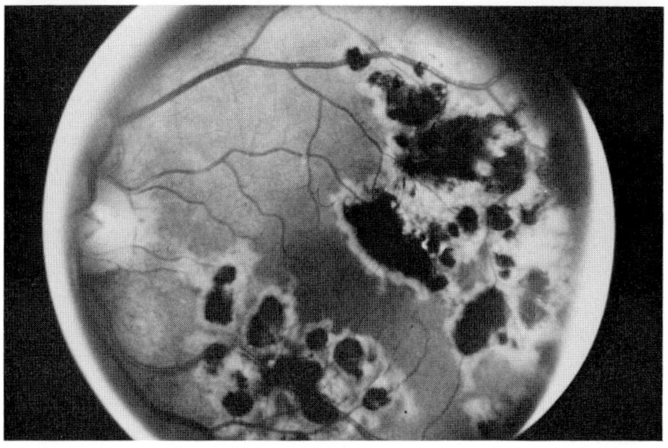
B

Figure 25.6 *Uveitis.* A, *multiple chorioretinal lesions in a sixty-one-year-old woman; etiology unknown.* B, *unusual densely pigmented bilateral lesions found incidentally in a 9-year-old boy. His acuity was 20/20 and an ERG was normal.* C, *punched-out miliary lesions in a 25-year-old male with a 3-year history of left upper lobe infiltrate, uveitis, and treatment for tuberculosis. This may represent miliary tuberculosis, but he was never severely ill.*

standard texts on ophthalmology or chorioretinal diseases.

Infections of the Retinal Pigment Epithelium

Most viruses and bacteria have their primary effects in the eye on tissues other than the retinal pigment epithelium, unless the infection is so rampant that indiscriminate tissue damage occurs. Possible exceptions are congenital rubella, which seems to affect the pigment epithelium directly, and a few infections such as measles and syphilis in which secondary damage to the pigment epithelium is a major effect of the pathologic process.

RUBELLA RETINOPATHY. Retinal pigmentary changes in rubella have been well documented since the congenital rubella syndrome was first recognized by Gregg in 1941. The retinopathy (or, more properly, pigment epitheliopathy) is characterized by a coarse "salt and pepper" fundus appearance (Fig. 25.7) without significant visual effects (unless other elements of the congenital rubella syndrome are present, such as cataracts or severe glaucoma). The retinopathy is often an isolated finding, and has been observed in 50 percent of children recognized to have congenital rubella (Francois 1976), although the significance of this figure is hard to assess since one does not know how many children have been exposed to the virus and not come to medical attention.

The visual acuity in congenital rubella retinopathy is usual normal or only minimally affected (Wolff 1973; Krill 1972). The ERG and EOG light responses are usually within normal limits (Krill 1967, 1972; Francois 1976; Franceschetti, Francois, and Babel 1974) although cases with severe retinopathy and ERG changes have been reported (Zetterström 1959). The most striking finding is that the EOG is normal in the face of an extensive pigment epithelial abnormality (Krill 1967). Apparently the pigmentary disturbance is not necessarily associated with a functional disturbance of the cells. However, local cellular damage may be severe enough in some cases to allow the complications of subretinal neovascularization (Gass 1977a; Deutman and Grizzard 1978).

Rubella retinopathy may be present at birth, or may develop during the first several years of life (Collis and Cohen 1970; Wolff 1973). Francois (1976) points out that, in contrast to the cataract and glaucomatous changes that depend upon exposure to the virus at a critical period of fetal development, the retinopathy has been documented after maternal exposures at different periods throughout pregnancy. Histopathology (Zimmerman 1968; Yanoff 1971; Wolff 1973; Franceschetti, Francois, and Babel 1974) shows an uneven distribution of pigment within the pigment epithelial layer, but no recognizable abnormality of either the choroid or the overlying neurosensory retina. Thus, the primary damage appears to be within the pigment epithelium.

In a child with good vision, the characteristic peppery retinopathy of congenital rubella will not ordinarily be confused with other disorders. The retinopathy of congenital syphilis (see below) may occasionally appear similar, but will usually show signs of optic nerve damage or of a vascular distribution of lesions that would be uncharacteristic of rubella, and the serology will be positive. Dystrophic diseases can be distinguished by their electrophysiologic characteristics and sometimes by a positive

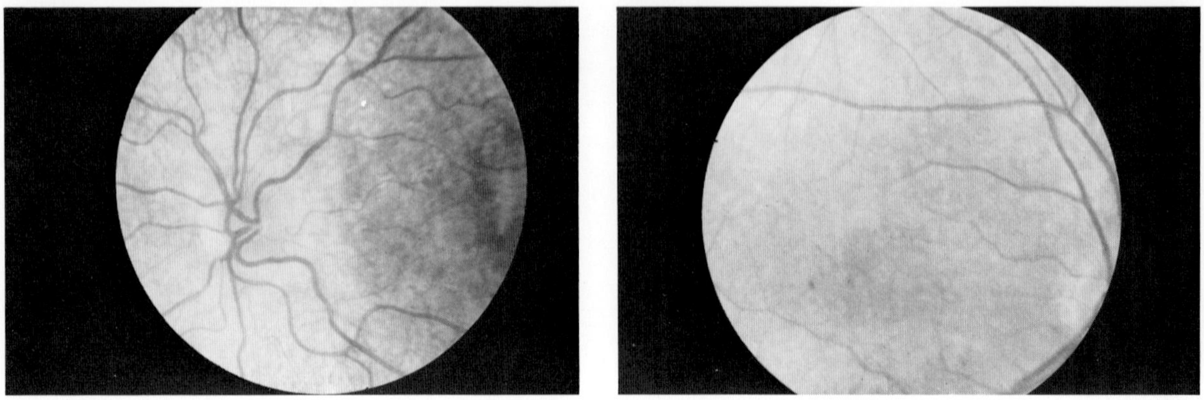

A B

Figure 25.7 *Rubella retinopathy.* A, *fundus photograph showing the characteristic irregular coarse granularity of the fundus.* B, *similar findings in a different patient.*

A

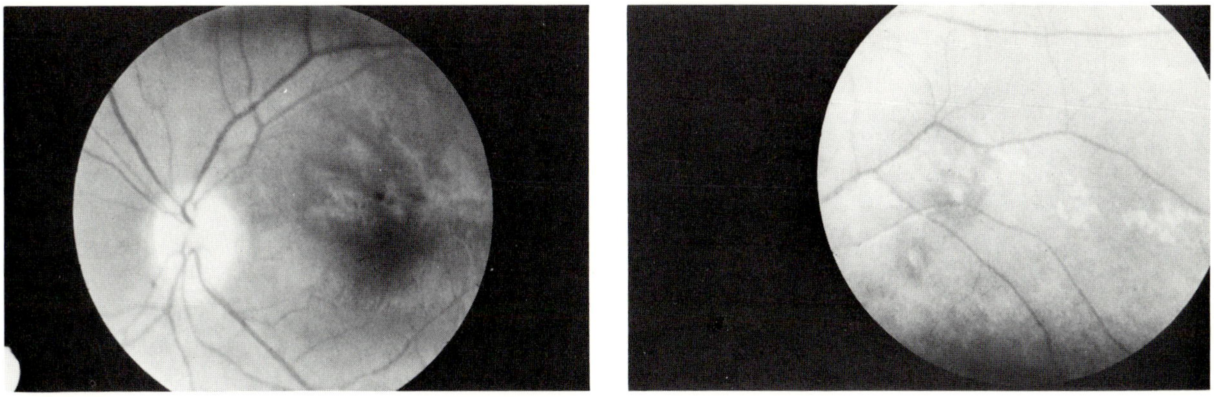

B

Figure 25.8 *Two cases of subacute sclerosing panencephalitis. (From Robb and Watters 1970; copyright 1970, American Medical Association.)*

family history. Postnatal infections with rubella (Gerstel and Zinn 1976) and other viruses (Franceschetti, Dieterle, and Schwartz 1958) usually produce a very different syndrome, probably vasculitic in nature, that results in extensive pigmentary scarring with severe effects on visual function.

MEASLES AND SUBACUTE SCLEROSING PANENCEPHALITIS. Measles (rubeola, morbilli), in contrast to rubella, is not typically a cause of congenital retinopathy; only rare cases have been reported (Guzzinati 1954; Metz and Harkey 1968). Acquired measles retinitis is also rare, but is better documented (Franceschetti, Dieterle, and Schwartz 1958; Hentsch and Külz 1969; Scheie and Morse 1972; Regensburg and Henkes 1976). The retinitis usually begins sometime in the second week of the disease, and is characterized by signs of arterial occlusive disease such as vascular attenuation, retinal edema, macular changes, and severe bilateral loss of acuity. A gradual recovery occurs, with the development of pigmentary retinopathy. The late fundus picture may resemble retinitis pigmentosa, with a pale optic disc, attenuated vessels, bone spicule figures, and an unrecordable ERG. Visual function may continue to deteriorate long after the acute stages of the disease, so the long-term prognosis must be guarded.

Another form of postnatal measles infection is subacute sclerosing panencephalitis (subacute sclerosing leukoencephalitis of van Bogaert, Dawson's inclusion body encephalitis). This disorder most commonly develops in childhood, with the insidious development of mental changes concomitant with, or preceded by, the appearance of focal edematous lesions of the posterior retina (Fig. 25.8), which evolve into relatively nonspecific pigmentary scars (Robb and Watters 1970; Obenour 1972; Green and Wirtschafter 1973). The disease relentlessly progresses, with the development of seizures, myoclonic jerks, cortical blindness, decorticate rigidity, and death within 1–2 years of onset. Measles virus is believed to be responsible, because virus-like particles are visible in brain tissue, and a virus antigenically similar to measles can be transmitted from infected brain tissue to animals (Obenour 1972; Green and Wirtschafter 1973). Inclusion bodies have been demonstrated in retinal neurons (Klouček and Otradovec 1969; Nelson et al. 1970), and the involvement of the pigment epithelium appears to be secondary to the retinitis.

SYPHILIS. Syphilis (lues) is a disease with protean manifestations as a congenital or acquired infection. Our concern here is with the propensity of syphilis to produce a chorioretinitis that can mimic the fundus appearance of rubella retinopathy or retinitis pigmentosa (Fig. 25.9). Syphilitic involvement of the choroid is not infrequent in congenital cases, occurring in anywhere from 5 to 70 percent depending upon the series (Franceschetti, Francois, and Babel 1974). Most descriptions have separated the congenital and acquired forms, but the pathology in both appears to be similar (Duke-Elder and Perkins 1966). The primary lesion is a choroiditis with nonspecific round cell infiltration and nodule formation at the level of the choriocapillaris. Depending upon the severity, there can be damage to the overlying Bruch's membrane, pigment epithelium, and retina. A curious finding in adult syphilitic chorioretinitis

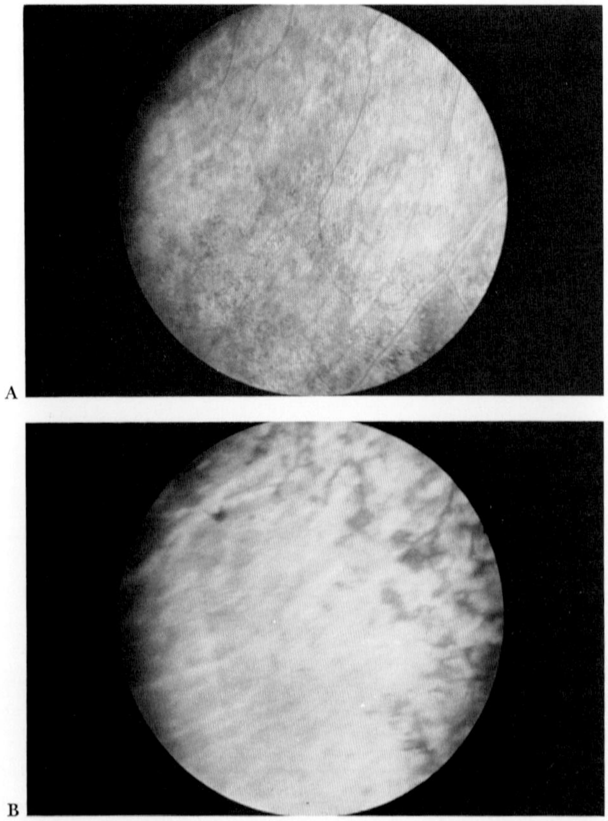

Figure 25.9 *Syphilitic chorioretinitis.* A, *salt and pepper fundus in a 33-year-old female.* B, *pseudoretinitis pigmentosa in a 53-year-old male with a 7-year history of night blindness. The VDRL was negative and the FTA-ABS borderline. Spirochetes were found in the cerebrospinal fluid. (Courtesy of J. L. Smith.)*

is invasion of the choroid by retinal elements, including pigment epithelium and glia (Blodi and Hervouet 1968).

Major reviews (Franceschetti, Francois, and Babel 1974; Ballantyne and Michaelson 1970; Duke-Elder and Perkins 1966) note that several patterns of pigmentation have been distinguished, including a coarse granular "salt and pepper" retinopathy, a more focal punched-out uveitic pattern, and a diffuse atrophic pigmentary degeneration that mimics retinitis pigmentosa. They caution that these forms frequently blend into one another, and all three types may be found in both the congenital or acquired patient populations. Nonetheless, the congenital form seems more likely to produce a diffuse pigmentary retinopathy, while the adult form tends to be more focal or show a vasculitic pattern (Savir and Kurtz 1976; Crouch and Goldberg 1975).

In the milder forms of syphilitic retinopathy, especially when there is only a salt-and-pepper pigmentation or a set of focal lesions, visual function may be relatively normal and the electrophysiologic data unremarkable. Differentiation from rubella retinopathy will be necessary and can usually be made on the basis of the maternal history and appropriate serologic tests. The similarity in fundus appearance is interesting, since the abnormality in rubella is confined to the pigment epithelium, whereas the abnormality in syphilis is an initial choroiditis. In the more severely atrophic forms of syphilitic retinopathy, in which there is a diffuse tapetoretinal degeneration, there is usually significant loss of visual function and a severely depressed ERG and EOG light response. Relative preservation of the ERG may sometimes be a clue suggesting syphilis, as opposed to retinitis pigmentosa, but severe cases with low signals provide a difficult differential diagnostic problem. One cannot always tell whether a positive serology (even an FTA-ABS) is the cause of longstanding retinal atrophy or is merely an incidental finding (cf. Heckenlively 1977).

OTHER INFECTIONS AND EMBRYOPATHIES

Viruses other than rubella and measles have been reported only rarely (and with marginal documentation) to cause a benign congenital pigmentary retinopathy (cf. Krill 1972; Francois 1976): varicella, mumps, and influenza have all been implicated in at least one case. More severe forms of congenital chorioretinitis, in which the damage is not localized to the pigment epithelium, have been reported with cytomegalic inclusion disease (Tarkkanen, Merenmies, and Holström 1972), herpes simplex retinitis (Cibis 1975), and varicella (Charles, Bennett, and Margolis 1977).

Another source of congenital damage to the developing eye is ionizing radiation (Franceschetti, Francois, and Babel 1974; Francois 1976), and various degrees of retinopathy have been observed. The pigmentary changes may in some cases appear similar to rubella retinopathy (Francois et al. 1962). Chromosomal defects may also produce abnormalities in the pigment epithelium, although they are usually associated with extensive deformation of other structures as well. An unusual distribution of pigment within the pigment epithelium was recently noted in a case of trisomy 18 (Rodrigues et al. 1973), and diffuse pigmentary changes have been observed in a child with familial microcephaly (Fisch et al. 1973).

Adult retinal infection with viruses can occur,

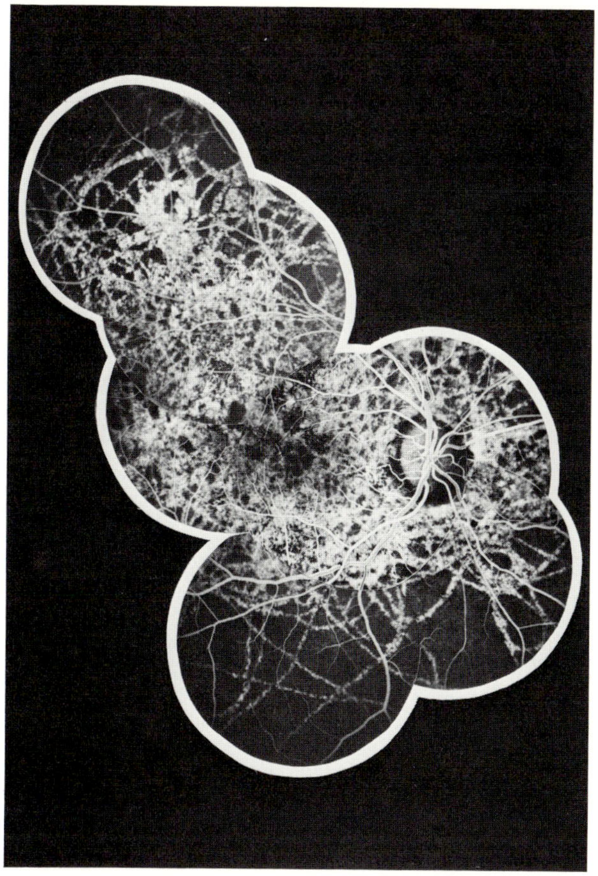

Figure 25.10. *Ophthalmomyiasis. Montage of the fluorescein angiogram, showing the pathognomonic tracks through the pigment epithelium. (From Gass and Lewis, 1976; copyright 1976, American Medical Association.)*

especially in immunologically compromised patients. A number of viruses may produce fundus changes similar to those described earlier for adult measles (Franceschetti, Dieterle, and Schwartz 1958). Some viruses, such as cytomegalovirus, can cause a more aggressive inflammation (Nicholson 1975), although the presentation may be quite variable (cf. Marmor et al. 1978). Involvement of the pigment epithelium in these acquired viral infections is secondary to neuronal involvement or vasculitis.

Various bacteria and fungi can also produce diffuse or multifocal chorioretinitis that involves the pigment epithelium secondarily (see Fig. 25.6*C*). In severe infection there is widespread destruction of ocular tissues, but in more limited cases the damage may be focal. For example, histoplasmosis may at times leave small scars scattered about the fundus (see below). Focal depigmented lesions may also be an occasional sequelae of miliary tuberculosis, candidiasis, or coccidiomycosis. Most parasitic infections of the fundus cause a severe inflammatory reaction, especially when the parasite dies in situ, and the result is a deep gliotic scar. An occasional exception is ophthalmomyiasis (infestation by fly larvae), in which a living maggot can enter the subretinal space and migrate extensively without significant inflammation (Fig. 25.10). Passage of the larva damages the pigment epithelium and leaves pathognomonic interlacing depigmented tracks throughout the fundus (Fitzgerald and Rubin 1974; Gass and Lewis 1976).

Focal Exudative Lesions Involving the Retinal Pigment Epithelium

Exudations of fluid under the retinal pigment epithelium characterize a group of acquired disorders that share common features in their secondary effects upon the eye. Central serous retinopathy, detachments of the pigment epithelium, and the presumed ocular histoplasmosis syndrome fall clearly within this group. Other conditions, such as multiple vitelliform cysts and foveo-macular dystrophy, should possibly be included, although whether these disorders are acquired or dystrophic has not been firmly established. The clinical diagnosis and management of these exudative conditions is treated at length in other sources (Gass 1977a; Zweng, Little, and Vassiliadis 1977), and the presentation here will emphasize the pathogenetic role of the pigment epithelium.

CENTRAL SEROUS RETINOPATHY. Central serous retinopathy (central serous choroidopathy, angiospastic retinopathy) is an acute disorder of young adults (predominantly males), which is characterized by serous detachment of the sensory retina in the macular region (Fig. 25.11; see also Fig. 25.5), often during a period of emotional stress (Gass 1967, 1977a). Depending on the location of the serous elevation, visual function may be minimally or moderately affected, there may be a distortion of objects (metamorphopsia and micropsia), and there may be an apparent hyperopia because of the foveal elevation. In most patients suffering a first attack, the lesion will spontaneously resolve over several weeks to several months, leaving little or no residual deficit. Occasionally, the lesion will persist and cause permanent degenerative changes. Resolution of the serous fluid is unaffected by drugs such as steroids, carbonic anhydrase inhibitors, and vasodilators, but can be enhanced by photocoagulation of the leaking

A B C D

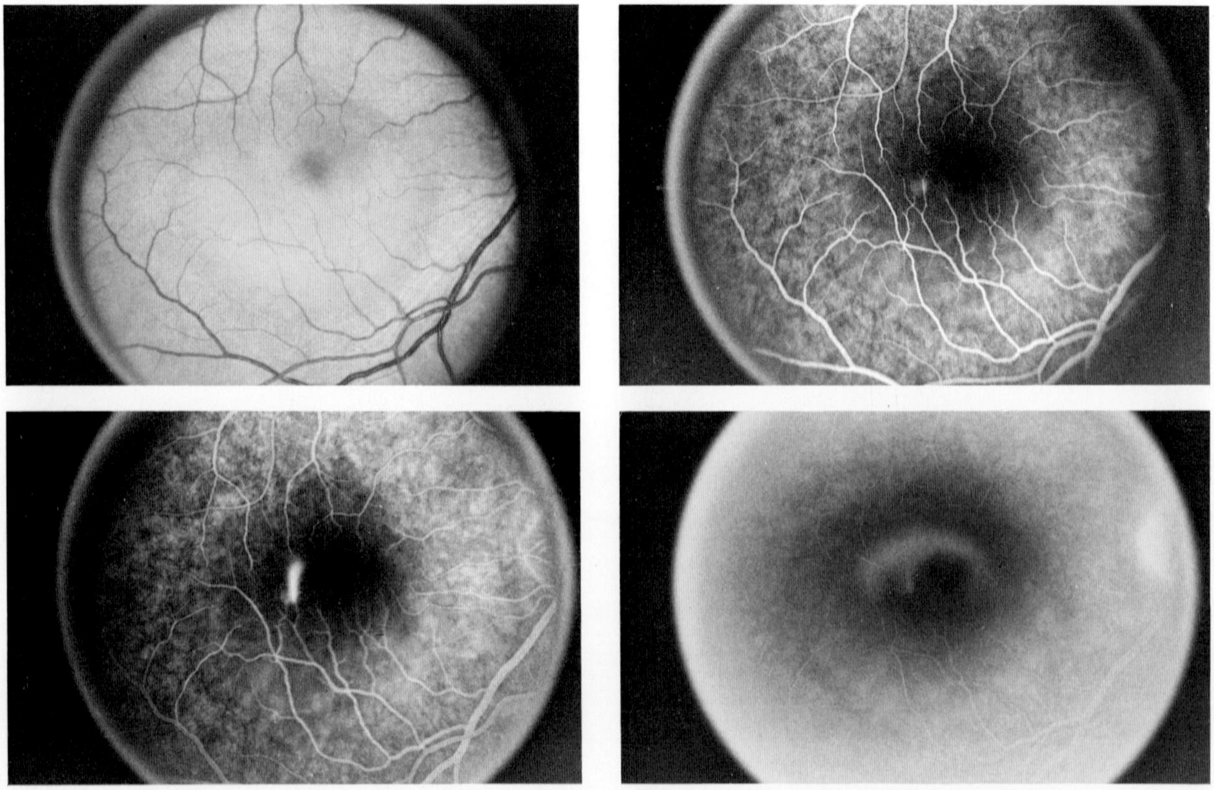

Figure 25.11 *Central serous retinopathy. Thirty-five-year-old male with 20/30 acuity (adding +0.50 diopters) and metamorphopsia.* A, *fundus photograph showing a large serous detachment.* B, *venous phase of the fluorescein angiogram; dye leakage is still not visible.* C, *later phase of the angiogram, showing dye rising (presumably by convection) into the detachment.* D, *very late angiogram, showing a faint umbrella pattern as dye slowly layers out against the superior wall of the detachment.*

spot, or, in fact, by photocoagulation of uninvolved areas within the detachment. However, because the lesions are usually self-limiting, most authorities favor a period of observation for initial attacks. Recurrences are not infrequent and the prognosis for recovery seems to diminish with repeated attacks. Multiple lesions may occasionally be present.

Although the most prominent feature of this condition is the detachment of the sensory retina, Gass (1977a, b) feels that the primary lesion is a small focal detachment of the pigment epithelium, which acts as a nidus for the leakage of fluid into the subretinal space. On performing fluorescein angiography, one characteristically finds a small lesion that lights up immediately during the early phases, and through which fluorescein appears to percolate during the later phases of the angiogram (Figs. 25.5, 25.11). Gass argues that the pigment epithelium normally functions to pump fluid out of the subretinal space, and that a small focal detachment of the pigment epithelium will not of itself produce a serous detachment of the retina. However, if this detached pigment epithelium suffers a gradual loss of its pumping abilities, then exudation into the subretinal space may occur. The effect of photocoagulation would be to destroy these malfunctioning pigment epithelial cells. Photocoagulation probably does not "seal" leaks, since Peyman and Bok (1972) have shown that laser scars are quite permeable to large molecules.

Gass's view of the pathogenesis of central serous retinopathy is widely believed, but other theories have also been proposed. For example, Behrendt (1977) argues that the source of the serous fluid is leakage from the retinal vessels, and that focal defects in the pigment epithelium may *help* the choroid to absorb the fluid. The characteristic spreading of fluorescein through these lesions during angiography can be accounted for by simple diffusion. This theory explains why photocoagulation can be effective regardless of where the lesions are placed. On the other hand, Friedman (1977) argues that obstruction of the choroidal venous system is a major factor in causing fluid to accumulate in the subretinal space.

DETACHMENT OF THE RETINAL PIGMENT EPITHELIUM. We have noted that small detachments of the retinal pigment epithelium can be part of the syndrome of central serous retinopathy. Larger and more prominent detachments of the pigment epithelium, without retinal detachments overlying, are characteristic of eyes with extensive drusen or early senile maculopathy (Gass, Norton, and Justice 1966; Gass 1973). These larger pigment epithelial detachments typically appear yellowish in color and show prompt hyperfluorescence during the early phases of angiography (Fig. 25.12). However, there are many exceptions to this description, particularly with long-standing detachments in which the fluid below the pigment epithelium has become inspissated and thick. These detachments can resemble the cysts of vitelliform dystrophy (see Fig. 24.6 in the preceding chapter) and may block fluorescein in the early stages of angiography because of their density. They often show characteristic branched pigment figures within the detachment (Fig. 25.13), which are accentuated by angiography. Pigment epithelial detachments may spontaneously resolve, leaving telltale branched pigment figures or areas of atrophic retina and pigment epithelium. Not infrequently, they are associated with a degeneration of Bruch's membrane, through which subretinal neovascularization can occur. Sometimes the development of a large pigment epithelial detachment may be associated with the fading of underlying drusen (Gass 1973; Bonnet, Pingault, and Tapissier 1976). The visual significance of pigment epithelial detachments depends upon their location (foveal or extrafoveal) and on their stage of evolution—but the prognosis is much less cheerful than for central serous retinopathy (Gass 1973, 1977a). Their significance will be discussed more fully in Chapter 26.

The etiology of pigment epithelial detachment, other than as a confluence of drusen, is unknown. The question of whether the pigment epithelial detachments of central serous retinopathy and senile macular degeneration are separate entities or forms

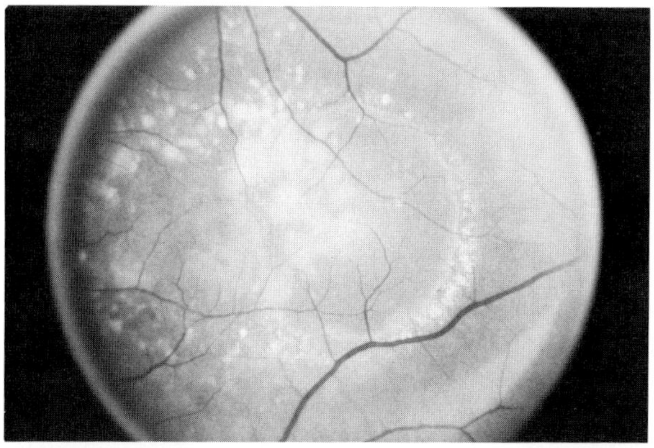

A

Figure 25.12 *Pigment epithelial detachment. Seventy-year-old male with 20/50 acuity in the left eye.* A, *large pigment epithelial detachment in the left eye, just temporal to the fovea.* B, *arteriovenous phase of the fluorescein angiogram, showing the prompt leakage of dye under the entire detachment (compare this to the slow focal leakage in Fig. 25.11).* C, *late phase of the angiogram, showing persistent staining of the detachment.*

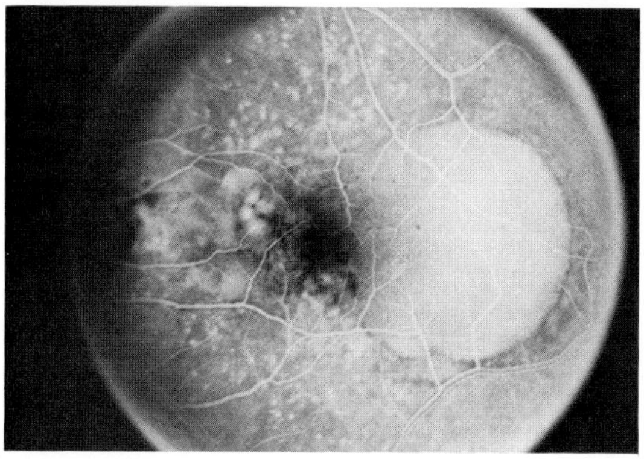

B

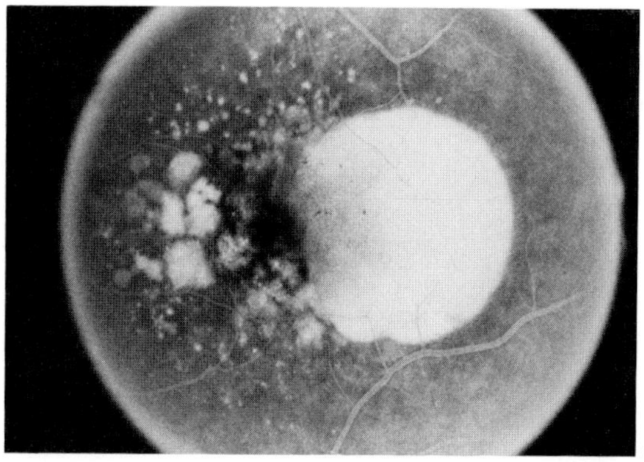

C

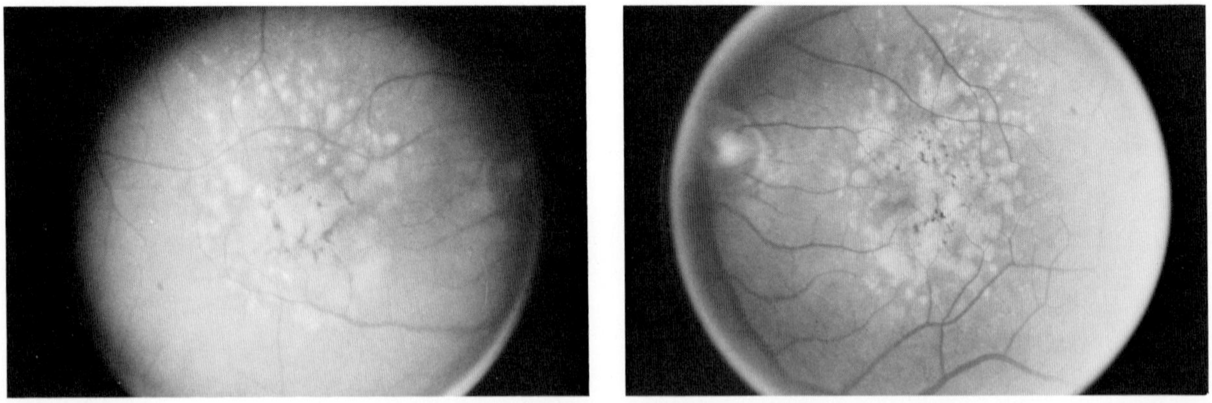

A B

Figure 25.13 *Pigment epithelial detachment. Confluent drusen and focal detachments of the pigment epithelium, containing characteristic branched pigment figures.* A, *right eye.* B, *left eye.*

of a common disease is not fully resolved (Faurschou, Rosenberg, and Nielsen 1977).

Pigment epithelial detachments may be confused with, or may underlie, two conditions that have been described as separate entities but which seem of uncertain definition. The first condition, "multiple vitelliform cysts" (Remky and Kolbl 1971; Noble, Levitzky, and Carr 1976), is characterized by multiple pigment epithelial detachments with somewhat variable electrophysiologic findings and effects upon vision. The EOG light response is not severely affected, as it is in vitelliform dystrophy. Others have reported "multiple serous detachments" of the pigment epithelium (Robinson and Gitter 1972; Gass 1977a) which appear typical of presenile degenerative lesions. The second condition is the foveomacular dystrophy described by Gass (1974), which is characterized by bilateral vitelliform lesions (see Fig. 24.5). Some of Gass's patients were related, but others had no family history of the disorder. He speculated that the lesions might be related to drusen, since the pathologic findings were similar in the one case examined histologically. This view is consistent with that of other authors (for example, Rosenberg, Faurschou, and Nielsen 1977) and with my own experience that the foveal lesions may evolve into scars having the typical branched pigmentary pattern of old pigment epithelial detachments (Marmor 1979).

PRESUMED OCULAR HISTOPLASMOSIS SYNDROME. The presumed ocular histoplasmosis syndrome (and its European affiliate, juvenile hemorrhagic macular choroidopathy) is a disorder characterized by multiple, small, old, pigment epithelial scars, by peripapillary atrophy, and by a tendency to develop subretinal neovascularization beneath existing scars in the macula (Fig. 25.14). The disorder typically affects individuals between ages 20 and 50. The initial complaint is usually metamorphopsia or a decrease in vision, and examination shows a focal exudation of fluid near an existing lesion in the macula, or a small grey-green ring (possibly surrounded by hemorrhages or exudate) signifying the presence of choroidal neovascularization (Gass 1977a). Gass states that the greenish color is not produced by either exudation or hemorrhage, as frequently described, but by a proliferation of the pigment epithelium and an accumulation of pigment-laden macrophages. Once neovascularization develops, the prognosis for central vision is relatively poor. Photocoagulation may occasionally be of benefit, but an end-stage disciform lesion (see Chapter 26) is the usual result.

The syndrome most often involves the reactivation or decompensation of old lesions (Smith, Knox, and Jensen 1973; Sawelson et al. 1976) to produce an exudative or hemorrhagic maculopathy. Thus, a macula free of lesions will generally be spared the disease. However, there have been several recent, well-documented reports (Ryan 1976; Miller, Stevens, and de Venecia 1976; Wilkinson 1976) demonstrating the appearance of new lesions and neovascularization in maculae that were ophthalmoscopically clear.

This syndrome has excited enormous debate in recent years about whether or not histoplasmosis is the primary etiologic agent, and whether the recurrences are triggered by vascular or inflammatory events (cf. Brockhurst et al. 1977; Meredith et al.

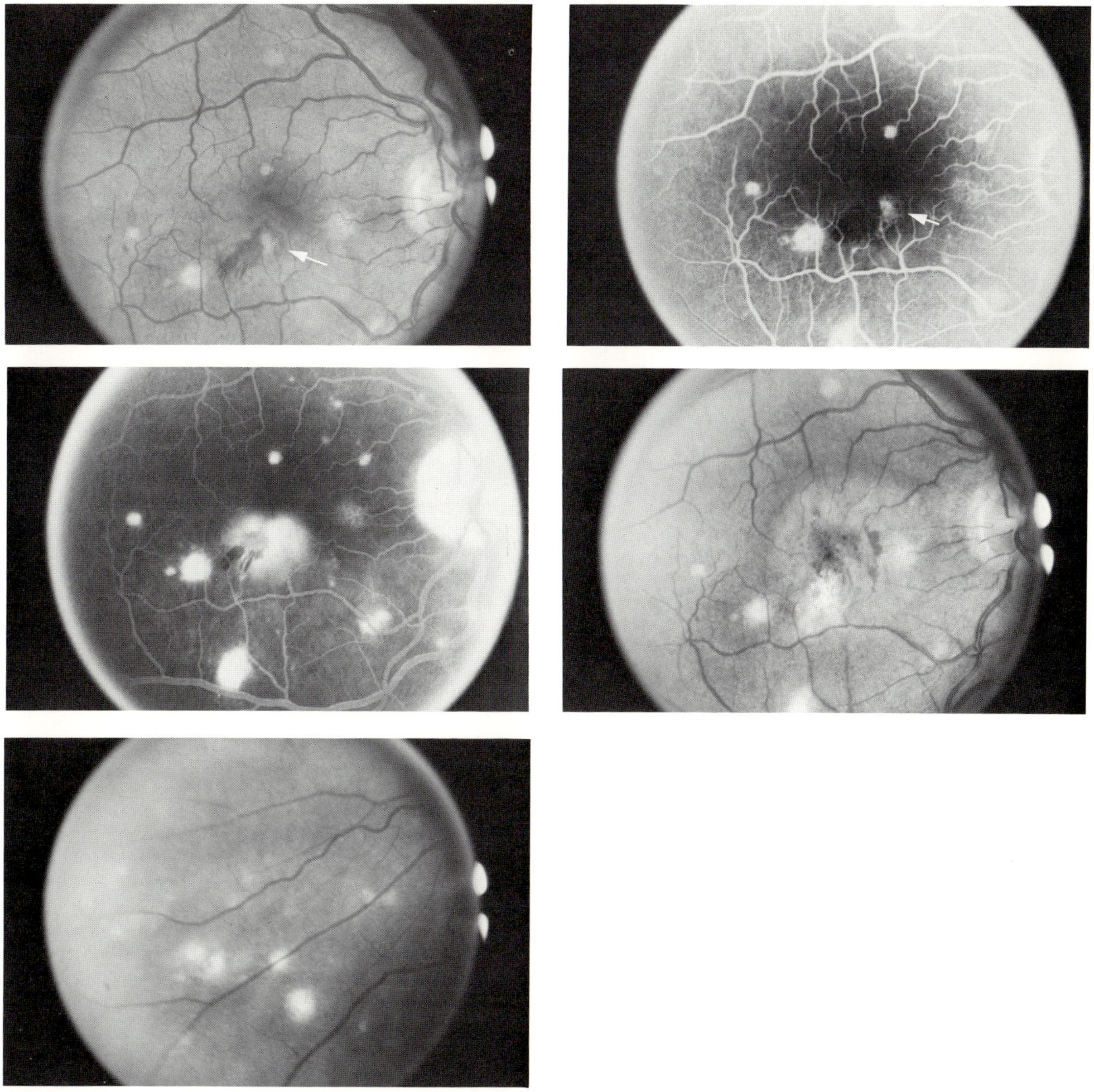

A B C D E

Figure 25.14 *Presumed ocular histoplasmosis syndrome. Forty-five-year-old female with 20/30 acuity.* A, *right macula showing characteristic punched out scars and a subretinal hemorrhage emanating from a pigmented ring (arrow).* B, *arterio-venous phase of the angiogram, showing subretinal neovascularization at the site of the ring lesion (arrow).* C, *late phase of the angiogram, showing leakage of dye from the neovascular tuft. The chorioretinal scars transmit fluorescence but do not show leakage.* D, *the same eye 2½ months later, showing extensive scarring in the macula despite focal laser photocoagulation; visual acuity had fallen to 20/200.* E, *peripheral punched-out lesions in the left eye of the same patient.*

1977). The major arguments for histoplasmosis being the etiologic agent are epidemiological, and are based upon a concentration of cases in regions of the United States where histoplasmosis is endemic and upon the high incidence of positive histoplasmin skin tests in patients with the syndrome. On the other hand, despite the prevalence of the syndrome, there has been only a single somewhat questionable case in which there has been histopathologic demonstration of histoplasma organisms (Roth 1977). Indeed, an identical syndrome called juvenile hemorrhagic macular choroidopathy is prevalent in Europe, where histoplasmosis is virtually unknown (Braunstein, Rosen, and Bird 1974; Francois, deLaey, and Dakir 1974; Flage, Sand, and Syrdalen 1977; Saari 1977a).

Reactivation of existing scars has been viewed as recrudescence of an allergic or infectious process (Schlaegel 1977), or, alternatively, as vascular ingrowth through a mechanically damaged Bruch's membrane (Key, Green, and Maumenee 1977; Gass 1977a). However, it is difficult for either of these theories to explain the de novo appearance of macular lesions long after the presumed infection took place. There seems little doubt that some cases of the syndrome in the United States are caused by histoplasmosis, but there are probably several other causes, either inflammatory or vascular in origin, which can produce this same spectrum of findings and may account for the European syndrome and the cases in which lesions develop progressively. For example, Saari (1977b) has proposed that intravascular coagulation in the choriocapillaris might initiate neovascularization.

An unanswered question is, Why do certain chorioretinal scars break down, while others do not? For example, subretinal neovascularization is uncommon after toxoplasmosis. Perhaps deeper scars, such as those in toxoplasmosis, destroy the choriocapillaris, which would be the source of neovascularization. The role of the pigment epithelium in the presumed histoplasmosis syndrome is obscure, although pigment epithelial changes are characteristic of the lesions. Gass (1977a) maintains that hemorrhage in this syndrome is typically subretinal, rather than sub–pigment-epithelial, so that a breakdown of the pigment epithelial barrier is postulated.

Degenerative Disorders of the Pigment Epithelium

Aging in the pigment epithelium, and the development of senile macular degeneration, are considered in Chapter 26. There are, however, degenerative conditions involving the pigment epithelium that do not fall conveniently under either of those headings; several examples will be considered in this section. Angioid streaks characterize a disorder whose primary pathology lies in Bruch's membrane, but in which there can be widespread disturbance of the pigment epithelium and a breakdown of its barrier functions. Photic maculopathy involves a degeneration of pigment epithelial cells as a result of photic or thermal injury. Senile reticular degeneration is associated with peripheral drusen and aging, but is of special interest because the gross pattern of pigmentary degeneration is similar to that in other pigment epithelial disorders (such as pattern dystrophies). In high myopia, damage to Bruch's membrane and the pigment epithelium are important to the development of hemorrhagic complications. Retinal detachment is often a disease of the vitreous and vitreoretinal junction, but may also involve the pigment epithelium, which participates in maintaining adhesion.

ANGIOID STREAKS. Angioid streaks (Figs. 25.15, 25.16) are bilateral, irregular, grey or brown bands, developing in adulthood, which can mimic the appearance of retinal vessels but are located at the level of Bruch's membrane and the pigment epithelium (cf. Paton 1972; Franceschetti, Francois, and Babel 1974; Archer and Logan 1977). Most angioid streaks radiate from the disc, but other patterns may be seen. They vary considerably in size and pigmentation, and the heavily pigmented ones are often surrounded by a narrow depigmented border. They are seen better by fluorescein angiography than by ophthalmoscopy, since they transmit early and late fluorescence through areas of thinned pigment epithelium (Fig. 25.15). Other fundus changes are often associated with angioid streaks. Most characteristic is a diffuse mottling of the posterior pole, sometimes called peau d'orange fundus (Fig. 25.16) which may (Krill, Klien, and Archer 1973) or may not (Hull and Aaberg 1974) represent an early state of the disorder. Also frequent are scattered punched-out lesions resembling those of presumed ocular histoplasmosis.

The basic pathology of angioid streaks is thought to be a degeneration and rupturing of Bruch's membrane, which, in the area of the lesions, shows deep basophilic staining and calcification (Archer and Logan 1977). The overlying pigment epithelium becomes thin, or occasionally hypertrophic in places, as a secondary change. Because of the breakdown in Bruch's membrane, a pathway is opened for vascular ingrowth under the pigment epithelium and retina —and neovasculation under the retina with resultant hemorrhage and disciform macular degeneration is the major complication of this disorder. The pathogenesis of the changes in Bruch's membrane is unknown, but an abnormality of elastic tissue is postulated because of the frequent association be-

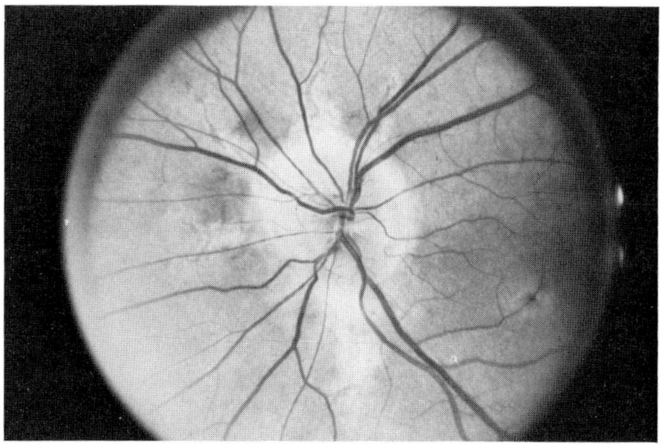

A

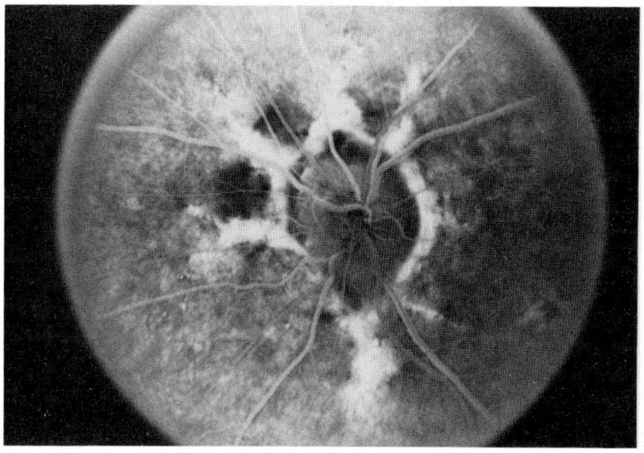

B

Figure 25.15 *Angioid streaks.* A, *fundus photograph. Small pigmented angioid streaks radiate from the disc.* B, *the regions of the streaks are accentuated by fluorescein angiography.*

tween angioid streaks and pseudoxanthoma elasticum. The initial localization of pathology to Bruch's membrane, rather than the retina or pigment epithelium, is confirmed by the normality of the ERG and EOG in uncomplicated cases (Weinstein and Reeser 1975; Archer and Logan 1977). These tests may be abnormal, however, when the disease is unusually extensive or the retina and pigment epithelium have been damaged secondarily.

Angioid streaks may be a sporadic finding, but are usually associated as a degenerative condition with one of several systemic diseases. The most prominent association is with pseudoxanthoma elasticum (Groenblad-Strandberg syndrome), a recessively inherited condition in which elastic tissue in the skin, vascular system, and presumably the eye degenerates and calcifies prematurely. Upwards of 85 percent of patients with pseudoxanthoma elas-

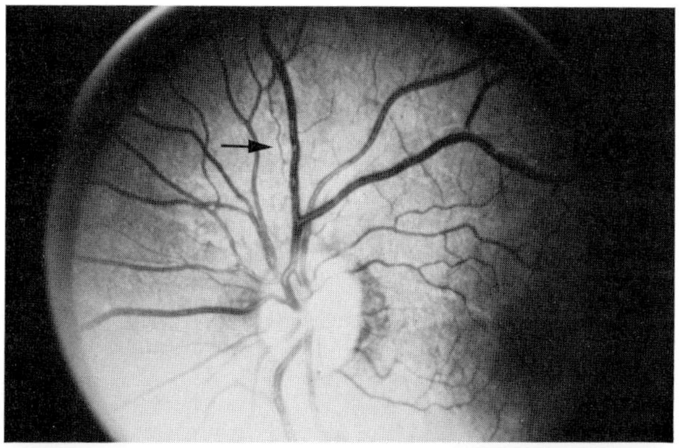

A

Figure 25.16 *Angioid streaks.* A, *peripapillary region showing a rather subtle angioid streak (arrow).* B, *macula showing characteristic mottling of the pigment epithelium (peau d'orange).* C, *coarser mottling temporal to the macula.*

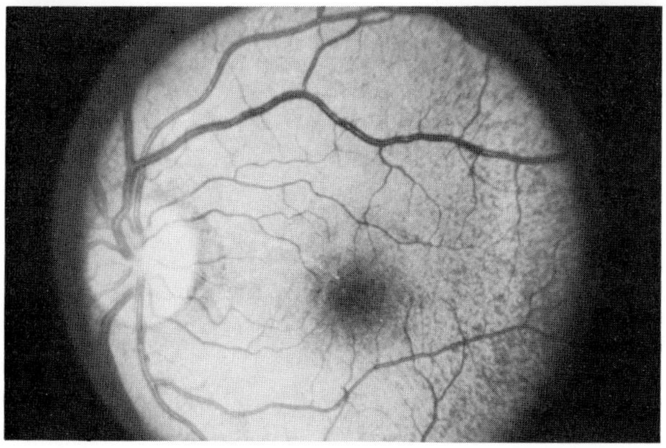

B

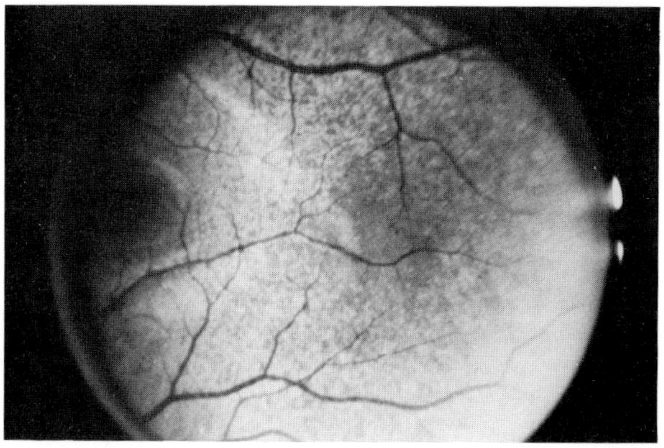

C

ticum have evidence of angioid streaks, and nearly that percent of patients with angioid streaks are found to have pseudoxanthoma elasticum when it is looked for (Paton 1972; Archer and Logan 1977). Thus, a diagnostic skin biopsy should be performed on patients with angioid streaks who do not have an obvious cause for the lesions. Angioid streaks are seen occasionally in Paget's disease of bone and in sickle-cell disorders, and rarely in a variety of other conditions (Paton 1972; Archer and Logan 1977). They may be associated with drusen of the optic disc (Paton 1972; Wobmann 1972; Shields et al. 1975).

The clinical course of angioid streaks is variable, because the lesions may range from minimal peau d'orange to extensive streaks in the posterior fundus. In most patients they are already widespread by the time they cause symptoms and are diagnosed (Shields et al. 1975). Shields et al. noted that about 70 percent of the lesions involved the macula, and that 80 percent of the patients with macular involvement showed signs of exudative maculopathy. Thus, the prognosis for central vision must be guarded. No form of therapy, including laser photocoagulation, has proven effective in stopping the progress of the lesions.

The role of the pigment epithelium in angioid streaks is presumed to be secondary, since damage to Bruch's membrane appears to be the primary event of the disease. Nonetheless, widespread pigment epithelial changes may be present (as peau d'orange fundus), and the propensity for exudation and hemorrhage under the macula indicates damage to, and breakdown of, the pigment epithelium. There may also be damage to the choriocapillaris (Federman, Shields, and Tomer 1975).

PHOTIC MACULOPATHY. Damage to the fovea from radiant energy is not a rare occurrence, and has been described under a variety of terms, including photic maculopathy, solar retinopathy, eclipse maculopathy, foveomacular retinitis, and welding-arc maculopathy. The source of exposure is sometimes obvious, as in arc welding or eclipse-watching, but the history may be obscure if the subject was under the influence of drugs (such as LSD—cf. Fuller 1976) or is reluctant to reveal what happened. There is still debate about whether the "epidemics" of foveomacular retinitis in military installations represent sun-gazing (possibly for secondary gain—cf. Ewald and Ritchey 1970) or some other etiology (Woldoff and Kilpatrick 1975; Marlor et al. 1973).

The primary symptom of photic maculopathy is a loss of central vision, which may be minimal or severe, depending upon the length and intensity of exposure. Examination of the fundus shortly after exposure shows a greyish-yellow lesion in each fovea, giving the appearance of retinal edema overlying a yellowish scar or exudation. The edema clears in a few days, leaving a characteristic yellow lesion near the level of the pigment epithelium, which usually does not leak or transmit fluorescein. Over the ensuing weeks this lesion may fade, give way to a small lamellar hole (Anaclerio and Wicker 1970), or evolve into a focal pigment epithelial defect that persists as a sign of the foveal insult (Fig. 25.17). The visual acuity usually improves over a period of several weeks or months after exposure, sometimes returning to normal, but sometimes leaving a significant degree of permanent loss (MacFaul 1969).

The question of whether the injury in photic maculopathy is thermal or photic remains a subject

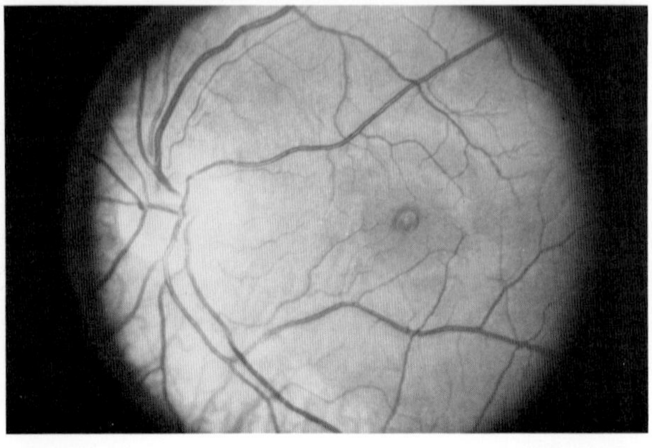

A

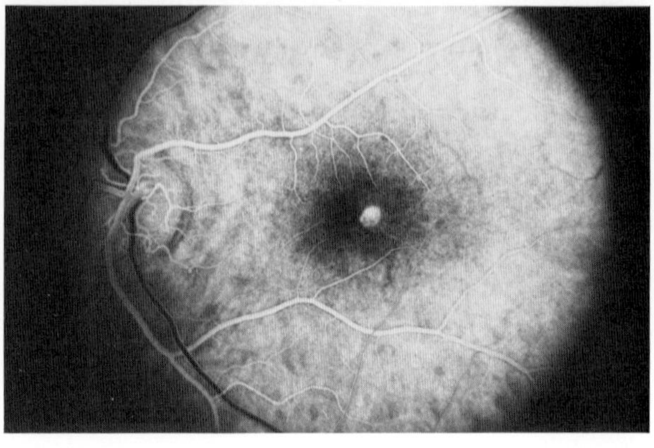

B

Figure 25.17 *Photic maculopathy. Twenty-year-old male with 20/30 acuity and a vague history of occasional looks at the sun.* A, *small depigmented lesion directly in the fovea.* B, *fluorescein angiogram.*

of debate. In general, thermal damage (from a therapeutic laser, for example) is inflicted primarily on the pigment epithelium, which contains the energy-absorbing melanin, whereas photic damage (such as from prolonged exposure to ordinary levels of light) is most evident in the photoreceptors (see Chapter 17). Primate studies by Tso and colleagues (Tso et al. 1972; Tso, Fine, and Zimmerman 1972; Tso 1973) have shown that injury from an indirect ophthalmoscope or low-intensity laser was most severe in the photoreceptor layer. However, a recent report by Tso and La Piana (1975) on humans who gazed at the sun for one hour with an eye scheduled to be enucleated for melanoma, showed damage confined primarily to the pigment epithelium. Surprisingly, the visual acuity in these subjects was only minimally affected, although the recovery time after a brief dazzle (photostress test) was significantly prolonged. The energy of a welding-arc beam has been estimated to have little thermal effect in the fundus (Naidoff and Sliney 1974), and histology on one case of welding-arc maculopathy (Gass 1977a) showed a focal loss of photoreceptor elements as well as some thinning and depigmentation of the pigment epithelium.

Another form of energy that may damage the pigment epithelium is ultrasound. Focused ultrasound can produce chorioretinal lesions without damage to the anterior structures in the eye (Purnell et al. 1964) and may even induce retinal detachment. Ultrasound energy is highly attenuated and absorbed by melanin (Marmor, Wickramasinghe, and Lemons 1977), so the pigment epithelium probably causes greater conversion of ultrasound energy to heat than does the retina. Ultrasound-induced hyperthermia is currently being used therapeutically to augment the treatment of focal cancers (Marmor et al. 1979), and medical exposures to ultrasound will probably increase in the future.

SENILE RETICULAR DEGENERATION. The general topic of peripheral retinal degenerations has been dealt with elsewhere (Rutnin and Schepens 1967; Kanski 1975). However, the entity of senile reticular degeneration deserves emphasis here, because it involves the pigment epithelium and has relevance for the differential diagnosis of other pigment epithelial disorders. This coarse reticular pattern of pigment clumps (Fig. 25.18) is found in the peripheral retina of many elderly people, some-

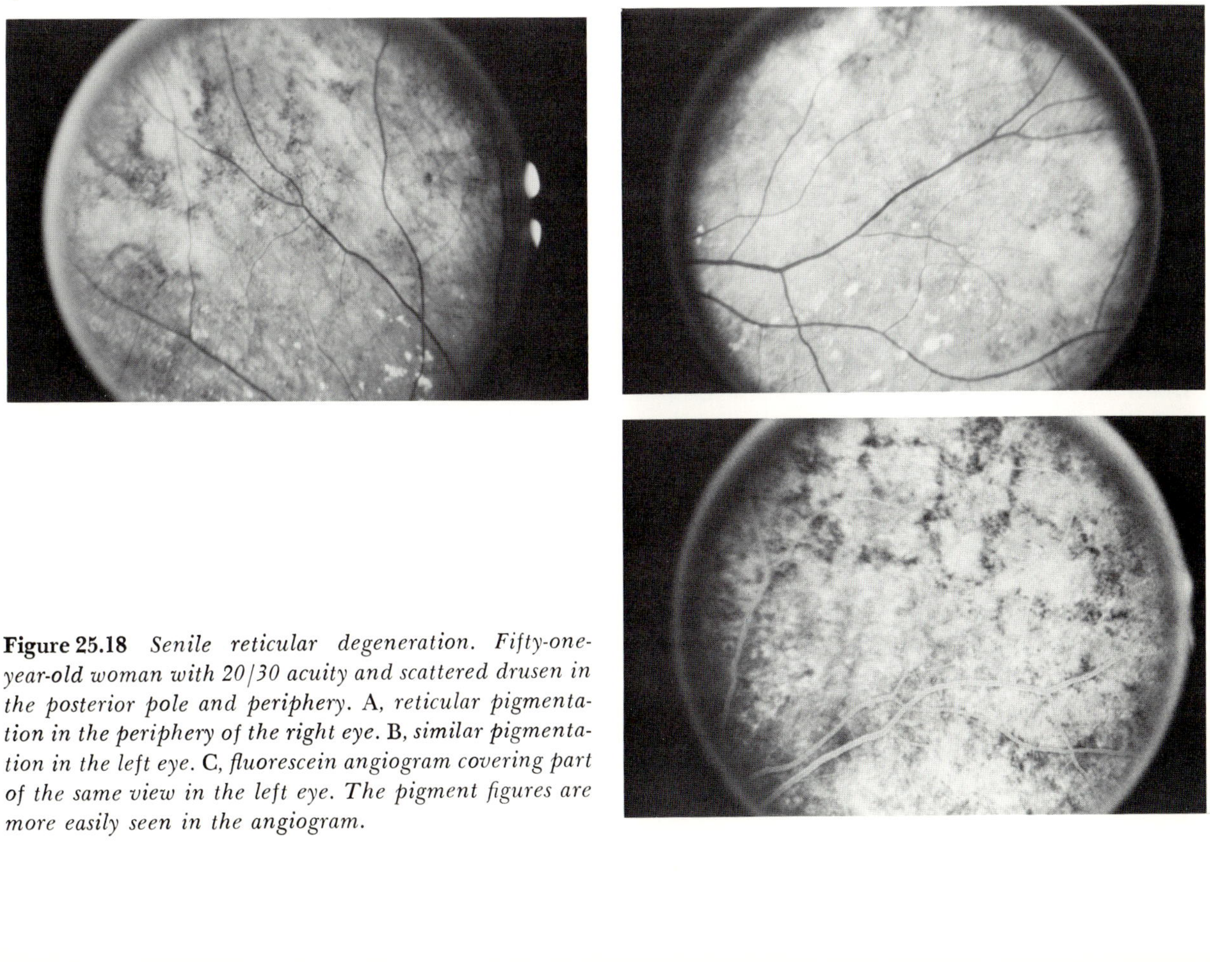

A B C

Figure 25.18 *Senile reticular degeneration. Fifty-one-year-old woman with 20/30 acuity and scattered drusen in the posterior pole and periphery.* A, *reticular pigmentation in the periphery of the right eye.* B, *similar pigmentation in the left eye.* C, *fluorescein angiogram covering part of the same view in the left eye. The pigment figures are more easily seen in the angiogram.*

times associated with extensive peripheral drusen (Deutman 1971; Daicker 1973; Gass 1973; Kanski 1975). It may be quite extensive, covering much of the equatorial fundus, and patients with the condition are occasionally referred for the differential diagnosis of retinitis pigmentosa. However, these lesions are usually benign and limited in extent, although they do signify a generalized aging of the fundus and may be found in individuals with varying degrees of senile changes in the macula. The ERG is often normal, but may be subnormal if extensive areas of the retina are involved or there are generalized effects of aging, such as ischemia from arteriosclerosis. Fishman et al. (1976) described similar lesions as a possible late stage of reticular dystrophy, but this seems doubtful since the lesions of reticular dystrophy tend to fade with age (Sjögren 1950).

Daicker (1973) and Gass (1973) have shown histologically that the senile reticular lesions are caused by bands of thickened and heaped-up pigment epithelial cells, interspersed with regions of thin and hypochromatic cells. Large drusen are frequently seen between the regions of hyperchromatic cells. No histologic material is available from the pattern dystrophies (see Chapter 24), so that we do not know whether their pigment pattern (which may also be reticular) has a similar anatomic basis. Of interest is the fact that reticular patterns appear in diverse disorders of the fundus, and thus may relate to a general anatomic substrate (such as the choriocapillary lobules), which can condition the expression of various diseases.

PATHOLOGIC MYOPIA. Damage to the pigment epithelium, and secondary migration of pigment, can occur in a variety of degenerative conditions in which focal or diffuse damage to the choroid or retina occur. An example is pathologic myopia.

Pathologic myopia (high myopia, degenerative myopia) is characterized by a severe and progressive enlargement of the eyeball, so that all coats of the eye become progressively stretched (Blach 1977). The pigment epithelium and retina become thinned, contributing to the pale appearance of the fundus, and may become regionally damaged and atrophic, with pigment migration occurring as a secondary reaction (Fig. 25.19). Bruch's membrane may rupture, producing "lacquer cracks" and providing a conduit for vascular ingrowth, particularly in the macula, where old hemorrhage may be evident as a Förster-Fuchs spot or disciform scar. The functional status of the retina and pigment epithelium depends upon the degree of diffuse cellular damage. Thus, in severely degenerated eyes Blach, Jay, and Kolb (1966) found that the EOG light response, ERG, and dark adaptation could all be significantly reduced. However, when the disease was less extensive, there was only a mild reduction in the EOG light response and the ERG (primarily in the b-wave rather than the a-wave). In general, if the ERG or dark adaptation are severely reduced in a myopic eye, one should strongly suspect that the cause is *not* myopia but one of the photoreceptor dystrophies or night-blinding disorders, many of which have a high association with myopia. The etiology of degenerative myopia is unknown, and no simple hereditary pattern has been demonstrated, except when it is associated with another ocular disease.

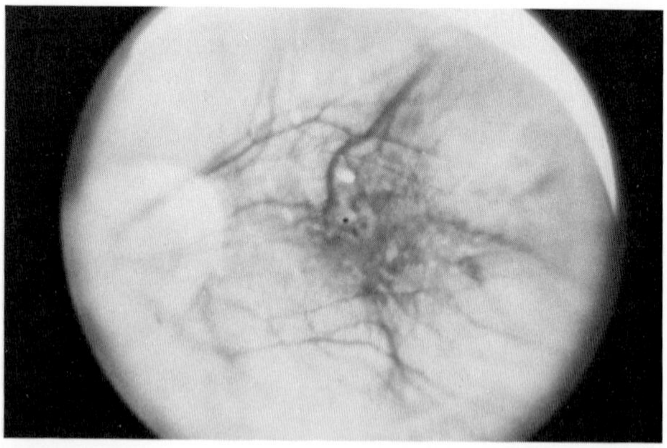

Figure 25.19 *Degenerative myopia. Thirty-year-old woman with 20/70 acuity and greater than 30 diopters of myopia.*

RETINAL DETACHMENT. Retinal detachment, so far as we know, is not primarily a pigment epithelial disease, although the pigment epithelium contributes to the adhesive forces that normally keep the retina apposed to the wall of the eye (see Chapter 11). The majority of clinical detachments occur in older individuals with degenerative changes in the retina and, most critically, with vitreous changes that include liquefaction and traction. Nonetheless, in some instances pathology in the pigment epithelium may weaken adhesive forces or the ability of the eye to resorb subretinal fluid, so that a detachment can easily occur.

Also of concern regarding retinal detachments are

the anatomic changes that take place in the underlying pigment epithelium. The pigment epithelial cells under a detached retina (Kroll and Machemer 1968; Kroll 1969) become taller, and melanin granules withdraw from apical processes, which become thickened and increased in number. Lamellar inclusion bodies (phagosomes), which signify phagocytosis of outer segment material (see Chapter 9), disappear. Of clinical concern is the fact that pigment epithelial cells may migrate under a detachment, transform into macrophages and fibrocytes, and sometimes grow through retinal tears into the vitreous (Machemer and Laqua 1975); this contributes to the formation of massive vitreous retraction. If the retina is reattached soon enough that the photoreceptors can regenerate, the pigment epithelium returns to its normal configuration and reestablishes an anatomic relationship with the outer segments (Kroll and Machemer 1969). However, the speed or accuracy with which these anatomic relationships are reformed has not been determined. If the detachment is long-standing, the photoreceptors may not regenerate and then the pigment epithelium reacts as it would in any degenerative disease, with the development of diffuse depigmentation and pigment dispersion. An important question to be resolved is whether the ability of the retina to recover from detachments of varying duration is a function solely of photoreceptor viability or also reflects the pigment epithelium's ability to reestablish and sustain the necessary interaction between the two cell layers.

REFERENCES

Anaclerio, A. M., and Wicker, H. S. 1970. Self-induced solar retinopathy by patients in a psychiatric hospital. *Am. J. Ophthalmol.* 69:731–736.

Annesley, W. H., Tomer, T. L., and Shields, J. A. 1973. Multifocal placoid pigment epitheliopathy. *Am. J. Ophthalmol.* 76:511–518.

Archer, D. B., and Logan, W. C. 1977. Bruch's membrane degenerations. II. Angioid streaks. In *Hereditary retinal and choroidal diseases,* ed. A. E. Krill. Vol. 2. *Clinical Characteristics.* Hagerstown: Harper and Row.

Azar, P., Jr., Gohd, R. S., Waltman, D., and Gitter, K. A. 1975. Acute posterior multifocal placoid pigment epitheliopathy associated with an adenovirus type 5 infection. *Am. J. Ophthalmol.* 80:1003–1005.

Baarsma, G. S., and Deutman, A. F. 1976. Serpiginous (geographic) choroiditis. *Doc. Ophthalmol.* 40:269–285.

Ballantyne, A. J., and Michaelson, I. C. 1970. *Textbook of the fundus of the eye,* 2d ed. Edinburgh: E & S Livingstone.

Behrendt, T. 1977. Central serous retinopathy (CSR). In *Controversy in ophthalmology,* ed. R. J. Brockhurst, S. A. Boruchoff, B. T. Hutchinson, and S. Lessell, pp. 698–705. Philadelphia: W. B. Saunders Co.

Bird, A. C., and Hamilton, A. M. 1972. Placoid pigment epitheliopathy. *Br. J. Ophthalmol.* 56:881–886.

Blach, R. K. 1977. Bruch's membrane degenerations. III. Degenerative myopia. In *Hereditary retinal and choroidal diseases,* ed. A. E. Krill. Vol. 2. *Clinical Characteristics.* Hagerstown: Harper and Row.

Blach, R. K., Jay, B., and Kolb, H. 1966. Electrical activity of the eye in high myopia. *Br. J. Ophthalmol.* 50:629–641.

Blodi, F. C., and Hervouet, F. 1968. Syphilitic chorioretinitis. *Arch. Ophthalmol.* 79:294–296.

Bonnet, M., Pingault, Cl., and Tapissier, J. 1976. Disparition progressive des verrucosités hyalines de la macula. *Arch. Ophtalmol.* (Paris) 36:465–474.

Braunstein, R. A., Rosen, D. A., and Bird, A. C. 1974. Ocular histoplasmosis syndrome in the United Kingdom. *Br. J. Ophthalmol.* 58:893–898.

Brockhurst, R. J., Boruchoff, S. A., Hutchinson, B. T., and Lessell, S., ed. 1977. *Controversy in ophthalmology.* Philadelphia: W. B. Saunders Co.

Brown, M., Eberdt, A., and Ladas, G. 1973. Pigment epitheliopathy in a patient with mycobacterial infection. *J. Pediatr. Ophthalmol.* 10:278–281.

Bullock, J. D., and Fletcher, R. L. 1977. Cerebrospinal fluid abnormalities in acute posterior multifocal placoid pigment epitheliopathy. *Am. J. Ophthalmol.* 84:45–49.

Charles, N. C., Bennett, T. W., and Margolis, S. 1977. Ocular pathology of the congenital varicella syndrome. *Arch. Ophthalmol.* 95:2034–2037.

Chisholm, I. H., Gass, J. D. M., and Hutton, W. L. 1976. The late stage of serpiginous (geographic) choroiditis. *Am. J. Ophthalmol.* 82:343–351.

Cibis, G. W. 1975. Neonatal herpes simplex retinitis. *Albrecht von Graefes Arch. Klin. Ophthalmol.* 196:39–47.

Collis, W. J., and Cohen, D. N. 1970. Rubella retinopathy: a progressive disorder. *Arch. Ophthalmol.* 84: 33–35.

Crouch, E. R., Jr., and Goldberg, M. F. 1975. Retinal periarteritis secondary to syphilis. *Arch. Ophthalmol.* 93: 384–387.

DAICKER, B. 1973. Lineare Degenerationen des peripheren retinalen Pigmentepithels. *Albrecht von Graefes Arch. Klin. Ophthalmol.* 186:1–12.

DEUTMAN, A. F. 1971. *The hereditary dystrophies of the posterior pole of the eye.* Assen: Van Gorcum and Co.

DEUTMAN, A. F. 1974. Acute retinal pigment epitheliitis. *Am. J. Ophthalmol.* 78:571–578.

DEUTMAN, A. F., and GRIZZARD, W. S. 1978. Rubella retinopathy and subretinal neovascularization. *Am. J. Ophthalmol.* 85:82–87.

DEUTMAN, A. F., and LION, F. 1977. Choriocapillaris nonperfusion in acute multifocal placoid pigment epitheliopathy. *Am. J. Ophthalmol.* 84:652–657.

DEUTMAN, A. F., OOSTERHUIS, J. A., BOEN-TAN, T. N., and AAN DE KERK, A. L. 1972. Acute posterior multifocal placoid pigment epitheliopathy. *Br. J. Ophthalmol.* 56: 863–874.

DUKE-ELDER, S., and PERKINS, E. S. 1966. Diseases of the uveal tract. In *System of ophthalmology*, ed. S. Duke-Elder, vol. 9, pp. 292–321. London: Henry Kimpton.

EIFRIG, D. E., KNOBLOCH, W. H., and MORAN, J. A. 1977. Retinal pigment epitheliitis. *Ann. Ophthalmol.* 9:639–642.

EWALD, R. A., and RITCHEY, C. L. 1970. Sun gazing as the cause of foveo-macular retinitis. *Am. J. Ophthalmol.* 70:491–497.

FAURSCHOU, S., ROSENBERG, T., and NIELSEN, N. 1977. Central serous retinopathy and presenile disciform exudative macular degeneration. *Acta Ophthalmol.* 55:515–524.

FEDERMAN, J. L., SHIELDS, J. A., and TOMER, T. L. 1975. Angioid streaks. II. Fluorescein angiographic features. *Arch. Ophthalmol.* 93:951–962.

FISCH, R. O., KETTERLING, W. C., SCHACHT, L. E., and LETSON, R. D. 1973. Ocular abnormalities of a child associated with familial microcephaly. *Am. J. Ophthalmol.* 76:260–264.

FISHMAN, G. A., BASKIN, M., and JEDNOCK, H. 1977. Spinal fluid pleocytosis in acute posterior multifocal placoid pigment epitheliopathy. *Ann. Ophthalmol.* 9:33–36.

FISHMAN, G. A., RABB, M. F., and KAPLAN, J. 1974. Acute posterior multifocal placoid pigment epitheliopathy. *Arch. Ophthalmol.* 92:173–177.

FISHMAN, G. A., WOOLF, M. B., GOLDBERG, M. F., and BUSSE, B. 1976. Reticular tapeto-retinal dystrophy as a possible late stage of Sjögren's reticular dystrophy. *Br. J. Ophthalmol.* 60:35–40.

FITZGERALD, C. R., and RUBIN, M. L. 1974. Intraocular parasite destroyed by photocoagulation. *Arch. Ophthalmol.* 91:162–164.

FITZPATRICK, P. J., and ROBERTSON, D. M. 1973. Acute posterior multifocal placoid pigment epitheliopathy. *Arch. Ophthalmol.* 89:373–376.

FLAGE, T., SAND, A. B., and SYRDALEN, P. 1977. Haemorrhagic maculopathy in young adults. *Acta Ophthalmol.* 55:489–496.

FRANCESCHETTI, A., DIETERLE, P., and SCHWARTZ, A. 1958. Rétinité pigmentaire à virus: relation entre tableau clinique et électrorétinogramme (ERG). *Ophthalmologica* 135:545–554.

FRANCESCHETTI, A., FRANÇOIS, J., and BABEL, J. 1974. *Chorioretinal heredodegenerations.* Springfield: Charles C. Thomas.

FRANÇOIS, J. 1976. Embryological pigment epithelial dystrophies. *Ophthalmologica* 172:417–433.

FRANÇOIS, J., DELAEY, J. J., and DAKIR, M. 1974. Juvenile hemorrhagic macular choroidopathy. *Bull. Soc. Belge. Ophtalmol.* 167:664–678.

FRANÇOIS, J., HOOFT, C., DE BLOND, R., and DE LOORE, F. 1962. Embryopathie par radiations ionisantes. *Ophthalmologica* 143:163–186.

FRIEDMAN, E. 1977. Central serous choroidopathy: pathogenesis and treatment. In *Controversy in ophthalmology*, ed. R. J. Brockhurst, S. A. Boruchoff, B. T. Hutchinson, and S. Lessell, pp. 706–709. Philadelphia: W. B. Saunders Co.

FRIEDMAN, M. W. 1975. Bilateral recurrent acute retinal pigment epitheliitis. *Am. J. Ophthalmol.* 79:567–570.

FULLER, D. B. 1976. Severe solar maculopathy associated with the use of lysergic acid diethylamide (LSD). *Am. J. Ophthalmol.* 81:413–416.

GASS, J. D. M. 1967. Pathogenesis of disciform detachment of the neuroepithelium. II. Idiopathic central serous choroidopathy. *Am. J. Ophthalmol.* 63:587–615.

GASS, J. D. M. 1968. Acute posterior multifocal placoid pigment epitheliopathy. *Arch. Ophthalmol.* 80:177–185.

GASS, J. D. M. 1973. Drusen and disciform macular detachment and degeneration. *Arch. Ophthalmol.* 90:206–217.

GASS, J. D. M. 1974. A clinicopathologic study of a peculiar foveomacular dystrophy. *Trans. Am. Ophthalmol. Soc.* 72:139–156.

GASS, J. D. M. 1977*a*. *Stereoscopic atlas of macular diseases,* 2d ed. St. Louis: C. V. Mosby Co.

GASS, J. D. M. 1977*b*. Idiopathic central serous choroidopathy (ICSC). In *Controversy in ophthalmology,* ed. R. J. Brockhurst, S. A. Boruchoff, B. T. Hutchinson, and S. Lessell, pp. 710–714. Philadelphia: W. B. Saunders Co.

Gass, J. D. M., and Lewis, R. A. 1976. Subretinal tracks in ophthalmomyiasis. *Arch. Ophthalmol.* 94:1500–1505.

Gass, J. D., Norton, E. W. D., and Justice, J., Jr. 1966. Serous detachment of the retinal pigment epithelium. *Trans. Am. Acad. Ophthalmol. Otolaryngol.* 70:990–1015.

Gerstle, C., and Zinn, K. M. 1976. Rubella-associated retinitis in an adult: report of a case. *Mt. Sinai J. Med.* 43:303–308.

Green, S. H., and Wirtschafter, J. D. 1973. Ophthalmoscopic findings in subacute sclerosing panencephalitis. *Br. J. Ophthalmol.* 57:780–787.

Gregg, N. M. 1941. Congenital cataract following German measles in the mother. *Trans. Ophthalmol. Soc. Aust.* 3:35–46.

Guzzinati, G. C. 1954. Sulla possibilita' di lesioni oculari congenite de morbillo e da epatite epidemica: osservazioni cliniche. *Boll. Oculist.* 33:833–841.

Hamilton, A. M., and Bird, A. C. 1974. Geographic choroidopathy. *Br. J. Ophthalmol.* 58:784–797.

Hayreh, S. S., and Baines, J. A. B. 1972. Occlusion of the posterior ciliary artery. II. Chorio-retinal lesions. *Br. J. Ophthalmol.* 56:736–753.

Heckenlively, J. 1977. Secondary retinitis pigmentosa (Syphilis). (XIV ISCERG Symposium, Louisville, 1976.) *Doc. Ophthalmol. Proc.* 13:245–255.

Henkes, H. E. 1977. Photic injury to the retina and the manifestation of acute posterior multifocal placoid pigment epitheliopathy. *Doc. Ophthalmol.* 44:113–120.

Hentsch, R., and Külz, J. 1969. Beitrag zur sekundaren Pigment-degeneration der Netzhaut nach Masern. *Klin. Monatsbl. Augenheilkd.* 154:706–712.

Holt, W. S., Regan, C. D. J., and Trempe, C. 1976. Acute posterior multifocal placoid pigment epitheliopathy. *Am. J. Ophthalmol.* 81:403–412.

Hull, D. S., and Aaberg, T. M. 1974. Fluorescein study of a family with angioid streaks and pseudoxanthoma elasticum. *Br. J. Ophthalmol.* 58:738–745.

Jacklin, H. N. 1977. Acute posterior multifocal placoid pigment epitheliopathy and thyroiditis. *Arch. Ophthalmol.* 95:995–997.

Kanski, J. J. 1975. Peripheral retinal degenerations. *Trans. Ophthalmol. Soc. U.K.* 95:173–179.

Key, S. H. III, Green, W. R., and Maumenee, A. E. 1977. Pathology of macular lesions of ocular histoplasmosis: its pathogenetic and therapeutic implications. In *Controversy in ophthalmology,* ed. R. J. Brockhurst, S. A. Boruchoff, B. T. Hutchinson, and S. Lessell, pp. 732–747. Philadelphia: W. B. Saunders Co.

Kirkham, T. H., Ffytche, T. J., and Sanders, M. D. 1972. Placoid pigment epitheliopathy with retinal vasculitis and papillitis. *Br. J. Ophthalmol.* 56:875–880.

Klouček, F., and Otradovec, J. 1969. Das histopathologische Bild der zentralen chorioretinalen Veränderungen bei subakuter sklerosierender Leukoenzephalitis (van Bogaert). *Ophthalmologica* 157:442–450.

Krill, A. E. 1967. The retinal disease of rubella. *Arch. Ophthalmol.* 77:445–449.

Krill, A. E. 1972. Retinopathy secondary to rubella. *Int. Ophthalmol. Clin.* 12:89–103.

Krill, A. E., and Deutman, A. F. 1972. Acute retinal pigment epitheliitis. *Am. J. Ophthalmol.* 74:193–205.

Krill, A. E., Klien, B. A., and Archer, D. B. 1973. Precursors of angioid streaks. *Am. J. Ophthalmol.* 76:875–879.

Kroll, A. J. 1969. Secondary retinal detachment. *Am. J. Ophthalmol.* 68:223–237.

Kroll, A. J., and Machemer, R. 1968. Experimental retinal detachment in the owl monkey. III. Electron microscopy of the retina and pigment epithelium. *Am. J. Ophthalmol.* 66:410–427.

Kroll, A. J., and Machemer, R. 1969. Experimental retinal detachment in the owl monkey. V. Electron microscopy of the reattached retina. *Am. J. Ophthalmol.* 67:117–130.

Laatikainen, L., and Erkkilä, H. 1974. Serpiginous choroiditis. *Br. J. Ophthalmol.* 58:777–783.

Lowes, M. 1977. Placoid pigment epitheliopathy presenting as an anterior uveitis. *Acta Ophthalmol.* 55:800–806.

MacFaul, P. 1969. Visual prognosis after solar retinopathy. *Br. J. Ophthalmol.* 53:534–541.

Machemer, R., and Laqua, H. 1975. Pigment epithelium proliferation in retinal detachment (massive periretinal proliferation). *Am. J. Ophthalmol.* 80:1–23.

Marlor, R. L., Blais, B. R., Preston, F. R., and Boyden, D. G. 1973. Foveo-macular retinitis, an important problem in military medicine: epidemiology. *Invest. Ophthalmol.* 12:5–16.

Marmor, J. B., Pounds, D., Postic, T. B., and Hahn, G. M. 1979. Treatment of superficial human neoplasms by local hyperthermia induced by ultrasound. *Cancer* 43:196–205.

Marmor, M. F. 1975. The retinal pigment epithelium and ocular disease. *Int. Ophthalmol. Clin.* 15:131–146.

Marmor, M. F. 1979. "Vitelliform" lesions in adults. *Ann. Ophthalmol.*, in press.

Marmor, M. F., Egbert, P. R., Egbert, B. M., and Marmor, J. B. 1978. Optic nerve head involvement with cytomegalovirus in an adult with lymphoma. *Arch. Ophthalmol.* 96:1252–1254.

Marmor, M. F., Wickramasinghe, H. K., and Lemons, R. A. 1977. Acoustic microscopy of the human retina and pigment epithelium. *Invest. Ophthalmol.* 16:660–666.

Meredith, T. A., Green, W. R., Key, S. N., III, Dolin, G. S., and Maumenee, A. E. 1977. Ocular histoplasmosis: clinicopathologic correlation of 3 cases. *Surv. Ophthalmol.* 22:189–205.

Metz, H. S., and Harkey, M. E. 1968. Pigmentary retinopathy following maternal measles (morbilli) infection. *Am. J. Ophthalmol.* 66:1107–1110.

Miller, S. A., Stevens, T. S., and de Venecia, G. 1976. De novo lesions in presumed ocular histoplasmosis-like syndrome. *Br. J. Ophthalmol.* 60:700–712.

Naidoff, M. A., and Sliney, D. H. 1974. Retinal injury from a welding arc. *Am. J. Ophthalmol.* 77:663–668.

Nelson, D. A., Weiner, A., Yanoff, M., and dePeralta, J. 1970. Retinal lesions in subacute sclerosing panencephalitis. *Arch. Ophthalmol.* 84:613–621.

Nicholson, D. H. 1975. Cytomegalovirus infection of the retina. *Int. Ophthalmol. Clin.* 15:151–162.

Noble, K. G., Levitzky, M. J., and Carr, R. E. 1976. Detachments of the retinal pigment epithelium at the posterior pole. *Am. J. Ophthalmol.* 82:246–251.

Obenour, L. C. 1972. Subacute sclerosing panencephalitis. *Int. Ophthalmol. Clin.* 12:215–223.

Pallin, S. L. 1977. Noninfectious posterior chorioretinopathies: curious multifocal and contiguous disorders. *Ann. Ophthalmol.* 9:713–718.

Paton, D. 1972. *The relationship of angioid streaks to systemic disease*. Springfield: Charles C. Thomas.

Peyman, G. A., and Bok, D. 1972. Peroxidase diffusion in the normal and laser-coagulated primate retina. *Invest. Ophthalmol.* 11:35–45.

Purnell, E. W., Sokollu, A., Torchia, R., and Taner, N. 1964. Focal chorioretinitis produced by ultrasound. *Invest. Ophthalmol.* 3:657–664.

Regensburg, N. I., and Henkes, H. E. 1976. Measles (morbilli) and ocular complications. *Doc. Ophthalmol.* 40:287–300.

Remky, H., and Kolbl, I. 1971. Multiple vitelliform cysts. *Klin. Monatsbl. Augenheilkd.* 159:322–330.

Robb, R. M., and Watters, G. V. 1970. Ophthalmic manifestations of subacute sclerosing panencephalitis. *Arch. Ophthalmol.* 83:426–435.

Robinson, T. R., and Gitter, K. A. 1972. Multiple serous detachments of the retinal pigment epithelium. *Arch. Ophthalmol.* 88:487–492.

Rodrigues, M. M., Purrnett, H. H., Valdes-Dapena, M., and Martyn, L. J. 1973. Retinal pigment epithelium in a case of trisomy 18. *Am. J. Ophthalmol.* 76:265–268.

Rosenberg, T., Faurschou, S., and Nielsen, N. 1977. The clinical classification of maculopathies in adults. *Acta Ophthalmol.* 55(suppl. 131):1–23.

Roth, A. M. 1977. Histoplasma capsulatum in the presumed ocular histoplasmosis syndrome. *Am. J. Ophthalmol.* 84:293–298.

Rutnin, U., and Schepens, C. L. 1967. Fundus appearance in normal eyes. III. Peripheral degenerations. *Am. J. Ophthalmol.* 64:1040–1062.

Ryan, S. J. 1976. De novo subretinal neovascularization in the histoplasmosis syndrome. *Arch. Ophthalmol.* 94: 321–327.

Ryan, S. J., and Maumenee, A. E. 1972. Acute posterior multifocal placoid pigment epitheliopathy. *Am. J. Ophthalmol.* 74:1066–1074.

Saari, M. 1977*a*. Disciform detachment of the macula. I. Juvenile haemorrhagic macular choroidopathy. *Acta Ophthalmol.* 55:525–529.

Saari, M. 1977*b*. Disciform detachment of the macula. II. Fluorescein and indocyanine green fluorescence angiographic findings in juvenile haemorrhagic macular choroidopathy. *Acta Ophthalmol.* 55:530–538.

Savino, P. J., Weinberg, R. J., Yassin, J. G., and Pilkerton, A. R. 1974. Diverse manifestations of acute posterior multifocal placoid pigment epitheliopathy. *Am. J. Ophthalmol.* 77:659–662.

Savir, H., and Kurz, O. 1976. Fluorescein angiography in syphilitic retinal vasculitis. *Ann. Ophthalmol.* 8:713–716.

Sawelson, H., Goldberg, R. E., Annesley, W. H., Jr., and Tomer, T. L. 1976. Presumed ocular histoplasmosis syndrome. *Arch. Ophthalmol.* 94:221–224.

Schatz, H., Maumenee, A. E., and Patz, A. 1974. Geographical helicoid peripapillary choroidopathy: clinical presentation and fluorescein angiographic findings. *Trans. Am. Acad. Ophthalmol. Otolaryngol.* 78:747–761.

SCHEIE, H. G., and MORSE, P. H. 1972. Rubeola retinopathy. *Arch. Ophthalmol.* 88:341–344.

SCHLAEGEL, T. F., JR. 1977. Controversy in ophthalmology: the presumed histoplasmosis syndrome. In *Controversy in ophthalmology,* ed. R. J. Brockhurst, S. A. Boruchoff, B. T. Hutchinson, and S. Lessell, pp. 720–731. Philadelphia: W. B. Saunders Co.

SHIELDS, J. A., FEDERMAN, J. L., TOMER, T. L., and ANNESLEY, W. H., JR. 1975. Angioid streaks. I. Ophthalmoscopic variations and diagnostic problems. *Br. J. Ophthalmol.* 59:257–266.

SJÖGREN, H. 1950. Dystrophia reticularis laminae pigmentosae retinae. *Acta Ophthalmol.* 28:279–295.

SMITH, R. E., KNOX, D. L., and JENSEN, A. D. 1973. Ocular histoplasmosis. *Arch. Ophthalmol.* 89:296–300.

SMITH, V. C., POKORNY, J., ERNEST, J. T., and STARR, S. J. 1978. Visual function in acute posterior multifocal placoid pigment epitheliopathy. *Am. J. Ophthalmol.* 85:192–199.

TAKEI, Y., and MIZUNO, K. 1976. Acute posterior multifocal placoid pigment epitheliopathy. *Jap. J. Ophthalmol.* 20:257–265.

TARKKANEN, A., MERENMIES, L., and HOLMSTRÖM, T. 1972. Ocular involvement in congenital cytomegalic inclusion disease. *J. Pediatr. Ophthalmol.* 9:82–86.

TSO, M. O. M. 1973. Photic maculopathy in rhesus monkey. *Invest. Ophthalmol.* 12:17–34.

TSO, M. O. M., FINE, B. S., and ZIMMERMAN, L. E. 1972. Photic maculopathy produced by the indirect ophthalmoscope. I. Clinical and histopathologic study. *Am. J. Ophthalmol.* 73:686–699.

TSO, M. O. M., and LA PIANA, F. G. 1975. The human fovea after sungazing. *Trans. Am. Acad. Ophthalmol. Otolaryngol.* 79:788–795.

TSO, M. O. M., WALLOW, I. H. L., POWELL, J. O., and ZIMMERMAN, L. E. 1972. Recovery of the rod and cone cells after photic injury. *Trans. Am. Acad. Ophthalmol. Otolaryngol.* 76:1247–1262.

VAN BUSKIRK, E. M., LESSELL, S., and FRIEDMAN, E. 1971. Pigmentary epitheliopathy and erythema nodosum. *Arch. Ophthalmol.* 85:369–372.

WEINSTEIN, G. W., and REESER, F. 1975. The electro-oculogram, angioid streaks and the R-membrane. *Eye Ear Nose Throat Mon.* 54:22–28.

WILKINSON, C. P. 1976. Presumed ocular histoplasmosis. *Am. J. Ophthalmol.* 82:140–142.

WOBMANN, P. 1972. Angioid streaks und Drusenpapille. *Klin. Monatsbl. Augenheilkd.* 161:191–197.

WOLDOFF, H. S., and KILPATRICK, W. R. J. 1975. Foveomacular retinitis: report of cases in three young females. *Ann. Ophthalmol.* 7:1035–1038.

WOLFF, S. M. 1973. The ocular manifestations of congenital rubella: a prospective study of 328 cases of congenital rubella. *J. Pediatr. Ophthalmol.* 10:101–141.

YANOFF, M. 1971. The retina in rubella. In *Retinal diseases in children,* ed. W. Tasman, pp. 223–232. New York: Harper and Row.

ZETTERSTRÖM, B. 1959. The electroretinogram in congenital rubella retinitis. XVIII Concilium Ophthalmologicum, Belgica, 1958. *Acta* 1:661–664.

ZIMMERMAN, L. E. 1968. Histopathologic basis for ocular manifestations of congenital rubella syndrome. *Am. J. Ophthalmol.* 65:837–862.

ZWENG, H. C., LITTLE, H. L., and VASSILIADIS, A. 1977. *Argon laser photocoagulation.* St. Louis: C. V. Mosby Co.

ACKNOWLEDGMENTS

My research is supported in part by National Eye Institute grant EY01678.

26 Aging of the Retinal Pigment Epithelium

Abraham L. Kornzweig

MUCH HAS BEEN written on the appearance of the aging pigment epithelium of the retina, as seen with the direct and indirect ophthalmoscope and the light microscope. Histochemical studies have added to our knowledge of the chemical changes involved in the aging process. The fluorescein angiogram has contributed its share of observed data. The electron microscope and autoradiographic techniques with labeled amino acids have expanded the boundaries of our knowledge of the aging process to a considerable degree.

It is always difficult to decide whether an observed change is due to the aging process alone or is the result of pathologic conditions caused by infections, general metabolic diseases, heredity, or other environmental trauma. Nevertheless, an attempt will be made to enumerate those conditions that occur most frequently in aged eyes that do not show evidence of disease, injury, or infection.

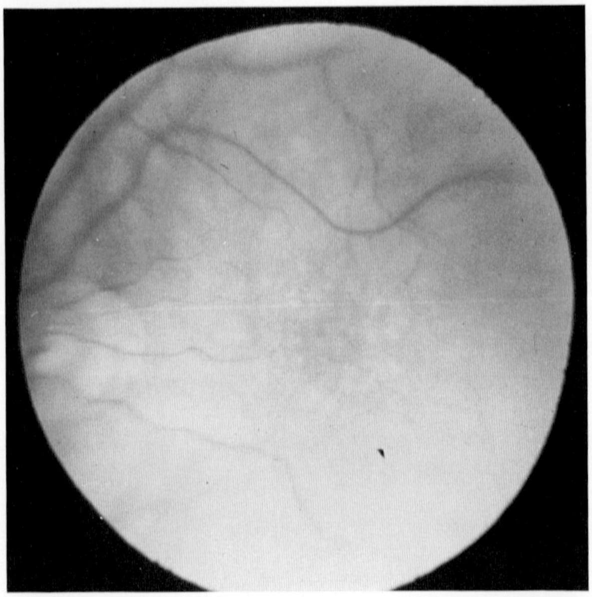

Figure 26.1 *Fundus photograph of an 83-year-old female, showing senile peripapillary halo, some drusen and early senile macular degeneration.*

Ophthalmoscopic Appearance

The pigmentary boundary of the optic disc becomes paler in aged eyes and often atrophies completely, giving rise to the senile halo or conus (Fig. 26.1). The macula, on the other hand, becomes darker, and the fovea may lose its central light reflex. This change is attributed to an increase in pigmentation of the pigment epithelial cells in this area. In the midperiphery and anterior to the equator, small yellow-white spots appear, irregular in outline with an occasional reddish border. These are areas of pigmentary atrophy involving small clumps of cells. A diffuse zone of decreased pigmentation may encircle the globe between the equator and the ora; this zone is seen best by indirect ophthalmoscopy or by Goldmann's three-mirror contact lens. It is also seen in post-mortem specimens. At the extreme periphery there is a zone of cystoid degeneration, especially on the temporal side, and there are large, punched-out areas of depigmentation. Histologically, the depigmented areas correspond to regions of atrophy, where there is an abrupt and complete disappearance of the outer layers of the retina, including the pigment epithelium. The choriocapillaris is atrophic or absent. Conversely, pigmentary proliferation and migration, resembling a pigmentary dystrophy clinically, may occur into the retina, appearing mainly at the equator and extending anteriorly to the ora serrata. The posterior pole is usually not involved in this type of hyperpigmentation, nor is there functional impairment in dark adaptation or visual fields. Some such pigment proliferation may occur around the optic disc (Duke-Elder and Dobree 1967).

A frequent sign of aging is the appearance of drusen around the posterior pole, in the macular area, and in the periphery. These are light brown spots, which may be few or numerous, single or confluent, usually with a well-defined boundary and occasionally with a slightly pigmented border. As will be shown later, the drusen are deposits on Bruch's membrane that elevate the pigment layer. If they are sufficiently large and located in the foveal region, they may interfere with vision, even causing a central scotoma (Fig. 26.1).

Aging changes occur much earlier in life anterior to the equator and at the ora serrata, in both the retina and its pigment epithelium. These changes include atrophic spots, pigmentary invasion of the retina, cystoid degeneration of the retina, especially on the temporal side, peripheral chorioretinopathy, and peripheral retinal holes. The enlargement of the eyeball from birth to maturity and the marked increase in surface area of the retina, especially anteriorly, is brought about mainly at the expense of thinning of the retinal tissue, the pigment epithelial layer, and the choroid. These changes in the anterior segment of the retina are more commonly seen in patients with axial myopia (Tso and Friedman 1968).

The fluorescein angiogram has also demonstrated multiple areas of atrophy of the pigment layer in the aged eye. These atrophic spots are most significant in patients with senile macular degeneration, and they indicate the extent of macular disease, which is usually greater than what one would expect from the appearance of the macula with the ophthalmoscope. The areas of atrophy light up as the fluorescein-stained blood plasma courses through the choriocapillaris. The appearance is distinctive and helps in the differential diagnosis of those cases that show leakage of dye-stained plasma from weakened capillaries and areas of hemorrhage. In multiple drusen, there is a similar lighting-up of the angiogram, but the distribution of such atrophic spots is more general, involving the periphery as well as the macula, and they are more discrete in appearance. This technique demonstrates one essential change in the retinal pigment epithelial cell with age, namely, a tendency to cellular atrophy (Fig. 26.2).

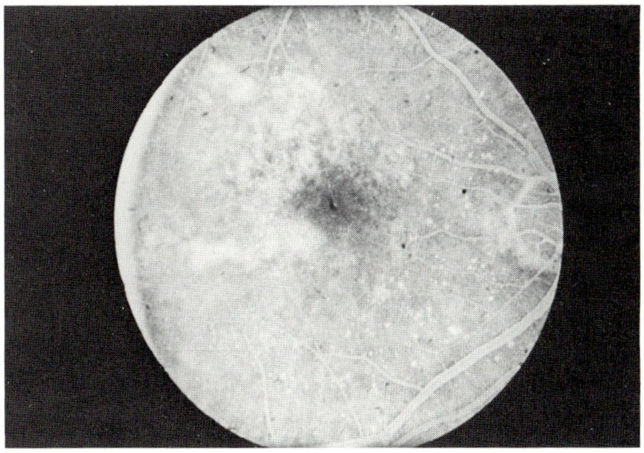

Figure 26.2 *Fluorescein angiogram showing extensive areas of atrophy in the macular region and numerous atrophic spots and drusen around the posterior poles in an elderly patient.*

Histology

The pigment epithelium of the retina shows the approach of old age more markedly than the visual retina does. It undergoes changes similar to those of the pigment epithelia of the ciliary body and iris, partly atrophic and partly hyperplastic.

Atrophy of the pigment layer as it approaches the optic nerve is common but not constant. Occasionally the layer may even become hyperpigmented. In the macular area the pigment cells are more heavily pigmented and more compressed and taller. Most of the cells have a single nucleus, but a small number, estimated at 3 percent, are binucleate.

The aging changes of the RPE in an aged eye that is relatively normal can be readily seen with the light microscope in most post-mortem eyes. In the peripapillary area, the retinal pigment epithelium is partly or completely atrophic for a short distance. This is a terminal area and probably not as well nourished as other areas. The choriocapillaris is missing in some aged eyes, the outer retinal layers are deficient in cells, and the photoreceptors are stunted or absent. As the macular region is approached, the RPE begins to appear more normal and soon reaches its most mature and aged stage—opposite the foveal region. Here the cells are fully pigmented and densely packed together. The cells are somewhat taller and narrower. The nuclei are not visible. This condition persists well beyond the macular region to the equator. Gradually, a change is noted. The cells become lighter in color, contain fewer pigment granules, and are not quite as tall. Further into the periphery and anterior to the equator, these changes become more pronounced. The pigment epithelial cells are flatter and missing in places. At higher magnification, extracellular pigment granules can be seen. Occasional drusen elevate the pigment layer. In some eyes there is a heaping-up or clumping of pigment cells. All these changes are increased in the region of the cystic degeneration of the retina and at the ora. Pigment granules can occasionally be seen in the retina (Fig. 26.3).

Tso and Friedman (1968) have contributed considerably to our knowledge of the retinal pigment epithelium by their study of the pigment epithelium laid out flat on a slide and bleached. They have noted that, with aging, there is a progressively increasing degree of pleomorphism in the size and shape of cells, nuclei, and pigment granules. A large number of senile eyes were found to contain pigment cells with small basophilic nuclei. In most cases these cells were otherwise normal in appearance. Occasionally, the nucleus was pyknotic, the cytoplasm depigmented, or the cell of unusual shape (Figs. 26.4–26.6).

As the eye develops from infancy to adulthood, there is a reversal of the normal growth gradient. In the macular area, the retinal pigment epithelial cells, which are broad and low in infants, become narrower and higher in adults. Conversely, the peripheral cells, which are narrow and high in infancy,

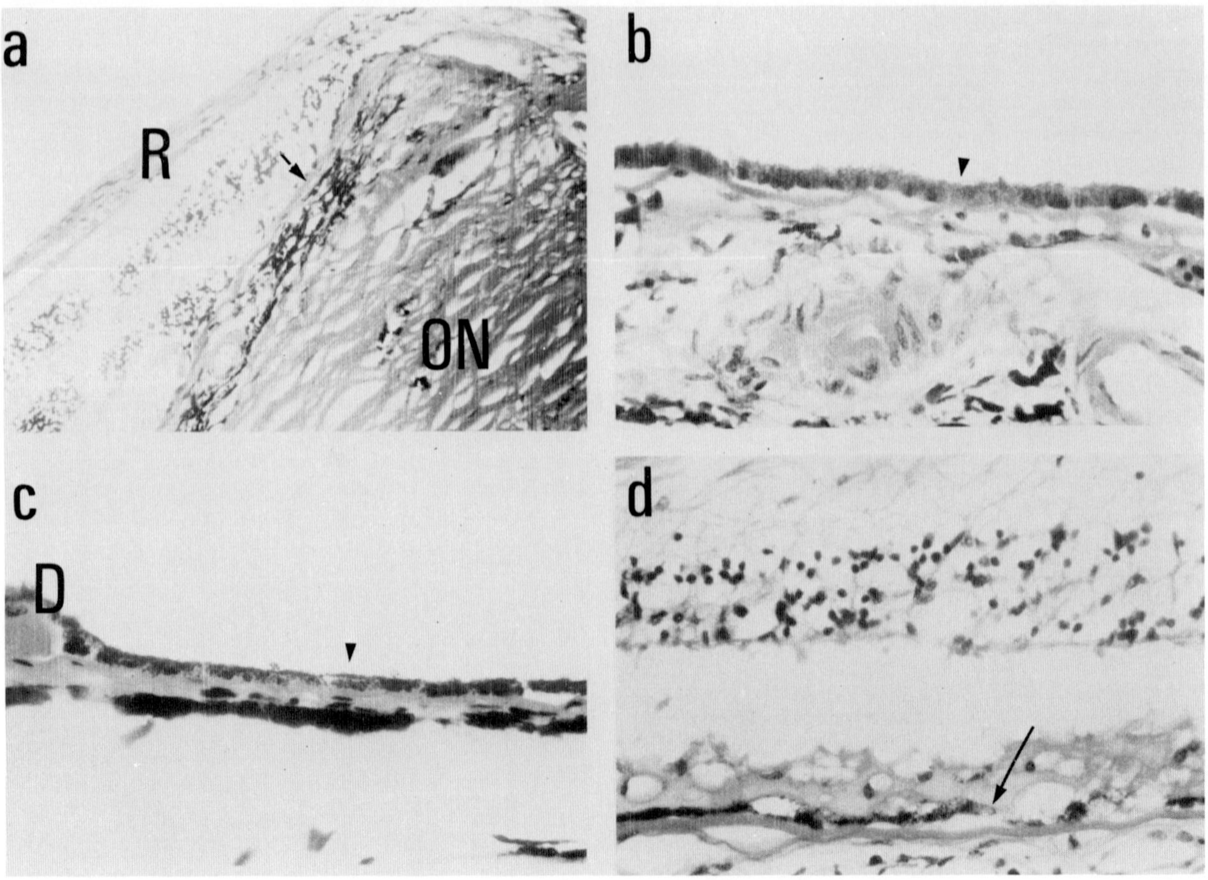

Figure 26.3 *The retinal pigment epithelium* (arrows) *in a relatively normal aged patient, showing variations in histological appearance.* a, *the disc, showing peripapillary atrophy of the RPE.* R, *retina;* ON, *optic nerve. (Stained with H&E. ×100.)* b, *in the macula area; the RPE is mature—height 14–16 microns.* c, *in the midperiphery; the RPE is flatter—height 10–12 microns,* D, *drusen* d, *near the ora serrata; the RPE is considerably flattened and atrophic in spots—height 6–8 microns. (Stained with H&E. ×268.)*

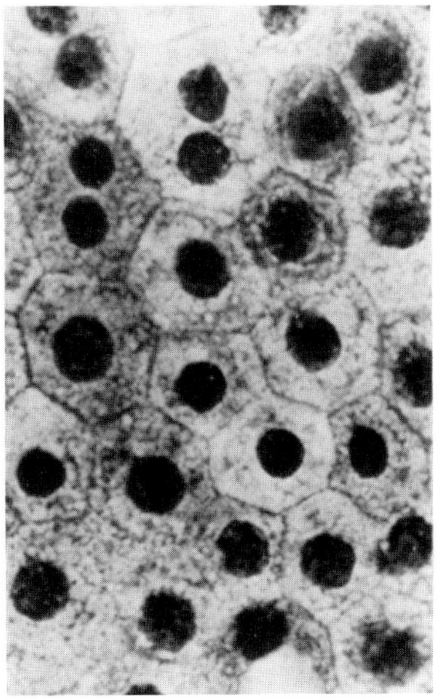

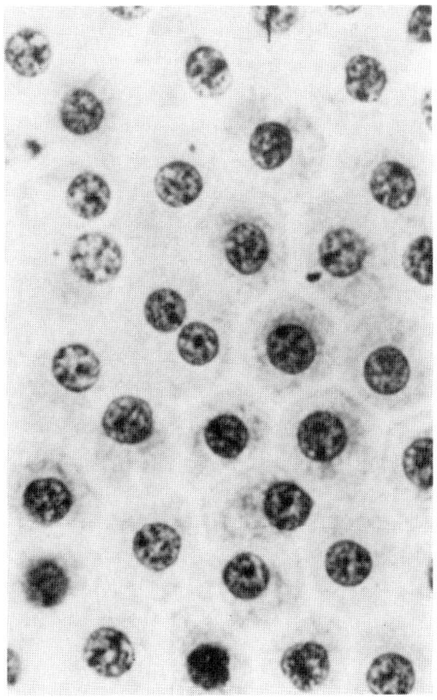

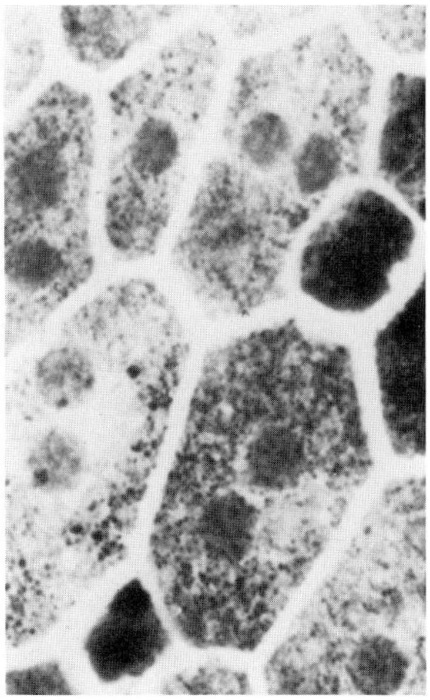

Figures 26.4–26.6 *Bleached specimens of RPE studied on the flat. (From Friedman and Tso 1968. Copyright 1968, American Medical Association.)*

Figure 26.4 *Macula RPE cells of adult (age 87). Cells show pleomorphism. Nuclei are smaller, more darkly staining, and irregular in size than in young eyes. Bleached flat preparation. (Stained with H&E. ×1000.)*

Figure 26.5 *Equatorial RPE cells of adult (age 87). Cells show moderate pleomorphism. Nuclei are not too different in terms of size or staining properties from those of young eyes. Bleached flat preparation. (Stained with H&E. ×1000.)*

Figure 26.6 *Peripheral RPE cells of adult (age 45). Cells are markedly pleomorphic in size, shape, and pigmentation. They are considerably larger than cells situated more anteriorly. Bleached flat preparation. (Stained with H&E. ×1000.)*

become broader and flatter in adulthood. In between, at the equator and midperiphery, there is no demonstrable change. There is a similar inversion in pigmentation. In infants and children the macular pigment cells in unbleached specimens are lighter than those in the periphery. In older eyes the macular cells become darker than those in the periphery. In senile eyes, pigmentation is irregular and distributed in no discernible pattern (Tso and Friedman 1968).

Height of the Pigment Cells in Eyes over Age 75

Peripapillary region	8–10 microns or atrophic
Macular region	14–16 microns
Equator	10–12 microns
Near and at ora serrata	6–8 microns or absent in spots

The most impressive senescent changes noted by Tso and Friedman were those associated with drusen. The pigment cells in contact with the smallest drusen were displaced and elevated by them, but were otherwise normal. Those cells in contact with larger or confluent drusen were abnormal in appearance, grossly degenerated, or absent (Fig. 26.7).

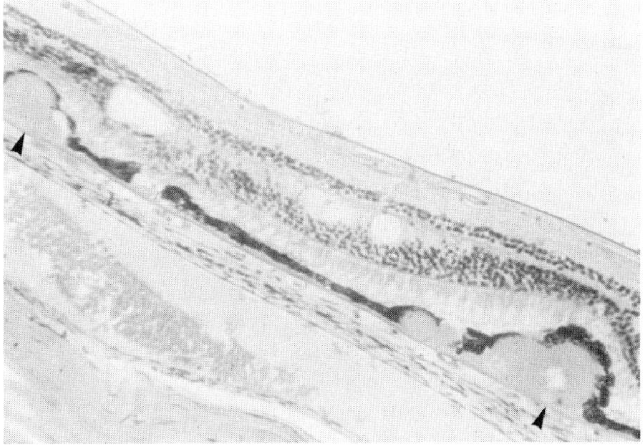

Figure 26.7 *Histologic appearance of confluent drusen* (arrows) *in male (age 88), showing elevation of pigment layer and degeneration and atrophy of some pigment cells. (Stained with Hemotoxylin and Eosin. ×154.)*

In the periphery, pleomorphism was especially noted, with many more binucleated cells and an occasional multinucleated cell. No mitotic figures were seen in any of Tso and Friedman's specimens. Pigment granules were changed in size and shape, being more spherical rather than the usual rod or oval shape. They were unevenly distributed, being heaped up in some cells and almost disappearing in others.

Further regressive changes include vacuolation and fatty degeneration, and eventually nucleolysis and the breaking up of the cell into an amorphous mass and the scattering of pigment granules into the retinal tissue (Duke-Elder and Dobree 1967). The retina may be so extensively invaded by pigment that the term senile pigmentary degeneration is merited. Alternatively, hyperplastic changes may occur, the cells proliferating and increasing their content of pigment, some containing two nuclei and attaining several times their normal size. Others become aggregated into piled-up groups of degenerated elements.

Lipofuscin is another pigment that has been described in the pigment epithelial cell (Streeten 1961; Feeney et al. 1965). This pigment is considered to be formed as a result of "wear and tear" and is like that formed in neural cells and other tissues. Lipofuscin granules probably are residual bodies, representing the end products of phagosomal activity. Using a light microscope, Streeten found that lipofuscin granules appear at the early age of 18 months. After age 20 they increase in number and size. The granules are sudanophilic and contain unsaturated fatty acids, probably in a phosphated complex. The sudanophilic granules and the rod outer segments show the same histochemical qualities. Stains for myelin-like substances, such as lipofuscin myelin and lipid of the retinal rod outer segments, are positive in these granules, suggesting that the granules may be derived from the rod discs.

The aging processes in the macular region—including the pigment epithelial layer, Bruch's membrane, and the choriocapillaris—were studied by Sarks (1976). This prospective study over ten years involved 376 eyes from 216 patients aged 43 to 97 years. These eyes were classified from Group I to Group VI, depending upon the degree of aging changes noted in the three tissues. Groups I and II had minimal and normal aging. Groups III and IV showed early distinct pathological degenerative changes, and Groups V and VI showed advanced degenerative changes, including neovascular and disciform formations and RPE and choroidal atrophy. Special consideration was given to the presence of a basilar linear deposit underneath the retinal pigment epithelial cells. The deposit was finely granular and eosinophilic; it was moderately positive to periodic acid–Schiff's reagent, but stained blue with picro-Mallory method. When observed with an electron-microscope it was seen to consist of banded fibers embedded in granular material lying between the plasma infoldings and the basement membrane of the retinal pigment epithelium. This deposit seemed to be a manifestation of the gradual failure of the pigment epithelium, and it proved to be the most suitable criterion by which to evaluate the pathogenesis of senile macular degeneration. This substance is now called a "basal lamina deposit," in view of the electron-microscopic studies, and in Sarks' view it is the primary factor in the ultimate development of degeneration in the macular region.

Electron Microscopy

Early observers found lamellar inclusions in the retinal pigment cell. They were first described by Bairati and Orzalesi (1963). Since the membranes of these lamellae greatly resembled those of the outer segments of the photoreceptors, they were believed to represent portions of photoreceptors. These cytoplasmic inclusions are now known as phagosomes (Young and Bok 1969). They are surrounded by the cell membrane of the photoreceptor. This in turn is surrounded by another cell membrane, which is derived from the pigment epithelial cell. The process that leads to these inclusions is a true example of phagocytosis. The phagosome is displaced deeper and deeper into the cytoplasm of the pigment cell and shows progressive disintegration (Fig. 26.8).

The phagocytic function of the pigment epithelial cell, as it involves the rod cell photoreceptor, was first described by Young and Bok (1969) and is fully discussed in Chapter 8. Spitznas and Hogan (1970) observed this process involving the cone cell photoreceptor in the human eye.

The electron-microscopic appearance of the aged RPE has been well described by Hogan, Alvarado, and Weddell (1971) and by Hogan (1972). Aging changes begin to appear at about age 30, are slowly progressive, and are of varying severity in different eyes. They are characterized by the following changes:

1. Formation of lipofuscin granules in the cytoplasm.

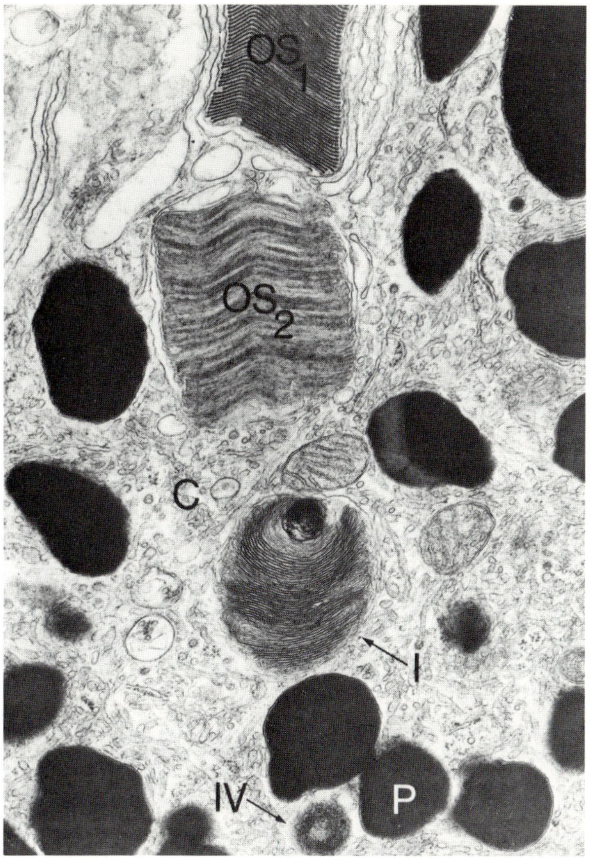

Figure 26.8 *Electron-microscopic photograph of a retinal pigment cell and an outer segment of a photoreceptor cell with progressive stages of disintegration of a phagosome.* OS_1 *and* OS_2 *are outer segments in the process of ingestion by a pigment cell* (C). *A recently formed phagosome is at* I, *and a phagosome in the end stages of hydrolytic digestion at* IV. *(From Hogan, Alvarado, and Weddell 1971. ×16,130.)*

2. Increasing amounts of lipid and numbers of residual bodies, especially in the basal portion of the cytoplasm.
3. Alteration in the character of the basal cytoplasmic infoldings.
4. Accumulation of residual bodies, granular material, and lipid external to the cell membrane.
5. Fiber formation between the basal infoldings.

Coated and uncoated vesicles, residual bodies, partially digested phagosomes, lipid, crystals, and mucoprotein all accumulate in Bruch's membrane during the aging process. The deposition in the inner collagenous zone may be diffuse or may accumulate in masses to form drusen. With advancing age the same cytoplasmic substances also accumulate in the outer collagenous zone of Bruch's membrane and in the intercapillary septa of the choriocapillaris. These changes can be seen in Figures 26.9–26.12.

Additional aging changes of RPE have been noted by Farkas, Sylvester, and Archer (1971):

1. Pigment granules become dispersed throughout the cell, not just in the upper two-thirds.
2. Mitochondria undergo degenerative changes.
3. Lysosomes are more numerous among degenerated pigment cells.
4. Mitochondria and lysosomes fuse to form cytolysosomes.
5. Homogenous fibrillar material is present in the cytoplasm and in the basal infoldings of the basal membrane.
6. Numerous single-membrane-bound vesicles appear in the RPE and inner collagenous zone of Bruch's membrane.
7. Widely-spaced collagen appears at the base of the RPE and inner collagenous layer of Bruch's membrane.
8. Drusen material accumulates in the inner collagenous layer of Bruch's membrane and elevates the RPE. Some cells of the RPE become incorporated into the drusen.
9. Remnants of phagosomes are numerous in the aging RPE.
10. Enlargement of the drusen causes displacement and eventual destruction of the photoreceptor outer segments.

The large number of vesicles bound by single membranes both in the RPE and in drusen was the most significant observation of this study. They are believed to be degenerated lysosomes.

Drusen

Possibly the most significant sign of aging in the pigment epithelial cell is the formation of drusen. These are deposits on the inner surface of Bruch's membrane that elevate the pigment layer. They are very numerous in aged eyes, some being single and others confluent. They have been described by many observers, including Salzmann, who called them warts and gave very accurate descriptions of them in his book on *The Anatomy and Histology of the Human Eyeball* (1912). He believed that drusen were secreted by the pigment epithelial cells, a belief to which many other authors have subscribed. This belief has now been materially strengthened by the

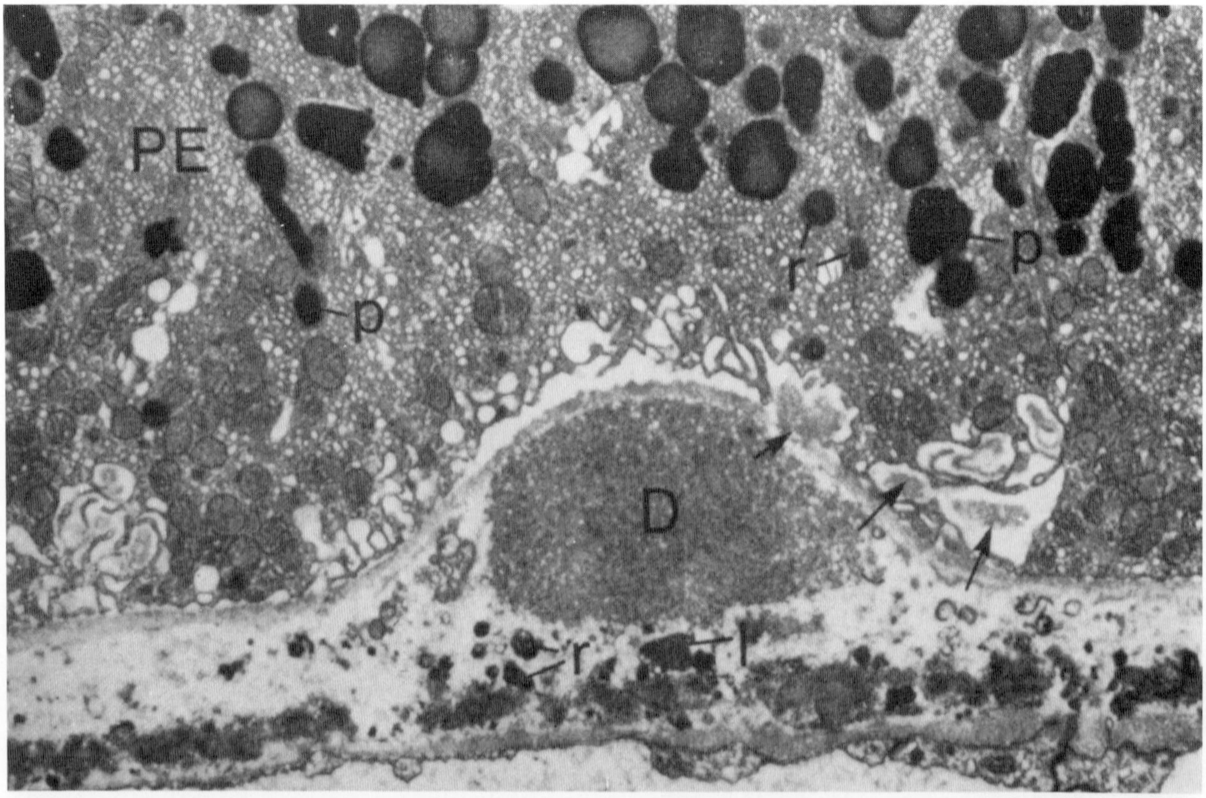
PE
p
r
p
D
r
l

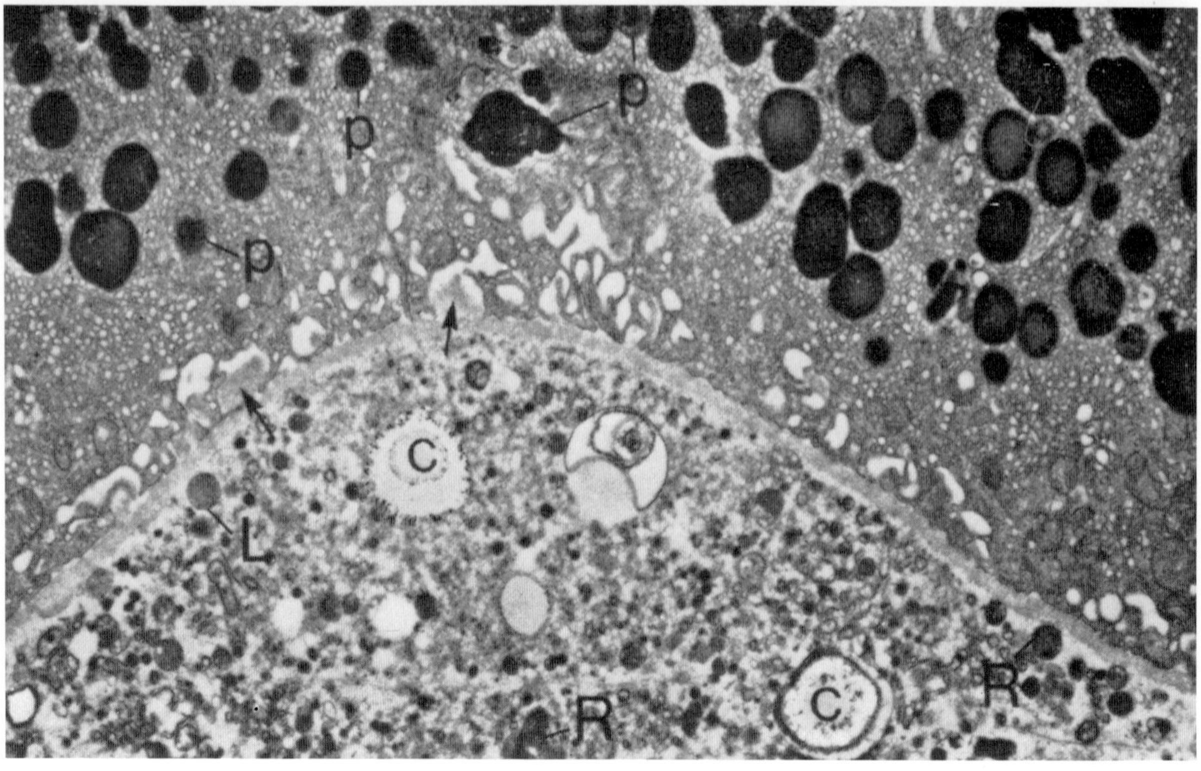
p
p
p
c
L
R
c
R

← **Figure 26.9** *Aging of the retinal pigment epithelium. The pigment epithelium* (PE) *contains phagosomes* (p), *residual bodies* (r), *and flattening and widening of basal infolding. Infoldings contain filaments and granular material* (arrows), *some of which are continuous with drusen* (D) *through breaks in the basement membrane. Beneath the drusen, Bruch's membrane contains residual bodies* (r), *lipid* (l), *and vesicular material. (From Hogan 1972. ×8,200.)*

↙ **Figure 26.10** *Aging of retinal pigment epithelium. The drusen contain crystals* (c), *many residual bodies* (R), *and lipid* (L). *Cytoplasm of RPE cell contains phagosomes* (p), *basal infolding filaments, and granular materials* (arrows). *(From Hogan 1972. ×8,600.)*

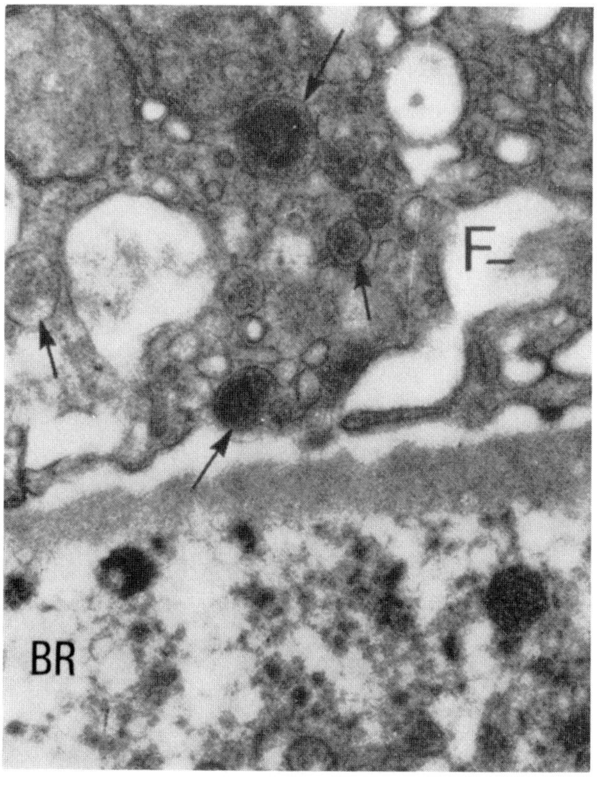

Figure 26.11 *Further degeneration of the RPE, with onset of senile maculopathies. The cytoplasm is filled with many membrane-bound particles* (arrows); *basal infoldings are attenuated; and extracellular fibrils* (F) *have formed. Bruch's membrane* (Br) *contains same material as is present in cytoplasm. (From Hogan 1972. ×16,280.)*

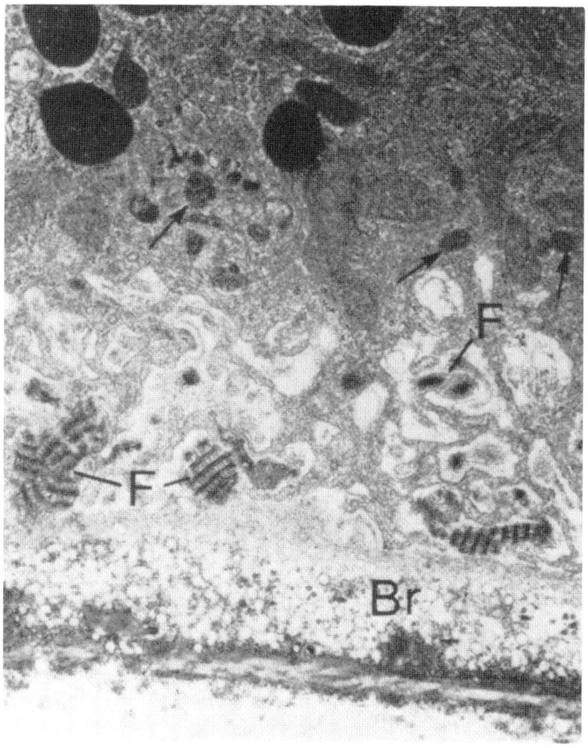

Figure 26.12 *Further degeneration of RPE and Bruch's membrane with onset of senile maculopathies. The markedly abnormal epithelial cytoplasm shows particulate material* (arrows), *and fibril formation. Banded fiber formation* (F) *is visible in the infoldings of cells. Marked accumulation of vesicular and particulate material is present in Bruch's membrane* (Br). *(From Hogan 1972. ×9,200.)*

newer knowledge that the tips of the photoreceptor outer processes are phagocytized by the pigment epithelial cells. It is conceivable that this function of the pigment cell may weaken in the aged eyes due to other metabolic factors. One such factor is suggested by the study of the aged choriocapillaris by Kornzweig (1977). He found that in patients with senile macular degeneration there is a marked increase in the thickness of the intercapillary septa and a narrowing of the lumen of the capillaries. This process—plus basophilia of Bruch's membrane and other aging changes of the lamina—could produce a sufficient hypoxia of the pigment epithelial cells and a gradual loss of function. The weakened pigment cell might then be unable to phagocytize completely the ingested phagosomes, which would then be excreted from the epithelial cell onto the inner collagenous layer of Bruch's membrane and thus contribute to drusen formation. The loss of function of the pigment epithelial cell could also lead to atrophy and to the other aging phenomena mentioned earlier. The toxic effects of the partially digested excretion plus the hypoxia could stimulate hyperplasia in some areas and migration of pigment cells in other areas, as well as rupture of pigment cells and dispersion of pigment granules into the retina.

Another theory of drusen formation has been described by Friedman et al. (1963). In a study of the choroid, using the trypsin digestion technique, they noted clumps of PAS-positive substances in the periphery of the choroid. These clumps were usually arranged along the walls of the capillaries in this area (Fig. 26.13). They were interpreted as plasmic substances extruded from the venous capillaries as a result of decompensation of the venous vortex system. After extrusion these substances would pass through the pores of Bruch's membrane and come to rest underneath the pigment epithelial layer, thus forming drusen. While this PAS-positive material could be readily seen in the midperiphery in aged post-mortem choroidal specimens processed by the trypsin digestion technique, it was not seen in the macular area by Kornzweig, who studied 107 aged eyes by the same technique. Hence this interesting theory could not account for drusen in the macular region, a common site. Also, the venous decompensation and partial obstruction mentioned by Friedman et al. has not been objectively demonstrated. Electron-microscopic studies would seem to favor the excretion hypothesis, that drusen are formed of undigested phagosomes and residual lipofuscin granules. It is also possible that plasmic substances, however derived, may contribute to the formation of some types of drusen in the periphery of the retina (Hogan 1972).

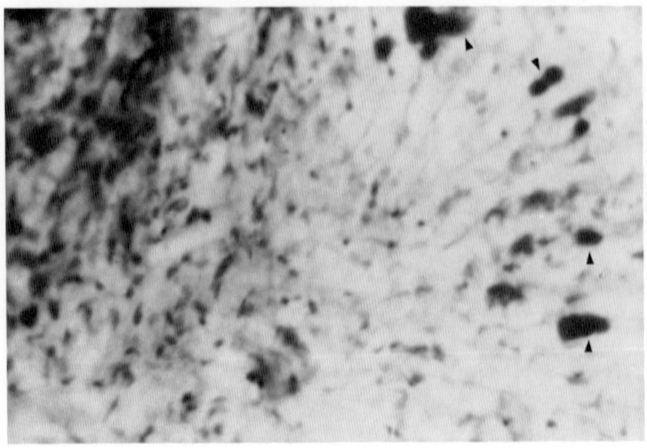

Figure 26.13 *Histological specimen of human choriocapillaris in the midperiphery after trypsin digestion, showing PAS-positive deposits alongside the capillaries* (arrows). *(Stained with PAS and giemsa. ×170.)*

There is a third theory that drusen are formed as a result of a primary hereditary dystrophy, usually transmitted as a regular dominant trait. Several distinct types of dystrophy are known, including Tay's guttate choroiditis, Doyne's honeycomb dystrophy, and Mallatia leventinese. For this group of clinical entities, the name "dominant drusen of Bruch's membrane" has been proposed by Deutman and Jansen (1970). A distinctive and virtually pathognomonic feature of this group is the finding of drusen on the nasal side of the disc.

Farkas, Sylvester, and Archer (1971) have added an additional theory of drusen formation. They believe drusen are the result of a pathologic autolysis of the RPE, caused by a breakdown of cellular lysozymes. Drusen, therefore, would be a conglomeration of denatured mitochondria, cytoplasmic debris, pigment granules, photoreceptor remnants, and residual bodies that have been extruded from the degenerated RPE into Bruch's membrane. Aging, irritation and genetic factors may have a role in triggering the breakdown of lysosomes. The aging pigment epithelial cell would be unable for some reason to function fully, retained phagosomes would not be completely digested by cellular lysosomes, residual bodies and lipofuscin granules would accumulate, and all would eventually be extruded into the inner collagenous layer of Bruch's membrane and form drusen. An exception to this theory is the fact that pigment granules have never been found in drusen, unless they are so altered by degeneration as to be unrecognizable as such. But pigment granules are a hardy tissue that can be scattered from broken pigment cells, migrate into retinal tissue, form clumps, and still retain their characteristic histologic appearance. Besides, pigment cells have been known to become atrophic individually or in groups without causing drusen formation. Until more is known of the aging process itself, the most satisfactory explanation of drusen formation is the excretion of undigested material from the RPE onto the inner layers of Bruch's membrane.

Farkas, Sylvester, Archer, and Altona (1971) have studied the histochemistry of drusen. Drusen are composed mainly of mucopolysaccharides and lipid components. The mucopolysaccharides are made up of sialic acid combined with mucin to form sialomucin. Sialic acid is also a component of lysosomes, and this may explain why this material is found in drusen. The lipid fraction is a cerebroside, since the pigment epithelium is of neuroectodermal origin. The lipofuscin granules in pigment epithelium may be another source of cerebroside in drusen. Sialomucin and cerebrosides may combine to form gangliosides.

Farkas et al. still believe that drusen are formed by the autophagic destruction of the RPE due to abnormal lysosomal activity. Further studies are needed to determine the stimulus that induces the RPE to embark on such a self-destructive course.

The Aging of Bruch's Membrane

Bruch's membrane (lamina vitrea of the choroid) is so intimately connected with the retinal pigment epithelium anatomically and physiologically, that they must be studied together.

With the electron microscope, filaments from the basal membrane of the RPE have been seen to penetrate the inner layers of Bruch's membrane, thus making a firmer union between these tissues than exists between the RPE and the outer layers of the retina. This would account for the observation that the RPE is more often adherent to the choroid than to the retina in histological and pathological specimens (Hogan, Alvarado, and Weddell 1971).

Bruch's membrane consists of five layers (Fig. 26.14). From the inside out, they are:

1. basal membrane of the RPE
2. inner collagenous layer
3. elastic layer
4. outer collagenous layer
5. basement membrane of the choriocapillaris

Bruch's membrane is about 2 μ thick in young eyes but increases in thickness with increasing age. In persons over age 50, the membrane shows areas of basophilia, especially in the peripapillary and the macular regions. PAS staining also becomes more intense with increasing age, because the membrane contains an increased amount of ground substance.

What was formerly thought to be the inner collagenous layer is now known to be partly the basement membrane of the retinal pigment epithelium and partly the inner collagenous layer. It measures 0.3 μ thick. It is PAS-positive and can be recognized on electron-micrographs as a fine line close to the pigment epithelial cells.

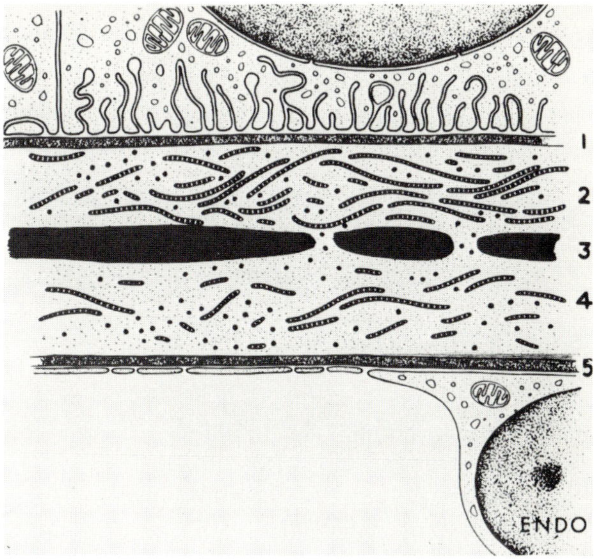

Figure 26.14 *Diagrammatic illustration showing the five layers of Bruch's membrane. (From Hogan, Alvarado, and Weddell 1971.)*

Bruch's membrane includes collagen and elastic fibers woven together like a lattice, running parallel to each other or in arcades or brushlike patterns (Feher and Valu 1967). Collagen fibers are seen to traverse the membrane in all directions in sections stained with Masson's stain. In the periphery, Bruch's membrane becomes thinner because of the larger choriocapillary vessels, but it is uniformly smooth where it is in contact with the basement membrane of the retinal pigment epithelium. The elastic layer lies in a single plane throughout the choroid and is about 0.8 μ thick. It is separated from the basement membrane of the RPE and the choriocapillaris by a narrow space. It is interrupted frequently by spaces, so that a netlike meshwork is formed along the whole inner choroid. The openings in the meshwork are triangular or irregularly polygonal and vary in size. Collagen fibers pass through the meshes of the elastic membrane from the basement membrane of the choriocapillaris to the basement membrane of the retinal pigment epithelium. The earliest studies of the tissue were by Sumita (1961), Lerche (1963–1965), Garron (1963), and Hogan (1961).

Indications of aging of Bruch's membrane are seen as early as age 20. They first appear in the inner collagenous zone and gradually extend to the elastic layer and the outer collagenous zone and the intercapillary connective tissue. These changes continue throughout life and are extensive in the eyes of many elderly persons, especially in the macular area and peripapillary zone. It is not yet clear whether these changes precede or follow the senescent changes in the retinal macula. Nor is it clear whether they result from the accumulation of material from the retinal pigment epithelium or the choriocapillaris, or whether they develop directly from the collagenous material and ground substances in situ (Hogan, Alvarado, and Weddell 1971).

The first indication of aging is the accumulation of vesicular, granular, and filamentous substances. Bruch's membrane becomes increasingly PAS-positive. Various forms of these substances are seen —granular, amorphous, ring-like, and linear—increasing in amount with aging, and gradually occupying larger areas of the collagenous zones. Some of these fibers are banded together, separated by a space measuring 0.12 μ. Long, dense osmophilic fibers appear, which are the same diameter as collagen, but lack continuous banding. These fibers may be altered collagen fibers. The elastic layer of fibers also undergoes changes with aging. The fibers develop a moth-eaten appearance, increased density of the central core, and deposits of long needle-like crystals. All these changes cause an increased basophilia of Bruch's membrane on routine histological preparations. These changes are more prominent in the macular and peripapillary regions, and less intense in the equatorial region (Figs. 26.15–26.16).

With the electron microscope it is possible to detect the early changes associated with drusen formation. Amorphous granular material accumulates external to the basement membrane of the RPE. This material contains round and oval globules of varying sizes. In time this material appears in the inner collagenous zone and elevates the overlying pigment epithelium internally. Crystals are also seen in the granular material. These substances are exactly similar to substances seen in the RPE. Other types of drusen are seen which contain fibrillar substances. When cut in cross section, they appear finely granular. They too are first seen internal to the basement membrane, but similar fibrillar substances are not seen in pigment cells.

The aging changes noted above are also present in the intercapillary extension of Bruch's membrane. There is an increased amount of vesicular material, clumps of widely spaced fibers, coiled osmophilic

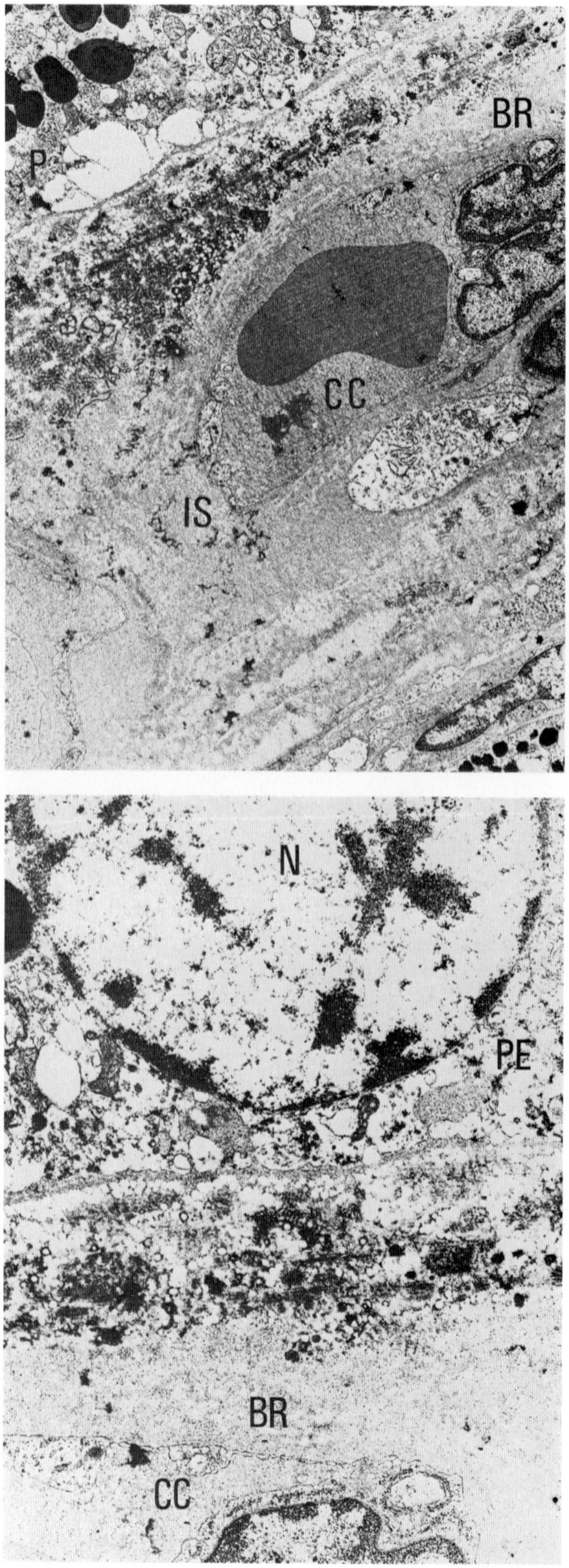

← **Figure 26.15** *Electron-microscopic study of Bruch's membrane* (Br) *and intercapillary septum* (IS) *of the choriocapillaris, male age 84. Aging changes in Bruch's membrane are continued on into intercapillary septum.* P, *pigment epithelial cells.* CC, *choriocapillaris with RBC.* (×5,440.)

↙ **Figure 26.16** *Electron-microscopic study of Bruch's membrane* (Br), *showing numerous vesicular changes especially in inner collagenous layer, banded elastic fibrils, and residual bodies. The basal portion of the RPE cell* (PE) *shows disorganization of basal infoldings and some residual bodies. The choriocapillaris* (cc) *has a fenestrated endothelial layer.* (×6,800.)

and dark granular ground substance. The intercapillary septa thus increase in thickness and impinge on the lumen of the capillaries, especially in the macular area (Kornzweig 1977) (Fig. 26.17).

Clinical Conditions Associated with the Aging of the RPE

The most important conditions resulting from the aging changes in the RPE and Bruch's membrane are those affecting central vision, the senile maculopathies. A statistical survey of over 1,000 aged residents at a home for the aged showed senile maculopathy in 30 percent of the patients over age 65. This percentage increases in each decade, reaching 38 percent in the over-80 age group (Kornzweig et al. 1957). Another survey, in a private practice, of 800 persons over age 65 came to a similar conclusion (Gordon 1960). Schatz (1975) has described two

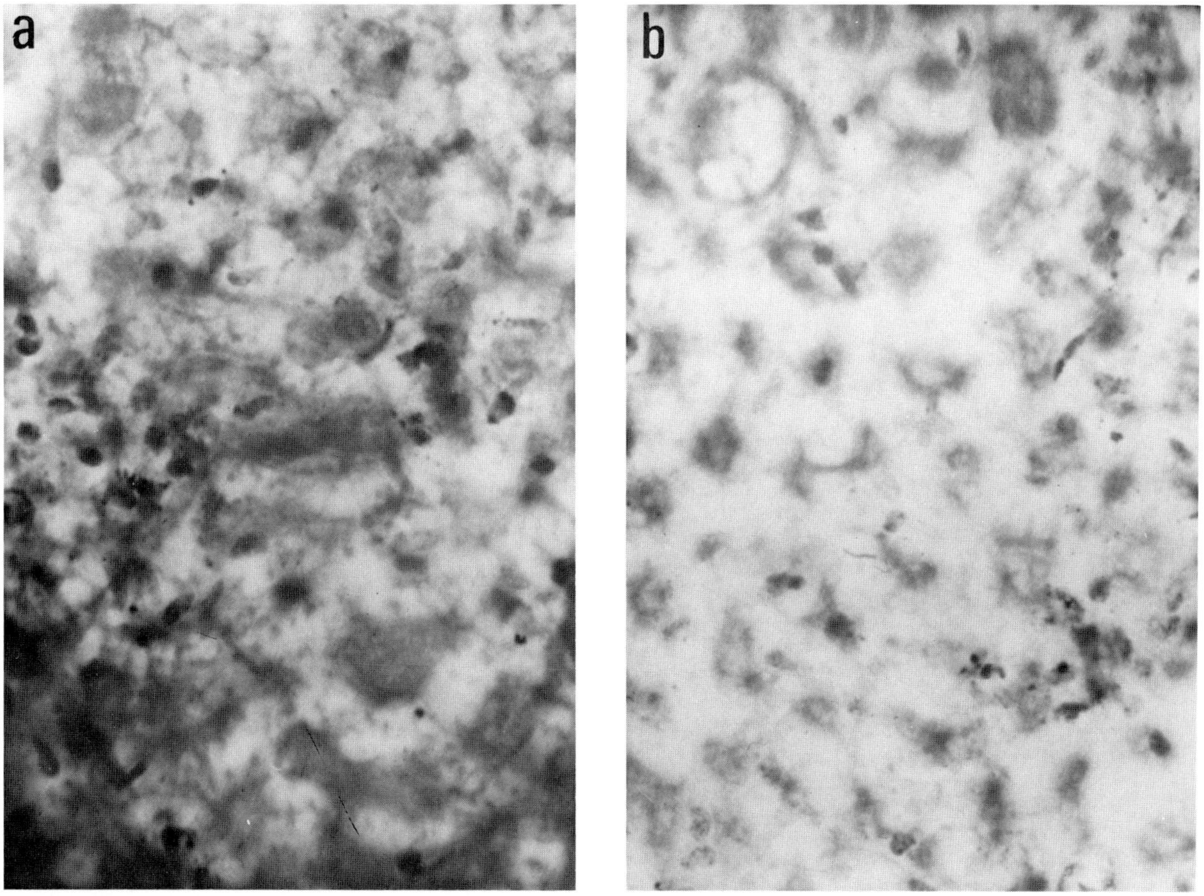

Figure 26.17 a, *Changes in choriocapillaris in aged patient with senile macular degeneration after trypsin digestion. There is an increase in thickness of the intercapillary septa and narrowing of the lumen of the capillaries. Compare with normal aged choriocapillaris* (b). *Clear spaces are capillaries. Dark areas are septa. (Stained with hemotoxylin and giemsa.* ×250.)

types of senile maculopathy: dry, or atrophic, and wet, or exudative. Both eyes are usually involved; one eye is typically more advanced than the other. Men and women are about equally affected. Senile maculopathy is relatively rare in dark-pigmented people.

ATROPHIC MACULOPATHY: SENILE MACULAR DEGENERATION. The earliest symptom noted by the patient or found by the ophthalmologist on routine eye examination is a gradual loss of central vision, usually starting in one eye. This will increase on succeeding examinations from 20/30 to 20/70 and eventually to 20/200 or worse, causing legal blindness. The fundus examination will show mild pigmentary changes in the macular region, described as mottling. The foveal light reflex is gone (Fig. 26.18). A central scotoma can be ascertained on central field testing with a one-millimeter white object. The fluorescein angiogram will show an area of pigmentary atrophy in the macular region greater than would be anticipated from the fundus appearance (Fig. 26.19). These are essential features of senile macular degeneration. Both eyes eventually become involved, causing considerable distress in the aged patient because of an inability to read ordinary print or sew a fine stitch, especially with black thread or on dark materials. Color perception is impaired, and there is difficulty identifying faces or objects at a distance. Objects appear to be distorted (metamorphosis) in size and shape when seen with the affected eye, especially in the later stages.

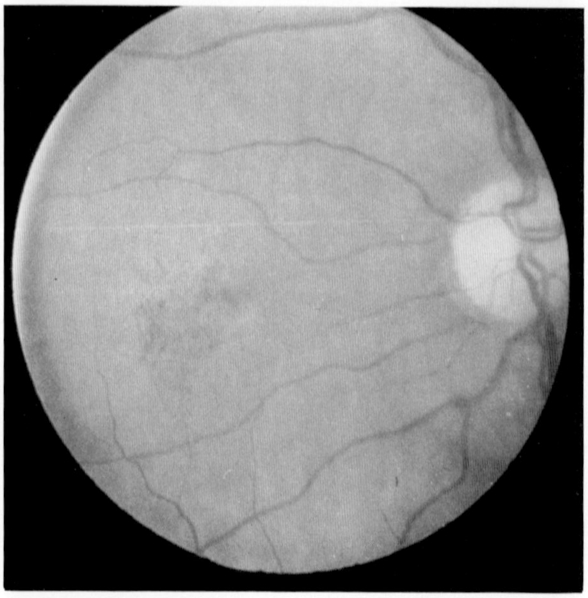

Figure 26.18 *Fundus photograph of 68-year-old female showing senile macular degeneration. Note pigmentary changes in macular area.*

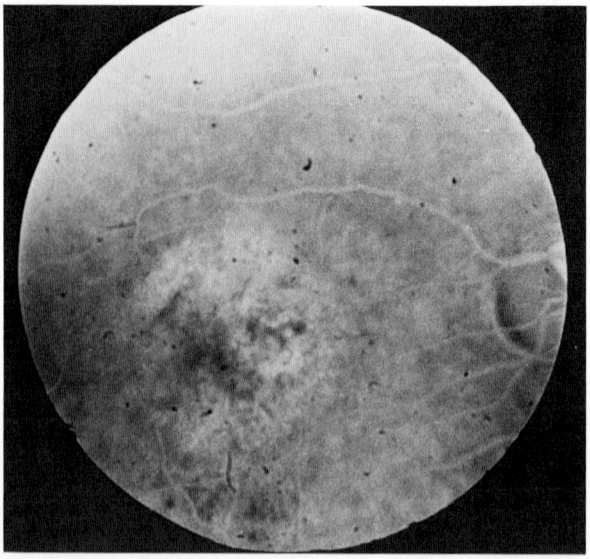

Figure 26.19 *Fluorescein angiogram of patient showing atrophic changes in RPE much greater than indicated by fundus photograph.*

The histopathological changes seen in postmortem eyes have been described by Hogan and Zimmerman (1962), Hogan (1967), and Kornzweig et al. (1959, 1965). Briefly, there are early changes in the retina, diminution of cone cells and processes, and early cystic changes in Henle's layer. The pigment epithelium shows some atrophy of pigment cells, migration of pigment cells, and clumping. On Bruch's membrane, drusen are seen, and areas of basophilia. In moderately advanced cases these changes are more pronounced (Fig. 26.20). Finally, in late cases there is an absence of cone cells and processes, marked cystic degeneration in Henle's layer, complete atrophy of the pigment layer, and frequently adhesion of retina to Bruch's membrane. Recently, as mentioned earlier in this report, changes in the choriocapillaris have been noted by Kornzweig (1977) (Fig. 26.17). These changes include a marked increase in the thickness of the intercapillary septa and narrowing of the lumen of the capillaries of the choriocapillaris, especially in the region of the macula, plus basophilia of Bruch's membrane.

Figure 26.20 *Light-micrograph of macular area of retina in a patient with moderately advanced senile macular degeneration, showing cystic degeneration of Henle's layer of the retina and atrophy of cone cells and processes. Pigment cells have atrophied, migrated into retina, and clumped in foveal region. (Stained with hemotoxylin and giemsa. ×68.)*

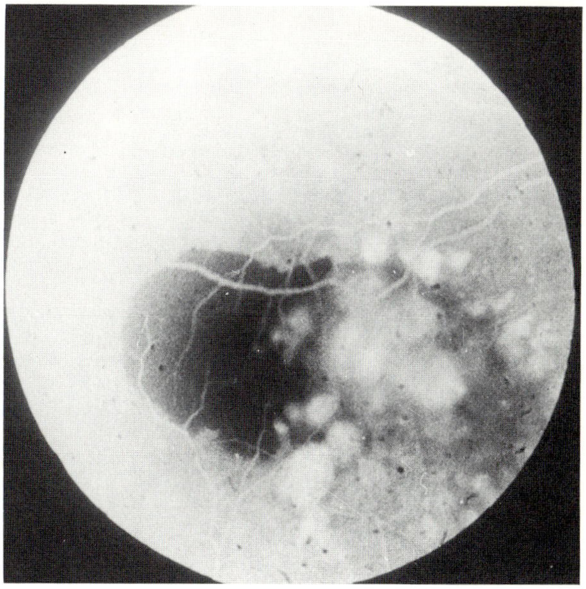

Figure 26.21 *Fluorescein angiogram of aged patient with Kühnt-Junius disciform macular degeneration. Note dark hemorrhagic areas.*

EXUDATIVE MACULOPATHY: KUHNT-JUNIUS DISCIFORM MACULAR DEGENERATION. This clinical entity has been well described by Kühnt and Junius and many others. Several authors believe that the atrophic type of senile macular degeneration may be a precursor to the disciform, exudative type. The clinical symptoms are similar, especially in the early stages, but the ophthalmoscopic appearance is different. In disciform degeneration, the macular area is elevated to some extent by subretinal serous fluid and appears grossly disciform. Pigmentary changes are noted, including areas of hyperpigmentation and other areas of depigmentation. An essential difference between the two types is the appearance of hemorrhagic areas in and around the macula in disciform degeneration. When this occurs the loss of central vision progresses rapidly.

The fluorescein angiogram will show early hyperfluorescence and late leakage into subretinal space. Cystoid leakage is noted in late cases. Dark areas of hemorrhage are seen in or around the macular region (Fig. 26.21), usually from subretinal neovascularization (Schatz 1975) (Fig. 26.22).

On pathological examination, hemorrhagic and serous fluid is seen subretinally—and even external to the pigment epithelium. The hemorrhage is poorly absorbed and is eventually replaced by fibrovascular tissue. Neovascularization also takes place in the same area, and the degenerated pigment layer is incorporated in the fibrovascular mass. The external retinal layers are destroyed or atrophied, and the retina is elevated in a disc-like area involving the macular region. Central vision is finally destroyed. Bruch's membrane is involved, showing basophilia, thickening, and the electron microscopic aging changes described earlier. Ruptures or breaks in Bruch's membrane, as originally described by Verhoeff and Grossman (1937), are seen in many specimens. These breaks may cause irritative reactions in the adjacent choriocapillaris, resulting in hemorrhages and serous exudation. More recently, several authors have found neovascularization in the subretinal tissue, which has been traced to retinal blood vessels (Teeters 1973). Both choroidal and retinal vessels may be involved at different stages of the disease (Fig. 26.23).

CENTRAL AREOLAR CHOROIDAL DYSTROPHY. In this condition—which primarily involves the macular area—the pigment epithelial layer and the outer layers of the retina are atrophied secondarily, due to selective atrophy of the choriocapillaris in the macular region (Ashton 1953; Sorsby 1939; Carr 1965). This rare disease is transmitted in an autosomal dominant fashion. It is an abiotrophy in which the involved tissue disappears in a genetically

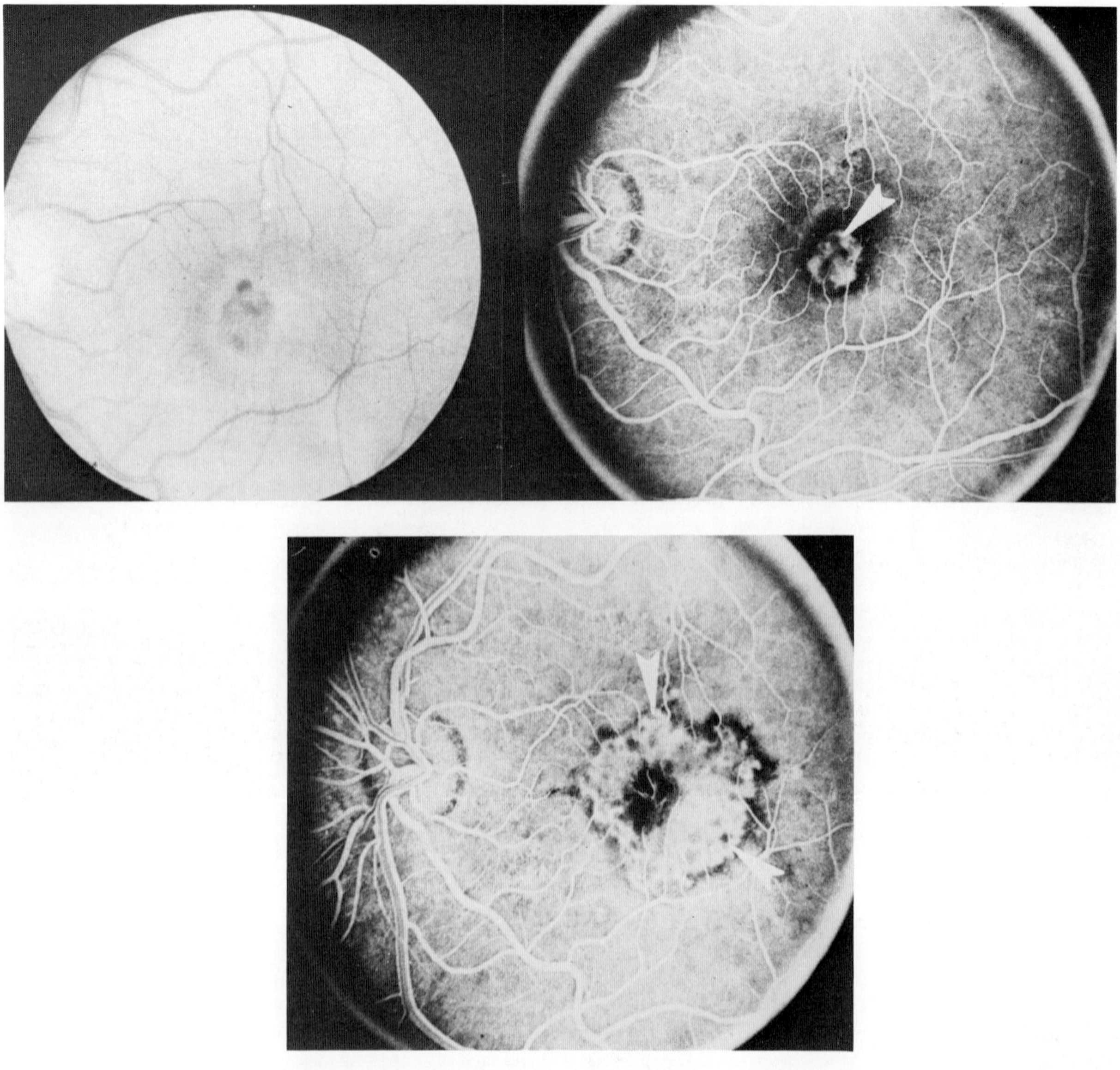

Figure 26.22 *Fundus photograph and fluorescein angiogram of patient with Kühnt-Junius disciform macular degeneration, showing development and progression of a subretinal neovascular membrane* (arrows). *(From Schatz 1975.)*

predetermined area. Only four cases have been examined pathologically, and the findings are similar in all reported cases. It is important to ophthalmologists because of its genetic character. When suspected cases are found in younger persons, a familial study should be initiated, especially of older relatives. Genetic counselling may then be important (Ferry, Llovea, and Shafer 1972).

CENTRAL SEROUS RETINOPATHY AND SEROUS DETACHMENT OF RETINAL PIGMENT EPITHELIUM. Ordinarily, central serous retinopathy and serous detachment of the retinal pigment epithelium would not be considered senile maculopathies. They occur most often in younger people, before the age of 60, and 80 percent of cases occur in males. They are relatively uncommon in

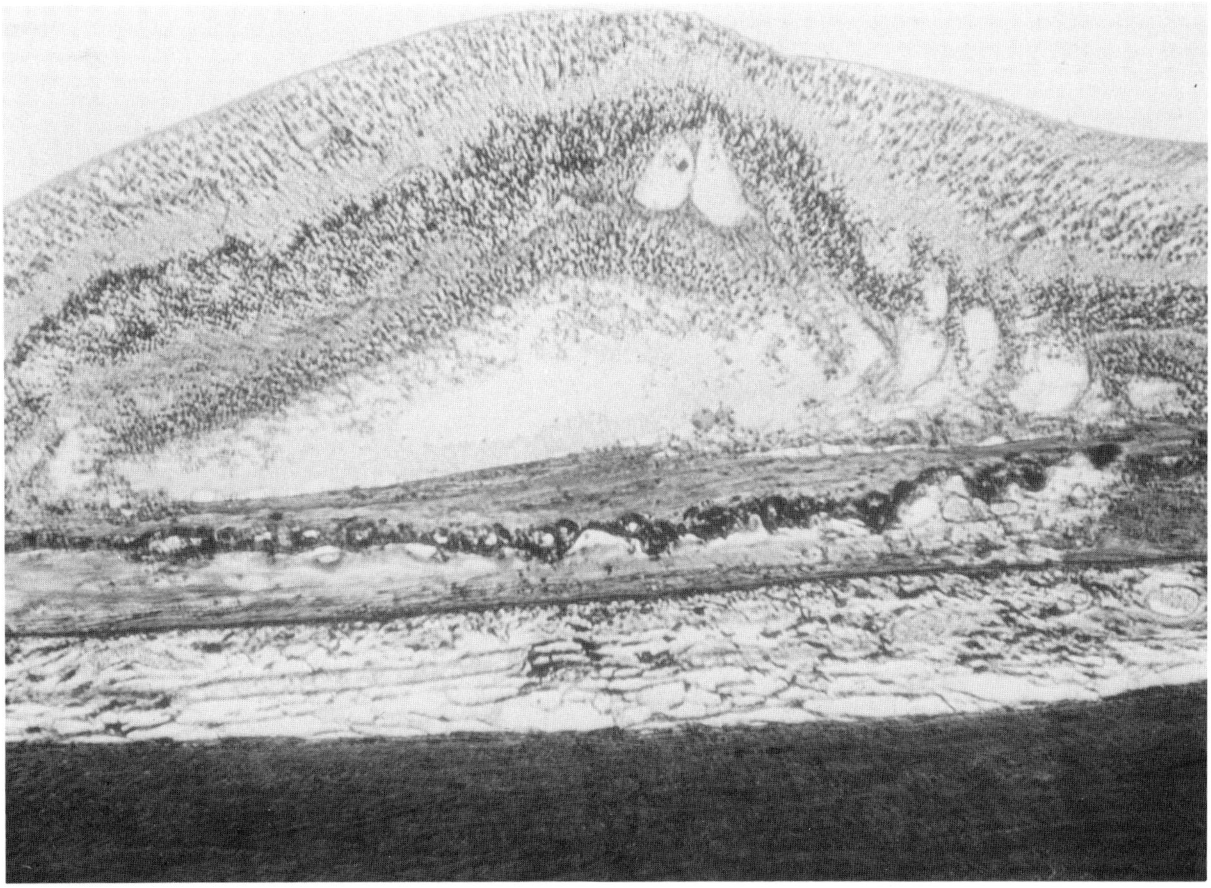

Figure 26.23 *Histopathology of senile disciform macular degeneration. The retina is elevated by a serous exudate in the subretinal area. There is neovascular, fibrous tissue in the subretinal space. Cone cells and processes are atrophied. The pigment layer is atrophied and incorporated in the neovascular fibrous tissue. (Stained with Masson Trichrome stain. ×100.)*

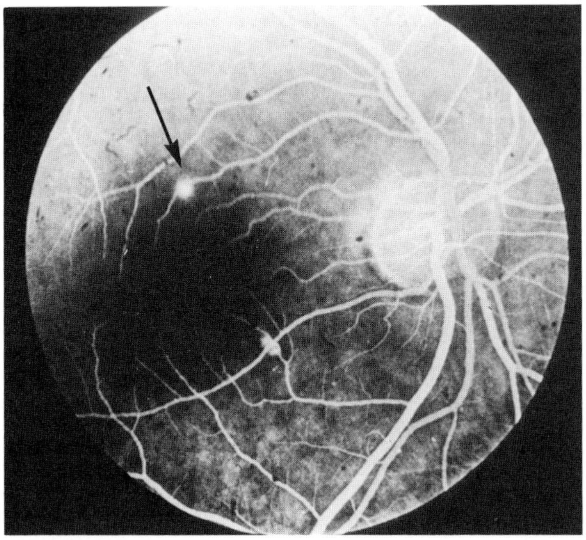

blacks. Smoking is believed to be a factor, and physical and emotional stress. Klien considered these conditions to be due to an instability of the sympathetic nervous system or an allergic reaction. These conditions may be precursors to the onset of the more recognized senile maculopathies. I was led to this belief by a case in a male, 48 years old, who had a typical attack of central serous retinopathy in one eye; the attack lasted several weeks. A fluorescein angiogram showed a leak in the paramacular area. Directly opposite, equidistant from the fovea, was an atrophic area in the pigment epithelium (Fig. 26.24). The leak was treated with a laser beam and the patient's vision improved. The atrophic area

Figure 26.24 *Fluorescein angiogram of a male, 48 years, with central serous retinopathy showing a leak in paramacular area* (arrow). *Directly opposite is an area of atrophy of the RPE.*

persisted and did not change. This situation has been seen several times in more recent cases, leading to the hypothesis that persons with these conditions are more prone to the onset of atrophy of the RPE in and around the macular area and probably to senile macular degeneration. This thought has also been expressed by G. Coscas (1972) in the exhaustive monograph on edema of the macula, with special reference to senile disciform macular degeneration.

REFERENCES

ASHTON, N. 1953. Central areolar choroidal sclerosis: a histopathologic study. *Br. J. Ophthalmol.* 37:140–147.

BAIRATI, A., JR., and ORZALESI, N. 1963. The ultrastructure of the pigment epithelium and of the photoreceptor pigment epithelium junction in the human retina. *J. Ultrastruct. Res.* 9:4–84.

CARR, R. E. 1965. Central areolar choroidal dystrophy. *Arch. Ophthalmol.* 73:32–35.

COSCAS, G., ed., with DHERMY, P., BERNARD, J. A., AUBRY, J. P., and LEGRAS, M. 1972. Maculopathies oedemateuses. *Bull. Soc. Ophtalmol. Fr.,* annual report, p. 70.

DEUTMAN, A. F., and JANSEN, L. M. A. A. 1970. Dominantly inherited drusen of Bruch's membrane. *Br. J. of Ophthalmol.* 54:373–382.

DUKE-ELDER, S., and DOBREE, J. H. 1967. *System of ophthalmology.* 10. *Diseases of the retina.* Chap. 5. Degenerations and dystrophies, p. 517. St. Louis, Missouri: C. V. Mosby.

DUKE-ELDER, S., and PERKINS, E. S. 1965. *System of Ophthalmology.* 9. *Diseases of the uveal tract,* pp. 673–674. St. Louis, Missouri: C. V. Mosby.

FARKAS, T. G., SYLVESTER, V., and ARCHER, D. 1971. The ultrastructure of drusen. *Am. J. Ophthalmol.* 71:1196–1205.

FARKAS, T. G., SYLVESTER, V., ARCHER, D., and ALTONA, M. 1971. *Histochemistry of drusen. Am. J. Ophthalmol.* 71:1206–1215.

FEENEY, L., GREISHABER, J., and HOGAN, M. J. 1965. Studies of human ocular pigment. In *The structure of the eye,* ed. J. W. Towen. Stuttgart: Shattauer-Verlag.

FEHER, J., and VALU, L. 1967. Aging changes of collagenous fibers of the choroid: investigation by polarization optics. *Graefes Arch. Klin. und Exper. Ophthalmol.* 172:249–253.

FERRY, A. P., LLOVEA, I., and SHAFER, D. M. 1972. Central areolar choroidal dystrophy. *Arch. Ophthalmol.* 88: 39–43.

FRIEDMAN, E., SMITH, T. R., and KUWABARA, T. 1963. Senile choroidal patterns and drusen. *Arch. Ophthalmol.* 69:220–230.

FRIEDMAN, E., and TSO, M. 1968. Retinal pigment epithelium. II. Histologic changes associated with age. *Arch. Ophthalmol.* 79:315–320.

GARRON, L. K. 1963. The ultrastructure of retinal pigment epithelium with observations on the choriocapillaris and Bruch's membrane. *Trans. Am. Ophthalmol. Soc.* 61:545.

GORDON, D. M. 1960. The eye: 65 and over. Exhibit at the annual meeting of the American Academy of Ophthalmology and Otolaryngology, Chicago.

HOGAN, M. J. 1967. Bruch's membrane and diseases of the macula: role of elastic tissue and collagen. *Trans. Ophthalmol. Soc. U.K.* 87:113–161.

HOGAN, M. J. 1961. Ultrastructure of the choroid: its role in the pathogenesis of choroidal disease. *Trans. Pac. Coast Otoophthalmol. Soc.* 43:61.

HOGAN, M. J. 1972. Role of the retinal pigment epithelium in macular disease. *Trans. Am. Acad. Ophthalmol. Otolaryngol.* 76:64–80.

HOGAN, M., ALVARADO, J. A., and WEDDELL, J. E. 1971. *Histology of the human eye: atlas and textbook.* Retinal pigment epithelium, p. 405. Philadelphia: W. B. Saunders.

HOGAN, M. J., and ZIMMERMAN, L. E. 1962. *Ophthalmic Pathology,* 2nd ed., p. 543. Philadelphia: W. B. Saunders.

JUNIUS, P., and KÜHNT, H. 1926. *Die Scheibenformige Entartung des Netzhautmitte.* Berlin: S. Karger.

KORNZWEIG, A. L. 1965. Symposium: macular diseases. *Pathogenesis:* diseases of the macula in the aged. *Trans. Am. Acad. Ophthalmol. Otolaryngol.* :668–682.

KORNZWEIG, A. L. 1977. Changes in choriocapillaris associated with senile macular degeneration. *Ann. Ophthalmol.* 9:753–764.

KORNZWEIG, A. L., FELDSTEIN, M., and SCHNEIDER, J. 1957. Eye in old age. IV. Ocular survey of over 1000 aged patients with special reference to disturbed visual function. *Am. J. Ophthalmol.* 44:29–37.

KORNZWEIG, A., FELDSTEIN, M., and SCHNEIDER, J. 1959. The pathogenesis of senile macular degeneration. *Am. J. Ophthalmol.* 48:22–28.

LERCHE, W. 1963–1965. Electron microscopic observations on Bruch's membrane of the human eye. *Ber. Dtsch. Ophthalmol. Ges.* 65:384.

NAKAIZUMI, Y. 1964. The ultrastructure of Bruch's membrane. I. Human, monkey, rabbit, guinea pig and rat's eyes. *Arch. Ophthalmol.* 72:380.

SALZMANN, M. 1912. *The anatomy and histology of the human eyeball: its development and senescence,* auth. trans. E. V. L. Brown. Chap. 11, The pigment epithelium. Chap. 14, The appearance of age in the eyeball. Leipzig and Wien: Franz Deuticke.

SARKS, S. H. 1976. Aging and degeneration in the macular region: a clinical-pathological study. *Br. J. Ophthalmol.* 60:324–341.

SCHÄTZ, H. 1975. *The senile maculopathies and the retinal pigment epithelium. (Int. Ophthalmol. Clin.* 15:169) Boston: Little, Brown.

SORSBY, A. 1939. Choroidal and angiosclerosis with special reference to its hereditary character. *Br. J. Ophthalmol.* 23:32–35.

SPITZNAS, M., and HOGAN, M. J. 1970. Outer segments of photoreceptors and the retinal pigment epithelium: interrelationship in the human eye. *Arch. Ophthalmol.* 84: 810–819.

SUMITA, R. 1961. The fine structure of Bruch's membrane in the choroid. *Acta. Soc. Ophthalmol. Jap.* 65:1188.

STREETEN, B. W. 1961. Sudanophilic granules of the human retinal pigment epithelium. *Arch. Ophthalmol.* 66:125–132 (391–398).

TSO, M., and FRIEDMAN, E. 1968. Retinal pigment epithelium. III. Growth and development. *Arch. Ophthalmol.* 80:214–216.

TEETERS, V. W., and BIRD, A. C. 1973. A clinical study of the vascularity of senile disciform macular degeneration. *Am. J. Ophthalmol.* 75:53–65.

VERHOEFF, F. H., and GROSSMAN, H. P. Pathogenesis of disciform degeneration of the macula. *Arch. Ophthalmol.* 18:561–585. Oct. 1937.

YOUNG, R., and BOK, D. 1969. Participation of the retinal pigment epithelium in the rod outer segment renewal process. *J. Cell Biol.* 42:392–403.

Name Index

Subject Index